Classics in Mathematics

David Gilbarg • Neil S. Trudinger

Elliptic Partial Differential Equations of Second Order

Springer
Berlin
Heidelberg
New York
Barcelona
Hong Kong
London
Milan
Paris
Singapore
Tokyo

David Gilbarg was born in New York in 1918, and was educated there through undergraduate school. He received his Ph.D. degree at Indiana University in 1941. His work in fluid dynamics during the war years motivated much of his later research on flows with free boundaries. He was on the Mathematics Faculty at Indiana University from 1946 to 1957 and at Stanford University from 1957 on. His principal interests and contributions have been in mathematical fluid dynamics and the theory of elliptic partial differential equations.

Neil S. Trudinger was born in Ballarat, Australia in 1942. After schooling and undergraduate education in Australia, he completed his Ph.D. at Stanford University, USA in 1966. He has been a Professor of Mathematics at the Australian National University, Canberra since 1973. His research contributions, while largely focused on non-linear elliptic partial differential equations, have also spread into geometry, functional analysis and computational mathematics. Among honours received are Fellowships of the Australian Academy of Science and of the Royal Society of London.

David Gilbarg • Neil S. Trudinger

Elliptic Partial Differential Equations of Second Order

Reprint of the 1998 Edition

David Gilbarg
Stanford University
Department of Mathematics
Stanford, CA 94305-2125
USA
e-mail: gilbarg@math.stanford.edu

Neil S. Trudinger
The Australian National University
School of Mathematical Sciences
Canberra ACT 0200
Australia
e-mail: neil.trudinger@anu.edu.au

Originally published as Vol. 224 of the
Grundlehren der mathematischen Wissenschaften

Cataloging-in-Publication Data applied for

Die Deutsche Bibliothek - CIP-Einheitsaufnahme

Gilbarg, David:
Elliptic partial differential equations of second order / David Gilbarg; Neil S. Trudinger. - Reprint of the 1998 ed. - Berlin; Heidelberg; New York; Barcelona; Hong Kong; London; Milan; Paris; Singapore; Tokyo: Springer, 2001
(Classics in mathematics)
ISBN 3-540-41160-7

Mathematics Subject Classification (2000): 35Jxx

ISSN 1431-0821
ISBN 3-540-41160-7 Springer-Verlag Berlin Heidelberg New York

Springer-Verlag Berlin Heidelberg New York
a member of Springer Science+Business Media

Printed in Germany

Printed on acid-free paper SPIN 10982136 41/3111ck-5 4 3 2 1

Preface to the Revised Third Printing

This revision of the 1983 second edition of "Elliptic Partial Differential Equations of Second Order" corresponds to the Russian edition, published in 1989, in which we essentially updated the previous version to 1984. The additional text relates to the boundary Hölder derivative estimates of Nikolai Krylov, which provided a fundamental component of the further development of the classical theory of elliptic (and parabolic), fully nonlinear equations in higher dimensions. In our presentation we adapted a simplification of Krylov's approach due to Luis Caffarelli.

The theory of nonlinear second order elliptic equations has continued to flourish during the last fifteen years and, in a brief epilogue to this volume, we signal some of the major advances. Although a proper treatment would necessitate at least another monograph, it is our hope that this book, most of whose text is now more than twenty years old, can continue to serve as background for these and future developments.

Since our first edition we have become indebted to numerous colleagues, all over the globe. It was particularly pleasant in recent years to make and renew friendships with our Russian colleagues, Olga Ladyzhenskaya, Nina Ural'tseva, Nina Ivochkina, Nikolai Krylov and Mikhail Safonov, who have contributed so much to this area. Sadly, we mourn the passing away in 1996 of Ennico De Giorgi, whose brilliant discovery forty years ago opened the door to the higher-dimensional nonlinear theory.

October 1997 *David Gilbarg · Neil S. Trudinger*

Preface to the First Edition

This volume is intended as an essentially self-contained exposition of portions of the theory of second order quasilinear elliptic partial differential equations, with emphasis on the Dirichlet problem in bounded domains. It grew out of lecture notes for graduate courses by the authors at Stanford University, the final material extending well beyond the scope of these courses. By including preparatory chapters on topics such as potential theory and functional analysis, we have attempted to make the work accessible to a broad spectrum of readers. Above all, we hope the readers of this book will gain an appreciation of the multitude of ingenious barehanded techniques that have been developed in the study of elliptic equations and have become part of the repertoire of analysis.

Many individuals have assisted us during the evolution of this work over the past several years. In particular, we are grateful for the valuable discussions with L. M. Simon and his contributions in Sections 15.4 to 15.8; for the helpful comments and corrections of J. M. Cross, A. S. Geue, J. Nash, P. Trudinger and B. Turkington; for the contributions of G. Williams in Section 10.5 and of A. S. Geue in Section 10.6; and for the impeccably typed manuscript which resulted from the dedicated efforts of Isolde Field at Stanford and Anna Zalucki at Canberra. The research of the authors connected with this volume was supported in part by the National Science Foundation.

August 1977

David Gilbarg
Stanford

Neil S. Trudinger
Canberra

Note: The Second Edition includes a new, additional Chapter 9. Consequently Chapters 10 and 15 referred to above have become Chapters 11 and 16.

Table of Contents

Chapter 1

Introduction

Summary

The principal objective of this work is the systematic development of the general theory of second order quasilinear elliptic equations and of the linear theory required in the process. This means we shall be concerned with the solvability of boundary value problems (primarily the Dirichlet problem) and related general properties of solutions of linear equations

$$Lu \equiv a^{ij}(x)D_{ij}u + b^i(x)D_iu + c(x)u = f(x), \quad i, j = 1, 2, \ldots, n, \tag{1.1}$$

and of quasilinear equations

$$Qu \equiv a^{ij}(x, u, Du)D_{ij}u + b(x, u, Du) = 0. \tag{1.2}$$

Here $Du = (D_1u, \ldots, D_nu)$, where $D_iu = \partial u/\partial x_i$, $D_{ij}u = \partial^2 u/\partial x_i\,\partial x_j$, etc., and the summation convention is understood. The ellipticity of these equations is expressed by the fact that the coefficient matrix $[a^{ij}]$ is (in each case) positive definite in the domain of the respective arguments. We refer to an equation as *uniformly elliptic* if the ratio γ of maximum to minimum eigenvalue of the matrix $[a^{ij}]$ is bounded. We shall be concerned with both non-uniformly and uniformly elliptic equations.

The classical prototypes of linear elliptic equations are of course Laplace's equation

$$\Delta u = \sum D_{ii}u = 0$$

and its inhomogeneous counterpart, Poisson's equation $\Delta u = f$. Probably the best known example of a quasilinear elliptic equation is the minimal surface equation

$$\sum D_i(D_iu/(1 + |Du|^2)^{1/2}) = 0,$$

which arises in the problem of least area. This equation is non-uniformly elliptic, with $\gamma = 1 + |Du|^2$. The properties of the differential operators in these examples motivate much of the theory of the general classes of equations discussed in this book.

The relevant linear theory is developed in Chapters 2–9 (and in part of Chapter 12). Although this material has independent interest, the emphasis here is on aspects needed for application to nonlinear problems. Thus the theory stresses weak hypotheses on the coefficients and passes over many of the important classical and modern results on linear elliptic equations.

Since we are ultimately interested in classical solutions of equation (1.2), what is required at some point is an underlying theory of classical solutions for a sufficiently large class of linear equations. This is provided by the Schauder theory in Chapter 6, which is an essentially complete theory for the class of equations (1.1) with Hölder continuous coefficients. Whereas such equations enjoy a definitive existence and regularity theory for classical solutions, corresponding results cease to be valid for equations in which the coefficients are assumed only continuous.

A natural starting point for the study of classical solutions is the theory of Laplace's and Poisson's equations. This is the content of Chapters 2 and 4. In anticipation of later developments the Dirichlet problem for harmonic functions with continuous boundary values is approached through the Perron method of subharmonic functions. This emphasizes the maximum principle, and with it the barrier concept for studying boundary behavior, in arguments that are readily extended to more general situations in later chapters. In Chapter 4 we derive the basic Hölder estimates for Poisson's equation from an analysis of the Newtonian potential. The principal result here (see Theorems 4.6, 4.8) states that all $C^2(\Omega)$ solutions of Poisson's equation, $\Delta u=f$, in a domain Ω of $\mathbb{R}^n$ satisfy a uniform estimate in any subset $\Omega'\subset\subset\Omega$

$$(1.3)\qquad \|u\|_{C^{2,\alpha}(\bar{\Omega}')}\leqslant C(\sup_{\Omega}|u|+\|f\|_{C^{\alpha}(\bar{\Omega})}),$$

where C is a constant depending only on α $(0<\alpha<1)$, the dimension n and dist $(\Omega', \partial\Omega)$; (for notation see Section 4.1). This *interior estimate* (interior since $\Omega'\subset\subset\Omega$) can be extended to a *global estimate* for solutions with sufficiently smooth boundary values provided the boundary $\partial\Omega$ is also sufficiently smooth. In Chapter 4 estimates up to the boundary are established only for hyperplane and spherical boundaries, but these suffice for the later applications.

The climax of the theory of classical solutions of linear second order elliptic equations is achieved in the Schauder theory, which is developed in modified and expanded form in Chapter 6. Essentially, this theory extends the results of potential theory to the class of equations (1.1) having Hölder continuous coefficients. This is accomplished by the simple but fundamental device of regarding the equation locally as a perturbation of the constant coefficient equation obtained by fixing the leading coefficients at their values at a single point. A careful calculation based on the above mentioned estimates for Poisson's equation yields the same inequality (1.3) for any $C^{2,\alpha}$ solution of (1.1), where the constant C now depends also on the bounds and Hölder constants of the coefficients and in addition on the minimum and maximum eigenvalues of the coefficient matrix $[a^{ij}]$ in Ω. These results are stated as interior estimates in terms of weighted interior norms (Theorem 6.2) and, in the case of sufficiently smooth boundary data, as global estimates in terms of global norms (Theorem 6.6). Here we meet the important and recurring concept of an *apriori* estimate; namely, an estimate (in terms of given data) valid for all *possible* solutions of a class of problems even if the hypotheses do not guarantee the existence of such solutions. A major part of this book is devoted to the establishment of apriori bounds for various problems. (We have taken the liberty of replacing the latin *a priori* with the single word *apriori*, which will be used throughout.)

The importance of such apriori estimates is visible in several applications in Chapter 6, among them in establishing the solvability of the Dirichlet problem by the method of continuity (Theorem 6.8) and in proving the higher order regularity of C^2 solutions under appropriate smoothness hypotheses (Theorems 6.17, 6.19). In both cases the estimates provide the necessary compactness properties for certain classes of solutions, from which the desired results are easily inferred.

We remark on several additional features of Chapter 6, which are not needed for the later developments but which broaden the scope of the basic Schauder theory. In Section 6.5 it is seen that for continuous boundary values and a suitably wide class of domains the proof of solvability of the Dirichlet problem for (1.1) can be achieved entirely with interior estimates, thereby simplifying the structure of the theory. The results of Section 6.6 extend the existence theory for the Dirichlet problem to certain classes of non-uniformly elliptic equations. Here we see how relations between geometric properties of the boundary and the degenerate ellipticity at the boundary determine the continuous assumption of boundary values. The methods are based on barrier arguments that foreshadow analogous (but deeper) results for nonlinear equations in Part II. In Section 6.7 we extend the theory of (1.1) to the regular oblique derivative problem. The method is basically an extrapolation to these boundary conditions of the earlier treatment of Poisson's equation and the Schauder theory (without barrier arguments, however).

In the preceding considerations, especially in the existence theory and barrier arguments, the maximum principle for the operator L (when $c \leqslant 0$) plays an essential part. This is a special feature of second order elliptic equations that simplifies and strengthens the theory. The basic facts concerning the maximum principle, as well as illustrative applications of comparison methods, are contained in Chapter 3. The maximum principle provides the earliest and simplest apriori estimates of the general theory. It is of considerable interest that all the estimates of Chapters 4 and 6 can be derived entirely from comparison arguments based on the maximum principle, without any mention of the Newtonian potential or integrals.

An alternative and more general approach to linear problems, without potential theory, can be achieved by Hilbert space methods based on *generalized* or *weak* solutions, as in Chapter 8. To be more specific, let L' be a second order differential operator, with principal part of *divergence form*, defined by

$$L'u \equiv D_i(a^{ij}(x)D_ju + b^i(x)u) + c^i(x)D_iu + d(x)u.$$

If the coefficients are sufficiently smooth, then clearly this operator falls within the class discussed in Chapter 6. However, even if the coefficients are in a much wider class and u is only weakly differentiable (in the sense of Chapter 7), one can still define weak or generalized solutions of $L'u=g$ in appropriate function classes. In particular, if the coefficients a^{ij}, b^i, c^i are bounded and measurable in Ω and g is an integrable function in Ω, let us call u a weak or generalized solution of $L'u=g$ in Ω if $u \in W^{1,2}(\Omega)$ (as defined in Chapter 7) and

$$\int_\Omega [(a^{ij}D_ju+b^iu)D_iv-(c^iD_iu+du)v]\,dx=-\int_\Omega gv\,dx \tag{1.4}$$

for all *test functions* $v \in C_0^1(\Omega)$. It is clear that if the coefficients and g are sufficiently smooth and $u \in C^2(\Omega)$, then u is also a classical solution.

We can now speak also of weak solutions u of the *generalized Dirichlet problem*,

$$L'u=g \text{ in } \Omega, \quad u=\varphi \text{ on } \partial\Omega,$$

if u is a weak solution satisfying $u-\varphi \in W_0^{1,2}(\Omega)$, where $\varphi \in W^{1,2}(\Omega)$. Assuming that the minimum eigenvalue of $[a^{ij}]$ is bounded away from zero in Ω, that

$$D_ib^i+d \leqslant 0 \tag{1.5}$$

in the weak sense, and that also $g \in L^2(\Omega)$, we find in Theorem 8.3 that the generalized Dirichlet problem has a unique solution $u \in W^{1,2}(\Omega)$. Condition (1.5), which is the analogue of $c \leqslant 0$ in (1.1), assures a maximum principle for weak solutions of $L'u \geqslant 0 (\leqslant 0)$ (Theorem 8.1) and hence uniqueness for the generalized Dirichlet problem. Existence of a solution then follows from the Fredholm alternative for the operator L' (Theorem 8.6), which is proved by an application of the Riesz representation theorem in the Hilbert space $W_0^{1,2}(\Omega)$.

The major part of Chapter 8 is taken up with the regularity theory for weak solutions. Additional regularity of the coefficients in (1.4) implies that the solutions belong to higher $W^{k,2}$ spaces (Theorems 8.8, 8.10). It follows from the Sobolev imbedding theorems in Chapter 7 that weak solutions are in fact classical solutions provided the coefficients are sufficiently regular. Global regularity of these solutions is inferred by extending interior regularity to the boundary when the boundary data are sufficiently smooth (Theorems 8.13, 8.14).

The regularity theory of weak solutions and the associated pointwise estimates are fundamental to the nonlinear theory. These results provide the starting point for the "bootstrap" arguments that are typical of nonlinear problems. Briefly, the idea here is to start with weak solutions of a quasilinear equation, regarding them as weak solutions of related *linear* equations obtained by inserting them into the coefficients, and then to proceed by establishing improved regularity of these solutions. Starting anew with the latter solutions and repeating the process, still further regularity is assured, and so on, until the original weak solutions are finally proved to be suitably smooth. This is the essence of the regularity proofs for the older variational problems and is implicit in the nonlinear theory presented here.

The Hölder estimates for weak solutions that are so vital for the nonlinear theory are derived in Chapter 8 from Harnack inequalities based on the Moser iteration technique (Theorems 8.17, 8.18, 8.20, 8.24). These results generalize the basic apriori Hölder estimate of De Giorgi, which provided the initial breakthrough in the theory of quasilinear equations in more than two independent variables. The arguments rest on integral estimates for weak solutions u derived from judicious choice of test functions v in (1.4). The test function technique is the dominant theme in the derivation of estimates throughout most of this work.

In this edition we have added new material to Chapter 8 covering the Wiener criterion for regular boundary points, eigenvalues and eigenfunctions, and Hölder estimates for first derivatives of solutions of linear divergence structure equations.

We conclude Part I of the present edition with a new chapter, Chapter 9, concerning strong solutions of linear elliptic equations. These are solutions which possess second derivatives, at least in a weak sense, and satisfy (1.1) almost everywhere. Two strands are interwoven in this chapter. First we derive a maximum principle of Aleksandrov, and a related apriori bound (Theorem 9.1) for solutions in the Sobolev space $W^{2,n}(\Omega)$, thereby extending certain basic results from Chapter 3 to nonclassical solutions. Later in the chapter, these results are applied to establish various pointwise estimates, including the recent Hölder and Harnack estimates of Krylov and Safonov (Theorems 9.20, 9.22; Corollaries 9.24, 9.25). The other strand in this chapter is the L^p theory of linear second-order elliptic equations that is analogous to the Schauder theory of Chapter 6. The basic estimate for Poisson's equation, namely the Calderon-Zygmund inequality (Theorem 9.9) is derived through the Marcinkiewicz interpolation theorem, although without the use of Fourier transform methods. Interior and global estimates in the Sobolev spaces $W^{2,p}(\Omega)$, $1 < p < \infty$, are established in Theorems 9.11, 9.13 and applied to the Dirichlet problem for strong solutions, in Theorem 9.15 and Corollary 9.18.

Part II of this book is devoted largely to the Dirichlet problem and related estimates for quasilinear equations. The results concern in part the general operator (1.2) while others apply especially to operators of divergence form

$$Qu \equiv \operatorname{div} \mathbf{A}(x, u, Du) + B(x, u, Du) \tag{1.6}$$

where $\mathbf{A}(x, z, p)$ and $B(x, z, p)$ are respectively vector and scalar functions defined on $\Omega \times \mathbb{R} \times \mathbb{R}^n$.

Chapter 10 extends maximum and comparison principles (analogous to results in Chapter 3) to solutions and subsolutions of quasilinear equations. We mention in particular apriori bounds for solutions of $Qu \geqslant 0\ (=0)$, where Q is a divergence form operator satisfying certain structure conditions more general than ellipticity (Theorem 10.9).

Chapter 11 provides the basic framework for the solution of the Dirichlet problem in the following chapters. We are concerned principally with classical solutions, and the equations may be uniformly or non-uniformly elliptic. Under suitable general hypotheses any globally smooth solution u of the boundary value problem for $Qu=0$ in a domain Ω with smooth boundary can be viewed as a fixed point, $u=Tu$, of a compact operator T from $C^{1,\alpha}(\bar{\Omega})$ to $C^{1,\alpha}(\bar{\Omega})$ for any $\alpha \in (0,1)$. In the applications the function defined by Tu, for any $u \in C^{1,\alpha}(\bar{\Omega})$, is the unique solution of the *linear* problem obtained by inserting u into the coefficients of Q. The Leray-Schauder fixed point theorem (proved in Chapter 11) then implies the existence of a solution of the boundary value problem provided an apriori bound, in $C^{1,\alpha}(\bar{\Omega})$, can be established for the solutions of a related continuous family of equations $u=T(u;\sigma)$, $0\leqslant\sigma\leqslant 1$, where $T(u;1)=Tu$ (Theorems 11.4, 11.6). The establishment of such bounds for certain broad classes of Dirichlet problems is the object of Chapters 13–15.

The general procedure for obtaining the required apriori bound for possible solutions u is a four-step process involving successive estimation of $\sup_\Omega |u|$, $\sup_{\partial\Omega} |Du|$, $\sup_\Omega |Du|$, and $\|u\|_{C^{1,\alpha}(\bar{\Omega})}$ for some $\alpha>0$. Each of these estimates presupposes the preceding ones and the final bound on $\|u\|_{C^{1,\alpha}(\bar{\Omega})}$ completes the existence proof based on the Leray-Schauder theorem.

As already observed, bounds on $\sup_\Omega |u|$ are discussed in Chapter 10. In the later chapters this bound is either assumed in the hypotheses or is implied by properties of the equation.

Equations in two variables (Chapter 12) occupy a special place in the theory. This is due in part to the distinctive methods that have been developed for them and also to the results, some of which have no counterpart for equations in more than two variables. The method of quasiconformal mappings and arguments based on divergence structure equations (cf. Chapter 11) are both applicable to equations in two variables and yield relatively easily the desired $C^{1,\alpha}$ apriori estimates, from which a solution of the Dirichlet problem follows readily.

Of particular interest is the fact that solutions of uniformly elliptic linear equations in two variables satisfy an apriori $C^{1,\alpha}$ estimate depending only on the ellipticity constants and bounds on the coefficients, without any regularity assumptions (Theorem 12.4). Such a $C^{1,\alpha}$ estimate, or even the existence of a gradient bound under the same general conditions is unknown for equations in more than two variables. Another special feature of the two-dimensional theory is the existence of an apriori C^1 bound $|Du|\leqslant K$ for solutions of *arbitrary* elliptic equations

$$(1.7) \qquad au_{xx}+2bu_{xy}+cu_{yy}=0,$$

where u is continuous on the closure of a bounded convex domain Ω and has boundary value φ on $\partial\Omega$ satisfying a bounded slope (or three-point) condition with constant K. This classical result, usually based on a theorem of Radó on saddle surfaces, is given an elementary proof in Lemma 12.6. The stated gradient bound, which is valid for all solutions u of the general quasilinear equation (1.7) in which $a=a(x, y, u, u_x, u_y)$, etc., and such that $u=\varphi$ on $\partial\Omega$, reduces this Dirichlet problem to the case of uniformly elliptic equations treated in Theorem 12.5. In Theorem 12.7 we obtain a solution of the general Dirichlet problem for (1.7), assuming local Hölder continuity of the coefficients and a bounded slope condition for the boundary data (without further smoothness restrictions on the data).

Chapters 13, 14 and 15 are devoted to the derivation of the gradient estimates involved in the existence procedure described above. In Chapter 13, we prove the fundamental results of Ladyzhenskaya and Ural'tseva on Hölder estimates of derivatives of elliptic quasilinear equations. In Chapter 14 we study the estimation of the gradient of solutions of elliptic quasilinear equations on the boundary.

After considering general and convex domains, we give an account of the theory of Serrin which associates generalized boundary curvature conditions with the solvability of the Dirichlet problem. In particular, we are able to conclude from the results of Chapters 11, 13 and 14 the Jenkins and Serrin criterion for solvability of the Dirichlet problem for the minimal surface equation, namely, that this problem is solvable for smooth domains and arbitrary smooth boundary values if and only if the mean curvature of the boundary (with respect to the inner normal) is non-negative at every point (Theorem 14.14).

Global and interior gradient bounds for solutions u of quasilinear equations are established in Chapter 15. Following a refinement of an old procedure of Bernstein we derive estimates for $\sup_{\Omega} |Du|$ in terms of $\sup_{\partial\Omega} |Du|$ for classes of equations that include both uniformly elliptic equations satisfying natural growth conditions and equations sharing common structural properties with the prescribed mean curvature equation (Theorem 15.2). A variant of our approach yields interior gradient estimates for a more restricted class of equations (Theorem 15.3). We also consider uniformly and non-uniformly elliptic equations in divergence form (Theorems 15.6, 15.7 and 15.8), in which cases, by appropriate test function arguments, we deduce gradient estimates under different types of coefficient conditions than in the general case. We conclude Chapter 15 with a selection of existence theorems, chosen to illustrate the scope of the theory. These theorems are all obtained by various combinations of the apriori estimates in Chapters 10, 14 and 15 and a judicious choice of a related family of problems to which Theorem 11.8 can be applied.

In Chapter 16, we concentrate on the prescribed mean curvature equation and derive an interior gradient bound (Theorem 16.5) thereby enabling us to deduce existence theorems for the Dirichlet problem when only continuous boundary values are assigned (Theorems 16.8, 16.10). We also consider a family of equations in two variables, which in a certain sense bear the same relationship to the prescribed mean curvature equation as the uniformly elliptic equations of Chapter 12 bear to Laplace's equation. Indeed, by means of a generalized notion of quasiconformal mapping, we derive interior estimates for first and second derivatives. The second derivative estimates provide a generalization of a well known curvature estimate of Heinz for solutions of the minimal surface equation (Theorem 16.20) and moreover, imply an extension of the famous result of Bernstein that entire solutions of the minimal surface equation in two variables must be linear (Corollary 16.19). However, perhaps the striking feature of Theorems 16.5 and 16.20 is the approach. Rather than working in the domain Ω, we work on the hypersurface S given by the graph of the solution u and exploit various relations between the tangential gradient and Laplacian operators on S and the mean curvature of S.

We have also added to the present edition a new final chapter. Chapter 17 is devoted to fully nonlinear elliptic equations, which incorporates recent work on equations of Monge-Ampère and Bellman-Pucci type. These are equations of the general form

$$F[u] = F(x, u, Du, D^2u) = 0 \tag{1.8}$$

and include linear and quasilinear equations of the forms (1.1) and (1.2) as special cases. The function F is defined for $(x, z, p, r) \in \Omega \times \mathbb{R} \times \mathbb{R}^n \times \mathbb{R}^{n \times n}$ where $\mathbb{R}^{n \times n}$ denotes the linear space of real symmetric $n \times n$ matrices. Equation (1.8) is elliptic when the derivative F_r is a positive definite matrix. The method of continuity (Theorem 17.8) reduces the solvability of the Dirichlet problem for (1.8) to the establishment of $C^{2,\alpha}(\bar{\Omega})$ estimates, for some $\alpha > 0$; that is, in addition to the first derivative estimation required for the quasilinear case, we need *second* derivative estimates for fully nonlinear equations. Such estimates are established for equations in two variables (Theorems 17.9, 17.10), uniformly elliptic equations (Theorems 17.14, 17.15) and equations of Monge-Ampère type (Theorems 17.19, 17.20, 17.26), yielding, in particular, recent results on the solvability of the Dirichlet problem for uniformly elliptic equations by Evans, Krylov and Lions (in Theorems 17.17, 17.18), and for equations of Monge-Ampère type by Krylov, and Caffarelli, Nirenberg and Spruck (in Theorem 17.23).

We conclude this summary with some guides to the reader. The material is not in strict logical order. Thus the theory of Poisson's equation (Chapter 4) would normally follow Laplace's equation (Chapter 2). However, the elementary character of the results on the maximum principle (Chapter 3) and the opportunity for the reader to meet early some general problems with variable coefficients recommends its insertion after Chapter 2. In fact, the general maximum principle is not used until the existence theory of Chapter 6. The basic material on functional analysis (Chapter 5) is needed in only a minor way for the Schauder theory: the contraction mapping principle and the basic concepts of Banach spaces suffice, except for the proof of the alternative in Theorem 6.15. For applications to nonlinear problems in Part II it is sufficient to know the results of Section 1–3 of Chapter 6. Depending on the reader's interests, it may be preferable to study the linear theory by starting directly with L^2 theory in Chapter 8; this assumes the preliminary material on functional analysis (Chapter 5) and on the calculus of weakly differentiable functions (Chapter 7). The Harnack inequalities and Hölder estimates in the regularity theory of Chapter 8 are not applied until Chapter 13.

The theory of quasilinear equations in two variables (Chapter 12) is essentially independent of Chapters 7–11 and can be read following Chapter 6 provided one assumes the Schauder fixed point theorem (Theorem 11.1). The method of quasiconformal mappings is met again in Chapter 16 but otherwise the remaining chapters are independent of Chapter 12. Accordingly, after the basic outline of the nonlinear theory in Chapter 11 the reader can proceed directly to the n-variable theory in Chapters 13–17. Chapter 16 is largely independent of Chapters 13–15. Chapters 6 and 9 are sufficient preparation for most of Chapter 17.

Further Remarks

Beyond the assumption of basic real analysis and linear algebra the material in this work is almost entirely self-contained. Thus, much of the preliminary development of potential theory and functional analysis, as well as results on Sobolev spaces and fixed point theorems, will be familiar to many readers, although the proof of the Leray-Schauder theorem without topological degree in Theorem 11.6 is probably not so well known. A number of well established auxiliary results, such as the interpolation inequalities and extension lemmas of Chapter 6, are proved for the sake of completeness.

There is substantial overlap with the monographs of Ladyzhenskaya and Ural'tséva [LU 4] and Morrey [MY 5]. This book differs from the former in some of the analytical techniques and in the emphasis on non-uniformly elliptic equations in the nonlinear theory; it differs from the latter in not being directly concerned with variational problems and methods. The present work also includes material developed since the publication of those books. On the other hand, it is much more limited in various ways. Among the topics not included are systems of equations, semilinear equations, the theory of monotone operators, and aspects of the subject based on geometric measure theory.

In a subject that is often quite technical we have not always striven for the greatest generality, especially with respect to the modulus of continuity, estimates, integral conditions, and the like. We have instead confined ourselves to conditions determined by *power* functions: for example, Hölder continuity rather than Dini continuity, L^p spaces in Chapter 8 rather than Orlicz spaces, structure conditions in terms of powers of $|p|$ rather than more general functions of $|p|$, etc. By suitable modification of the proofs the reader will usually be able to supply the appropriate generalizations.

Historical material and bibliographical references are discussed primarily in the Notes at the end of the chapters. These are not intended to be complete but rather to supplement the text and place it in better perspective. A much more extensive survey of the literature until 1968 is contained in Miranda [MR 2]. The problems attached to the chapters are also intended to supplement the text; hopefully they will be useful exercises for the reader.

Basic Notation

$\mathbb{R}^n$: Euclidean n-space, $n \geqslant 2$, with points $x=(x_1, \ldots, x_n)$, $x_i \in \mathbb{R}$ (real numbers); $|x|=(\sum x_i^2)^{1/2}$; if $\mathbf{b}=(b_1, \ldots, b_n)$ is an ordered n-tuple, then $|\mathbf{b}|=(\sum b_i^2)^{1/2}$.

$\mathbb{R}^n_+$: half-space in $\mathbb{R}^n = \{x \in \mathbb{R}^n \,|\, x_n > 0\}$.

∂S: boundary of the point set S; $\bar{S}$ = closure of $S = S \cup \partial S$.

$S - S'$: $\{x \in S \,|\, x \notin S'\}$.

$S' \subset\subset S$: S' has compact closure in S; S' is *strictly contained* in S.

Ω: a proper open subset of $\mathbb{R}^n$, not necessarily bounded; Ω is a *domain* if it is also connected; $|\Omega|$ = volume of Ω.

$B(y)$: a ball in $\mathbb{R}^n$ with center y; $B_r(y)$ is the open ball of radius r centered at y.

ω_n: volume of unit ball in $\mathbb{R}^n = \dfrac{2\pi^{n/2}}{n\Gamma(n/2)}$.

$D_i u = \partial u/\partial x_i$, $D_{ij} u = \partial^2 u/\partial x_i\, \partial x_j$, etc.; $Du = (D_1 u, \ldots, D_n u)$ = gradient of u; $D^2 u = [D_{ij} u]$ = Hessian matrix of second derivatives $D_{ij} u$, $i, j = 1, 2, \ldots, n$.

$\beta=(\beta_1, \ldots, \beta_n)$, $\beta_i=$ integer $\geqslant 0$, with $|\beta| = \sum \beta_i$, is a *multi-index*; we define

$$D^\beta u = \frac{\partial^{|\beta|} u}{\partial x_1^{\beta_1} \cdots \partial x_n^{\beta_n}}$$

$C^0(\Omega)$ $(C^0(\bar{\Omega}))$: the set of continuous functions on Ω $(\bar{\Omega})$.

$C^k(\Omega)$: the set of functions having all derivatives of order $\leqslant k$ continuous in Ω ($k=$ integer $\geqslant 0$ or $k=\infty$).

$C^k(\bar{\Omega})$: the set of functions in $C^k(\Omega)$ all of whose derivatives of order $\leqslant k$ have continuous extensions to $\bar{\Omega}$.

supp u: the *support* of u, the closure of the set on which $u \neq 0$.

$C_0^k(\Omega)$: the set of functions in $C^k(\Omega)$ with compact support in Ω.

$C = C(*, \ldots, *)$ denotes a constant depending only on the quantities appearing in parentheses. In a given context, the same letter C will (generally) be used to denote different constants depending on the same set of arguments.

Part I

Linear Equations

Chapter 2

Laplace's Equation

Let Ω be a domain in $\mathbb{R}^n$ and u a $C^2(\Omega)$ function. The Laplacian of u, denoted Δu, is defined by

$$(2.1) \qquad \Delta u = \sum_{i=1}^{n} D_{ii}u = \operatorname{div} Du.$$

The function u is called *harmonic* (*subharmonic*, *superharmonic*) in Ω if it satisfies there

$$(2.2) \qquad \Delta u = 0 \quad (\geqslant 0, \leqslant 0).$$

In this chapter we develop some basic properties of harmonic, subharmonic and superharmonic functions which we use to study the solvability of the classical Dirichlet problem for *Laplace's equation*, $\Delta u = 0$. As mentioned in Chapter 1, Laplace's equation and its inhomogeneous form, Poisson's equation, are basic models of linear elliptic equations.

Our starting point here will be the well known divergence theorem in $\mathbb{R}^n$. Let Ω_0 be a bounded domain with C^1 boundary $\partial\Omega_0$ and let ν denote the unit outward normal to $\partial\Omega_0$. For any vector field $\mathbf{w}$ in $C^1(\bar{\Omega}_0)$, we then have

$$(2.3) \qquad \int_{\Omega_0} \operatorname{div} \mathbf{w}\, dx = \int_{\partial\Omega_0} \mathbf{w} \cdot \nu\, ds$$

where ds indicates the $(n-1)$-dimensional area element in $\partial\Omega_0$. In particular if u is a $C^2(\bar{\Omega}_0)$ function we have, by taking $\mathbf{w} = Du$ in (2.3),

$$(2.4) \qquad \int_{\Omega_0} \Delta u\, dx = \int_{\partial\Omega_0} Du \cdot \nu\, ds = \int_{\partial\Omega_0} \frac{\partial u}{\partial \nu}\, ds.$$

(For a more general formulation of the divergence theorem, see [KE 2].)

2.1. The Mean Value Inequalities

Our first theorem, which is a consequence of the identity (2.4), comprises the well known mean value properties of harmonic, subharmonic and superharmonic functions.

Theorem 2.1. *Let $u \in C^2(\Omega)$ satisfy $\Delta u = 0$ ($\geqslant 0$, $\leqslant 0$) in Ω. Then for any ball $B = B_R(y) \subset\subset \Omega$, we have*

$$(2.5)\qquad u(y) = (\leqslant, \geqslant)\, \frac{1}{n\omega_n R^{n-1}} \int_{\partial B} u\, ds,$$

$$(2.6)\qquad u(y) = (\leqslant, \geqslant)\, \frac{1}{\omega_n R^n} \int_B u\, dx.$$

For harmonic functions, Theorem 2.1 thus asserts that the function value at the center of the ball B is equal to the integral mean values over both the surface ∂B and B itself. These results, known as the *mean value theorems*, in fact also characterize harmonic functions; (see Theorem 2.7).

Proof of Theorem 2.1. Let $\rho \in (0, R)$ and apply the identity (2.4) to the ball $B_\rho = B_\rho(y)$. We obtain

$$\int_{\partial B_\rho} \frac{\partial u}{\partial \nu}\, ds = \int_{B_\rho} \Delta u\, dx = (\geqslant, \leqslant)\, 0.$$

Introducing radial and angular coordinates $r = |x - y|$, $\omega = \dfrac{x-y}{r}$, and writing $u(x) = u(y + r\omega)$, we have

$$\begin{aligned}\int_{\partial B_\rho} \frac{\partial u}{\partial \nu}\, ds &= \int_{\partial B_\rho} \frac{\partial u}{\partial r}(y + \rho\omega)\, ds = \rho^{n-1} \int_{|\omega|=1} \frac{\partial u}{\partial r}(y + \rho\omega)\, d\omega \\ &= \rho^{n-1} \frac{\partial}{\partial \rho} \int_{|\omega|=1} u(y + \rho\omega)\, d\omega = \rho^{n-1} \frac{\partial}{\partial \rho}\left[\rho^{1-n} \int_{\partial B_\rho} u\, ds\right] \\ &= (\geqslant, \leqslant)\, 0.\end{aligned}$$

Consequently for any $\rho \in (0, R)$,

$$\rho^{1-n} \int_{\partial B_\rho} u\, ds = (\leqslant, \geqslant)\, R^{1-n} \int_{\partial B_R} u\, ds$$

and since

$$\lim_{\rho \to 0} \rho^{1-n} \int_{\partial B_\rho} u\, ds = n\omega_n u(y)$$

relations (2.5) follow. To get the solid mean value inequalities, that is, relations (2.6), we write (2.5) in the form

$$n\omega_n \rho^{n-1} u(y) = (\leqslant, \geqslant) \int_{\partial B_\rho} u\, ds, \quad \rho \leqslant R$$

and integrate with respect to ρ from 0 to R. The relations (2.6) follow immediately. □

2.2. Maximum and Minimum Principle

With the aid of Theorem 2.1 the *strong maximum principle* for subharmonic functions and the *strong minimum principle* for superharmonic functions may be derived.

Theorem 2.2. *Let $\Delta u \geqslant 0$ ($\leqslant 0$) in Ω and suppose there exists a point $y \in \Omega$ for which $u(y) = \sup_\Omega u$ ($\inf_\Omega u$). Then u is constant. Consequently a harmonic function cannot assume an interior maximum or minimum value unless it is constant.*

Proof. Let $\Delta u \geqslant 0$ in Ω, $M = \sup_\Omega u$ and define $\Omega_M = \{x \in \Omega \mid u(x) = M\}$. By assumption Ω_M is not empty. Furthermore since u is continuous, Ω_M is closed relative to Ω. Let z be any point in Ω_M and apply the mean value inequality (2.6) to the subharmonic function $u - M$ in a ball $B = B_R(z) \subset\subset \Omega$. We therefore obtain

$$0 = u(z) - M \leqslant \frac{1}{\omega_n R^n} \int_B (u - M)\, dx \leqslant 0,$$

so that $u = M$ in $B_R(z)$. Consequently Ω_M is also open relative to Ω. Hence $\Omega_M = \Omega$. The result for superharmonic functions follows by replacement of u by $-u$. □

The strong maximum and minimum principles immediately imply global estimates, namely the following *weak maximum* and *minimum principles*.

Theorem 2.3. *Let $u \in C^2(\Omega) \cap C^0(\bar{\Omega})$ with $\Delta u \geqslant 0$ ($\leqslant 0$) in Ω. Then, provided Ω is bounded,*

$$\sup_\Omega u = \sup_{\partial\Omega} u \quad (\inf_\Omega u = \inf_{\partial\Omega} u). \tag{2.7}$$

Consequently, for harmonic u

$$\inf_{\partial\Omega} u \leqslant u(x) \leqslant \sup_{\partial\Omega} u, \quad x \in \Omega.$$

A uniqueness theorem for the classical Dirichlet problem for Laplace's and Poisson's equation in bounded domains now follows from Theorem 2.3.

Theorem 2.4. *Let $u, v \in C^2(\Omega) \cap C^0(\bar{\Omega})$ satisfy $\Delta u = \Delta v$ in Ω, $u = v$ on $\partial\Omega$. Then $u = v$ in Ω.*

Proof. Let $w = u - v$. Then $\Delta w = 0$ in Ω and $w = 0$ on $\partial\Omega$. It follows from Theorem 2.3 that $w = 0$ in Ω. □

Note that also by Theorem 2.3, we have that if u and v are harmonic and subharmonic functions respectively, agreeing on the boundary $\partial\Omega$, then $v \leqslant u$ in Ω.

Hence the term subharmonic. A corresponding remark is true for superharmonic functions. Later in this chapter, we employ this property of $C^2(\Omega)$ subharmonic and superharmonic functions to expand their definition to larger classes of functions. In the next chapter, an alternate method of proof for Theorems 2.2, 2.3 and 2.4 will be supplied when we treat maximum principles for general elliptic equations; (see also Problem 2.1).

2.3. The Harnack Inequality

A further consequence of Theorem 2.1 is the following Harnack inequality for harmonic functions.

Theorem 2.5. *Let u be a non-negative harmonic function in Ω. Then for any bounded subdomain $\Omega' \subset\subset \Omega$, there exists a constant C depending only on n, Ω' and Ω such that*

$$\sup_{\Omega'} u \leq C \inf_{\Omega'} u. \tag{2.8}$$

Proof. Let $y \in \Omega$, $B_{4R}(y) \subset \Omega$. Then for any two points $x_1, x_2 \in B_R(y)$, we have by the inequalities (2.6)

$$u(x_1) = \frac{1}{\omega_n R^n} \int_{B_R(x_1)} u\,dx \leq \frac{1}{\omega_n R^n} \int_{B_{2R}(y)} u\,dx.$$

$$u(x_2) = \frac{1}{\omega_n (3R)^n} \int_{B_{3R}(x_2)} u\,dx \geq \frac{1}{\omega_n (3R)^n} \int_{B_{2R}(y)} u\,dx.$$

Consequently we obtain

$$\sup_{B_R(y)} u \leq 3^n \inf_{B_R(y)} u. \tag{2.9}$$

Now let $\Omega' \subset\subset \Omega$ and choose $x_1, x_2 \in \bar{\Omega}'$ so that $u(x_1) = \sup_{\Omega'} u$, $u(x_2) = \inf_{\Omega'} u$. Let $\Gamma \subset \bar{\Omega}'$ be a closed arc joining x_1 and x_2 and choose R so that $4R < \operatorname{dist}(\Gamma, \partial\Omega)$. By virtue of the Heine-Borel theorem, Γ can be covered by a finite number N (depending only on Ω' and Ω) of balls of radius R. Applying the estimate (2.9) in each ball and combining the resulting inequalities, we obtain.

$$u(x_1) \leq 3^{nN} u(x_2).$$

Hence the estimate (2.8) holds with $C = 3^{nN}$. □

Note that the constant in (2.8) is invariant under similarity and orthogonal transformations. A Harnack inequality for weak solutions of homogeneous elliptic equations will be established in Chapter 8.

2.4. Green's Representation

As a prelude to *existence* considerations we derive now some further consequences of the divergence theorem, namely, the Green identities. Let Ω be a domain for which the divergence theorem holds and let u and v be $C^2(\bar{\Omega})$ functions. We select $\mathbf{w} = vDu$ in the identity (2.3) to obtain *Green's first identity*:

$$\int_\Omega v\,\Delta u\,dx + \int_\Omega Du\cdot Dv\,dx = \int_{\partial\Omega} v\frac{\partial u}{\partial\nu}\,ds. \tag{2.10}$$

Interchanging u and v in (2.10) and subtracting, we obtain *Green's second identity*:

$$\int_\Omega (v\,\Delta u - u\,\Delta v)\,dx = \int_{\partial\Omega}\left(v\frac{\partial u}{\partial\nu} - u\frac{\partial v}{\partial\nu}\right)ds. \tag{2.11}$$

Laplace's equation has the radially symmetric solution r^{2-n} for $n>2$ and $\log r$ for $n=2$, r being radial distance from a fixed point. To proceed further from (2.11), we fix a point y in Ω and introduce the normalized *fundamental solution* of Laplace's equation:

$$\Gamma(x-y) = \Gamma(|x-y|) = \begin{cases} \dfrac{1}{n(2-n)\omega_n}|x-y|^{2-n}, & n>2 \\[2ex] \dfrac{1}{2\pi}\log|x-y|, & n=2. \end{cases} \tag{2.12}$$

By simple computation we have

$$\begin{aligned} D_i\Gamma(x-y) &= \frac{1}{n\omega_n}(x_i-y_i)|x-y|^{-n}; \\ D_{ij}\Gamma(x-y) &= \frac{1}{n\omega_n}\{|x-y|^2\delta_{ij} - n(x_i-y_i)(x_j-y_j)\}\,|x-y|^{-n-2}. \end{aligned} \tag{2.13}$$

Clearly Γ is harmonic for $x\neq y$. For later purposes we note the following derivative estimates:

$$\begin{aligned} |D_i\Gamma(x-y)| &\leqslant \frac{1}{n\omega_n}|x-y|^{1-n}; \\ |D_{ij}\Gamma(x-y)| &\leqslant \frac{1}{\omega_n}|x-y|^{-n}. \\ |D^\beta\Gamma(x-y)| &\leqslant C|x-y|^{2-n-|\beta|}, \qquad C = C(n,|\beta|). \end{aligned} \tag{2.14}$$

The singularity at $x=y$ prevents us from using Γ in place of v in Green's second identity (2.11). One way of overcoming this difficulty is to replace Ω by $\Omega - \bar{B}_\rho$

where $B_\rho = B_\rho(y)$ for sufficiently small ρ. We can then conclude from (2.11) that

$$\int_{\Omega - B_\rho} \Gamma\, \Delta u\, dx = \int_{\partial\Omega} \left(\Gamma \frac{\partial u}{\partial \nu} - u \frac{\partial \Gamma}{\partial \nu}\right) ds + \int_{\partial B_\rho} \left(\Gamma \frac{\partial u}{\partial \nu} - u \frac{\partial \Gamma}{\partial \nu}\right) ds. \tag{2.15}$$

Now

$$\begin{aligned}\int_{\partial B_\rho} \Gamma \frac{\partial u}{\partial \nu}\, ds &= \Gamma(\rho) \int_{\partial B_\rho} \frac{\partial u}{\partial \nu}\, ds \\ &\leqslant n\omega_n \rho^{n-1} \Gamma(\rho) \sup_{B_\rho} |Du| \to 0 \quad \text{as } \rho \to 0\end{aligned}$$

and

$$\begin{aligned}\int_{\partial B_\rho} u \frac{\partial \Gamma}{\partial \nu}\, ds &= -\Gamma'(\rho) \int_{\partial B_\rho} u\, ds \quad \text{(recall that } \nu \text{ is } \textit{outer} \text{ normal to } \Omega - B_\rho) \\ &= \frac{-1}{n\omega_n \rho^{n-1}} \int_{\partial B_\rho} u\, ds \to -u(y) \quad \text{as } \rho \to 0.\end{aligned}$$

Hence letting ρ tend to zero in (2.15) we arrive at *Green's representation formula*:

$$u(y) = \int_{\partial\Omega} \left(u \frac{\partial \Gamma}{\partial \nu}(x-y) - \Gamma(x-y) \frac{\partial u}{\partial \nu}\right) ds + \int_\Omega \Gamma(x-y)\, \Delta u\, dx, \quad (y \in \Omega). \tag{2.16}$$

For an integrable function f, the integral $\int_\Omega \Gamma(x-y) f(x)\, dx$ is called the *Newtonian potential* with density f. If u has compact support in $\mathbb{R}^n$, then (2.16) yields the frequently useful representation formula,

$$u(y) = \int \Gamma(x-y)\, \Delta u(x)\, dx. \tag{2.17}$$

For harmonic u, we also obtain the representation

$$u(y) = \int_{\partial\Omega} \left(u \frac{\partial \Gamma}{\partial \nu}(x-y) - \Gamma(x-y) \frac{\partial u}{\partial \nu}\right) ds, \quad (y \in \Omega). \tag{2.18}$$

Since the integrand above is infinitely differentiable and, in fact, also analytic with respect to y, it follows that u is also analytic in Ω. Thus harmonic functions are analytic throughout their domain of definition and therefore uniquely determined by their values in any open subset.

Now suppose that $h \in C^1(\bar{\Omega}) \cap C^2(\Omega)$ satisfies $\Delta h = 0$ in Ω. Then again by Green's second identity (2.11) we obtain

$$-\int_{\partial\Omega} \left(u \frac{\partial h}{\partial \nu} - h \frac{\partial u}{\partial \nu}\right) ds = \int_\Omega h\, \Delta u\, dx. \tag{2.19}$$

Writing $G=\Gamma+h$ and adding (2.16) and (2.19) we then obtain a more general version of Green's representation formula:

$$u(y)=\int_{\partial\Omega}\left(u\frac{\partial G}{\partial\nu}-G\frac{\partial u}{\partial\nu}\right)ds+\int_{\Omega}G\,\Delta u\,dx. \tag{2.20}$$

If in addition $G=0$ on $\partial\Omega$ we have

$$u(y)=\int_{\partial\Omega}u\frac{\partial G}{\partial\nu}\,ds+\int_{\Omega}G\,\Delta u\,dx \tag{2.21}$$

and the function $G=G(x,y)$ is called the (Dirichlet) *Green's function* for the domain Ω, sometimes also called the *Green's function of the first kind* for Ω. By Theorem 2.4, the Green's function is unique and from the formula (2.21) its existence implies a representation for a $C^1(\bar{\Omega})\cap C^2(\Omega)$ harmonic function in terms of its boundary values.

2.5. The Poisson Integral

When the domain Ω is a ball the Green's function can be explicitly determined by the method of images and leads to the well known Poisson integral representation for harmonic functions in a ball. Namely, let $B_R=B_R(0)$ and for $x\in B_R$, $x\neq0$ let

$$\bar{x}=\frac{R^2}{|x|^2}x \tag{2.22}$$

denote its inverse point with respect to B_R; if $x=0$, take $\bar{x}=\infty$. It is then easily verified that the Green's function for B_R is given by

$$G(x,y)=\begin{cases}\Gamma(|x-y|)-\Gamma\left(\dfrac{|y|}{R}|x-\bar{y}|\right), & y\neq0\\ \Gamma(|x|)-\Gamma(R), & y=0.\end{cases} \tag{2.23}$$

$$=\Gamma(\sqrt{|x|^2+|y|^2-2x\cdot y})-\Gamma\left(\sqrt{\left(\frac{|x|\,|y|}{R}\right)^2+R^2-2x\cdot y}\right)$$

for all $x,y\in B_R$, $x\neq y$.

The function G defined by (2.23) has the properties

$$G(x,y)=G(y,x),\quad G(x,y)\leqslant0\quad\text{for } x,y\in\bar{B}_R. \tag{2.24}$$

Furthermore, direct calculation shows that at $x \in \partial B_R$ the normal derivative of G is given by

$$\frac{\partial G}{\partial \nu} = \frac{\partial G}{\partial |x|} = \frac{R^2 - |y|^2}{n\omega_n R} |x-y|^{-n} \geqslant 0. \tag{2.25}$$

Hence if $u \in C^2(B_R) \cap C^1(\bar{B}_R)$ is harmonic, we have by (2.21) the *Poisson integral formula*:

$$u(y) = \frac{R^2 - |y|^2}{n\omega_n R} \int_{\partial B_R} \frac{u \, ds_x}{|x-y|^n}. \tag{2.26}$$

The right hand side of formula (2.26) is called the Poisson integral of u. A simple approximation argument shows that the Poisson integral formula continues to hold for $u \in C^2(B_R) \cap C^0(\bar{B}_R)$. Note that by taking $y=0$, we recover the mean value theorem for harmonic functions. In fact all the previous theorems of this chapter could have been derived as consequences of the representation (2.21) with $\Omega = B_R(0)$.

To establish the existence of solutions of the classical Dirichlet problem for balls we need the converse result to the representation (2.26), and we prove this now.

Theorem 2.6. *Let $B = B_R(0)$ and φ be a continuous function on ∂B. Then the function u defined by*

$$u(x) = \begin{cases} \dfrac{R^2 - |x|^2}{n\omega_n R} \displaystyle\int_{\partial B} \frac{\varphi(y) \, ds_y}{|x-y|^n} & \text{for } x \in B \\ \varphi(x) & \text{for } x \in \partial B \end{cases} \tag{2.27}$$

belongs to $C^2(B) \cap C^0(\bar{B})$ and satisfies $\Delta u = 0$ in B.

Proof. That u defined by (2.27) is harmonic in B is evident from the fact that G, and hence $\partial G/\partial \nu$, is harmonic in x, or it may be verified by direct calculation. To establish the continuity of u on ∂B, we use the Poisson formula (2.26) for the special case $u = 1$ to obtain the identity.

$$\int_{\partial B} K(x, y) \, ds_y = 1 \quad \text{for all } x \in B \tag{2.28}$$

where K is the *Poisson kernel*

$$K(x, y) = \frac{R^2 - |x|^2}{n\omega_n R |x-y|^n}; \quad x \in B, \ y \in \partial B. \tag{2.29}$$

Of course the integral in (2.28) may be evaluated directly but this is a complicated calculation. Now let $x_0 \in \partial B$ and ε be an arbitrary positive number. Choose $\delta > 0$

so that $|\varphi(x)-\varphi(x_0)|<\varepsilon$ if $|x-x_0|<\delta$ and let $|\varphi|\leqslant M$ on ∂B. Then if $|x-x_0|<\delta/2$, we have by (2.27) and (2.28)

$$\begin{aligned}|u(x)-u(x_0)| &= \left|\int_{\partial B} K(x,y)(\varphi(y)-\varphi(x_0))\,ds_y\right| \\ &\leqslant \int_{|y-x_0|\leqslant\delta} K(x,y)|\varphi(y)-\varphi(x_0)|\,ds_y \\ &\quad+\int_{|y-x_0|>\delta} K(x,y)|\varphi(y)-\varphi(x_0)|\,ds_y \\ &\leqslant \varepsilon+\frac{2M(R^2-|x|^2)R^{n-2}}{(\delta/2)^n}.\end{aligned}$$

If now $|x-x_0|$ is sufficiently small it is clear that $|u(x)-u(x_0)|<2\varepsilon$ and hence u is continuous at x_0. Consequently $u\in C^0(\bar{B})$ as required. □

We note that the preceding argument is local; that is, if φ is only bounded and integrable on ∂B, and continuous at x_0, then $u(x)\to\varphi(x_0)$ as $x\to x_0$.

2.6. Convergence Theorems

We consider now some immediate consequences of the Poisson integral formula. The following three theorems will not however be required for the later development. We show first that harmonic functions can in fact be characterized by their mean value property.

Theorem 2.7. *A $C^0(\Omega)$ function u is harmonic if and only if for every ball $B=B_R(y)\subset\subset\Omega$ it satisfies the mean value property,*

$$u(y)=\frac{1}{n\omega_n R^{n-1}}\int_{\partial B} u\,ds. \tag{2.30}$$

Proof. By Theorem 2.6, there exists for any ball $B\subset\subset\Omega$ a harmonic function h such that $h=u$ on ∂B. The difference $w=u-h$ will then be a function satisfying the mean value property on any ball in B. Consequently the maximum principle and uniqueness results of Theorems 2.2, 2.3 and 2.4 apply to w since the mean value inequalities were the only properties of harmonic functions used in their derivation. Hence $w=0$ in B and consequently u must be harmonic in Ω. □

As an immediate consequence of the preceding theorem we have:

Theorem 2.8. *The limit of a uniformly convergent sequence of harmonic functions is harmonic.*

It follows from Theorem 2.8, that if $\{u_n\}$ is a sequence of harmonic functions in a bounded domain Ω, with continuous boundary values $\{\varphi_n\}$ which converge uniformly on $\partial\Omega$ to a function φ, then the sequence $\{u_n\}$ converges uniformly (by the maximum principle) to a harmonic function having the boundary values φ on $\partial\Omega$. By means of Harnack's inequality, Theorem 2.5, we can also derive, from Theorem 2.8, *Harnack's convergence theorem.*

Theorem 2.9. *Let $\{u_n\}$ be a monotone increasing sequence of harmonic functions in a domain Ω and suppose that for some point $y \in \Omega$, the sequence $\{u_n(y)\}$ is bounded. Then the sequence converges uniformly on any bounded subdomain $\Omega' \subset\subset \Omega$ to a harmonic function.*

Proof. The sequence $\{u_n(y)\}$ will converge, so that for arbitrary $\varepsilon > 0$ there is a number N such that $0 \leqslant u_m(y) - u_n(y) < \varepsilon$ for all $m \geqslant n > N$. But then by Theorem 2.5, we must have

$$\sup_{\Omega'} |u_m(x) - u_n(x)| < C\varepsilon$$

for some constant C depending on Ω' and Ω. Consequently $\{u_n\}$ converges uniformly and by virtue of Theorem 2.8, the limit function is harmonic. □

2.7. Interior Estimates of Derivatives

By direct differentiation of the Poisson integral it is possible to obtain interior derivative estimates for harmonic functions. Alternatively, such estimates also follow from the mean value theorem. For let u be harmonic in Ω and $B = B_R(y) \subset\subset \Omega$. Since the gradient Du is also harmonic in Ω it follows by the mean value and divergence theorems that

$$Du(y) = \frac{1}{\omega_n R^n} \int_B Du\, dx = \frac{1}{\omega_n R^n} \int_{\partial B} u\nu\, ds,$$

$$|Du(y)| \leqslant \frac{n}{R} \sup_{\partial B} |u|$$

and hence

$$|Du(y)| \leqslant \frac{n}{d_y} \sup_{\Omega} |u|, \tag{2.31}$$

where $d_y = \operatorname{dist}(y, \partial\Omega)$. By successive application of the estimate (2.31) in equally spaced nested balls we obtain an estimate for higher order derivatives:

Theorem 2.10. *Let u be harmonic in Ω and let Ω' be any compact subset of Ω. Then for any multi-index α we have*

$$\sup_{\Omega'} |D^\alpha u| \leqslant \left(\frac{n|\alpha|}{d}\right)^{|\alpha|} \sup_{\Omega} |u| \tag{2.32}$$

where $d = \operatorname{dist}(\Omega', \partial\Omega)$.

An immediate consequence of the bound (2.32) is the equicontinuity on compact subdomains of the derivatives of any bounded set of harmonic functions. Consequently by Arzela's theorem, we see that any bounded set of harmonic functions forms a *normal family*; that is, we have:

Theorem 2.11. *Any bounded sequence of harmonic functions on a domain Ω contains a subsequence converging uniformly on compact subdomains of Ω to a harmonic function.*

The previous convergence theorem, Theorem 2.8, would also follow immediately from Theorem 2.11.

2.8. The Dirichlet Problem; the Method of Subharmonic Functions

We are in a position now to approach the question of existence of solutions of the classical Dirichlet problem in arbitrary bounded domains. The treatment here will be accomplished by *Perron's method of subharmonic functions* [PE] which relies heavily on the maximum principle and the solvability of the Dirichlet problem in balls. The method has a number of attractive features in that it is elementary, it separates the interior existence problem from that of the boundary behaviour of solutions, and it is easily extended to more general classes of second order elliptic equations. There are other well known approaches to existence theorems such as the method of integral equations, treated for example in the books [KE 2] [GU], and the variational or Hilbert space approach which we describe in a more general context in Chapter 8.

The definition of $C^2(\Omega)$ subharmonic and superharmonic function is generalized as follows. A $C^0(\Omega)$ function u will be called *subharmonic* (*superharmonic*) in Ω if for every ball $B \subset\subset \Omega$ and every function h harmonic in B satisfying $u \leqslant (\geqslant) h$ on ∂B, we also have $u \leqslant (\geqslant) h$ in B. The following properties of $C^0(\Omega)$ subharmonic functions are readily established:

(i) If u is subharmonic in a domain Ω, it satisfies the strong maximum principle in Ω; and if v is superharmonic in a bounded domain Ω with $v \geqslant u$ on $\partial\Omega$, then either $v > u$ throughout Ω or $v \equiv u$. To prove the latter assertion, suppose the contrary. Then at some point $x_0 \in \Omega$ we have

$$(u-v)(x_0) = \sup_{\Omega} (u-v) = M \geqslant 0,$$

and we may assume there is a ball $B=B(x_0)$ such that $u-v\not\equiv M$ on ∂B. Letting $\bar{u}, \bar{v}$ denote the harmonic functions respectively equal to u, v on ∂B (Theorem 2.6), one sees that

$$M \geq \sup_{\partial B} (\bar{u}-\bar{v}) \geq (\bar{u}-\bar{v})(x_0) \geq (u-v)(x_0)=M,$$

and hence the equality holds throughout. By the strong maximum principle for harmonic functions (Theorem 2.2) it follows that $\bar{u}-\bar{v}\equiv M$ in B and hence $u-v\equiv M$ on ∂B, which contradicts the choice of B.

(ii) Let u be subharmonic in Ω and B be a ball strictly contained in Ω. Denote by $\bar{u}$ the harmonic function in B (given by the Poisson integral of u on ∂B) satisfying $\bar{u}=u$ on ∂B. We define in Ω the *harmonic lifting* of u (in B) by

$$U(x)=\begin{cases}\bar{u}(x), & x\in B\\ u(x), & x\in \Omega-B.\end{cases} \tag{2.33}$$

Then the function U is also subharmonic in Ω. For consider an arbitrary ball $B'\subset\subset\Omega$ and let h be a harmonic function in B' satisfying $h\geq U$ on $\partial B'$. Since $u\leq U$ in B' we have $u\leq h$ in B' and hence $U\leq h$ in $B'-B$. Also since U is harmonic in B, we have by the maximum principle $U\leq h$ in $B\cap B'$. Consequently $U\leq h$ in B' and U is subharmonic in Ω.

(iii) Let $u_1, u_2, \ldots, u_N$ be subharmonic in Ω. Then the function $u(x)=\max\{u_1(x), \ldots, u_N(x)\}$ is also subharmonic in Ω. This is a trivial consequence of the definition of subharmonicity. Corresponding results for superharmonic functions are obtained by replacing u by $-u$ in properties (i), (ii) and (iii).

Now let Ω be bounded and φ be a bounded function on $\partial\Omega$. A $C^0(\bar{\Omega})$ subharmonic function u is called a *subfunction* relative to φ if it satisfies $u\leq\varphi$ on $\partial\Omega$. Similarly a $C^0(\bar{\Omega})$ superharmonic function is called a *superfunction* relative to φ if it satisfies $u\geq\varphi$ on $\partial\Omega$. By the maximum principle every subfunction is less than or equal to every superfunction. In particular, constant functions $\leq \inf_{\partial\Omega}\varphi$ ($\geq \sup_{\partial\Omega}\varphi$) are subfunctions (superfunctions). Let S_φ denote the set of subfunctions relative to φ. The basic result of the Perron method is contained in the following theorem.

Theorem 2.12. *The function* $u(x)=\sup_{v\in S_\varphi} v(x)$ *is harmonic in* Ω.

Proof. By the maximum principle any function $v\in S_\varphi$ satisfies $v\leq\sup\varphi$, so that u is well defined. Let y be an arbitrary fixed point of Ω. By the definition of u, there exists a sequence $\{v_n\}\subset S_\varphi$ such that $v_n(y)\to u(y)$. By replacing v_n with $\max(v_n, \inf\varphi)$, we may assume that the sequence $\{v_n\}$ is bounded. Now choose R so that the ball $B=B_R(y)\subset\subset\Omega$ and define V_n to be the harmonic lifting of v_n in B according to (2.33). Then $V_n\in S_\varphi$, $V_n(y)\to u(y)$ and by Theorem 2.11 the sequence $\{V_n\}$ contains a subsequence $\{V_{n_k}\}$ converging uniformly in any ball $B_\rho(y)$ with $\rho<R$ to a function v that is harmonic in B. Clearly $v\leq u$ in B and $v(y)=u(y)$. We claim now that in fact $v=u$ in B. For suppose $v(z)<u(z)$ at some $z\in B$. Then there exists

a function $\bar{u} \in S_\varphi$ such that $v(z) < \bar{u}(z)$. Defining $w_k = \max(\bar{u}, V_{n_k})$ and also the harmonic liftings W_k as in (2.33), we obtain as before a subsequence of the sequence $\{W_k\}$ converging to a harmonic function w satisfying $v \leqslant w \leqslant u$ in B and $v(y) = w(y) = u(y)$. But then by the maximum principle we must have $v = w$ in B. This contradicts the definition of $\bar{u}$ and hence u is harmonic in Ω. □

The preceding result exhibits a harmonic function which is a prospective solution (called the *Perron solution*) of the classical Dirichlet problem: $\Delta u = 0$, $u = \varphi$ on $\partial\Omega$. Indeed, if the Dirichlet problem is solvable, its solution is identical with the Perron solution. For let w be the presumed solution. Then clearly $w \in S_\varphi$ and by the maximum principle $w \geqslant u$ for all $u \in S_\varphi$. We note here also that the proof of Theorem 2.12 could have been based on the Harnack convergence theorem, Theorem 2.9, instead of the compactness theorem, Theorem 2.11; (see Problem 2.10).

In the Perron method the study of boundary behaviour of the solution is essentially separate from the existence problem. The continuous assumption of boundary values is connected to the geometric properties of the boundary through the concept of *barrier* function. Let ξ be a point of $\partial\Omega$. Then a $C^0(\bar{\Omega})$ function $w = w_\xi$ is called a *barrier* at ξ relative to Ω if:

(i) w is superharmonic in Ω;
(ii) $w > 0$ in $\bar{\Omega} - \xi$; $w(\xi) = 0$.

A more general definition of barrier requires only that the superharmonic function w be continuous and positive in Ω, and that $w(x) \to 0$ as $x \to \xi$. The results of this section are valid for these *weak* barriers as well (see [HL, p. 168], for example). An important feature of the barrier concept is that it is a local property of the boundary $\partial\Omega$. Namely, let us define w to be a *local barrier* at $\xi \in \partial\Omega$ if there is a neighborhood N of ξ such that w satisfies the above definition in $\Omega \cap N$. Then a barrier at ξ relative to Ω can be defined as follows. Let B be a ball satisfying $\xi \in B \subset\subset N$ and $m = \inf_{N-B} w > 0$. The function

$$\bar{w}(x) = \begin{cases} \min(m, w(x)), & x \in \bar{\Omega} \cap B \\ m, & x \in \bar{\Omega} - B \end{cases}$$

is then a barrier at ξ relative to Ω, as one sees by confirming properties (i) and (ii). Indeed, $\bar{w}$ is continuous in $\bar{\Omega}$ and is superharmonic in Ω by property (iii) of subharmonic functions; property (ii) is immediate.

A boundary point will be called *regular* (with respect to the Laplacian) if there exists a barrier at that point.

The connection between the barrier and boundary behavior of solutions is contained in the following.

Lemma 2.13. *Let u be the harmonic function defined in Ω by the Perron method (Theorem 2.12). If ξ is a regular boundary point of Ω and φ is continuous at ξ, then $u(x) \to \varphi(\xi)$ as $x \to \xi$.*

Proof. Choose $\varepsilon > 0$, and let $M = \sup|\varphi|$. Since ξ is a regular boundary point, there is a barrier w at ξ and, by virtue of the continuity of φ, there are constants δ and k such that $|\varphi(x) - \varphi(\xi)| < \varepsilon$ if $|x - \xi| < \delta$, and $kw(x) \geqslant 2M$ if $|x - \xi| \geqslant \delta$. The functions $\varphi(\xi) + \varepsilon + kw$, $\varphi(\xi) - \varepsilon - kw$ are respectively superfunction and subfunction relative to φ. Hence from the definition of u and the fact that every superfunction dominates every subfunction, we have in Ω,

$$\varphi(\xi) - \varepsilon - kw(x) \leqslant u(x) \leqslant \varphi(\xi) + \varepsilon + kw(x)$$

or

$$|u(x) - \varphi(\xi)| \leqslant \varepsilon + kw(x).$$

Since $w(x) \to 0$ as $x \to \xi$, we obtain $u(x) \to \varphi(\xi)$ as $x \to \xi$. □

This leads immediately to

Theorem 2.14. *The classical Dirichlet problem in a bounded domain is solvable for arbitrary continuous boundary values if and only if the boundary points are all regular.*

Proof. If the boundary values φ are continuous and the boundary $\partial\Omega$ consists of regular points, the preceding lemma states that the harmonic function provided by the Perron method solves the Dirichlet problem. Conversely, suppose that the Dirichlet problem is solvable for all continuous boundary values. Let $\xi \in \partial\Omega$. Then the function $\varphi(x) = |x - \xi|$ is continuous on $\partial\Omega$ and the harmonic function solving the Dirichlet problem in Ω with boundary values φ is obviously a barrier at ξ. Hence ξ is regular, as are all points of $\partial\Omega$. □

The important question remains: For what domains are the boundary points regular? It turns out that general sufficient conditions can be stated in terms of local geometric properties of the boundary. We mention some of these conditions below.

If $n = 2$, consider a boundary point z_0 of a bounded domain Ω and take the origin at z_0 with polar coordinates r, θ. Suppose there is a neighborhood N of z_0 such that a single valued branch of θ is defined in $\Omega \cap N$, or in a component of $\Omega \cap N$ having z_0 on its boundary. One sees that

$$w = -\mathrm{Re}\,\frac{1}{\log z} = -\frac{\log r}{\log^2 r + \theta^2}$$

is a (weak) local barrier at z_0 and hence z_0 is a regular point. In particular, z_0 is a regular boundary point if it is the endpoint of a simple arc lying in the exterior of Ω. Thus the Dirichlet problem in the plane is always solvable for continuous boundary values in a (bounded) domain whose boundary points are each accessible from the exterior by a simple arc. More generally, the same barrier shows that the boundary value problem is solvable if every component of the complement of the domain consists of more than a single point. Examples of such domains are domains bounded by a finite number of simple closed curves. Another is the unit disc slit along an arc; in this case the boundary values can be assigned on opposite sides of the slit.

For higher dimensions the situation is substantially different and the Dirichlet problem cannot be solved in corresponding generality. Thus, an example due to Lebesgue shows that a closed surface in three dimensions with a sufficiently sharp inward directed cusp has a non-regular point at the tip of the cusp; (see for example [CH]).

A simple sufficient condition for solvability in a bounded domain $\Omega\subset\mathbb{R}^n$ is that Ω satisfy the *exterior sphere condition*; that is, for every point $\xi\in\partial\Omega$, there exists a ball $B=B_R(y)$ satisfying $\bar{B}\cap\bar{\Omega}=\xi$. If such a condition is fulfilled, then the function w given by

$$w(x)=\begin{cases} R^{2-n}-|x-y|^{2-n} & \text{for } n\geqslant 3\\ \log\dfrac{|x-y|}{R} & \text{for } n=2\end{cases} \tag{2.34}$$

will be a barrier at ξ. Consequently the boundary points of a domain with C^2 boundary are all regular points; (see Problem 2.11).

2.9. Capacity

The physical concept of *capacity* provides another means of characterizing regular and exceptional boundary points. Let Ω be a bounded domain in $\mathbb{R}^n (n\geqslant 3)$ with smooth boundary $\partial\Omega$, and let u be the harmonic function (often called the *conductor potential*) defined in the complement of $\bar{\Omega}$ and satisfying the boundary conditions $u=1$ on $\partial\Omega$ and $u=0$ at infinity. The existence of u is easily established as the (unique) limit of harmonic functions u' in an expanding sequence of bounded domains having $\partial\Omega$ as an inner boundary (on which $u'=1$) and with outer boundaries (on which $u'=0$) tending to infinity. If Σ denotes $\partial\Omega$ or any smooth closed surface enclosing Ω, then the quantity

$$\operatorname{cap}\Omega=-\int_\Sigma\frac{\partial u}{\partial\nu}\,ds=\int_{\mathbb{R}^n-\Omega}|Du|^2\,dx\qquad \nu=\text{outer normal} \tag{2.35}$$

is defined to be the capacity of Ω. In electrostatics, cap Ω is within a constant factor the total electric charge on the conductor $\partial\Omega$ held at unit potential (relative to infinity).

Capacity can also be defined for domains with nonsmooth boundaries and for any compact set as the (unique) limit of the capacities of a nested sequence of approximating smoothly bounded domains. Equivalent definitions of capacity can be given directly without use of approximating domains (e.g., see [LK]). In particular, we have the variational characterization

$$\operatorname{cap}\Omega=\inf_{v\in K}\int|Dv|^2. \tag{2.36}$$

where

$$K = \{v \in C_0^1(\mathbb{R}^n) \,|\, v = 1 \text{ on } \Omega\}.$$

To investigate the regularity of a point $x_0 \in \partial\Omega$, consider for any fixed $\lambda \in (0, 1)$ the capacity

$$C_j = \text{cap}\,\{x \notin \Omega \,|\, |x - x_0| \leqslant \lambda^j\}.$$

The *Wiener criterion* states that x_0 is a regular boundary point of Ω if and only if the series

$$\sum_{j=0}^{\infty} C_j/\lambda^{j(n-2)} \tag{2.37}$$

diverges.

For a discussion of capacity and proof of the Wiener criterion we refer to the literature, e.g., [KE 2, LK]. In Chapter 8 this condition for regularity will be proved for a general class of elliptic operators in divergence form.

Problems

2.1. Derive the weak maximum principle for $C^2(\Omega)$ subharmonic functions from a consideration of necessary conditions for a relative maximum.

2.2. Prove that if $\Delta u = 0$ in Ω and $u = \partial u/\partial \nu = 0$ on an open, smooth portion of $\partial\Omega$, then u is identically zero.

2.3. Let G be the Green's function for a bounded domain Ω. Prove

a) $G(x, y) = G(y, x)$ for all $x, y \in \Omega$, $x \neq y$;

b) $G(x, y) < 0$ for all $x, y \in \Omega$, $x \neq y$;

c) $\int_\Omega G(x, y) f(y)\, dy \to 0$ as $x \to \partial\Omega$, if f is bounded and integrable on Ω.

2.4. (*Schwarz reflection principle.*) Let Ω^+ be a subdomain of the half-space $x_n > 0$ having as part of its boundary an open section T of the hyperplane $x_n = 0$. Suppose that u is harmonic in Ω^+, continuous in $\Omega^+ \cup T$, and that $u = 0$ on T. Show that the function U defined by

$$U(x_1, \ldots, x_n) = \begin{cases} u(x_1, \ldots, x_n), & x_n \geqslant 0 \\ -u(x_1, \ldots, -x_n), & x_n < 0 \end{cases}$$

is harmonic in the domain $\Omega^+ \cup T \cup \Omega^-$, where Ω^- is the reflection of Ω^+ in $x_n=0$ (*i.e.*, $\Omega^-=\{(x_1,\ldots,x_n)\in\mathbb{R}^n \mid (x_1,\ldots,-x_n)\in\Omega^+\}$).

2.5. Determine the Green's function for the annular region bounded by two concentric spheres in $\mathbb{R}^n$.

2.6. Let u be a non-negative harmonic function in a ball $B_R(0)$. Deduce from the Poisson integral formula, the following version of *Harnack's inequality*

$$\frac{R^{n-2}(R-|x|)}{(R+|x|)^{n-1}}u(0)\leqslant u(x)\leqslant\frac{R^{n-2}(R+|x|)}{(R-|x|)^{n-1}}u(0).$$

2.7. Show that a $C^0(\Omega)$ function u is subharmonic in Ω if and only if it satisfies the mean value inequality locally; that is, for every $y\in\Omega$ there exists $\delta=\delta(y)>0$ such that

$$u(y)\leqslant\frac{1}{n\omega_n R^{n-1}}\int_{\partial B_R(y)} u\,ds \quad \text{for all } R\leqslant\delta.$$

2.8. An integrable function u in a domain Ω is called *weakly harmonic* (*subharmonic, superharmonic*) in Ω if

$$\int_\Omega u\,\Delta\varphi\,dx=(\geqslant,\leqslant)\,0$$

for all functions $\varphi\geqslant0$ in $C^2(\Omega)$ having compact support in Ω. Show that a $C^0(\Omega)$ weakly harmonic (subharmonic, superharmonic) function is harmonic (subharmonic, superharmonic).

2.9. Show that for $C^2(\Omega)$ functions u, the conditions: (i) $\Delta u\geqslant0$ in Ω; (ii) u is subharmonic in Ω; (iii) u is weakly subharmonic in Ω, are equivalent.

2.10. Prove Theorem 2.12 using Theorem 2.9 instead of Theorem 2.11.

2.11. Show that a domain Ω with C^2 boundary $\partial\Omega$ satisfies an exterior sphere condition.

2.12. Show that the Dirichlet problem is solvable for any domain Ω satisfying an *exterior cone condition*; that is, for every point $\xi\in\partial\Omega$ there exists a finite right circular cone K, with vertex ξ, satisfying $\bar{K}\cap\bar{\Omega}=\xi$. At each point $\xi\in\partial\Omega$ taken as origin, show that a suitable local barrier can be chosen in the form $w=r^\lambda f(\theta)$ where θ is the polar angle.

2.13. Let u be harmonic in $\Omega\subset\mathbb{R}^n$. Use the argument leading to (2.31) to prove the interior gradient bound,

$$|Du(x_0)|\leqslant\frac{n}{d_0}[\sup_\Omega u-u(x_0)],\quad d_0=\operatorname{dist}(x_0,\partial\Omega).$$

If $u \geqslant 0$ in Ω infer that

$$|Du(x_0)| \leqslant \frac{n}{d_0} u(x_0).$$

2.14. (a) Prove *Liouville's theorem*: A harmonic function defined over $\mathbb{R}^n$ and bounded above is constant.

(b) If $n=2$ prove that the Liouville theorem in part (a) is valid for *subharmonic* functions.

(c) If $n>2$ show that a bounded subharmonic function defined over $\mathbb{R}^n$ need not be constant.

2.15. Let $u \in C^2(\bar{\Omega})$, $u = 0$ on $\partial\Omega \in C^1$. Prove the interpolation inequality: For every $\varepsilon > 0$,

$$\int_\Omega |Du|^2 \, dx \leqslant \varepsilon \int_\Omega (\Delta u)^2 \, dx + \frac{1}{4\varepsilon} \int_\Omega u^2 \, dx.$$

2.16. Prove Theorem 2.12 by finding in every ball $B \subset\subset \Omega$ a monotone increasing sequence of harmonic functions that are restrictions of subfunctions on B and that converge uniformly to u on a dense set of points in B. Hence show that Theorems 2.12 and 2.14 can be proved without use of the strong maximum principle.

2.17. Show that the volume integral in (2.35) is defined, and prove the equivalence of the capacity definitions (2.35) and (2.36).

2.18. Let u be harmonic in (open, connected) $\Omega \subset \mathbb{R}^n$, and suppose $B_c(x_0) \subset\subset \Omega$. If $a \leqslant b \leqslant c$, where $b^2 = ac$, show that

$$\int_{|\omega|=1} u(x_0 + a\omega)u(x_0 + c\omega) \, d\omega = \int_{|\omega|=1} u^2(x_0 + b\omega) \, d\omega.$$

Hence, conclude that if u is constant in a neighbourhood it is identically constant. (Cf. [GN].)

Chapter 3

The Classical Maximum Principle

The purpose of this chapter is to extend the classical maximum principles for the Laplace operator, derived in Chapter 2, to linear elliptic differential operators of the form

$$Lu = a^{ij}(x)D_{ij}u + b^i(x)D_iu + c(x)u, \quad a^{ij} = a^{ji}, \tag{3.1}$$

where $x = (x_1, \ldots, x_n)$ lies in a domain Ω of $\mathbb{R}^n$, $n \geqslant 2$. It will be assumed, unless otherwise stated, that u belongs to $C^2(\Omega)$. The summation convention that repeated indices indicate summation from 1 to n is followed here as it will be throughout. L will always denote the operator (3.1).

We adopt the following definitions: L is *elliptic* at a point $x \in \Omega$ if the coefficient matrix $[a^{ij}(x)]$ is positive; that is, if $\lambda(x)$, $\Lambda(x)$ denote respectively the minimum and maximum eigenvalues of $[a^{ij}(x)]$, then

$$0 < \lambda(x)|\xi|^2 \leqslant a^{ij}(x)\xi_i\xi_j \leqslant \Lambda(x)|\xi|^2 \tag{3.2}$$

for all $\xi = (\xi_1, \ldots, \xi_n) \in \mathbb{R}^n - \{0\}$. If $\lambda > 0$ in Ω, then L is elliptic in Ω, and *strictly elliptic* if $\lambda \geqslant \lambda_0 > 0$ for some constant λ_0. If Λ/λ is bounded in Ω, we shall call L *uniformly elliptic* in Ω. Thus the operator $D_{11} + x_1 D_{22}$ is elliptic but not uniformly elliptic in the half plane, $x_1 > 0$, while it is uniformly elliptic in strips of the form $(\alpha, \beta) \times \mathbb{R}$ where $0 < \alpha < \beta < \infty$.

Most results concerning elliptic operators of the form (3.1) require additional conditions limiting the relative importance of the lower order terms b^iD_iu, cu with respect to the principal term $a^{ij}D_{ij}u$. The condition

$$\frac{|b^i(x)|}{\lambda(x)} \leqslant \text{const} < \infty, \quad i = 1, \ldots, n, \quad x \in \Omega \tag{3.3}$$

will be assumed throughout this chapter. By then considering $L' = \lambda^{-1} L$ in place of L we can reduce to the case where $\lambda = 1$ and the b^i are bounded. If, in addition, L is uniformly elliptic, we can also take the a^{ij} to be bounded. Note that if the coefficients a^{ij}, b^i are continuous in Ω, then on any bounded subdomain $\Omega' \subset\subset \Omega$, L is uniformly elliptic and (3.3) holds. The coefficient c will also be subject to restrictions but these will vary and consequently be indicated in the appropriate hypotheses.

The maximum principle is an important feature of second order elliptic equations that distinguishes them from equations of higher order and systems of equations. In addition to its many applications, the maximum principle provides pointwise estimates that lead to a more definitive theory than would be otherwise available. In this chapter, most of the results will be based solely on the ellipticity of L and not on other special properties of the coefficients (such as smoothness). It is this generality which makes the maximum principle useful in apriori estimation of solutions, especially in nonlinear problems.

3.1. The Weak Maximum Principle

For many purposes it suffices to have the following *weak maximum principle.*

Theorem 3.1. *Let L be elliptic in the bounded domain Ω. Suppose that*

$$Lu \geqslant 0 \ (\leqslant 0) \text{ in } \Omega, \qquad c=0 \text{ in } \Omega, \tag{3.4}$$

with $u \in C^2(\Omega) \cap C^0(\bar{\Omega})$. Then the maximum (minimum) of u in $\bar{\Omega}$ is achieved on $\partial\Omega$, that is,

$$\sup_{\Omega} u = \sup_{\partial\Omega} u \ (\inf_{\Omega} u = \inf_{\partial\Omega} u). \tag{3.5}$$

It is apparent that the conclusion remains valid if $|b^i|/\lambda$ is only locally bounded in Ω, for example if $a^{ij}, b^i \in C^0(\Omega)$. Also, if u is not assumed continuous in $\bar{\Omega}$, the conclusion (3.5) can be replaced by

$$\sup_{\Omega} u = \limsup_{x\to\partial\Omega} u(x) \ (\inf_{\Omega} u = \liminf_{x\to\partial\Omega} u(x)) \tag{3.6}$$

Proof. It is readily seen that if $Lu>0$ in Ω, then a strong maximum principle holds; that is, u cannot achieve an interior maximum in $\bar{\Omega}$. For at such a point x_0, $Du(x_0)=0$ and the matrix $D^2u(x_0)=[D_{ij}u(x_0)]$ is nonpositive. But the matrix $[a^{ij}(x_0)]$ is positive since L is elliptic. Consequently $Lu(x_0)=a^{ij}(x_0)D_{ij}u(x_0)\leqslant 0$, contradicting $Lu>0$. (Note that only the semi-definiteness of the coefficient matrix $[a_{ij}]$ is needed in this argument.)

By hypothesis (3.3), $|b^i|/\lambda \leqslant b_0=$constant. Then since $a^{11}\geqslant\lambda$, there is a sufficiently large constant γ for which

$$L\, e^{\gamma x_1}=(\gamma^2 a^{11}+\gamma b^1)\, e^{\gamma x_1}\geqslant \lambda(\gamma^2-\gamma b_0)\, e^{\gamma x_1}>0.$$

Hence for any $\varepsilon>0$, $L(u+\varepsilon\, e^{\gamma x_1})>0$ in Ω so that

$$\sup_{\Omega}(u+\varepsilon\, e^{\gamma x_1})=\sup_{\partial\Omega}(u+\varepsilon\, e^{\gamma x_1})$$

by the above. Letting $\varepsilon\to 0$, we see that $\sup_{\Omega} u=\sup_{\partial\Omega} u$, as asserted in the theorem. □

Remark. It is clear from the proof that the theorem holds under the weaker hypothesis that the coefficient matrix $[a^{ij}]$ is non-negative and that for some k the ratio $|b^k|/a^{kk}$ is locally bounded.

It is convenient to introduce the following terminology suggested by the maximum principle: a function satisfying $Lu=0$ ($\geqslant 0$, $\leqslant 0$) in Ω is a *solution* (*subsolution*, *supersolution*) of $Lu=0$ in Ω. When L is the Laplacian, these terms correspond respectively to harmonic, subharmonic and superharmonic functions.

Let us suppose more generally that $c \leqslant 0$ in Ω. By considering the subset $\Omega^+ \subset \Omega$ in which $u>0$, one sees that if $Lu \geqslant 0$ in Ω, then $L_0 u = a^{ij}D_{ij}u + b^i D_i u \geqslant -cu \geqslant 0$ in Ω^+ and hence the maximum of u on $\bar{\Omega}^+$ must be achieved on $\partial\bar{\Omega}^+$ and hence also on $\partial\Omega$. Thus, writing $u^+ = \max(u, 0)$, $u^- = \min(u, 0)$ we obtain:

Corollary 3.2. *Let L be elliptic in the bounded domain Ω. Suppose that in Ω*

$$Lu \geqslant 0 \ (\leqslant 0), \qquad c \leqslant 0, \tag{3.7}$$

with $u \in C^0(\bar{\Omega})$. Then

$$\sup_{\Omega} u \leqslant \sup_{\partial\Omega} u^+ \quad (\inf_{\Omega} u \geqslant \inf_{\partial\Omega} u^-). \tag{3.8}$$

If $Lu=0$ in Ω, then

$$\sup_{\Omega} |u| = \sup_{\partial\Omega} |u|. \tag{3.9}$$

In this corollary, the condition $c \leqslant 0$ cannot be relaxed in general to allow $c>0$, as is evident from the existence of positive eigenvalues κ for the problem: $\Delta u + \kappa u = 0$, $u=0$ on $\partial\Omega$. An immediate and important application of the weak maximum principle is to the problem of uniqueness and continuous dependence of solutions on their boundary values. From Corollary 3.2 follows automatically a uniqueness result for the classical Dirichlet problem for operators L, and a comparison principle, which is the typical form of application of the corollary.

Theorem 3.3. *Let L be elliptic in Ω with $c \leqslant 0$ in Ω. Suppose that u and v are functions in $C^2(\Omega) \cap C^0(\bar{\Omega})$ satisfying $Lu = Lv$ in Ω, $u = v$ on $\partial\Omega$. Then $u = v$ in Ω. If $Lu \geqslant Lv$ in Ω and $u \leqslant v$ on $\partial\Omega$, then $u \leqslant v$ in Ω.*

3.2. The Strong Maximum Principle

Although the weak maximum principle suffices for most applications, it is often necessary to have the strong form which excludes the assumption of a non-trivial interior maximum. We shall obtain such a result for locally uniformly elliptic operators by means of the following frequently useful boundary point lemma. The domain Ω is said to satisfy an *interior sphere condition* at $x_0 \in \partial\Omega$ if there exists a ball $B \subset \Omega$ with $x_0 \in \partial B$, (that is, the complement of Ω satisfies an exterior sphere condition at x_0).

Lemma 3.4. *Suppose that L is uniformly elliptic, $c=0$ and $Lu \geqslant 0$ in Ω. Let $x_0 \in \partial\Omega$ be such that*

(i) *u is continuous at x_0;*
(ii) *$u(x_0) > u(x)$ for all $x \in \Omega$;*
(iii) *$\partial\Omega$ satisfies an interior sphere condition at x_0.*

Then the outer normal derivative of u at x_0, if it exists, satisfies the strict inequality

$$\frac{\partial u}{\partial \nu}(x_0) > 0. \tag{3.10}$$

If $c \leqslant 0$ and c/λ is bounded, the same conclusion holds provided $u(x_0) \geqslant 0$, and if $u(x_0) = 0$ the same conclusion holds irrespective of the sign of c.

Proof. Since Ω satisfies an interior sphere condition at x_0, there exists a ball $B = B_R(y) \subset \Omega$ with $x_0 \in \partial B$. For $0 < \rho < R$, we introduce an auxiliary function v by defining

$$v(x) = e^{-\alpha r^2} - e^{-\alpha R^2}$$

where $r = |x - y| > \rho$ and α is a positive constant yet to be determined. Direct calculation gives for $c \leqslant 0$

$$\begin{aligned} Lv(x) &= e^{-\alpha r^2}\,[4\alpha^2 a^{ij}(x_i - y_i)(x_j - y_j) - 2\alpha(a^{ii} + b^i(x_i - y_i))] + cv \\ &\geqslant e^{-\alpha r^2}\,[4\alpha^2 \lambda(x) r^2 - 2\alpha(a^{ii} + |\mathbf{b}| r) + c], \quad \mathbf{b} = (b^1, \ldots, b^n). \end{aligned}$$

By assumption a^{ii}/λ, $|\mathbf{b}|/\lambda$, and c/λ are bounded. Hence α may be chosen large enough so that $Lv \geqslant 0$ throughout the annular region $A = B_R(y) - B_\rho(y)$. Since $u - u(x_0) < 0$ on $\partial B_\rho(y)$ there is a constant $\varepsilon > 0$ for which $u - u(x_0) + \varepsilon v \leqslant 0$ on $\partial B_\rho(y)$. This inequality is also satisfied on $\partial B_R(y)$ where $v = 0$. Thus we have $L(u - u(x_0) + \varepsilon v) \geqslant -cu(x_0) \geqslant 0$ in A, and $u - u(x_0) + \varepsilon v \leqslant 0$ on ∂A. The weak maximum principle (Corollary 3.2) now implies that $u - u(x_0) + \varepsilon v \leqslant 0$ throughout A. Taking the normal derivative at x_0, we obtain, as required,

$$\frac{\partial u}{\partial \nu}(x_0) \geqslant -\varepsilon \frac{\partial v}{\partial \nu}(x_0) = -\varepsilon\, v'(R) > 0.$$

For c of arbitrary sign, if $u(x_0) = 0$ the preceding argument remains valid if L is replaced everywhere by $L - c^+$. □

More generally, whether or not the normal derivative exists, we get

$$\liminf_{x \to x_0} \frac{u(x_0) - u(x)}{|x - x_0|} > 0, \tag{3.11}$$

where the angle between the vector $x_0 - x$ and the normal at x_0 is less than $\pi/2 - \delta$ for some fixed $\delta > 0$.

Although the interior sphere condition can be relaxed somewhat, it is not possible to assert (3.11) without suitable smoothness of $\partial\Omega$ at x_0. For example, let $L = \Delta$ and $\Omega \subset \mathbb{R}^2$ be the region in the right-half-plane in which $u = \text{Re}\,(z/\log z) < 0$. An elementary calculation shows that $\partial\Omega \subset C^1$ near $z = 0$ and $u_x(0, 0) = 0$, so that (3.11) is false.

We are now in a position to derive the following *strong maximum principle* of E. Hopf [HO 1].

Theorem 3.5. *Let L be uniformly elliptic, $c=0$ and $Lu \geqslant 0 (\leqslant 0)$ in a domain Ω (not necessarily bounded). Then if u achieves its maximum (minimum) in the interior of Ω, u is a constant. If $c \leqslant 0$ and c/λ is bounded, then u cannot achieve a non-negative maximum (non-positive minimum) in the interior of Ω unless it is constant.*

The conclusion obviously remains valid if L is only locally uniformly elliptic and $|b^i|/\lambda$, c/λ are only locally bounded.

Proof. If we assume, contrary to the theorem, that u is non-constant and achieves its maximum $M \geqslant 0$ in the interior of Ω, then the set Ω^- on which $u < M$ satisfies $\Omega^- \subset \Omega$ and $\partial\Omega^- \cap \Omega \neq \phi$. Let x_0 be a point in Ω^- that is closer to $\partial\Omega^-$ than to $\partial\Omega$, and consider the largest ball $B \subset \Omega^-$ having x_0 as center. Then $u(y) = M$ for some point $y \in \partial B$ while $u < M$ in B. The preceding lemma implies that $Du(y) \neq 0$, which is impossible at the interior maximum y. □

If $c < 0$ at some point, then the constant of the theorem is obviously zero. Also, if $u = 0$ at an interior maximum (minimum), then it follows from the proof of the theorem that $u \equiv 0$, irrespective of the sign of c.

It is of course possible to prove the strong maximum principle directly without going through Theorem 3.1 and Lemma 3.4; (see [MR 2] for example).

Uniqueness theorems for other types of boundary value problems are consequences of Lemma 3.4 and Theorem 3.5. In particular, we have the following uniqueness theorem for the classical Neumann problem.

Theorem 3.6. *Let $u \in C^2(\Omega) \cap C^0(\bar{\Omega})$ be a solution of $Lu = 0$ in the bounded domain Ω, where L is uniformly elliptic, $c \leqslant 0$, c/λ is bounded and Ω satisfies an interior sphere condition at each point of $\partial\Omega$. If the normal derivative is defined everywhere on $\partial\Omega$ and $\partial u/\partial\nu = 0$ on $\partial\Omega$, then u is constant in Ω. If, also, $c < 0$ at some point in Ω, then $u \equiv 0$.*

Proof. If $u \neq$ const., we may assume that either of the functions u or $-u$ achieves a non-negative maximum M at a point x_0 on $\partial\Omega$ and is less than M in Ω (by the strong maximum principle). Applying Lemma 3.4 at x_0 we infer that $\partial u/\partial\nu(x_0) \neq 0$, contradicting the hypothesis. □

The result of Theorem 3.6 may also be generalized to mixed boundary value and oblique derivative problems; (see Problem 3.1). When $\partial\Omega$ has corners or edges where the derivatives of u are not defined, these results are false in the stated generality, even if u is assumed continuous on $\bar{\Omega}$; (see Problem 3.8(a)).

3.3. Apriori Bounds

The maximum principle also provides a simple pointwise estimate for solutions of the inhomogeneous equation $Lu=f$ in bounded domains. We remark that only the ellipticity and bounds on the coefficients are involved. This proves to be an important consideration in nonlinear problems.

Theorem 3.7. *Let $Lu \geqslant f\ (=f)$ in a bounded domain Ω, where L is elliptic, $c \leqslant 0$, and $u \in C^0(\bar{\Omega}) \cap C^2(\Omega)$. Then*

$$\sup_{\Omega} u(|u|) \leqslant \sup_{\partial\Omega} u^+(|u|) + C \sup_{\Omega} \frac{|f^-|}{\lambda}\left(\frac{|f|}{\lambda}\right), \tag{3.12}$$

where C is a constant depending only on diam Ω *and $\beta=\sup|\mathbf{b}|/\lambda$. In particular, if Ω lies between two parallel planes a distance d apart, then* (3.12) *is satisfied with $C=e^{(\beta+1)d}-1$.*

Proof. Let Ω lie in the slab $0<x_1<d$, and set $L_0=a^{ij}D_{ij}+b^iD_i$. For $\alpha \geqslant \beta+1$ we have

$$L_0\, e^{\alpha x_1}=(\alpha^2 a^{11}+\alpha b^1)\, e^{\alpha x_1} \geqslant \lambda(\alpha^2-\alpha\beta)\, e^{\alpha x_1} \geqslant \lambda.$$

Let

$$v=\sup_{\partial\Omega} u^+ + (e^{\alpha d}-e^{\alpha x_1}) \sup_{\Omega} \frac{|f^-|}{\lambda}.$$

Then, since $Lv=L_0 v+cv \leqslant -\lambda \sup_{\partial\Omega}(|f^-|/\lambda)$,

$$L(v-u) \leqslant -\lambda\left(\sup_{\Omega}\frac{|f^-|}{\lambda}+\frac{f}{\lambda}\right) \leqslant 0 \text{ in } \Omega,$$

and $v-u \geqslant 0$ on $\partial\Omega$. Hence, for $C=e^{\alpha d}-1$ and $\alpha \geqslant \beta+1$, we obtain the desired result for the case $Lu \geqslant f$,

$$\sup_{\Omega} u \leqslant \sup_{\Omega} v \leqslant \sup_{\partial\Omega} u^+ + C \sup_{\Omega} \frac{|f^-|}{\lambda}.$$

Replacing u by $-u$, we obtain (3.12) for the case $Lu=f$. □

Theorem 3.7 will be strengthened in Chapters 8 and 9 to provide analogous estimates for sup u in terms of integral norms of f.

When the condition $c \leqslant 0$ is not satisfied, it is still possible to assert an apriori bound analogous to (3.12) provided the domain Ω lies between sufficiently close parallel planes.

Corollary 3.8. *Let $Lu=f$ in a bounded domain Ω, where L is elliptic and $u \in C^0(\bar{\Omega}) \cap C^2(\Omega)$. Let C be the constant of Theorem 3.7 and suppose that*

$$C_1 = 1 - C \sup_{\Omega} \frac{c^+}{\lambda} > 0. \tag{3.13}$$

Then

$$\sup_{\Omega} |u| \leqslant \frac{1}{C_1}\left(\sup_{\partial\Omega} |u| + C \sup_{\Omega} \frac{|f|}{\lambda}\right). \tag{3.14}$$

Remark. Since $C = e^{(\beta+1)d} - 1$ is a possible value of the constant in (3.12), where d is the width of any slab containing Ω, condition (3.13) will be satisfied in any sufficiently narrow domain in which the quantities $|\mathbf{b}|/\lambda$ and c/λ are bounded above. If $c^+ \equiv 0$ (i.e., $c \leqslant 0$), then $C_1 = 1$ and (3.14) reduces to (3.12).

Proof of Corollary 3.8. Let us rewrite $Lu=(L_0+c)u=f$ in the form

$$(L_0+c^-)u = f' \equiv f + (c^- - c)u = f - c^+ u.$$

From (3.12) we obtain

$$\begin{aligned}\sup_{\Omega} |u| &\leqslant \sup_{\partial\Omega} |u| + C \sup_{\Omega} \frac{|f'|}{\lambda}\\ &\leqslant \sup_{\partial\Omega} |u| + C\left(\sup_{\Omega} \frac{|f|}{\lambda} + \sup_{\Omega} |u| \sup_{\Omega} \frac{c^+}{\lambda}\right).\end{aligned}$$

This inequality and (3.13) imply (3.14). □

An immediate consequence of Corollary 3.8 is uniqueness for solutions of the Dirichlet problem in sufficiently small domains (assuming of course fixed upper bounds on the quantities $|\mathbf{b}|/\lambda$ and c/λ).

3.4. Gradient Estimates for Poisson's Equation

The maximum principle can also be used to derive estimates on derivatives of solutions provided additional conditions are placed on the equation. To illustrate the method we obtain such estimates for Poisson's equation. The results derived here will not be required for later developments.

Let $\Delta u = f$ in the cube $Q = \{x = (x_1, \ldots, x_n) \in \mathbb{R}^n \mid |x_i| < d,\ i = 1, \ldots, n\}$, with $u \in C^2(Q) \cap C^0(\bar{Q})$ and f bounded in Q. By means of a comparison argument we shall derive the estimate

$$|D_i u(0)| \leqslant \frac{n}{d} \sup_{\partial Q} |u| + \frac{d}{2} \sup_{Q} |f|, \quad i = 1, \ldots, n. \tag{3.15}$$

In the half-cube

$$Q'=\{(x_1,\ldots,x_n) \mid |x_i|<d,\ i=1,\ldots,n-1,\ 0<x_n<d\},$$

consider the function

$$\varphi(x',x_n)=\tfrac{1}{2}[u(x',x_n)-u(x',-x_n)],$$

where we write $x'=(x_1,\ldots,x_{n-1})$ and $x=(x',x_n)$. One sees that $\varphi(x',0)=0$, $\sup_{\partial Q'}|\varphi|\leqslant M=\sup_{\partial Q}|u|$, and $|\Delta\varphi|\leqslant N=\sup_{Q}|f|$ in Q'. Consider also the function

$$\psi(x',x_n)=\frac{M}{d^2}[|x'|^2+x_n(nd-(n-1)x_n)]+N\frac{x_n}{2}(d-x_n).$$

Obviously $\psi(x',x_n)\geqslant 0$ on $x_n=0$ and $\psi\geqslant M$ on the remaining portion of $\partial Q'$; also $\Delta\psi=-N$. Hence $\Delta(\psi\pm\varphi)\leqslant 0$ in Q' and $\psi\pm\varphi\geqslant 0$ on $\partial Q'$, from which it follows by the maximum principle that $|\varphi(x',x_n)|\leqslant\psi(x',x_n)$ in Q'. Letting $x'=0$ in the expressions for φ and ψ, then dividing by x_n and letting x_n tend to zero, we obtain

$$|D_n u(0)|=\lim_{x_n\to 0}\left|\frac{\varphi(0,x_n)}{x_n}\right|\leqslant\frac{n}{d}M+\frac{d}{2}N,$$

which is the asserted estimate (3.15) for $i=n$. The result follows in the corresponding way for $i=1,\ldots,n-1$. If $f=0$, (3.15) provides an independent proof of (essentially) the gradient bound (2.31) for harmonic functions.

From (3.15) we infer that in any domain Ω a bounded solution u of $\Delta u=f$ satisfies an estimate

$$\sup_{\Omega} d_x|Du(x)|\leqslant C\,(\sup_{\Omega}|u|+\sup_{\Omega} d_x^2|f(x)|), \tag{3.16}$$

where $d_x=\operatorname{dist}(x,\partial\Omega)$ and $C=C(n)$. For if $x\in\Omega$ and Q is a cube of side $d=d_x/\sqrt{n}$ with its center at x, we have from (3.15) the inequality

$$\begin{aligned} d_x|Du(x)| &\leqslant C(\sup_{\partial Q}|u|+d^2\sup_{Q}|f|)\\ &\leqslant C(\sup_{\Omega}|u|+\sup_{\Omega} d_y^2|f(y)|). \end{aligned}$$

(Here we have used the same letter C to denote constants depending only on n.)

In the same generality as above, we now derive by a similar comparison argument an estimate of the modulus of continuity of the gradient of solutions of Poisson's equation.

Again let $u \in C^2(Q) \cap C^0(\bar{Q})$ be a solution of $\Delta u = f$ in the cube Q, and set $M = \sup_Q |u|$, $N = \sup_Q |f|$. Let Q' be the domain in $\mathbb{R}^{n+1}$ given by

$$Q' = \{(x_1, \ldots, x_{n-1}, y, z) \mid |x_i| < d/2,\ i = 1, \ldots, n-1,\ 0 < y, z < d/4\},$$

and let us define in Q' the function

$$\varphi(x', y, z) = \tfrac{1}{4}[u(x', y+z) - u(x', y-z) - u(x', -y+z) + u(x', -y-z)].$$

Introducing the elliptic operator

$$L \equiv \sum_{i=1}^{n-1} \frac{\partial^2}{\partial x_i^2} + \frac{1}{2}\frac{\partial^2}{\partial y^2} + \frac{1}{2}\frac{\partial^2}{\partial z^2}$$

in the $n+1$ variables $x_1, \ldots, x_{n-1}, y, z$, we see that $|L\varphi| \leqslant N$ in Q'. Also, on Q' we have: (i) $\varphi(x', 0, z) = \varphi(x', y, 0) = 0$; (ii) $|\varphi| \leqslant M$ on $|x_i| = d/2$, $i = 1, \ldots, n-1$; (iii) $|\varphi(x', d/4, z)| \leqslant \mu z$ and $|\varphi(x', y, d/4)| \leqslant \mu y$, where $|Du| \leqslant \mu$ in Q', μ being given in terms of M and N by (3.16). We now choose a comparison function in Q' of the form,

$$\psi(x', y, z) = \frac{4M|x'|^2}{d^2} + \frac{4\mu}{d} yz + kyz \log \frac{2d}{y+z}, \tag{3.17}$$

where k is a positive constant yet to be determined. We observe first that $|\varphi| \leqslant \psi$ on $\partial Q'$. Since

$$L\psi = \frac{8(n-1)}{d^2} M + k\left(-1 + \frac{yz}{(y+z)^2}\right) \leqslant \frac{8(n-1)M}{d^2} - \frac{3}{4}k,$$

we see that $L\psi \leqslant -N$ provided

$$k \geqslant \tfrac{4}{3}(N + 8(n-1)M/d^2).$$

With such a choice of k, the function

$$\psi(x', y, z) = \frac{4M|x'|^2}{d^2} + yz\left(\frac{4\mu}{d} + k \log \frac{2d}{y+z}\right)$$

satisfies the conditions, $L(\psi \pm \varphi) \leqslant 0$ in Q', $\psi \pm \varphi \geqslant 0$ on $\partial Q'$. Accordingly, $|\varphi| \leqslant \psi$ in Q'. Letting $x' = 0$ in this inequality, then dividing by z and letting z tend to zero, we obtain

$$\tfrac{1}{2}|u_y(0, y) - u_y(0, -y)| = \lim_{z \to 0} \frac{|\varphi(0, y, z)|}{z} \leqslant \frac{4\mu}{d} y + ky \log \frac{2d}{y}. \tag{3.18}$$

With a slight modification of the argument an analogous estimate can be derived for $|D_i u(0, x_n) - D_i u(0, -x_n)|$ (where $D_i = \partial/\partial x_i$, $i = 1, \ldots, n-1$). Let us define

$$\varphi(\hat{x}, y, z) = \tfrac{1}{4}[u(\hat{x}, y, z) - u(\hat{x}, -y, z) - u(\hat{x}, y, -z) + u(\hat{x}, -y, -z)]$$

where $\hat{x} = (x_1, \ldots, x_{n-2})$. In the domain

$$Q' = \{(x_1, \ldots, x_{n-2}, y, z) \mid |x_i| < d/2, i = 1, \ldots, n-2, 0 < y, z < d/2\}$$

we choose a comparison function similar to (3.17) of the form

$$\psi(\hat{x}, y, z) = \frac{4M|\hat{x}|^2}{d^2} + yz\left(\frac{4\mu}{d} + \bar{k} \log \frac{2d}{y+z}\right),$$

where μ and $\bar{k}$ are constants such that $|Du| \leqslant \mu$ in Q' and

$$\bar{k} \geqslant \tfrac{2}{3}[N + 8(n-2)M/d^2].$$

One verifies that $\Delta(\psi \pm \varphi) \leqslant 0$ in Q' and $\psi \pm \varphi \geqslant 0$ on $\partial Q'$, from which it follows $|\varphi| \leqslant \psi$ in Q'. As above, if we set $\hat{x} = 0$ in this inequality, then divide by y and let y tend to zero, we obtain

$$\tfrac{1}{2}|D_{n-1}u(0, z) - D_{n-1}u(0, -z)| = \lim_{y \to 0} \frac{|\varphi(0, y, z)|}{y} \leqslant \frac{4\mu}{d} z + \bar{k} z \log \frac{2d}{z}. \tag{3.19}$$

Obviously the same result is obtained if D_{n-1} is replaced by D_i, $i = 1, \ldots, n-2$. We note that unlike the argument for (3.18), the proof of (3.19) did not require the introduction of an operator in $\mathbb{R}^{n+1}$.

If now $\Delta u = f$ in a domain Ω of $\mathbb{R}^n$, we can obtain from (3.18) and (3.19) an estimate for $|Du(x) - Du(y)|$, where x and y are any two points of Ω. Let $d_x = \text{dist}(x, \partial\Omega)$, $d_y = \text{dist}(y, \partial\Omega)$ and $d_{x,y} = \min(d_x, d_y)$. Assume $d_x \leqslant d_y$, so that $d_x = d_{x,y}$. Suppose first that $|x - y| \leqslant d = d_x/2\sqrt{n}$, and consider the segment joining x and y. We choose the center of this segment as origin and rotate coordinates so that x and y lie on the x_n axis with $x = (0, x_n)$, $y = (0, -x_n)$ in the new coordinates. The cube $Q = \{(x_1, \ldots, x_n) \mid |x_i| < d, i = 1, \ldots, n\}$ lies in Ω at a distance greater than $d_x/2$ from $\partial\Omega$. We may apply (3.16), (3.18) and (3.19) directly in Q to obtain

$$d^2 \frac{|Du(x) - Du(y)|}{|x - y|} \leqslant C\left(\sup_Q |u| + d^2 \sup_Q |f|\right) \log \frac{2d}{|x - y|},$$

for some constant $C = C(n)$. Hence

$$d_{x,y}^2 \frac{|Du(x) - Du(y)|}{|x - y|} \leqslant C\left(\sup_\Omega |u| + \sup_\Omega d_x^2 |f(x)|\right) \log \frac{d_{x,y}}{|x - y|}.$$

If x and y are now points in Ω such that $|x-y|>d$, we have from (3.16),

$$d_{x,y}^2 \frac{|Du(x)-Du(y)|}{|x-y|} \leqslant C\,(\sup_\Omega |u| + \sup_\Omega d_x^2 |f(x)|).$$

Combining these results, we obtain

$$d_{x,y}^2 \frac{|Du(x)-Du(y)|}{|x-y|} \leqslant C\,(\sup_\Omega |u| + \sup_\Omega d_x^2 |f(x)|)\left(\left|\log \frac{d_{x,y}}{|x-y|}\right|+1\right)$$

where C is a constant depending only on n.

The preceding results are collected in the following.

Theorem 3.9. *Let $u \in C^2(\Omega)$ satisfy Poisson's equation, $\Delta u = f$, in Ω. Then*

$$\sup_\Omega d_x |Du(x)| \leqslant C\,(\sup_\Omega |u| + \sup_\Omega d_x^2 |f(x)|),$$

and for all x, y in Ω, $x \neq y$,

$$d_{x,y}^2 \frac{|Du(x)-Du(y)|}{|x-y|} \leqslant C\,(\sup_\Omega |u| + \sup_\Omega d_x^2 |f(x)|)\left(\left|\log \frac{d_{x,y}}{|x-y|}\right|+1\right), \tag{3.20}$$

where $C=C(n)$. Here $d_x = \mathrm{dist}\,(x, \partial\Omega)$, $d_{x,y} = \min\,(d_x, d_y)$.

Despite the elementary character of its proof, this theorem is essentially sharp and the estimate (3.20) cannot be improved without further continuity assumptions on f. Theorem 3.9 will also hold for weak solutions in the sense of Chapter 8 provided f is bounded; (see Problem 8.4).

Extensions of the above results for the case of Hölder continuous f are treated by other methods in Chapter 4, although the comparison methods of this section can be used to obtain these extensions as well; (see [BR 1, 2]).

3.5. A Harnack Inequality

The maximum principle provides an elementary proof of a general Harnack inequality for uniformly elliptic equations in two independent variables. Letting $D_\rho = D_\rho(0)$ denote the open disk of radius ρ centered at the origin, we state the result in the following form.

Theorem 3.10. *Let u be a non-negative C^2 solution of*

$$Lu = au_{xx} + 2bu_{xy} + cu_{yy} = 0$$

in the disk D_R, and suppose that L is uniformly elliptic in D_R. Then at all points $z=(x, y) \in D_{R/4}$ we have the inequality

$$Ku(0) \leqslant u(z) \leqslant K^{-1}u(0), \tag{3.21}$$

where K is a constant depending only on the ellipticity modulus $\mu = \sup_D \Lambda/\lambda$.

Proof. We note to begin with that since the equation $Lu=0$ and the modulus μ are invariant under similarity transformation, it suffices to prove the theorem in the unit disk $D=D_1$. Since $u \geqslant 0$ in D, the strong maximum principle (Theorem 3.5) implies that either $u \equiv 0$ or $u>0$ in D, so it suffices to assume the latter. Consider the set G in D where $u>u(0)/2$, and let $G' \subset G$ be the component containing 0. One sees from the maximum principle that $\partial G' \cap \partial D$ is non-empty, and hence there is no loss of generality in assuming that the point $Q=(0, 1)$ is in $\partial G'$. We define the functions v_+ and v_- by

$$v_\pm(x, y) = \pm x + \tfrac{3}{4} - k(y-\tfrac{1}{2})^2,$$

where k is a positive constant. The parabolas, $\Gamma_\pm : v_\pm = 0$, have vertices $(\mp\frac{3}{4}, \frac{1}{2})$ in D and common axis $y=\frac{1}{2}$. If k is sufficiently large (it suffices that $k \geqslant 3$), the domains $P_\pm$ in D in which $v_\pm > 0$ have an intersection $P_+ \cap P_-$ bounded by arcs of Γ_+, Γ_- and lying entirely in the upper half of D; (see Figure 1). In $P_\pm$, the functions $v_\pm$ obviously satisfy the inequality $0 < v_\pm < \frac{7}{4}$.

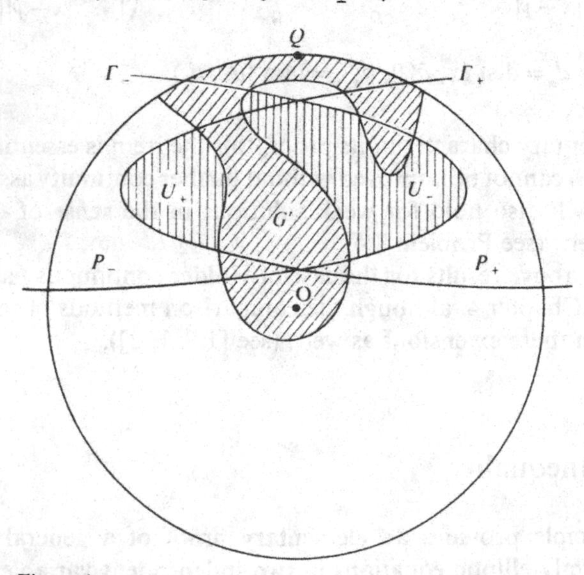

Figure 1

Setting $E_\pm = \exp(\alpha v_\pm)$, where α is a positive constant yet to be chosen, we find by direct calculation

$$\begin{aligned} LE_\pm &= E_\pm\{\alpha^2[a \mp 4bk(y-\tfrac{1}{2}) + 4ck^2(y-\tfrac{1}{2})^2] - 2\alpha kc\} \\ &\geqslant E_\pm(\alpha^2\lambda - 2\alpha k\Lambda) \\ &\geqslant 0 \text{ in } D, \quad \text{if } \alpha \geqslant 2k\mu. \end{aligned}$$

Consequently, with such a choice of α, the functions

(3.22) $$w_{\pm}=(E_{\pm}-1)/(e^{7\alpha/4}-1)$$

have the properties:

$$Lw_{\pm}\geqslant 0 \text{ in } D; \qquad w_{\pm}=0 \text{ on } \Gamma_{\pm}; \qquad 0<w_{\pm}<1 \text{ in } P_{\pm}.$$

Now let z be any point in $P_{+}\cap P_{-}$. Then either: (i) $u\geqslant u(0)/2$ and $z\in \bar{G}$; or (ii) z lies in a component U_{+} of $P_{+}-\bar{G}$ such that $\partial U_{+}\subset \Gamma_{+}\cup \partial G$; or (iii) z lies in a component U_{-} of $P_{-}-\bar{G}$ such that $\partial U_{-}\subset \Gamma_{-}\cup \partial G$; (see Figure 1). These are the only alternatives, since either $P_{+}\cap P_{-}\subset G'$ or $\partial G'$ separates $P_{+}\cup P_{-}$. (The two dimensionality is used here in an essential way.) In cases (ii) and (iii) we have

$$u-\tfrac{1}{2}u(0)w_{\pm}=\tfrac{1}{2}u(0)(1-w_{\pm})>0 \quad \text{on } \partial G\cap \partial U_{\pm},$$
$$u-\tfrac{1}{2}u(0)w_{\pm}=u>0 \quad \text{on } \Gamma_{\pm}\cap \partial U_{\pm}.$$

Thus $u-\frac{1}{2}u(0)w_{\pm}>0$ on $\partial U_{\pm}$. Since $L(u-\frac{1}{2}u(0)w_{\pm})\leqslant 0$, we infer that

$$u(z)>\tfrac{1}{2}u(0)\min(w_{+}(z), w_{-}(z)) \quad \forall z\in P_{+}\cap P_{-}.$$

In particular, on the segment, $y=\frac{1}{2}$, $|x|\leqslant\frac{1}{2}$, we have

(3.23) $$u(x,\tfrac{1}{2})>K_1u(0) \qquad \forall x\in[-\tfrac{1}{2},\tfrac{1}{2}],$$

where

$$K_1=\tfrac{1}{2}(e^{\alpha/4}-1)/(e^{7\alpha/4}-1)=\tfrac{1}{2}\inf_{|x|\leqslant 1/2}[w_{+}(x,\tfrac{1}{2}), w_{-}(x,\tfrac{1}{2})].$$

We now define another comparison function, similar to (3.22). Namely, setting $v=y+1-6x^2$, we consider the domain

$$P=\{(x,y)\in D\mid v(x,y)>0,\ y<\tfrac{1}{2}\}.$$

P is bounded by the segment, $y=\frac{1}{2}$, $|x|\leqslant\frac{1}{2}$, and the arc Γ of the parabola $v=0$, with vertex at $(0,-1)$ and passing through the points $(\pm\frac{1}{2},\frac{1}{2})$. As before, for a suitable choice of $\beta>0$ depending only on μ, the function

$$w=(e^{\beta v}-1)/(e^{3\beta/2}-1)$$

has the properties:

$$Lw\geqslant 0 \text{ in } D; \qquad w=0 \text{ on } \Gamma; \qquad 0<w<1 \text{ in } P.$$

From (3.23) we have that $u-K_1u(0)w>0$ on ∂P, and since $L(u-K_1u(0)w)\leqslant 0$, it follows from the maximum principle that

$$u(z)>K_1u(0)w(z) \quad \forall z\in P.$$

Noting that $D_{1/3} \subset P$, and setting $K_2 = \inf_{D_{1/3}} w$, we obtain

$$u(z) > K_1 K_2 u(0) = K u(0) \quad \forall z \in D_{1/3}. \tag{3.24}$$

Clearly K depends only on μ.

If now $z \in D_{1/4}$ the disk $D_{3/4}(z)$ is contained in D and the inequality (3.24), applied in the disk $D_{1/4}(z)$, implies

$$u(0) > K u(z) \quad \forall z \in D_{1/4}.$$

Combining this inequality with (3.24), we obtain

$$K u(0) < u(z) < K^{-1} u(0) \quad \forall z \in D_{1/4}. \quad \square$$

It follows immediately from (3.21) that

$$\sup_{D_{R/4}} u \leqslant \kappa \inf_{D_{R/4}} u, \tag{3.25}$$

where $\kappa = 1/K^2$. By the same chaining argument as in Theorem 2.5, we obtain the following Harnack inequality for arbitrary domains in $\mathbb{R}^2$.

Corollary 3.11. *Let the hypotheses of Theorem* 3.10 *hold in a domain $\Omega \subset \mathbb{R}^2$. Then for any bounded subdomain $\Omega' \subset\subset \Omega$, there is a constant κ depending only on Ω, Ω' and μ such that*

$$\sup_{\Omega'} u \leqslant \kappa \inf_{\Omega'} u. \tag{3.26}$$

If we consider the more general elliptic equation

$$Lu = a^{ij} D_{ij} u + b^i D_i u + cu = 0, \quad c \leqslant 0, \quad i, j = 1, 2, \tag{3.27}$$

where the coefficients of the operator L are bounded and $\lambda \geqslant \lambda_0 > 0$, the proof of Theorem 3.10 in the unit disk D is still valid (with slight modification); and the conclusion remains the same, but the constant K now depends on the bounds for the coefficients in D as well as on μ. In stating the analogous result for a disk of radius R, the constant K will therefore depend on R in addition to the other quantities; (see Problem 3.4).

The Harnack inequality (3.21) has as consequence the following Liouville theorem.

Corollary 3.12. *If the equation $Lu = au_{xx} + 2bu_{xy} + cu_{yy} = 0$ is uniformly elliptic in $\mathbb{R}^2$ and u is a solution bounded below (or above) and defined over the entire plane, then u is constant.*

Proof. We may assume that $\inf u = 0$, and, hence, for any $\varepsilon > 0$, $u(z_0) < \varepsilon$ for some z_0. In every disc $D_{2R}(z_0)$, we have from (3.21) that $u(z) < K\varepsilon$ for all $z \in D_R(z_0)$. Since K is a constant independent of R, it follows that $u(z) < K\varepsilon$ for all $z \in \mathbb{R}^2$, and the conclusion is immediate by letting $\varepsilon \to 0$. □

A proof of the extension of the Harnack inequality (Theorem 3.10) and of Corollary 3.12 to higher dimensions appears in Chapter 9. Other Harnack inequalities, for equations in divergence form, together with some important applications, are contained in Chapters 8 and 13.

3.6. Operators in Divergence Form

We conclude this chapter with a brief look at the situation for operators in *divergence form*. In many situations it is more natural to consider these than operators of the form (3.1). The simplest such case is

$$Lu = D_j(a^{ij}D_iu). \tag{3.28}$$

Later it will be necessary to consider more general operators whose principal part is in divergence form. L will be called elliptic in Ω if the coefficient matrix $[a^{ij}(x)]$ is positive for all $x \in \Omega$.

Evidently the results concerning the maximum principle apply equally well to the operator (3.28) when the coefficients a^{ij} are sufficiently smooth. However, when this is not the case, or, as in nonlinear problems, when it is often inappropriate to make quantitative assumptions concerning the smoothness of the coefficients (e.g., bounds on their derivatives), the essentially algebraic methods of the earlier part of this chapter cease to be applicable and must be replaced by integral methods, which are more natural for the divergence structure of L.

The relations $Lu = 0$ ($\geqslant 0$, $\leqslant 0$) satisfied by solutions (subsolutions, supersolutions) of $Lu = 0$ can be defined for broader classes of coefficients and functions u than those formally permitted in (3.28). Thus, if the coefficients a^{ij} are bounded and measurable and $u \in C^1(\Omega)$, then, in a generalized sense, u is said to satisfy $Lu = 0$ ($\geqslant 0$, $\leqslant 0$) respectively in Ω, according as

$$\int_\Omega a^{ij}(x)D_iuD_j\varphi\,dx = 0 \ (\leqslant 0, \ \geqslant 0) \tag{3.29}$$

for all non-negative functions $\varphi \in C_0^1(\Omega)$. By application of the divergence theorem this is easily seen to be equivalent to $Lu = 0$ ($\geqslant 0$, $\leqslant 0$) when $a^{ij} \in C^1(\Omega)$ and $u \in C^2(\Omega)$. In later chapters generalized solutions will be defined in wider and more appropriate function spaces.

The weak maximum principle is an immediate consequence of (3.29). For let u satisfy

$$\int_\Omega a^{ij} D_i u D_j \varphi \, dx \leqslant 0 \quad \text{for all } \varphi \in C_0^1(\Omega),\ \varphi \geqslant 0; \tag{3.30}$$

and suppose, contrary to our assertion, that $\sup_\Omega u > \sup_{\partial\Omega} u = u_0$. Then for some constant $c > 0$, there is a subdomain $\Omega' \subset\subset \Omega$ in which $v = u - u_0 - c > 0$ and $v = 0$ on $\partial\Omega'$. The relation (3.30) remains true with u replaced by v and with $\varphi = v$ in Ω', $= 0$ elsewhere. (As thus defined $\varphi \notin C_0^1(\Omega)$, but (3.30) can be seen to hold by approximating this φ with functions in $C_0^1(\Omega)$.) It follows that

$$\int_{\Omega'} a^{ij} D_i v D_j v \, dx \leqslant 0,$$

and hence since $[a^{ij}]$ is positive, we infer that $Dv = 0$ in Ω'. Since $v = 0$ on $\partial\Omega'$, we have $v = 0$ in Ω', which contradicts the definition of v. This establishes the weak maximum principle.

Stronger and more general maximum principles for divergence structure operators will be presented in later chapters. Aside from the already noted difference in methods in treating the two classes of operators, it should be remarked also that results relating to the maximum principle are often different for the operators (3.1) and (3.28) when there are weak smoothness conditions on the coefficients. For example, Lemma 3.4 is not necessarily true for the uniformly elliptic operator of divergence form (3.28) even when the coefficients are arbitrarily smooth in the interior and continuous up to the boundary; (see Problem 3.9).

Notes

The boundary point lemma (Lemma 3.4) as proved here is due to Hopf [HO5]; an independent proof, differing only in the choice of comparison function, was obtained by Oleinik [OL]. The result remains valid, under the same hypotheses on the coefficients, if $\partial\Omega$ has a Dini continuous normal [KH]. A further extension, valid for a class of domains including Lipschitz domains, provides a proof of uniqueness for the Neumann problem in such domains [ND]. Lemma 3.4 is false in general for strictly and uniformly elliptic equations of divergence form even if the coefficients are continuous at the boundary point (see Problem 3.9), but is true if the coefficients are Hölder continuous in a neighborhood (Finn-Gilbarg [FG 1]).

Results analogous to Lemma 3.4 for domains satisfying an interior cone condition, in place of the interior sphere condition, have been obtained by Oddson [OD] and Miller [ML 1, 3]. They prove (3.11) and more precise results, with $|x - x_0|^\mu$ in place of $|x - x_0|$, the exponent μ depending only on the cone angle and the ellipticity constant; (here the vector $x - x_0$ lies within a fixed subcone of the assumed interior cone at x_0). These essentially sharp results are based on the extremal elliptic operators of Pucci [PU 2].

The maximum principle in the generality of Theorem 3.5 was first proved by Hopf [HO 1]. For earlier results, under more restrictive hypotheses, see references in [PW], p. 156, where there is also a discussion of various extensions of the maximum principle. Some of these are considered in Chapters 8 and 9.

Section 3.4 is based on the ideas of Brandt [BR 1, 2], who has shown that much of the linear theory of classical solutions of second order elliptic and parabolic equations, including the deeper estimates of Chapters 4 and 6, can be derived from comparison arguments using the maximum principle. As in Section 3.4, the method requires appropriate (and generally not obvious) choices of comparison functions, which are used to estimate difference quotients and hence derivatives.

The Harnack inequality (Theorem 3.10) and some extensions are due to Serrin [SE 1]. This seems to be the first proof of a Harnack inequality by the maximum principle. Bers and Nirenberg [BN] derive a very similar result by altogether different and deeper methods.

The Liouville theorem (Corollary 3.12) is related to Bernstein's geometric theorem on surfaces of non-positive curvature (see [HO 4]) which implies that an entire solution u of any elliptic equation $au_{xx}+2bu_{xy}+cu_{yy}=0$ such that $u=o(r)$ as $r\to\infty$ must be constant. Of particular interest is the fact that the equation need only be pointwise elliptic. In this generality Corollary 3.12 ceases to be valid, as counterexamples show. Bernstein's result is also based on the maximum principle but the argument is quite different and is more geometric.

Problems

3.1. Let L satisfy the conditions of Theorem 3.6 in a bounded domain Ω and suppose $Lu=0$ in Ω.

(a) Let $\partial\Omega=S_1\cup S_2$ (S_1 non-empty) and assume an interior sphere condition at each point of S_2. Suppose $u\in C^2(\Omega)\cap C^1(\Omega\cup S_2)\cap C^0(\bar{\Omega})$ satisfies the *mixed* boundary condition

$$u=0 \text{ on } S_1, \qquad \sum \beta_i D_i u=0 \text{ on } S_2$$

where the vector $\boldsymbol{\beta}(x)=(\beta_1(x),\ldots,\beta_n(x))$ has a non-zero normal component (to the interior sphere) at each point $x\in S_2$. Then $u\equiv 0$.

(b) Let $\partial\Omega$ satisfy an interior sphere condition, and assume that $u\in C^2(\Omega)\cap C^1(\bar{\Omega})$ satisfies the *regular oblique derivative* boundary condition

$$\alpha(x)u+\sum \beta_i(x)D_i u=0 \text{ on } \partial\Omega,$$

where $\alpha(\boldsymbol{\beta}\cdot\boldsymbol{\nu})>0$, $\boldsymbol{\nu}=$outward normal. Then $u\equiv 0$.

3.2. (a) If L is elliptic, $Lu\geqslant 0$ ($\leqslant 0$) and $c<0$ in a domain Ω, then u cannot achieve an interior positive maximum (negative minimum). (No assumption is made concerning the coefficients b^i.)

(b) If L is elliptic with $c<0$ in a bounded domain Ω, and $u \in C^2(\Omega) \cap C^0(\bar{\Omega})$ satisfies $Lu=f$ in Ω, then

$$\sup_{\Omega} |u| \leqslant \sup_{\partial\Omega} |u| + \sup_{\Omega} |f/c|.$$

3.3. Let $Lu = au_{xx} + 2bu_{xy} + cu_{yy} = 0$ in an exterior domain $r > r_0$, L being uniformly elliptic. Prove that if u is bounded on one side then u has a limit (possibly infinite) as $r \to \infty$; (cf. [GS]). [Apply the Harnack inequality in suitable annuli extending to infinity.] Use this result to prove the Liouville theorem, Corollary 3.12.

3.4. Let u be a non-negative solution of

$$Lu \equiv a^{ij} D_{ij} u + b^i D_i u + cu = 0, \quad c \leqslant 0, \quad i, j = 1, 2,$$

where the coefficients of L satisfy the inequalities

$$\Lambda/\lambda \leqslant \mu, \quad |b^i/\lambda|, |c/\lambda| \leqslant \nu \quad (\mu, \nu = \text{const.}).$$

Prove the Harnack inequality (3.21) with $K = K(\mu, \nu)$ and Corollary 3.11 with $\kappa = \kappa(\mu, \nu, \Omega, \Omega')$.

3.5. Assume the conditions on L in Problem 3.4 are satisfied in the punctured disk $D_0: 0 < r < r_0$, and let $Lu = 0$ in D_0. Prove that if u is bounded on one side, then u has a limit (possibly infinite) as $r \to 0$; (cf. [GS]).

3.6. Let $u \in C^2(\Omega) \cap C^0(\bar{\Omega})$ be a solution of

$$Lu \equiv a^{ij} D_{ij} u + b^i D_i u + cu = f, \; c \leqslant 0$$

in a bounded C^1 domain Ω of $\mathbb{R}^n$ satisfying an exterior sphere condition at $x_0 \in \partial\Omega$, with $\bar{B}_R(y) \cap \bar{\Omega} = x_0$, and let λ, Λ be positive constants such that

$$a^{ij}(x)\xi_i\xi_j \geqslant \lambda|\xi|^2 \quad \forall x \in \Omega, \; \xi \in \mathbb{R}^n$$
$$|a^{ij}|, |b^i|, |c| \leqslant \Lambda.$$

If $\varphi \in C^2(\bar{\Omega})$ and $u = \varphi$ on $\partial\Omega$, show that u satisfies a Lipschitz condition at x_0,

$$|u(x) - u(x_0)| \leqslant K|x - x_0|, \; x \in \Omega,$$

where $K = K(\lambda, \Lambda, R, \operatorname{diam} \Omega, \sup_{\Omega} |f|, |\varphi|_{2;\Omega})$. Hence conclude that K provides a gradient bound for u on $\partial\Omega$ when $u \in C^1(\bar{\Omega})$ and $\partial\Omega$ is sufficiently smooth; (cf. [CH], p. 343). If the sign of c is unrestricted, show that the same result holds provided K depends also on $\sup_{\Omega} |u|$.

3.7. (a) Let the operator L in the preceding problem have Hölder continuous coefficients a^{ij} at the origin: $|a^{ij}(x)-a^{ij}(0)|\leqslant K|x|^{\alpha}$, $\alpha>0$, in $|x|<r_0$ for some constant K. Suppose $Lu\geqslant 0$ ($c\equiv 0$) in the punctured ball $0<r\leqslant r_0$, and assume

$$u=\begin{cases} o(\log r), & n=2 \\ o(r^{2-n}), & n>2 \end{cases} \quad \text{as } r\to 0.$$

Show that

$$\limsup_{x\to 0} u(x)\leqslant \sup_{|x|=r_0} u(x) \tag{3.31}$$

and that equality holds only if u is constant.

(b) If $n>2$ show that the same conclusion holds as in part (a) if the coefficients a^{ij} are continuous at $x=0$ and $u=0(r^{2-n+\delta})$ as $r\to 0$ for some $\delta>0$; (cf. [GS]).

3.8. Consider the equation

$$L_n u\equiv a^{ij}D_{ij}u=0, \qquad a^{ij}=\delta^{ij}+g(r)\frac{x_i x_j}{r^2}, \quad i,j=1,\ldots,n. \tag{3.32}$$

Show that $L_n u=0$ has a radially symmetric solution $u=u(r)$, $r\neq 0$, satisfying the ordinary differential equation

$$\frac{u''}{u'}=\frac{1-n}{r(1+g)}.$$

(a) If $n=2$ and $g(r)=-2/(2+\log r)$, show that equation (3.32) is uniformly elliptic in the disk $D:0\leqslant r\leqslant r_0=e^{-3}$, with continuous coefficients at the origin, and has bounded solutions $a+b/\log r$ in the punctured disk $D-\{0\}$ that do not satisfy (3.31).

(b) If $n>2$ and $g(r)=-[1+(n-1)\log r]^{-1}$, show that (3.32) is uniformly elliptic in $0\leqslant r\leqslant r_0=e^{-1}$ and has continuous coefficients at the origin. Show that the corresponding solution $u=u(r)$ satisfies the condition $u=o(r^{2-n})$ as $r\to 0$ but does not satisfy (3.31).

(c) If $n>2$, determine a function $g(r)$ such that (3.32) is uniformly elliptic and has a bounded solution $u=u(r)$ continuous at $r=0$ that does not satisfy (3.31).

3.9. Let $w=z\exp[-(\log(1/|z|))^{1/2}]$. By considering the relation $w_{\bar{z}}=v(z)w_z$, where

$$\frac{\partial}{\partial z}=\frac{1}{2}\left(\frac{\partial}{\partial x}-i\frac{\partial}{\partial y}\right), \qquad \frac{\partial}{\partial \bar{z}}=\frac{1}{2}\left(\frac{\partial}{\partial x}+i\frac{\partial}{\partial y}\right),$$

show that $u = \text{Re}\, w = x \exp[-(\log(1/r))^{1/2}]$ satisfies a uniformly elliptic equation of divergence form

$$(au_x + bu_y)_x + (bu_x + cu_y)_y = 0,$$

in which the coefficients $a \to 1$, $b \to 0$, $c \to 1$ at the origin and are regular in $0 < r < 1$. Observe that $u(0, 0) = 0$, $u(x, y) > 0$ for $x > 0$ and $u_x(0, 0) = 0$. Compare with Lemma 3.4.

3.10. Let L be the operator of Problem 3.6, but without any condition on the sign of the coefficient c. Assume there is a function v such that $v > 0$ and $Lv \leqslant 0$ in Ω. Then, if $Lu \geqslant 0$, show that the function $w = u/v$ cannot achieve a non-negative maximum in the interior of Ω unless it is constant.

Chapter 4

Poisson's Equation and the Newtonian Potential

In Chapter 2 we introduced the fundamental solution Γ of Laplace's equation given by

$$(4.1) \qquad \Gamma(x-y)=\Gamma(|x-y|)=\begin{cases} \dfrac{1}{n(2-n)\omega_n}|x-y|^{2-n}, & n>2 \\ \dfrac{1}{2\pi}\log|x-y|, & n=2. \end{cases}$$

For an integrable function f on a domain Ω, the *Newtonian potential of f* is the function w defined on $\mathbb{R}^n$ by

$$(4.2) \qquad w(x)=\int_{\Omega}\Gamma(x-y)f(y)\,dy.$$

From Green's representation formula (2.16), we see that when $\partial\Omega$ is sufficiently smooth a $C^2(\bar{\Omega})$ function may be expressed as the sum of a harmonic function and the Newtonian potential of its Laplacian. It is not surprising therefore that the study of *Poisson's equation* $\Delta u = f$ can largely be effected through the study of the Newtonian potential of f. This chapter is primarily devoted to the estimation of derivatives of the Newtonian potential. As well as leading to existence theorems for the classical Dirichlet problem for Poisson's equation, these estimates form the basis for the Schauder or potential theoretic approach to linear elliptic equations treated in Chapter 6.

4.1. Hölder Continuity

If the function f in (4.2) belongs to $C_0^\infty(\Omega)$, we see by writing

$$w(x)=\int_{\Omega}\Gamma(x-y)f(y)\,dy=\int_{\mathbb{R}^n}\Gamma(x-y)f(y)\,dy$$

$$=\int_{\mathbb{R}^n}\Gamma(z)f(x-z)\,dz$$

that the function w will belong to $C^\infty(\bar{\Omega})$. If, on the other hand, f is merely assumed continuous, the Newtonian potential w is not necessarily twice differentiable. It turns out that a convenient class of functions f to work with is the class of Hölder continuous functions which we introduce now.

Let x_0 be a point in $\mathbb{R}^n$ and f a function defined on a bounded set D containing x_0. If $0<\alpha<1$, we say that f is *Hölder continuous with exponent* α *at* x_0 if the quantity

$$[f]_{\alpha;x_0}=\sup_D \frac{|f(x)-f(x_0)|}{|x-x_0|^\alpha} \tag{4.3}$$

is finite. We call $[f]_{\alpha;x_0}$ the *α-Hölder coefficient* of f at x_0 with respect to D. Clearly if f is Hölder continuous at x_0, then f is continuous at x_0. When (4.3) is finite for $\alpha=1$, f is said to be *Lipschitz continuous* at x_0.

Example. The function f on $B_1(0)$ given by $f(x)=|x|^\beta$, $0<\beta<1$, is Hölder continuous with exponent β at $x=0$, and is Lipschitz continuous when $\beta=1$.

The notion of Hölder continuity is readily extended to the whole of D (not necessarily bounded). We call f *uniformly Hölder continuous with exponent* α *in* D if the quantity

$$[f]_{\alpha;D}=\sup_{\substack{x,y\in D\\ x\neq y}} \frac{|f(x)-f(y)|}{|x-y|^\alpha}, \quad 0<\alpha\leq 1, \tag{4.4}$$

is finite; and *locally Hölder continuous with exponent* α *in* D if f is uniformly Hölder continuous with exponent α on compact subsets of D. These two concepts obviously coincide when D is compact. Furthermore note that local Hölder continuity is a stronger property than pointwise Hölder continuity in compact subsets. A locally Hölder continuous function will be pointwise Hölder continuous in D provided it is also bounded in D.

Hölder continuity proves to be a quantitative measure of continuity that is especially well suited to the study of partial differential equations. In a certain sense, it may also be viewed as a fractional differentiability. This suggests a natural widening of the well known spaces of differentiable functions. Let Ω be an open set in $\mathbb{R}^n$ and k a non-negative integer. The *Hölder spaces* $C^{k,\alpha}(\bar{\Omega})$ ($C^{k,\alpha}(\Omega)$) are defined as the subspaces of $C^k(\bar{\Omega})$ ($C^k(\Omega)$) consisting of functions whose k-th order partial derivatives are uniformly Hölder continuous (locally Hölder continuous) with exponent α in Ω. For simplicity we write

$$C^{0,\alpha}(\Omega)=C^\alpha(\Omega), \qquad C^{0,\alpha}(\bar{\Omega})=C^\alpha(\bar{\Omega}),$$

with the understanding that $0<\alpha<1$ whenever this notation is used, unless otherwise stated.

Also, by setting

$$C^{k,0}(\Omega)=C^k(\Omega), \qquad C^{k,0}(\bar{\Omega})=C^k(\bar{\Omega}),$$

we may include the $C^k(\Omega)$ ($C^k(\bar{\Omega})$) spaces among the $C^{k,\alpha}(\Omega)$ ($C^{k,\alpha}(\bar{\Omega})$) spaces for $0 \leqslant \alpha \leqslant 1$. We also designate by $C_0^{k,\alpha}(\Omega)$ the space of functions on $C^{k,\alpha}(\Omega)$ having compact support in Ω.

Let us set

$$(4.5)\qquad \begin{aligned} [u]_{k,0;\Omega} &= |D^k u|_{0;\Omega} = \sup_{|\beta|=k} \sup_{\Omega} |D^\beta u|, \quad k=0,1,2,\dots. \\ [u]_{k,\alpha;\Omega} &= [D^k u]_{\alpha;\Omega} = \sup_{|\beta|=k} [D^\beta u]_{\alpha;\Omega}. \end{aligned}$$

With these seminorms, we can define the related norms

$$(4.6)\qquad \begin{aligned} \|u\|_{C^k(\bar{\Omega})} &= |u|_{k;\Omega} = |u|_{k,0;\Omega} = \sum_{j=0}^{k} [u]_{j,0;\Omega} = \sum_{j=0}^{k} |D^j u|_{0;\Omega}, \\ \|u\|_{C^{k,\alpha}(\bar{\Omega})} &= |u|_{k,\alpha;\Omega} = |u|_{k;\Omega} + [u]_{k,\alpha;\Omega} = |u|_{k;\Omega} + [D^k u]_{\alpha;\Omega}, \end{aligned}$$

on the spaces $C^k(\bar{\Omega})$, $C^{k,\alpha}(\bar{\Omega})$, respectively. It is sometimes useful, especially in this chapter, to introduce non-dimensional norms on $C^k(\bar{\Omega})$, $C^{k,\alpha}(\bar{\Omega})$. If Ω is bounded, with $d = \operatorname{diam} \Omega$, we set

$$(4.6)'\qquad \begin{aligned} \|u\|'_{C^k(\bar{\Omega})} &= |u|'_{k;\Omega} = \sum_{j=0}^{k} d^j [u]_{j,0;\Omega} = \sum_{j=0}^{k} d^j |D^j u|_{0;\Omega}; \\ \|u\|'_{C^{k,\alpha}(\bar{\Omega})} &= |u|'_{k,\alpha;\Omega} = |u|'_{k;\Omega} + d^{k+\alpha}[u]_{k,\alpha;\Omega} = |u|'_{k;\Omega} + d^{k+\alpha}[D^k u]_{\alpha;\Omega}. \end{aligned}$$

The spaces $C^k(\bar{\Omega})$, $C^{k,\alpha}(\bar{\Omega})$, equipped with their respective norms, are Banach spaces; (see Chapter 5).

We note here that the product of Hölder continuous functions is again Hölder continuous. In fact if $u \in C^\alpha(\bar{\Omega})$, $v \in C^\beta(\bar{\Omega})$, we have $uv \in C^\gamma(\bar{\Omega})$ where $\gamma = \min(\alpha, \beta)$, and

$$(4.7)\qquad \begin{aligned} \|uv\|_{C^\gamma(\bar{\Omega})} &\leqslant \max(1, d^{\alpha+\beta-2\gamma}) \|u\|_{C^\alpha(\bar{\Omega})} \|v\|_{C^\beta(\bar{\Omega})}; \\ \|uv\|'_{C^\gamma(\bar{\Omega})} &\leqslant \|u\|'_{C^\alpha(\bar{\Omega})} \|v\|'_{C^\beta(\bar{\Omega})}. \end{aligned}$$

For the domains Ω of interest in this work the inclusion relation $C^{k',\alpha'}(\bar{\Omega}) \subset C^{k,\alpha}(\bar{\Omega})$ will hold whenever $k+\alpha < k'+\alpha'$. It should be noted, however, that such a relation will not be true in general. For example, consider the cusped domain

$$\Omega = \{(x, y) \in \mathbb{R}^2 \mid y < |x|^{1/2},\ x^2 + y^2 < 1\};$$

and for some β, $1 < \beta < 2$, let $u(x, y) = (\operatorname{sgn} x) y^\beta$ if $y > 0$, $= 0$ if $y \leqslant 0$. Clearly $u \in C^1(\bar{\Omega})$. However, if $1 > \alpha > \beta/2$, it is easily seen that $u \notin C^\alpha(\bar{\Omega})$, and hence $C^1(\bar{\Omega}) \not\subset C^\alpha(\bar{\Omega})$.

4.2. The Dirichlet Problem for Poisson's Equation

We show that if f is bounded and Hölder continuous in the bounded domain Ω, the classical Dirichlet problem for Poisson's equation may be solved under the same boundary conditions for which Laplace's equation is solvable (Theorem 2.14). First we require some differentiability results for the Newtonian potential in bounded domains.

In the following the D operator is always taken with respect to the x variable.

Lemma 4.1. *Let f be bounded and integrable in Ω, and let w be the Newtonian potential of f. Then $w \in C^1(\mathbb{R}^n)$ and for any $x \in \Omega$,*

$$D_i w(x) = \int_\Omega D_i \Gamma(x-y) f(y)\, dy, \quad i=1,\ldots,n. \tag{4.8}$$

Proof. By virtue of the estimate (2.14) for $D\Gamma$, the function

$$v(x) = \int_\Omega D_i \Gamma(x-y) f(y)\, dy$$

is well defined. To show that $v = D_i w$, we fix a function η in $C^1(\mathbb{R})$ satisfying $0 \leqslant \eta \leqslant 1$, $0 \leqslant \eta' \leqslant 2$, $\eta(t) = 0$ for $t \leqslant 1$, $\eta(t) = 1$ for $t \geqslant 2$ and define for $\varepsilon > 0$,

$$w_\varepsilon(x) = \int_\Omega \Gamma \eta_\varepsilon f(y)\, dy, \qquad \Gamma = \Gamma(x-y), \quad \eta_\varepsilon = \eta(|x-y|/\varepsilon).$$

Clearly, $w_\varepsilon \in C^1(\mathbb{R}^n)$ and

$$v(x) - D_i w_\varepsilon(x) = \int_{|x-y| \leqslant 2\varepsilon} D_i\{(1-\eta_\varepsilon)\Gamma\} f(y)\, dy$$

so that

$$\begin{aligned} |v(x) - D_i w_\varepsilon(x)| &\leqslant \sup|f| \int_{|x-y| \leqslant 2\varepsilon} \left(|D_i \Gamma| + \frac{2}{\varepsilon}|\Gamma|\right) dy \\ &\leqslant \sup|f| \begin{cases} \dfrac{2n\varepsilon}{n-2} & \text{for } n > 2 \\ 4\varepsilon(1 + |\log 2\varepsilon|) & \text{for } n = 2. \end{cases} \end{aligned}$$

Consequently, w_ε and $D_i w_\varepsilon$ converge uniformly in compact subsets of $\mathbb{R}^n$ to w and v respectively as $\varepsilon \to 0$. Hence, $w \in C^1(\mathbb{R}^n)$ and $D_i w = v$. □

Lemma 4.2. *Let f be bounded and locally Hölder continuous (with exponent $\alpha \leqslant 1$) in Ω, and let w be the Newtonian potential of f. Then $w \in C^2(\Omega)$, $\Delta w = f$ in Ω, and for any $x \in \Omega$,*

$$(4.9) \qquad D_{ij}w(x) = \int_{\Omega_0} D_{ij}\Gamma(x-y)(f(y)-f(x))\,dy - f(x)\int_{\partial\Omega_0} D_i\Gamma(x-y)\nu_j(y)\,ds_y, \quad i,j=1,\ldots,n;$$

here Ω_0 is any domain containing Ω for which the divergence theorem holds and f is extended to vanish outside Ω.

Proof. By virtue of the estimate (2.14) for $D^2\Gamma$ and the pointwise Hölder continuity of f in Ω the function

$$u(x) = \int_{\Omega_0} D_{ij}\Gamma(f(y) - f(x))\,dy - f(x)\int_{\partial\Omega_0} D_i\Gamma\nu_j(y)\,ds_y$$

is well-defined. Let $v = D_i w$, and define for $\varepsilon > 0$

$$v_\varepsilon(x) = \int_{\Omega} D_i\Gamma\eta_\varepsilon f(y)\,dy,$$

where η_ε is the function introduced in the preceding lemma. Clearly, $v_\varepsilon \in C^1(\Omega)$, and differentiating, we obtain

$$\begin{aligned} D_j v_\varepsilon(x) &= \int_{\Omega} D_j(D_i\Gamma\eta_\varepsilon)f(y)\,dy \\ &= \int_{\Omega} D_j(D_i\Gamma\eta_\varepsilon)(f(y) - f(x))\,dy \\ &\quad + f(x)\int_{\Omega_0} D_j(D_i\Gamma\eta_\varepsilon)\,dy \\ &= \int_{\Omega_0} D_j(D_i\Gamma\eta_\varepsilon)(f(y) - f(x))\,dy \\ &\quad - f(x)\int_{\partial\Omega_0} D_i\Gamma\nu_j(y)\,ds_y \end{aligned}$$

provided ε is sufficiently small. Hence, by subtraction

$$\begin{aligned}|u(x) - D_j v_\varepsilon(x)| &= \left| \int_{|x-y|\leqslant 2\varepsilon} D_j\{(1-\eta_\varepsilon)D_i\Gamma\}(f(y)-f(x))\,dy \right| \\ &\leqslant [f]_{\alpha;x} \int_{|x-y|\leqslant 2\varepsilon} \left(|D_{ij}\Gamma| + \frac{2}{\varepsilon}|D_i\Gamma|\right)|x-y|^\alpha\,dy \\ &\leqslant \left(\frac{n}{\alpha}+4\right)[f]_{\alpha;x}(2\varepsilon)^\alpha\end{aligned}$$

provided $2\varepsilon < \operatorname{dist}(x, \partial\Omega)$. Consequently $D_j v_\varepsilon$ converges to u uniformly on compact subsets of Ω as $\varepsilon \to 0$, and since v_ε converges uniformly to $v = D_i w$ in Ω, we obtain $w \in C^2(\Omega)$ and $u = D_{ij}w$. Finally, setting $\Omega_0 = B_R(x)$ in (4.9), we have for sufficiently large R,

$$\Delta w(x) = \frac{1}{n\omega_n R^{n-1}} f(x) \int_{|x-y|=R} \nu_i(y)\nu_i(y)\,ds_y = f(x).$$

This completes the proof of Lemma 4.2. □

From Lemmas 4.1, 4.2 and Theorem 2.14 we can now conclude:

Theorem 4.3. *Let Ω be a bounded domain and suppose that each point of $\partial\Omega$ is regular (with respect to the Laplacian). Then if f is a bounded, locally Hölder continuous function in Ω, the classical Dirichlet problem: $\Delta u = f$ in Ω, $u = \varphi$ on $\partial\Omega$, is uniquely solvable for any continuous boundary values φ.*

Proof. We define w to be the Newtonian potential of f and set $v = u - w$. Then the problem $\Delta u = f$ in Ω, $u = \varphi$ on $\partial\Omega$ is equivalent to the problem $\Delta v = 0$ in Ω, $v = \varphi - w$ on $\partial\Omega$, and its unique solvability follows by Theorem 2.14. □

In the case where Ω is a ball, $B = B_R(0)$ say, Theorem 4.3 follows from the Poisson integral formula (Theorem 2.6) and Lemmas 4.1, 4.2. Moreover we have the explicit formula for the solution:

$$u(x) = \int_{\partial B} K(x,y)\varphi(y)\,ds_y + \int_B G(x,y)f(y)\,dy \tag{4.10}$$

where K is the Poisson kernel (2.29) and G is the Green's function (2.23).

4.3. Hölder Estimates for the Second Derivatives

The following lemma provides the basic estimate in the subsequent development of the theory.

Lemma 4.4. *Let $B_1 = B_R(x_0)$, $B_2 = B_{2R}(x_0)$ be concentric balls in $\mathbb{R}^n$. Suppose $f \in C^\alpha(\bar{B}_2)$, $0<\alpha<1$, and let w be the Newtonian potential of f in B_2. Then $w \in C^{2,\alpha}(\bar{B}_1)$ and*

$$(4.11) \qquad |D^2 w|'_{0,\alpha;B_1} \leqslant C|f|'_{0,\alpha;B_2},$$

i.e.,

$$|D^2 w|_{0;B_1} + R^\alpha [D^2 w]_{\alpha;B_1} \leqslant C(|f|_{0;B_2} + R^\alpha [f]_{\alpha;B_2})$$

where $C = C(n, \alpha)$.

Proof. For any $x \in B_1$, we have by formula (4.9),

$$D_{ij}w(x) = \int_{B_2} D_{ij}\Gamma(x-y)(f(y)-f(x))\,dy - f(x)\int_{\partial B_2} D_i\Gamma(x-y)\nu_j(y)\,ds_y$$

so that by (2.14)

$$(4.12) \qquad \begin{aligned} |D_{ij}w(x)| &\leqslant \frac{|f(x)|}{n\omega_n} R^{1-n} \int_{\partial B_2} ds_y + \frac{[f]_{\alpha;x}}{\omega_n}\int_{B_2} |x-y|^{\alpha-n}\,dy \\ &\leqslant 2^{n-1}|f(x)| + \frac{n}{\alpha}(3R)^\alpha [f]_{\alpha;x} \\ &\leqslant C_1(|f(x)| + R^\alpha [f]_{\alpha;x}) \end{aligned}$$

where $C_1 = C_1(n, \alpha)$.

Next, for any other point $\bar{x} \in B_1$ we have again by formula (4.9),

$$D_{ij}w(\bar{x}) = \int_{B_2} D_{ij}\Gamma(\bar{x}-y)(f(y)-f(\bar{x}))\,dy - f(\bar{x})\int_{\partial B_2} D_i\Gamma(\bar{x}-y)\nu_j(y)\,ds_y.$$

Writing $\delta = |x-\bar{x}|$, $\xi = \frac{1}{2}(x+\bar{x})$, we consequently obtain by subtraction

$$D_{ij}w(\bar{x}) - D_{ij}w(x) = f(x)I_1 + (f(x)-f(\bar{x}))I_2 + I_3 + I_4 + (f(x)-f(\bar{x}))I_5 + I_6,$$

where the integrals I_1, I_2, I_3, I_4, I_5 and I_6 are given by

$$I_1 = \int_{\partial B_2} (D_i\Gamma(x-y) - D_i\Gamma(\bar{x}-y))\nu_j(y)\,ds_y$$

$$I_2 = \int_{\partial B_2} D_i\Gamma(\bar{x}-y)\nu_j(y)\,ds_y$$

$$I_3 = \int_{B_\delta(\xi)} D_{ij}\Gamma(x-y)(f(x)-f(y))\,dy$$

$$I_4 = \int_{B_\delta(\xi)} D_{ij}\Gamma(\bar{x}-y)(f(y)-f(\bar{x}))\,dy$$

$$I_5 = \int_{B_2 - B_\delta(\xi)} D_{ij}\Gamma(x-y)\,dy$$

$$I_6 = \int_{B_2 - B_\delta(\xi)} (D_{ij}\Gamma(x-y) - D_{ij}\Gamma(\bar{x}-y))(f(\bar{x})-f(y))\,dy.$$

The estimation of these integrals can be achieved as follows:

$$|I_1| \leqslant |x-\bar{x}| \int_{\partial B_2} |DD_i\Gamma(\hat{x}-y)|\,ds_y \quad \text{for some point } \hat{x} \text{ between } x \text{ and } \bar{x},$$

$$\leqslant \frac{n^2 2^{n-1}|x-\bar{x}|}{R}, \quad \text{since } |\hat{x}-y| \geqslant R \text{ for } y \in \partial B_2,$$

$$\leqslant n^2 2^{n-\alpha}\left(\frac{\delta}{R}\right)^\alpha, \quad \text{since } \delta = |x-\bar{x}| < 2R.$$

$$|I_2| \leqslant \frac{1}{n\omega_n} R^{1-n} \int_{\partial B_2} ds_y = 2^{n-1}.$$

$$|I_3| \leqslant \int_{B_\delta(\xi)} |D_{ij}\Gamma(x-y)||f(x)-f(y)|\,dy$$

$$\leqslant \frac{1}{\omega_n}[f]_{\alpha;x} \int_{B_{3\delta/2}(x)} |x-y|^{\alpha-n}\,dy$$

$$= \frac{n}{\alpha}\left(\frac{3\delta}{2}\right)^\alpha [f]_{\alpha;x}$$

$$|I_4| \leqslant \frac{n}{\alpha}\left(\frac{3\delta}{2}\right)^\alpha [f]_{\alpha;\bar{x}}, \quad \text{as in the estimation of } I_3.$$

Integration by parts gives

$$|I_5| = \left| \int_{\partial(B_2 - B_\delta(\xi))} D_i\Gamma(x-y)\nu_j(y)\, ds_y \right|$$

$$\leqslant \left| \int_{\partial B_2} D_i\Gamma(x-y)\nu_j(y)\, ds_y \right| + \left| \int_{\partial B_\delta(\xi)} D_i\Gamma(x-y)\nu_j(y)\, ds_y \right|$$

$$\leqslant 2^{n-1} + \frac{1}{n\omega_n}\left(\frac{\delta}{2}\right)^{1-n} \int_{\partial B_\delta(\xi)} ds_y = 2^n.$$

$$|I_6| \leqslant |x-\bar{x}| \int_{B_2 - B_\delta(\xi)} |DD_{ij}\Gamma(\hat{x}-y)|\,|f(\bar{x})-f(y)|\, dy$$

for some $\hat{x}$ between x and $\bar{x}$,

$$\leqslant c\delta \int_{|y-\xi|\geqslant\delta} \frac{|f(\bar{x})-f(y)|}{|\hat{x}-y|^{n+1}}\, dy, \quad c = n(n+5)/\omega_n$$

$$\leqslant c\delta[f]_{\alpha;\bar{x}} \int_{|y-\xi|\geqslant\delta} \frac{|\bar{x}-y|^\alpha}{|\hat{x}-y|^{n+1}}\, dy,$$

$$\leqslant c\left(\frac{3}{2}\right)^\alpha 2^{n+1}\delta[f]_{\alpha;\bar{x}} \int_{|y-\xi|\geqslant\delta} |\xi-y|^{\alpha-n-1}\, dy$$

since $|\bar{x}-y| \leqslant \frac{3}{2}|\xi-y| \leqslant 3|\hat{x}-y|$,

$$\leqslant \frac{c'}{1-\alpha} 2^{n+1}\left(\frac{3}{2}\right)^\alpha \delta^\alpha [f]_{\alpha;\bar{x}}, \quad c' = n^2(n+5).$$

Collecting terms, we thus have

$$(4.13) \qquad |D_{ij}w(\bar{x}) - D_{ij}w(x)| \leqslant C_2(R^{-\alpha}|f(x)| + [f]_{\alpha;x} + [f]_{\alpha;\bar{x}})|x-\bar{x}|^\alpha,$$

where C_2 is a constant depending only on n and α. The required estimate then follows by combining (4.12) and (4.13). □

Remark. If Ω_1, Ω_2 are domains such that $\Omega_1 \subset B_1$, $\Omega_2 \supset B_2$, and $f \in C^\alpha(\bar{\Omega}_2)$, and if w is the Newtonian potential of f over Ω_2, then evidently Lemma 4.4 remains true with Ω_1, Ω_2 replacing B_1, B_2, respectively, in (4.11); that is,

$$|D^2 w|'_{0,\alpha;\Omega_1} \leqslant C|f|'_{0,\alpha;\Omega_2}.$$

Hölder estimates for solutions of Poisson's equation follow immediately from Lemma 4.4.

Theorem 4.5. *Let $u \in C_0^2(\mathbb{R}^n)$, $f \in C_0^\alpha(\mathbb{R}^n)$, satisfy Poisson's equation $\Delta u = f$ in $\mathbb{R}^n$. Then $u \in C_0^{2,\alpha}(\mathbb{R}^n)$ and, if $B = B_R(x_0)$ is any ball containing the support of u, we have*

$$
\begin{aligned}
&|D^2u|'_{0,\alpha;B} \leqslant C|f|'_{0,\alpha;B}, \quad C = C(n, \alpha) \\
&|u|'_{1;B} \leqslant CR^2|f|_{0;B}, \quad C = C(n)
\end{aligned} \tag{4.14}
$$

Proof. By virtue of the representation (2.17),

$$
u(x) = \int_{\mathbb{R}^n} \Gamma(x-y)f(y)\,dy, \tag{4.15}
$$

so that the estimates for Du and D^2u follow respectively from Lemmas 4.1 and 4.4 and the fact that f has compact support in B. The estimate for $|u|_{0;B}$ follows at once from that for Du. □

The restriction that u has compact support can be removed, by various means, in order to achieve the following interior Hölder estimate for solutions of Poisson's equation. (See also Problem 4.4.)

Theorem 4.6. *Let Ω be a domain in $\mathbb{R}^n$ and let $u \in C^2(\Omega)$, $f \in C^\alpha(\Omega)$, satisfy Poisson's equation $\Delta u = f$ in Ω. Then $u \in C^{2,\alpha}(\Omega)$ and for any two concentric balls $B_1 = B_R(x_0)$, $B_2 = B_{2R}(x_0) \subset\subset \Omega$ we have*

$$
|u|'_{2,\alpha;B_1} \leqslant C(|u|_{0;B_2} + R^2|f|'_{0,\alpha;B_2}) \tag{4.16}
$$

where $C = C(n, \alpha)$.

Proof. By either Green's representation (2.16) or Lemma 4.2 we can write for $x \in B_2$, $u(x) = v(x) + w(x)$, where v is harmonic in B_2 and w is the Newtonian potential of f in B_2. By Theorem 2.10 and Lemmas 4.1 and 4.4, we have

$$
\begin{aligned}
&R|Dw|_{0;B_1} + R^2|D^2w|'_{0,\alpha;B_1} \leqslant CR^2|f|'_{0,\alpha;B_2} \\
&R|Dv|_{0;B_1} + R^2|D^2v|'_{0,\alpha;B_1} \leqslant C|v|_{0;B_2} \leqslant C(|u|_{0;B_2} + R^2|f|_{0;B_2}).
\end{aligned}
$$

The last inequality is immediate from $v = u - w$ when $n > 2$. For $n = 2$, by writing $u(x_1, x_2, x_3) = u(x_1, x_2)$, we may consider u to be a solution of Poisson's equation in a ball in $\mathbb{R}^3$ and the inequality follows in the same way. The desired estimate for u is obtained by combining these inequalities. □

An immediate consequence of the interior estimate (4.16) is the equicontinuity on compact subdomains of the second derivatives of any bounded set of solutions of Poisson's equation $\Delta u = f$. Consequently by Arzela's theorem we obtain an extension of the compactness result, Theorem 2.11, to solutions of Poisson's equation.

Corollary 4.7. *Any bounded sequence of solutions of Poisson's equation $\Delta u=f$ in a domain Ω with $f\in C^{\alpha}(\Omega)$ contains a subsequence converging uniformly on compact subdomains to a solution.*

It is sometimes preferable to work with an alternative (but equivalent) formulation of the interior estimate (4.16) in terms of certain interior norms which will be useful later. For $x, y\in\Omega$, which may be any proper open subset of $\mathbb{R}^n$, let us write $d_x=\text{dist}\,(x,\partial\Omega)$, $d_{x,y}=\min\,(d_x, d_y)$. We define for $u\in C^k(\Omega)$, $C^{k,\alpha}(\Omega)$ the following quantities, which are the analogues of the global seminorms and norms (4.5), (4.6).

$$[u]^*_{k,0;\Omega}=[u]^*_{k;\Omega}=\sup_{\substack{x\in\Omega\\|\beta|=k}} d_x^k|D^\beta u(x)|,\quad k=0,1,2,\ldots;$$

$$|u|^*_{k;\Omega}=|u|^*_{k,0;\Omega}=\sum_{j=0}^{k}[u]^*_{j,\Omega};$$

$$[u]^*_{k,\alpha;\Omega}=\sup_{\substack{x,y\in\Omega\\|\beta|=k}} d_{x,y}^{k+\alpha}\frac{|D^\beta u(x)-D^\beta u(y)|}{|x-y|^\alpha},\quad 0<\alpha\leqslant 1; \tag{4.17}$$

$$|u|^*_{k,\alpha;\Omega}=|u|^*_{k;\Omega}+[u]^*_{k,\alpha;\Omega}.$$

In this notation,

$$[u]^*_{0;\Omega}=|u|^*_{0;\Omega}=|u|_{0;\Omega}.$$

We note that $|u|^*_{k;\Omega}$ and $|u|^*_{k,\alpha;\Omega}$ are norms on the subspaces of $C^k(\Omega)$ and $C^{k,\alpha}(\Omega)$ respectively for which they are finite. If Ω is bounded and $d=\text{diam}\,\Omega$, then obviously these interior norms and the global norms (4.6) are related by

$$|u|^*_{k,\alpha;\Omega}\leqslant\max\,(1, d^{k+\alpha})|u|_{k,\alpha;\Omega}. \tag{4.17)$'$($$

If $\Omega'\subset\subset\Omega$ and $\sigma=\text{dist}\,(\Omega',\partial\Omega)$, then

$$\min\,(1,\sigma^{k+\alpha})|u|_{k,\alpha;\Omega'}\leqslant|u|^*_{k,\alpha;\Omega}. \tag{4.17)$''$($$

It is convenient here to also introduce the quantity

$$|f|^{(k)}_{0,\alpha;\Omega}=\sup_{x\in\Omega} d_x^k|f(x)|+\sup_{x,y\in\Omega} d_{x,y}^{k+\alpha}\frac{|f(x)-f(y)|}{|x-y|^\alpha}. \tag{4.18}$$

This is a special case of certain norms to be defined later.

From Theorem 4.6, we can now derive an interior estimate for arbitrary domains which will be generalized in very similar form to elliptic equations in Chapter 6.

Theorem 4.8. *Let $u \in C^2(\Omega)$, $f \in C^\alpha(\Omega)$ satisfy $\Delta u = f$ in an open set Ω of $\mathbb{R}^n$. Then*

$$(4.19) \qquad |u|^*_{2,\alpha;\Omega} \leqslant C(|u|_{0;\Omega} + |f|^{(2)}_{0,\alpha;\Omega}),$$

where $C = C(n, \alpha)$.

Proof. If either $|u|_{0,\Omega}$ or $|f|^{(2)}_{0,\alpha;\Omega}$ is infinite, the estimate (4.19) is trivial. Otherwise for $x \in \Omega$, $R = \frac{1}{3} d_x$, $B_1 = B_R(x)$, $B_2 = B_{2R}(x)$, we have for any first derivative Du and second derivative D^2u

$$\begin{aligned} d_x|Du(x)| + d_x^2|D^2u(x)| &\leqslant (3R)|Du|_{0;B_1} + (3R)^2|D^2u|_{0;B_1} \\ &\leqslant C(|u|_{0;B_2} + R^2|f|'_{0,\alpha;B_2}) \quad \text{by (4.16)} \\ &\leqslant C(|u|_{0;\Omega} + |f|^{(2)}_{0,\alpha;\Omega}). \end{aligned}$$

Hence we obtain

$$(4.20) \qquad |u|^*_{2;\Omega} \leqslant C(|u|_{0;\Omega} + |f|^{(2)}_{0,\alpha;\Omega}).$$

To estimate $[u]^*_{2,\alpha;\Omega}$ we let $x, y \in \Omega$ with $d_x \leqslant d_y$. Then

$$\begin{aligned} d^{2+\alpha}_{x,y} \frac{|D^2u(x) - D^2u(y)|}{|x-y|^\alpha} &\leqslant (3R)^{2+\alpha}[D^2u]_{\alpha;B_1} \\ &\quad + 3^\alpha(3R)^2(|D^2u(x)| + |D^2u(y)|) \\ &\leqslant C(|u|_{0;B_2} + R^2|f|'_{0,\alpha;B_2}) + 6[u]^*_{2;\Omega} \quad \text{by (4.16)} \\ &\leqslant C(|u|_{0;\Omega} + |f|^{(2)}_{0,\alpha;\Omega}) \quad \text{by (4.20)}. \end{aligned}$$

The estimate (4.19) then follows. □

The preceding result provides bounds in compact subsets for Du, D^2u and the Hölder coefficients of D^2u in terms of a bound on the right member of (4.19), and hence it is the basis of compactness results for solutions of Poisson's equation. In particular, Corollary 4.7 is also an immediate consequence of Theorem 4.8, after noting that the latter implies the equicontinuity of solutions and of their first and second derivatives on compact subsets.

By means of the compactness result, Corollary 4.7, we can now derive an existence theorem for Poisson's equation $\Delta u = f$ for unbounded f.

Theorem 4.9. *Let B be a ball in $\mathbb{R}^n$ and f a function in $C^\alpha(B)$ for which $\sup_{x \in B} d_x^{2-\beta}|f(x)| \leqslant N < \infty$ for some β, $0 < \beta < 1$. Then there exists a unique function $u \in C^2(B) \cap C^0(\bar{B})$ satisfying $\Delta u = f$ in B, $u = 0$ on ∂B. Furthermore, u satisfies an estimate*

$$(4.21) \qquad \sup_{x \in B} d_x^{-\beta}|u(x)| \leqslant CN,$$

where the constant C depends only on β.

Proof. The estimate (4.21) follows from a simple barrier argument. Namely, let $B = B_R(x_0)$, $r = |x - x_0|$ and set

$$w(x) = (R^2 - r^2)^\beta.$$

By direct calculation, we have for $r < R$

$$\begin{aligned}\Delta w(x) &= -2\beta(R^2 - r^2)^{\beta-2}[n(R^2 - r^2) + 2(1-\beta)r^2] \\ &\leqslant -4\beta(1-\beta)R^2(R^2-r^2)^{\beta-2} \leqslant -\beta(1-\beta)R^\beta(R-r)^{\beta-2}.\end{aligned}$$

Now suppose that $\Delta u = f$ in B, $u = 0$ on ∂B. Since $d_x = R - r$, we have by hypothesis

$$\begin{aligned}|f(x)| &\leqslant N\, d_x^{\beta-2} = N(R-r)^{\beta-2} \\ &\leqslant -C_0 N \Delta w, \quad \text{where } C_0 = [\beta(1-\beta)R^\beta]^{-1},\end{aligned}$$

so that

$$\Delta(C_0 N w \pm u) \leqslant 0 \text{ in } B \quad \text{and} \quad C_0 N w \pm u = 0 \text{ on } \partial B.$$

Consequently, by the maximum principle,

$$|u(x)| \leqslant C_0 N w(x) \leqslant C N d_x^\beta \quad \text{for } x \in B, \tag{4.22}$$

which implies (4.21) with the constant $C = 2/\beta(1-\beta)$.

Finally to show the existence of u, we let

$$f_m = \begin{cases} m & \text{if } f \geqslant m \\ f & \text{if } |f| \leqslant m \\ -m & \text{if } f \leqslant -m \end{cases}$$

and let $\{B_k\}$ be a sequence of concentric balls exhausting B such that $|f| \leqslant k$ in B_k. We define u_m by $\Delta u_m = f_m$ in B, $u_m = 0$ on ∂B.

By (4.21), we have

$$\sup_{x\in B} d_x^{-\beta}|u_m(x)| \leqslant C \sup_{x \in B} d_x^{2-\beta}|f_m(x)| \leqslant CN,$$

so that the sequence $\{u_m\}$ is uniformly bounded and $\Delta u_m = f$ in B_k, for $m \geqslant k$. Hence by Corollary 4.7, applied successively to the sequence of balls B_k, a subsequence of $\{u_m\}$ converges in B to a $C^2(B)$ function u satisfying $\Delta u = f$ in B. It follows that $|u(x)| \leqslant CNd_x^\beta$ and hence $u = 0$ on ∂B. □

It is easy to show by counterexample that Theorem 4.9 is false if $\beta \leqslant 0$. We remark that the theorem may be extended to more general domains than balls; (see Problem 4.6). Also, for arbitrary domains with regular boundary points, the classical Dirichlet problem for Poisson's equation, $\Delta u = f$, is solvable for unbounded f satisfying certain integrability conditions; (see Problem 4.3).

4.4. Estimates at the Boundary

Theorem 4.8 will be applied in Chapter 6 to the derivation of interior Hölder estimates for linear elliptic equations. However in order to establish global estimates, which are required for the existence theory, we need a version of Theorem 4.8 applicable to the intersection of a domain Ω and a half-space. Let us first derive the appropriate extension of the Hölder estimate for the Newtonian potential, Lemma 4.4. In what follows, $\mathbb{R}^n_+$ will denote the half-space, $x_n>0$, and T the hyperplane, $x_n=0$; $B_2=B_{2R}(x_0)$, $B_1=B_R(x_0)$ will be balls with center $x_0\in\bar{\mathbb{R}}^n_+$ and we let $B_2^+=B_2\cap\mathbb{R}^n_+$, $B_1^+=B_1\cap\mathbb{R}^n_+$.

Lemma 4.10. *Let $f\in C^\alpha(\bar{B}_2^+)$, and let w be the Newtonian potential of f in B_2^+. Then $w\in C^{2,\alpha}(\bar{B}_1^+)$ and*

$$|D^2w|'_{0,\alpha;B_1^+}\leqslant C|f|'_{0,\alpha;B_2^+} \tag{4.23}$$

where $C=C(n,\alpha)$.

Proof. We assume that B_2 intersects T since otherwise the result is already contained in Lemma 4.4. The representation (4.9) holds for $D_{ij}w$ with $\Omega_0=B_2^+$. If either i or $j\neq n$, then the portion of the boundary integral

$$\int_{\partial B_2^+} D_i\Gamma(x-y)\nu_j(y)\,ds_y\left(=\int_{\partial B_2^+} D_j\Gamma(x-y)\nu_i(y)\,ds_y\right)$$

on T vanishes since ν_i or $\nu_j=0$ there. The estimates in Lemma 4.4 for $D_{ij}w$ (i or $j\neq n$) then proceed *exactly* as before with B_2 replaced by B_2^+, $B_\delta(\xi)$ replaced by $B_\delta(\xi)\cap B_2^+$ and ∂B_2 replaced by ∂B_2^+-T. Finally $D_{nn}w$ can be estimated from the equation $\Delta w=f$ and the estimates on $D_{kk}w$ for $k=1,\dots,n-1$. □

Theorem 4.11. *Let $u\in C^2(B_2^+)\cap C^0(\bar{B}_2^+)$, $f\in C^\alpha(\bar{B}_2^+)$, satisfy $\Delta u=f$ in B_2^+, $u=0$ on T. Then $u\in C^{2,\alpha}(\bar{B}_1^+)$ and we have*

$$|u|'_{2,\alpha;B_1^+}\leqslant C(|u|_{0;B_2^+}+R^2|f|'_{0,\alpha;B_2^+}) \tag{4.24}$$

where $C=C(n,\alpha)$.

Proof. Let $x'=(x_1,\dots,x_{n-1})$, $x^*=(x',-x_n)$ and define

$$f^*(x)=f^*(x',x_n)=\begin{cases} f(x',x_n) & \text{if } x_n\geqslant 0\\ f(x',-x_n) & \text{if } x_n\leqslant 0.\end{cases}$$

We assume that B_2 intersects T; otherwise Theorem 4.6 implies (4.24). We set $B_2^-=\{x\in\mathbb{R}^n \mid x^*\in B_2^+\}$ and $D=B_2^+\cup B_2^-\cup(B_2\cap T)$. Then $f^*\in C^\alpha(\bar{D})$ and $|f^*|'_{0,\alpha;D}\leqslant 2|f|'_{0,\alpha;B_2^+}$. Now, defining

$$(4.25)\qquad w(x)=\int_{B_2^+}[\Gamma(x-y)-\Gamma(x^*-y)]f(y)\,dy$$

$$=\int_{B_2^+}[\Gamma(x-y)-\Gamma(x-y^*)]f(y)\,dy,$$

we have $w(x',0)=0$ (see Problem 2.3c) and $\Delta w=f$ in B_2^+. Noting that

$$\int_{B_2^+}\Gamma(x-y^*)f(y)\,dy=\int_{B_2^-}\Gamma(x-y)f^*(y)\,dy,$$

we then obtain

$$w(x)=2\int_{B_2^+}\Gamma(x-y)f(y)\,dy-\int_D\Gamma(x-y)f^*(y)\,dy.$$

Letting $w^*(x)=\int_D\Gamma(x-y)f^*(y)\,dy$, we have by the Remark following Lemma 4.4 (in which we set $\Omega_1=B_1^+$, $\Omega_2=D$)

$$|D^2w^*|'_{0,\alpha;B_1^+}\leqslant C|f^*|'_{0,\alpha;D}\leqslant 2C|f|'_{0,\alpha;B_2^+}.$$

Combining this with Lemma 4.10, we obtain

$$(4.26)\qquad |D^2w|'_{0,\alpha;B_1^+}\leqslant C|f|'_{0,\alpha;B_2^+}.$$

Now let $v=u-w$. Then $\Delta v=0$ in B_2^+ and $v=0$ on T. By reflection v may be extended to a harmonic function in B_2 (Problem 2.4) and hence the estimate (4.24) follows from the interior derivative estimate for harmonic functions, Theorem 2.10. □

Remark. If in addition to the hypotheses of Theorem 4.11, u has compact support in $B_2^+\cup T$, we obtain from (4.26) the simpler estimate (extending (4.14))

$$(4.27)\qquad |D^2u|'_{0,\alpha;B_2^+}\leqslant C|f|'_{0,\alpha;B_2^+}.$$

In this case we have the representation

$$(4.28)\qquad u(x)=w(x)=\int_{B_2^+}[\Gamma(x-y)-\Gamma(x^*-y)]f(y)\,dy.$$

It will be useful to have an analogue of Theorem 4.8 in which the estimates are valid up to a hyperplane boundary piece. For this purpose we introduce certain partially interior norms and seminorms, analogous to (4.17) and (4.18). Let Ω be a proper open subset of $\mathbb{R}^n_+$ with open boundary portion T on $x_n=0$. For $x, y \in \Omega$ let us write

$$\bar{d}_x = \text{dist}\,(x, \partial\Omega - T), \; \bar{d}_{x,y} = \min\,(\bar{d}_x, \bar{d}_y).$$

We define the following quantities:

$$[u]^*_{k,0;\Omega\cup T} = [u]^*_{k;\Omega\cup T} = \sup_{\substack{x\in\Omega \\ |\beta|=k}} \bar{d}^k_x |D^\beta u(x)|, \quad k=0,1,2,\ldots;$$

$$(4.29) \qquad |u|^*_{k;\Omega\cup T} = |u|^*_{k,0;\Omega\cup T} = \sum_{j=0}^{k} [u]^*_{j;\Omega\cup T};$$

$$[u]^*_{k,\alpha;\Omega\cup T} = \sup_{\substack{x,y\in\Omega \\ |\beta|=k}} \bar{d}^{k+\alpha}_{x,y} \frac{|D^\beta u(x) - D^\beta u(y)|}{|x-y|^\alpha}, \quad 0<\alpha\leqslant 1;$$

$$|u|^*_{k,\alpha;\Omega\cup T} = |u|^*_{k;\Omega\cup T} + [u]^*_{k,\alpha;\Omega\cup T};$$

$$|u|^{(k)}_{0,\alpha;\Omega\cup T} = \sup_{x\in\Omega} \bar{d}^k_x |u(x)| + \sup_{x,y\in\Omega} \bar{d}^{k+\alpha}_{x,y} \frac{|u(x)-u(y)|}{|x-y|^\alpha}.$$

We can now state:

Theorem 4.12. *Let Ω be an open set in $\mathbb{R}^n_+$ with a boundary portion T on $x_n=0$, and let $u \in C^2(\Omega) \cap C^0(\Omega \cup T)$, $f \in C^\alpha(\Omega \cup T)$ satisfy $\Delta u = f$ in Ω, $u=0$ on T. Then*

$$(4.30) \qquad |u|^*_{2,\alpha;\Omega\cup T} \leqslant C(|u|_{0;\Omega} + |f|^{(2)}_{0,\alpha;\Omega\cup T}),$$

where $C = C(n, \alpha)$.

This result follows from Theorem 4.11 in the same way that Theorem 4.8 follows from Theorem 4.6; the details of proof are therefore omitted.

Theorems 4.11 and 4.12 provide a regularity result for solutions of Poisson's equation at a hyperplane portion of the boundary. More generally, if Ω is a bounded domain, $f \in C^\alpha(\bar{\Omega})$, $u \in C^2(\Omega) \cap C^0(\bar{\Omega})$, $\Delta u = f$ in Ω, and if $\partial\Omega$ and the boundary values of u are sufficiently smooth, it follows that $u \in C^{2,\alpha}(\bar{\Omega})$. This result, essentially *Kellogg's theorem* [KE 1], will be established as a byproduct of our treatment of linear elliptic equations in Chapter 6. The case when Ω is a ball however is directly derivable from Theorem 4.11.

Theorem 4.13. *Let B be a ball in $\mathbb{R}^n$ and u and f functions on $\bar{B}$ satisfying $u \in C^2(B) \cap C^0(\bar{B})$, $f \in C^\alpha(\bar{B})$, $\Delta u = f$ in B, $u=0$ on ∂B. Then $u \in C^{2,\alpha}(\bar{B})$.*

Proof. By a translation we may assume ∂B passes through the origin. The inversion mapping $x \to x^* = x/|x|^2$ is a bicontinuous, smooth mapping of $\mathbb{R}^n - \{0\}$ onto itself which maps B onto a half-space, B^*. Furthermore if $u \in C^2(B) \cap C^0(\bar{B})$, the *Kelvin transform* of u, defined by

$$(4.31) \qquad v(x) = |x|^{2-n} u\left(\frac{x}{|x|^2}\right)$$

belongs to $C^2(B^*) \cap C^0(\bar{B}^*)$ and satisfies (see Problem 4.7).

$$(4.32) \qquad \begin{aligned} \Delta_{x^*} v(x^*) &= |x^*|^{-n-2} \Delta_x u(x), \quad x^* \in B^*, x \in B, \\ &= |x^*|^{-n-2} f\left(\frac{x^*}{|x^*|^2}\right), \quad x^* \in B^*. \end{aligned}$$

Hence Theorem 4.11 is applicable to the Kelvin transform v and since by translation any point of ∂B may be taken for the origin we obtain $u \in C^{2,\alpha}(\bar{B})$. □

Corollary 4.14. *Let $\varphi \in C^{2,\alpha}(\bar{B})$, $f \in C^\alpha(\bar{B})$. Then the Dirichlet problem, $\Delta u = f$ in B, $u = \varphi$ on ∂B, is uniquely solvable for a function $u \in C^{2,\alpha}(\bar{B})$.*

Proof. Writing $v = u - \varphi$, the problem is reduced to the problem $\Delta v = f - \Delta\varphi$ in B, $v = 0$ on ∂B, which is solvable for $v \in C^2(B) \cap C^0(\bar{B})$ by Theorem 4.3 and consequently for $v \in C^{2,\alpha}(\bar{B})$ by Theorem 4.13. □

As a byproduct of the proof of Theorem 4.13, we see that Lemma 4.4 may be improved in the sense that if $f \in C^\alpha(\bar{B})$, its Newtonian potential in B will belong to $C^{2,\alpha}(\bar{B})$.

4.5. Hölder Estimates for the First Derivatives

Poisson's equation often appears in the form

$$(4.33) \qquad \Delta u = \operatorname{div} \mathbf{f} = D_i f^i, \qquad \mathbf{f} = (f^1, \ldots, f^n)$$

where the density function is a divergence. The corresponding estimates of solutions can be reduced to those of the preceding sections, with certain generalizations that will be useful later.

If $\mathbf{f} \in C^{1,\alpha}(\Omega)$, then, obviously, the estimates for the Newtonian potential of $\operatorname{div} \mathbf{f}$ and for solutions of (4.33) are the same as before provided $\operatorname{div} \mathbf{f}$ replaces f throughout. If Ω is sufficiently smooth, we have

$$\int_\Omega \Gamma(x-y) \operatorname{div} \mathbf{f}(y)\, dy = \int_\Omega D\Gamma(x-y) \cdot \mathbf{f}(y)\, dy + \int_{\partial\Omega} \Gamma(x-y) \mathbf{f} \cdot \mathbf{v}\, ds_y,$$

and, thus, the Newtonian potential of div **f** in Ω is within a harmonic function given by

$$(4.34)\qquad w(x) = D\int_{\Omega}\Gamma(x-y)\mathbf{f}(y)\,dy = D_j\int_{\Omega}\Gamma(x-y)f^j(y)\,dy.$$

This expression is identical with the Newtonian potential when **f** has compact support in Ω. We see that w is still defined when **f** is only integrable, in which case (4.34) can be taken as the definition of a generalized Newtonian potential of div **f** in Ω. If, in addition, **f** is Hölder continuous, the first derivatives of w in Ω are (as in Lemma 4.2) given by

$$(4.35)\qquad D_i w(x) = \int_{\Omega} D_{ij}\Gamma(x-y)(f^j(y)-f^j(x))\,dy - f^j(x)\int_{\partial\Omega} D_i\Gamma(x-y)\nu_j\,ds_y,$$

which can be estimated as in Lemma 4.4. Thus we have in the notation of Lemma 4.4

$$(4.36)\qquad |Dw|'_{0,\alpha;B_1} \leqslant C|\mathbf{f}|'_{0,\alpha;B_2},\ C = C(n,\alpha).$$

We can, therefore, assert a $C^{1,\alpha}$ interior estimate:

Theorem 4.15. *Let Ω be a domain in $\mathbb{R}^n$, and let u satisfy Poisson's equation* (4.33), *where $\mathbf{f}\in C^{\alpha}(\Omega)$, $0<\alpha<1$. Then for any two concentric balls $B_1 = B_R(x_0)$, $B_2 = B_{2R}(x_0)\subset\subset\Omega$ we have*

$$(4.37)\qquad |u|'_{1,\alpha;B_1} \leqslant C(|u|_{0;B_2} + R|\mathbf{f}|'_{0,\alpha;B_2}),\qquad C = C(n,\alpha).$$

The proof is the same as that of Theorem 4.6 if (4.36) replaces (4.11) in the argument.

Corresponding boundary estimates can be derived in a similar way. If B_1 and B_2 are as in Lemma 4.10, then the potential (4.34) has first derivatives given by (4.35), with $\Omega = B_2^+$, and we obtain the estimate

$$(4.38)\qquad |Dw|'_{0,\alpha;B_1^+} \leqslant C|\mathbf{f}|'_{0,\alpha;B_2^+},\qquad C = C(n,\alpha).$$

To obtain a $C^{1,\alpha}$ analogue of Theorem 4.11 for solutions of (4.33) vanishing on $x_n = 0$, we proceed as in that theorem with a method based on reflection. Let

$$G(x,y) = \Gamma(x-y) - \Gamma(x-y^*) = \Gamma(x-y) - \Gamma(x^*-y)$$

denote the Green's function of the half-space $\mathbb{R}^n_+$, and consider

$$(4.39)\qquad v(x) = -\int_{B_2^+} D_y G(x,y)\cdot\mathbf{f}(y)\,dy \qquad D_y = (D_{y_1},\ldots,D_{y_n})$$

$$= \int_{B_2^+} D\Gamma(x-y)\cdot f(y)\,dy + \int_{B_2^+} D_y\Gamma(x^*-y)\cdot\mathbf{f}(y)\,dy.$$

For each $i = 1, \dots, n$, let v_i denote the component of v given by

$$v_i(x) = \int_{B_2^+} D_i \Gamma(x-y) f^i(y)\, dy + \int_{B_2^+} D_{y_i} \Gamma(x^*-y) f^i(y)\, dy.$$

We see that v and v_i vanish on $T = B_2 \cap \{x_n = 0\}$. Now suppose $\mathbf{f} \in C^\alpha(\bar{B}_2^+)$, and let $\mathbf{f}$ be extended by even reflection in $x_n = 0$, and denote the extended function again by $\mathbf{f}$. Then, for $i = 1, \dots, n-1$, we have as in the proof of Theorem 4.11,

$$v_i(x) = D_i \left[2 \int_{B_2^+} \Gamma(x-y) f^i(y)\, dy - \int_{B_2^+ \cup B_2^-} \Gamma(x-y) f^i(y)\, dy \right]. \tag{4.40}$$

And when $i = n$, since

$$\int_{B_2^+} D_{y_n} \Gamma(x^*-y) f^n(y)\, dy = \int_{B_2^-} D_n \Gamma(x-y) f^n(y)\, dy,$$

we obtain

$$v_n(x) = D_n \int_{B_2^+ \cup B_2^-} \Gamma(x-y) f^n(y)\, dy. \tag{4.41}$$

It now follows from (4.36) and (4.38), applied to (4.40) and (4.41),

$$|Dv|'_{0,\alpha;B_1^+} \leqslant C|\mathbf{f}|'_{0,\alpha;B_2^+}, \qquad C = C(n, \alpha). \tag{4.42}$$

Theorem 4.16. *Let $u \in C^0(\bar{B}_2^+)$ satisfy Poisson's equation* (4.33), *where $f \in C^\alpha(\bar{B}_2^+)$, and suppose $u = 0$ on $B_2 \cap \{x_n = 0\}$. Then*

$$|u|'_{1,\alpha;B_1^+} \leqslant C(|u|_{0;B_2^+} + R|\mathbf{f}|'_{0,\alpha;B_2^+}), \qquad C = C(n, \alpha). \tag{4.43}$$

The proof is obtained from (4.42) by the concluding argument in Theorem 4.11.

It is also possible to state analogues of Theorems 4.3, 4.13 and Corollary 4.14 for solutions of (4.33). The details are left to the reader.

The preceding results can also be extended to equations of the form

$$\Delta u = g + \operatorname{div} \mathbf{f} \tag{4.44}$$

when $\mathbf{f}$ is Hölder continuous and g is bounded and integrable. We observe that the first derivatives of the Newtonian potential of g satisfy an α-Hölder estimate for every $\alpha < 1$ (see Problem 4.8(a); also Theorem 3.9). From this estimate for the

Newtonian potential, it follows that the $C^{1,\alpha}$ estimates for solutions of (4.44) corresponding to (4.37) and (4.43) take the form

$$|u|'_{1,\alpha;B_1} \leqslant C(|u|_{0;B_2} + R^2|g|_{0;B_2} + R|\mathbf{f}|'_{0,\alpha;B_2}) \tag{4.45}$$

$$|u|'_{1,\alpha;B_1^+} \leqslant C(|u|_{0;B_2^+} + R^2|g|_{0;B_2^+} + R|\mathbf{f}|'_{0,\alpha;B_2^+}) \tag{4.46}$$

where $C = C(n, \alpha)$.

Remark. If $g \in L^p(\Omega)$, where $p = n/(1-\alpha)$, then the terms containing g on the right-hand side in (4.45), (4.46) can be replaced by $R^{1+\alpha}\|g\|_p$ (see Problem 4.8(b)).

Notes

The Hölder estimates of this chapter are essentially due to Korn [KR 1].

In Lemma 4.2, Hölder continuity can be replaced by Dini continuity, so that the Newtonian potential of f is a C^2 solution of Poisson's equation $\Delta u = f$ if

$$|f(x)-f(y)| \leqslant \varphi(|x-y|), \quad \text{where} \int_{0^+} \varphi(r)r^{-1}\,dr < \infty; \tag{4.47}$$

(see Problem 4.2). However, if f is only continuous the Newtonian potential need not be twice differentiable; (see Problem 4.9).

The weighted interior norms and seminorms (4.17), (4.29) are adapted from Douglis and Nirenberg [DN]. The primary function of the partially interior norms and seminorms (4.29) is to facilitate the derivation of boundary estimates by direct imitation of the proofs of interior estimates; (see, for example, Theorem 4.12 and Lemma 6.4).

Problems

4.1. (a) Prove (4.7). (b) If $f \in C^\alpha(\mathbb{R})$ and $g \in C^\beta(\Omega)$, show that $f \circ g \in C^{\alpha\beta}(\Omega)$.

4.2. Prove Lemma 4.2 if f is Dini continuous in Ω (i.e., f satisfies (4.47)).

4.3. Show that Theorem 4.3 continues to hold if the boundedness of f is replaced by $f \in L^p(\Omega)$ for some $p > n/2$; (see Lemma 7.12).

4.4. Derive Theorem 4.6 from Theorem 4.5 by applying the representation formula (2.17) to $v(x) = u(x)\eta(|x-x_0|/R)$, where η is a cut-off function such that $\eta \in C^2(\mathbb{R})$, $\eta(r) = 1$ for $r \leqslant \frac{3}{2}$, $\eta(r) = 0$ for $r \geqslant 2$.

4.5. Prove the following extensions to Poisson's equation of the solid mean value inequalities (2.6) for Laplace's equation. Let $u \in C^2(\Omega) \cap C^0(\bar{\Omega})$ satisfy $\Delta u = (\geqslant, \leqslant)\, f$ in Ω. Then for any ball $B = B_R(y) \subset \Omega$, we have

$$u(y) = (\leqslant, \geqslant) \left\{ \frac{1}{|B|} \int_B u\, dx - \frac{1}{n\omega_n} \int_B f(x)\Theta(r, R)\, dx \right\}, \quad r = |x - y|,$$

where

$$\Theta(r, R) = \begin{cases} \dfrac{1}{n-2}(r^{2-n} - R^{2-n}) - (R^2 - r^2)/2R^n, & n > 2, \\ \log(R/r) - \frac{1}{2}(1 - r^2/R^2), & n = 2. \end{cases}$$

4.6. Prove Theorem 4.9 if the ball B is replaced by an arbitrary bounded C^2 domain (use Lemma 14.16 and a comparison function $d^\beta \eta$, where η is a suitable cut-off function and d is the distance function).

4.7. Let $\Delta u = f$ in $\Omega \subset \mathbb{R}^n$. Show that the Kelvin transform of u, defined by

$$v(x) = |x|^{2-n} u(x/|x|^2) \quad \text{for } x/|x|^2 \in \Omega$$

satisfies

$$\Delta v(x) = |x|^{-n-2} f(x/|x|^2).$$

4.8. Let w be the Newtonian potential of f in $B = B_R(x_0)$.

(a) If $f \in L^\infty(B)$, show that $Dw \in C^\alpha(\mathbb{R}^n)$ for every $\alpha \in (0, 1)$ and

$$[Dw]_{\alpha; B} \leqslant C(n, \alpha) R^{1-\alpha} \| f \|_{\infty; B}$$

(b) If $f \in L^p(B)$, where $p = n/(1 - \alpha)$, $0 < \alpha < 1$, show that $Dw \in C^\alpha(\mathbb{R}^n)$ and

$$[Dw]_{\alpha; B} \leqslant C(n, \alpha) \| f \|_{p; B}.$$

4.9. (a) For α with $|\alpha| = 2$ let P be a homogeneous harmonic polynomial of degree 2 with $D^\alpha P \neq 0$ (e.g., $P = x_1 x_2$, $D_{12} P = 1$). Choose $\eta \in C_0^\infty(\{x \,|\, |x| < 2\})$ with $\eta = 1$ when $|x| < 1$, set $t_k = 2^k$, and let $c_k \to 0$ as $k \to \infty$, with $\sum c_k$ divergent. Define

$$f(x) = \sum_0^\infty c_k \Delta(\eta P)(t_k x).$$

Show that f is continuous but that $\Delta u = f$ does not have a C^2 solution in any neighbourhood of the origin.

(b) For α with $|\alpha| = 3$ choose a homogeneous harmonic polynomial Q of degree 3 with $D^\alpha Q \neq 0$. With η, t_k and c_k as in part (a), define

$$u(x) = \sum_0^\infty c_k \eta(t_k x) Q(x) = \sum_0^\infty c_k (\eta Q)(t_k x)/t_k^3.$$

Then

$$\Delta u = g(x) = \sum_0^\infty c_k \Delta(\eta Q)(t_k x)/t_k.$$

Show that $g \in C^1$ but that $u \notin C^{2,1}$ in any neighbourhood of the origin. Hence Lemma 4.4 is not valid for $\alpha = 1$.

4.10. Let $u \in C_0^2(B)$ satisfy $\Delta u = f$ in $B = B_R(x_0)$. Show that

(a) $|u|_0 \leqslant \dfrac{R^2}{2n} |f|_0$; (b) $|D_i u|_0 \leqslant R|f|_0$, $i = 1, \ldots, n$.

Hence in (4.14), $|u|'_{1;B} \leqslant 3R^2 |f|_{0;B}$.

Chapter 5

Banach and Hilbert Spaces

This chapter supplies the functional analytic material required for our study of existence of solutions of linear elliptic equations in Chapters 6 and 8. This material will be familiar to a reader already versed in basic functional analysis but we shall assume some acquaintance with elementary linear algebra and the theory of metric spaces. Unless otherwise indicated, all linear spaces used in this book are assumed to be defined over the real number field. The theory of this chapter, however, carries over almost unchanged if the real numbers are replaced by the complex numbers. Let $\mathscr{V}$ be a linear space over $\mathbb{R}$. A *norm* on $\mathscr{V}$ is a mapping $p: \mathscr{V} \to \mathbb{R}$ (henceforth we write $p(x) = \|x\| = \|x\|_{\mathscr{V}}$, $x \in \mathscr{V}$) satisfying

(i) $\|x\| \geqslant 0$ for all $x \in \mathscr{V}$, $\|x\| = 0$ if and only if $x = 0$;
(ii) $\|\alpha x\| = |\alpha|\ \|x\|$ for all $\alpha \in \mathbb{R}$, $x \in \mathscr{V}$;
(iii) $\|x+y\| \leqslant \|x\| + \|y\|$ for all $x, y \in \mathscr{V}$ (triangle inequality).

A linear space $\mathscr{V}$ equipped with a norm is called a *normed linear space*. A normed linear space $\mathscr{V}$ is a metric space under the metric ρ defined by

$$\rho(x, y) = \|x-y\|, \quad x, y \in \mathscr{V}.$$

Consequently a sequence $\{x_n\} \subset \mathscr{V}$ converges to an element $x \in \mathscr{V}$ if $\|x_n - x\| \to 0$. Also $\{x_n\}$ is a Cauchy sequence if $\|x_n - x_m\| \to 0$ as $m, n \to \infty$. If $\mathscr{V}$ is complete, that is every Cauchy sequence converges, then $\mathscr{V}$ is called a *Banach space*.

Examples. (i) Euclidean space $\mathbb{R}^n$ is a Banach space under the standard norm:

$$\|x\| = \left(\sum_{i=1}^{n} x_i^2\right)^{1/2}, \quad x = (x_1, \ldots, x_n).$$

(ii) For a bounded domain $\Omega \subset \mathbb{R}^n$, the *Hölder spaces* $C^{k,\alpha}(\bar{\Omega})$ are Banach spaces under either of the equivalent norms (4.6) or (4.6)′ introduced in Chapter 4; (see Problems 5.1, 5.2).

(iii) The Sobolev spaces $W^{k,p}(\Omega)$, $W_0^{k,p}(\Omega)$ (see Chapter 7).

Existence theorems in partial differential equations are often reducible to the solvability of equations in appropriate function spaces. For the Schauder theory of linear elliptic equations we will employ two basic existence theorems for operator equations in Banach spaces, namely the Contraction Mapping Principle and the Fredholm alternative.

5.1. The Contraction Mapping Principle

A mapping T from a normed linear space $\mathscr{V}$ into itself is called a *contraction* mapping if there exists a number $\theta<1$ such that

$$\|Tx-Ty\|\leqslant\theta\|x-y\| \quad \text{for all } x, y\in\mathscr{V}. \tag{5.1}$$

Theorem 5.1. *A contraction mapping T in a Banach space $\mathscr{B}$ has a unique fixed point, that is there exists a unique solution $x\in\mathscr{B}$ of the equation $Tx=x$.*

Proof. (*Method of successive approximations.*) Let $x_0\in\mathscr{B}$ and define a sequence $\{x_n\}\subset\mathscr{B}$ by $x_n=T^n x_0$, $n=1,2,\ldots$ Then if $n\geqslant m$, we have

$$\begin{aligned}\|x_n-x_m\|&\leqslant\sum_{j=m+1}^{n}\|x_j-x_{j-1}\| \quad \text{by the triangle inequality}\\ &=\sum_{j=m+1}^{n}\|T^{j-1}x_1-T^{j-1}x_0\|\\ &\leqslant\sum_{j=m+1}^{n}\theta^{j-1}\|x_1-x_0\| \quad \text{by (5.1)}\\ &\leqslant\frac{\|x_1-x_0\|\theta^m}{1-\theta}\to 0 \quad \text{as } m\to\infty.\end{aligned}$$

Consequently $\{x_n\}$ is a Cauchy sequence and, since $\mathscr{B}$ is complete, converges to an element $x\in\mathscr{B}$. Clearly T is also a continuous mapping and hence we have

$$Tx=\lim Tx_n=\lim x_{n+1}=x$$

so that x is a fixed point of T. The uniqueness of x follows immediately from (5.1). □

In the statement of Theorem 5.1, the space $\mathscr{B}$ can obviously be replaced by any closed subset.

5.2. The Method of Continuity

Let $\mathscr{V}_1$ and $\mathscr{V}_2$ be normed linear spaces. A linear mapping $T:\mathscr{V}_1\to\mathscr{V}_2$ is *bounded* if the quantity

$$\|T\|=\sup_{x\in\mathscr{V}_1,\,x\neq 0}\frac{\|Tx\|_{\mathscr{V}_2}}{\|x\|_{\mathscr{V}_1}} \tag{5.2}$$

is finite. It is easy to show that a linear mapping T is bounded if and only if it is continuous. The invertibility of a bounded linear mapping may sometimes be deduced from the invertibility of a similar mapping through the following theorem, which is known in applications as the *method of continuity*.

Theorem 5.2. *Let $\mathscr{B}$ be a Banach space, $\mathscr{V}$ a normed linear space and let L_0, L_1 be bounded linear operators from $\mathscr{B}$ into $\mathscr{V}$. For each $t \in [0, 1]$, set*

$$L_t = (1-t)L_0 + tL_1$$

and suppose that there is a constant C such that

$$\|x\|_{\mathscr{B}} \leqslant C\|L_t x\|_{\mathscr{V}} \tag{5.3}$$

for $t \in [0, 1]$. Then L_1 maps $\mathscr{B}$ onto $\mathscr{V}$ if and only if L_0 maps $\mathscr{B}$ onto $\mathscr{V}$.

Proof. Suppose that L_s is onto for some $s \in [0, 1]$. By (5.3), L_s is one-to-one and hence the inverse mapping $L_s^{-1}: \mathscr{V} \to \mathscr{B}$ exists. For $t \in [0, 1]$ and $y \in \mathscr{V}$, the equation $L_t x = y$ is equivalent to the equation

$$\begin{aligned} L_s(x) &= y + (L_s - L_t)x \\ &= y + (t-s)L_0 x - (t-s)L_1 x \end{aligned}$$

which in turn, is equivalent to the equation

$$x = L_s^{-1}y + (t-s)L_s^{-1}(L_0 - L_1)x$$

The mapping T from $\mathscr{B}$ into itself given by $Tx = L_s^{-1}y + (t-s)L_s^{-1}(L_0 - L_1)x$ is clearly a contraction mapping if

$$|s-t| < \delta = [C(\|L_0\| + \|L_1\|)]^{-1}$$

and hence the mapping L_t is onto for all $t \in [0, 1]$, satisfying $|s-t| < \delta$. By dividing the interval $[0, 1]$ into subintervals of length less than δ, we see that the mapping L_t is onto for all $t \in [0, 1]$ provided it is onto for any fixed $t \in [0, 1]$, in particular for $t=0$ or $t=1$. □

5.3. The Fredholm Alternative

Let $\mathscr{V}_1$ and $\mathscr{V}_2$ be normed linear spaces. A mapping $T: \mathscr{V}_1 \to \mathscr{V}_2$ is called *compact* (or *completely continuous*) if T maps bounded sets in $\mathscr{V}_1$ into relatively compact sets in $\mathscr{V}_2$ or equivalently T maps bounded sequences in $\mathscr{V}_1$ into sequences in $\mathscr{V}_2$ which contain convergent subsequences. It follows that a compact linear mapping is also continuous but the converse is not true in general unless $\mathscr{V}_2$ is finite dimensional.

The Fredholm alternative (or Riesz-Schauder theory) concerns compact linear operators from a space $\mathscr{V}$ into itself and is an extension of the theory of linear mappings in finite dimensional spaces.

Theorem 5.3. *Let T be a compact linear mapping of a normed linear space $\mathscr{V}$ into itself. Then either (i) the homogeneous equation*

$$x - Tx = 0$$

has a nontrivial solution $x \in \mathscr{V}$ or (ii) *for each $y \in \mathscr{V}$ the equation*

$$x - Tx = y$$

has a uniquely determined solution $x \in \mathscr{V}$. Furthermore, in case (ii), *the operator $(I-T)^{-1}$ whose existence is asserted there is also bounded.*

The proof of Theorem 5.3 depends upon the following simple result of Riesz.

Lemma 5.4. *Let $\mathscr{V}$ be a normed linear space and $\mathscr{M}$ a proper closed subspace of $\mathscr{V}$. Then for any $\theta < 1$, there exists an element $x_\theta \in \mathscr{V}$ satisfying $\|x_\theta\| = 1$ and* $\operatorname{dist}(x_\theta, \mathscr{M}) \geqslant \theta$.

Proof. Let $x \in \mathscr{V} - \mathscr{M}$. Since $\mathscr{M}$ is closed, we have

$$\operatorname{dist}(x, \mathscr{M}) = \inf_{y \in M} \|x - y\| = d > 0.$$

Consequently there exists an element $y_\theta \in \mathscr{M}$ such that

$$\|x - y_\theta\| \leqslant \frac{d}{\theta},$$

so that, defining

$$x_\theta = \frac{x - y_\theta}{\|x - y_\theta\|},$$

we have $\|x_\theta\| = 1$ and for any $y \in \mathscr{M}$,

$$\|x_\theta - y\| = \frac{\|x - y_\theta - \|y_\theta - x\|\, y\|}{\|y_\theta - x\|}$$

$$\geqslant \frac{d}{\|y_\theta - x\|} \geqslant \theta.$$

The lemma is thus proved. □

If $\mathscr{V}=\mathbb{R}^n$, it is clear that one can take $\theta=1$ by choosing x_θ orthogonal to $\mathscr{M}$. This will also be possible in any Hilbert space but in general Lemma 5.4, which asserts the existence of a "nearly orthogonal" element to $\mathscr{M}$, cannot be improved to allow $\theta=1$.

Proof of Theorem 5.3. It is convenient to split our proof into four stages.

(1) *Let $S=I-T$ where I is the identity mapping and let $\mathscr{N}=S^{-1}(0)=\{x\in\mathscr{V}\mid Sx=0\}$ be the null space of S. Then there exists a constant K such that*

$$\text{dist}\,(x,\mathscr{N})\leqslant K\|Sx\| \quad \textit{for all } x\in\mathscr{V}. \tag{5.4}$$

Proof. Suppose the result is not true. Then there exists a sequence $\{x_n\}\subset\mathscr{V}$ satisfying $\|Sx_n\|=1$ and $d_n=\text{dist}\,(x_n,\mathscr{N})\to\infty$. Choose a sequence $\{y_n\}\subset\mathscr{N}$ such that $d_n\leqslant\|x_n-y_n\|\leqslant 2d_n$. Then if

$$z_n=\frac{x_n-y_n}{\|x_n-y_n\|}$$

we have $\|z_n\|=1$, and $\|Sz_n\|\leqslant d_n^{-1}\to 0$ so that the sequence $\{Sz_n\}$ converges to 0. But since T is compact, by passing to a subsequence if necessary, we may assume that the sequence $\{Tz_n\}$ converges to an element $y_0\in\mathscr{V}$. Since $z_n=(S+T)z_n$, we then also have $\{z_n\}$ converging to y_0 and consequently $y_0\in\mathscr{N}$. However this leads to a contradiction as

$$\begin{aligned}\text{dist}\,(z_n,\mathscr{N})&=\inf_{y\in\mathscr{N}}\|z_n-y\|\\&=\|x_n-y_n\|^{-1}\inf_{y\in\mathscr{N}}\|x_n-y_n-\|x_n-y_n\|\,y\|\\&=\|x_n-y_n\|^{-1}\,\text{dist}\,(x_n,\mathscr{N})\geqslant\tfrac{1}{2}.\quad\square\end{aligned}$$

(2) *Let $\mathscr{R}=S(\mathscr{V})$ be the range of S. Then $\mathscr{R}$ is a closed subspace of $\mathscr{V}$.*

Proof. Let $\{x_n\}$ be a sequence in $\mathscr{V}$ whose image $\{Sx_n\}$ converges to an element $y\in\mathscr{V}$. To show that $\mathscr{R}$ is closed we must show that $y=Sx$ for some element $x\in\mathscr{V}$. By our previous result the sequence $\{d_n\}$ where $d_n=\text{dist}\,(x_n,\mathscr{N})$ is bounded. Choosing $y_n\in\mathscr{N}$ as before and writing $w_n=x_n-y_n$, we consequently have that the sequence $\{w_n\}$ is bounded while the sequence $\{Sw_n\}$ converges to y. Since T is compact, by passing to a subsequence if necessary we may assume that the sequence $\{Tw_n\}$ converges to an element $w_0\in\mathscr{V}$. Hence the sequence $\{w_n\}$ itself converges to $y+w_0$ and by the continuity of S, we have $S(y+w_0)=y$. Consequently $\mathscr{R}$ is closed. $\square$

(3) *If $\mathscr{N}=\{0\}$, then $\mathscr{R}=\mathscr{V}$. That is, if case* (i) *of Theorem* 5.3 *does not hold, then case* (ii) *is true.*

Proof. By our previous result the sets $\mathscr{R}_j$ defined by $\mathscr{R}_j=S^j(\mathscr{V})$, $j=1, 2, \ldots$ form a non-increasing sequence of closed subspaces of $\mathscr{V}$. Suppose that no two of these spaces coincide. Then each is a proper subspace of its predecessor. Hence by Lemma 5.4, there exists a sequence $\{y_n\} \subset \mathscr{V}$ such that $y_n \in \mathscr{R}_n$, $\|y_n\| = 1$ and dist $(y_n, \mathscr{R}_{n+1}) \geqslant \frac{1}{2}$. Thus if $n > m$,

$$\begin{aligned} Ty_m - Ty_n &= y_m + (-y_n - Sy_m + Sy_n) \\ &= y_m - y \quad \text{for some } y \in \mathscr{R}_{m+1}. \end{aligned}$$

Hence $\|Ty_m - Ty_n\| \geqslant \frac{1}{2}$ contrary to the compactness of T. Consequently there exists an integer k such that $\mathscr{R}_j = \mathscr{R}_k$ for all $j \geqslant k$. Up to this point we have not used the condition: $\mathscr{N} = \{0\}$. Now let y be an arbitrary element of $\mathscr{V}$. Then $S^k y \in \mathscr{R}_k = \mathscr{R}_{k+1}$ and so $S^k y = S^{k+1} x$ for some $x \in \mathscr{V}$. Therefore $S^k(y - Sx) = 0$ and so $y = Sx$ since $S^{-k}(0) = S^{-1}(0) = 0$. Consequently $\mathscr{R} = \mathscr{R}_j = \mathscr{V}$ for all j. □

(4) *If* $\mathscr{R} = \mathscr{V}$, *then* $\mathscr{N} = \{0\}$. *Consequently either case* (i) *or case* (ii) *holds.*

Proof. This time we define a non-decreasing sequence of closed subspaces $\{\mathscr{N}_j\}$ by setting $\mathscr{N}_j = S^{-j}(0)$. The closure of $\mathscr{N}_j$ follows from the continuity of S. By employing an analogous argument based on Lemma 5.4 to that used in step (3), we obtain that $\mathscr{N}_j = \mathscr{N}_l$ for all $j \geqslant$ some integer l. Then if $\mathscr{R} = \mathscr{V}$, any element $y \in \mathscr{N}_l$ satisfies $y = S^l x$ for some $x \in \mathscr{V}$. Consequently $S^{2l} x = 0$ so that $x \in \mathscr{N}_{2l} = \mathscr{N}_l$ whence $y = S^l x = 0$. Step (4) is thus proved. □

The boundedness of the operator $S^{-1} = (I - T)^{-1}$ in case (ii) follows from step (1) with $\mathscr{N} = \{0\}$. Note that a slight simplification could be achieved by taking $\mathscr{N} = \{0\}$ at the outset in steps (1) and (2) and that step (4) is independent of the previous steps. Theorem 5.3 is thus completely proved. □

Certain aspects of the spectral behaviour of compact linear operators follow from Theorem 5.3 and Lemma 5.4. A number λ is called an *eigenvalue* of T if there exists a non-zero element x in $\mathscr{V}$ (called an *eigenvector*) satisfying $Tx = \lambda x$. It is clear that eigenvectors belonging to different eigenvalues must be linearly independent. Also the dimension of the null space of the operator $S_\lambda = \lambda I - T$ is called the *multiplicity* of λ. If $\lambda \neq 0, \in \mathbb{R}$ is not an eigenvalue of T, it follows from Theorem 5.3 that the *resolvent* operator $R_\lambda = (\lambda I - T)^{-1}$ is a well defined, bounded linear mapping of $\mathscr{V}$ onto itself. From Lemma 5.4 we may deduce the following result.

Theorem 5.5. *A compact linear mapping T of a normed linear space into itself possesses a countable set of eigenvalues having no limit points except possibly $\lambda = 0$. Each non-zero eigenvalue has finite multiplicity.*

Proof. Suppose that there exists a sequence $\{\lambda_n\}$ of not necessarily distinct eigenvalues and a sequence of corresponding linearly independent eigenvectors $\{x_n\}$ satisfying $\lambda_n \to \lambda \neq 0$. Let $\mathscr{M}_n$ be the closed subspace spanned by $\{x_1, \ldots x_n\}$.

By Lemma 5.4, there exists a sequence $\{y_n\}$ such that $y_n \in \mathcal{M}_n$, $\|y_n\|=1$ and $\operatorname{dist}(y_n, \mathcal{M}_{n-1}) \geq \frac{1}{2}$, $(n=2, 3\ldots)$. If $n>m$, we have

$$\begin{aligned}\lambda_n^{-1}Ty_n-\lambda_m^{-1}Ty_m &= y_n+(-y_m-\lambda_n^{-1}S_{\lambda_n}y_n+\lambda_m^{-1}S_{\lambda_m}y_m)\\ &= y_n-z \quad \text{where } z\in\mathcal{M}_{n-1}.\end{aligned}$$

For, if $y_n=\sum_{j=1}^{n}\beta_j x_j$ then $y_n-\lambda_n^{-1}Ty_n=\sum_{j=1}^{n}\beta_j(1-\lambda_n^{-1}\lambda_j)x_j\in\mathcal{M}_{n-1}$ and similarly $S_{\lambda_m}y_m\in\mathcal{M}_m$. Therefore we have

$$\|\lambda_n^{-1}Ty_n-\lambda_m^{-1}Ty_m\|\geq\tfrac{1}{2}$$

which contradicts the compactness of T combined with the hypothesis $\lambda_n\to\lambda\neq 0$. Hence our initial supposition is false and this implies the validity of the theorem. □

5.4. Dual Spaces and Adjoints

For the sake of completeness we mention a few results here that will be proved and applied in this book only in Hilbert spaces. Let $\mathscr{V}$ be a normed linear space. A functional on $\mathscr{V}$ is a mapping from $\mathscr{V}$ into $\mathbb{R}$. The space of all bounded linear functionals on $\mathscr{V}$ is called the *dual space* of $\mathscr{V}$ and is denoted by $\mathscr{V}^*$. It can be shown easily that $\mathscr{V}^*$ is a Banach space under the norm:

$$\|f\|_{\mathscr{V}^*}=\sup_{x\neq 0}\frac{|f(x)|}{\|x\|} \tag{5.5}$$

Example. The dual space of $\mathbb{R}^n$ is isomorphic to $\mathbb{R}^n$ itself.

The dual space of $\mathscr{V}^*$, denoted $\mathscr{V}^{**}$, is called the second dual of $\mathscr{V}$. Clearly the mapping $J:\mathscr{V}\to\mathscr{V}^{**}$ given by $Jx(f)=f(x)$ for $f\in\mathscr{V}^*$ is a norm preserving, linear, one-to-one mapping of $\mathscr{V}$ into $\mathscr{V}^{**}$. If $J\mathscr{V}=\mathscr{V}^{**}$, then we call $\mathscr{V}$ *reflexive*. Reflexive Banach spaces have certain properties that make them more amenable to applications to differential equations than Banach spaces in general. The Sobolev spaces $W^{k,p}(\Omega)$ introduced in Chapter 7 are reflexive for $p>1$ but the Hölder spaces $C^{k,\alpha}(\bar{\Omega})$ of Chapter 4 are nonreflexive.

Let T be a bounded linear mapping between two Banach spaces $\mathscr{B}_1$ and $\mathscr{B}_2$. The *adjoint* of T, denoted T^*, is a bounded linear mapping between $\mathscr{B}_2^*$ and $\mathscr{B}_1^*$ defined by

$$(T^*g)(x)=g(Tx) \quad \text{for } g\in\mathscr{B}_2^*,\ x\in\mathscr{B}_1. \tag{5.6}$$

Letting $\mathcal{N}$, $\mathcal{R}$, $\mathcal{N}^*$, $\mathcal{R}^*$ denote the null spaces and ranges of T, T^* respectively, the following relations hold provided $\mathcal{R}$ is closed,

$$\mathcal{R}=\mathcal{N}^{*\perp}=\{y\in\mathcal{B}_2 \mid g(y)=0 \quad \text{for all } g\in\mathcal{N}^*\},$$

$$\mathcal{R}^*=\mathcal{N}^{\perp}=\{f\in\mathcal{B}_1^* \mid f(x)=0 \quad \text{for all } x\in\mathcal{N}\}.$$

Also the compactness of T implies the compactness of T^*. These two results are proved for example in [YO]. Consequently we see that if case (i) of the Fredholm alternative holds for a Banach space $\mathcal{B}$, then the equation $x-Tx=y$ is solvable for $x\in\mathcal{B}$ if and only if $g(y)=0$ for all $g\in\mathcal{B}^*$ satisfying $T^*g=g$. This last result will be established directly in Hilbert spaces.

5.5. Hilbert Spaces

We develop here the Hilbert space theory required for our treatment of linear elliptic operators in Chapter 8. A *scalar* (or *inner*) *product* on a linear space $\mathscr{V}$ is a mapping $q: \mathscr{V}\times\mathscr{V}\to\mathbb{R}$ (henceforth we write $q(x, y)=(x, y)$ or $(x, y)_{\mathscr{V}}$, $x, y\in\mathscr{V}$) satisfying

(i) $(x, y)=(y, x)$ for all $x, y\in\mathscr{V}$,

(ii) $(\lambda_1 x_1+\lambda_2 x_2, y)=\lambda_1(x_1, y)+\lambda_2(x_2, y)$ for all $\lambda_1, \lambda_2\in\mathbb{R}$, $x_1, x_2, y\in\mathscr{V}$,

(iii) $(x, x)>0$ for all $x\neq 0, \in\mathscr{V}$.

A linear space $\mathscr{V}$ equipped with an inner product is called an *inner product space* or a *pre-Hilbert space*. Writing $\|x\|=(x, x)^{1/2}$ for $x\in\mathscr{V}$, we have the following inequalities:

Schwarz inequality

$$(5.7) \qquad |(x, y)|\leqslant\|x\|\,\|y\|;$$

Triangle inequality

$$(5.8) \qquad \|x+y\|\leqslant\|x\|+\|y\|;$$

Parallelogram law

$$(5.9) \qquad \|x+y\|^2+\|x-y\|^2=2(\|x\|^2+\|y\|^2).$$

In particular an inner product space $\mathscr{V}$ is a normed linear space. A *Hilbert space* is defined to be a complete inner product space.

Examples. (i) Euclidean space $\mathbb{R}^n$ is a Hilbert space under the inner product

$$(x, y)=\sum x_i y_i, \quad x=(x_1, \ldots, x_n), \quad y=(y_1, \ldots, y_n).$$

(ii) The Sobolev spaces $W^{k,2}(\Omega)$; (see Chapter 7).

5.6. The Projection Theorem

Two elements x and y in an inner product space are called *orthogonal* (or *perpendicular*) if $(x, y)=0$. Given a subset $\mathcal{M}$ of an inner product space we denote by $\mathcal{M}^\perp$ the set of elements orthogonal to every element of $\mathcal{M}$. The following theorem asserts the existence of an orthogonal projection of any element in a Hilbert space onto a closed subspace.

Theorem 5.6. *Let $\mathcal{M}$ be a closed subspace of a Hilbert space $\mathcal{H}$. Then for every $x \in \mathcal{H}$ we have $x=y+z$ where $y \in \mathcal{M}$ and $z \in \mathcal{M}^\perp$.*

Proof. If $x \in \mathcal{M}$, we set $y=x$, $z=0$. Hence we may assume $\mathcal{M} \neq \mathcal{H}$ and $x \notin \mathcal{M}$. Define

$$d=\operatorname{dist}(x, \mathcal{M})=\inf_{y \in \mathcal{M}} \|x-y\|>0$$

and let $\{y_n\} \subset \mathcal{M}$ be a *minimizing sequence*, that is $\|x-y_n\| \to d$. Using the parallelogram law we obtain

$$4\|x-\tfrac{1}{2}(y_m+y_n)\|^2+\|y_m-y_n\|^2=2(\|x-y_m\|^2+\|x-y_n\|^2)$$

so that, since $\frac{1}{2}(y_m+y_n) \in \mathcal{M}$, also we have $\|y_m-y_n\| \to 0$ as $m, n \to \infty$; that is the sequence $\{y_n\}$ converges since $\mathcal{H}$ is complete. Also, since $\mathcal{M}$ is closed, $y=\lim y_n \in \mathcal{M}$ and $\|x-y\|=d$.

Now write $x=y+z$ where $z=x-y$. To complete the proof we must show $z \in \mathcal{M}^\perp$. For any $y' \in \mathcal{M}$ and $\alpha \in \mathbb{R}$ we have $y+\alpha y' \in \mathcal{M}$ and so

$$\begin{aligned} d^2 \leqslant \|x-y-\alpha y'\|^2 &= (z-\alpha y', z-\alpha y') \\ &= \|z\|^2-2\alpha(y', z)+\alpha^2\|y'^2\|. \end{aligned}$$

Therefore, since $\|z\|=d$, we obtain for all $\alpha>0$

$$|(y', z)| \leqslant \frac{\alpha}{2}\|y'^2\|$$

so that $(y', z)=0$ for all $y' \in \mathcal{M}$. Hence $z \in \mathcal{M}^\perp$. □

The element y is called the *orthogonal projection* of x on $\mathcal{M}$. Theorem 5.6 also shows that any closed proper subspace of $\mathcal{H}$ is orthogonal to some element of $\mathcal{H}$.

5.7. The Riesz Representation Theorem

The Riesz representation theorem provides an extremely useful characterization of the bounded linear functionals on a Hilbert space as inner products.

Theorem 5.7. *For every bounded linear functional F on a Hilbert space $\mathcal{H}$, there is a uniquely determined element $f \in \mathcal{H}$ such that $F(x)=(x,f)$ for all $x \in \mathcal{H}$ and $\|F\|=\|f\|$.*

Proof. Let $\mathcal{N}=\{x|F(x)=0\}$ be the null space of F. If $\mathcal{N}=\mathcal{H}$, the result is proved by taking $f=0$. Otherwise, since $\mathcal{N}$ is a closed subspace of $\mathcal{H}$, there exists by Theorem 5.6 an element $z\neq 0, \in \mathcal{H}$ such that $(x, z)=0$ for all $x \in \mathcal{N}$. Hence $F(z)\neq 0$ and moreover for any $x \in \mathcal{H}$,

$$F\left(x-\frac{F(x)}{F(z)}z\right)=F(x)-\frac{F(x)}{F(z)}F(z)=0$$

so that the element $x-\dfrac{F(x)}{F(z)}z \in \mathcal{N}$. This means that

$$\left(x-\frac{F(x)}{F(z)}z, z\right)=0,$$

that is, that

$$(x, z)=\frac{F(x)}{F(z)}\|z\|^2$$

and hence $F(x)=(f, x)$ where $f=zF(z)/\|z\|^2$. The uniqueness of f is easily proved and is left to the reader. To show that $\|F\|=\|f\|$, we have first, by the Schwarz inequality,

$$\|F\|=\sup_{x\neq 0}\frac{|(x,f)|}{\|x\|}\leqslant \sup_{x\neq 0}\frac{\|x\|\ \|f\|}{\|x\|}=\|f\|;$$

and secondly,

$$\|f\|^2=(f,f)=F(f)\leqslant\|F\|\ \|f\|,$$

so that $\|f\|\leqslant\|F\|$, and hence $\|F\|=\|f\|$. □

Theorem 5.7 shows that the dual space of a Hilbert space may be identified with the space itself and consequently that Hilbert spaces are reflexive.

5.8. The Lax-Milgram Theorem

The Riesz representation theorem suffices for the treatment of linear elliptic equations that are variational, that is, they are the Euler-Lagrange equations of certain multiple integrals. For general divergence structure equations we will require a slight extension of Theorem 5.7 due to Lax and Milgram. A bilinear form $\mathbf{B}$ on a Hilbert space $\mathscr{H}$ is called *bounded* if there exists a constant K such that

$$|\mathbf{B}(x, y)| \leqslant K\|x\| \, \|y\| \quad \text{for all } x, y \in \mathscr{H} \tag{5.10}$$

and *coercive* if there exists a number $\nu > 0$ such that

$$\mathbf{B}(x, x) \geqslant \nu\|x\|^2 \quad \text{for all } x \in \mathscr{H}. \tag{5.11}$$

A particular example of a bounded, coercive bilinear form is the inner product itself.

Theorem 5.8. *Let $\mathbf{B}$ be a bounded, coercive bilinear form on a Hilbert space $\mathscr{H}$. Then for every bounded linear functional $F \in \mathscr{H}^*$, there exists a unique element $f \in \mathscr{H}$ such that*

$$\mathbf{B}(x, f) = F(x) \quad \textit{for all } x \in \mathscr{H}.$$

Proof. By virtue of Theorem 5.7, there exists a linear mapping $T\colon \mathscr{H} \to \mathscr{H}$ defined by $\mathbf{B}(x, f) = (x, Tf)$ for all $x \in \mathscr{H}$. Furthermore $\|Tf\| \leqslant K\|f\|$ by (5.10) so that T is bounded. By (5.11) we obtain $\nu\|f\|^2 \leqslant \mathbf{B}(f, f) = (f, Tf) \leqslant \|f\| \, \|Tf\|$, so that

$$\nu\|f\| \leqslant \|Tf\| \leqslant K\|f\| \quad \text{for all } f \in \mathscr{H}.$$

This estimate implies that T is one-to-one, has closed range (see Problem 5.3) and that T^{-1} is bounded. Suppose that T is not onto $\mathscr{H}$. Then there exists an element $z \neq 0$ satisfying $(z, Tf) = 0$ for all $f \in \mathscr{H}$. Choosing $f = z$, we obtain $(z, Tz) = \mathbf{B}(z, z) = 0$ implying $z = 0$ by (5.11). Consequently T^{-1} is a bounded linear mapping on $\mathscr{H}$. We then have $F(x) = (x, g) = \mathbf{B}(x, T^{-1}g)$ for all $x \in \mathscr{H}$ and some unique $g \in \mathscr{H}$ and the result is proved with $f = T^{-1}g$. □

5.9. The Fredholm Alternative in Hilbert Spaces

Theorems 5.3 and 5.5 are of course applicable to compact operators in Hilbert spaces. Let us derive now for Hilbert spaces our earlier remarks concerning adjoints in Banach spaces. In light of Theorem 5.7, we define the adjoint slightly differently.

If T is a bounded linear operator in a Hilbert space $\mathscr{H}$, its adjoint T^* is also a bounded linear mapping in $\mathscr{H}$ defined by

(5.12) $\quad (T^*y, x)=(y, Tx) \quad$ for all $x, y\in \mathscr{H}$.

Clearly $\|T^*\|=\|T\|$, where $\|T\|=\sup_{x\neq 0}\|Tx\|/\|x\|$.

Lemma 5.9. *If T is compact, then T^* is also compact.*

Proof. Let $\{x_n\}$ be a sequence in $\mathscr{H}$ satisfying $\|x_n\|\leqslant M$.
Then

$$\begin{aligned}\|T^*x_n\|^2=(T^*x_n, T^*x_n)&=(x_n, TT^*x_n)\\ &\leqslant\|x_n\|\,\|TT^*x_n\|\\ &\leqslant M\,\|T\|\,\|T^*x_n\|,\end{aligned}$$

so that $\|T^*x_n\|\leqslant M\,\|T\|$; that is, the sequence $\{T^*x_n\}$ is also bounded. Hence, since T is compact, by passing to a subsequence if necessary, we may assume that the sequence $\{TT^*x_n\}$ converges. But then

$$\begin{aligned}\|T^*(x_n-x_m)\|^2&=(T^*(x_n-x_m), T^*(x_n-x_m))\\ &=(x_n-x_m, TT^*(x_n-x_m))\\ &\leqslant 2M\,\|TT^*(x_n-x_m)\|\to 0 \quad \text{as } m, n\to\infty.\end{aligned}$$

Since $\mathscr{H}$ is complete, the sequence $\{T^*x_n\}$ is convergent and hence T^* is compact. □

Lemma 5.10. *The closure of the range of T is the orthogonal complement of the null space of T^*.*

Proof. Let $\mathscr{R}=$ the range of T, $\mathscr{N}^*=$ the null space of T^*. If $y=Tx$, we have $(y,f)=(Tx,f)=(x,T^*f)=0$ for all $f\in\mathscr{N}^*$ so that $\mathscr{R}\subset\mathscr{N}^{*\perp}$, and since $\mathscr{N}^{*\perp}$ is closed, $\bar{\mathscr{R}}\subset\mathscr{N}^{*\perp}$. Now suppose that $y\notin\bar{\mathscr{R}}$. By the projection theorem, Theorem 5.7, $y=y_1+y_2$ where $y_1\in\bar{\mathscr{R}}$, $y_2\in\bar{\mathscr{R}}^{\perp}-\{0\}$. Consequently $(y_2, Tx)=(T^*y_2, x)=0$ for all $x\in\mathscr{H}$, so that $y_2\in\mathscr{N}^*$. Therefore $(y_2, y)=(y_2, y_1)+\|y_2\|^2=\|y_2\|^2$ and hence $y\notin\mathscr{N}^{*\perp}$. □

Note that Lemma 5.10 is valid whether or not T is compact. By combining Lemmas 5.9 and 5.10 with Theorems 5.3 and 5.5, we then obtain the following Fredholm alternative for compact operators in Hilbert spaces.

Theorem 5.11. *Let $\mathscr{H}$ be a Hilbert space and T a compact mapping of $\mathscr{H}$ into itself. Then there exists a countable set $\Lambda\subset\mathbb{R}$ having no limit points except possibly $\lambda=0$, such that if $\lambda\neq 0$, $\lambda\notin\Lambda$ the equations*

(5.13) $\quad \lambda x-Tx=y, \; \lambda x-T^*x=y$

have uniquely determined solutions $x \in \mathscr{H}$ for every $y \in \mathscr{H}$, and the inverse mappings $(\lambda I - T)^{-1}, (\lambda I - T^*)^{-1}$ are bounded. If $\lambda \in \Lambda$, the null spaces of the mappings $\lambda I - T$, $\lambda I - T^*$ have positive finite dimension and the equations (5.13) are solvable if and only if y is orthogonal to the null space of $\lambda I - T^*$ in the first case and $\lambda I - T$ in the other.

5.10. Weak Compactness

Let $\mathscr{V}$ be a normed linear space. A sequence $\{x_m\}$ *converges weakly* to an element $x \in \mathscr{V}$ if $f(x_n) \to f(x)$ for all f in the dual space $\mathscr{V}^*$. By the Riesz representation theorem, Theorem 5.7, a sequence $\{x_n\}$ in a Hilbert space $\mathscr{H}$ will converge weakly to $x \in \mathscr{H}$ if $(x_n, y) \to (x, y)$ for all $y \in \mathscr{H}$. The following result is useful in the Hilbert space approach to differential equations.

Theorem 5.12. *A bounded sequence in a Hilbert space contains a weakly convergent subsequence.*

Proof. Let us assume initially that $\mathscr{H}$ is separable and suppose that the sequence $\{x_n\} \subset \mathscr{H}$ satisfies $\|x_n\| \leq M$. Let $\{y_m\}$ be a dense subset of $\mathscr{H}$. By the Cantor diagonal process we obtain a subsequence $\{x_{n_k}\}$ of our original sequence satisfying $(x_{n_k}, y_m) \to \alpha_m \in \mathbb{R}$ as $k \to \infty$. The mapping $f: \{y_m\} \to \mathbb{R}$ defined by $f(y_m) = \alpha_m$ may consequently be extended to a bounded linear functional f on $\mathscr{H}$ and hence by the Riesz representation theorem, there exists an element $x \in \mathscr{H}$ satisfying $(x_{n_k}, y) \to f(y) = (x, y)$ as $k \to \infty$, for all $y \in \mathscr{H}$. Hence the subsequence $\{x_{n_k}\}$ converges weakly to x.

To extend the result to an arbitrary Hilbert space $\mathscr{H}$, we let $\mathscr{H}_0$ be the closure of the linear hull of the sequence $\{x_n\}$. Then by our previous argument there exists a subsequence $\{x_{n_k}\} \subset \mathscr{H}_0$ and an element $x \in \mathscr{H}_0$ satisfying $(x_{n_k}, y) \to (x, y)$ for all $y \in \mathscr{H}_0$. But by Theorem 5.5, we have for arbitrary $y \in \mathscr{H}$, $y = y_0 + y_1$, where $y_0 \in \mathscr{H}_0$, $y_1 \in \mathscr{H}_0^\perp$. Hence $(x_{n_k}, y) = (x_{n_k}, y_0) \to (x, y_0) = (x, y)$ for all $y \in \mathscr{H}$ so that $\{x_{n_k}\}$ converges weakly to x, as required. □

The first part of the proof of Theorem 5.12 extends automatically to reflexive Banach spaces with separable dual spaces (see Problem 5.4). The result is true however for arbitrary reflexive Banach spaces (see [YO]).

Notes

The material in this chapter is standard and can be found in texts on functional analysis such as [DS], [EW] and [YO].

Problems

5.1. Prove that the Hölder spaces $C^{k,\alpha}(\bar{\Omega})$, introduced in Chapter 4, are Banach spaces under either of the equivalent norms (4.6) or (4.6)′.

5.2. Prove that the interior Hölder spaces $C_*^{k,\alpha}(\Omega)$ defined by

$$C_*^{k,\alpha}(\Omega)=\{u\in C^{k,\alpha}(\Omega)|\,|u|_{k,\alpha;\Omega}^*<\infty\}$$

are Banach spaces under the interior norms given by (4.17).

5.3. Let $\mathscr{B}$ be a Banach space and T be a bounded linear mapping of $\mathscr{B}$ into itself satisfying

$$\|x\|\leqslant K\|Tx\| \quad \text{for all } x\in\mathscr{B},$$

for some $K\in\mathbb{R}$. Prove that the range of T is closed.

5.4. Prove that a bounded sequence in a separable, reflexive Banach space contains a weakly convergent subsequence.

Chapter 6

Classical Solutions; the Schauder Approach

This chapter develops a theory of second order linear elliptic equations that is essentially an extension of potential theory. It is based on the fundamental observation that equations with Hölder continuous coefficients can be treated locally as a perturbation of constant coefficient equations. From this fact Schauder [SC 4, 5] was able to construct a global theory, an extension of which is presented here. Basic to this approach are apriori estimates of solutions, extending those of potential theory to equations with Hölder continuous coefficients. These estimates provide compactness results that are essential for the existence and regularity theory, and since they apply to classical solutions under relatively weak hypotheses on the coefficients, they play an important part in the subsequent nonlinear theory.

Throughout this chapter we shall denote by $Lu=f$ the equation

$$Lu=a^{ij}(x)D_{ij}u+b^{i}(x)D_{i}u+c(x)u=f(x), \quad a^{ij}=a^{ji}, \tag{6.1}$$

where the coefficients and f are defined in an open set $\Omega\subset\mathbb{R}^n$ and, unless otherwise stated, the operator L is *strictly elliptic*; that is,

$$a^{ij}(x)\xi_i\xi_j\geqslant\lambda|\xi|^2, \quad \forall x\in\Omega, \quad \xi\in\mathbb{R}^n, \tag{6.2}$$

for some positive constant λ.

Equations with constant coefficients. Before treating equation (6.1) with variable coefficients, we establish a necessary preliminary result that extends Theorems 4.8 and 4.12 from Poisson's equation to other elliptic equations with constant coefficients. We state these extensions in the following lemma, recalling the interior and partially interior norms defined in (4.17), (4.18) and (4.29). Here and throughout this chapter all Hölder exponents will be assumed to lie in (0, 1) unless otherwise stated.

Lemma 6.1. *In the equation*

$$L_0u=A^{ij}D_{ij}u=f(x), \quad A^{ij}=A^{ji}, \tag{6.3}$$

let $[A^{ij}]$ be a constant matrix such that

$$\lambda|\xi|^2 \leqslant A^{ij}\xi_i\xi_j \leqslant \Lambda|\xi|^2, \quad \forall \xi \in \mathbb{R}^n$$

for positive constants λ, Λ.

(a) *Let $u \in C^2(\Omega)$, $f \in C^\alpha(\Omega)$ satisfy $L_0 u = f$ in an open set Ω of $\mathbb{R}^n$. Then*

$$|u|^*_{2,\alpha;\Omega} \leqslant C(|u|_{0;\Omega} + |f|^{(2)}_{0,\alpha;\Omega}) \tag{6.4}$$

where $C = C(n, \alpha, \lambda, \Lambda)$.

(b) *Let Ω be an open subset of $\mathbb{R}^n_+$ with a boundary portion T on $x_n = 0$, and let $u \in C^2(\Omega) \cap C^0(\Omega \cup T)$, $f \in C^\alpha(\Omega \cup T)$ satisfy $L_0 u = f$ in Ω, $u = 0$ on T. Then*

$$|u|^*_{2,\alpha;\Omega\cup T} \leqslant C(|u|_{0;\Omega} + |f|^{(2)}_{0,\alpha;\Omega\cup T}), \tag{6.5}$$

where $C = C(n, \alpha, \lambda, \Lambda)$.

Proof. Let $\mathbf{P}$ be a constant matrix which defines a nonsingular linear transformation $y = x\mathbf{P}$ from $\mathbb{R}^n$ onto $\mathbb{R}^n$. Letting $u(x) \to \tilde{u}(y)$ under this transformation one verifies easily that

$$A^{ij}D_{ij}u(x) = \tilde{A}^{ij}D_{ij}\tilde{u}(y),$$

where $\tilde{\mathbf{A}} = \mathbf{P}^t\mathbf{A}\mathbf{P}$ ($\mathbf{P}^t = \mathbf{P}$ transpose). For a suitable orthogonal matrix $\mathbf{P}$, $\tilde{\mathbf{A}}$ is a diagonal matrix whose diagonal elements are the eigenvalues $\lambda_1, \ldots, \lambda_n$ of $\mathbf{A}$. If, further, $\mathbf{Q} = \mathbf{PD}$, where $\mathbf{D}$ is the diagonal matrix $[\lambda_i^{-1/2}\delta_{ij}]$, then the transformation $y = x\mathbf{Q}$ takes $L_0 u = f(x)$ into the Poisson equation $\Delta\tilde{u}(y) = \tilde{f}(y)$, where $u(x) \to \tilde{u}(y)$, $f(x) \to \tilde{f}(y)$ under the transformation. By a further rotation we may assume that $\mathbf{Q}$ takes the half-space $x_n > 0$ onto the half-space $y_n > 0$.

Since the orthogonal matrix $\mathbf{P}$ preserves length, we have

$$\Lambda^{-1/2}|x| \leqslant |x\mathbf{Q}| \leqslant \lambda^{-1/2}|x|.$$

It follows that if $\Omega \to \tilde{\Omega}$, $v(x) \to \tilde{v}(y)$ under the transformation $y = x\mathbf{Q}$ then the norms (4.17), (4.18) defined on Ω and $\tilde{\Omega}$ are related by the inequalities

$$\begin{aligned} c^{-1}|v|^*_{k,\alpha;\Omega} &\leqslant |\tilde{v}|^*_{k,\alpha;\tilde{\Omega}} \leqslant c|v|^*_{k,\alpha;\Omega}, \\ c^{-1}|v|^{(k)}_{0,\alpha;\Omega} &\leqslant |\tilde{v}|^{(k)}_{0,\alpha;\tilde{\Omega}} \leqslant c|v|^{(k)}_{0,\alpha;\Omega} \end{aligned} \qquad k = 0, 1, 2, \ldots, \quad 0 \leqslant \alpha \leqslant 1, \tag{6.6}$$

where $c = c(k, n, \lambda, \Lambda)$.

Similarly if Ω is an open subset of $\mathbb{R}^n_+$ with a boundary portion T on $x_n = 0$ which is taken by $y = x\mathbf{Q}$ into the set $\tilde{\Omega}$ in $\mathbb{R}^n_+$ with boundary portion $\tilde{T}$, the norms

(4.29) in Ω and $\tilde{\Omega}$ satisfy the inequalities

$$c^{-1}|v|^*_{k,\alpha;\Omega\cup T}\leqslant|\tilde{v}|^*_{k,\alpha;\tilde{\Omega}\cup\tilde{T}}\leqslant c|v|^*_{k,\alpha;\Omega\cup T},$$
(6.7)
$$c^{-1}|v|^{(k)}_{0,\alpha;\Omega\cup T}\leqslant|\tilde{v}|^{(k)}_{0,\alpha;\tilde{\Omega}\cup\tilde{T}}\leqslant c|v|^{(k)}_{0,\alpha;\Omega\cup T},$$

where c is the same constant as in (6.6).

To prove part (a) of the lemma we apply Theorem 4.8 in $\tilde{\Omega}$ and the inequalities (6.6) to obtain

$$|u|^*_{2,\alpha;\Omega}\leqslant C|\tilde{u}|^*_{2,\alpha;\tilde{\Omega}}\leqslant C(|\tilde{u}|_{0;\tilde{\Omega}}+|\tilde{f}|^{(2)}_{0,\alpha;\tilde{\Omega}})\leqslant C(|u|_{0;\Omega}+|f|^{(2)}_{0,\alpha;\Omega}),$$

which is the desired conclusion (6.4). (Here we have used the same letter C to denote constants depending on n, α, λ, Λ.)

Part (b) of the lemma is proved in the same way, using Theorem 4.12 and the inequalities (6.7). □

Lemma 6.1 provides an immediate extension of the estimates on balls in Theorems 4.6 and 4.11 from Poisson s equation to the more general equation with constant coefficients (6.3). Of course in the latter case the constant C depends on λ, Λ as well as n, α.

6.1. The Schauder Interior Estimates

Our first objective in the study of the equation $Lu=f$ is the derivation of the Schauder interior estimates, which play an essential part in the ensuing treatment of the existence and regularity theory. These estimates are based on the same kind of result already obtained in (6.4) for solutions of $L_0u=f$.

To obtain estimates for the interior norm $|u|^*_{2,\alpha;\Omega}$ of solutions of $Lu=f$ in Ω, it suffices to bound only $|u|_{0;\Omega}$ and the seminorm $[u]^*_{2,\alpha;\Omega}$ (defined in (4.17)). That this is so is a consequence of the following *interpolation inequalities.*

Let $u\in C^{2,\alpha}(\Omega)$, where Ω is an open subset of $\mathbb{R}^n$. Then for any $\varepsilon>0$ there is a constant $C=C(\varepsilon)$ such that

$$[u]^*_{j,\beta;\Omega}\leqslant C|u|_{0;\Omega}+\varepsilon[u]^*_{2,\alpha;\Omega}, \tag{6.8}$$

$$j=0,1,2;\quad 0\leqslant\alpha,\beta\leqslant 1, j+\beta<2+\alpha;$$

$$|u|^*_{j,\beta;\Omega}\leqslant C|u|_{0;\Omega}+\varepsilon[u]^*_{2,\alpha;\Omega}. \tag{6.9}$$

These inequalities are proved in Lemma 6.32 of Appendix 1 to this chapter.

In order to state the Schauder estimates in a sharp form, and also for subsequent applications, we introduce the following additional interior seminorms and norms on the spaces $C^k(\Omega)$, $C^{k,\alpha}(\Omega)$. For σ a real number and k a non-negative integer we define

$$
\begin{aligned}
&[f]^{(\sigma)}_{k,0;\Omega}=[f]^{(\sigma)}_{k;\Omega}=\sup_{\substack{x\in\Omega\\|\beta|=k}} d_x^{k+\sigma}|D^\beta f(x)|;\\
&[f]^{(\sigma)}_{k,\alpha;\Omega}=\sup_{\substack{x,y\in\Omega\\|\beta|=k}} d_{x,y}^{k+\alpha+\sigma}\frac{|D^\beta f(x)-D^\beta f(y)|}{|x-y|^\alpha},\quad 0<\alpha\leqslant 1;\\
(6.10)\qquad &|f|^{(\sigma)}_{k;\Omega}=\sum_{j=0}^{k}[f]^{(\sigma)}_{j;\Omega};\\
&|f|^{(\sigma)}_{k,\alpha;\Omega}=|f|^{(\sigma)}_{k;\Omega}+[f]^{(\sigma)}_{k,\alpha;\Omega}.
\end{aligned}
$$

In this notation, when $\sigma=0$ these quantities are identical with those defined in (4.17), so that $[\cdot]^{(0)}=[\cdot]^*$ and $|\cdot|^{(0)}=|\cdot|^*$.

It is easy to verify that

$$
(6.11)\qquad |fg|^{(\sigma+\tau)}_{0,\alpha;\Omega}\leqslant|f|^{(\sigma)}_{0,\alpha;\Omega}|g|^{(\tau)}_{0,\alpha;\Omega}\quad\text{for } \sigma+\tau\geqslant 0.
$$

We now establish the basic *Schauder interior estimates*.

Theorem 6.2. *Let Ω be an open subset of $\mathbb{R}^n$, and let $u\in C^{2,\alpha}(\Omega)$ be a bounded solution in Ω of the equation*

$$
Lu=a^{ij}D_{ij}u+b^iD_iu+cu=f,
$$

where $f\in C^\alpha(\Omega)$ and there are positive constants λ, Λ such that the coefficients satisfy

$$
(6.12)\qquad a^{ij}\xi_i\xi_j\geqslant\lambda|\xi|^2\quad\forall x\in\Omega,\ \xi\in\mathbb{R}^n
$$

and

$$
(6.13)\qquad |a^{ij}|^{(0)}_{0,\alpha;\Omega},\ |b^i|^{(1)}_{0,\alpha;\Omega},\ |c|^{(2)}_{0,\alpha;\Omega}\leqslant\Lambda.
$$

Then

$$
(6.14)\qquad |u|^*_{2,\alpha;\Omega}\leqslant C(|u|_{0;\Omega}+|f|^{(2)}_{0,\alpha;\Omega})
$$

where $C=C(n,\alpha,\lambda,\Lambda)$.

Proof. By virtue of (6.9) it suffices to prove the inequality (6.14) for $[u]^*_{2,\alpha;\Omega}$, and it suffices to prove the latter for compact subsets of Ω. Namely, let $\{\Omega_i\}$ be a sequence of open subsets of Ω such that $\Omega_i\subset\subset\Omega_{i+1}\subset\subset\Omega$ and $\cup\,\Omega_i=\Omega$. We have that

$[u]^*_{2,\alpha;\Omega_i}$ is finite for each i and if (6.14) holds in Ω_i, we may assert for any pair of points $x, y \in \Omega$, for all sufficiently large i, and for any second derivative D^2u,

$$(d^{(i)}_{x,y})^{2+\alpha}\frac{|D^2u(x)-D^2u(y)|}{|x-y|^\alpha}\leqslant[u]^*_{2,\alpha;\Omega_i}\leqslant C(|u|_{0;\Omega_i}+|f|^{(2)}_{0,\alpha;\Omega_i})$$
$$\leqslant C(|u|_{0;\Omega}+|f|^{(2)}_{0,\alpha;\Omega})$$

where $d^{(i)}_{x,y}=\min\,[\operatorname{dist}(x,\partial\Omega_i), \operatorname{dist}(y,\partial\Omega_i)]$. Letting $i\to\infty$, we obtain the inequality

$$d^{2+\alpha}_{x,y}\frac{|D^2u(x)-D^2u(y)|}{|x-y|^\alpha}\leqslant C(|u|_{0;\Omega}+|f|^{(2)}_{0,\alpha;\Omega})\quad \forall x,y\in\Omega,$$

which implies the same bound for $[u]^*_{2,\alpha;\Omega}$. We may therefore assume in the following that $[u]^*_{2,\alpha;\Omega}$ is finite.

For notational convenience the same letter C will be used to denote constants depending on n, α, λ, Λ.

Let x_0, y_0 be any two distinct points in Ω and suppose $d_{x_0}=d_{x_0,y_0}=\min\,(d_{x_0}, d_{y_0})$. Let $\mu\leqslant\frac{1}{2}$ be a positive constant to be specified later, and set $d=\mu d_{x_0}$, $B=B_d(x_0)$. We rewrite $Lu=f$ in the form

$$a^{ij}(x_0)D_{ij}u=(a^{ij}(x_0)-a^{ij}(x))D_{ij}u-b^iD_iu-cu+f\equiv F(x),\tag{6.15}$$

and we consider this as an equation in B with constant coefficients $a^{ij}(x_0)$. Lemma 6.1(a), applied to this equation, asserts that if $y_0\in B_{d/2}(x_0)$, then for any second derivative D^2u

$$\left(\frac{d}{2}\right)^{2+\alpha}\frac{|D^2u(x_0)-D^2u(y_0)|}{|x_0-y_0|^\alpha}\leqslant C(|u|_{0;B}+|F|^{(2)}_{0,\alpha;B});$$

and thus

$$d^{2+\alpha}_{x_0}\frac{|D^2u(x_0)-D^2u(y_0)|}{|x_0-y_0|^\alpha}\leqslant\frac{C}{\mu^{2+\alpha}}(|u|_{0;B}+|F|^{(2)}_{0,\alpha;B}).$$

On the other hand, if $|x_0-y_0|\geqslant d/2$,

$$d^{2+\alpha}_{x_0}\frac{|D^2u(x_0)-D^2u(y_0)|}{|x_0-y_0|^\alpha}\leqslant\left(\frac{2}{\mu}\right)^\alpha[d^2_{x_0}|D^2u(x_0)|+d^2_{y_0}|D^2u(y_0)|]$$
$$\leqslant\frac{4}{\mu^\alpha}[u]^*_{2;\Omega},$$

so that, combining these two inequalities, we obtain

$$d^{2+\alpha}_{x_0}\frac{|D^2u(x_0)-D^2u(y_0)|}{|x_0-y_0|^\alpha}\leqslant\frac{C}{\mu^{2+\alpha}}(|u|_{0;\Omega}+|F|^{(2)}_{0,\alpha;B})+\frac{4}{\mu^\alpha}[u]^*_{2;\Omega}.\tag{6.16}$$

We proceed to estimate $|F|^{(2)}_{0,\alpha;B}$ in terms of $|u|_{0;\Omega}$ and $[u]^*_{2,\alpha;\Omega}$. We have

$$|F|^{(2)}_{0,\alpha;B} \leqslant \sum_{i,j} |(a^{ij}(x_0)-a^{ij}(x))D_{ij}u|^{(2)}_{0,\alpha;B} + \sum_i |b^iD_iu|^{(2)}_{0,\alpha;B} + |cu|^{(2)}_{0,\alpha;B} + |f|^{(2)}_{0,\alpha;B}. \tag{6.17}$$

It will be useful in estimating these terms to have the following inequality. Recalling that for all $x \in B$, $d_x(=\text{dist}\,(x,\partial\Omega)) > (1-\mu)d_{x_0} \geqslant \frac{1}{2}d_{x_0}$, we have for $g \in C^\alpha(\Omega)$

$$\begin{aligned}|g|^{(2)}_{0,\alpha;B} &\leqslant d^2|g|_{0;B} + d^{2+\alpha}[g]_{\alpha;B} \leqslant \frac{\mu^2}{(1-\mu)^2}[g]^{(2)}_{0;\Omega} + \frac{\mu^{2+\alpha}}{(1-\mu)^{2+\alpha}}[g]^{(2)}_{0,\alpha;\Omega} \\ &\leqslant 4\mu^2[g]^{(2)}_{0;\Omega} + 8\mu^{2+\alpha}[g]^{(2)}_{0,\alpha;\Omega} \leqslant 8\mu^2|g|^{(2)}_{0,\alpha;\Omega}.\end{aligned} \tag{6.18}$$

Writing $(a(x_0)-a(x))D^2u = (a^{ij}(x_0)-a^{ij}(x))D_{ij}u$ for each pair i, j, we obtain from (6.11) and (6.18)

$$\begin{aligned}|(a(x_0)-a(x))D^2u|^{(2)}_{0,\alpha;B} &\leqslant |a(x_0)-a(x)|^{(0)}_{0,\alpha;B}|D^2u|^{(2)}_{0,\alpha;B} \\ &\leqslant |a(x_0)-a(x)|^{(0)}_{0,\alpha;B}(4\mu^2[u]^*_{2;\Omega} + 8\mu^{2+\alpha}[u]^*_{2,\alpha;\Omega}).\end{aligned}$$

Since

$$\begin{aligned}|a(x_0)-a(x)|^{(0)}_{0,\alpha;B} &\leqslant \sup_{x\in B}|a(x_0)-a(x)| + d^\alpha[a]_{\alpha;B} \leqslant 2d^\alpha[a]_{\alpha;B} \\ &\leqslant 2^{1+\alpha}\mu^\alpha[a]^*_{0,\alpha;\Omega} \leqslant 4\Lambda\mu^\alpha,\end{aligned}$$

we arrive at the following estimate for the principal term in (6.17),

$$\begin{aligned}\sum_{i,j}|(a^{ij}(x_0)-a^{ij}(x))D_{ij}u|^{(2)}_{0,\alpha;B} &\leqslant 32n^2\Lambda\mu^{2+\alpha}([u]^*_{2;\Omega} + \mu^\alpha[u]^*_{2,\alpha;\Omega}) \\ &\leqslant 32n^2\Lambda\mu^{2+\alpha}(C(\mu)|u|_{0;\Omega} + 2\mu^\alpha[u]^*_{2,\alpha;\Omega}).\end{aligned} \tag{6.19}$$

The last inequality is obtained by setting $\varepsilon=\mu^\alpha$ in the interpolation inequality (6.8).

Letting $bDu = b^iD_iu$ for each i, we obtain from (6.18) and (6.13)

$$\begin{aligned}|bDu|^{(2)}_{0,\alpha;B} &\leqslant 8\mu^2|bDu|^{(2)}_{0,\alpha;\Omega} \leqslant 8\mu^2|b|^{(1)}_{0,\alpha;\Omega}|Du|^{(1)}_{0,\alpha;\Omega} \\ &\leqslant 8\mu^2\Lambda|u|^*_{1,\alpha;\Omega} \leqslant 8\mu^2\Lambda(C(\mu)|u|_{0;\Omega} + \mu^{2\alpha}[u]^*_{2,\alpha;\Omega}).\end{aligned}$$

The last inequality is obtained by setting $\varepsilon=\mu^{2\alpha}$ in (6.9). Thus we have

$$|b^iD_iu|^{(2)}_{0,\alpha;B} \leqslant 8n\Lambda\mu^2(C(\mu)|u|_{0;\Omega} + \mu^{2\alpha}[u]^*_{2,\alpha;\Omega}). \tag{6.20}$$

Similarly, from (6.18), (6.11) and (6.9), we obtain

$$|cu|^{(2)}_{0,\alpha;B} \leqslant 8\mu^2|c|^{(2)}_{0,\alpha;\Omega}|u|^{(0)}_{0,\alpha;\Omega} \leqslant 8\Lambda\mu^2(C(\mu)|u|_{0;\Omega} + \mu^{2\alpha}[u]^*_{2,\alpha;\Omega}). \tag{6.21}$$

Finally,

$$|f|^{(2)}_{0,\alpha;B} \leqslant 8\mu^2 |f|^{(2)}_{0,\alpha;\Omega}. \tag{6.22}$$

Letting C denote constants depending only on n, α, λ, Λ, and $C(\mu)$ constants depending also on μ, we find after combining (6.19)–(6.22),

$$|F|^{(2)}_{0,\alpha;B} \leqslant C\mu^{2+2\alpha}[u]^*_{2,\alpha;\Omega} + C(\mu)(|u|_{0;\Omega} + |f|^{(2)}_{0,\alpha;\Omega}).$$

Inserting this into the right member of (6.16), and using (6.8) with $\varepsilon = \mu^{2\alpha}$ to estimate $[u]^*_{2;\Omega}$, we obtain from (6.16)

$$d^{2+\alpha}_{x_0,y_0} \frac{|D^2u(x_0) - D^2u(y_0)|}{|x_0 - y_0|^\alpha} \leqslant C\mu^\alpha[u]^*_{2,\alpha;\Omega} + C(\mu)(|u|_{0;\Omega} + |f|^{(2)}_{0,\alpha;\Omega}).$$

The right member of this inequality is independent of x_0, y_0. Taking the supremum over all $x_0, y_0 \in \Omega$, we obtain

$$[u]^*_{2,\alpha;\Omega} \leqslant C\mu^\alpha[u]^*_{2,\alpha;\Omega} + C(\mu)(|u|_{0;\Omega} + |f|^{(2)}_{0,\alpha;\Omega}).$$

We now choose and fix $\mu = \mu_0$ so that $C\mu_0^\alpha \leqslant \frac{1}{2}$. Then we arrive at the desired estimate

$$[u]^*_{2,\alpha;\Omega} \leqslant C(|u|_{0;\Omega} + |f|^{(2)}_{0,\alpha;\Omega}). \quad \square$$

The preceding form of the interior estimates for $Lu = f$ permits the coefficients and f to be unbounded subject to the condition (6.13). In typical applications of the interior estimates to convergence results, it suffices to know equicontinuity of solutions and their derivatives up to second order on compact subsets. For this purpose the following corollary is usually adequate.

Corollary 6.3. *Let $u \in C^{2,\alpha}(\Omega)$, $f \in C^\alpha(\bar{\Omega})$ satisfy $Lu = f$ in a bounded domain Ω where L satisfies (6.2) and its coefficients are in $C^\alpha(\bar{\Omega})$. Then if $\Omega' \subset\subset \Omega$ with* $\operatorname{dist}(\Omega', \partial\Omega) \geqslant d$, *there is a constant C such that*

$$d|Du|_{0;\Omega'} + d^2|D^2u|_{0;\Omega'} + d^{2+\alpha}[D^2u]_{\alpha;\Omega'} \leqslant C(|u|_{0;\Omega} + |f|_{0,\alpha;\Omega}) \tag{6.23}$$

where C depends only on the ellipticity constant λ and the $C^\alpha(\bar{\Omega})$ norms of the coefficients of L (as well as on n, α and the diameter of Ω).

Remark. An immediate consequence of this result is that uniformly bounded solutions of elliptic equations $Lu = f$ with locally Hölder continuous coefficients and locally Hölder continuous f are equicontinuous with their first and second derivatives on compact subsets. This is true as well for the solutions of any *family* of such equations with ellipticity constant λ uniformly bounded away from zero in compact subsets $\Omega' \subset\subset \Omega$ and with coefficients and inhomogeneous term f having uniformly bounded $C^{\alpha}(\bar{\Omega}')$ norms.

6.2. Boundary and Global Estimates

To extend the preceding interior estimates to the entire domain it is necessary to have estimates that are meaningful near the boundary. These can be obtained provided the boundary values of the solution and the boundary itself are sufficiently smooth. In many respects the proof of these boundary estimates follows closely that of the interior estimates.

The global estimates which are the principal goal of this section will be established in *domains of class* $C^{2,\alpha}$.

Definition. A bounded domain Ω in $\mathbb{R}^n$ and its boundary are of class $C^{k,\alpha}$, $0 \leqslant \alpha \leqslant 1$, if at each point $x_0 \in \partial\Omega$ there is a ball $B = B(x_0)$ and a one-to-one mapping ψ of B onto $D \subset \mathbb{R}^n$ such that:

(i) $\psi(B \cap \Omega) \subset \mathbb{R}^n_+$; (ii) $\psi(B \cap \partial\Omega) \subset \partial\mathbb{R}^n_+$; (iii) $\psi \in C^{k,\alpha}(B)$, $\psi^{-1} \in C^{k,\alpha}(D)$.

A domain Ω will be said to have a *boundary portion* $T \subset \partial\Omega$ of class $C^{k,\alpha}$ if at each point $x_0 \in T$ there is a ball $B = B(x_0)$ in which the above conditions are satisfied and such that $B \cap \partial\Omega \subset T$. We shall say that the diffeomorphism ψ *straightens the boundary* near x_0.

We note in particular that Ω is a $C^{k,\alpha}$ domain if each point of $\partial\Omega$ has a neighborhood in which $\partial\Omega$ is the graph of a $C^{k,\alpha}$ function of $n-1$ of the coordinates $x_1, \ldots, x_n$. The converse is also true if $k \geqslant 1$.

It follows from the above definition that a domain of class $C^{k,\alpha}$ is also of class $C^{j,\beta}$ provided $j + \beta < k + \alpha$, $0 \leqslant \alpha, \beta \leqslant 1$.

A function φ defined on a $C^{k,\alpha}$ boundary portion T of a domain Ω will be said to be in class $C^{k,\alpha}(T)$ if $\varphi \circ \psi^{-1} \in C^{k,\alpha}(D \cap \partial\mathbb{R}^n_+)$ for each $x_0 \in T$. It is important to note that if $\partial\Omega$ is in $C^{k,\alpha}(k \geqslant 1)$, then a function $\varphi \in C^{k,\alpha}(\partial\Omega)$ can be extended to a function in $C^{k,\alpha}(\bar{\Omega})$, and conversely, a function in $C^{k,\alpha}(\bar{\Omega})$ has boundary values in $C^{k,\alpha}(\partial\Omega)$ (see Lemma 6.38). Hence, in what follows it is immaterial whether we consider boundary values φ belonging to $C^{k,\alpha}(\partial\Omega)$ or $C^{k,\alpha}(\bar{\Omega})$.

It is also possible to define a boundary norm on $C^{k,\alpha}(\partial\Omega)$, in various ways. For example, if $\varphi \in C^{k,\alpha}(\partial\Omega)$ let Φ denote an extension of φ to $\bar{\Omega}$ and define $\|\varphi\|_{C^{k,\alpha}(\partial\Omega)} = \inf_{\Phi} \|\Phi\|_{C^{k,\alpha}(\bar{\Omega})}$, where the infimum is taken over the set of all global extensions Φ. Equipped with this norm, $C^{k,\alpha}(\partial\Omega)$ becomes a Banach space. In this work we shall not use boundary norms for functions on curved boundaries but instead we generally consider a boundary function as the restriction of a globally defined function with its appropriate norm.

In obtaining boundary estimates for $Lu = f$ in domains with a $C^{2,\alpha}(\alpha > 0)$ boundary portion we first establish such an estimate in domains with a hyperplane boundary portion. For this purpose we use the following interpolation inequality, analogous to (6.8), (6.9). For its statement we recall the partially interior norms and seminorms defined in (4.29).

Let Ω be an open subset of $\mathbb{R}^n_+$ with a boundary portion T on $x_n=0$, and assume $u \in C^{2,\alpha}(\Omega \cup T)$. Then for any $\varepsilon>0$ and some constant $C(\varepsilon)$ we have

$$(6.24)\qquad [u]^*_{j,\beta;\Omega\cup T} \leqslant C|u|_{0;\Omega} + \varepsilon[u]^*_{2,\alpha;\Omega\cup T},$$

$$j=0,1,2;\quad 0\leqslant \alpha,\beta\leqslant 1,\ j+\beta<2+\alpha;$$

$$(6.25)\qquad |u|^*_{j,\beta;\Omega\cup T} \leqslant C|u|_{0;\Omega} + \varepsilon[u]^*_{2,\alpha;\Omega\cup T}.$$

These inequalities are proved in Lemma 6.34 of Appendix 1 to this chapter. We can now assert the following essential local boundary estimate.

Lemma 6.4. *Let Ω be an open subset of $\mathbb{R}^n_+$, with a boundary portion T on $x_n=0$. Suppose that $u \in C^{2,\alpha}(\Omega \cup T)$ is a bounded solution in Ω of $Lu=f$ satisfying the boundary condition $u=0$ on T. In addition to (6.2) it is assumed that*

$$(6.26)\qquad |a^{ij}|^{(0)}_{0,\alpha;\Omega\cup T},\ |b^i|^{(1)}_{0,\alpha;\Omega\cup T},\ |c|^{(2)}_{0,\alpha;\Omega\cup T} \leqslant \Lambda;\quad |f|^{(2)}_{0,\alpha;\Omega\cup T}<\infty.$$

Then

$$(6.27)\qquad |u|^*_{2,\alpha;\Omega\cup T} \leqslant C(|u|_{0;\Omega} + |f|^{(2)}_{0,\alpha;\Omega\cup T})$$

where $C=C(n,\alpha,\lambda,\Lambda)$.

Proof. The proof is identical with that of Theorem 6.2 if in the latter d_x is replaced by $\bar{d}_x$ and Lemma 6.1(a) and the inequalities (6.8), (6.9) are replaced when necessary by Lemma 6.1(b) and the inequalities (6.24), (6.25). □

This lemma provides a bound on the first and second derivatives of u and the Hölder coefficients of its second derivatives in any subset $\Omega' \subset \Omega$ for which dist $(\Omega', \partial\Omega - T)>0$. In particular $\partial\Omega'$ may contain any portion of T at a non-zero distance from $\partial\Omega - T$.

In order to extend the preceding lemma to domains with a curved boundary portion, we introduce the relevant seminorms and norms, in obvious generalization of (4.29). Let Ω be an open set in $\mathbb{R}^n$ with $C^{k,\alpha}$ boundary portion T. For $x, y \in \Omega$ we set

$$\bar{d}_x = \text{dist}\,(x, \partial\Omega - T),\qquad \bar{d}_{x,y} = \min\,(\bar{d}_x, \bar{d}_y)$$

and for functions $u \in C^{k,\alpha}(\Omega \cup T)$ we define the following quantities:

$$
\begin{aligned}
&[u]^*_{k,0;\Omega\cup T} = [u]^*_{k;\Omega\cup T} = \sup_{\substack{x\in\Omega\\ |\beta|=k}} d_x^k |D^\beta u(x)|, \quad k=0,1,2,\ldots;\\
&[u]^*_{k,\alpha;\Omega\cup T} = \sup_{\substack{x,y\in\Omega\\ |\beta|=k}} d_{x,y}^{k+\alpha} \frac{|D^\beta u(x)-D^\beta u(y)|}{|x-y|^\alpha}, \quad 0<\alpha\leqslant 1;\\
(6.28)\qquad &|u|^*_{k,0;\Omega\cup T} = |u|^*_{k;\Omega\cup T} = \sum_{j=0}^{k} [u]^*_{j;\Omega\cup T};\\
&|u|^*_{k,\alpha;\Omega\cup T} = |u|^*_{k;\Omega\cup T} + [u]^*_{k,\alpha;\Omega\cup T};\\
&|u|^{(k)}_{0,\alpha;\Omega\cup T} = \sup_{x\in\Omega} d_x^k |u(x)| + \sup_{x,y\in\Omega} d_{x,y}^{k+\alpha} \frac{|u(x)-u(y)|}{|x-y|^\alpha}.
\end{aligned}
$$

When $T=\phi$ and $\Omega \cup T=\Omega$, these quantities reduce to the interior seminorms and norms already defined in (4.17) and (4.18).

Let Ω be a bounded domain with $C^{k,\alpha}$ boundary portion T, $k\geqslant 1$, $0\leqslant\alpha\leqslant 1$. Suppose that $\Omega \subset\subset D$, where D is a domain that is mapped by a $C^{k,\alpha}$ diffeomorphism ψ onto D'. Letting $\psi(\Omega)=\Omega'$ and $\psi(T)=T'$, we can define the quantities in (6.28) with respect to Ω' and T'. If $x'=\psi(x)$, $y'=\psi(y)$, one sees that

$$
(6.29)\qquad K^{-1}|x-y| \leqslant |x'-y'| \leqslant K|x-y|
$$

for all points $x, y \in \Omega$, where K is a constant depending on ψ and Ω. Letting $u(x) \to \tilde{u}(x')$ under the mapping $x \to x'$, we find after a calculation using (6.29): for $0\leqslant j\leqslant k$, $0\leqslant\beta\leqslant 1$, $j+\beta\leqslant k+\alpha$,

$$
\begin{aligned}
&K^{-1}|u(x)|_{j,\beta;\Omega} \leqslant |\tilde{u}(x')|_{j,\beta;\Omega'} \leqslant K|u(x)|_{j,\beta;\Omega};\\
(6.30)\qquad &K^{-1}|u(x)|^*_{j,\beta;\Omega\cup T} \leqslant |\tilde{u}(x')|^*_{j,\beta;\Omega'\cup T'} \leqslant K|u(x)|^*_{j,\beta;\Omega\cup T};\\
&K^{-1}|u(x)|^{(\sigma)}_{0,\beta;\Omega\cup T} \leqslant |\tilde{u}(x')|^{(\sigma)}_{0,\beta;\Omega'\cup T'} \leqslant K|u(x)|^{(\sigma)}_{0,\beta;\Omega\cup T}.
\end{aligned}
$$

In these inequalities K denotes constants depending on the mapping ψ and the domain Ω.

Lemma 6.4 can now be applied together with (6.30) to obtain a local boundary estimate for curved boundaries. For this it is convenient to use the global norms (4.6).

Lemma 6.5. *Let Ω be a $C^{2,\alpha}$ domain in $\mathbb{R}^n$, and let $u \in C^{2,\alpha}(\bar{\Omega})$ be a solution of $Lu=f$ in Ω, $u=0$ on $\partial\Omega$, where $f \in C^\alpha(\bar{\Omega})$. It is assumed that the coefficients of L satisfy* (6.2) *and*

$$
(6.31)\qquad |a^{ij}|_{0,\alpha;\Omega},\ |b^i|_{0,\alpha;\Omega},\ |c|_{0,\alpha;\Omega} \leqslant \Lambda.
$$

Then for some δ there is a ball $B=B_\delta(x_0)$ at each point $x_0 \in \partial\Omega$ such that

$$|u|_{2,\alpha;B\cap\Omega} \leqslant C(|u|_{0;\Omega}+|f|_{0,\alpha;\Omega}) \tag{6.32}$$

where $C=C(n, \alpha, \lambda, \Lambda, \Omega)$.

Proof. By the definition of a $C^{2,\alpha}$ domain, at each point $x_0 \in \partial\Omega$ there is a neighborhood N of x_0 and a $C^{2,\alpha}$ diffeomorphism that straightens the boundary in N. Let $B_\rho(x_0) \subset\subset N$ and set $B'=B_\rho(x_0)\cap\Omega$, $D'=\psi(B')$, $T=B_\rho(x_0)\cap\partial\Omega\subset\partial B'$ and $T'=\psi(T)\subset\partial D'$ (T' is a hyperplane portion of $\partial D'$). Under the mapping $y=\psi(x)=(\psi_1(x), \ldots, \psi_n(x))$, let $\tilde{u}(y)=u(x)$ and $\tilde{L}\tilde{u}(y)=Lu(x)$, where

$$\tilde{L}\tilde{u} \equiv \tilde{a}^{ij}D_{ij}\tilde{u}+\tilde{b}^iD_i\tilde{u}+\tilde{c}\tilde{u}=\tilde{f}(y)$$

and

$$\tilde{a}^{ij}(y)=\frac{\partial\psi_i}{\partial x_r}\frac{\partial\psi_j}{\partial x_s}a^{rs}(x), \qquad \tilde{b}^i(y)=\frac{\partial^2\psi_i}{\partial x_r\partial x_s}a^{rs}(x)+\frac{\partial\psi_i}{\partial x_r}b^r(x),$$

$$\tilde{c}(y)=c(x), \qquad \tilde{f}(y)=f(x).$$

We observe that in D'

$$\tilde{\lambda}|\xi|^2 \leqslant \tilde{a}^{ij}\xi_i\xi_j \quad \forall\xi\in\mathbb{R}^n,$$

where

$$\tilde{\lambda}=\lambda/K \tag{6.33}$$

for a suitable positive constant K depending only on the mapping ψ on B'. By virtue of (6.30) we have also (for appropriate choice of K in (6.33))

$$|\tilde{a}^{ij}|_{0,\alpha;D'}, |\tilde{b}^i|_{0,\alpha;D'}, |\tilde{c}|_{0,\alpha;D'} \leqslant \tilde{\Lambda}=K\Lambda; \quad |\tilde{f}|_{0,\alpha;D'}<\infty. \tag{6.34}$$

Thus the conditions of Lemma 6.4 are satisfied for the equation $\tilde{L}\tilde{u}=\tilde{f}$ in D' with the hyperplane portion T'. We can therefore assert

$$|\tilde{u}|^*_{2,\alpha;D'\cup T'} \leqslant C(|\tilde{u}|_{0;D'}+|\tilde{f}|^{(2)}_{0,\alpha;D'\cup T'})$$

where the constant $C=C(n, \alpha, \tilde{\lambda}, \tilde{\Lambda})$. It follows from (6.30) that

$$\begin{aligned}|u|^*_{2,\alpha;B'\cup T} &\leqslant C(|u|_{0,B'}+|f|^{(2)}_{0,\alpha;B'\cup T}) \leqslant C(|u|_{0;B'}+|f|_{0,\alpha;B'})\\ &\leqslant C(|u|_{0;\Omega}+|f|_{0,\alpha;\Omega}),\end{aligned}$$

where C now depends on $n, \alpha, \lambda, \Lambda$ and B'. Letting $B''=B_{\rho/2}(x_0)\cap\Omega$ and observing that

$$\min(1, (\rho/2)^{2+\alpha})|u|_{2,\alpha;B''} \leqslant |u|^*_{2,\alpha;B'\cup T},$$

we obtain

$$(6.35)\qquad |u|_{2,\alpha;B''}\leqslant C(|u|_{0;\Omega}+|f|_{0,\alpha;\Omega}).$$

The radius ρ appearing in the previous estimates depends in general on the point $x_0\in\partial\Omega$. Consider now the collection of balls $B_{\rho/4}(x)$ for all $x\in\partial\Omega$. A finite subset $B_{\rho_i/4}(x_i)$, $i=1,\ldots,N$, of this collection covers $\partial\Omega$. Let $\delta=\min\rho_i/4$ be the minimum radius of the balls in this finite covering. We assert that for this δ the conclusion of the lemma is true. Namely, let C_i be the constant in (6.35) corresponding to x_i, and let $C=\max C_i$. Consider any point $x_0\in\partial\Omega$ and the ball $B_\delta(x_0)$. For some i we must have $x_0\in B_{\rho_i/4}(x_i)$ and hence $B=B_\delta(x_0)\subset B_{\rho_i/2}(x_i)=B_i$. From (6.35) we obtain the desired conclusion

$$|u|_{2,\alpha;B\cap\Omega}\leqslant|u|_{2,\alpha;B_i\cap\Omega}\leqslant C(|u|_{0,\Omega}+|f|_{0,\alpha;\Omega})$$

where C depends on n, α, λ, Λ, and Ω. □

We remark that in the preceding lemma the dependence of the constant C on the domain Ω is through the constant K in (6.30), (6.33) and (6.34), and K in turn depends only on the $C^{2,\alpha}$ bounds on the mapping ψ which defines the local representation of the boundary $\partial\Omega$. If the bounds on the mapping ψ can be stated uniformly over the boundary (which is always possible for $C^{2,\alpha}$ domains), then K can replace Ω in the statement of the estimate (6.32) and the domain Ω may also be unbounded.

The principal result of this section is the following apriori global estimate for solutions with $C^{2,\alpha}$ boundary values defined on $C^{2,\alpha}$ domains.

Theorem 6.6. *Let Ω be a $C^{2,\alpha}$ domain in $\mathbb{R}^n$ and let $u\in C^{2,\alpha}(\bar\Omega)$ be a solution of $Lu=f$ in Ω, where $f\in C^\alpha(\bar\Omega)$ and the coefficients of L satisfy, for positive constants λ, Λ,*

$$a^{ij}\xi_i\xi_j\geqslant\lambda|\xi|^2\quad\forall x\in\Omega,\ \xi\in\mathbb{R}^n,$$

and

$$|a^{ij}|_{0,\alpha;\Omega},\ |b^i|_{0,\alpha;\Omega},\ |c|_{0,\alpha;\Omega}\leqslant\Lambda.$$

Let $\varphi(x)\in C^{2,\alpha}(\bar\Omega)$, and suppose $u=\varphi$ on $\partial\Omega$. Then

$$(6.36)\qquad |u|_{2,\alpha;\Omega}\leqslant C(|u|_{0;\Omega}+|\varphi|_{2,\alpha;\Omega}+|f|_{0,\alpha;\Omega})$$

where $C=C(n,\alpha,\lambda,\Lambda,\Omega)$.

Proof. It suffices to prove the theorem for the case $u=0$ on $\partial\Omega$ and $\varphi=0$. Namely, if we set $v=u-\varphi$, then $v=0$ on $\partial\Omega$ and $Lv=f-L\varphi\equiv f'\in C^\alpha(\bar\Omega)$. The conclusion (6.36), applied to v with $\varphi=0$, asserts

$$|v|_{2,\alpha;\Omega}\leqslant C(|v|_{0;\Omega}+|f'|_{0,\alpha;\Omega}).$$

Since $|L\varphi|_{0,\alpha;\Omega}\leqslant C|\varphi|_{2,\alpha;\Omega}$, it follows

$$|u|_{2,\alpha;\Omega}\leqslant|v|_{2,\alpha;\Omega}+|\varphi|_{2,\alpha;\Omega}\leqslant C(|u|_{0;\Omega}+|\varphi|_{2,\alpha;\Omega}+|f|_{0,\alpha;\Omega})$$

as asserted in the theorem. In the remainder of the proof we assume $u=0$ on $\partial\Omega$.

Let $x\in\Omega$. We consider the two possibilities: (i) $x\in B_0=B_{2\sigma}(x_0)\cap\Omega$ for some $x_0\in\partial\Omega$, where $\delta=2\sigma$ is the radius in Lemma 6.5; (ii) $x\in\Omega_\sigma=\{x\in\Omega\mid \operatorname{dist}(x,\partial\Omega)>\sigma\}$. In case (i) Lemma 6.5 implies that

$$|Du(x)|+|D^2u(x)|\leqslant C(|u|_0+|f|_{0,\alpha}). \tag{6.37}$$

(Here and in the following we omit the subscript Ω without ambiguity.) In case (ii) we obtain the same inequality, with a different constant C, from Corollary 6.3 after setting $d=\sigma$ in (6.23). Choosing the larger of the two constants C, we may assume (6.37) for any point x in Ω, and hence for $|u|_2$.

Now let x, y be distinct points in Ω and consider the three possibilities: (i) $x, y\in B_0$ for some x_0; (ii) $x, y\in\Omega_\sigma$; (iii) x or y is in $\Omega-\Omega_\sigma$ but not both x and y are in the same ball B_0 for any x_0. These exhaust all possibilities. We consider the Hölder quotient $|D^2u(x)-D^2u(y)|/|x-y|^\alpha$. In case (i), Lemma 6.5 gives the inequality

$$\frac{|D^2u(x)-D^2u(y)|}{|x-y|^\alpha}\leqslant C_1(|u|_0+|f|_{0,\alpha}).$$

In case (ii) we obtain the same inequality, with a different constant C_2, from Corollary 6.3. In case (iii), $\operatorname{dist}(x, y)>\sigma$, so that

$$\begin{aligned}\frac{|D^2u(x)-D^2u(y)|}{|x-y|^\alpha}&\leqslant\sigma^{-\alpha}(|D^2u(x)|+|D^2u(y)|)\\&\leqslant C_3(|u|_0+|f|_{0,\alpha})\quad\text{by (6.37).}\end{aligned}$$

Letting $C=\max(C_1, C_2, C_3)$, and taking the supremum over all $x, y\in\Omega$, we obtain

$$[D^2u]_\alpha\leqslant C(|u|_0+|f|_{0,\alpha}).$$

Combining this result with the bound for $|u|_2$ given by (6.37), we get

$$|u|_{2,\alpha}\leqslant C(|u|_0+|f|_{0,\alpha}),$$

which establishes the theorem. □

Remark. The typical application of Theorem 6.6 concerns a set of solutions of an equation or family of equations, each solution satisfying a uniform estimate (6.36). The boundedness of the set of solutions in $C^{2,\alpha}(\bar{\Omega})$ then assures their precompactness in $C^2(\bar{\Omega})$; (see Lemma 6.36).

A simple modification of the argument in Theorem 6.6 yields the following local estimate in domains with a $C^{2,\alpha}$ boundary portion.

Corollary 6.7. *Let $\Omega \subset \mathbb{R}^n$ be a domain with a $C^{2,\alpha}$ boundary portion $T \subset \partial\Omega$. Let $u \in C^{2,\alpha}(\Omega \cup T)$ be a solution of $Lu=f$ in Ω, $u=\varphi$ on T, where L and f satisfy the conditions of Theorem 6.6, and $\varphi \in C^{2,\alpha}(\bar{\Omega})$. Then, if $x_0 \in T$ and $B=B_\rho(x_0)$ is a ball with radius $\rho < \text{dist}\,(x_0, \partial\Omega - T)$, we have*

$$|u|_{2,\alpha;B\cap\Omega} \leqslant C(|u|_{0;\Omega} + |\varphi|_{2,\alpha;\Omega} + |f|_{0,\alpha;\Omega}), \tag{6.38}$$

where $C = C(n, \alpha, \lambda, \Lambda, B \cap \Omega)$.

It is a straightforward matter to extend the estimates of this and the preceding section to equations whose coefficients and inhomogeneous term are in $C^{k,\alpha}$, where $k>0$; (see Problems 6.1, 6.2).

6.3. The Dirichlet Problem

We consider the Dirichlet problem for $Lu=f$ in a bounded domain Ω of $\mathbb{R}^n$. Our procedure for solving this variable coefficient equation is to reduce it to the case of constant coefficients by the *method of continuity* (Theorem 5.2). Briefly summarized, this method as applied here starts with the solution for Poisson's equation $\Delta u=f$ and then arrives at a solution for $Lu=f$ through solutions for a continuous family of equations connecting $\Delta u=f$ and $Lu=f$.

We treat first the Dirichlet problem for sufficiently smooth domains and boundary values. In this case the connection between the solvability of Poisson's equation and of $Lu=f$ when $c \leqslant 0$ is contained in the following theorem.

Theorem 6.8. *Let Ω be a $C^{2,\alpha}$ domain in $\mathbb{R}^n$, and let the operator L be strictly elliptic in Ω with coefficients in $C^\alpha(\bar{\Omega})$ and with $c \leqslant 0$. Then if the Dirichlet problem for Poisson's equation, $\Delta u=f$ in Ω, $u=\varphi$ on $\partial\Omega$, has a $C^{2,\alpha}(\bar{\Omega})$ solution for all $f \in C^\alpha(\bar{\Omega})$ and all $\varphi \in C^{2,\alpha}(\bar{\Omega})$, the problem,*

$$Lu=f \text{ in } \Omega, \qquad u=\varphi \text{ on } \partial\Omega, \tag{6.39}$$

also has a (unique) $C^{2,\alpha}(\bar{\Omega})$ solution for all such f and φ.

Proof. By hypothesis we may assume that the coefficients of L satisfy the conditions

$$\begin{aligned} &\lambda|\xi|^2 \leqslant a^{ij}\xi_i\xi_j \quad \forall x \in \Omega,\ \xi \in \mathbb{R}^n, \\ &|a^{ij}|_{0,\alpha}, |b^i|_{0,\alpha}, |c|_{0,\alpha} \leqslant \Lambda, \end{aligned} \tag{6.40}$$

with positive constants λ, Λ. (In writing norms, we omit the subscript Ω, which will be implicitly understood.) It suffices to restrict consideration to zero boundary values, since the problem (6.39) is equivalent to $Lv=f-L\varphi\equiv f'$ in Ω, $v=0$ on $\partial\Omega$.

We consider the family of equations,

$$L_t u \equiv tLu+(1-t)\Delta u=f, \quad 0\leqslant t\leqslant 1. \tag{6.41}$$

We note that $L_0=\Delta$, $L_1=L$, and that the coefficients of L_t satisfy (6.40) with

$$\lambda_t=\min(1,\lambda), \qquad \Lambda_t=\max(1,\Lambda).$$

The operator L_t may be considered a bounded linear operator from the Banach space $\mathfrak{B}_1=\{u\in C^{2,\alpha}(\bar\Omega)\mid u=0 \text{ on } \partial\Omega\}$ into the Banach space $\mathfrak{B}_2=C^\alpha(\bar\Omega)$. The solvability of the Dirichlet problem, $L_t u=f$ in Ω, $u=0$ on $\partial\Omega$, for arbitrary $f\in C^\alpha(\bar\Omega)$ is then equivalent to the invertibility of the mapping L_t. Let u_t denote a solution of this problem. By virtue of Theorem 3.7, we have the bound

$$|u_t|_0\leqslant C\sup_\Omega|f|\leqslant C|f|_{0,\alpha},$$

where C depends only on λ, Λ and the diameter of Ω. Hence from (6.36), we have

$$|u_t|_{2,\alpha}\leqslant C|f|_{0,\alpha}, \tag{6.42}$$

that is,

$$\|u\|_{\mathfrak{B}_1}\leqslant C\|L_t u\|_{\mathfrak{B}_2},$$

the constant C being independent of t. Since, by hypothesis, $L_0=\Delta$ maps $\mathfrak{B}_1$ onto $\mathfrak{B}_2$, the method of continuity (Theorem 5.2) is applicable and the theorem follows. □

The preceding theorem presupposes the solvability in $C^{2,\alpha}(\bar\Omega)$ of the Dirichlet problem for Poisson's equation when Ω and the boundary values are in class $C^{2,\alpha}$. Although this result—Kellogg's theorem—can be established independently by potential theoretic methods, we shall not assume it or prove it here, but rather derive it later as a consequence of elliptic theory. However, in the special case that Ω is a ball, Kellogg's theorem has already been proved, in Corollary 4.14. This provides the following existence theorem for balls.

Corollary 6.9. *In Theorem 6.8 let Ω be a ball B, and let the operator L satisfy the same conditions as in that theorem. Then if $f\in C^\alpha(\bar B)$ and $\varphi\in C^{2,\alpha}(\bar B)$, the Dirichlet problem, $Lu=f$ in B, $u=\varphi$ on ∂B, has a (unique) solution $u\in C^{2,\alpha}(\bar B)$.*

The conditions on the boundary data can be weakened to give the following generalization, which will be useful later.

Lemma 6.10. *Let T be a boundary portion (possibly empty) of a ball B in $\mathbb{R}^n$, and let $\varphi \in C^0(\partial B) \cap C^{2,\alpha}(T)$. Then if L satisfies the conditions of Theorem 6.8 in B and $f \in C^\alpha(\bar{B})$, the Dirichlet problem, $Lu=f$ in B, $u=\varphi$ on ∂B, has a (unique) solution $u \in C^{2,\alpha}(B \cup T) \cap C^0(\bar{B})$.*

Proof. Let $x_0 \in T$ if T is non-empty. We can assume that the boundary function φ is continued by radial extension to a function $\varphi \in C^0(B') \cap C^{2,\alpha}(\bar{G})$, where B' is a ball containing $\bar{B}$ and $G = B_\rho(x_0) \subset\subset B'$; (see Remark 2 after Lemma 6.38). Let $\{\varphi_k\}$ be a sequence of sufficiently smooth (say C^3) functions in B' such that

$$(6.43) \qquad |\varphi_k - \varphi|_{0;B} \to 0 \quad \text{and} \quad |\varphi_k|_{2,\alpha;G} \leqslant C|\varphi|_{2,\alpha;G} \quad \text{as } k \to \infty,$$

where C is a constant independent of k. (For the existence of such an approximation see the discussion after Lemma 7.1.) For each k let u_k be the corresponding solution of the Dirichlet problem, $Lu=f$ in B, $u=\varphi_k$ on ∂B. By Corollary 6.9 the functions u_k are known to exist in $C^{2,\alpha}(\bar{B})$. From the maximum principle it follows that the sequence $\{u_k\}$ converges uniformly to a function $u \in C^0(\bar{B})$ such that $u=\varphi$ on $\partial \bar{B}$. The compactness of $\{u_k\}$ provided by Corollary 6.3 assures that this sequence converges to a solution of $Lu=f$ on compact subsets of B and hence the limit function u is a solution in B lying in $C^0(\bar{B})$. Furthermore, by Corollary 6.7, in $D = B_{\rho/2}(x_0) \cap B$ the functions u_k satisfy the estimate

$$|u_k|_{2,\alpha;D} \leqslant C(|u_k|_{0;B} + |\varphi_k|_{2,\alpha;G} + |f|_{0,\alpha;B}).$$

It follows from (6.43) and Arzela's theorem that u satisfies such an estimate in D (with φ replacing φ_k) and in particular that $u \in C^{2,\alpha}(\bar{D})$. Thus $u \in C^{2,\alpha}(B \cup T)$ and the lemma is proved. □

We note especially that the preceding lemma provides a solution of the Dirichlet problem in balls when the boundary values are only continuous; the solution is then in class $C^0(\bar{B}) \cap C^{2,\alpha}(B)$.

It is now possible to imitate the Perron method of subharmonic functions of Chapter 2 and to extend the results obtained there for harmonic functions to the Dirichlet problem for $Lu = f$. A function $u \in C^0(\Omega)$ will be called a *subsolution* (*supersolution*) of $Lu = f$ in Ω if for every ball $B \subset\subset \Omega$ and every solution v such that $Lv = f$ in B, the inequality $u \leqslant v$ ($u \geqslant v$) on ∂B implies also $u \leqslant v$ ($u \geqslant v$) in B. If we suppose that L satisfies the strong maximum principle and that the Dirichlet problem for $Lu = f$ is solvable in balls for continuous boundary values, then subsolutions and subharmonic functions are seen to share many of the same properties. In particular, we assert the following propositions without proof, the details being essentially the same as for subharmonic functions. When f and the coefficients of L are in $C^\alpha(\Omega)$, with $c \leqslant 0$, we need only observe that in the proofs Theorem 3.5 and Lemma 6.10 replace Theorems 2.2 and 2.6, respectively.

(i) A function $u \in C^2(\Omega)$ is a subsolution if and only if $Lu \geqslant f$.

(ii) If u is a subsolution in a bounded domain Ω and v is a supersolution such that $v \geqslant u$ on $\partial\Omega$, then either $v > u$ throughout Ω or $v \equiv u$.

(iii) Let u be a subsolution in Ω and B a ball such that $\bar{B} \subset \Omega$. We denote by $\bar{u}$ the solution of $Lu = f$ in B satisfying the condition $\bar{u} = u$ on ∂B. Then the function U defined by

$$U(x) = \begin{cases} \bar{u}(x), & x \in B \\ u(x), & x \in \Omega - B \end{cases}$$

is a subsolution in Ω.

(iv) Let $u_1, u_2, \ldots, u_N$ be subsolutions in Ω, then the function $u(x) = \max\{u_1(x), u_2(x), \ldots, u_N(x)\}$ is also a subsolution in Ω.

In (i), (iii), and (iv) corresponding statements can obviously be made for supersolutions.

Now let Ω be a bounded domain and φ a bounded function on $\partial\Omega$. A function $u \in C^0(\bar{\Omega})$ will be called a *subfunction* (*superfunction*) relative to φ if u is a subsolution (supersolution) in Ω and $u \leqslant \varphi$ ($u \geqslant \varphi$) on $\partial\Omega$. By (ii) above every subfunction is less than or equal to every superfunction. We denote by S_φ the set of subfunctions in Ω relative to φ. Let us assume that S_φ is non-empty and is bounded from above. This is the case, for example, when L is strictly elliptic in Ω and its coefficients and f are bounded. Namely, if Ω lies in the slab $0 < x_1 < d$, then the functions

$$\begin{aligned} v^+ &= \sup|\varphi| + (e^{\gamma d} - e^{\gamma x_1}) \frac{\sup|f|}{\lambda} \\ v^- &= -\sup|\varphi| - (e^{\gamma d} - e^{\gamma x_1}) \frac{\sup|f|}{\lambda} \end{aligned} \tag{6.44}$$

are respectively superfunction and subfunction if the positive constant γ is sufficiently large; (see Theorem 3.7). The superfunction v^+ provides an upper bound for the functions of S_φ, and the existence of the subfunction v^- assures that S_φ is non-empty.

We can now assert the basic existence result of the Perron process for $Lu = f$, assuming that f and the coefficients of L are in $C^\alpha(\Omega)$ and that $c \leqslant 0$.

Theorem 6.11. *The function* $u(x) = \sup_{v \in S_\varphi} v(x)$ *lies in* $C^{2,\alpha}(\Omega)$ *and satisfies* $Lu = f$ *in* Ω *provided* u *is bounded.*

The proof differs only in minor details from that of Theorem 2.12 and is left to the reader. We call attention to the fact that the compactness of solutions required in the proof is provided by the interior estimate of Corollary 6.3, and that the appropriate form of the maximum principle in the argument is given by Theorem 3.5.

We wish now to determine conditions that the solution defined in Theorem 6.11 assumes the boundary values φ continuously. As in the case of harmonic functions, this problem can be treated by means of the barrier concept, which we define for $Lu=f$ with $c\leqslant 0$ in a bounded domain Ω. Let φ be a bounded function on $\partial\Omega$, continuous at $x_0\in\partial\Omega$. Then a sequence of functions $\{w_i^+(x)\}$ ($\{w_i^-(x)\}$) in $C^0(\bar{\Omega})$ is an *upper (lower) barrier* in Ω relative to L, f, and φ at x_0 if:

(i) $w_i^+(w_i^-)$ is a super(sub)function relative to φ in Ω;
(ii) $w_i^\pm(x_0)\to\varphi(x_0)$ as $i\to\infty$.

If both an upper and lower barrier exist at a point, it will be convenient to speak simply of a *barrier* at the point.

The basic property of barriers is contained in the following:

Lemma 6.12. *Let u be the solution of $Lu=f$ in Ω defined by Theorem* 6.11, *where φ is a bounded function on $\partial\Omega$, continuous at x_0. If there exists a barrier at x_0, then $u(x)\to\varphi(x_0)$ as $x\to x_0$.*

Proof. From the definition of u and the fact that every subfunction is dominated by every superfunction, we have for all i,

$$w_i^-(x)\leqslant u(x)\leqslant w_i^+(x)\quad\text{in }\Omega.$$

For any $\varepsilon>0$ and all sufficiently large i, condition (ii) above implies

$$\lim_{x\to x_0} w_i^-(x)>\varphi(x_0)-\varepsilon,\qquad \lim_{x\to x_0} w_i^+(x)<\varphi(x_0)+\varepsilon.$$

It follows that

$$\lim_{x\to x_0}\sup|u(x)-\varphi(x_0)|<\varepsilon,$$

and hence we conclude $u(x)\to\varphi(x_0)$ as $x\to x_0$. □

We amplify on the barrier concept with the following remarks.

Remark 1. In many cases of interest the special structure of the equation simplifies the determination of a barrier. Thus, if $c\equiv 0$, $f\equiv 0$ in the equation $Lu=f$, the situation is the same as for Laplace's equation in that a barrier at x_0 is determined by a single supersolution $w\in C^0(\bar{\Omega})$ with the property that $w>0$ on $\partial\Omega-x_0$, $w(x_0)=0$. To see this, let $\varepsilon>0$; then by the boundedness of φ and its continuity at x_0, there is a positive constant k_ε such that

$$w_\varepsilon^+\equiv\varphi(x_0)+\varepsilon+k_\varepsilon w,\qquad w_\varepsilon^-\equiv\varphi(x_0)-\varepsilon-k_\varepsilon w$$

are respectively superfunction and subfunction in Ω with respect to φ, and obviously $w_\varepsilon^\pm(x_0)\to\varphi(x_0)$ as $\varepsilon\to 0$. Thus the family $w_\varepsilon^\pm(x)$ determines a barrier.

To consider another class of equations, let f and the coefficients of L be bounded in Ω. Again a single function $w \in C^0(\bar{\Omega}) \cap C^2(\Omega)$ determines a barrier at x_0 if it satisfies the conditions: (a) $Lw \leqslant -1$ in Ω; (b) $w>0$ on $\partial\Omega - x_0$, $w(x_0)=0$. Namely, given $\varepsilon>0$, then as above there is a positive constant k_ε such that

$$\varphi(x_0)+\varepsilon+k_\varepsilon w(x) \geqslant \varphi(x), \qquad \varphi(x_0)-\varepsilon-k_\varepsilon w(x) \leqslant \varphi(x) \quad \text{on } \partial\Omega.$$

If now we set $k'_\varepsilon = \max\,(k_\varepsilon, \sup_\Omega |f-c\varphi(x_0)|)$, the functions

$$w_\varepsilon^+ \equiv \varphi(x_0)+\varepsilon+k'_\varepsilon w, \qquad w_\varepsilon^- \equiv \varphi(x_0)-\varepsilon-k'_\varepsilon w$$

are respectively superfunction and subfunction and define a barrier at x_0. In fact, $w_\varepsilon^+ \geqslant \varphi$, $w_\varepsilon^- \leqslant \varphi$ on $\partial\Omega$, and since $Lw \leqslant -1$,

$$L[\varphi(x_0)+\varepsilon+k'_\varepsilon w] \leqslant c(\varphi(x_0)+\varepsilon)-k'_\varepsilon \leqslant c\varphi(x_0)-k'_\varepsilon \leqslant f;$$

similarly, $L[\varphi(x_0)-\varepsilon-k'_\varepsilon w] \geqslant f$, and hence the family $w_\varepsilon^\pm$ determines a barrier with respect to φ at x_0.

When, as in the above cases, a barrier at x_0 is constructed from a fixed supersolution w of $Lu=0$ depending only on L and the domain, we shall say that w *determines a barrier* at x_0.

Remark 2. The above definition of barrier is often difficult to apply, since it requires the construction of global sub- and supersolutions defined over all of Ω. It may then become necessary to find *local barriers* to achieve the desired results. To motivate the definition of this concept, let $M^+(M^-)$ be an upper (lower) bound for a solution in Ω whose boundary behavior is being studied at a point $x_0 \in \partial\Omega$. Then a sequence of functions $\{w_i^+(x)\}$ ($\{w_i^-(x)\}$) is a *local* upper (lower) barrier relative to L, f, φ and $M^+(M^-)$ at x_0 if there is an open neighborhood $\mathcal{N}$ of x_0 such that:

(i) $w_i^+(w_i^-)$ is a super (sub) solution in $\mathcal{N} \cap \Omega$;
(ii) $w_i^+ \geqslant \varphi \quad (w_i^- \leqslant \varphi) \quad$ on $\mathcal{N} \cap \partial\Omega$;
(iii) $w_i^+ \geqslant M^+$ $(w_i^- \leqslant M^-)$ on $\Omega \cap \partial\mathcal{N}$;
(iv) $w_i^\pm(x_0) \to \varphi(x_0)$ as $i \to \infty$.

(In particular, if $\bar{\Omega} \subset \mathcal{N}$, the functions $w_i^\pm$ define a barrier at x_0 in the previous global sense and condition (iii) can be omitted.) One sees immediately that if the solution u defined in Theorem 6.11 satisfies $|u| \leqslant M$ in Ω, then Lemma 6.12 remains valid if there exists a local barrier at x_0 with respect to the bounds $\pm M$. Later in this work local barriers will play an important part in the study of boundary behavior of solutions.

Remark 3. The same argument as in Lemma 6.12 shows that a barrier determines a modulus of continuity at the boundary for any solution assuming its boundary values continuously. Thus in the cases considered in Remark 1, if u is a bounded solution of $Lu=f$ such that $u(x) \to \varphi(x_0)$ as $x \to x_0$, we have for any $\varepsilon>0$ and a suitable positive constant k_ε

$$|u(x)-\varphi(x_0)| \leqslant \varepsilon + k_\varepsilon w(x) \quad \text{in } \Omega.$$

If w determines a local barrier at x_0 the same inequality holds in a fixed neighborhood of x_0 (independent of ε).

For the equation $Lu=f$, as for Laplace's equation, the existence and construction of barriers is closely connected to the local properties of the boundary. To illustrate by an example of interest in later application, let L be strictly elliptic in a bounded domain Ω, with $c \leqslant 0$, and let f and the coefficients of L be bounded. We suppose that Ω satisfies an exterior sphere condition at $x_0 \in \partial\Omega$, so that $\bar{B} \cap \bar{\Omega} = x_0$ for some ball $B = B_R(y)$. We show that the function

$$w(x) = \tau(R^{-\sigma} - r^{-\sigma}), \quad r = |x-y|, \tag{6.45}$$

satisfies $Lw \leqslant -1$ in Ω for suitable positive constants τ and σ and hence (by Remark 1 above) w determines a barrier at x_0. Taking $y=0$ for convenience, we have from (6.40) and the fact that $c \leqslant 0$, for $x \in \Omega$

$$\begin{aligned} L(R^{-\sigma} - r^{-\sigma}) &\leqslant \sigma r^{-\sigma-4}[-(\sigma+2)a^{ij}x_i x_j + r^2(\Sigma a^{ii} + b^i x_i)] \\ &\leqslant \sigma r^{-\sigma-2}[-(\sigma+2)\lambda + (\Sigma a^{ii} + b^i x_i)]. \end{aligned}$$

Since Ω and the coefficients of L are bounded, the right member is negative and bounded away from zero provided σ is sufficiently large. Hence, for suitably large τ and σ, $Lw \leqslant -1$, as asserted.

Thus, under the above assumptions on the equation $Lu=f$, a bounded domain Ω satisfying an exterior sphere condition at every point $x_0 \in \partial\Omega$, (e.g. any C^2 domain), has a barrier at every boundary point, and Lemma 6.12 is applicable whenever the prescribed boundary values are continuous. Combining this fact with Theorem 6.11, we are led to the following general existence theorem.

Theorem 6.13. *Let L be strictly elliptic in a bounded domain Ω, with $c \leqslant 0$, and let f and the coefficients of L be bounded and belong to $C^\alpha(\Omega)$. Suppose that Ω satisfies an exterior sphere condition at every boundary point. Then, if φ is continuous on $\partial\Omega$, the Dirichlet problem,*

$$Lu = f \text{ in } \Omega, \quad u = \varphi \text{ on } \partial\Omega,$$

has a (unique) solution $u \in C^0(\bar{\Omega}) \cap C^{2,\alpha}(\Omega)$.

The preceding theorem can be extended to more general domains. In particular, under the same hypotheses on L and f, it can be proved for domains satisfying an exterior cone condition (see Problem 6.3).

If the hypotheses of this theorem are strengthened so that f and the coefficients of L are in $C^{\alpha}(\bar{\Omega})$, it can be shown that the domains for which the Dirichlet problem is solvable for continuous boundary values are precisely the same for both the Laplace operator and L (see Notes).

We turn now to the question of global regularity of the above solutions when the boundary data are sufficiently smooth. We have seen that under the hypotheses of Theorem 6.8 the solution of the Dirichlet problem for $Lu=f$ lies in $C^{2,\alpha}(\bar{\Omega})$ provided the same result (Kellogg's theorem) is true for Poisson's equation. We proceed now to prove this regularity theorem directly from the results of this section.

Theorem 6.14. *Let L be strictly elliptic in a bounded domain Ω, with $c \leqslant 0$, and let f and the coefficients of L belong to $C^{\alpha}(\bar{\Omega})$. Suppose that Ω is a $C^{2,\alpha}$ domain and that $\varphi \in C^{2,\alpha}(\bar{\Omega})$. Then the Dirichlet problem,*

$$Lu=f \text{ in } \Omega, \quad u=\varphi \text{ on } \partial\Omega,$$

has a (unique) solution lying in $C^{2,\alpha}(\bar{\Omega})$.

Proof. Since the hypotheses of Theorem 6.13 are satisfied, let u be the corresponding solution of the Dirichlet problem. We know from Theorem 6.13 that $u \in C^0(\bar{\Omega}) \cap C^{2,\alpha}(\Omega)$, so it remains only to show that at each point $x_0 \in \partial\Omega$, we have $u \in C^{2,\alpha}(D \cap \bar{\Omega})$, where D is some neighbourhood of x_0. Since Ω is a $C^{2,\alpha}$ domain, there is a neighborhood N of x_0 that can be taken by a $C^{2,\alpha}$ mapping $y=\psi(x)$ with $C^{2,\alpha}$ inverse into a neighborhood $\tilde{N}$ in such a way that $\psi(N \cap \bar{\Omega})$ contains the closure of a ball B, and a portion T of $N \cap \partial\Omega$ containing x_0 is mapped by ψ into a boundary portion $\tilde{T}$ of B. Under this mapping the equation $Lu(x)=f(x)$ transforms into an equation $\tilde{L}\tilde{u}(y)=\tilde{f}(y)$ defined in B. Because of the $C^{2,\alpha}$ character of the mapping, we have $\varphi \to \tilde{\varphi} \in C^{2,\alpha}(\bar{B})$, and $\tilde{L}$ and $\tilde{f}$ satisfy the same hypotheses in B as L and f do in Ω; that is, $\tilde{L}$ is strictly elliptic in B, with $\tilde{c} \leqslant 0$, and $\tilde{f}$ and the coefficients of $\tilde{L}$ are in $C^{\alpha}(\bar{B})$; (see Lemma 6.5.). Consider then the solution v of the Dirichlet problem, $\tilde{L}v=\tilde{f}$ in B, $v=\tilde{u}$ on ∂B. Since $\tilde{u}=\tilde{\varphi}$ on $\tilde{T} \subset \partial B$, we have that $\tilde{u} \in C^0(\partial B) \cap C^{2,\alpha}(\tilde{T})$ on ∂B. Hence, by uniqueness for the Dirichlet problem and by Lemma 6.10 it follows $\tilde{u}=v \in C^0(\bar{B}) \cap C^{2,\alpha}(B \cup \tilde{T})$. Returning to Ω, let $D'=\psi^{-1}(B)$. We see that $u \in C^{2,\alpha}(D' \cup T)$, and since x_0 was arbitrary on $\partial\Omega$, we conclude that $u \in C^{2,\alpha}(\bar{\Omega})$. □

The preceding result can be extended to conditions of lower regularity of the coefficients, domain and boundary values (see Notes).

If the operator L does not satisfy the condition $c \leqslant 0$, then, as is well known from simple examples, the Dirichlet problem for $Lu=f$ no longer has a solution in general. However, it is still possible to assert a Fredholm alternative, which we formulate as follows.

Theorem 6.15. *Let $L \equiv a^{ij}D_{ij}+b^iD_i+c$ be strictly elliptic with coefficients in $C^{\alpha}(\bar{\Omega})$ in a $C^{2,\alpha}$ domain Ω. Then either: (a) the homogeneous problem, $Lu=0$ in Ω, $u=0$ on $\partial\Omega$, has only the trivial solution, in which case the inhomogeneous problem,*

$Lu=f$ in Ω, $u=\varphi$ in $\partial\Omega$, has a unique $C^{2,\alpha}(\bar{\Omega})$ solution for all $f\in C^\alpha(\bar{\Omega})$, $\varphi\in C^{2,\alpha}(\bar{\Omega})$; or (b) *the homogeneous problem has nontrivial solutions, which form a finite dimensional subspace of $C^{2,\alpha}(\bar{\Omega})$.*

Proof. We have seen that under the stated assumptions concerning f and φ, the inhomogeneous problem, $Lu=f$ in Ω, $u=\varphi$ on $\partial\Omega$, is equivalent to the problem $Lv=f-L\varphi$, $v=0$ on $\partial\Omega$. We shall therefore consider the Dirichlet problem with only the homogeneous boundary condition, $u=0$ on $\partial\Omega$, and it will suffice to restrict the operator L to the linear space

$$\mathfrak{B}=\{u\in C^{2,\alpha}(\bar{\Omega})\mid u=0 \text{ on } \partial\Omega\}.$$

Let σ be any constant such that $\sigma\geqslant\sup_\Omega c$, and define the operator $L_\sigma\equiv L-\sigma$. Then, by Theorem 6.14, the mapping

$$L_\sigma:\mathfrak{B}\to C^\alpha(\bar{\Omega})$$

is invertible. Furthermore, the inverse mapping L_σ^{-1}, by the estimate (6.36) and Theorem 3.7, is a compact mapping from $C^\alpha(\bar{\Omega})$ into $C^2(\bar{\Omega})$ and hence is also compact as a mapping from $C^\alpha(\bar{\Omega})$ into $C^\alpha(\bar{\Omega})$. Consider then the equation

$$(6.46)\qquad u+\sigma L_\sigma^{-1}u=L_\sigma^{-1}f,\quad f\in C^\alpha(\bar{\Omega}),$$

in which $L_\sigma^{-1}:C^\alpha(\bar{\Omega})\to C^\alpha(\bar{\Omega})$ is compact. By Theorem 5.3 on compact operators in a Banach space, the alternative applies and (6.46) always has a solution $u\in C^\alpha(\bar{\Omega})$ provided the homogeneous equation $u+\sigma L_\sigma^{-1}u=0$ has only the trivial solution $u=0$. When this condition is not satisfied, the null space of the operator $I+\sigma L_\sigma^{-1}$ (I=identity) is a finite dimensional subspace of $C^\alpha(\bar{\Omega})$ (Theorem 5.5).

To interpret these statements in terms of the Dirichlet problem for $Lu=f$, we observe first that since L_σ^{-1} maps $C^\alpha(\bar{\Omega})$ onto $\mathfrak{B}$, any solution $u\in C^\alpha(\bar{\Omega})$ of (6.46) must also belong to $\mathfrak{B}$. Hence, operating on (6.46) with L_σ we obtain

$$(6.47)\qquad Lu=L_\sigma(u+\sigma L_\sigma^{-1}u)=f,\quad u\in\mathfrak{B}.$$

Thus, the solutions of (6.46) are in one-to-one correspondence with the solutions of the boundary value problem (6.47), and we can therefore conclude the alternative stated in the theorem. □

The importance of the alternative for the Dirichlet problem is that it shows uniqueness to be a sufficient condition for existence. We remark that by virtue of Lemma 6.18 (to be proved in the next section), the $C^2(\bar{\Omega})$ solutions of $Lu=f$ are also in $C^{2,\alpha}(\bar{\Omega})$, and hence the null space of L in $C^2(\bar{\Omega})$ is also finite dimensional. We note also that Theorem 5.5 implies that the set Σ of real values σ for which the homogeneous problem, $Lu-\sigma u=0$, $u=0$ on $\partial\Omega$, has nontrivial solutions is at most countable and discrete. Furthermore (by Theorem 5.3), if $\sigma\notin\Sigma$, any solution of the Dirichlet problem $L_\sigma u=f$ in Ω, $u=\varphi$ on $\partial\Omega$, satisfies an estimate

$$|u|_{2,\alpha}\leqslant C(|\varphi|_{2,\alpha}+|f|_{0,\alpha})$$

where the constant C is independent of u, f and φ.

6.4. Interior and Boundary Regularity

In the preceding sections the inhomogeneous term f and the coefficients of L were assumed to be C^α functions and the corresponding solutions of the equation $Lu=f$ were in class $C^{2,\alpha}$. We now investigate the higher regularity properties of the solutions in their dependence on the smoothness of f and the coefficients of L. The global regularity of the solutions will depend as well on the smoothness of the boundary and the boundary values.

First we show that any C^2 solution of $Lu=f$ must also be in class $C^{2,\alpha}$ if f and the coefficients of L are in C^α.

Lemma 6.16. *Let u be a $C^2(\Omega)$ solution of the equation $Lu=f$ in an open set Ω, where f and the coefficients of the elliptic operator L are in $C^\alpha(\Omega)$. Then $u\in C^{2,\alpha}(\Omega)$.*

Proof. It obviously suffices to prove $u\in C^{2,\alpha}(B)$ for arbitrary balls $B\subset\subset\Omega$. Let B be such a ball and in B consider the Dirichlet problem for v:

$$L_0v=a^{ij}D_{ij}v+b^iD_iv=f'\equiv f-cu,\quad v=u \text{ on } \partial B. \tag{6.48}$$

Since, by hypothesis, $u\in C^2(\bar{B})$, we have that f' and the coefficients of L_0 are in $C^\alpha(\bar{B})$. Hence, by Lemma 6.10, there is a solution v of (6.48) lying in $C^{2,\alpha}(B)\cap C^0(\bar{B})$. By uniqueness we infer that the solutions u and v of (6.48) are identical in B and thus $u\in C^{2,\alpha}(B)$. □

We note that the preceding result and those that follow in this section make no assumption concerning the sign of the coefficient c.

The above lemma and the Schauder interior estimates yield the following interior regularity theorem.

Theorem 6.17. *Let u be a $C^2(\Omega)$ solution of the equation $Lu=f$ in an open set Ω, where f and the coefficients of the elliptic operator L are in $C^{k,\alpha}(\Omega)$. Then $u\in C^{k+2,\alpha}(\Omega)$. If f and the coefficients of L lie in $C^\infty(\Omega)$, then $u\in C^\infty(\Omega)$.*

Proof. By Lemma 6.16 the theorem is established for $k=0$. We now prove it for $k=1$. Let v be a function on Ω and denote by e_l $(l=1,\ldots,n)$ the unit coordinate vector in the x_l direction. We define the difference quotient of v at x in the direction e_l by

$$\Delta^h v(x)=\Delta_l^h v(x)=\frac{v(x+he_l)-v(x)}{h}.$$

Taking the difference quotients of both sides of the equation

$$Lu = a^{ij}D_{ij}u + b^iD_iu + cu = f$$

we obtain

$$\begin{aligned} (6.49) \qquad L(\Delta^h u) &= a^{ij}D_{ij}\Delta^h u + b^iD_i\Delta^h u + c\Delta^h u = F_h(x) \\ &\equiv \Delta^h f - (\Delta^h a^{ij})D_{ij}\bar{u} - (\Delta^h b^i)D_i\bar{u} - (\Delta^h c)\bar{u}, \quad \bar{u} = u(x + he_l). \end{aligned}$$

All the difference quotients in this equation are assumed taken at $x \in \Omega$ in the direction e_l for some $l = 1, \ldots, n$. Since $f \in C^{1,\alpha}(\Omega)$ and

$$\Delta^h f(x) = \frac{1}{h}\int_0^1 \frac{d}{dt} f(x + the_l)\, dt = \int_0^1 D_l f(x + the_l)\, dt,$$

one sees that $\Delta^h f \in C^\alpha(\Omega')$ in any subset $\Omega' \subset\subset \Omega$ for which $|h| < \operatorname{dist}(\Omega', \partial\Omega)$. In particular, if B and B' are balls in Ω such that $B' \subset B \subset\subset \Omega$ and $\operatorname{dist}(B', \partial B) = h_0 > 0$, then $\Delta^h f \in C^\alpha(\bar{B}')$ for $0 < |h| < h_0$, and there is a uniform bound: $|\Delta^h f|_{0,\alpha;B'} \leqslant$ const independent of h. Similar bounds hold for the difference quotients $\Delta^h a^{ij}$, $\Delta^h b^i$, $\Delta^h c$, which also belong to $C^\alpha(\bar{B}')$. Since $u \in C^{2,\alpha}(\Omega)$ (by Lemma 6.16), it follows that $F_h \in C^\alpha(\bar{B}')$ for $|h| < h_0$ and, furthermore, $|F_h|_{0,\alpha;B'} \leqslant$ const independent of h. Observing also the bound, $\sup_{B'} |\Delta^h u| \leqslant \sup_B |Du|$, we can infer from the interior estimates of Corollary 6.3 that the set of functions $\Delta^h u$ and their first and second derivatives $D_i\Delta^h u$, $D_{ij}\Delta^h u$, $(i, j = 1, \ldots, n)$, are bounded and equicontinuous on any ball $B'' \subset\subset B'$, and hence every sequence in these sets contains a uniformly convergent subsequence on B''. Since $\Delta^h_l u \to D_l u$ as $h \to 0$, we may therefore assert that $D_{ij}\Delta^h_l u \to D_{ijl}u$ as $h \to 0$ and that $u \in C^{3,\alpha}(B'')$. Since B'' could be an arbitrary ball having its closure in Ω, we infer that $u \in C^{3,\alpha}(\Omega)$, thereby proving the theorem for $k = 1$.

To prove the theorem for $k > 1$, we proceed by induction on k. Under the stated hypotheses on L and f we may accordingly assume that $u \in C^{k+1,\alpha}(\Omega)$, and we wish to show that $u \in C^{k+2,\alpha}(\Omega)$. Since f and the coefficients of L are in $C^{k,\alpha}(\Omega)$, the equation $Lu = f$ may be differentiated $k-1$ times, and this yields an equation $L\hat{u} = \hat{f}$, where $\hat{u} = D^\beta u$ for some index β such that $|\beta| = k-1$, and where $\hat{f}$ equals $D^\beta f$ plus a sum of terms which are products of derivatives of the coefficients of order $\leqslant k-1$ and of derivatives of u of order $\leqslant k$. Thus $\hat{f} \in C^{1,\alpha}(\Omega)$. By the same argument as above for $k = 1$, we see that $\hat{u} \in C^{3,\alpha}(\Omega)$ and hence $u \in C^{k+2,\alpha}(\Omega)$, as claimed in the theorem. The final assertion concerning solutions in $C^\infty(\Omega)$ follows immediately. □

It is also the case that if f and the coefficients of L are real analytic functions, then any solution of $Lu = f$ is likewise analytic. For the proof we refer to the literature (e.g. [HO 3]).

To be able to assert analogous regularity properties up to the boundary, it is obviously necessary that the boundary itself and the boundary values of the solution be sufficiently smooth. To establish the appropriate results on regularity up to the boundary we first prove an analogue of Lemma 6.16.

Lemma 6.18. *Let Ω be a domain with a $C^{2,\alpha}$ boundary portion T, and let $\varphi \in C^{2,\alpha}(\bar{\Omega})$. Suppose that u is a $C^0(\bar{\Omega}) \cap C^2(\Omega)$ function satisfying $Lu=f$ in Ω, $u=\varphi$ on T, where f and the coefficients of the strictly elliptic operator L belong to $C^\alpha(\bar{\Omega})$. Then $u \in C^{2,\alpha}(\Omega \cup T)$.*

Proof. Since T is a $C^{2,\alpha}$ portion of $\partial\Omega$, it is possible to find at each point x_0 of T a boundary neighborhood $T' \subset\subset T$ and a $C^{2,\alpha}$ domain D in Ω such that $x_0 \in T' \subset \partial D$. In addition, D can be chosen so small that Corollary 3.8 is applicable and hence the Dirichlet problem for $Lu=f$ in D has at most one solution in $C^0(\bar{D}) \cap C^2(D)$.

The argument is now very similar to that in Lemma 6.10. Since $u \in C^0(\partial D) \cap C^{2,\alpha}(T')$, we can extend the boundary values of u on ∂D to a function $v \in C^0(D') \cap C^{2,\alpha}(\bar{B})$, where $D \subset\subset D'$ and $B = B_\rho(x_0) \subset\subset D'$. (See Remark 1 after Lemma 6.38 for the construction of v.) Let $\{v_k\}$ be a sequence of functions in $C^3(D')$ such that

$$|v_k - v|_{0;D} \to 0 \quad \text{and} \quad |v_k|_{2,\alpha;B} \leqslant C|v|_{2,\alpha;B} \quad \text{as } k \to \infty.$$

By virtue of the Fredholm alternative (Theorem 6.15), the Dirichlet problem, $Lu=f$ in D, $u=v_k$ on ∂D, has a unique solution u_k in $C^{2,\alpha}(\bar{D})$ for each k. As a consequence of Corollaries 6.3 and 3.8, the sequence $\{u_k\}$ converges to the solution u in D; and by Corollary 6.7 we have $u \in C^{2,\alpha}(B' \cap \bar{D})$, where $B' = B_{\rho/2}(x_0)$. Since x_0 was an arbitrary point on T, it follows that $u \in C^{2,\alpha}(\Omega \cup T)$. □

If $c \leqslant 0$, the preceding result is essentially contained in the proof of Theorem 6.14. However, when there are no restrictions on the sign of the coefficient c, the argument in Theorem 6.14 has to be suitably modified, as in the above proof. Another proof, along different lines, is contained in the Notes.

Lemma 6.18 remains valid if $C^{2,\alpha}$ is replaced by $C^{1,\alpha}$ in the hypotheses and conclusion, and under more general circumstances as well (see [GH]).

With Lemma 6.18 as a starting point we can establish the following global regularity theorem.

Theorem 6.19. *Let Ω be a $C^{k+2,\alpha}$ domain $(k \geqslant 0)$ and let $\varphi \in C^{k+2,\alpha}(\bar{\Omega})$. Suppose that u is a $C^0(\bar{\Omega}) \cap C^2(\Omega)$ function satisfying $Lu=f$ in Ω, $u=\varphi$ on $\partial\Omega$, where f and the coefficients of the strictly elliptic operator L belong to $C^{k,\alpha}(\bar{\Omega})$. Then $u \in C^{k+2,\alpha}(\bar{\Omega})$.*

Proof. For $k=0$ the theorem is implied by Lemma 6.18. We now prove the result for $k=1$. Let x_0 be an arbitrary boundary point of Ω, and consider a suitable $C^{3,\alpha}$ diffeomorphism ψ that straightens the boundary near x_0. As a consequence of this mapping we may consider the equation $Lu=f$ as defined in a domain G with a hyperplane portion T on $x_n=0$, while the remaining hypotheses of the theorem are unchanged (see p. 91). By considering the function $u-\varphi$ in place of u and noting that $L\varphi \in C^{1,\alpha}(\bar{G})$, we may assume that $\varphi=0$ in the statement of the theorem.

As in Theorem 6.17, we take the difference quotient of the equation $Lu=f$ in the direction e_l for any $l=1, \ldots, n-1$, obtaining thereby an equation of the form (6.49), which is satisfied by the difference quotient $\Delta^h u\ (=\Delta_l^h u)$. If $0<|h|<h_0$, then this equation is valid in the set $G'=\{x \in G|\ \mathrm{dist}\,(x, \partial G-T)>h_0\}$, which has a hyperplane boundary portion $T' \subset T$. Under the assumptions on L and f, and since $u=0$ on T and $u \in C^{2,\alpha}(\bar{G})$, the conditions of Lemma 6.4 are satisfied in G' by equation (6.49) and the solution $\Delta^h u$. It follows from Lemma 6.4 that the function families $\Delta^h u, D_i \Delta^h u, D_{ij} \Delta^h u, (i, j=1, \ldots, n)$, are bounded and equicontinuous on compact subsets of $G' \cup T'$. Since $\Delta_l^h u \to D_l u$ as $h \to 0$, we can assert that $D_{ij} \Delta_l^h u \to D_{ijl} u$ for $i, j=1, \ldots, n$ and $l=1, \ldots, n-1$, and, in addition, $D_l u \in C^{2,\alpha}(G' \cup T')$ for $l=1, \ldots, n-1$. It remains only to show that $D_n u \in C^{2,\alpha}(G' \cup T')$ as well. This follows at once by writing

$$D_{nn} u=(1/a^{nn})(f-(L-a^{nn} D_{nn})u)$$

and observing from the preceding results that the right hand side is contained in $C^{1,\alpha}(G' \cup T')$. Since x_0 was an arbitrary point on $\partial\Omega$, we conclude that $u \in C^{3,\alpha}(\bar{\Omega})$.

The proof of the theorem for $k>1$ proceeds by induction on k, as in Theorem 6.17, by considering the equation satisfied by any derivative of order $k-1$ and thereby reducing to the case $k=1$ treated above. □

It is clear from the preceding argument, which is essentially local, that the regularity result remains true up to any $C^{k+2,\alpha}$ boundary portion T provided the solution u is continuous up to T and takes on $C^{k+2,\alpha}$ boundary values on T.

6.5. An Alternative Approach

An examination of the proof of Theorem 6.13 shows that this existence theorem follows by the Perron process from the solvability of the Dirichlet problem in balls for arbitrary continuous boundary values. The latter result (contained in Lemma 6.10) depended in an essential way for its proof on the boundary estimates of the Schauder theory. However, as we shall see below, it is possible to develop a theory of the Dirichlet problem for continuous boundary values that is based entirely on the Schauder *interior* estimates, without any use of boundary estimates.

The developments of this section will rest on the following extension of the interior estimates in Theorem 6.2. For its statement we make use of the seminorms and norms defined in (6.10).

Lemma 6.20. *Let $u \in C^{2,\alpha}(\Omega)$ satisfy the equation $Lu=f$ in an open set Ω of $\mathbb{R}^n$, where the coefficients of L satisfy* (6.12) *and* (6.13). *Suppose that $|u|_{0;\Omega}^{(-\beta)}<\infty$ and $|f|_{0,\alpha;\Omega}^{(2-\beta)}<\infty$ for some $\beta \in \mathbb{R}$. Then we have*

$$|u|_{2,\alpha;\Omega}^{(-\beta)} \leq C(|u|_{0;\Omega}^{(-\beta)}+|f|_{0,\alpha;\Omega}^{(2-\beta)}), \tag{6.50}$$

where $C=C(n, \alpha, \lambda, \Lambda, \beta)$.

Proof. Let x be any point in Ω, d_x its distance from $\partial\Omega$ and $d = d_x/2$. Then by (6.14), applied to the ball $B = B_d(x)$, we have

$$\begin{aligned} d^{1-\beta}|Du(x)| + d^{2-\beta}|D^2u(x)| &\leq Cd^{-\beta}(|u|_{0;B} + |f|^{(2)}_{0,\alpha;B}) \\ &\leq C\Big[\sup_{y\in B} d_y^{-\beta}|u(y)| + \sup_{y\in B} d_y^{2-\beta}|f(y)| \\ &\quad + \sup_{y\in B} d_{x,y}^{2+\alpha-\beta}\frac{|f(x)-f(y)|}{|x-y|^\alpha}\Big] \\ &\leq C(|u|^{(-\beta)}_{0;\Omega} + |f|^{(2-\beta)}_{0,\alpha;\Omega}). \end{aligned}$$

Hence

$$|u|^{(-\beta)}_{2;\Omega} \leq C(|u|^{(-\beta)}_{0;\Omega} + |f|^{(2-\beta)}_{0,\alpha;\Omega}). \tag{6.51}$$

To estimate $[u]^{(-\beta)}_{2,\alpha;\Omega}$, let x, y be distinct points in Ω, with $d_x \leq d_y$ and $B = B_d(x)$ as above. Then, by considering the two cases $|x-y| \leq d/2$ and $|x-y| > d/2$, we have for any second derivative D^2u.

$$\begin{aligned} d_{x,y}^{2+\alpha-\beta}\frac{|D^2u(x) - D^2u(y)|}{|x-y|^\alpha} &\leq Cd^{-\beta}(|u|_{0;B} + |f|^{(2)}_{0,\alpha;B}) \\ &\quad + d_x^{2+\alpha-\beta}\frac{(|D^2u(x)| + |D^2u(y)|)}{(d/2)^\alpha} \\ &\leq C(|u|^{(-\beta)}_{0;\Omega} + |f|^{(2-\beta)}_{0,\alpha;\Omega}) + 8[u]^{(-\beta)}_{2;\Omega}. \end{aligned}$$

Taking the supremum with respect to x, y, and applying (6.51), we obtain

$$[u]^{(-\beta)}_{2,\alpha;\Omega} \leq C(|u|^{(-\beta)}_{0,\Omega} + |f|^{(2-\beta)}_{0,\alpha;\Omega}).$$

Combining this with (6.51), we get the desired estimate (6.50). □

We observe that for $\beta = 0$ the preceding result reduces to the previous estimate (6.14). When $\beta > 0$, the hypothesis that $|u|^{(-\beta)}_{0;\Omega}$ is finite obviously requires that $u = 0$ on $\partial\Omega$.

In this section we shall solve the Dirichlet problem for $Lu = f$ in balls using the method of continuity. This will require an apriori estimate of solutions for suitably unbounded f, given by the following lemma. The corresponding result for Poisson's equation is contained in Theorem 4.9.

Lemma 6.21. *Let L be strictly elliptic (satisfying (6.2)), with $c \leq 0$ and with coefficients bounded in magnitude by Λ in a ball $B = B_R(x_0)$. Suppose that $u \in C^0(\bar{B}) \cap C^2(B)$ is a solution of $Lu = f$ in B, $u = 0$ on ∂B. Then, for any $\beta \in (0, 1)$ we have*

$$\sup_{x\in B} d_x^{-\beta}|u(x)| \leq C \sup_{x\in B} d_x^{2-\beta}|f(x)|, \tag{6.52}$$

where $C = C(\beta, n, R, \lambda, \Lambda)$.

Proof. Let β be fixed and assume that $\sup\limits_{x\in B} d_x^{2-\beta}|f(x)|=N<\infty$. The required estimate (6.52) will be obtained by the construction of a suitable comparison function bounding u. For convenience we take $x_0=0$, and let us set

$$w_1(x)=(R^2-r^2)^\beta, \quad r=|x|.$$

Then

$$\begin{aligned} Lw_1(x)&=\beta(R^2-r^2)^{\beta-2}[4(\beta-1)a^{ij}x_ix_j-2(R^2-r^2)(\Sigma a^{ii}+b^ix_i)\\ &\quad+(c/\beta)(R^2-r^2)^2]\\ &\leqslant-\beta(R^2-r^2)^{\beta-2}[4(1-\beta)\lambda r^2+2(R^2-r^2)(n\lambda-\sqrt{n}\,\Lambda r)]. \end{aligned}$$

It is clear that for some R_0, $0\leqslant R_0<R$, the expression in brackets is positive if $R_0\leqslant r\leqslant R$. Hence

$$\begin{aligned} Lw_1(x)&\leqslant-c_1(R-r)^{\beta-2} && \text{in } R_0\leqslant r<R,\\ &\leqslant c_2(R-r)^{\beta-2} && \text{in } 0\leqslant r<R_0, \end{aligned}$$

where c_1 and c_2 are positive constants depending only on β, n, R, λ, and Λ. (If $R_0=0$, the second inequality is of course superflous.)

Now let $w_2(x)=e^{\alpha R}-e^{\alpha x_1}$, where $\alpha\geqslant 1+\sup\limits_B|\mathbf{b}|/\lambda$. Then, as in Theorem 3.7, we have $Lw_2(x)\leqslant-\lambda\, e^{-\alpha R}$ in B, and hence

$$\begin{aligned} Lw_2(x)&\leqslant-c_3(R-r)^{\beta-2} && \text{in } 0\leqslant r<R_0,\\ &\leqslant 0 && \text{in } R_0\leqslant r<R, \end{aligned}$$

where $c_3=\lambda\, e^{-\alpha R}(R-R_0)^{2-\beta}$. Since, by assumption, $|f(x)|\leqslant Nd_x^{\beta-2}$ and $d_x=R-r$, it follows that for the positive constants $\gamma_1=1/c_1$ and $\gamma_2=(1+c_2/c_1)/c_3$,

$$L(\gamma_1w_1+\gamma_2w_2)\leqslant-(R-r)^{\beta-2}\leqslant-|f(x)|/N \quad \text{in } 0\leqslant|x|<R.$$

Letting $w=\gamma_1w_1+\gamma_2w_2$, one sees that $w(x)\geqslant0$ on ∂B, $w(x)=0$ at $x=(R,0,\ldots,0)$, and

$$L(Nw\pm u)\leqslant0 \quad \text{in } B, \qquad Nw\pm u\geqslant0 \quad \text{on } \partial B.$$

From the maximum principle (Corollary 3.2) we infer

$$|u(x)|\leqslant Nw(x) \quad \text{in } B. \tag{6.53}$$

Consider now any point $x\in B$, which we assume without loss of generality to lie on the x_1 axis. Then (6.53) implies the inequality

$$|u(x)|\leqslant CN(R-r)^\beta=CNd_x^\beta$$

for some constant $C=C(\beta,n,R,\lambda,\Lambda)$, and the lemma is proved. □

The preceding result can be extended by similar methods to more general domains, for example, to C^2 domains (see Problem 6.5).

With the help of the preceding two lemmas we can now prove the following extension of Theorem 4.9 to the equation $Lu=f$. We observe that the proof makes no use of boundary estimates.

Theorem 6.22. *Let B be a ball in $\mathbb{R}^n$ and f a function in $C^\alpha(B)$ such that $|f|^{(2-\beta)}_{0,\alpha;B}<\infty$ for some $\beta\in(0,1)$. Assume L to be strictly elliptic in B, with $c\leqslant 0$ and coefficients satisfying* (6.2) *and* (6.31). *Then there exists a (unique) solution $u\in C^0(\bar B)\cap C^{2,\alpha}(B)$ of the Dirichlet problem, $Lu=f$ in B, $u=0$ on ∂B. In addition, $|u|^{(-\beta)}_{0;B}<\infty$ and hence u satisfies the estimate* (6.50) *in B.*

Proof. The argument is based on the method of continuity. As in Theorem 6.8, we consider the family of equations,

$$L_t u\equiv tLu+(1-t)\,\Delta u=f,\quad 0\leqslant t\leqslant 1,$$

and observe that the coefficients of L_t also satisfy (6.2) and (6.31) in B, with $\lambda_t=\min(1,\lambda)$, $\Lambda_t=\max(1,\Lambda)$ replacing λ and Λ respectively. By virtue of (6.11) we have

$$|a^{ij}D_{ij}u|^{(2-\beta)}_{0,\alpha},\ |b^iD_iu|^{(2-\beta)}_{0,\alpha},\ |cu|^{(2-\beta)}_{0,\alpha}\leqslant C|u|^{(-\beta)}_{2,\alpha},$$

and hence the operator L_t, for each t, is a bounded linear operator from the Banach space

$$\mathfrak{B}_1=\{u\in C^{2,\alpha}(B)\mid |u|^{(-\beta)}_{2,\alpha;B}<\infty\}$$

into the Banach space

$$\mathfrak{B}_2=\{f\in C^\alpha(B)\mid |f|^{(2-\beta)}_{0,\alpha;B}<\infty\}.$$

The solvability of the Dirichlet problem, $L_tu=f$ in B, $u=0$ on ∂B, for arbitrary $f\in\mathfrak{B}_2$ is equivalent to the invertibility of the mapping $\mathfrak{B}_1\to\mathfrak{B}_2$ defined by $u\to L_tu$. Let u_t denote a solution of this problem for some $t\in[0,1]$. Then from (6.52) we have

$$|u_t|^{(-\beta)}_0\leqslant C|f|^{(2-\beta)}_0\leqslant C|f|^{(2-\beta)}_{0,\alpha}.$$

It follows from (6.50) that

$$|u_t|^{(-\beta)}_{2,\alpha}\leqslant C|f|^{(2-\beta)}_{0,\alpha},$$

or, equivalently,

$$\|u\|_{\mathfrak{B}_1}\leqslant C\|L_tu\|_{\mathfrak{B}_2},$$

the constant C being independent of t. The fact that L_0 is onto is already contained in Theorem 4.9. The method of continuity (Theorem 5.2) is now applicable and the theorem is proved. □

The preceding theorem can be extended to yield the following generalization of Lemma 6.10 for the case of continuous boundary values.

Corollary 6.23. *Under the hypotheses of Theorem 6.22, if $\varphi \in C^0(\bar{B})$ there exists a (unique) solution $u \in C^0(\bar{B}) \cap C^{2,\alpha}(B)$ of the Dirichlet problem, $Lu=f$ in B, $u=\varphi$ on ∂B.*

Proof. Let $\{\varphi_k\}$ be a sequence of functions in $C^3(\bar{B})$ converging uniformly to φ on $\bar{B}$. By Theorem 6.22, the Dirichlet problem, $Lv_k=f-L\varphi_k$ in B, $v_k=0$ on ∂B, is uniquely solvable for each k, and defines a solution $u_k=v_k+\varphi_k$ of the corresponding problem with inhomogeneous boundary values, $Lu_k=f$ in B, $u_k=\varphi_k$ on ∂B. It follows in the usual way from the maximum principle that u_k converges uniformly on $\bar{B}$ to a function $u \in C^0(\bar{B})$ such that $u=\varphi$ on ∂B. From the compactness provided by the interior estimates (Corollary 6.3), it follows also that $Lu=f$ in B, and hence u is the required solution. □

Starting with this existence theorem in balls, we can proceed as before to apply the Perron process in more general domains, obtaining in particular Theorem 6.13.

6.6. Non-Uniformly Elliptic Equations

The existence results of the preceding sections, which were valid in arbitrary smooth bounded domains, were derived under the assumption of uniform ellipticity of the differential operator L. When the equation ceases to be uniformly elliptic, the conditions for solvability are greatly circumscribed and will in general require limitations on the geometry of the domain or connections between the differential operator and the geometry.

It is instructive to consider an example for which the Dirichlet problem is not solvable. Let us consider solutions $u(x, y)$ of the equation

$$u_{xx}+y^2u_{yy}=0 \tag{6.54}$$

in the rectangle R, $0<x<\pi$, $0<y<Y$, such that $u \in C^0(\bar{R}) \cap C^2(R)$ satisfies the boundary conditions $u(0, y)=u(\pi, y)=0$, $0 \leqslant y \leqslant Y$. Any such solution $u(x, y)$ has a Fourier series expansion $\Sigma f_n(y) \sin nx$, in which the coefficients $f_n(y)$ satisfy the ordinary differential equation $y^2 f_n''-n^2 f_n=0$. This equation has the independent solutions y^{β_n}, y^{γ_n}, where $\beta_n=\frac{1}{2}(1+\sqrt{1+4n^2})>0$, $\gamma_n=\frac{1}{2}(1-\sqrt{1+4n^2})<0$. The fact that $u(x, y)$ is bounded at $y=0$ requires that $f_n(y)=\text{const } y^{\beta_n}$, and hence $f_n(0)=0$. It follows that $u(x, 0)=0$, and accordingly the only continuous solutions on $\bar{R}$ satisfying the prescribed boundary conditions must have zero boundary values on $y=0$ and therefore cannot take on prescribed non-zero boundary data.

The arguments leading to Theorem 6.13 can be extended to non-uniformly elliptic equations under suitable hypotheses on the coefficients and the domain. We observe to begin with that if f and the coefficients of L in the equation $Lu=f$ $(c\leqslant 0)$ are locally Hölder continuous in the domain Ω, and if the set of subfunctions of the Dirichlet problem with respect to the prescribed boundary function φ is non-empty and bounded above, then the Perron process defines a bounded solution of $Lu=f$ in Ω (Theorem 6.11). In particular, this is the case if $|\mathbf{b}|/\lambda$, f/λ and the domain Ω are bounded, where $\lambda(x)$ is the minimum eigenvalue of the coefficient matrix $\mathbf{A}(x)=[a^{ij}(x)]$; (see Theorem 3.7). This assumption will be understood in the following considerations.

To study whether $u(x)\to\varphi(x_0)$ at a specific boundary point $x_0\in\partial\Omega$ where φ is continuous, let us suppose, as in the discussion preceding Theorem 6.13, that Ω satisfies an exterior sphere condition at x_0, and let $B=B_R(y)$ be a ball such that $\bar{B}\cap\bar{\Omega}=x_0$. While no longer assuming that L is uniformly elliptic near x_0, we make the additional (but less restrictive) hypothesis that $|\mathbf{A}(x)\cdot(x-y)|\geqslant\delta>0$ for all x in the intersection $N\cap\Omega$ of Ω with some neighborhood N of x_0. (In particular, if the coefficient matrix $\mathbf{A}(x)$ is continuous at x_0, this condition will be satisfied if $\mathbf{A}(x_0)\cdot(x_0-y)\neq 0$; that is, if the normal vector to $\partial\Omega$ is not in the null space of $\mathbf{A}$ at x_0.) It then follows that $a^{ij}(x_i-y_i)(x_j-y_j)\geqslant\lambda'|x-y|^2$ for all $x\in N\cap\Omega$, where λ' is a suitable positive constant, although the minimum eigenvalue λ may tend to zero at x_0. Let us assume also that all coefficients of L are bounded. The barrier argument preceding Theorem 6.13 now proceeds as before and we conclude that the function $w(x)=\tau(R^{-\sigma}-r^{-\sigma})$ defined in (6.45) determines a local barrier at x_0 for appropriate choices of τ and σ, and hence $u(x)\to\varphi(x_0)$. We note that in the boundary value problem considered for equation (6.54), the normal vector at each point of the boundary segment, $y=0$, $0<x<\pi$, is in the null-space of the coefficient matrix $[1,0/0,0]$, and thus the above hypothesis is not fulfilled.

Alternatively, let $[a^{ij}(x)]$ be an arbitrary positive matrix and assume that the functions $|\mathbf{b}|/\lambda$, c/λ, f/λ, are bounded. By dividing by the minimum eigenvalue λ we may assume without loss of generality that the operator L is strictly elliptic in Ω with $\lambda=1$. The limit behaviour $u(x)\to\varphi(x_0)$ is now guaranteed if Ω satisfies a *strict exterior plane condition* at x_0. By this we shall mean that in some neighborhood of x_0 there is a hyperplane intersecting $\bar{\Omega}$ in the single point x_0. Such a condition will be satisfied, for example, if $\partial\Omega$ is strictly convex near x_0. To prove the assertion $u(x)\to\varphi(x_0)$, let us for convenience choose x_0 to be the origin and let the normal to the assumed exterior plane at x_0 be the x_1 axis, with $x_1>0$ in Ω near x_0. Under the stated conditions there is a slab $0<x_1<d$ whose intersection D with Ω near x_0 is such that $x_1>0$ on $\bar{D}-x_0$. As in the proof of Theorem 3.7 we see that the function

$$w(x)=e^{\gamma d}(1-e^{-\gamma x_1})$$

satisfies $Lw\leqslant-\lambda$ in D provided $\gamma\geqslant 1+\sup\limits_D(|\mathbf{b}|/\lambda)$, and hence $Lw\leqslant-1$ if we take $\lambda=1$. It follows as in Remark 1 following Lemma 6.12 that for a suitable constant $k=k(\varepsilon)$, the functions

$$w_\varepsilon^+\equiv\varphi(x_0)+\varepsilon+kw,\qquad w_\varepsilon^-\equiv\varphi(x_0)-\varepsilon-kw$$

determine a local barrier at x_0 with respect to the upper and lower bounds on u in Ω.

We observe that if the boundary function φ is *constant* near x_0, then $u(x) \to \varphi(x_0)$ follows even if Ω satisfies a non-strict exterior plane condition at x_0. In this connection one notes that in the boundary value problem considered previously for (6.54), the boundary segment, $y=0$, $0<x<\pi$, is convex but not strictly convex, and the stated boundary value problem is solvable only for zero data on the interval.

The preceding remarks immediately yield the following simple extension of Theorem 6.13 to non-uniformly elliptic equations.

Theorem 6.24. *Let L be strictly elliptic (satisfying (6.2)) in a bounded domain Ω, with $c \leqslant 0$, a^{ij}, b^i, c, $f \in C^\alpha(\Omega)$, and assume b^i, c, f are bounded. Suppose that Ω satisfies the exterior sphere condition and, in addition, a strict exterior plane condition at those boundary points where any of the coefficients a^{ij} are unbounded. Then, if φ is continuous on $\partial\Omega$, the Dirichlet problem, $Lu=f$ in Ω, $u=\varphi$ on $\partial\Omega$, has a (unique) solution $u \in C^0(\bar{\Omega}) \cap C^{2,\alpha}(\Omega)$.*

It is clear from the above arguments that this result can be modified in various ways (see, for example, Problem 6.4). When the equation is homogeneous and the lower order terms of L are not present, we obtain the following:

Corollary 6.24′. *Let the coefficients a^{ij} of the elliptic equation $a^{ij}D_{ij}u = 0$ belong to $C^\alpha(\Omega)$, where Ω is a bounded strictly convex domain. Then the Dirichlet problem is solvable in $C^0(\bar{\Omega}) \cap C^{2,\alpha}(\Omega)$ for arbitrary continuous boundary values.*

Although this result is an immediate consequence of Theorem 6.24 its proof can be obtained more directly by observing that because of the strict convexity of Ω and the special form of the equation, a *linear* function determines a barrier at each boundary point.

A closer examination of the relation between the coefficients of L and the local curvature properties of the boundary makes it possible to derive other general sufficient conditions for the existence of barriers. Let f and the coefficients of L (with $c \leqslant 0$) be bounded and assume Ω is a C^2 domain. The minimum eigenvalue of the principal coefficient matrix $[a^{ij}(x)]$ may approach zero on $\partial\Omega$. We shall seek conditions for the existence of a local barrier at $x_0 \in \partial\Omega$ with respect to the continuous boundary function φ and the bound M.

Let B denote a ball centered at x_0 and set $G = B \cap \Omega$; B will be specified later. Let $\psi \in C^2(\bar{G})$ be a fixed function such that $\psi(x) > 0$ on $\bar{G} - x_0$ and $\psi(x_0) = 0$. For every $\varepsilon > 0$ and a suitable constant $k = k(\varepsilon)$, we can satisfy the inequalities

$$\begin{aligned} \varepsilon + k\psi(x) &\geqslant |\varphi(x) - \varphi(x_0)| \quad \text{on } \partial\Omega \cap B, \\ &\geqslant M \quad \text{on } \partial B \cap \Omega. \end{aligned}$$

Let us now define the distance function (see Appendix to Chapter 14).

$$d(x) = \operatorname{dist}(x, \partial\Omega), \; x \in \Omega,$$

which is in class C^2 in some neighborhood

$$\mathcal{N} = \{x \in \Omega \mid d(x) < d_0\}.$$

It can be assumed that $G \subset \mathcal{N}$. We propose to find conditions under which functions $w^{\pm}(x)$ of the form

$$w^+ = \varphi(x_0) + \varepsilon + k\psi + Kd, \qquad w^- = \varphi(x_0) - \varepsilon - k\psi - Kd$$

define a barrier at x_0 in G, where $K = K(\varepsilon)$ is an appropriate positive constant. This will be the case if $Lw^+ \leqslant f$ and $Lw^- \geqslant f$ in G. Thus, the functions $w^{\pm}$ define a barrier if the inequality

$$K(a^{ij}D_{ij}d + b^iD_id) \leqslant -k|L\psi| - |f - c\varphi(x_0)| \tag{6.55}$$

holds in G for some K. This provides a sufficient condition for the existence of a barrier which, in principle, can be verified by inspection of the given equation and domain.

To realize the condition (6.55) more concretely, let us, for example, assume the coefficients $a^{ij}(x)$ to be continuous at x_0. Choose a coordinate system with origin at x_0 and the x_n axis coincident with the inner normal Dd at x_0. A rotation $\mathbf{P}$ of the coordinate system about the x_n axis, in which the new axes are the principal directions of $\partial\Omega$ at x_0, diagonalizes the Hessian matrix $[D_{ij}d(x_0)]$, so that

$$\mathbf{P}^t[D_{ij}d(x_0)]\mathbf{P} = \operatorname{diag}\,[-\kappa_1, -\kappa_2, \ldots, -\kappa_{n-1}, 0],$$

where $\kappa_1, \kappa_2, \ldots, \kappa_{n-1}$ are the principal curvatures of $\partial\Omega$ with respect to the inner normal at x_0; (see Appendix to Chapter 14). If $a_1, a_2, \ldots, a_n$ denote the corresponding diagonal elements of the matrix $\mathbf{P}^t[a^{ij}(x_0)]\mathbf{P}$, we see that

$$(a^{ij}D_{ij}d)_{x=x_0} = -\sum_{i=1}^{n-1} a_i\kappa_i. \tag{6.56}$$

Hence, if

$$\sum_{i=1}^{n-1} a_i\kappa_i > \sup_{\Omega} |\mathbf{b}| \tag{6.57}$$

it follows by continuity that the inequality (6.55) is satisfied in $G = B \cap \Omega$ for some ball B about x_0, provided K is chosen large enough. If in addition, the coefficients b^i are continuous at x_0, and if b_ν denotes the normal component of the vector $\mathbf{b}(x_0) = (b^1(x_0), \ldots, b^n(x_0))$ with respect to the inner normal, condition (6.55) is satisfied in a suitable G provided

$$\sum_{i=1}^{n-1} a_i\kappa_i - b_\nu > 0. \tag{6.58}$$

Under the stated hypotheses, (6.57) and (6.58) are therefore sufficient conditions for the existence of a local barrier at x_0. If $\mathbf{b}(x_0)=0$, (6.58) reduces to the simple condition,

$$\sum_{i=1}^{n-1} a_i \kappa_i > 0,$$

which involves only the leading coefficients and the principal curvatures and directions of the boundary. It is not difficult to remove the continuity assumptions on the coefficients at the boundary and to reformulate (6.57) and (6.58) appropriately.

We remark that when $\partial\Omega$ is not in C^2, the preceding considerations are still applicable provided a C^2 domain $\tilde{\Omega}$ can be found such that $x_0 \in \partial\Omega \cap \partial\tilde{\Omega}$ and $B \cap \Omega \subset B \cap \tilde{\Omega}$ for some ball B containing x_0. The above conditions for the existence of a barrier at x_0 then remain valid if the distance function $d(x)$ is replaced by $\tilde{d}(x) = \text{dist}(x, \partial\tilde{\Omega})$.

The preceding remarks and those earlier in this section concerning the exterior sphere condition show the existence of a barrier at the following sets of boundary points of a C^2 domain Ω where the coefficients a^{ij}, b^i are continuous:

$$(6.59)\qquad \begin{aligned} \Sigma_1 &= \{x_0 \in \partial\Omega \,|\, \text{inequality (6.58) holds}\} \\ \Sigma_2 &= \{x_0 \in \partial\Omega \,|\, a^{ij}\nu_i(x_0)\nu_j(x_0) \neq 0, \text{ where } \boldsymbol{\nu}(x_0) \text{ is normal to } \partial\Omega\}. \end{aligned}$$

Taken together with Theorem 6.11 these results yield:

Theorem 6.25. *Let the operator L in (6.1) be elliptic in a C^2 domain Ω with $c \leq 0$ in Ω and a^{ij}, b^i/λ, c, $f/\lambda \in C^\alpha(\Omega) \cap C^0(\bar{\Omega})$. Assume $\Sigma_1 \cup \Sigma_2 = \partial\Omega$. Then the problem*

$$Lu = f \text{ in } \Omega, \quad u = \varphi \text{ on } \partial\Omega,$$

has a unique solution $u \in C^2(\Omega) \cap C^0(\bar{\Omega})$ for all $\varphi \in C^0(\partial\Omega)$.

We note that under certain conditions on L and $\partial\Omega$, a solution is uniquely determined by the values of φ on $\Sigma_1 \cup \Sigma_2$ even if $\Sigma_1 \cup \Sigma_2 \neq \partial\Omega$, as in the example (6.54). A maximum principle related to such a boundary value problem is contained in Problem 6.10.

6.7. Other Boundary Conditions; the Oblique Derivative Problem

Until now we have been concerned only with Dirichlet boundary conditions. We now develop an analogue of the Schauder theory for the regular oblique derivative problem.

Poisson's Equation

In extending the Schauder theory to other linear boundary value problems our point of departure is the theory of Poisson's equation in the half-space $\mathbb{R}^n_+ = \{x \mid x_n > 0\}$ with the oblique derivative boundary condition,

$$Nu \equiv au + \sum_{i=1}^{n} b_i D_i u = \varphi \quad \text{on } x_n = 0, \tag{6.60}$$

in which the coefficients a, b_i are constants. We also write the boundary operator N in the equivalent forms

$$Nu \equiv au + \mathbf{b} \cdot Du = au + b_t D_t u + b_n D_n u$$

where $\mathbf{b} = (b_1, \ldots, b_n) = (b_t, b_n)$ and $D = (D_1, \ldots, D_n) = (D_t, D_n)$, D_t denoting the tangential gradient. We assume throughout the *regular* oblique derivative condition $b_n \neq 0$, and to fix our ideas we take at first

$$b_n > 0, \qquad |\mathbf{b}| = (|b_t|^2 + b_n^2)^{1/2} = 1. \tag{6.61}$$

The latter condition is an inessential normalization which allows us to write

$$Nu = au + D_s u$$

where $D_s u = \partial u / \partial s$ is the directional derivative in the direction of the vector $\mathbf{b}$.

We consider initially the homogeneous boundary condition $Nu = 0$ on $x_n = 0$ and construct a harmonic Green's function in $\mathbb{R}^n_+$ satisfying this boundary condition. Letting Γ denote the fundamental solution (4.1) of Laplace's equation, we write for $n \geqslant 3$ and $a \leqslant 0$

$$G(x, y) = \Gamma(x - y) - \Gamma(x - y^*) - 2b_n \int_0^\infty e^{as} D_n \Gamma(x - y^* + \mathbf{b}s)\, ds, \tag{6.62}$$

where $x, y \in \mathbb{R}^n_+$, $D_n = \partial/\partial x_n$, and $y^* = (y_1, \ldots, y_{n-1}, -y_n) = (y', -y_n)$. Clearly G is harmonic in x and y for $x \neq y$ and by direct calculation one finds

$$NG(x, y) = 0 \quad \text{on } x_n = 0. \tag{6.63}$$

(Here N operates on G as a function of x for fixed y.) Thus G has the required properties of a Green's function for the boundary condition (6.63).

The choice of G in (6.62) is motivated by the following considerations. If $G(x, y) = \Gamma(x - y) + h(x, y)$ is the desired harmonic Green's function satisfying (6.63), then NG is also harmonic in x (for $y \neq x$) and vanishes on $x_n = 0$. It follows from the Schwarz reflection principle (Problem 2.4) that NG is regular in $\mathbb{R}^n$ except for the singularity

$$a\Gamma(x - y) + b_t D_t \Gamma(x - y) + b_n D_n \Gamma(x - y) - a\Gamma(x - y^*)$$
$$- b_t D_t \Gamma(x - y^*) + b_n D_n \Gamma(x - y^*).$$

Here we have used the fact that $D_i\Gamma(x^*-y^*)=D_i\Gamma(x-y)$ for $i=1,\ldots,n-1$, while $D_n\Gamma(x^*-y^*)=-D_n\Gamma(x-y)$. Thus if we impose the condition that NG vanish at infinity, it follows from the Liouville theorem

$$\begin{aligned} D_s h + ah &= -a\Gamma(x-y^*) - b_i D_i\Gamma(x-y^*) + b_n D_n(x-y^*) \\ &= -a\Gamma(x-y^*) - D_s\Gamma(x-y^*) + 2b_n D_n\Gamma(x-y^*). \end{aligned}$$

This implies

$$\begin{aligned} D_s[e^{as}\,h(x+\mathbf{b}s,\,y)] = &-[a\Gamma(x-y^*+\mathbf{b}s) + D_s\Gamma(x-y^*+\mathbf{b}s)]\,e^{as} \\ &+ 2b_n D_n\Gamma(x-y^*+\mathbf{b}s)\,e^{as}. \end{aligned}$$

Integrating with respect to s from $s=0$ to $s=\infty$ and then integrating by parts, we obtain

$$h(x,y) = -\Gamma(x-y^*) - 2b_n\int_0^\infty e^{as}\,D_n\Gamma(x-y^*+\mathbf{b}s)\,ds.$$

This expression for h gives us (6.62), which is valid also when $b_n=0$.

We now examine $G(x,y)$ more closely with the objective of deriving estimates analogous to those in Chapter 4 for the Newtonian potential and solutions of Poisson's equation.

Letting $\boldsymbol{\xi}=(x-y^*)/|x-y^*|$, we have

$$G(x,y) = \Gamma(x-y) - \Gamma(x-y^*) + \Theta(x,y)$$

where

$$\begin{aligned} \Theta(x,y) &\equiv -2b_n\int_0^\infty e^{as}D_n\Gamma(x-y^*+\mathbf{b}s)\,ds, \quad a\leqslant 0, \\ &= -|x-y^*|^{2-n}\left[\frac{2b_n}{n\omega_n}\int_0^\infty e^{a|x-y^*|s}\,\frac{(\xi_n+b_n s)\,ds}{(1+2(\boldsymbol{\xi}\cdot\mathbf{b})s+s^2)^{n/2}}\right] \\ &= |x-y^*|^{2-n}\,g(\boldsymbol{\xi},|x-y^*|). \end{aligned}$$

One sees that g is regular in its arguments since (by (6.61)) $\boldsymbol{\xi}\cdot\mathbf{b}>-|b_t|>-1$ for all $x,y\in\mathbb{R}^n_+$ and hence the denominator of the integrand is bounded away from zero. The function $\Theta(x,y)=\Theta(x-y^*)$ satisfies the relations

$$\begin{aligned} &D_{x_i}\Theta(x,y) = -D_{y_i}\Theta(x,y), \qquad i=1,\ldots,n-1; \\ (6.64)\qquad &D_{x_n}\Theta(x,y) = D_{y_n}\Theta(x,y); \\ &|D^\beta\Theta(x,y)| \leqslant C|x-y^*|^{2-n-|\beta|}, \qquad C=C(n,|\beta|,b_n). \end{aligned}$$

These relations will suffice in extending the details of Lemma 4.10 for the Newtonian potential to the analogous integral $\int \Theta(x, y)f(y)\,dy$. If $|\mathbf{b}| \neq 1$ we replace a by $a/|\mathbf{b}|$ and b_n by $b_n/|\mathbf{b}|$ in the preceding. We note that since $a \leqslant 0$ the constant C in (6.64) can be taken independent of a.

Theorem 6.26. *Let $B_1 = B_R(x_0)$, $B_2 = B_{2R}(x_0)$ denote balls with center $x_0 \in \bar{\mathbb{R}}^n_+$, and let $B_1^+ = B_1 \cap \mathbb{R}^n_+$, $B_2^+ = B_2 \cap \mathbb{R}^n_+$, $T = B_2 \cap \{x_n = 0\}$. Suppose that $u \in C^2(B_2^+) \cap C^1(B_2^+ \cup T)$ satisfies $\Delta u = f$ in B_2^+, $f \in C^\alpha(\bar{B}_2^+)$, and the boundary condition (6.60), $Nu = \varphi$ on T, with $a \leqslant 0, b_n > 0$ and $\varphi \in C^{1,\alpha}(T)$. Then $u \in C^{2,\alpha}(\bar{B}_1^+)$ and*

$$(6.65) \qquad |u|'_{2,\alpha;B_1^+} \leqslant C(|u|_{0;B_2^+} + R|\varphi|'_{1,\alpha;T} + R^2|f|'_{0,\alpha;B_2^+}),$$

where $C = C(n, \alpha, b_n/|\mathbf{b}|)$.

(Here $|\ \ |'$ denotes the weighted norms (4.6)$'$ defined with respect to R.)

Proof. It is assumed that T is non-empty for otherwise the stated result is already contained in Theorem 4.6. Suppose first that $\varphi = 0$, $|\mathbf{b}| = 1$ and $n > 2$. Consider the function

$$w(x) = \int_{B_2^+} G(x, y) f(y)\,dy = w_1(x) + w_2(x),$$

where

$$w_1(x) = \int_{B_2^+} [\Gamma(x-y) - \Gamma(x-y^*)] f(y)\,dy$$

$$w_2(x) = \int_{B_2^+} \Theta(x, y) f(y)\,dy.$$

We have already seen in (4.26) that w_1 satisfies the estimate

$$(6.66) \qquad |D^2 w_1|'_{0,\alpha;B_1^+} \leqslant C|f|'_{0,\alpha;B_2^+}, \quad C = C(n, \alpha).$$

The estimation of w_2 is essentially the same as that for the Newtonian potential in Lemma 4.10. Let $f(x)$ be defined by even reflection with respect to x_n, so that $f(x', -x_n) = f(x', x_n)$. Then a representation analogous to (4.9) is valid for the second derivatives of w_2; namely, for $x \in B_2^+$ and $i, j = 1, \ldots, n$, we have

$$D_{ij} w_2(x) = \int_{B_2^+} D_{ij}\Theta(x - y^*)(f(y^*) - f(x))\,dy$$

$$-f(x) \int_{\partial B_2^+} D_i\Theta(x - y^*)\nu_j(y)\,ds_y,$$

where $\boldsymbol{\nu}=(\nu_1,\ldots,\nu_n)$ is the outward unit normal to ∂B_2^+. By virtue of (6.64) the arguments of Lemmas 4.4 and 4.10 carry over without essential change to give the estimate

$$|D^2w_2|'_{0,\alpha;B_1^+}\leqslant C|f|'_{0,\alpha;B_2^+},\quad C=C(n,\alpha,b_n).$$

Combining this inequality with (6.66), we obtain

$$|D^2w|'_{0,\alpha;B_1^+}\leqslant C|f|'_{0,\alpha;B_2^+},\quad C=C(n,\alpha,b_n). \tag{6.67}$$

If $|\mathbf{b}|\neq 1$ we replace b_n by $b_n/|\mathbf{b}|$ in this estimate.

To obtain estimates for u from the preceding, let $\eta\in C_0^2(B_2)$ be a cutoff function such that $\eta(x)=1$ for $|x-x_0|\leqslant(\frac{3}{2})R$ and $|D^\beta\eta|\leqslant C/R^{|\beta|}$ for $|\beta|\leqslant 2$. Then, we have

$$u(x)\eta(x)=\int_{B_2^+}G(x,y)\Delta[u(y)\eta(y)]\,dy,\qquad x\in B_2^+.$$

If $x\in B_1^+$, so that $|x-y|>R/2$ where $D\eta\neq 0$, we obtain

$$\begin{aligned}u(x)&=\int_{B_2^+}(\eta Gf+u\Delta\eta)\,dy+2\int_{B_2^+}GDu\cdot D\eta\,dy,\\&=\int_{B_2^+}(\eta Gf+u\Delta\eta)\,dy-2\int_{B_2^+}u(DG\cdot D\eta+G\Delta\eta)\,dy.\end{aligned}$$

The required estimate (6.65) when $\varphi=0$ now follows from (6.67), the bounds $|D^\beta\eta|\leqslant C/R^{|\beta|}$, and, by (2.14) and (6.64),

$$|D^\beta G(x,y)|\leqslant C|x-y|^{2-n-|\beta|}\leqslant CR^{2-n-|\beta|},$$

where the indicated derivatives may be with respect to both x and y variables.

We now remove the restriction that $\varphi=0$. For this purpose we seek a function $\psi\in C^{2,\alpha}(\bar{B}_2^+)$ satisfying $N\psi=\varphi$ on T. It can be assumed that φ is suitably extended outside T so that $\varphi\in C_0^{1,\alpha}(\mathbb{R}^{n-1})$ on $x_n=0$; (see Lemma 6.38). Choosing a non-negative function $\eta\in C_0^2(\mathbb{R}^{n-1})$ such that $\int\eta(y')\,dy'=1$, $y'=(y_1,\ldots,y_{n-1})$, we define

$$\psi(x)=\psi(x',x_n)=b_n^{-1}x_n\int_{\mathbb{R}^{n-1}}\varphi(x'-x_ny')\eta(y')\,dy'. \tag{6.68}$$

One verifies easily that $\psi(x',0)=0$, $D_n\psi(x',0)=b_n^{-1}\varphi(x')$, and hence

$$N\psi=\varphi\quad\text{on}\quad x_n=0. \tag{6.69}$$

That $\psi \in C^{2,\alpha}(\mathbb{R}^n_+)$ is seen from the relations

$$b_n D_{ij}\psi(x) = \int D_i\varphi(x' - x_n y')D_j\eta(y')\,dy', \quad i, j \neq n;$$

$$b_n D_{in}\psi(x) = -\int y' \cdot D\varphi(x' - x_n y')D_i\eta(y')\,dy', \quad i \neq n;$$

$$b_n D_{nn}\psi(x) = \int y' \cdot D\varphi(x' - x_n y')[(n-2)\eta(y') + y' \cdot D\eta(y')]\,dy'.$$

We observe also that

$$(6.70)\qquad b_n|\psi|'_{2,\alpha;B_2^+} \leqslant C(R|\varphi|_{0;T} + R^2|D\varphi|_{0;T} + R^{2+\alpha}[D\varphi]_{\alpha;T}) = CR|\varphi|'_{1,\alpha;T},$$

where the constant C depends only on n and the choice of η.

We are now able to make the reduction to the case $\varphi = 0$. Letting $v = u - \psi$, it follows that $\Delta v = f - \Delta\psi \in C^\alpha(\bar{B}_2^+)$, and by virtue of (6.69) we have $Nv = 0$ on T. From (6.70) and the proven estimate for v we obtain (6.65). □

We remark that in general the constant C in (6.65) becomes unbounded as $b_n \to 0$.

The following estimate is a consequence of the preceding theorem and is stated without proof. The details are similar to the proof of Theorem 4.8.

Lemma 6.27. *Let Ω be a bounded open set in $\mathbb{R}^n_+$ with boundary portion T on $x_n = 0$, and let $u \in C^2(\Omega) \cap C^1(\Omega \cup T)$ satisfy $\Delta u = f$ in Ω, $f \in C^\alpha(\bar{\Omega})$, and the boundary condition (6.60), $Nu = \varphi$ on T, with $a \leqslant 0$, $b_n > 0$ and $\varphi \in C^{1,\alpha}(T)$. Then*

$$|u|^*_{2,\alpha;\Omega\cup T} \leqslant C(|u|_{0;\Omega} + |\varphi|_{1,\alpha;T} + |f|_{0,\alpha;\Omega}),$$

where $C = C(n, \alpha, b_n/|\mathbf{b}|, \operatorname{diam}\Omega)$.

The same argument as in Lemma 6.1 provides the following extension to the oblique derivative problem.

Lemma 6.28. *Under the same hypotheses as in Lemma 6.27 let u satisfy $L_0 u = f$ (in place of $\Delta u = f$), where L_0 is the constant coefficient operator defined in Lemma 6.1. Then*

$$(6.71)\qquad |u|^*_{2,\alpha;\Omega\cup T} \leqslant C(|u|_{0;\Omega} + |\varphi|_{1,\alpha;T} + |f|_{0,\alpha;\Omega})$$

where $C = C(n, \alpha, \lambda, \Lambda, b_n/|\mathbf{b}|, \operatorname{diam}\Omega)$.

Variable Coefficients

We consider now equations with variable coefficients and the corresponding oblique derivative problem in domains with curved boundaries. The extension of the estimates of the Schauder theory to these boundary conditions follows closely the ideas in Sections 6.1 and 6.2. We derive first an analogue of Lemma 6.4.

Lemma 6.29. *Let Ω be a bounded open set in $\mathbb{R}^n_+$ with a boundary portion T on $x_n=0$. Suppose that $u\in C^{2,\alpha}(\Omega\cup T)$ is a solution in Ω of $Lu=f$ (equation* (6.1)) *satisfying the boundary condition*

$$N(x')u\equiv\gamma(x')u+\sum_{i=1}^{n}\beta_i(x')D_iu=\varphi(x'),\quad x'\in T, \tag{6.72}$$

where $|\beta_n|\geqslant\kappa>0$ for some constant κ. It is assumed that L satisfies (6.2) *and that $f\in C^\alpha(\bar\Omega)$, $\varphi\in C^{1,\alpha}(\bar T)$, a^{ij}, b^i, $c\in C^\alpha(\bar\Omega)$ and γ, $\beta_i\in C^{1,\alpha}(\bar T)$ with*

$$|a^{ij}, b^i, c|_{0,\alpha;\Omega},\ |\gamma,\beta_i|_{1,\alpha;T}\leqslant\Lambda,\quad i,j=1,\dots,n.$$

Then

$$|u|^*_{2,\alpha;\Omega\cup T}\leqslant C(|u|_{0;\Omega}+|\varphi|_{1,\alpha;T}+|f|_{0,\alpha;\Omega}), \tag{6.73}$$

where $C=C(n,\alpha,\lambda,\Lambda,\kappa,\operatorname{diam}\Omega)$.

Proof. We remark first that it can be assumed $\gamma\leqslant0$ and $\beta_n>0$. For by setting $v=ue^{kx_n}$, where $k\geqslant\sup|\gamma|/\kappa$, the boundary condition (6.72) is taken into

$$N'v=(\gamma-k\beta_n)v+\sum\beta_iD_iv=\varphi\quad\text{on } T$$

with $\beta_n(\gamma-k\beta_n)\leqslant0$. At the same time $Lu=f$ is transformed into an equation $L'v=f'$ satisfying the same hypotheses as in the theorem. The desired estimate (6.73) is obviously equivalent to the corresponding estimate for v.

The technique in Theorem 6.2 and Lemma 6.4 of "freezing" the coefficients is once again applicable, with certain modifications due to the boundary condition (6.72). Thus, let x_0, y_0 be any two distinct points in Ω and suppose $\bar d_{x_0}=\min(\bar d_{x_0},\bar d_{y_0})$, where $\bar d_x=\operatorname{dist}(x,\partial\Omega-T)$. Let $\mu\leqslant1/4$ be a positive constant (to be specified later), and set $d=\mu\bar d_{x_0}$, $B_d=B_d(x_0)$. If $B_d\cap T\neq\phi$ let x_0' denote the projection of x_0 on T. As in Theorem 6.2 we rewrite the equation $Lu=f$ in the form (6.15) and the boundary condition (6.72) as

$$N(x_0')u=[N(x_0')-N(x')]u(x')+\varphi(x')\equiv\Phi(x'),\quad x'\in T, \tag{6.74}$$

and we consider (6.15), (6.74) as a problem in $B_d\cap\Omega$ with constant coefficients, disregarding (6.74) if $B_d\cap T=\phi$. The argument in Theorem 6.2 and the analogous one indicated in Lemma 6.4 proceed in essentially the same way, except that Lemma 6.28 now replaces Lemma 6.1 in the details. Thus in place of (6.16) we now obtain

$$\bar d^{2+\alpha}_{x_0}\frac{|D^2u(x_0)-D^2u(y_0)|}{|x_0-y_0|^\alpha}\leqslant\frac{C}{\mu^{2+\alpha}}(|u|_{0;\Omega}+|\Phi|_{1,\alpha;B\cap T}+|F|_{0,\alpha;B\cap\Omega})$$
$$+\frac{4}{\mu^\alpha}[u]^*_{2;\Omega\cup T}. \tag{6.75}$$

The estimation of the right number is essentially as in Theorem 6.2 with the exception of the additional term $|\Phi|_{1,\alpha;B\cap T}$. Concerning this term, one sees that

$$|[N(x_0') - N(x')]u(x')|_{1,\alpha;B\cap T} \leqslant C\mu^{2+\alpha}(C(\mu)|u|_{0;\Omega} + \mu^\alpha[u]^*_{2,\alpha;\Omega\cup T}).$$

The details are similar to those leading to (6.19) and are not carried out here. Combining this estimate with that for the other terms in (6.75) we arrive at the desired estimate (6.73). □

The preceding lemma can now be extended to domains with curved boundaries. A repetition of the arguments in Lemma 6.5 and Theorem 6.6 leads to the following global estimate for solutions of the oblique derivative problem.

Theorem 6.30. *Let Ω be a $C^{2,\alpha}$ domain in $\mathbb{R}^n$, and let $u \in C^{2,\alpha}(\bar\Omega)$ be a solution in Ω of $Lu=f$ satisfying the boundary condition*

$$N(x)u \equiv \gamma(x)u + \sum_{i=1}^n \beta_i(x)D_iu = \varphi(x), \quad x \in \partial\Omega,$$

where the normal component β_ν of the vector $\boldsymbol\beta = (\beta_1, \ldots, \beta_n)$ is non-zero and

$$(6.76) \qquad |\beta_\nu| \geqslant \kappa > 0 \text{ on } \partial\Omega \quad (\kappa = \text{const}).$$

It is assumed that L satisfies (6.2) and that $f \in C^\alpha(\bar\Omega)$, $\varphi \in C^{1,\alpha}(\bar\Omega)$, $a^{ij}, b^i, c \in C^\alpha(\bar\Omega)$ and $\gamma, \beta_i \in C^{1,\alpha}(\bar\Omega)$ with

$$|a^{ij}, b^i, c|_{0,\alpha;\Omega}, |\gamma, \beta_i|_{1,\alpha;\Omega} \leqslant \Lambda, \quad i, j = 1, \ldots, n.$$

Then

$$(6.77) \qquad |u|_{2,\alpha;\Omega} \leqslant C(|u|_{0;\Omega} + |\varphi|_{1,\alpha;\Omega} + |f|_{0,\alpha;\Omega})$$

where $C = C(n, \alpha, \lambda, \Lambda, \kappa, \Omega)$.

Remarks. 1) Condition (6.76) implies that the directional derivative $\boldsymbol\beta \cdot Du$ is nowhere tangential to $\partial\Omega$. This hypothesis is essential in the present considerations. 2) In the statement of the theorem it is convenient and involves no loss of generality to assume that φ, γ and β_i are defined globally (rather than on $\partial\Omega$), so that the norm $|\ |_{1,\alpha;\Omega}$ is well defined for these functions. In Theorem 6.26 and Lemmas 6.27–6.29 where T was a hyperplane boundary portion, a global extension of φ (and of γ, β_i in Lemma 6.29) was not used since the norm $|\ |_{1,\alpha;T}$ was naturally defined. 3) That $|\varphi|_{1,\alpha}$ appears in the estimate (6.77) is to be expected since Nu involves first order differentiation of the $C^{2,\alpha}$ function u. This is in contrast with the corresponding global estimate (6.36) for the Dirichlet boundary condition which requires that $\varphi \in C^{2,\alpha}$.

We have thus far been concerned only with estimates for the oblique derivative problem. The actual solution of the problem for $Lu=f$ can be reduced to that for Poisson's equation by the method of continuity (as in Theorem 6.8), but the method now involves a continuous family of both differential operators and boundary operators. Unique solvability of the oblique derivative problem, under appropriate additional restrictions on the operators L and N, is provided in the following theorem.

Theorem 6.31. *Let L be a strictly elliptic operator with $c\leqslant 0$ and coefficients in $C^{\alpha}(\bar{\Omega})$ in a $C^{2,\alpha}$ domain Ω. Let $Nu\equiv\gamma u+\boldsymbol{\beta}\cdot Du$ define a boundary operator on $\partial\Omega$ such that $\gamma(\boldsymbol{\beta}\cdot\boldsymbol{\nu})>0$ on $\partial\Omega$ if $\boldsymbol{\nu}$ is the outward unit normal on $\partial\Omega$. Assume that γ, $\boldsymbol{\beta}\in C^{1,\alpha}(\partial\Omega)$. Then the oblique derivative problem*

$$Lu=f \text{ in } \Omega, \qquad Nu=\varphi \text{ on } \partial\Omega \tag{6.78}$$

has a unique $C^{2,\alpha}(\bar{\Omega})$ solution for all $f\in C^{\alpha}(\bar{\Omega})$ and $\varphi\in C^{1,\alpha}(\partial\Omega)$.

Proof. We assume without loss of generality that $\gamma>0$ and $\boldsymbol{\beta}\cdot\boldsymbol{\nu}>0$ on $\partial\Omega$ and also that φ and $\boldsymbol{\beta}$ are extended to all of $\bar{\Omega}$ and are contained in $C^{1,\alpha}(\bar{\Omega})$. Consider the family of problems for $0\leqslant t\leqslant 1$:

$$\begin{aligned} L_t u &\equiv tLu+(1-t)\Delta u=f \text{ in } \Omega, \\ N_t u &\equiv tNu+(1-t)\left(\frac{\partial u}{\partial\nu}+u\right)=\varphi \text{ on } \partial\Omega. \end{aligned} \tag{6.79}$$

We note that $L_1=L$, $L_0=\Delta$, $N_1=N$, $N_0=\partial/\partial\nu+\text{identity}$, and that for suitable positive constants λ, Λ the coefficients a_t^{ij}, b_t^i, c_t of L_t satisfy

$$a_t^{ij}\xi_i\xi_j\geqslant\lambda_t|\xi|^2=\min(1,\lambda)|\xi|^2 \quad \forall x\in\Omega,\ \xi\in\mathbb{R}^n$$

and

$$|a_t^{ij}, b_t^i, c_t|_{0,\alpha}\leqslant\Lambda_t=\max(1,\Lambda).$$

Also, since $\boldsymbol{\beta}\cdot\boldsymbol{\nu}\geqslant\beta'>0$ and $\gamma\geqslant\gamma'>0$ for some constants β', γ', we have

$$\gamma_t=(1-t)+t\gamma\geqslant\min(1,\gamma')>0$$
$$\boldsymbol{\beta}_t\cdot\boldsymbol{\nu}=(1-t)+t\boldsymbol{\beta}\cdot\boldsymbol{\nu}\geqslant\min(1,\beta')>0 \text{ on } \partial\Omega$$

while $|\boldsymbol{\beta}|_{1,\alpha}$ is also bounded independently of t.

Consider any solution $u\in C^{2,\alpha}(\bar{\Omega})$ of the problem (6.79) (for some t). $|u|_{2,\alpha}$ satisfies the estimate (6.77) with a constant C independent of t. By estimating $|u|_0$ in terms of φ and f we shall obtain in addition a bound

$$|u|_{2,\alpha}\leqslant C(|\varphi|_{1,\alpha}+|f|_{0,\alpha}) \tag{6.80}$$

valid for all $C^{2,\alpha}(\bar{\Omega})$ solutions of (6.79).

To estimate $|u|_0$ we first make a substitution $v=u/\omega$ where ω is a fixed $C^0(\bar{\Omega}) \cap C^2(\Omega)$ function (independent of t) satisfying the conditions: (i) $\omega \geqslant \bar{\omega} > 0$; (ii) $L_t\omega \leqslant \bar{c} < 0$; (iii) $\gamma_t + \boldsymbol{\beta}_t \cdot D\omega/\omega \geqslant \bar{\gamma} > 0$ on $\partial\Omega$, where $\bar{\omega}$, $\bar{c}$, $\bar{\gamma}$ are constants. Such a function ω can be chosen in the form $\omega(x) = c_1 - c_2\, e^{\mu x_1}$ if μ is sufficiently large and c_1, c_2 are suitable positive constants. The substitution $v=u/\omega$ transforms (6.79) into another problem: $\tilde{L}_t v = \tilde{f} = f/\omega$ in Ω, $\tilde{N}_t v = \tilde{\varphi} = \varphi/\omega$ on $\partial\Omega$, in which the coefficient $L_t\omega/\omega$ of v in $\tilde{L}_t v$ satisfies $L_t\omega/\omega \leqslant \bar{c}/\omega < 0$ and the coefficient $\tilde{\gamma}_t$ of v in $\tilde{N}_t v$ satisfies $\tilde{\gamma}_t = \gamma_t + \boldsymbol{\beta}_t \cdot D\omega/\omega \geqslant \bar{\gamma} > 0$. If now $\sup_\Omega |v| = |v(x_0)|$ at some $x_0 \in \Omega$, then

$$\sup_\Omega |v| = |v(x_0)| \leqslant |\tilde{f}(x_0)/\bar{c}| \leqslant \sup_\Omega |\tilde{f}|/|\bar{c}|,$$

and hence

$$|u|_0 \leqslant \sup_\Omega \omega \sup_\Omega |v| \leqslant C|f|_0,$$

where C is a constant independent of t. On the other hand, if $\sup_\Omega |v| = |v(x_0)|$ for some $x_0 \in \partial\Omega$, then either

$$\sup_\Omega |v| = v(x_0) \leqslant \bar{\gamma}^{-1}(\tilde{\varphi} - \boldsymbol{\beta}_t \cdot Dv)_{x=x_0} \leqslant \tilde{\varphi}(x_0)/\bar{\gamma}$$

or

$$\sup_\Omega |v| = -v(x_0) \leqslant \bar{\gamma}^{-1}(-\tilde{\varphi} + \boldsymbol{\beta}_t \cdot Dv)_{x=x_0} \leqslant -\tilde{\varphi}(x_0)/\bar{\gamma}$$

and thus $|u|_0 \leqslant C \sup_{\partial\Omega} |\varphi|$. The estimate (6.80) follows at once from (6.77).

The argument proceeds now essentially as in Theorem 6.8. Let

$$\mathfrak{B}_1 = C^{2,\alpha}(\bar{\Omega}), \qquad \mathfrak{B}_2 = C^{\alpha}(\bar{\Omega}) \times C^{1,\alpha}(\partial\Omega),$$

with

$$\|(f, \varphi)\|_{\mathfrak{B}_2} = |f|_{0,\alpha;\Omega} + |\varphi|_{1,\alpha;\partial\Omega},$$

and consider the operator

$$\mathfrak{L}_t = (L_t, N_t)\colon \mathfrak{B}_1 \to \mathfrak{B}_2.$$

The solvability of the problem (6.79) for arbitrary $f \in C^{\alpha}(\bar{\Omega})$, $\varphi \in C^{1,\alpha}(\partial\Omega)$ is equivalent to showing that $\mathfrak{L}_t$ is one-to-one *and* onto. Let u_t denote a solution of this problem for given f, φ. It is unique (see Problem 3.1) and from (6.80) we have the bound

$$|u_t|_{2,\alpha} \leqslant C(|f|_{0,\alpha} + |\varphi|_{1,\alpha})$$

or, equivalently,

(6.81) $$\|u\|_{\mathfrak{B}_1} \leqslant C\|\mathfrak{L}_t u\|_{\mathfrak{B}_2},$$

the constant C being independent of t. The fact that $\mathfrak{L}_0$ is invertible is a consequence of the solvability in $C^{2,\alpha}(\bar{\Omega})$ of the third boundary value problem:

$$\Delta u = f \text{ in } \Omega, \qquad u + \frac{\partial u}{\partial \nu} = \varphi \text{ on } \partial\Omega.$$

We refer to the literature of potential theory for this result; (see [GU], for example). Assuming this result, we conclude from (6.81) and the method of continuity (Theorem 5.2) that the theorem is proved. □

If either of the conditions $\gamma > 0$ or $c \leqslant 0$ is not satisfied in the preceding theorem, unique solvability is no longer assured, and one can assert instead the Fredholm alternative as in Theorem 6.15. The method of proof is basically the same. An immediate consequence of the alternative is solvability when $c \leqslant 0$, $\gamma \geqslant 0$ and either $c \not\equiv 0$ or $\gamma \not\equiv 0$, since uniqueness holds under these conditions.

6.8. Appendix 1: Interpolation Inequalities

We prove here the interpolation inequalities quoted in the course of Chapter 6. We begin with the inequalities for interior norms and seminorms.

Lemma 6.32. *Suppose $j+\beta < k+\alpha$, where j, $k = 0, 1, 2, \ldots$, and $0 \leqslant \alpha$, $\beta \leqslant 1$. Let Ω be an open subset of $\mathbb{R}^n$ and assume $u \in C^{k,\alpha}(\Omega)$. Then for any $\varepsilon > 0$ and some constant $C = C(\varepsilon, k, j)$ we have*

(6.82)
$$[u]^*_{j,\beta;\Omega} \leqslant C|u|_{0;\Omega} + \varepsilon[u]^*_{k,\alpha;\Omega},$$
$$|u|^*_{j,\beta;\Omega} \leqslant C|u|_{0;\Omega} + \varepsilon[u]^*_{k,\alpha;\Omega}.$$

Proof. We shall establish (6.82) for the cases j, $k = 0, 1, 2$ needed in the present work. A direct extension of the ideas and a suitable induction yield the stated result for arbitrary j, k.

It is assumed that the right member of (6.82) is finite since otherwise the assertion is trivial. For notational convenience we omit the subscript Ω, the domain Ω being implicitly understood. We consider several cases:

(i) $j = 1$, $k = 2$; $\alpha = \beta = 0$. We wish to show

(6.83) $$[u]^*_1 \leqslant C(\varepsilon)|u|_0 + \varepsilon[u]^*_2$$

for any $\varepsilon>0$. Let x be any point in Ω, d_x its distance from $\partial\Omega$, and $\mu\leqslant\frac{1}{2}$ a positive constant to be specified later. We set $d=\mu d_x$ and $B=B_d(x)$. For any $i=1, 2, \ldots, n$, let x', x'' be the endpoints of the segment of length $2d$ parallel to the x_i axis with its center at x. Then for some $\bar{x}$ in this segment we have

$$|D_iu(\bar{x})|=\frac{|u(x')-u(x'')|}{2d}\leqslant\frac{1}{d}|u|_0$$

and

$$|D_iu(x)|=|D_iu(\bar{x})+\int_{\bar{x}}^{x}D_{ii}u\,dx_i|\leqslant\frac{1}{d}|u|_0+d\sup_B|D_{ii}u|$$

$$\leqslant\frac{1}{d}|u|_0+d\sup_{y\in B}d_y^{-2}\sup_{y\in B}d_y^2|D_{ii}u(y)|.$$

Since $d_y>d_x-d=(1-\mu)\,d_x\geqslant d_x/2$ for all $y\in B$, it follows that

$$d_x|D_iu(x)|\leqslant\mu^{-1}|u|_0+4\mu\sup_{y\in\Omega}d_y^2|D_{ii}u(y)|\leqslant\mu^{-1}|u|_0+4\mu[u]_2^*.$$

Hence

$$[u]_1^*=\sup_{\substack{x\in\Omega\\ i=1,\ldots,n}}d_x|D_iu(x)|\leqslant\mu^{-1}|u|_0+4\mu[u]_2^*.$$

If now μ is chosen so that $\mu\leqslant\varepsilon/4$ we conclude (6.82) with $C=\mu^{-1}$.

(ii) $j\leqslant k$; $\beta=0$, $\alpha>0$. We proceed in a similar manner. As before let $x\in\Omega$, $0<\mu\leqslant\frac{1}{2}$, $d=\mu d_x$, $B=B_d(x)$, and let x', x'' be the endpoints of the segment of length $2d$ parallel to the x_l axis with its center at x. For some $\bar{x}$ on this segment we have

$$|D_{il}u(\bar{x})|=\frac{|D_iu(x')-D_iu(x'')|}{2d}\leqslant\frac{1}{d}\sup_B|D_iu| \tag{6.84}$$

and

$$|D_{il}u(x)|\leqslant|D_{il}u(\bar{x})|+|D_{il}u(x)-D_{il}u(\bar{x})|$$

$$\leqslant\frac{1}{d}\sup_{y\in B}d_y^{-1}\sup_{y\in B}d_y|D_iu(y)|$$

$$+d^\alpha\sup_{y\in B}d_{x,y}^{-2-\alpha}\sup_{y\in B}d_{x,y}^{2+\alpha}\frac{|D_{il}u(x)-D_{il}u(y)|}{|x-y|^\alpha}.$$

Again, since d_y, $d_{x,y}>d_x/2$ for all $y\in B$, it follows that

$$d_x^2|D_{il}u(x)|\leqslant\frac{2}{\mu}[u]_1^*+2^{2+\alpha}\mu^\alpha[u]_{2,\alpha}^*.$$

Taking the supremum over i, l and $x \in \Omega$, choosing μ so that $8\mu^\alpha \leqslant \varepsilon$ and setting $C = 2/\mu$, we obtain the inequality

$$(6.85) \qquad [u]_2^* \leqslant C(\varepsilon)[u]_1^* + \varepsilon[u]_{2,\alpha}^*.$$

If u replaces $D_i u$ in (6.84) and the obvious modifications are made in the ensuing details, we obtain (6.82) for $j = k = 1$, $\beta = 0$, $\alpha > 0$:

$$(6.86) \qquad [u]_1^* \leqslant C(\varepsilon)|u|_0 + \varepsilon[u]_{1,\alpha}^*.$$

Combining (6.85) with (6.83) after appropriate choice of ε in each of these inequalities, we arrive at (6.82) for $k = 2$ and $j = 1, 2$.

(iii) $j < k$; $\beta > 0$, $\alpha = 0$. Let $x, y \in \Omega$ with $d_x \leqslant d_y$, so that $d_x = d_{x,y}$. Let μ, d, and B be defined as before. We prove (6.82) for the case $j = 0$, first establishing the interpolation inequality

$$[u]_{0,\beta}^* \leqslant C(\varepsilon)|u|_0 + \varepsilon[u]_1^*,$$

where $\varepsilon > 0$ may be arbitrary for $0 < \beta < 1$. If $y \in B$, we obtain from the theorem of the mean, for $0 < \beta \leqslant 1$,

$$d_x^\beta \frac{|u(x) - u(y)|}{|x-y|^\beta} \leqslant \mu^{1-\beta}\, d_x |Du|_{0;B} \leqslant 2\mu^{1-\beta}[u]_1^*;$$

and if $y \notin B$, we have

$$(6.87) \qquad d_x^\beta \frac{|u(x) - u(y)|}{|x-y|^\beta} \leqslant 2\mu^{-\beta}|u|_0.$$

Combining these inequalities, we obtain for $0 < \beta \leqslant 1$

$$(6.88) \qquad [u]_{0,\beta}^* = \sup_{x,y \in \Omega} d_{x,y}^\beta \frac{|u(x) - u(y)|}{|x-y|^\beta} \leqslant 2\mu^{-\beta}|u|_0 + 2\mu^{1-\beta}[u]_1^*.$$

This implies (6.82) when $\beta < 1$ and $2\mu^{1-\beta} \leqslant \varepsilon$. Applying (6.83) to the right member of (6.88) and choosing μ appropriately, we obtain (6.82) for $j = 0$, $k = 2$, $\alpha = 0$, $0 < \beta \leqslant 1$. The proof for $j = 1$, $k = 2$ proceeds in much the same way after replacing u with $D_i u$. There is the following difference, however. In place of (6.87) we now have the inequality,

$$d_x^{1+\beta} \frac{|D_i u(x) - D_i u(y)|}{|x-y|^\beta} \leqslant \mu^{-\beta}[d_x|D_i u(x)| + d_y|D_i u(y)|] \leqslant 2\mu^{-\beta}[u]_1^*.$$

The conclusion again follows by application of (6.83).

(iv) $j \leqslant k$; $\alpha, \beta > 0$. It suffices to take $j=k$ and hence $\alpha > \beta$. With the same notation as above, we have for $y \in B$

$$d_x^\beta \frac{|u(x)-u(y)|}{|x-y|^\beta} \leqslant \mu^{\alpha-\beta}\, d_x^\alpha \frac{|u(x)-u(y)|}{|x-y|^\alpha},$$

while if $y \notin B$,

$$d_x^\beta \frac{|u(x)-u(y)|}{|x-y|^\beta} \leqslant 2\mu^{-\beta}|u|_0.$$

Combining these inequalities and taking the supremum over $x, y \in \Omega$, we obtain (6.82) for $j=k=0$ with $\varepsilon = \mu^{\alpha-\beta}$ and $C = 2/\mu^\beta$. The remaining cases when $j=k=1, 2$ follow in a similar way with the use of results from case (ii).

The interpolation inequality (6.82) for seminorms immediately implies the norm inequality

$$|u|^*_{j,\beta;\Omega} \leqslant C|u|_{0;\Omega} + \varepsilon[u]^*_{k,\alpha;\Omega}. \quad \square$$

An application of Lemma 6.32 yields the following compactness result.

Lemma 6.33. *Let Ω be a bounded open set in $\mathbb{R}^n$ and let S be a bounded subset of the Banach space*

$$C^{k,\alpha}_* = \{u \in C^{k,\alpha}(\Omega) \mid |u|^*_{k,\alpha;\Omega} < \infty\}, \quad k=0,1,2,\ldots, 0 \leqslant \alpha \leqslant 1.$$

Suppose the functions of S are also equicontinuous on $\bar{\Omega}$. Then if $k+\alpha > j+\beta$, it follows that S is precompact in $C^{j,\beta}_*(\Omega)$.

Proof. Since S is equicontinuous on $\bar{\Omega}$ and bounded in $C^{k,\alpha}_*$, it contains a sequence $\{u_m\}$ that converges uniformly on $\bar{\Omega}$ to a function $u \in C^{k,\alpha}_*$. By hypothesis we may assume $|u_m|^*_{k,\alpha} \leqslant M$ (independent of m). From (6.82) we have for any $\varepsilon > 0$ and some constant $C = C(\varepsilon)$

$$|u_m - u|^*_{j,\beta} \leqslant C|u_m - u|_0 + \varepsilon|u_m - u|^*_{k,\alpha}.$$

If now N is so large that $|u_m - u|_0 \leqslant \varepsilon/C$ for all $m > N$, it follows that $|u_m - u|^*_{j,\beta} \leqslant \varepsilon(1+2M)$ for $m > N$. Thus $\{u_m\}$ converges to u in $C^{j,\beta}_*$, which proves the assertion of the lemma. $\square$

We now extend Lemma 6.32 to partially interior norms and seminorms in domains with a hyperplane boundary portion.

Lemma 6.34. *Suppose $j+\beta<k+\alpha$, where $j, k=0, 1, 2, \ldots$, and $0\leqslant\alpha,\beta\leqslant 1$. Let Ω be an open subset of $\mathbb{R}^n_+$ with a boundary portion T on $x_n=0$, and assume $u\in C^{k,\alpha}(\Omega\cup T)$. Then for any $\varepsilon>0$ and some constant $C=C(\varepsilon, j, k)$ we have*

$$(6.89)\qquad \begin{aligned}[][u]^*_{j,\beta;\Omega\cup T}&\leqslant C|u|_{0;\Omega}+\varepsilon[u]^*_{k,\alpha;\Omega\cup T},\\ |u|^*_{j,\beta;\Omega\cup T}&\leqslant C|u|_{0;\Omega}+\varepsilon[u]^*_{k,\alpha;\Omega\cup T}.\end{aligned}$$

Proof. Again we suppose that the right members are finite. The proof is patterned after that of Lemma 6.32 and we emphasize only the details in which the proofs differ. In the following we omit the subscript $\Omega\cup T$, which will be implicitly understood.

We consider first the cases $1\leqslant j\leqslant k\leqslant 2$, $\beta=0$, $\alpha\geqslant 0$, starting with the inequality

$$(6.90)\qquad [u]^*_2\leqslant C(\varepsilon)|u|_0+\varepsilon[u]^*_{2,\alpha},\ \alpha>0.$$

Let x be any point in Ω, $\bar{d}_x$ its distance from $\partial\Omega-T$, and $d=\mu\bar{d}_x$ where $\mu\leqslant\frac{1}{4}$ is a constant to be specified later. If dist $(x, T)\geqslant d$, then the ball $B_d(x)\subset\Omega$ and the argument proceeds as in Lemma 6.32, leading to the inequality

$$\bar{d}_x^2|D_{il}u(x)|\leqslant C(\varepsilon)[u]^*_1+\varepsilon[u]^*_{2,\alpha}$$

provided $\mu=\mu(\varepsilon)$ is chosen sufficiently small. If dist $(x, T)<d$ we consider the ball $B=B_d(x_0)\subset\Omega$, where x_0 is on the perpendicular to T passing through x and dist $(x, x_0)=d$. Let x', x'' be the endpoints of the diameter of B parallel to the x_l axis. Then we have for some $\bar{x}$ on this diameter

$$\begin{aligned}|D_{il}u(\bar{x})|&=\frac{|D_iu(x')-D_iu(x'')|}{2d}\leqslant\frac{1}{d}\sup_B|D_iu|\leqslant\frac{2}{\mu}\bar{d}_x^{-2}\sup_{y\in B}\bar{d}_y|D_iu(y)|\\ &\leqslant\frac{2}{\mu}\bar{d}_x^{-2}[u]^*_1,\quad\text{since }\bar{d}_y>\bar{d}_x/2\text{ for all }y\in B;\end{aligned}$$

and

$$\begin{aligned}|D_{il}u(x)|&\leqslant|D_{il}u(\bar{x})|+|D_{il}u(x)-D_{il}u(\bar{x})|\\ &\leqslant\frac{2}{\mu}\bar{d}_x^{-2}[u]^*_1+2d^\alpha\sup_{y\in B}\bar{d}_{x,y}^{-2-\alpha}\sup_{y\in B}\bar{d}_{x,y}^{2+\alpha}\frac{|D_{il}u(x)-D_{il}u(y)|}{|x-y|^\alpha};\end{aligned}$$

hence

$$\begin{aligned}\bar{d}_x^2|D_{il}u(x)|&\leqslant\frac{2}{\mu}[u]^*_1+16\mu^\alpha[u]^*_{2,\alpha}\\ &\leqslant C[u]^*_1+\varepsilon[u]^*_{2,\alpha}\end{aligned}$$

provided $16\mu^\alpha\leqslant\varepsilon$, $C=2/\mu$. Choosing the smaller value of μ, corresponding to the two cases dist $(x, T)\geqslant d$ and dist $(x, T)<d$, and taking the supremum over all

$x \in \Omega$ and $i, l=1, \ldots, n$, we obtain (6.90). If u replaces $D_i u$ in the preceding and the details are modified accordingly, we obtain (6.90) for $j=k=1$.

To prove (6.89) for $j=1$, $k=2$, $\alpha=\beta=0$, we proceed as in Lemma 6.32 with the modifications suggested by the above proof of (6.90). Together with the preceding cases, this gives us (6.89) for $1 \leqslant j \leqslant k \leqslant 2$, $\beta=0$, $\alpha \geqslant 0$.

The proof of (6.89) for $\beta>0$ follows closely that in cases (iii) and (iv) of Lemma 6.32. The principal difference is that the argument for $\beta>0$, $\alpha=0$ now requires application of the theorem of the mean in the truncated ball $B_d(x) \cap \Omega$ for points x such that dist $(x, T)<d$. □

We conclude this section with the proof of *global interpolation inequalities* in smooth domains.

Lemma 6.35. *Suppose $j+\beta<k+\alpha$, where $j=0, 1, 2, \ldots, k=1, 2, \ldots$, and $0 \leqslant \alpha, \beta \leqslant 1$. Let Ω be a $C^{k,\alpha}$ domain in $\mathbb{R}^n$, and assume $u \in C^{k,\alpha}(\bar{\Omega})$. Then for any $\varepsilon>0$ and some constant $C=C(\varepsilon, j, k, \Omega)$ we have*

$$|u|_{j,\beta;\Omega} \leqslant C|u|_{0;\Omega}+\varepsilon|u|_{k,\alpha;\Omega}. \tag{6.91}$$

Proof. The proof is based on a reduction to Lemma 6.34 by means of an argument very similar to that in Lemma 6.5. As in that lemma, at each point $x_0 \in \partial\Omega$ let $B_\rho(x_0)$ be a ball and ψ be a $C^{k,\alpha}$ diffeomorphism that straightens the boundary in a neighborhood containing $B'=B_\rho(x_0) \cap \Omega$ and $T=B_\rho(x_0) \cap \partial\Omega$. Let $\psi(B')= D' \subset \mathbb{R}^n_+$, $\psi(T)=T' \subset \partial\mathbb{R}^n_+$. Since T' is a hyperplane portion of $\partial D'$, we may apply the interpolation inequality (6.89) in D' to the function $\tilde{u}=u \circ \psi^{-1}$ to obtain

$$|\tilde{u}|^*_{j,\beta;D' \cup T'} \leqslant C(\varepsilon)|\tilde{u}|_{0;D'}+\varepsilon|\tilde{u}|^*_{k,\alpha;D' \cup T'}.$$

From (6.30) it follows

$$|u|^*_{j,\beta;B' \cup T} \leqslant C(\varepsilon)|u|_{0;B'}+\varepsilon|u|^*_{k,\alpha;B' \cup T}.$$

(We recall that the same notation $C(\varepsilon)$ is being used for different functions of ε.) Letting $B''=B_{\rho/2}(x_0) \cap \Omega$, we infer from (4.17)', (4.17)''

$$\begin{aligned} |u|_{j,\beta;B''} &\leqslant C(\varepsilon)|u|_{0,B'}+\varepsilon|u|_{k,\alpha;B'} \\ &\leqslant C(\varepsilon)|u|_{0;\Omega}+\varepsilon|u|_{k,\alpha;\Omega}. \end{aligned} \tag{6.92}$$

Let $B_{\rho_i/4}(x_i)$, $x_i \in \partial\Omega$, $i=1, \ldots, N$, be a finite collection of balls covering $\partial\Omega$, such that the inequality (6.90) holds in each set $B''_i=B_{\rho_i/2}(x_i) \cap \Omega$, with a constant $C_i(\varepsilon)$. Let $\delta=\min \rho_i/4$ and $C=C(\varepsilon)=\max C_i(\varepsilon)$. Then at every point $x_0 \in \partial\Omega$, we have $B=B_\delta(x_0) \subset B_{\rho_i/2}(x_i)$ for some i and hence

$$|u|_{j,\beta;B \cap \Omega} \leqslant C|u|_{0;\Omega}+\varepsilon|u|_{k,\alpha;\Omega}. \tag{6.93}$$

The remainder of the argument is analogous to that in Theorem 6.6 and is left to the reader. □

The global interpolation inequality (6.91) is valid in more general domains, for example in $C^{0,1}$ domains; (see Problem 6.7). However, as shown in the example on page 53, suitable regularity of the domain Ω is required to insure the inclusion relation $C^{k,\alpha}(\bar{\Omega})\subset C^{j,\beta}(\bar{\Omega})$ when $k+\alpha>j+\beta$, and hence the global interpolation inequalities are not true in arbitrary domains.

Lemma 6.35 implies the following compactness result.

Lemma 6.36. *Let Ω be a $C^{k,\alpha}$ domain in $\mathbb{R}^n$ (with $k\geqslant 1$) and let S be a bounded set in $C^{k,\alpha}(\bar{\Omega})$. Then S is precompact in $C^{j,\beta}(\bar{\Omega})$ if $j+\beta<k+\alpha$.*

The proof is essentially the same as that of Lemma 6.33 and is therefore omitted. The result is obviously valid for domains in which the global interpolation inequality (6.91) holds, hence in $C^{0,1}$ domains.

6.9. Appendix 2: Extension Lemmas

This section establishes some results needed earlier in this chapter and elsewhere in this work concerning the extension of globally defined functions into larger domains and the extension of functions defined on the boundary to globally defined functions.

We shall use the concept of a partition of unity. Let Ω be an open set in $\mathbb{R}^n$ covered by a countable collection $\{\Omega_j\}$ of open sets Ω_j. A countable set of functions $\{\eta_i\}$ is a *locally finite partition of unity subordinate to the covering* $\{\Omega_j\}$ if: (i) $\eta_i\in C_0^\infty(\Omega_j)$ for some $j=j(i)$; (ii) $\eta_i\geqslant 0$, $\Sigma\eta_i=1$ in Ω; (iii) at each point of Ω there is a neighborhood in which only a finite number of the η_i are non-zero. For the proof of existence of such a partition we refer to the literature; (for example, see [YO], also Problem 6.8). In the following applications the construction of the partition is relatively simple.

Lemma 6.37. *Let Ω be a $C^{k,\alpha}$ domain in $\mathbb{R}^n$ (with $k\geqslant 1$) and let Ω' be an open set containing $\bar{\Omega}$. Suppose $u\in C^{k,\alpha}(\bar{\Omega})$. Then there exists a function $w\in C_0^{k,\alpha}(\Omega')$ such that $w=u$ in Ω and*

$$|w|_{k,\alpha;\Omega'}\leqslant C|u|_{k,\alpha;\Omega},\tag{6.94}$$

where $C=C(k,\Omega,\Omega')$.

Proof. Let $y=\psi(x)$ define a $C^{k,\alpha}$ diffeomorphism that straightens the boundary near $x_0\in\partial\Omega$, and let G and $G^+=G\cap\mathbb{R}^n_+$ be respectively a ball and half-ball in the image of ψ such that $\psi(x_0)\in G$. Setting $\tilde{u}(y)=u\circ\psi^{-1}(y)$ and $y=(y_1,\ldots,y_{n-1},y_n)=(y',y_n)$, we define an extension of $\tilde{u}(y)$ into $y_n<0$ by

$$\tilde{u}(y',y_n)=\sum_{i=1}^{k+1}c_i\tilde{u}(y',-y_n/i),\quad y_n<0,$$

where $c_1, \ldots, c_{k+1}$ are constants determined by the system of equations

$$\sum_{i=1}^{k+1} c_i(-1/i)^m = 1, \quad m=0, 1, \ldots, k.$$

One verifies readily that the extended function $\tilde{u}$ is continuous with all derivatives up to order k in G and that $\tilde{u} \in C^{k,\alpha}(G)$. Thus $w = \tilde{u} \circ \psi \in C^{k,\alpha}(\bar{B})$ for some ball $B = B(x_0)$ and $w = u$ in $B \cap \Omega$, so that w provides a $C^{k,\alpha}$ extension of u into $\Omega \cup B$. By (6.30) the inequality (6.94) holds (with $\Omega \cup B$ in place of Ω').

Now consider a finite covering of $\partial\Omega$ by balls B_i, $i=1, \ldots, N$, such as B in the preceding, and let $\{w_i\}$ be the corresponding $C^{k,\alpha}$ extensions. We may assume the balls B_i are so small that their union with Ω is contained in Ω'. Let $\Omega_0 \subset\subset \Omega$ be an open subset of Ω such that Ω_0 and the balls B_i provide a finite open covering of Ω. Let $\{\eta_i\}$, $i=0, 1, \ldots, N$, be a partition of unity subordinate to this covering, and set

$$w = u\eta_0 + \sum w_i \eta_i$$

with the understanding that $w_i \eta_i = 0$ if $\eta_i = 0$. One verifies from the above discussion that w is an extension of u into Ω' and has the properties asserted in the lemma. □

The following result provides an extension of a boundary function to a globally defined function in the same regularity class.

Lemma 6.38. *Let Ω be a $C^{k,\alpha}$ domain in $\mathbb{R}^n (k \geq 1)$ and let Ω' be an open set containing $\bar{\Omega}$. Suppose $\varphi \in C^{k,\alpha}(\partial\Omega)$. Then there exists a function $\Phi \in C_0^{k,\alpha}(\Omega')$ such that $\Phi = \varphi$ on $\partial\Omega$.*

Proof. At any point $x_0 \in \partial\Omega$ let the mapping ψ and the ball G be defined as in the preceding lemma, and assume that $\tilde{\varphi} = \varphi \circ \psi^{-1} \in C^{k,\alpha}(G \cap \partial\mathbb{R}^n_+)$. We define $\tilde{\Phi}(y', y_n) = \tilde{\varphi}(y')$ in G and set $\Phi(x) = \tilde{\Phi} \circ \psi(x)$ for $x \in \psi^{-1}(G)$. Clearly $\Phi \in C^{k,\alpha}(\bar{B})$ for some ball $B = B(x_0)$ and $\Phi = \varphi$ on $B \cap \partial\Omega$. Now let $\{B_i\}$ be a finite covering of $\partial\Omega$ by balls such as B, and let Φ_i be the corresponding $C^{k,\alpha}$ functions defined on B_i. The proof can now be completed as in the preceding lemma by use of an appropriate partition of unity. □

Remarks. 1) In the preceding lemma, if $\varphi \in C^0(\partial\Omega) \cap C^{k,\alpha}(T)$ where $T \subset \partial\Omega$, then the same argument leads to an extension $\Phi \in C^0(\Omega') \cap C^{k,\alpha}(G)$, where G is an open set containing T. A simple modification of the above proof shows that if Ω is any domain with $C^{k,\alpha}$ boundary portion T and if $\varphi \in C^{k,\alpha}(T)$, then φ can be extended to a function $\Phi \in C^{k,\alpha}(G)$, where G is an open set containing T and $\Phi = \varphi$ on T. A countable covering of T by balls is required for the argument. If $\varphi \in C^0(\partial\Omega) \cap C^{k,\alpha}(T)$ then the extension Φ can be determined so that $\Phi \in C^0(\bar{\Omega}) \cap C^{k,\alpha}(G)$. 2) For domains with a simple geometry the construction of an extended function can often be made directly and easily. Thus if $B = B_R(x_0)$ is a ball in $\mathbb{R}^n$ and $\varphi \in C^0(\partial\Omega) \cap C^{k,\alpha}(T)$, $T \subset \partial B$, then an extension of φ into $\mathbb{R}^n$ can be obtained by setting

$$\Phi(x) = \Phi(x_0 + r\omega) = \varphi(R\omega)\eta(r),$$

where $r=|x-x_0|$, $\omega=(x-x_0)/r$, and $\eta(r)$ is a C^∞ cut-off function such that $\eta(r)=0$ for $0\leqslant r\leqslant R/4$, $\eta(r)=1$ for $r\geqslant R/2$. The function $\Phi(x)$ obviously coincides with φ on ∂B, is in $C^0(\mathbb{R}^n)$ and is of class $C^{k,\alpha}$ in the conical region determined by the rays from the origin through the points of T.

Notes

The apriori estimates and existence theory in Sections 1–3 are, in modified form, the contributions of Schauder [SC 4, 5]. At about the same time, Caccioppoli [CA 1] stated without details similar results, which were elaborated by Miranda [MR 2]. Closely related ideas are contained in the work of Hopf [HO 3] who earlier established the interior regularity theorems of Section 4. The existence theory and general properties of solutions for essentially the same class of problems were previously obtained by Giraud [GR 1–3], who used the method of integral equations based on representing solutions as surface potentials. Further details amplifying on the respective contributions are discussed by Miranda [MR 2]. A development of the Schauder estimates based on methods of Fourier analysis, for equations of arbitrary order, is contained in Hörmander [HM 3].

The formulation of the interior estimates of Section 1 in terms of interior norms and the method of derivation are patterned after Douglis and Nirenberg [DN], who also extended the interior estimates to elliptic systems. The details of proof of Theorem 6.2 can be simplified somewhat if the estimates are first carried out in balls, using Theorem 4.6, and at the end are converted into the bound (6.14) for the $C^{2,\alpha}$ interior norm. See, for example, the proof of Theorem 9.11.

The global estimates of Section 6.2 and the proof of Theorem 6.8 based on these estimates assume $C^{2,\alpha}$ boundary data. Under weaker regularity hypotheses an existence proof for say $C^{1,\alpha}(\bar{\Omega})\cap C^2(\Omega)$ solutions does not follow from the Schauder theory in its usual form. Such an existence theorem is implied by regularity results of Widman [WI 1]. Gilbarg and Hörmander [GH] have extended the global Schauder theory to include conditions of lower regularity of the coefficients, domain and boundary values. We summarize their results, which apply to conditions of higher regularity as well:

If

$$0\leqslant k<a=k+\alpha\leqslant k+1,$$

let $H_a(\Omega)$ denote the Hölder space of functions with finite norm $|u|_{a,\Omega}=|u|_{k,\alpha;\Omega}$ (the latter being in the notation of this book; thus $H_a(\Omega)=C^{k,\alpha}(\bar{\Omega})$). Setting

$$\Omega_\delta=\{x\in\Omega\,|\,\mathrm{dist}\,(x,\partial\Omega)>\delta\},$$

let $H_a^{(b)}(\Omega)$ denote the set of functions on Ω belonging to $H_a(\Omega_\delta)$ for all $\delta>0$ and with finite norm

$$|u|_{a,\Omega}^{(b)}=|u|_a^{(b)}=\sup_{\delta>0}\delta^{a+b}|u|_{a,\Omega_\delta},$$

where $a + b \geqslant 0$. Since $H_a^{(-b)} \subset H_b^{(-b)} = H_b$ for $a \geqslant b > 0$, b a noninteger, the upper and lower indices in the norm $|u|_a^{(-b)}$ describe, respectively, the global and interior regularity of u. Define also $H_a^{(b-0)}(\Omega)$ to be the set of functions in $H_a^{(b)}$ such that $\delta^{a+b}|u|_{a,\Omega_\delta} \to 0$ as $\delta \to 0$. Having these spaces, now let Ω be a bounded C^γ domain for some $\gamma \geqslant 1$, and a, b be nonintegers such that $0 < b \leqslant a$, $a > 2$, $b \leqslant \gamma$. Let

$$P = \sum_{|\beta| \leqslant 2} p_\beta(x) D^\beta$$

be an elliptic second-order differential operator on $\overline{\Omega}$ such that

$$\begin{aligned} &p_\beta \in H_{a-2}^{(2-b)}(\Omega) \quad \text{if } |\beta| \leqslant 2, \\ &p_\beta \in C^0(\overline{\Omega}) \quad \text{if } |\beta| = 2, \\ &p_\beta \in H_{a-2}^{(2-|\beta|-0)}(\Omega) \quad \text{if } b < |\beta|. \end{aligned}$$

(Thus the lower-order coefficients may be unbounded if $b < 2$.) Then, if $u \in C^2(\Omega) \cap C^0(\overline{\Omega})$ is a solution of

$$Pu = f \text{ in } \Omega, \quad u = \varphi \text{ on } \partial\Omega \tag{6.95}$$

where $f \in H_{a-2}^{(2-b)}(\Omega)$, $\varphi \in H_b(\partial\Omega)$, it follows that $u \in H_a^{(-b)}(\Omega)$ and satisfies an estimate

$$|u|_a^{(-b)} \leqslant C(|u|_0 + |\varphi|_{b,\partial\Omega} + |f|_{a-2}^{(2-b)}),$$

where C depends on Ω, a, b, the norms of the coefficients and their minimum eigenvalue. If $p_0 \leqslant 0$ the Dirichlet problem (6.95) has a unique solution in $H_a^{(-b)}$, and a corresponding Fredholm-type theorem holds in general. The case $2 + \alpha \leqslant a = b \leqslant \gamma$ is the one treated in this chapter. If Ω is a Lipschitz domain, analogous results hold for values $b < 1$ that depend on the exterior cone condition satisfied at the boundary; and it suffices that $p_\beta \in H_{a-2}^{(0)}$ for $|\beta| = 2$, so that the principal coefficients need not be continuous at the boundary.

The conditions under which a regular boundary point for the Laplacian is also a regular point for an elliptic operator, and conversely, have been studied by several authors. The equivalence has been proved for strictly elliptic operators L, as in Theorem 6.13, whose coefficients near the boundary are Lipschitz continuous [HR] or Dini continuous [KV 1], [NO 2]; and also for certain classes of discontinuous coefficients [AK] and degenerate elliptic operators [MM], [NO 3]. Capacity and the Wiener criterion (Section 2.9) play an important part in the arguments. If the coefficients of L are only continuous, the equivalence no longer holds in general (see the example in Problem 3.8(a), also [ML 4]). However, in the case of divergence structure equations, there is equivalence when the coefficients are only bounded and measurable [LSW] (see also Chapter 8). For additional results on regular boundary points, see [NO 1], [MZ], [LN 2], [ML 2, 4].

Hopf [HO 3] proves directly the interior regularity result, Lemma 6.16, without an existence theorem. His method, based on Korn's device (in [KR 2]) of perturbation about the constant coefficient equation, provides an extension of his results in [HO 2] on the regularity of solutions of variational problems and anticipates important aspects of the Schauder theory. A simple direct proof of Lemma 6.16 (and of more general results), based on regularization and interior estimates, is contained in [ADN 1], p. 723. For another approach see the proof of Lemma 9.16; also [MY 5], Section 5.6.

A basically simpler proof of the boundary regularity result, Lemma 6.18, can be obtained as follows. It suffices to prove that $u \in C^{\alpha}(\Omega \cup T)$, after which the argument proceeds essentially as in Lemma 6.16. By considering $u-\varphi$ in place of u, we may assume $\varphi \equiv 0$. Let $\Omega' \subset \Omega$ with $\partial\Omega' \cap \partial\Omega = T' \subset\subset T$, and let $\delta = \text{dist}\,(\Omega', \partial\Omega - T) < 1$. For any $x' \in \Omega'$ suppose first

$$d = \text{dist}\,(x', \partial\Omega) = |x' - x_0| \leqslant \delta, \quad x_0 \in \partial\Omega.$$

Then by Problem 3.6 we have

$$|u(x)| \leqslant C|x - x_0| \quad \text{for } x \in \Omega,$$

and hence $|u(x)| \leqslant Cd$ for $x \in B_d(x')$. From (6.23), in which we set $\Omega' = B_{d/2}(x')$ and $\Omega = B_d(x')$, it follows that for all $x \in B_{d/2}(x')$,

$$d|Du(x)| \leqslant C(\sup_{B_d} |u| + d^2|f|_{0,\alpha;B_d}),$$

and hence

$$|Du(x')| \leqslant C(1 + |f|_{0,\alpha;\Omega}) \leqslant C,$$

where the constant C depends only on δ and the given data. If $d > \delta$, then the same inequality holds, with the constant C now depending on δ^{-1}. We thus have a bound on $|Du|$ in Ω' and consequently $u \in C^{0,1}(\Omega \cup T)$.

Section 6.5 is a modification of the ideas of Michael [MI 1] who has shown that a general existence theory for continuous boundary values can be developed from interior estimates only. His results apply as well to certain classes of equations with unbounded coefficients near the boundary; (see Problems 6.5, 6.6).

Section 6.6 considers some cases of nonuniformly elliptic operators for which the ellipticity degenerates on the boundary. The theory of elliptic operators that degenerate in the interior is based on essentially different methods from those of this chapter. For the relevant results the reader is directed to the literature on hypoellipticity, e.g., [HM 2], [OR], [KJ].

The Schauder theory of the oblique derivative problem in Section 6.7 differs in some respects from earlier versions. In particular, Fiorenza [FI 1] based his approach on the representation by surface potentials of solutions of the boundary value problem (6.60) for Poisson's equation in a half-space, to which he applied some of the results of Giraud [GR 3]. His Schauder-type estimates for the case of variable coefficients exhibit a quite precise dependence on the bounds and Hölder constants of the coefficients. This dependence is used in [LU 4], Chapter 10, and by

Fiorenza [FI 2] and Ural'tseva [UR] to treat quasilinear equations with nonlinear boundary conditions. An extension of the Schauder theory to other boundary conditions for higher order equations and systems appears in Agmon, Douglis and Nirenberg [ADN 1, 2]. Their method is based on explicit integral representations of solutions of the constant coefficient problem in a half-space in terms of appropriate Poisson kernels for the given boundary conditions. Although the details are quite different, the development in Section 6.7 can be viewed as a special case of these results. See also Bouligand [BGD]. The nonregular oblique derivative problem, in which the directional derivative in the boundary condition becomes tangential ($\beta_\nu=0$ in (6.76)), is essentially deeper than the regular case and the results are different; see, for example, [HM 1], [EK], [SJ] and [WZ 1, 2].

The solution of exterior boundary value problems can be inferred readily from the results of this chapter. Meyers and Serrin [MS 1] treat boundary value problems for the equation $Lu=f$ in an exterior domain Ω (containing the exterior of some ball), with $c\leqslant 0$ and coefficients and f Hölder continuous in bounded subsets of Ω. Under suitable general hypotheses on the behavior of the coefficients at infinity they prove the existence of solutions in Ω with a limit at infinity from the convergence of solutions in expanding domains. They obtain the result, among others, that if $f=0$, $b^i=0$, $a^{ij}\to a_0^{ij}$ at ∞ and the matrix $[a_0^{ij}]$ has rank $\geqslant 3$, then there is a unique solution in Ω of the Dirichlet (and other) problems vanishing at infinity. Under these conditions, if $n>3$ the operator L may be non-uniformly elliptic at infinity and the boundary value problem is still well posed. An extension of the Schauder theory to exterior domains for $n\geqslant 3$, including Hölder estimates at infinity and a corresponding treatment of the exterior Dirichlet and Neumann problems, has been given by Oskolkov [OS 1]. An exterior Neumann problem for a class of quasilinear equations when $n\geqslant 3$ is treated in [FG 2].

The interpolation inequalities of Appendix 1 can easily be derived from the general convexity property of Hölder norms:

$$|u|_{k,\alpha}\leqslant C(|u|_{k_1,\alpha_1})^t(|u|_{k_2,\alpha_2})^{1-t},$$

where $0<t<1$,

$$k+\alpha=t(k_1+\alpha_1)+(1-t)(k_2+\alpha_2),$$

and the norms may be interior or global. For proof of this inequality, see Hörmander [HM 3].

Problems

6.1. (a) Let $u\in C^{k+2,\alpha}(\Omega)$, $k\geqslant 0$, be a solution of $Lu=f$ in a bounded open set $\Omega\subset\mathbb{R}^n$, and assume the coefficients of L satisfy (6.2) and $|a^{ij}, b^i, c|_{k,\alpha;\Omega}\leqslant\Lambda$. If $\Omega'\subset\subset\Omega$, show that

$$|u|_{k+2,\alpha;\Omega'}\leqslant C(|u|_{0;\Omega}+|f|_{k,\alpha;\Omega})$$

where $C=C(n,k,\alpha,\lambda,\Lambda,d)$, $d=\operatorname{dist}(\Omega',\partial\Omega)$.

(b) In Theorem 6.2 assume $u \in C^{k+2,\alpha}(\Omega)$, $k \geqslant 0$, and replace (6.13) by the conditions

$$|a^{ij}|^{(0)}_{k,\alpha},\ |b^i|^{(1)}_{k,\alpha},\ |c|^{(2)}_{k,\alpha} \leqslant \Lambda.$$

Prove the interior estimate

$$|u|^*_{k+2,\alpha} \leqslant C(|u|_0 + |f|^{(2)}_{k,\alpha})$$

where $C = C(n, k, \alpha, \lambda, \Lambda)$.

6.2. In Theorem 6.6 let Ω be a $C^{k+2,\alpha}$ domain, $k \geqslant 0$, and assume $u \in C^{k+2,\alpha}(\bar{\Omega})$, $\varphi \in C^{k+2,\alpha}(\bar{\Omega})$, $f \in C^{k,\alpha}(\bar{\Omega})$, $|a^{ij}, b^i, c|_{k,\alpha;\Omega} \leqslant \Lambda$. Prove the global estimate

$$|u|_{k+2,\alpha} \leqslant C(|u|_0 + |\varphi|_{k+2,\alpha} + |f|_{k,\alpha})$$

where $C = C(n, k, \alpha, \lambda, \Lambda, \Omega)$.

6.3. Prove Theorem 6.13 for a bounded domain Ω satisfying an exterior cone condition. Show that at each point $x_0 \in \partial\Omega$ there is a local barrier determined by a function of the form $r^\mu f(\theta)$, where $r = |x - x_0|$ and θ is the angle between the vector $x - x_0$ and the axis of the exterior cone; (cf. [ML 1, 3]).

6.4. Prove the following extension of Corollary 6.24′. Let Ω be a bounded strictly convex domain in $\mathbb{R}^n$ and let the equation

$$Lu \equiv a^{ij}D_{ij}u + b^iD_iu = 0$$

be elliptic in Ω with coefficients in $C^\alpha(\Omega)$. For $x_0 \in \partial\Omega$ let $\nu = \nu(x_0)$ denote the unit normal to a supporting plane (directed outward from Ω). At each x_0 suppose $\mathbf{b} \cdot \nu > 0$ in $B(x_0) \cap \Omega$ for some ball $B(x_0)$. Then the Dirichlet problem for $Lu = 0$ is (uniquely) solvable in $C^{2,\alpha}(\Omega) \cap C^0(\bar{\Omega})$ for arbitrary continuous boundary values.

6.5. (a) Prove Lemma 6.21 if the ball B is replaced by an arbitrary C^2 domain Ω; (see Problem 4.6).

(b) Extend part (a) to admit coefficients b^i satisfying $\sup_\Omega d_x^{1-\gamma}|b^i(x)| < \infty$ for some $\gamma \in (0, 1)$ and coefficients $c \leqslant 0$.

6.6. (a) Use the argument in Problem 6.5 to construct a barrier and prove the solvability of the Dirichlet problem, $Lu = f$ $(c \leqslant 0)$ in Ω, $u = 0$ on $\partial\Omega$, where L is strictly elliptic in the C^2 domain Ω, and the following conditions are satisfied: the coefficients a^{ij} are bounded; a^{ij}, b^i, c, $f \in C^\alpha(\Omega)$; $\sup_\Omega d_x^{1-\beta}|b^i(x)| < \infty$ and $\sup_\Omega d_x^{2-\beta}|f(x)| < \infty$ for some $\beta \in (0, 1)$.

(b) Under the additional hypothesis, $\sup_{\Omega} d_x^{2-\beta}|c(x)| < \infty$, extend part (a) to include the boundary condition $u = \varphi$ on $\partial\Omega$, where φ is continuous.

6.7. Prove the global interpolation inequality (6.91) if Ω is a $C^{0,1}$ (Lipschitz) domain. The result can be obtained as follows:

(i) Prove there is a constant K depending only on Ω such that every pair of points x, y in Ω can be connected by an arc $\gamma(x, y)$ in Ω whose length $|\gamma(x, y)| \leqslant K|x - y|$.

(ii) Show that for some constants ρ_0 and L depending only on Ω, if $y \in \Omega$ and dist $(y, \partial\Omega) < \rho_0$, then for all $\rho < \rho_0$ there is a point $x \in B_\rho(y)$ such that $B_{\rho/L^2}(x) \subset \Omega$.

(iii) Use parts (i) and (ii) in a modification of the proof of Lemma 6.34 to establish (6.91).

6.8. Let $\{\Omega_i\}$ be a countable open covering of an open set Ω in $\mathbb{R}^n$. If either (a) Ω is bounded and $\bar{\Omega} \subset \cup\, \Omega_i$, or (b) Ω_i is bounded and $\bar{\Omega}_i \subset \Omega$, $i = 1, 2, \ldots$, prove the existence of a partition of unity $\{\eta_i\}$ such that $\eta_i \in C_0^\infty(\Omega_i)$.

6.9. (a) Use a partition of unity and the definition of $C^{k,\alpha}$ domains in Section 6.2 to show that any such domain Ω with $k \geqslant 1$ can be defined by a function $F \in C^{k,\alpha}(\bar{\Omega})$, such that $F > 0$ in Ω, $F = 0$ on $\partial\Omega$ and grad $F \neq 0$ on $\partial\Omega$.

(b) Use part (a) and approximation by smooth functions to show that any $C^{k,\alpha}$ domain $\Omega(k \geqslant 1)$, defined by $F > 0$, can be exhausted by arbitrarily smooth domains Ω_ν, defined by $F_\nu > 0$, such that

$$\partial\Omega_\nu \to \partial\Omega_\nu \quad \text{and} \quad |F_\nu|_{k,\alpha;\Omega_\nu} \leqslant C|F|_{k,\alpha;\Omega} \qquad \text{as} \quad \nu \to \infty,$$

where C is a constant independent of k.

6.10. Let L be elliptic in a C^2 domain Ω, with $c \leqslant 0$ and $a^{ij}, b^i \in C^0(\bar{\Omega})$. Let $\Sigma = \Sigma_1 \cup \Sigma_2 \subset \partial\Omega$ be defined by (6.59). Suppose $u \in C^0(\bar{\Omega}) \cap C^2(\bar{\Omega} - \Sigma)$ satisfies $Lu \geqslant 0$ in Ω. Then, if either $c < 0$, or $a^{ii} > 0$, for some i, on $\partial\Omega - \Sigma$, prove that

$$\sup_{\Omega} u \leqslant \sup_{\Sigma} u^+.$$

(Use the distance function to flatten the boundary near an assumed maximum on $\partial\Omega - \Sigma$, and consider the differential equation there. Cf. [OR].)

6.11. Under the hypotheses of Theorem 6.30, but assuming $|\beta_i|_{0,\alpha} \leqslant \Lambda$ in place of $|\beta_i|_{1,\alpha} \leqslant \Lambda$, prove the estimate

$$|u|_{2,\alpha} \leqslant C(|u|_0 + |\varphi|_{1,\alpha} + |f|_{0,\alpha} + |Du|_0 \cdot [D\beta]_\alpha),$$

where $C = C(n, \alpha, \lambda, \Lambda, \kappa, \Omega)$.

Chapter 7

Sobolev Spaces

To motivate the theory of this chapter we now consider a different approach to Poisson's equation from that of Chapter 4. By the divergence theorem (equation (2.3)) a $C^2(\Omega)$ solution of $\Delta u = f$ satisfies the integral identity

$$\int_\Omega Du \cdot D\varphi \, dx = -\int_\Omega f\varphi \, dx \tag{7.1}$$

for all $\varphi \in C_0^1(\Omega)$. The bilinear form

$$(u, \varphi) = \int_\Omega Du \cdot D\varphi \, dx \tag{7.2}$$

is an inner product on the space $C_0^1(\Omega)$ and the completion of $C_0^1(\Omega)$ under the metric induced by (7.2) is consequently a Hilbert space, which we call $W_0^{1,2}(\Omega)$. Furthermore, for appropriate f the linear functional F defined by $F(\varphi) = -\int_\Omega f\varphi \, dx$ may be extended to a bounded linear functional on $W_0^{1,2}(\Omega)$. Hence, by the Riesz representation theorem (Theorem 5.7), there exists an element $u \in W_0^{1,2}(\Omega)$ satisfying $(u, \varphi) = F(\varphi)$ for all $\varphi \in C_0^1(\Omega)$. Thus, the existence of a *generalized solution* to the Dirichlet problem, $\Delta u = f$, $u = 0$ on $\partial\Omega$, is readily established. The question of classical existence is accordingly transformed into the question of regularity of generalized solutions under appropriately smooth boundary conditions. In the next chapter, the Lax-Milgram theorem (Theorem 5.8) will be applied to linear elliptic equations in divergence form in a similar manner to the above application of the Riesz representation theorem and by means of various arguments based on integral identities, regularity results will be established. But before we can so proceed we need to examine the class of Sobolev spaces, that is, the $W^{k,p}(\Omega)$ and $W_0^{k,p}(\Omega)$ spaces of which the space $W_0^{1,2}(\Omega)$ is a member. Some of the inequalities we treat will also be necessary for the development of the theory of quasilinear equations in Part II.

7.1. L^p Spaces

Throughout this chapter Ω will denote a bounded domain in $\mathbb{R}^n$. By a measurable function on Ω we shall mean an equivalence class of measurable functions on Ω which differ only on a subset of measure zero. Any pointwise property attributed to a measurable function will thus be understood to hold in the usual sense for some function in the same equivalence class. The supremum and infimum of a measurable function will then be understood as the essential supremum and infimum.

For $p \geqslant 1$, we let $L^p(\Omega)$ denote the classical Banach space consisting of measurable functions on Ω that are p-integrable. The norm in $L^p(\Omega)$ is defined by

$$(7.3) \qquad \|u\|_{p;\Omega} = \|u\|_{L^p(\Omega)} = \left(\int_\Omega |u|^p\,dx\right)^{1/p}.$$

When u is a vector or matrix function the same notation will be used, the norm $|u|$ denoting the usual Euclidean norm. For $p = \infty$, $L^\infty(\Omega)$ denotes the Banach space of bounded functions on Ω with the norm

$$(7.4) \qquad \|u\|_{\infty;\Omega} = \|u\|_{L^\infty(\Omega)} = \sup_\Omega |u|.$$

In the following we shall use $\|u\|_p$ for $\|u\|_{L^p(\Omega)}$ when there is no ambiguity.

We shall need the following inequalities in dealing with integral estimates: *Young's inequality*.

$$(7.5) \qquad ab \leqslant \frac{a^p}{p} + \frac{b^q}{q};$$

this holds for positive real numbers a, b, p, q satisfying

$$\frac{1}{p} + \frac{1}{q} = 1.$$

The case $p = q = 2$ of inequality (7.5) is known as Cauchy's inequality. Replacing a by $\varepsilon^{1/p}a$, b by $\varepsilon^{-1/p}b$ for positive ε, we obtain a useful interpolation inequality

$$(7.6) \qquad ab \leqslant \frac{\varepsilon a^p}{p} + \frac{\varepsilon^{-q/p} b^q}{q} \leqslant \varepsilon a^p + \varepsilon^{-q/p} b^q.$$

Hölder's inequality.

$$(7.7) \qquad \int_\Omega uv\,dx \leqslant \|u\|_p \|v\|_q;$$

this holds for functions $u \in L^p(\Omega)$, $v \in L^q(\Omega)$, $1/p + 1/q = 1$ and is a consequence of Young's inequality. When $p = q = 2$, Hölder's inequality reduces to the well known Schwarz inequality. That the expression (7.3) defines a norm on $L^p(\Omega)$ is a consequence of Hölder's inequality. Let us note some other simple consequences of Hölder's inequality.

$$(7.8)\qquad |\Omega|^{-1/p}\|u\|_p \leqslant |\Omega|^{-1/q}\|u\|_q \quad \text{for } u \in L^q(\Omega),\quad p \leqslant q.$$

$$(7.9)\qquad \|u\|_q \leqslant \|u\|_p^{\lambda}\|u\|_r^{1-\lambda} \quad \text{for } u \in L^r(\Omega),$$

where $p \leqslant q \leqslant r$ and $1/q = \lambda/p + (1-\lambda)/r$.
Combining inequalities (7.6) and (7.9), we obtain an interpolation inequality for L^p norms, namely,

$$(7.10)\qquad \|u\|_q \leqslant \varepsilon\|u\|_r + \varepsilon^{-\mu}\|u\|_p,$$

where

$$\mu = \left(\frac{1}{p} - \frac{1}{q}\right) \bigg/ \left(\frac{1}{q} - \frac{1}{r}\right).$$

We shall also have occasion to use a generalization of Hölder's inequality to m functions, $u_1, \ldots u_m$, lying respectively in spaces $L^{p_1}, \ldots L^{p_m}$, where

$$\frac{1}{p_1} + \cdots + \frac{1}{p_m} = 1.$$

The resulting inequality, obtainable from the case $m=2$ by an induction argument, is then

$$(7.11)\qquad \int_\Omega u_1 \cdots u_m \, dx \leqslant \|u_1\|_{p_1} \cdots \|u_m\|_{p_m}.$$

It is also of interest to study the L^p norm as a function of p. Writing

$$(7.12)\qquad \Phi_p(u) = \left(\frac{1}{|\Omega|}\int_\Omega |u|^p \, dx\right)^{1/p}$$

for $p>0$, we see that Φ is non-decreasing in p for fixed u, by inequality (7.8), while the inequality (7.9) shows that Φ is logarithmically convex in p^{-1}. Note that $\Phi_p(u) = |\Omega|^{-1/p}\|u\|_p$ for $p \geqslant 1$. Although the functional Φ does not extend the L^p norm as a norm for values of p less than one, it will nevertheless be useful for later purposes (see Chapter 8).

We also note here some of the well known functional analytic properties of the L^p spaces; (see for example Royden [RY]). The space $L^p(\Omega)$ is separable for $p<\infty$, $C^0(\bar{\Omega})$ being in particular a dense subspace. The dual space of $L^p(\Omega)$ is isomorphic to $L^q(\Omega)$ provided $1/p + 1/q = 1$ and $p < \infty$. Hence $L^p(\Omega)$ is reflexive for $1<p<\infty$. The number q, the *Hölder conjugate* of p, will often be denoted p'. Finally, $L^2(\Omega)$ is a *Hilbert space* under the scalar product

$$(u, v)=\int_\Omega uv\, dx.$$

7.2. Regularization and Approximation by Smooth Functions

The spaces $C^{k,\alpha}(\Omega)$ which were introduced in Chapter 4 are local spaces. Let us define local analogues of the $L^p(\Omega)$ spaces by letting $L^p_{\text{loc}}(\Omega)$ denote the linear space of measurable functions locally p-integrable in Ω. Although they are not normed spaces, the $L^p_{\text{loc}}(\Omega)$ spaces are readily topologized. Namely, a sequence $\{u_m\}$ converges to u in the sense of $L^p_{\text{loc}}(\Omega)$ if $\{u_m\}$ converges to u in $L^p(\Omega')$ for each $\Omega'\subset\subset\Omega$.

Let ρ be a non-negative function in $C^\infty(\mathbb{R}^n)$ vanishing outside the unit ball $B_1(0)$ and satisfying $\int \rho\, dx=1$. Such a function is often called a *mollifier*. A typical example is the function ρ given by

$$\rho(x)=\begin{cases} c\exp\left(\dfrac{1}{|x|^2-1}\right) & \text{for } |x|\leqslant 1\\ 0 & \text{for } |x|\geqslant 1\end{cases}$$

where c is chosen so that $\int \rho\, dx=1$ and whose graph has the familiar bell shape. For $u\in L^1_{\text{loc}}(\Omega)$ and $h>0$, the *regularization* of u, denoted by u_h, is then defined by the convolution

$$u_h(x)=h^{-n}\int_\Omega \rho\left(\frac{x-y}{h}\right)u(y)\, dy \tag{7.13}$$

provided $h<\text{dist}\,(x, \partial\Omega)$. It is clear that u_h belongs to $C^\infty(\Omega')$ for any $\Omega'\subset\subset\Omega$ provided $h<\text{dist}\,(\Omega', \partial\Omega)$. Furthermore, if u belongs to $L^1(\Omega)$, Ω bounded, then u_h lies in $C_0^\infty(\mathbb{R}^n)$ for arbitrary $h>0$. As h tends to zero, the function $y\mapsto h^{-n}\rho(x-y/h)$ tends to the Dirac delta distribution at the point x. The significant feature of regularization, which we partly explore now, is the sense in which u_h approximates u as h tends to zero. It turns out, roughly stated, that if u lies in a local space, then u_h approximates u in the natural topology of that space.

Lemma 7.1. *Let $u\in C^0(\Omega)$. Then u_h converges to u uniformly on any domain $\Omega'\subset\subset\Omega$.*

Proof. We have

$$u_h(x) = h^{-n} \int_{|x-y| \leqslant h} \rho\left(\frac{x-y}{h}\right) u(y)\, dy$$

$$= \int_{|z| \leqslant 1} \rho(z) u(x-hz)\, dz \quad \left(\text{putting } z = \frac{x-y}{h}\right);$$

hence if $\Omega' \subset\subset \Omega$ and $2h < \text{dist}\,(\Omega', \partial\Omega)$,

$$\sup_{\Omega'} |u - u_h| \leqslant \sup_{x \in \Omega'} \int_{|z| \leqslant 1} \rho(z) |u(x) - u(x-hz)|\, dz$$

$$\leqslant \sup_{x \in \Omega'} \sup_{|z| \leqslant 1} |u(x) - u(x-hz)|.$$

Since u is uniformly continuous over the set

$$B_h(\Omega') = \{x \mid \text{dist}\,(x, \Omega') < h\},$$

u_h tends to u uniformly on Ω'. □

The convergence in Lemma 7.1 would be uniform over all of Ω if u vanished continuously on $\partial\Omega$. More generally if $u \in C^0(\bar{\Omega})$ we can define an extension $\tilde{u}$ of u such that $\tilde{u} = u$ in Ω and $\tilde{u} \in C^0(\tilde{\Omega})$ for some $\tilde{\Omega} \supset\supset \Omega$. Then $\tilde{u}_h$, the regularization of $\tilde{u}$ in $\tilde{\Omega}$, converges to u uniformly in Ω as $h \to 0$.

The process of regularization can also be used to approximate Hölder continuous functions. In particular if $u \in C^\alpha(\Omega)$, $0 \leqslant \alpha \leqslant 1$, then

$$[u_h]_{\alpha;\Omega'} \leqslant [u]_{\alpha;\Omega''} \tag{7.14}$$

where $\Omega'' = B_h(\Omega')$, and consequently u_h tends to u in the sense of $C^{\alpha'}(\Omega')$ for every $\alpha' < \alpha$ and $\Omega' \subset\subset \Omega$, as $h \to 0$. Using Lemma 6.37 and Lemma 7.3 of the following section we can then conclude approximation results for $C^{k,\alpha}(\bar{\Omega})$ functions; (see Section 6.3).

We turn our attention now to the approximation of functions in the $L^p_{\text{loc}}(\Omega)$ spaces.

Lemma 7.2. *Let $u \in L^p_{\text{loc}}(\Omega)(L^p(\Omega))$, $p < \infty$. Then u_h converges to u in the sense of $L^p_{\text{loc}}(\Omega)(L^p(\Omega))$.*

Proof. Using Hölder's inequality, we obtain from (7.13)

$$|u_h(x)|^p \leqslant \int_{|z| \leqslant 1} \rho(z) |u(x-hz)|^p\, dz,$$

so that if $\Omega' \subset\subset \Omega$ and $2h < \text{dist}(\Omega', \partial\Omega)$,

$$\int_{\Omega'} |u_h|^p \, dx \leqslant \int_{\Omega'} \int_{|z| \leqslant 1} \rho(z) |u(x-hz)|^p \, dz \, dx$$

$$= \int_{|z| \leqslant 1} \rho(z) \, dz \int_{\Omega'} |u(x-hz)|^p \, dx$$

$$\leqslant \int_{B_h(\Omega')} |u|^p \, dx,$$

where $B_h(\Omega') = \{x \mid \text{dist}(x, \Omega') < h\}$. Consequently

$$\|u_h\|_{L^p(\Omega')} \leqslant \|u\|_{L^p(\Omega'')}, \quad \Omega'' = B_h(\Omega'). \tag{7.15}$$

The proof can now be completed by approximation based on Lemma 7.1. Choose $\varepsilon > 0$ together with a $C^0(\Omega)$ function w satisfying

$$\|u - w\|_{L^p(\Omega'')} \leqslant \varepsilon$$

where $\Omega'' = B_{h'}(\Omega')$ and $2h' < \text{dist}(\Omega', \partial\Omega)$. By virtue of Lemma 7.1, we have for sufficiently small h

$$\|w - w_h\|_{L^p(\Omega')} \leqslant \varepsilon.$$

Applying the estimate (7.15) to the difference $u - w$, we therefore obtain

$$\|u - u_h\|_{L^p(\Omega')} \leqslant \|u - w\|_{L^p(\Omega')} + \|w - w_h\|_{L^p(\Omega')} + \|u_h - w_h\|_{L^p(\Omega')}$$

$$\leqslant 2\varepsilon + \|u - w\|_{L^p(\Omega'')} \leqslant 3\varepsilon$$

for small enough $h \leqslant h'$. Hence u_h converges to u in $L^p_{\text{loc}}(\Omega)$. The result for $u \in L^p(\Omega)$ can then be obtained by extending u to be zero outside Ω and applying the result for $L^p_{\text{loc}}(\mathbb{R}^n)$. □

7.3. Weak Derivatives

Let u be locally integrable in Ω and α any multi-index. Then a locally integrable function v is called the αth *weak derivative* of u if it satisfies

$$\int_\Omega \varphi v \, dx = (-1)^{|\alpha|} \int_\Omega u D^\alpha \varphi \, dx \quad \text{for all } \varphi \in C_0^{|\alpha|}(\Omega). \tag{7.16}$$

We write $v=D^\alpha u$ and note that $D^\alpha u$ is uniquely determined up to sets of measure zero. Pointwise relations involving weak derivatives will be accordingly understood to hold almost everywhere. We call a function *weakly differentiable* if all its weak derivatives of first order exist and *k times weakly differentiable* if all its weak derivatives exist for orders up to and including k. Let us denote the linear space of k times weakly differentiable functions by $W^k(\Omega)$. Clearly $C^k(\Omega)\subset W^k(\Omega)$. The concept of weak derivative is thus an extension of the classical concept which maintains the validity of integration by parts (formula (7.16)).

We proceed to consider some basic properties of weakly differentiable functions. The first lemma describes the interaction of weak derivatives and mollifiers.

Lemma 7.3. *Let $u\in L^1_{\text{loc}}(\Omega)$, α a multi-index, and suppose that $D^\alpha u$ exists. Then if* dist $(x, \partial\Omega)>h$, *we have*

$$D^\alpha u_h(x)=(D^\alpha u)_h(x). \tag{7.17}$$

Proof. By differentiating under the integral sign, we obtain

$$\begin{aligned} D^\alpha u_h(x) &= h^{-n}\int_\Omega D^\alpha_x\rho\left(\frac{x-y}{h}\right)u(y)\,dy \\ &= (-1)^{|\alpha|}h^{-n}\int_\Omega D^\alpha_y\rho\left(\frac{x-y}{h}\right)u(y)\,dy \\ &= h^{-n}\int_\Omega \rho\left(\frac{x-y}{h}\right)D^\alpha u(y)\,dy \quad \text{by (7.16)} \\ &= (D^\alpha u)_h(x). \quad \square \end{aligned}$$

From Lemmas 7.1, 7.3 and the definition (7.16), now follows automatically a basic approximation theorem for weak derivatives, the explicit verification of which is left to the reader.

Theorem 7.4. *Let u and v be locally integrable in Ω. Then $v=D^\alpha u$ if and only if there exists a sequence of $C^\infty(\Omega)$ functions $\{u_m\}$ converging to u in $L^1_{\text{loc}}(\Omega)$ whose derivatives $D^\alpha u_m$ converge to v in $L^1_{\text{loc}}(\Omega)$.*

This equivalent characterization of weak derivatives can also be used as their definition, as is often the case. The resulting derivatives then are usually called *strong derivatives* so that Theorem 7.4 establishes the equivalence of weak and strong derivatives. Through Theorem 7.4, many results from the classical differential calculus may be extended to weak derivatives simply by approximation. In particular we have the *product formula*

$$D(uv)=uDv+vDu; \tag{7.18}$$

this holds for all $u, v \in W^1(\Omega)$ such that $uv, uDv+vDu \in L^1_{loc}(\Omega)$; (see Problem 7.4). Also if ψ maps Ω onto a domain $\tilde{\Omega} \subset \mathbb{R}^n$ with $\psi \in C^1(\Omega)$, $\psi^{-1} \in C^1(\tilde{\Omega})$ and if $u \in W^1(\Omega)$, $v = u \circ \psi^{-1}$, then $v \in W^1(\tilde{\Omega})$ and the usual change of variables formula applies, that is

$$D_i u(x) = \frac{\partial y_j}{\partial x_i} D_{y_j} v(y) \tag{7.19}$$

for almost all $x \in \Omega$, $y \in \tilde{\Omega}$, $y = \psi(x)$ (see Problem 7.5).

It is important to note that locally uniformly Lipschitz continuous functions are weakly differentiable, that is, $C^{0,1}(\Omega) \subset W^1(\Omega)$. This assertion follows since a function in $C^{0,1}(\Omega)$ will be absolutely continuous on any line segment in Ω. Consequently its partial derivatives (which exist almost everywhere) satisfy (7.16) and hence coincide almost everywhere with the weak derivatives. By means of regularization, we can in fact prove that a function is weakly differentiable if and only if it is equivalent to a function that is absolutely continuous on almost all line segments in Ω parallel to the coordinate directions and whose partial derivatives are locally integrable; (see Problem 7.8). The basic properties of weak differentiation treated in this and the following section can be alternatively derived from this characterization.

7.4. The Chain Rule

To complete our basic calculus of weak differentiation, we consider now a simple type of chain rule.

Lemma 7.5. *Let $f \in C^1(\mathbb{R})$, $f' \in L^\infty(\mathbb{R})$ and $u \in W^1(\Omega)$. Then the composite function $f \circ u \in W^1(\Omega)$ and $D(f \circ u) = f'(u)\, Du$.*

Proof. Let u_m, $m = 1, 2, \ldots \in C^1(\Omega)$, and let $\{u_m\}$, $\{Du_m\}$ converge to u, Du respectively in $L^1_{loc}(\Omega)$. Then for $\Omega' \subset\subset \Omega$, we have

$$\int_{\Omega'} |f(u_m) - f(u)|\, dx \leqslant \sup |f'| \int_{\Omega'} |u_m - u|\, dx \to 0 \quad \text{as } m \to \infty$$

$$\int_{\Omega'} |f'(u_m)Du_m - f'(u)Du|\, dx \leqslant \sup |f'| \int_{\Omega'} |Du_m - Du|\, dx$$

$$+ \int_{\Omega'} |f'(u_m) - f'(u)|\, |Du|\, dx.$$

A subsequence of $\{u_m\}$, which we renumber $\{u_m\}$, must converge a.e. (Ω') to u. Since f' is continuous, $\{f'(u_m)\}$ converges to $f'(u)$ a.e. (Ω'). Hence the last integral tends to zero by the dominated convergence theorem. Consequently the sequences $\{f(u_m)\}$, $\{f'(u_m)Du_m\}$ tend to $f(u)$, $f'(u)Du$ respectively, and therefore $Df(u)=f'(u)Du$. □

The positive and negative parts of a function u are defined by

$$u^+=\max\{u, 0\}, \qquad u^-=\min\{u, 0\}.$$

Clearly $u=u^+ + u^-$ and $|u|=u^+ - u^-$. From Lemma 7.5 we can derive the following chain rule for these functions.

Lemma 7.6. *Let $u \in W^1(\Omega)$; then u^+, u^-, $|u| \in W^1(\Omega)$ and*

$$Du^+=\begin{cases} Du & \text{if } u>0 \\ 0 & \text{if } u\leqslant 0 \end{cases}$$

$$Du^-=\begin{cases} 0 & \text{if } u\geqslant 0 \\ Du & \text{if } u<0 \end{cases} \tag{7.20}$$

$$D|u|=\begin{cases} Du & \text{if } u>0 \\ 0 & \text{if } u=0 \\ -Du & \text{if } u<0. \end{cases}$$

Proof. For $\varepsilon>0$, define

$$f_\varepsilon(u)=\begin{cases} (u^2+\varepsilon^2)^{1/2}-\varepsilon & \text{if } u>0 \\ 0 & \text{if } u\leqslant 0. \end{cases}$$

Applying Lemma 7.5 we then have, for any $\varphi \in C_0^1(\Omega)$,

$$\int_\Omega f_\varepsilon(u)D\varphi\, dx = -\int_{u>0} \varphi \frac{uDu}{(u^2+\varepsilon^2)^{1/2}}\, dx$$

and on letting $\varepsilon \to 0$, we obtain

$$\int_\Omega u^+ D\varphi\, dx = -\int_{u>0} \varphi Du\, dx$$

so that (7.20) is established for u^+. The other results follow since $u^-=-(-u)^+$ and $|u|=u^+-u^-$. □

Lemma 7.7. *Let $u \in W^1(\Omega)$. Then $Du=0$ a.e. on any set where u is constant.*

Proof. Without loss of generality we may take the constant to be zero. The result follows immediately from (7.20) since $Du=Du^+ + Du^-$. □

We call a function piecewise smooth if it is continuous and has piecewise continuous first derivatives. The following chain rule then generalizes Lemmas 7.5 and 7.6.

Theorem 7.8. *Let f be a piecewise smooth function on $\mathbb{R}$ with $f' \in L^\infty(\mathbb{R})$. Then if $u \in W^1(\Omega)$, we have $f \circ u \in W^1(\Omega)$. Furthermore, letting L denote the set of corner points of f, we have*

$$D(f \circ u) = \begin{cases} f'(u)Du & \text{if } u \notin L \\ 0 & \text{if } u \in L. \end{cases} \tag{7.21}$$

Proof. By an induction argument the proof is reduced to the case of one corner which we may take without loss of generality at the origin. Let $f_1, f_2 \in C^1(\mathbb{R})$ satisfy $f_1', f_2' \in L^\infty(\mathbb{R})$, $f_1(u) = f(u)$ for $u \geqslant 0$, $f_2(u) = f(u)$ for $u \leqslant 0$. Then since $f(u) = f_1(u^+) + f_2(u^-)$, the result follows by Lemmas 7.5 and 7.6. □

Combining Lemma 7.7 and Theorem 7.8, we see that if h is a finite valued function on $\mathbb{R}$, satisfying $h(u) = f'(u)$ for $u \notin L$, then $Df(u) = h(u)Du$. The chain rule in this form may be extended to Lipschitz continuous f and $u \in W^1(\Omega)$ for which $h(u)Du \in L^1_{\text{loc}}(\Omega)$. The proof of this assertion requires somewhat more measure theory than we have used; it is however a consequence of the characterization of weakly differentiable functions given in Problem 7.8.

7.5. The $W^{k,p}$ Spaces

The $W^{k,p}(\Omega)$ spaces are Banach spaces analogous in a certain sense to the $C^{k,\alpha}(\bar{\Omega})$ spaces. In the $W^{k,p}(\Omega)$ spaces, continuous differentiability is replaced by weak differentiability and Hölder continuity by p-integrability. For $p \geqslant 1$ and k a non-negative integer, we let

$$W^{k,p}(\Omega) = \{u \in W^k(\Omega);\ D^\alpha u \in L^p(\Omega) \text{ for all } |\alpha| \leqslant k\}.$$

The space $W^{k,p}(\Omega)$ is clearly linear. A norm is introduced by defining

$$\|u\|_{k,p;\Omega} = \|u\|_{W^{k,p}(\Omega)} = \left(\int_\Omega \sum_{|\alpha| \leqslant k} |D^\alpha u|^p \, dx \right)^{1/p}. \tag{7.22}$$

We shall also use $\|u\|_{k,p}$ for $\|u\|_{k,p;\Omega}$ when there is no ambiguity. An equivalent norm would be

$$\|u\|_{W^{k,p}(\Omega)} = \sum_{|\alpha| \leqslant k} \|D^\alpha u\|_p. \tag{7.23}$$

The verification that $W^{k,p}(\Omega)$ is a Banach space under (7.22) is left to the reader (Problem 7.10).

Another Banach space $W_0^{k,p}(\Omega)$ arises by taking the closure of $C_0^k(\Omega)$ in $W^{k,p}(\Omega)$. The spaces $W^{k,p}(\Omega)$, $W_0^{k,p}(\Omega)$ do not coincide for bounded Ω. The case $p=2$ is special, since the spaces $W^{k,2}(\Omega)$, $W_0^{k,2}(\Omega)$ (sometimes written $H^k(\Omega)$, $H_0^k(\Omega)$) will be Hilbert spaces under the scalar product

$$(7.24) \qquad (u, v)_k = \int_\Omega \sum_{|\alpha| \leqslant k} D^\alpha u D^\alpha v \, dx.$$

Further functional analytic properties of $W^{k,p}(\Omega)$ and $W_0^{k,p}(\Omega)$ follow by considering their natural imbedding into the product of N_k copies of $L^p(\Omega)$ where N_k is the number of multi-indices α satisfying $|\alpha| \leqslant k$. Using the facts that finite products and closed subspaces of separable (reflexive) Banach spaces are again separable (reflexive) [DS], we obtain accordingly that the spaces $W^{k,p}(\Omega)$, $W_0^{k,p}(\Omega)$ are separable for $1 \leqslant p < \infty$ (reflexive for $1 < p < \infty$).

The chain rule of Theorem 7.8 also extends to the spaces $W^{1,p}(\Omega)$, $W_0^{1,p}(\Omega)$. In fact as a consequence of Theorem 7.8 and the definitions of these spaces we have immediately that the space $W^1(\Omega)$ in the statement of Theorem 7.8 may be replaced by $W^{1,p}(\Omega)$, and by $W_0^{1,p}(\Omega)$ if also $f(0)=0$.

Local spaces $W_{\text{loc}}^{k,p}(\Omega)$ can be defined to consist of functions belonging to $W^{k,p}(\Omega')$ for all $\Omega' \subset\subset \Omega$. Theorem 7.4 shows that functions in $W_{\text{loc}}^{k,p}(\Omega)$ with compact support will in fact belong to $W_0^{k,p}(\Omega)$. Also, functions in $W^{1,p}(\Omega)$ which vanish continuously on $\partial\Omega$ will belong to $W_0^{1,p}(\Omega)$, since they can be approximated by functions with compact support.

In the case $p = \infty$, the Sobolev and Lipschitz spaces are related. In particular, $W_{\text{loc}}^{k,\infty}(\Omega) = C^{k-1,1}(\Omega)$ for arbitrary Ω, and $W^{k,\infty}(\Omega) = C^{k-1,1}(\bar{\Omega})$ for sufficiently smooth Ω, e.g., for Lipschitz Ω; (see Problem 7.7).

7.6. Density Theorems

It is clear from Lemmas 7.2 and 7.3, that if u lies in $W^{k,p}(\Omega)$, then $D^\alpha u_h$ tends to $D^\alpha u$ in the sense of $L_{\text{loc}}^p(\Omega)$ as h approaches zero, for all multi-indices α satisfying $|\alpha| \leqslant k$. Using this fact we shall derive a global approximation result.

Theorem 7.9. *The subspace $C^\infty(\Omega) \cap W^{k,p}(\Omega)$ is dense in $W^{k,p}(\Omega)$.*

Proof. Let Ω_j, $j=1, 2, \ldots$, be strictly contained subdomains of Ω satisfying $\Omega_j \subset\subset \Omega_{j+1}$ and $\cup\, \Omega_j = \Omega$, and let $\{\psi_j\}$, $j=0, 1, 2, \ldots$, be a partition of unity (see Problem 6.8) subordinate to the covering $\{\Omega_{j+1} - \Omega_{j-1}\}$, Ω_0 and Ω_{-1} being defined as empty sets. Then for arbitrary $u \in W^{k,p}(\Omega)$ and $\varepsilon > 0$, we can choose h_j, $j=1, 2, \ldots$, satisfying

$$(7.25) \qquad \begin{aligned} &h_j \leqslant \operatorname{dist}(\Omega_j, \partial\Omega_{j+1}), \quad j \geqslant 1 \\ &\|(\psi_j u)_{h_j} - \psi_j u\|_{W^{k,p}(\Omega)} \leqslant \frac{\varepsilon}{2^j}. \end{aligned}$$

Writing $v_j=(\psi_j u)_{h_j}$, we obtain from (7.25) that only a finite number of v_j are non-vanishing on any given $\Omega'\subset\subset\Omega$. Consequently the function $v=\sum v_j$ belongs to $C^\infty(\Omega)$. Furthermore

$$\|u-v\|_{W^{k,p}(\Omega)}\leqslant\sum\|v_j-\psi_j u\|_{W^{k,p}(\Omega)}\leqslant\varepsilon.$$

This completes the proof. □

Theorem 7.9 shows that $W^{k,p}(\Omega)$ could have been characterized as the completion of $C^\infty(\Omega)$ under the norm (7.22). In many instances this is a convenient definition.

In the case of arbitrary Ω we cannot replace $C^\infty(\Omega)$ by $C^\infty(\bar\Omega)$ in Theorem 7.9. However, $C^\infty(\bar\Omega)$ is dense in $W^{k,p}(\Omega)$ for a large class of domains Ω which includes for example C^1 domains (see Problem 7.11). More generally, if Ω satisfies a *segment condition* (that is, there exists a locally finite open covering $\{\mathscr{U}_i\}$ of $\partial\Omega$ and corresponding vectors y^i such that $x+ty^i\in\Omega$ for all $x\in\bar\Omega\cap\mathscr{U}_i$, $t\in(0,1)$), then $C^\infty(\bar\Omega)$ is dense in $W^{k,p}(\Omega)$. (See [AD]).

7.7. Imbedding Theorems

This and the following section are concerned with the connection between pointwise and integrability properties of weakly differentiable functions and the integrability properties of their derivatives. One of the simplest results in this direction is that weakly differentiable functions of one variable must be absolutely continuous. In this section we prove the well known *Sobolev inequalities* for functions in $W_0^{1,p}(\Omega)$.

Theorem 7.10.

$$W_0^{1,p}(\Omega)\subset\begin{cases}L^{np/(n-p)}(\Omega) & \text{for } p<n\\ C^0(\bar\Omega) & \text{for } p>n.\end{cases}$$

Furthermore, there exists a constant $C=C(n,p)$ such that, for any $u\in W_0^{1,p}(\Omega)$,

$$\|u\|_{np/(n-p)}\leqslant C\|Du\|_p \quad \textit{for } p<n,$$

(7.26)

$$\sup_\Omega|u|\leqslant C|\Omega|^{1/n-1/p}\|Du\|_p \quad \textit{for } p>n.$$

Proof. Let us first establish the estimates (7.26) for $C_0^1(\Omega)$ functions. We proceed from the case $p=1$. Clearly for any $u\in C_0^1(\Omega)$ and any i, $1\leqslant i\leqslant n$,

$$|u(x)|\leqslant\int_{-\infty}^{x_i}|D_i u|\,dx_i,$$

so that

$$(7.27)\qquad |u(x)|^{n/(n-1)}\leqslant\left(\prod_{i=1}^{n}\int_{-\infty}^{\infty}|D_iu|\,dx_i\right)^{1/(n-1)}.$$

The inequality (7.27) is now integrated successively over each variable x_i, $i=1,\ldots,n$, the generalized Hölder inequality (7.11) for $m=p_1=\cdots=p_m=n-1$ then being applied after each integration. Accordingly we obtain

$$\|u\|_{n/(n-1)}\leqslant\left(\prod_{i=1}^{n}\int_{\Omega}|D_iu|\,dx\right)^{1/n}$$

$$(7.28)\qquad \leqslant\frac{1}{n}\int_{\Omega}\sum_{i=1}^{n}|D_iu|\,dx$$

$$\leqslant\frac{1}{\sqrt{n}}\|Du\|_1.$$

Thus inequality (7.26) is established for the case $p=1$. The remaining cases can now be obtained by replacing u in the estimate (7.28) by powers of $|u|$. In this way we get for $\gamma>1$,

$$\|\,|u|^{\gamma}\|_{n/(n-1)}\leqslant\frac{\gamma}{\sqrt{n}}\int_{\Omega}|u|^{\gamma-1}|Du|\,dx$$

$$\leqslant\frac{\gamma}{\sqrt{n}}\|\,|u|^{\gamma-1}\|_{p'}\|Du\|_p$$

by Hölder's inequality. Now for $p<n$ we may choose γ to satisfy

$$\frac{\gamma n}{n-1}=\frac{(\gamma-1)p}{p-1},\quad \text{i.e. } \gamma=\frac{(n-1)p}{n-p},$$

and consequently obtain

$$\|u\|_{np/(n-p)}\leqslant\frac{\gamma}{\sqrt{n}}\|Du\|_p,$$

as required.

The case $p>n$ follows immediately by combining inequalities (7.34), with $q=\infty$, $\mu=1/n$, and (7.37) of the following section. We insert here an alternative proof which is based on the case $p=1$.

For $p>n$, let us write

$$\tilde{u}=\frac{\sqrt{n}|u|}{\|Du\|_p}$$

and assume that $|\Omega|=1$. We obtain then

$$\|\tilde{u}^\gamma\|_{n'} \leqslant \gamma \|\tilde{u}^{\gamma-1}\|_{p'}, \quad n'=\frac{n}{n-1}, p'=\frac{p}{p-1},$$

so that

$$\|\tilde{u}\|_{\gamma n'} \leqslant \gamma^{1/\gamma} \|\tilde{u}\|_{p'(\gamma-1)}^{1-1/\gamma}$$
$$\leqslant \gamma^{1/\gamma} \|\tilde{u}\|_{\gamma p'}^{1-1/\gamma} \quad \text{since } |\Omega|=1.$$

Let us substitute for γ the values δ^ν, $\nu=1, 2, \ldots$, where

$$\delta=\frac{n'}{p'}>1.$$

We obtain thus

$$\|\tilde{u}\|_{n'\delta^\nu} \leqslant \delta^{\nu\delta^{-\nu}} \|\tilde{u}\|_{n'\delta^{\nu-1}}^{1-\delta^{-\nu}}, \quad \nu=1, 2, \ldots$$

Iterating from $\nu=1$ and using (7.28), we get for any ν

$$\|\tilde{u}\|_{\delta^\nu} \leqslant \delta^{\Sigma \nu\delta^{-\nu}} \equiv \chi.$$

Consequently as $\nu \to \infty$, we obtain by Problem 7.1,

$$\sup_\Omega \tilde{u} \leqslant \chi,$$

and hence

$$\sup_\Omega |u| \leqslant \frac{\chi}{\sqrt{n}} \|Du\|_p.$$

To eliminate the restriction $|\Omega|=1$, we consider a transformation: $y_i=|\Omega|^{1/n}x_i$. We obtain thus

$$\sup_\Omega |u| \leqslant \frac{\chi}{\sqrt{n}} |\Omega|^{1/n-1/p} \|Du\|_p$$

as required.

To extend the estimates (7.26) to arbitrary $u \in W_0^{1,p}(\Omega)$, we let $\{u_m\}$ be a sequence of $C_0^1(\Omega)$ functions tending to u in $W^{1,p}(\Omega)$. Applying the estimates (7.26) to differences $u_{m_1}-u_{m_2}$, we see that $\{u_m\}$ will be a Cauchy sequence in $L^{np/(n-p)}(\Omega)$ for $p<n$ and in $C^0(\bar{\Omega})$ for $p>n$. Consequently the limit function u will lie in the desired spaces and satisfy (7.26). □

Remark. The best constant C satisfying (7.26) for the case $p<n$ was calculated by Rodemich [RO], see also [BL], [TA 2], who showed that

$$C=\frac{1}{n\sqrt{\pi}}\left(\frac{n!\,\Gamma(n/2)}{2\Gamma(n/p)\Gamma(n+1-n/p)}\right)^{1/n}\gamma^{1-1/p},\ \gamma=\frac{n(p-1)}{n-p}.$$

When $p=1$, the above number reduces to the well known isoperimetric constant $n^{-1}(\omega_n)^{-1/n}$.

A Banach space $\mathscr{B}_1$ is said to be *continuously imbedded* in a Banach space $\mathscr{B}_2$ (notation: $\mathscr{B}_1\to\mathscr{B}_2$) if there exists a bounded, linear, one-to-one mapping: $\mathscr{B}_1\to\mathscr{B}_2$. Theorem 7.10 may be thus expressed as $W_0^{1,p}(\Omega)\to L^{np/(n-p)}(\Omega)$ if $p<n$, $\to C^0(\bar{\Omega})$ if $p>n$. By iterating the result of Theorem 7.10 k times we arrive at an extension to the spaces $W_0^{k,p}(\Omega)$.

Corollary 7.11.

$$W_0^{k,p}(\Omega)\begin{cases}\to L^{np/(n-kp)}(\Omega) & \text{for } kp<n\\ \to C^m(\bar{\Omega}) & \text{for } 0\leqslant m<k-\dfrac{n}{p}.\end{cases}$$

The second case is a consequence of the first, together with the case $p>n$ in Theorem 7.10.

The estimates (7.26) and their extension to the spaces $W_0^{k,p}(\Omega)$ also show that a norm on $W_0^{k,p}(\Omega)$ equivalent to (7.22) may be defined by

$$\|u\|_{W_0^{k,p}(\Omega)}=\left(\int_\Omega\sum_{|\alpha|=k}|D^\alpha u|^p\,dx\right)^{1/p}. \tag{7.29}$$

In general, $W_0^{k,p}(\Omega)$ cannot be replaced by $W^{k,p}(\Omega)$ in Corollary 7.11. However, this replacement can be made for a large class of domains Ω, which includes for example domains with Lipschitz continuous boundaries. (See Theorem 7.26). More generally, if Ω satisfies a *uniform interior cone condition*, (that is, there exists a fixed cone K_Ω such that each $x\in\Omega$ is the vertex of a cone $K_\Omega(x)\subset\bar{\Omega}$ and congruent to K_Ω), then there is an imbedding

$$W^{k,p}(\Omega)\begin{cases}\to L^{np/(n-kp)}(\Omega) & \text{for } kp<n\\ \to C_B^m(\Omega) & \text{for } 0\leqslant m<k-\dfrac{n}{p}.\end{cases} \tag{7.30}$$

where $C_B^m(\Omega)=\{u\in C^m(\Omega)\,|\,D^\alpha u\in L^\infty(\Omega) \text{ for } |\alpha|\leqslant m\}$.

7.8. Potential Estimates and Imbedding Theorems

The imbedding results of the preceding section can be alternatively derived and also improved through the use of certain potential estimates. Let $\mu \in (0, 1]$ and define the operator V_μ on $L^1(\Omega)$ by the Riesz potential

$$(V_\mu f)(x) = \int_\Omega |x-y|^{n(\mu-1)} f(y)\, dy. \tag{7.31}$$

That V_μ is in fact well defined and maps $L^1(\Omega)$ into itself will appear as an incidental consequence of the next lemma. First we observe, by setting $f \equiv 1$ in (7.31),

$$V_\mu 1 \leqslant \mu^{-1} \omega_n^{1-\mu} |\Omega|^\mu. \tag{7.32}$$

For, choose $R > 0$ so that $|\Omega| = |B_R(x)| = \omega_n R^n$. Then

$$\begin{aligned}\int_\Omega |x-y|^{n(\mu-1)}\, dy &\leqslant \int_{B_R(x)} |x-y|^{n(\mu-1)}\, dy \\ &= \mu^{-1} \omega_n R^{n\mu} \\ &= \mu^{-1} \omega_n^{1-\mu} |\Omega|^\mu.\end{aligned}$$

Lemma 7.12. *The operator V_μ maps $L^p(\Omega)$ continuously into $L^q(\Omega)$ for any q, $1 \leqslant q \leqslant \infty$ satisfying*

$$0 \leqslant \delta = \delta(p, q) = p^{-1} - q^{-1} < \mu. \tag{7.33}$$

Furthermore, for any $f \in L^p(\Omega)$,

$$\| V_\mu f \|_q \leqslant \left(\frac{1-\delta}{\mu-\delta} \right)^{1-\delta} \omega_n^{1-\mu} |\Omega|^{\mu-\delta} \| f \|_p. \tag{7.34}$$

Proof. Choose $r \geqslant 1$ so that

$$r^{-1} = 1 + q^{-1} - p^{-1} = 1 - \delta.$$

Then it follows that $h(x-y) = |x-y|^{n(\mu-1)} \in L^r(\Omega)$, and by (7.32) one obtains

$$\|h\|_r \leqslant \left(\frac{1-\delta}{\mu-\delta} \right)^{1-\delta} \omega_n^{1-\mu} |\Omega|^{\mu-\delta}.$$

The estimate (7.34) can now be derived by adapting the usual proof of the Young inequality for convolutions in $\mathbb{R}^n$. Writing

$$h|f| = h^{r/q}\, h^{r(1-1/p)}\, |f|^{p/q} |f|^{p\delta},$$

we may estimate by the Hölder inequality (7.11)

$$|V_\mu f(x)| \leqslant \left\{\int_\Omega h^r(x-y)|f(y)|^p\,dy\right\}^{1/q} \left\{\int_\Omega h^r(x-y)\,dy\right\}^{1-1/p} \left\{\int_\Omega |f(y)|^p\,dy\right\}^{\delta},$$

so that

$$\|V_\mu f\|_q \leqslant \sup_\Omega \left\{\int h^r(x-y)\,dy\right\}^{1/r} \|f\|_p$$

$$\leqslant \left(\frac{1-\delta}{\mu-\delta}\right)^{1-\delta} \omega_n^{1-\mu}|\Omega|^{\mu-\delta}\|f\|_p. \quad \square$$

We mention here that Lemma 7.12 may be strengthened in the sense that V_μ maps $L^p(\Omega)$ continuously into $L^q(\Omega)$ provided $p>1$ and $\delta \leqslant \mu$. The proof requires a well known integral inequality of Hardy and Littlewood (see [HL]). However, Lemma 7.12 is adequate for our purposes here. Observe that when $p>\mu^{-1}$, V_μ maps $L^p(\Omega)$ continuously into $L^\infty(\Omega)$. Let us examine now the intermediate case $p=\mu^{-1}$.

Lemma 7.13. *Let $f \in L^p(\Omega)$ and $g=V_{1/p}f$. Then there exist constants c_1 and c_2 depending only on n and p such that*

$$\int_\Omega \exp\left[\frac{g}{c_1\|f\|_p}\right]^{p'} dx \leqslant c_2|\Omega|, \quad p'=p/(p-1). \tag{7.35}$$

Proof. From Lemma 7.12, we get for any $q \geqslant p$

$$\|g\|_q \leqslant q^{1-1/p+1/q}\omega_n^{1-1/p}|\Omega|^{1/q}\|f\|_p,$$

so that

$$\int_\Omega |g|^q\,dx \leqslant q^{1+q/p'}\omega_n^{q/p'}|\Omega|\,\|f\|_p^q$$

and hence for $q \geqslant p-1$

$$\int_\Omega |g|^{p'q}\,dx \leqslant p'q(\omega_n p'q\|f\|_p^{p'})^q|\Omega|.$$

Consequently

$$\int_\Omega \sum_{N_0}^{N} \frac{1}{k!}\left(\frac{|g|}{c_1\|f\|_p}\right)^{p'k} dx \leqslant p'|\Omega| \sum \left(\frac{p'\omega_n}{c_1^{p'}}\right)^k \frac{k^k}{(k-1)!}, \quad N_0=[p]$$

The series on the right hand side converges provided $c_1^{p'} > e\omega_n p'$, whence by the monotone convergence theorem and (7.8) the desired estimate (7.35) follows. □

The next lemmas serve to clarify the connection between weak derivatives and potentials of the above type.

Lemma 7.14. *Let $u \in W_0^{1,1}(\Omega)$. Then*

$$(7.36) \qquad u(x) = \frac{1}{n\omega_n}\int_\Omega \frac{(x_i - y_i)D_i u(y)}{|x-y|^n}\, dy \quad \text{a.e. } (\Omega).$$

Proof. Suppose that $u \in C_0^1(\Omega)$ and extend u to be zero outside Ω. Then, for any ω with $|\omega| = 1$,

$$u(x) = -\int_0^\infty D_r u(x + r\omega)\, dr.$$

Integrating with respect to ω, we obtain

$$u(x) = -\frac{1}{n\omega_n}\int_0^\infty \int_{|\omega|=1} D_r u(x + r\omega)\, dr\, d\omega$$

$$= \frac{1}{n\omega_n}\int_\Omega \frac{(x_i - y_i)D_i u(y)}{|x-y|^n}\, dy$$

and (7.36) follows from Lemma 7.12 and the fact that $C_0^1(\Omega)$ is dense in $W_0^{1,1}(\Omega)$. □

Note that by means of the formula (7.16) for integration by parts, the Newtonian potential representation for $C_0^2(\Omega)$ functions, equation (2.17), is deducible from formula (7.36). Also we obtain for $u \in W_0^{1,1}(\Omega)$

$$(7.37) \qquad |u| \leqslant \frac{1}{n\omega_n} V_{1/n}|Du|.$$

Combining Lemma 7.12 and inequality (7.37) we obtain immediately the imbeddings $W_0^{1,p}(\Omega) \to L^q(\Omega)$ for $p^{-1}-q^{-1}<n^{-1}$, which is almost the conclusion of Theorem 7.10. In fact, this weaker version would be adequate for the purposes of this book. But also combining Lemma 7.13 and (7.37), we obtain a sharpening of the case $p=n$ expressed by the following theorem.

Theorem 7.15. *Let $u \in W_0^{1,n}(\Omega)$. Then there exist constants c_1 and c_2 depending only on n, such that*

$$\int_\Omega \exp\left(\frac{|u|}{c_1\|Du\|_n}\right)^{n/(n-1)} dx \leqslant c_2|\Omega|. \tag{7.38}$$

Remark. The estimate (7.37) is readily generalized to higher order weak derivatives. One obtains then for $u \in W_0^{k,1}(\Omega)$,

$$|u| \leqslant \frac{1}{(k-1)!\,n\omega_n} V_{k/n}|D^k u|, \tag{7.39}$$

and using Lemma 7.13 we have an extension of Theorem 7.15. Namely there exist constants c_1 and c_2 depending only on n and k such that if $u \in W_0^{k,p}(\Omega)$ with $n=kp$, then

$$\int_\Omega \exp\left(\frac{|u|}{c_1\|D^k u\|_p}\right)^{p/(p-1)} dx \leqslant c_2|\Omega|. \tag{7.40}$$

The case $p>n$ of the Sobolev imbedding theorem may be sharpened through the following lemma.

Lemma 7.16. *Let Ω be convex and $u \in W^{1,1}(\Omega)$. Then*

$$|u(x) - u_S| \leqslant \frac{d^n}{n|S|} \int_\Omega |x-y|^{1-n}|Du(y)|\,dy \quad \text{a.e. } (\Omega), \tag{7.41}$$

where

$$u_S = \frac{1}{|S|}\int_S u\,dx, \quad d = \operatorname{diam}\Omega,$$

and S is any measurable subset of Ω.

Proof. By Theorem 7.9, it is enough to establish (7.41) for $u \in C^1(\Omega)$. We then have for $x, y \in \Omega$,

$$u(x)-u(y) = -\int_0^{|x-y|} D_r u(x+r\omega)\,dr, \quad \omega = \frac{y-x}{|y-x|}.$$

Integrating with respect to y over S, we obtain

$$|S|(u(x)-u_S)=-\int_S dy \int_0^{|x-y|} D_r u(x+r\omega)\,dr.$$

Writing

$$V(x)=\begin{cases}|D_r u(x)|, & x\in\Omega\\ 0, & x\notin\Omega\end{cases}$$

we thus have

$$|u(x)-u_S|\leqslant\frac{1}{|S|}\int_{|x-y|<d} dy\int_0^\infty V(x+r\omega)\,dr$$

$$=\frac{1}{|S|}\int_0^\infty\int_{|\omega|=1}\int_0^d V(x+r\omega)\rho^{n-1}\,d\rho\,d\omega\,dr$$

$$=\frac{d^n}{n|S|}\int_0^\infty\int_{|\omega|=1} V(x+r\omega)\,d\omega\,dr$$

$$=\frac{d^n}{n|S|}\int_\Omega |x-y|^{1-n}|D_r u(y)|\,dy. \quad \square$$

We can now prove the imbedding theorem of Morrey.

Theorem 7.17. *Let $u\in W_0^{1,p}(\Omega)$, $p>n$. Then $u\in C^\gamma(\bar{\Omega})$, where $\gamma=1-n/p$. Furthermore, for any ball $B=B_R$,*

$$\operatorname*{osc}_{\Omega\cap B_R} u\leqslant CR^\gamma\|Du\|_p, \tag{7.42}$$

where $C=C(n,p)$.

Proof. Coupling the estimates (7.41) and (7.34) for $S=\Omega=B$, $q=\infty$ and $\mu=n^{-1}$, we have

$$|u(x)-u_B|\leqslant C(n,p)R^\gamma\|Du\|_p \quad \text{a.e. } (\Omega\cap B).$$

The result then follows since

$$\begin{aligned}|u(x)-u(y)|&\leqslant|u(x)-u_B|+|u(y)-u_B|\\&\leqslant 2C(n,p)R^\gamma\|Du\|_p \quad \text{a.e. } (\Omega\cap B). \quad \square\end{aligned}$$

Combining Theorems 7.10 and 7.17, we have for $u \in W_0^{1,p}(\Omega)$ and $p>n$ the estimate

$$(7.43) \qquad |u|_{0,\gamma} \leqslant C[1+(\operatorname{diam}\Omega)^\gamma]\|Du\|_p.$$

Further, the results of Theorems 7.10, 7.15, 7.17 may be summarized by the following diagram

$$\begin{array}{lll} & L^{np/(n-p)}(\Omega), & p<n \\ & \nearrow & \\ W_0^{1,p}(\Omega) & \rightarrow L^\varphi(\Omega), & \varphi=\exp(|t|^{n/(n-1)})-1,\ p=n \\ & \searrow & \\ & C^\lambda(\bar\Omega), & \lambda=1-\dfrac{n}{p},\ p>n \end{array}$$

where $L^\varphi(\Omega)$ denotes the Orlicz space with defining function φ. (See [TR 2] for a more explicit definition of $L^\varphi(\Omega)$.)

For the derivation of many of the apriori estimates in this book weaker forms of the Sobolev inequalities known as the *Poincaré inequalities* are sufficient. From Lemmas 7.12 and 7.14 we have for $u \in W_0^{1,p}(\Omega)$, $1 \leqslant p < \infty$

$$(7.44) \qquad \|u\|_p \leqslant \left(\frac{1}{\omega_n}|\Omega|\right)^{1/n}\|Du\|_p;$$

while from Lemmas 7.12 and 7.16 we have, for $u \in W^{1,p}(\Omega)$ and convex Ω,

$$(7.45) \qquad \|u-u_S\|_p \leqslant \left(\frac{\omega_n}{|S|}\right)^{1-1/n} d^n\|Du\|_p, \quad d=\operatorname{diam}\Omega.$$

7.9. The Morrey and John-Nirenberg Estimates

We proceed now to a consideration of the potential operators V_μ on a different class of spaces in order to prove useful imbedding results due to Morrey (Theorem 7.19) and John and Nirenberg (Theorem 7.21). Namely, the integrable function f is said to belong to $M^p(\Omega)$, $1 \leqslant p \leqslant \infty$, if there exists a constant K such that

$$(7.46) \qquad \int_{\Omega \cap B_R} |f|\,dx \leqslant KR^{n(1-1/p)}$$

for all balls B_R. We define the p norm $\|f\|_{M^p(\Omega)}$ to be the infimum of the constants K satisfying (7.46). It is easy to see that $L^p(\Omega) \subset M^p(\Omega)$, $L^1(\Omega)=M^1(\Omega)$, $L^\infty(\Omega)=M^\infty(\Omega)$. Instead of considering in detail the action of the operators V_μ on arbitrary $M^p(\Omega)$ spaces, it will be enough to limit ourselves to the cases $p \geqslant \mu^{-1}$.

Lemma 7.18. *Let $f \in M^p(\Omega)$, $\delta = p^{-1} < \mu$. Then*

$$|V_\mu f(x)| \leqslant \frac{1-\delta}{\mu-\delta} (\text{diam } \Omega)^{n(\mu-\delta)} \|f\|_{M^p(\Omega)} \quad \text{a.e. } (\Omega). \tag{7.47}$$

Proof. Extend f to be zero outside Ω and write

$$v(\rho) = \int_{B_\rho(x)} |f(y)| \, dy.$$

Then

$$\begin{aligned} |V_\mu f(x)| &\leqslant \int_\Omega \rho^{n(\mu-1)} |f(y)| \, dy, \quad \rho = |x-y| \\ &= \int_0^d \rho^{n(\mu-1)} v'(\rho) \, d\rho, \qquad d = \text{diam } \Omega \\ &= d^{n(\mu-1)} v(d) + n(1-\mu) \int_0^d \rho^{n(\mu-1)-1} v(\rho) \, d\rho \\ &\leqslant \frac{1-\delta}{\mu-\delta} d^{n(\mu-\delta)} K \quad \text{by (7.46).} \quad \square \end{aligned}$$

The following theorem now generalizes Theorem 7.17.

Theorem 7.19. *Let $u \in W^{1,1}(\Omega)$, and suppose there exist positive constants K, α ($\alpha \leqslant 1$) such that*

$$\int_{B_R} |Du| \, dx \leqslant K R^{n-1+\alpha} \quad \textit{for all balls } B_R \subset \Omega. \tag{7.48}$$

Then $u \in C^{0,\alpha}(\Omega)$, and for any ball $B_R \subset \Omega$

$$\operatorname*{osc}_{B_R} u \leqslant C K R^\alpha, \tag{7.49}$$

where $C = C(n, \alpha)$. If $\Omega = \tilde{\Omega} \cap \mathbb{R}^n_+ = \{x \in \tilde{\Omega} \mid x_n > 0\}$ for some domain $\tilde{\Omega} \subset \mathbb{R}^n$ and (7.48) holds for all balls $B_R \subset \tilde{\Omega}$, then $u \in C^{0,\alpha}(\bar{\Omega} \cap \tilde{\Omega})$ and (7.49) holds for all $B_R \subset \tilde{\Omega}$.

Theorem 7.19 is obtained by combining Lemma 7.16 ($S = \Omega$) with Lemma 7.18. As a further consequence of Lemma 7.18 we have

Lemma 7.20. *Let $f \in M^p(\Omega)$ ($p > 1$) and $g = V_\mu f$, $\mu = p^{-1}$. Then there exist constants c_1 and c_2 depending only on n and p such that*

$$\int_\Omega \exp\left(\frac{|g|}{c_1 K}\right) dx \leqslant c_2 \, (\text{diam } \Omega)^n \tag{7.50}$$

where $K = \|f\|_{M^p(\Omega)}$.

Proof. Writing for any $q \geq 1$

$$|x-y|^{n(\mu-1)} = |x-y|^{(\mu/q-1)n/q}|x-y|^{n(1-1/q)(\mu/q+\mu-1)}$$

we have by Hölder's inequality

$$|g(x)| \leq (V_{\mu/q}|f|)^{1/q}(V_{\mu+\mu/q}|f|)^{1-1/q}.$$

By Lemma 7.18

$$\begin{aligned} V_{\mu+\mu/q}|f| &\leq \frac{(1-\mu)q}{\mu} d^{n/pq}K, \quad d = \text{diam } \Omega \\ &\leq (p-1)qd^{n/pq}K. \end{aligned}$$

Also by Lemma 7.12

$$\begin{aligned} \int_\Omega V_{\mu/q}|f|\, dx &\leq pq\omega_n^{1-1/pq}|\Omega|^{1/pq}\|f\|_1 \\ &\leq pq\omega_n K d^{n(1-1/p+1/pq)}. \end{aligned}$$

Hence

$$\begin{aligned} \int_\Omega |g|^q\, dx &\leq p(p-1)^{q-1}\omega_n q^q\, d^n K^q \\ &\leq p'\omega_n\{(p-1)qK\}^q\, d^n, \quad p' = p/(p-1). \end{aligned}$$

Consequently

$$\begin{aligned} \int_\Omega \sum_{m=0}^N \frac{|g|^m}{m!(c_1 K)^m}\, dx &\leq p'\omega_n\, d^n \sum_{m=0}^N \left(\frac{p-1}{c_1}\right)^m \frac{m^m}{m!} \\ &\leq c_2 d^n \quad \text{if } (p-1)e < c_1. \end{aligned}$$

Letting $N \to \infty$, we thus obtain (7.50). □

Combining Lemmas 7.16 and 7.20 we then get

Theorem 7.21. *Let $u \in W^{1,1}(\Omega)$ where Ω is convex, and suppose there exists a constant K such that*

$$\int_{\Omega \cap B_R} |Du|\, dx \leq KR^{n-1} \quad \textit{for all balls } B_R. \tag{7.51}$$

Then there exist positive constants σ_0 and C depending only on n such that

$$\int_\Omega \exp\left(\frac{\sigma}{K}|u-u_\Omega|\right) dx \leq C\,(\text{diam } \Omega)^n \tag{7.52}$$

where $\sigma = \sigma_0|\Omega|\,(\text{diam } \Omega)^{-n}$.

7.10. Compactness Results

Let $\mathscr{B}_1$ be a Banach space continuously imbedded in a Banach space $\mathscr{B}_2$. Then $\mathscr{B}_1$ is *compactly imbedded* in $\mathscr{B}_2$ if the imbedding operator $I: \mathscr{B}_1 \to \mathscr{B}_2$ is compact, that is, if the images of bounded sets in $\mathscr{B}_1$ are precompact in $\mathscr{B}_2$. Let us now prove the *Kondrachov compactness theorem* for the spaces $W_0^{1,p}(\Omega)$.

Theorem 7.22. *The spaces $W_0^{1,p}(\Omega)$ are compactly imbedded* (i) *in the spaces $L^q(\Omega)$ for any $q<np/(n-p)$, if $p<n$, and* (ii) *in $C^0(\bar{\Omega})$, if $p>n$.*

Proof. Part (ii) is a consequence of Morrey's theorem (Theorem 7.17) and Arzela's theorem on equicontinuous families of functions. Let us thus concentrate on part (i) and prove it initially for the case $q=1$. Let A be a bounded set in $W_0^{1,p}(\Omega)$. Without loss of generality we may assume that $A \subset C_0^1(\Omega)$ and that $\|u\|_{1,p;\Omega} \leqslant 1$ for all $u \in A$. For $h>0$, we define $A_h = \{u_h \mid u \in A\}$ where u_h is the regularization of u (see formula (7.13)). It then follows that the set A_h is precompact in $L^1(\Omega)$. For if $u \in A$ we have

$$|u_h(x)| \leqslant \int_{|z|\leqslant 1} \rho(z)|u(x-hz)|\,dz \leqslant h^{-n} \sup \rho \|u\|_1$$

and

$$|Du_h(x)| \leqslant h^{-1} \int_{|z|\leqslant 1} |D\rho(z)||u(x-hz)|\,dz \leqslant h^{-n-1} \sup |D\rho|\,\|u\|_1$$

so that A_h is a bounded, equicontinuous subset of $C^0(\bar{\Omega})$ and hence precompact in $C^0(\bar{\Omega})$ by Arzela's theorem, and consequently also precompact in $L^1(\Omega)$. Next we may estimate for $u \in A$

$$\begin{aligned} |u(x)-u_h(x)| &\leqslant \int_{|z|\leqslant 1} \rho(z)|u(x)-u(x-hz)|\,dz \\ &\leqslant \int_{|z|\leqslant 1} \rho(z) \int_0^{h|z|} |D_r u(x-r\omega)|\,dr\,dz, \quad \omega = \frac{z}{|z|}; \end{aligned}$$

hence integrating over x we obtain

$$\int_\Omega |u(x)-u_h(x)|\,dx \leqslant h \int_\Omega |Du|\,dx \leqslant h\,|\Omega|^{1-1/p}.$$

Consequently u_h is uniformly close to u in $L^1(\Omega)$ (relative to A). Since we have shown above that A_h is totally bounded in $L^1(\Omega)$ for all $h>0$, it follows that A

is also totally bounded in $L^1(\Omega)$ and hence precompact. The case $q=1$ is thus established. To extend the result to arbitrary $q<np/(n-p)$, we estimate by (7.9)

$$\|u\|_q \leqslant \|u\|_1^\lambda \|u\|_{np/(n-p)}^{1-\lambda} \quad \text{where } \lambda+(1-\lambda)\left(\frac{1}{p}-\frac{1}{n}\right)=\frac{1}{q}$$
$$\leqslant \|u\|_1^\lambda (C\|Du\|_p)^{1-\lambda} \quad \text{by Theorem 7.10.}$$

Consequently a bounded set in $W_0^{1,p}(\Omega)$ must be precompact in $L^q(\Omega)$ for $q>1$ and the theorem is proved. □

A simple extension of Theorem 7.22 shows that the imbeddings

$$W_0^{k,p}(\Omega) \begin{array}{l} \nearrow L^q(\Omega) \quad \text{for } kp<n,\ q<\dfrac{np}{n-kp} \\ \searrow C^m(\bar{\Omega}) \quad \text{for } 0\leqslant m<k-\dfrac{n}{p} \end{array}$$

are compact and that $W_0^{k,p}(\Omega)$ may be replaced by $W^{k,p}(\Omega)$ for certain Ω; see Theorem 7.26; Problem 7.14.

7.11. Difference Quotients

In partial differential equations, the weak or classical differentiability of functions may often be deduced through a consideration of their difference quotients. Let u be a function on a domain Ω in $\mathbb{R}^n$ and denote by e_i the unit coordinate vector in the x_i direction. As in Chapter 6, we define the difference quotient in the direction e_i by

$$(7.53) \qquad \Delta^h u(x)=\Delta_i^h u(x)=\frac{u(x+he_i)-u(x)}{h}, \quad h\neq 0.$$

The following basic lemmas pertain to difference quotients of functions in Sobolev spaces.

Lemma 7.23. *Let $u\in W^{1,p}(\Omega)$. Then $\Delta^h u\in L^p(\Omega')$ for any $\Omega'\subset\subset\Omega$ satisfying $h<\operatorname{dist}(\Omega',\partial\Omega)$, and we have*

$$\|\Delta^h u\|_{L^p(\Omega')}\leqslant \|D_i u\|_{L^p(\Omega)}.$$

Proof. Let us suppose initially that $u\in C^1(\Omega)\cap W^{1,p}(\Omega)$. Then

$$\Delta^h u(x)=\frac{u(x+he_i)-u(x)}{h}$$
$$=\frac{1}{h}\int_0^h D_i u(x_1,\ldots,x_{i-1},x_i+\xi,x_{i+1},\ldots,x_n)\,d\xi$$

so that by Hölder's inequality

$$|\Delta^h u(x)|^p \leqslant \frac{1}{h} \int_0^h |D_i u(x_1, \ldots, x_{i-1}, x_i + \xi, x_{i+1}, \ldots, x_n)|^p \, d\xi,$$

and hence

$$\int_{\Omega'} |\Delta^h u|^p \, dx \leqslant \frac{1}{h} \int_0^h \int_{B_h(\Omega')} |D_i u|^p \, dx \, d\xi \leqslant \int_{\Omega} |D_i u|^p \, dx.$$

The extension to arbitrary functions in $W^{1,p}(\Omega)$ follows by a straight-forward approximation argument using Theorem 7.9. □

Lemma 7.24. *Let $u \in L^p(\Omega)$, $1<p<\infty$, and suppose there exists a constant K such that $\Delta^h u \in L^p(\Omega')$ and $\|\Delta^h u\|_{L^p(\Omega')} \leqslant K$ for all $h>0$ and $\Omega' \subset\subset \Omega$ satisfying $h < \operatorname{dist}(\Omega', \partial\Omega)$. Then the weak derivative $D_i u$ exists and satisfies $\|D_i u\|_{L^p(\Omega)} \leqslant K$.*

Proof. By the weak compactness of bounded sets in $L^p(\Omega')$, (Problem 5.4), there exists a sequence $\{h_m\}$ tending to zero and a function $v \in L^p(\Omega)$ with $\|v\|_p \leqslant K$ satisfying for all $\varphi \in C_0^1(\Omega)$

$$\int_{\Omega} \varphi \, \Delta^{h_m} u \, dx \to \int_{\Omega} \varphi v \, dx.$$

Now for $h_m < \operatorname{dist}(\operatorname{supp} \varphi, \partial\Omega)$, we have

$$\int_{\Omega} \varphi \, \Delta^{h_m} u \, dx = -\int_{\Omega} u \, \Delta^{-h_m} \varphi \, dx \to -\int_{\Omega} u \, D_i \varphi \, dx.$$

Hence

$$\int_{\Omega} \varphi v \, dx = -\int_{\Omega} u D_i \varphi \, dx$$

whence $v = D_i u$. □

7.12. Extension and Interpolation

Under certain hypotheses on the domain Ω, functions in Sobolev spaces $W^{k,p}(\Omega)$ may be extended as functions in $W^{k,p}(\mathbb{R}^n)$. We commence this section with a basic extension result, analogous to Lemma 6.37, which will be used both to improve previous imbedding results and to establish interpolation inequalities for Sobolev space norms.

Theorem 7.25. *Let Ω be a $C^{k-1,1}$ domain in $\mathbb{R}^n$, $k \geqslant 1$. Then (i) $C^\infty(\bar{\Omega})$ is dense in $W^{k,p}(\Omega)$, $1 \leqslant p < \infty$, and (ii) for any open set $\Omega' \supset\supset \Omega$ there exists a bounded linear extension operator E from $W^{k,p}(\Omega)$ into $W_0^{k,p}(\Omega')$ such that $Eu = u$ in Ω and*

$$\|Eu\|_{k,p;\Omega'} \leqslant C\|u\|_{k,p;\Omega} \tag{7.54}$$

for all $u \in W^{k,p}(\Omega)$ where $C = C(k, \Omega, \Omega')$.

Proof. We observe, by virtue of Lemmas 6.37 and 7.4, that assertions (i) and (ii) are equivalent. Let us first consider the density result (i) for the half-space $\mathbb{R}^n_+ = \{x \in \mathbb{R}^n \mid x_n > 0\}$. In this case it is readily shown that the translated mollifications of u, given by

$$\begin{aligned} v_h(x) &= u_h(x + 2he_n) \\ &= h^{-n} \int\limits_{y_n > 0} u(y)\rho\left(\frac{x + 2he_n - y}{h}\right) dy, \quad h > 0, \end{aligned} \tag{7.55}$$

converge to u in $W^{k,p}(\mathbb{R}^n_+)$ as $h \to 0$. Accordingly an extension $E_0 u$ of u to all of $\mathbb{R}^n$ may be defined by the formula in Lemma 6.37, namely,

$$E_0 u(x) = \begin{cases} u(x) & \text{for } x_n > 0, \\ \sum_{i=1}^k c_i u(x', -x_n/i) & \text{for } x_n < 0 \end{cases} \tag{7.56}$$

where $c_1, \ldots, c_k$ are constants determined by the system of equations

$$\sum_{i=1}^k c_i(-1/i)^m = 1, \qquad m = 0, \ldots, k-1.$$

If $u \in C^\infty(\bar{\mathbb{R}}^n_+) \cap W^{k,p}(\mathbb{R}^n_+)$ it follows that $E_0 u \in C^{k-1,1}(\mathbb{R}^n) \cap W^{k,p}(\mathbb{R}^n)$ and, moreover,

$$\|E_0 u\|_{k,p;\mathbb{R}^n} \leqslant C\|u\|_{k,p;\mathbb{R}^n_+}, \tag{7.57}$$

where $C = C(k)$. Therefore, by approximation we obtain that E_0 maps $W^{k,p}(\mathbb{R}^n_+)$ into $W^{k,p}(\mathbb{R}^n)$ and satisfies (7.57) for all $u \in W^{k,p}(\mathbb{R}^n_+)$.

Having treated the half-space case, let us now suppose that Ω is a $C^{k-1,1}$ domain in $\mathbb{R}^n$. According to the definition in Section 6.2, there exist a finite number of open sets $\Omega_j \subset \Omega'$, $j = 1, \ldots, N$, which cover $\partial\Omega$, and corresponding mappings ψ_j of Ω_j onto the unit ball $B = B_1(0)$ in $\mathbb{R}^n$ such that

(i) $\psi_j(\Omega_j \cap \Omega) = B^+ = B \cap \mathbb{R}^n_+$;
(ii) $\psi_j(\Omega_j \cap \partial\Omega) = B \cap \partial\mathbb{R}^n_+$;
(iii) $\psi_j \in C^{k-1,1}(\Omega_j)$, $\quad \psi_j^{-1} \in C^{k-1,1}(B)$.

We let $\Omega_0 \subset\subset \Omega$ be a subdomain of Ω such that $\{\Omega_j\}, j = 0, \ldots, N$, is a finite covering of Ω, and let $\eta_j, j = 0, \ldots, N$ be a partition of unity subordinate to this covering. Then $(\eta_j u) \circ \psi_j^{-1} \in W^{k,p}(\mathbb{R}^n_+)$ (Problem 7.5) and hence $E_0[(\eta_j u) \circ \psi_j^{-1}] \in W^{k,p}(\mathbb{R}^n)$, whence $E_0[(\eta_j u) \circ \psi_j^{-1}] \circ \psi_j \in W_0^{k,p}(\Omega_j), j = 1, \ldots, N$, since $\operatorname{supp} \eta_j \subset \Omega_j$, Thus the mapping E defined for $u \in W^{k,p}(\Omega)$ by

$$Eu = u\eta_0 + \sum_{j=1}^{N} E_0[(\eta_j u) \circ \psi_j^{-1}] \circ \psi_j \tag{7.58}$$

satisfies $Eu \in W_0^{k,p}(\Omega')$, $Eu = u$ in Ω and

$$\|Eu\|_{k,p;\Omega'} \leqslant C\|u\|_{k,p;\Omega}$$

where $C = C(k, N, \psi_j, \eta_j) = C(k, \Omega, \Omega')$. Furthermore $(Eu)_h \to u$ in $W^{k,p}(\Omega)$ as $h \to 0$. □

By combining the case $k = 1$ in Theorem 7.25 with our previous imbedding results, Theorems 7.10, 7.12 and 7.22, we obtain corresponding imbedding results for the Sobolev spaces $W^{1,p}(\Omega)$ for Lipschitz domains Ω. By iteration we then have the following general imbedding theorem for $W^{k,p}(\Omega)$.

Theorem 7.26. *Let Ω be a $C^{0,1}$ domain in $\mathbb{R}^n$. Then,*

(i) if $kp < n$, the space $W^{k,p}(\Omega)$ is continuously imbedded in $L^{p^*}(\Omega)$, $p^* = np/(n - kp)$, and compactly imbedded in $L^q(\Omega)$ for any $q < p^*$;

(ii) if $0 \leqslant m < k - \dfrac{n}{p} < m + 1$, the space $W^{k,p}(\Omega)$ is continuously imbedded in $C^{m,\alpha}(\overline{\Omega})$, $\alpha = k - n/p - m$, and compactly imbedded in $C^{m,\beta}(\overline{\Omega})$ for any $\beta < \alpha$.

We turn now to interpolation inequalities which we treat initially for the spaces $W_0^{k,p}(\Omega)$.

Theorem 7.27. *Let $u \in W_0^{k,p}(\Omega)$. Then for any $\varepsilon > 0$, $0 < |\beta| < k$,*

$$\|D^\beta u\|_{p;\Omega} \leqslant \varepsilon\|u\|_{k,p;\Omega} + C\varepsilon^{|\beta|/(|\beta|-k)}\|u\|_{p;\Omega}, \tag{7.59}$$

where $C = C(k)$.

Proof. We establish (7.59) for the case $|\beta| = 1, k = 2$ which is needed in Chapter 9. A suitable induction argument yields the stated result for arbitrary β, k.

Let us first suppose $u \in C_0^2(\mathbb{R})$ and consider an interval (a, b) of length $b - a = \varepsilon$. For $x' \in (a, a + \varepsilon/3)$, $x'' \in (b - (\varepsilon/3), b)$, we have, by the mean valuc theorem,

$$\begin{aligned} |u'(\bar{x})| &= \left|\frac{u(x') - u(x'')}{x' - x''}\right| \\ &\leqslant \frac{3}{\varepsilon}(|u(x')| + |u(x'')|) \end{aligned}$$

for some $\bar{x} \in (a, b)$. Consequently for any $x \in (a, b)$,

$$|u'(x)| \leqslant \frac{3}{\varepsilon}(|u(x')| + |u(x'')|) + \int_a^b |u''|.$$

Integrating with respect to x' and x'', over the intervals $(a, a + \varepsilon/3)$, $(b - \varepsilon/3, b)$, respectively, we then obtain

$$|u'(x)| \leqslant \int_a^b |u''| + \frac{18}{\varepsilon^2} \int_a^b |u|,$$

so that by Hölder's inequality

$$|u'(x)|^p \leqslant 2^{p-1}\left\{\varepsilon^{p-1} \int_a^b |u''|^p + \frac{(18)^p}{\varepsilon^{p+1}} \int_a^b |u|^p\right\}.$$

Hence, integrating with respect to x over (a, b) we have

$$\int_a^b |u'(x)|^p \leqslant 2^{p-1}\left\{\varepsilon^p \int_a^b |u''|^p + \left(\frac{18}{\varepsilon}\right)^p \int_a^b |u|^p\right\}.$$

Consequently if we subdivide $\mathbb{R}$ into intervals of length ε, we obtain by adding all such inequalities

$$\int |u'|^p \leqslant 2^{p-1}\left\{\varepsilon^p \int |u''|^p + \left(\frac{18}{\varepsilon}\right)^p \int |u|^p\right\} \tag{7.60}$$

which is the desired result in the one-dimensional case. To extend to higher dimensions we fix i, $1 \leqslant i \leqslant n$, and apply (7.60) to $u \in C_0^2(\Omega)$ regarded as a function of x_i only. By successive integration over the remaining variables we thus obtain

$$\int |D_i u|^p \leqslant 2^{p-1}\left\{\varepsilon^p \int |D_{ii} u|^p + \left(\frac{18}{\varepsilon}\right)^p \int |u|^p\right\},$$

so that

$$\|D_i u\|_p \leqslant \varepsilon \|D_{ii} u\|_p + \frac{C}{\varepsilon} \|u\|_p$$

for $C = 36$. □

By combining Theorems 7.25 and 7.27, we obtain interpolation inequalities for the Sobolev spaces $W^{k,p}(\Omega)$.

Theorem 7.28. *Let Ω be a $C^{1,1}$ domain in $\mathbb{R}^n$ and $u \in W^{k,p}(\Omega)$. Then for any $\varepsilon > 0$, $0 < |\beta| < k$,*

$$\|D^\beta u\|_{p;\Omega} \leqslant \varepsilon \|u\|_{k,p;\Omega} + C\varepsilon^{|\beta|/(|\beta|-k)}\|u\|_{p;\Omega} \tag{7.61}$$

where $C = C(k, \Omega)$.

Alternative derivations of interpolation inequalities are treated in Problems 2.15, 7.18 and 7.19. The density, extension, imbedding, and interpolation results of Theorems 7.25, 7.26 and 7.28 are all valid under less restrictive hypotheses on the domains Ω; (see [AD]).

Notes

For related material on Sobolev spaces the reader is referred to the books [AD], [FR], [MY 5] and [NE]. We have followed the custom of referring to the spaces of this chapter as Sobolev spaces although various notions of spaces of weakly differentiable functions were used prior to Sobolev's work [SO 1]; (in this regard see [MY 1] and [MY 5]). The process of mollification or regularization appeared in Friedrich's work [FD 1]. The density theorem, Theorem 7.9, is due to Meyers and Serrin [MS 2]. The Sobolev inequalities, Theorem 7.10, were essentially proved by Sobolev [SO 1, 2]; we have followed the proof of Nirenberg [NI 3] for the case $p < n$. The Hölder estimates, Theorems 7.17 and 7.19 were derived by Morrey [MY 1]. Theorem 7.21 is due to John and Nirenberg [JN]; our proof is taken from [TR 2] where also the estimate Theorem 7.15 appeared. The compactness result, Theorem 7.22, is due to Rellich [RE] in the case $p = 2$ and to Kondrachov [KN] for the general case.

Problems

7.1. Let Ω be a bounded domain in $\mathbb{R}^n$. If u is a measurable function on Ω such that $|u|^p \in L^1(\Omega)$ for some $p \in \mathbb{R}$, we define

$$\Phi_p(u) = \left[\frac{1}{|\Omega|}\int_\Omega |u|^p\,dx\right]^{1/p}.$$

Show that: (i) $\lim_{p\to\infty} \Phi_p(u) = \sup_\Omega |u|$;

(ii) $\lim_{p\to-\infty} \Phi_p(u) = \inf_\Omega |u|$;

(iii) $\lim_{p\to 0} \Phi_p(u) = \exp\left[\frac{1}{|\Omega|}\int_\Omega \log|u|\,dx\right]$.

7.2. Show that a function u is weakly differentiable in a domain Ω if and only if it is weakly differentiable in a neighborhood of every point in Ω.

7.3. Let α, β be multi-indices and u be a locally integrable function on a domain Ω. Show that provided any two of the weak derivatives $D^{\alpha+\beta}u$, $D^{\alpha}(D^{\beta}u)$, $D^{\beta}(D^{\alpha}u)$ exist, they all exist and coincide a.e. (Ω).

7.4. Derive the product formula (7.18). (Hint: consider first the case, $u \in W^1(\Omega)$, $v \in C^1(\Omega)$).

7.5. Derive the formula (7.19), and show that it remains valid if we assume only $\psi \in C^{0,1}(\Omega)$, $\psi^{-1} \in C^{0,1}(\tilde{\Omega})$.

7.6. Let Ω be a domain in $\mathbb{R}^n$ containing the origin. Show that the function γ given by $\gamma(x)=|x|^{-\alpha}$ belongs to $W^k(\Omega)$ provided $k+\alpha<n$.

7.7. Let Ω be a domain in $\mathbb{R}^n$. Show that a function $u \in C^{0,1}(\Omega)$ if and only if u is weakly differentiable with locally bounded weak derivatives.

7.8. Let Ω be a domain in $\mathbb{R}^n$. Show that a function u is weakly differentiable in Ω if and only if it is equivalent to a function $\bar{u}$ that is absolutely continuous on almost all line segments in Ω parallel to the coordinate axes and whose partial derivatives, (which consequently exist a.e. (Ω)), are locally integrable in Ω. (See [MY 5], p. 66). Derive from this characterization the product formula and chain rule for weak differentiation.

7.9. Show that the norms (7.22) and (7.23) are equivalent norms on $W^{k,p}(\Omega)$.

7.10. Prove that the space $W^{k,p}(\Omega)$ is complete under either of the norms (7.22), (7.23).

7.11. Let Ω be a domain whose boundary can be locally represented as the graph of a Lipschitz continuous function. Show that $C^{\infty}(\bar{\Omega})$ is dense in $W^{k,p}(\Omega)$ for $1 \leqslant p < \infty, k \geqslant 1$, and compare this result with the density result in Theorem 7.25.

7.12. Let Ω be a $C^{0,1}$ domain. For any function $u \in W^{1,p}(\Omega)$ and $1 \leqslant p < n$, derive the Sobolev–Poincaré inequality

$$\|u - u_{\Omega}\|_{np/(n-p);\Omega} \leqslant C\|Du\|_{p;\Omega}$$

(where C is independent of u) by a contradiction argument based on the compactness result of Theorem 7.26.

7.13. Deduce from Theorem 7.19 the corresponding global result. Namely let $u \in W^1(\Omega)$, $\partial\Omega \in C^{0,1}$ and suppose there exist positive constants K, α $(\alpha<1)$ such that

$$\int_{B_R} |Du|\,dx \leqslant KR^{n-1+\alpha} \quad \text{for all balls } B_R \subset \mathbb{R}^n.$$

Then $u \in C^{0,\alpha}(\bar{\Omega})$ and

$$[u]_{\alpha;\Omega} \leq CK$$

where $C = C(n, \alpha, \Omega)$.

7.14. Let Ω be a bounded domain for which an imbedding

$$W^{1,p}(\Omega) \to L^{p^*}(\Omega), \quad 1 \leq p < \infty,$$

is valid. Show that the imbedding

$$W^{1,p}(\Omega) \to L^{q}(\Omega)$$

is compact for any $q < p^*$.

7.15. Let Ω be a domain in $\mathbb{R}^n$. The *total variation* of a function $u \in L^1(\Omega)$ is defined by

$$\int_\Omega |Du| = \sup\left\{\int_\Omega u \operatorname{div} \mathbf{v} \mid \mathbf{v} \in C_0^1(\Omega), |\mathbf{v}| \leq 1\right\}.$$

Show that the space $BV(\Omega)$ of functions of finite total variation is a Banach space under the norm

$$\|u\|_{BV(\Omega)} = \|u\|_1 + \int_\Omega |Du|,$$

and that $W^{1,1}(\Omega)$ is a closed subspace.

7.16. Let $u \in BV(\Omega)$. By invoking the regularization of u and appropriately modifying the proof of Theorem 7.9, show that there exists a sequence $\{u_m\} \subset C^\infty(\Omega) \cap W^{1,1}(\Omega)$ such that $u_m \to u$ in $L^1(\Omega)$ and

$$\int_\Omega |Du_m| \to \int_\Omega |Du|.$$

7.17 Let Ω be a bounded domain for which the Sobolev imbedding

$$W^{1,1}(\Omega) \to L^{n/(n-1)}(\Omega)$$

is valid. Show that also

$$BV(\Omega) \to L^{n/(n-1)}(\Omega)$$

and furthermore that the imbedding

$$BV(\Omega) \to L^q(\Omega)$$

is compact for any $q < n/(n-1)$.

7.18. Derive Theorem 7.27 for $p \geqslant 2$ from Green's first identity (2.10); (see Problem 2.15).

7.19. Let Ω be a $C^{0,1}$ domain. Derive the interpolation inequality (7.61) in the weaker form

$$\|D^\beta u\|_{p;\Omega} \leqslant \varepsilon\|u\|_{k,p;\Omega} + C_\varepsilon\|u\|_{p;\Omega},$$

(C_ε independent of u), by means of a contradiction argument based on the compactness result of Theorem 7.26.

7.20. Using regularization, show that locally integrable solutions of Laplace's equation (in the sense of Problem 2.8) are smooth and hence deduce the validity of the interior estimates in Chapter 4 for such solutions of Poisson's equation.

7.21. Using Morrey's inequality (7.42), prove that functions in the Sobolev space $W^{1,p}(\Omega)$, where $p > n$, are classically differentiable almost everywhere in Ω.

Chapter 8

Generalized Solutions and Regularity

This chapter treats linear elliptic operators having principal part in divergence form under relatively weak smoothness assumptions on the coefficients. We consider operators L of the form

$$Lu = D_i(a^{ij}(x)D_ju + b^i(x)u) + c^i(x)D_iu + d(x)u \tag{8.1}$$

whose coefficients a^{ij}, b^i, c^i, d $(i, j=1, \ldots, n)$ are assumed to be measurable functions on a domain $\Omega \subset \mathbb{R}^n$. An operator L of the general form (3.1) may be written in the form (8.1) provided its principal coefficients a^{ij} are differentiable. The Hilbert space approach developed here can then be viewed as providing an alternative existence theory to that of Chapter 6. On the other hand, if in (8.1) the coefficients a^{ij} and b^i are differentiable and the function $u \in C^2(\Omega)$, then L may be written in the general form (3.1) so that the theory of Chapter 6 would apply. The divergence form however has the advantage that the operator L may be defined for significantly broader classes of functions than the class $C^2(\Omega)$. Indeed, if we assume that the function u is only weakly differentiable and that the functions $a^{ij}D_ju + b^iu$, $c^iD_iu + du$, $i=1, \ldots, n$ are locally integrable, then, in a *weak* or *generalized* sense, u is said to satisfy $Lu=0$ $(\geqslant 0, \leqslant 0)$ respectively in Ω according as

$$\mathfrak{L}(u, v) = \int_\Omega \{(a^{ij}D_ju + b^iu)D_iv - (c^iD_iu + du)v\}\,dx = 0 \; (\leqslant 0, \geqslant 0) \tag{8.2}$$

for all non-negative functions $v \in C_0^1(\Omega)$. Provided the coefficients of L are locally integrable, it follows from the divergence theorem (2.3) that a function $u \in C^2(\Omega)$ satisfying $Lu=0$ $(\geqslant 0, \leqslant 0)$ in the classical sense also satisfies these relations in the generalized sense. Moreover, if the coefficients a^{ij}, b^i have locally integrable derivatives, then a generalized solution $u \in C^2(\Omega)$ is also a classical solution.

Let f^i, g, $i=1, \ldots, n$ be locally integrable functions in Ω. Then a weakly differentiable function u will be called a *weak* or *generalized* solution of the inhomogeneous equation

$$Lu = g + D_if^i \tag{8.3}$$

in Ω if

(8.4) $$\mathfrak{L}(u, v) = F(v) = \int_{\Omega} (f^i D_i v - gv)\, dx \quad \forall v \in C_0^1(\Omega).$$

As above we see that classical solutions of (8.3) are also generalized solutions and that a $C^2(\Omega)$ generalized solution is also a classical solution when the coefficients of L are sufficiently smooth.

Our plan is to study the generalized Dirichlet problem for the equation (8.3). The sense in which this problem is naturally posed depends on the coefficients of L. We shall assume throughout that L is strictly elliptic in Ω; that is, there exists a positive number λ such that

(8.5) $$a^{ij}(x)\xi_i\xi_j \geq \lambda|\xi|^2, \quad \forall x \in \Omega,\ \xi \in \mathbb{R}^n.$$

We also assume (unless stated otherwise) that L has bounded coefficients; that is for some constants Λ and $\nu \geq 0$ we have for all $x \in \Omega$

(8.6) $$\sum |a^{ij}(x)|^2 \leq \Lambda^2, \qquad \lambda^{-2} \sum (|b^i(x)|^2 + |c^i(x)|^2) + \lambda^{-1}|d(x)| \leq \nu^2.$$

We point out however that a satisfactory theory can still be developed if these conditions are relaxed [TR 7]. A function u belonging to the Sobolev space $W^{1,2}(\Omega)$ will then be called a solution of the *generalized Dirichlet problem:* $Lu = g + D_i f^i$, $u = \varphi$ on $\partial\Omega$, if u is a generalized solution of equation (8.3), $\varphi \in W^{1,2}(\Omega)$ and $u - \varphi \in W_0^{1,2}(\Omega)$.

The functions $v \in C_0^1(\Omega)$ that occur in the formulations (8.2) and (8.4) are often referred to as *test functions*. Note that by condition (8.6) we have

(8.7) $$|\mathfrak{L}(u, v)| \leq \int_{\Omega} \{|a^{ij}D_j u D_i v| + |b^i u D_i v| + |c^i v D_i u| + |duv|\}\, dx$$

$$\leq C\, \|u\|_{W^{1,2}(\Omega)} \|v\|_{W^{1,2}(\Omega)} \quad \text{by Schwarz's inequality}$$

Hence for fixed $u \in W^{1,2}(\Omega)$, the mapping $v \to \mathfrak{L}(u, v)$ is a bounded linear functional on $W_0^{1,2}(\Omega)$. Consequently the validity of the relations (8.2) for $v \in C_0^1(\Omega)$ implies their validity for $v \in W_0^{1,2}(\Omega)$.

The estimate (8.7) is also significant from the point of view of the existence theory for (8.3) as it shows that the operator L defines through (8.2) a bounded bilinear form on each of the Hilbert spaces $W^{1,2}(\Omega)$, $W_0^{1,2}(\Omega)$. For fixed $u \in W^{1,2}(\Omega)$, Lu may be defined as an element of the dual space of $W_0^{1,2}(\Omega)$ by setting $Lu(v) = \mathfrak{L}(u, v)$, $v \in W_0^{1,2}(\Omega)$. By virtue of the Riesz representation theorem, $W_0^{1,2}(\Omega)$ may be identified with its dual, and consequently the operator L induces a mapping $W^{1,2}(\Omega) \to W_0^{1,2}(\Omega)$. As we shall show presently, the solvability of the Dirichlet problem for equation (8.3) is readily reduced to the invertibility of this mapping.

The alternative approach to the linear Dirichlet problem described above is by no means the only important contribution of this chapter. The pointwise estimates developed in Sections 8.6, 8.9 and 8.10 are crucial for the subsequent development of the theory of quasilinear equations in Part II. For the purposes of this application, the reader need only consider $C^1(\bar{\Omega})$ subsolutions or supersolutions of equation (8.3) and moreover take $b^i = c^i = d = 0$ in (8.1), that is $\nu = 0$ in (8.6).

8.1. The Weak Maximum Principle

The classical weak maximum principle, Theorem 3.1, has a natural extension to operators in divergence form. In order to formulate it, we require a notion of inequality at the boundary for functions in the Sobolev space $W^{1,2}(\Omega)$. Namely, let us say that $u \in W^{1,2}(\Omega)$ satisfies $u \leqslant 0$ on $\partial\Omega$ if its positive part $u^+ = \max\{u, 0\} \in W_0^{1,2}(\Omega)$. If u is continuous in a neighborhood of $\partial\Omega$, then u satisfies $u \leqslant 0$ on $\partial\Omega$ if the inequality holds in the classical pointwise sense. Other definitions of inequality at $\partial\Omega$ follow naturally. For example: $u \geqslant 0$ on $\partial\Omega$ if $-u \leqslant 0$ on $\partial\Omega$; $u \leqslant v \in W^{1,2}(\Omega)$ on $\partial\Omega$ if $u - v \leqslant 0$ on $\partial\Omega$;

$$\sup_{\partial\Omega} u = \inf\{k \mid u \leqslant k \text{ on } \partial\Omega, k \in \mathbb{R}\}; \qquad \inf_{\partial\Omega} u = -\sup_{\partial\Omega}(-u).$$

For the classical weak maximum principle of Corollary 3.2, we imposed the condition that the coefficient of u in (3.1) is non-positive. The corresponding quantity in (8.1) is $D_i b^i + d$ but since the derivatives $D_j b^i$ need not exist as functions, the non-positivity of this term must be interpreted in a generalized sense, that is, we assume

$$\int_\Omega (dv - b^i D_i v)\, dx \leqslant 0 \quad \forall v \geqslant 0,\ v \in C_0^1(\Omega). \tag{8.8}$$

Since b^i and d are bounded, inequality (8.8) will continue to hold for all non-negative $v \in W_0^{1,1}(\Omega)$.

We can now state the following weak maximum principle.

Theorem 8.1. *Let $u \in W^{1,2}(\Omega)$ satisfy $Lu \geqslant 0$ $(\leqslant 0)$ in Ω. Then*

$$\sup_\Omega u \leqslant \sup_{\partial\Omega} u^+ \quad (\inf_\Omega u \geqslant \inf_{\partial\Omega} u^-). \tag{8.9}$$

Proof. If $u \in W^{1,2}(\Omega)$, $v \in W_0^{1,2}(\Omega)$ we have $uv \in W_0^{1,1}(\Omega)$ and $Duv = vDu + uDv$ (Problem 7.4). We may then write the inequality $\mathfrak{L}(u, v) \leqslant 0$ in the form

$$\int_\Omega \{a^{ij} D_j u D_i v - (b^i + c^i) v D_i u\}\, dx \leqslant \int_\Omega \{duv - b^i D_i(uv)\}\, dx \leqslant 0$$

for all $v \geqslant 0$ such that $uv \geqslant 0$, (by (8.8)). Hence, by the coefficient bounds (8.6), we have

$$\int_\Omega a^{ij} D_j u D_i v\, dx \leqslant 2\lambda\nu \int_\Omega v|Du|\, dx \tag{8.10}$$

for all $v \geqslant 0$ such that $uv \geqslant 0$. In the special case $b^i + c^i = 0$, the proof is immediate by taking $v = \max\{u - l, 0\}$ where $l = \sup_{\partial\Omega} u^+$. For the general case, we choose k to satisfy $l \leqslant k < \sup_{\Omega} u$, and we set $v = (u-k)^+$. (If no such k exists we are done.) By the chain rule, Theorem 7.8, we have $v \in W_0^{1,2}(\Omega)$ and

$$Dv = \begin{cases} Du & \text{for } u > k \quad (\text{i.e. for } v \neq 0), \\ 0 & \text{for } u \leqslant k \quad (\text{i.e. for } v = 0). \end{cases}$$

Consequently we obtain from (8.10)

$$\int_{\Omega} a^{ij} D_j v D_i v \, dx \leqslant 2\lambda\nu \int_{\Gamma} v|Dv| \, dx, \qquad \Gamma = \operatorname{supp} Dv \subset \operatorname{supp} v,$$

and hence by the strict ellipticity of L, (8.5),

$$\int_{\Omega} |Dv|^2 \, dx \leqslant 2\nu \int_{\Gamma} v|Dv| \, dx \leqslant 2\nu \, \|v\|_{2;\Gamma} \, \|Dv\|_2,$$

so that

$$\|Dv\|_2 \leqslant 2\nu \, \|v\|_{2;\Gamma}$$

Let us now apply the Sobolev inequality, Theorem 7.10, for $n \geqslant 3$, to obtain

$$\|v\|_{2n/(n-2)} \leqslant C\|v\|_{2;\Gamma} \leqslant C|\operatorname{supp} Dv|^{1/n}\|v\|_{2n/(n-2)}$$

where $C = C(n, \nu)$, so that

$$|\operatorname{supp} Dv| \geqslant C^{-n}.$$

In the case $n=2$, an inequality of the same form with $C = C(n, \nu, |\Omega|)$ also follows from the Sobolev inequality by replacing $2n/(n-2)$ by any number greater than 2. Since these inequalities are independent of k they must hold as k tends to $\sup_{\Omega} u$. That is, the function u must attain its supremum in Ω on a set of positive measure, where at the same time $Du = 0$ (by Lemma 7.7). This contradiction of the preceding inequality implies $\sup_{\Omega} u \leqslant l$. □

The uniqueness of solutions of the generalized Dirichlet problem for equation (8.3) is an immediate consequence of Theorem 8.1.

Corollary 8.2. *Let $u \in W_0^{1,2}(\Omega)$ satisfy $Lu = 0$ in Ω. Then $u = 0$ in Ω.*

For alternative conditions to inequality (8.8), the reader is referred to Problem 8.1; see also [TR 11].

8.2. Solvability of the Dirichlet Problem

The main objective of this section is the following existence result.

Theorem 8.3. *Let the operator L satisfy conditions* (8.5), (8.6) *and* (8.8). *Then for $\varphi \in W^{1,2}(\Omega)$ and $g, f^i \in L^2(\Omega)$, $i=1,\ldots,n$, the generalized Dirichlet problem, $Lu=g+D_if^i$ in Ω, $u=\varphi$ on $\partial\Omega$ is uniquely solvable.*

Proof. Theorem 8.3 will be derived as a byproduct of a Fredholm alternative for the operator L. Let us first reduce the Dirichlet problem to the case of zero boundary values. Setting $w=u-\varphi$, we obtain from (8.3)

$$\begin{aligned} Lw &= Lu - L\varphi \\ &= g - c^iD_i\varphi - d\varphi + D_i(f^i - a^{ij}D_j\varphi - b^i\varphi) \\ &= \hat{g} + D_i\hat{f}^i \end{aligned}$$

and from our conditions on L and φ, we clearly have $\hat{g}, \hat{f}^i \in L^2(\Omega)$, $i=1,\ldots,n$ and $w \in W_0^{1,2}(\Omega)$. Therefore it suffices to prove Theorem 8.3 for the case $\varphi \equiv 0$.

Let us write $\mathscr{H} = W_0^{1,2}(\Omega)$, $\mathbf{g}=(g, f^1, \ldots, f^n)$ and $F(v) = -\int_\Omega (gv - f^iD_iv)\,dx$ for $v \in \mathscr{H}$. Then since

$$|F(v)| \leqslant \|\mathbf{g}\|_2 \|v\|_{W^{1,2}(\Omega)}$$

we have $F \in \mathscr{H}^*$. If the bilinear form $\mathfrak{L}$ defined by (8.2) were coercive on $\mathscr{H}$ as well as bounded, we could conclude immediately the unique solvability of the Dirichlet problem for L from Theorem 5.8. Related to the coercivety of $\mathfrak{L}$ is the following.

Lemma 8.4. *Let L satisfy conditions* (8.5) *and* (8.6). *Then*

$$\mathfrak{L}(u,u) \geqslant \frac{\lambda}{2}\int_\Omega |Du|^2\,dx - \lambda\nu^2 \int_\Omega u^2\,dx. \tag{8.11}$$

Proof.

$$\begin{aligned} \mathfrak{L}(u,u) &= \int_\Omega (a^{ij}D_iuD_ju + (b^i - c^i)uD_iu - du^2)\,dx \\ &\geqslant \int_\Omega \left(\lambda|Du|^2 - \frac{\lambda}{2}|Du|^2 - \lambda\nu^2u^2\right)dx \quad \text{by Schwarz's inequality,} \\ &= \frac{\lambda}{2}\int_\Omega |Du|^2\,dx - \lambda\nu^2\int_\Omega u^2\,dx. \quad \square \end{aligned}$$

For $\sigma \in \mathbb{R}$, let us now define the operators L_σ by $L_\sigma u = Lu - \sigma u$. By Lemma 8.4 we see that the associated forms $\mathfrak{L}_\sigma$ will be coercive if either σ is sufficiently large or $|\Omega|$ is sufficiently small. To proceed further, we define an imbedding $I: \mathscr{H} \to \mathscr{H}^*$ by

$$(8.12) \qquad Iu(v) = \int_\Omega uv\,dx, \quad v \in \mathscr{H}.$$

Then we have

Lemma 8.5. *The mapping I is compact.*

Proof. We may write $I = I_1 I_2$ where $I_2 : \mathscr{H} \to L^2(\Omega)$ is the natural imbedding and $I_1 : L^2(\Omega) \to \mathscr{H}^*$ is given by (8.12). By the compactness result, Theorem 7.22, I_2 is compact (also if $p = n = 2$) and, since I_1 is clearly continuous, it follows that I is compact. □

To proceed further we choose σ_0 so that the form $\mathfrak{L}_{\sigma_0}$ is bounded and coercive on the Hilbert space $\mathscr{H}$. The equation $Lu = F$ for $u \in \mathscr{H}$, $F \in \mathscr{H}^*$ is then equivalent to the equation

$$L_{\sigma_0} u + \sigma_0 Iu = F.$$

By Theorem 5.8, $L_{\sigma_0}^{-1}$ is a continuous, one-to-one mapping of $\mathscr{H}^*$ onto $\mathscr{H}$ and so, applying it to the above equation, we obtain the equivalent equation

$$(8.13) \qquad u + \sigma_0 L_{\sigma_0}^{-1} Iu = L_{\sigma_0}^{-1} F.$$

The mapping $T = -\sigma_0 L_{\sigma_0}^{-1} I$ is compact by Lemma 8.5 and hence by the Fredholm alternative, Theorem 5.3, the existence of a function $u \in \mathscr{H}$ satisfying equation (8.13) is a consequence of the uniqueness in $\mathscr{H}$ of the trivial solution of the equation $Lu = 0$. Theorem 8.3 thus follows by the uniqueness result, Corollary 8.2. □

A description of the spectral behavior of the operator L follows from Theorem 5.11. For let us define the *formal adjoint* L^* of L by

$$(8.14) \qquad L^* u = D_i(a^{ji} D_j u - c^i u) - b^i D_i u + du.$$

Since $\mathfrak{L}^*(u, v) = \mathfrak{L}(v, u)$ for $u, v \in \mathscr{H} = W_0^{1,2}(\Omega)$ it follows that L^* is also the adjoint of L in the Hilbert space $\mathscr{H}$. By replacing L with L_σ in the above argument, we see that the equation $L_\sigma u = F$ will be equivalent to the equation $u + (\sigma_0 - \sigma) L_{\sigma_0}^{-1} Iu = L_{\sigma_0}^{-1} F$ and that the adjoint T_σ^* of the compact mapping $T_\sigma = (\sigma_0 - \sigma) L_{\sigma_0}^{-1} I$ is given by $T_\sigma^* = (\sigma_0 - \sigma)(L_{\sigma_0}^*)^{-1} I$. We can then apply Theorem 5.11 to obtain the following result.

Theorem 8.6. *Let the operator L satisfy conditions* (8.5) *and* (8.6). *Then there exists a countable, discrete set* $\Sigma \subset \mathbb{R}$ *such that if* $\sigma \notin \Sigma$, *the Dirichlet problems,* $L_\sigma u$,

$L_\sigma^* u = g + D_i f^i$, $u = \varphi$ on $\partial\Omega$, are uniquely solvable for arbitrary $g, f^i \in L^2(\Omega)$ and $\varphi \in W^{1,2}(\Omega)$. If $\sigma \in \Sigma$, then the subspaces of solutions of the homogeneous problems, $L_\sigma u$, $L_\sigma^* u = 0$, $u = 0$ on $\partial\Omega$ are of positive, finite dimension and the problem $L_\sigma u = g + D_i f^i$, $u = \varphi$ on $\partial\Omega$ is solvable if and only if

$$\int_\Omega \{(g - c^i D_i\varphi - d\varphi + \sigma\varphi)v - (f^i - a^{ij}D_j\varphi - b^i\varphi)D_i v\}\, dx = 0 \tag{8.15}$$

for all v satisfying $L_\sigma^* v = 0$, $v = 0$ on $\partial\Omega$. Furthermore if condition (8.8) *holds, then $\Sigma \subset (-\infty, 0)$.*

The operator $G_\sigma : \mathscr{H}^* \to \mathscr{H}$ given by $G_\sigma = L_\sigma^{-1}$ for $\sigma \notin \Sigma$ is called the *Green's operator* for the Dirichlet problem for L_σ. By Theorem 5.3, G_σ is a bounded linear operator on $\mathscr{H}^*$, consequently we have the following apriori estimate.

Corollary 8.7. *Let $u \in W^{1,2}(\Omega)$ satisfy $L_\sigma u = g + D_i f^i$, $u = \varphi$ on $\partial\Omega$ with $\sigma \notin \Sigma$. Then there exists a constant C depending only on L, σ and Ω such that*

$$\|u\|_{W^{1,2}(\Omega)} \leqslant C(\|\mathbf{g}\|_2 + \|\varphi\|_{W^{1,2}(\Omega)}). \tag{8.16}$$

It follows from Theorem 8.6 that Theorem 8.3 remains valid if we replace b^i by $-c^i$ in the condition (8.8).

8.3. Differentiability of Weak Solutions

The rest of this chapter is largely devoted to regularity considerations. We shall study in this section the existence of higher order weak derivatives of weak solutions of equation (8.3). With the aid of the differentiability results derived below, we shall deduce existence theorems for the classical Dirichlet problem from Theorem 8.3. In later sections we shall treat pointwise properties of weak solutions, such as the strong maximum principle and Hölder continuity. Our first regularity result provides conditions under which weak solutions of the equation $Lu = f$ are twice weakly differentiable.

Theorem 8.8. *Let $u \in W^{1,2}(\Omega)$ be a weak solution of the equation $Lu = f$ in Ω where L is strictly elliptic in Ω, the coefficients a^{ij}, b^i, $i, j = 1, \ldots, n$ are uniformly Lipschitz continuous in Ω, the coefficients c^i, d, $i = 1, \ldots, n$ are essentially bounded in Ω and the function f is in $L^2(\Omega)$. Then for any subdomain $\Omega' \subset\subset \Omega$, we have $u \in W^{2,2}(\Omega')$ and*

$$\|u\|_{W^{2,2}(\Omega')} \leqslant C(\|u\|_{W^{1,2}(\Omega)} + \|f\|_{L^2(\Omega)}) \tag{8.17}$$

for $C = C(n, \lambda, K, d')$, where λ is given by (8.5),

$$K = \max\{\|a^{ij}, b^i\|_{C^{0,1}(\bar\Omega)}, \|c^i, d\|_{L^\infty(\Omega)}\} \quad \textit{and} \quad d' = \operatorname{dist}(\Omega', \partial\Omega).$$

Furthermore u satisfies the equation

$$Lu = a^{ij}D_{ij}u + (D_j a^{ji} + b^i + c^i)D_i u + (D_i b^i + d)u = f \tag{8.18}$$

almost everywhere in Ω.

Proof. From the integral identity (8.4) we have

$$\int_\Omega a^{ij}D_j u D_i v \, dx = \int_\Omega gv \, dx \quad \forall v \in C_0^1(\Omega) \tag{8.19}$$

where $g \in L^2(\Omega)$ is given by

$$g = (b^i + c^i)D_i u + (D_i b^i + d)u - f. \tag{8.20}$$

For $|2h| < \text{dist}(\text{supp } v, \partial\Omega)$, let us replace v by its difference quotient $\Delta^{-h}v = \Delta_k^{-h}v$ for some k, $1 \leqslant k \leqslant n$. We then obtain

$$\begin{aligned} \int_\Omega \Delta^h(a^{ij}D_j u)D_i v \, dx &= -\int_\Omega a^{ij}D_j u D_i \, \Delta^{-h} v \, dx \\ &= -\int_\Omega g \, \Delta^{-h} v \, dx. \end{aligned}$$

Since

$$\Delta^h(a^{ij}D_j u)(x) = a^{ij}(x + he_k) \, \Delta^h D_j u(x) + \Delta^h a^{ij}(x) D_j u(x)$$

we then have

$$\int_\Omega a^{ij}(x + he_k)D_j \, \Delta^h u D_i v \, dx = -\int_\Omega (\bar{\mathbf{g}} \cdot Dv + g \, \Delta^{-h} v) \, dx$$

where $\bar{\mathbf{g}} = (\bar{g}^1, \ldots, \bar{g}^n)$ and $\bar{g}^i = \Delta^h a^{ij} D_j u$. Using (8.20) and Lemma 7.23, we can then estimate

$$\begin{aligned} \int_\Omega a^{ij}(x + he_k)D_j \, \Delta^h u D_i v \, dx &\leqslant (\|\bar{\mathbf{g}}\|_2 + \|g\|_2)\|Dv\|_2 \\ &\leqslant (C(n)K\|u\|_{W^{1,2}(\Omega)} + \|f\|_2)\|Dv\|_2. \end{aligned}$$

To proceed further let us take a function $\eta \in C_0^1(\Omega)$, satisfying $0 \leqslant \eta \leqslant 1$, and set $v = \eta^2 \Delta^h u$. We then obtain, using (8.5) and the Schwarz inequality,

$$\begin{aligned}
\lambda \int_\Omega |\eta D\Delta^h u|^2 \, dx &\leqslant \int_\Omega \eta^2 a^{ij}(x+he_k)\, \Delta^h D_i u\, \Delta^h D_j u \, dx \\
&= \int_\Omega a^{ij}(x+he_k) D_j\, \Delta^h u (D_i v - 2\, \Delta^h u \eta D_i \eta)\, dx \\
&\leqslant (C(n)K\|u\|_{W^{1,2}(\Omega)} + \|f\|_2)(\|\eta D\Delta^h u\|_2 + 2\|\Delta^h u D\eta\|_2) \\
&\quad + C(n)K\|\eta D\, \Delta^h u\|_2\, \|\Delta^h u D\eta\|_2 .
\end{aligned}$$

It then follows (with the help of Young's inequality (7.6)) that

$$\begin{aligned}
\|\eta\, \Delta^h Du\|_2 &\leqslant C(\|u\|_{W^{1,2}(\Omega)} + \|f\|_2 + \|\Delta^h u D\eta\|_2) \\
&\leqslant C(1 + \sup_\Omega |D\eta|)(\|u\|_{W^{1,2}(\Omega)} + \|f\|_2)
\end{aligned}$$

by Lemma 7.23, where $C = C(n, \lambda, K)$. The function η may now be chosen as a cut-off function such that $\eta = 1$ on $\Omega' \subset\subset \Omega$ and $|D\eta| \leqslant 2/d'$, where $d' = \text{dist}(\partial\Omega, \Omega')$. By Lemma 7.24 we obtain $Du \in W^{1,2}(\Omega')$ for any $\Omega' \subset\subset \Omega$, so that $u \in W^2(\Omega)$ and the estimate (8.17) holds. Finally, we have $Lu \in L^2_{\text{loc}}(\Omega)$ and clearly the integral identity (8.4) implies that $Lu = f$ almost everywhere in Ω. □

We note here (see Problem 8.2) that in the estimate (8.17), the quantity $\|u\|_{W^{1,2}(\Omega)}$ may be replaced by $\|u\|_{L^2(\Omega)}$.

The following general existence result for the Dirichlet problem for elliptic equations of the form

$$Lu \equiv a^{ij}(x) D_{ij} u + b^i(x) D_i u + c(x) u = f \tag{8.21}$$

can now be concluded from Theorems 8.3 and 8.8.

Theorem 8.9. *Let the operator L be strictly elliptic in Ω and have coefficients $a^{ij} \in C^{0,1}(\bar{\Omega})$, b^i, $c \in L^\infty(\Omega)$, $c \leqslant 0$. Then for arbitrary $f \in L^2(\Omega)$ and $\varphi \in W^{1,2}(\Omega)$, there exists a unique function $u \in W^{1,2}(\Omega) \cap W^{2,2}_{\text{loc}}(\Omega)$ satisfying $Lu = f$ in Ω and $u - \varphi \in W_0^{1,2}(\Omega)$.*

Theorem 8.9 continues to hold for sufficiently smooth $\partial\Omega$ with $\varphi \in W^{2,2}(\Omega)$ if we assume only that the principal coefficients a^{ij} are in $C^0(\bar{\Omega})$ (see Theorem 9.15). However, the uniqueness result will break down if the hypotheses are further weakened to allow discontinuous $a^{ij} \in L^\infty(\Omega)$, as is evidenced by the equation

$$\Delta u + b \frac{x_i x_j}{|x|^2} D_{ij} u = 0, \quad b = -1 + \frac{n-1}{1-\lambda}, \quad 0 < \lambda < 1. \tag{8.22}$$

which has for $n>2(2-\lambda)>2$ the two solutions $u_1(x)=1$, $u_2(x)=|x|^\lambda \in W^{2,2}(B)$ and agreeing on ∂B, where B is the unit ball, $B_1(0)$.

Further differentiability of weak solutions can be deduced readily from the proof of Theorem 8.8. For, suppose that we strengthen the smoothness conditions on the coefficients by assuming a^{ij}, $b^i \in C^{1,1}(\bar{\Omega})$, c^i, $d \in C^{0,1}(\bar{\Omega})$, together with $f \in W^{1,2}(\Omega)$. Then, replacing v by $D_k v$ for some k, $1 \leqslant k \leqslant n$, in the identity (8.19), we obtain on integration by parts

$$\int_\Omega a^{ij} D_{jk} u D_i v \, dx = \int_\Omega D_k \hat{g} v \, dx \quad \forall v \in C_0^1(\Omega), \tag{8.23}$$

and since $u \in W^{2,2}_{\text{loc}}(\Omega)$, we have $D_k \hat{g} \in L^2_{\text{loc}}(\Omega)$. Hence $D_k u \in W^{2,2}_{\text{loc}}(\Omega)$. By a straightforward induction argument, we can then conclude the following extension of Theorem 8.8.

Theorem 8.10. *Let $u \in W^{1,2}(\Omega)$ be a weak solution of the equation $Lu=f$ in Ω where L is strictly elliptic in Ω, the coefficients a^{ij}, $b^i \in C^{k,1}(\bar{\Omega})$, the coefficients c^i, $d \in C^{k-1,1}(\bar{\Omega})$ and the function $f \in W^{k,2}(\Omega)$, $k \geqslant 1$. Then for any subdomain $\Omega' \subset\subset \Omega$, we have $u \in W^{k+2,2}(\Omega')$ and*

$$\|u\|_{W^{k+2,2}(\Omega')} \leqslant C(\|u\|_{W^{1,2}(\Omega)} + \|f\|_{W^{k,2}(\Omega)}) \tag{8.24}$$

for $C=C(n, \lambda, K, d', k)$, where $K=\max\{\|a^{ij}, b^i\|_{C^{k,1}(\bar{\Omega})}, \|c^i, d\|_{C^{k-1,1}(\bar{\Omega})}\}$.

By the Sobolev imbedding theorem, Corollary 7.11, we now obtain from Theorem 8.10,

Corollary 8.11. *Let $u \in W^{1,2}(\Omega)$ be a weak solution of the strictly elliptic equation $Lu=f$ in Ω and suppose that the functions a^{ij}, b^i, c^i, d, f are in $C^\infty(\Omega)$. Then also $u \in C^\infty(\Omega)$.*

8.4. Global Regularity

Under appropriate smoothness conditions on the boundary $\partial\Omega$ the preceding interior regularity results can be extended to all of Ω. We first derive the global analogue of Theorem 8.8.

Theorem 8.12. *Let us assume, in addition to the hypotheses of Theorem 8.8, that $\partial\Omega$ is of class C^2 and that there exists a function $\varphi \in W^{2,2}(\Omega)$ for which $u-\varphi \in W_0^{1,2}(\Omega)$. Then we have also $u \in W^{2,2}(\Omega)$ and*

$$\|u\|_{W^{2,2}(\Omega)} \leqslant C(\|u\|_{L^2(\Omega)} + \|f\|_{L^2(\Omega)} + \|\varphi\|_{W^{2,2}(\Omega)}) \tag{8.25}$$

where $C=C(n, \lambda, K, \partial\Omega)$.

Proof. Replacing u by $u-\varphi$, we see that there is no loss of generality in assuming $\varphi\equiv 0$ and hence $u\in W_0^{1,2}(\Omega)$. Also by Lemma 8.4 we can estimate

$$(8.26)\qquad \|u\|_{W^{1,2}(\Omega)}\leqslant C(\|u\|_2+\|f\|_2)$$

where $C=C(n,\lambda,K)$. Since $\partial\Omega\in C^2$, there exists for each point $x_0\in\partial\Omega$, a ball $B=B(x_0)$ and a one-to-one mapping ψ from B onto an open set $D\subset\mathbb{R}^n$ such that $\psi(B\cap\Omega)\subset\mathbb{R}^n_+=\{x\in\mathbb{R}^n|x_n>0\}$, $\psi(B\cap\partial\Omega)\subset\partial\mathbb{R}^n_+$ and $\psi\in C^2(B)$, $\psi^{-1}\in C^2(D)$. Let $B_R(x_0)\subset\subset B$ and set $B^+=B_R(x_0)\cap\Omega$, $D'=\psi(B_R(x_0))$, $D^+=\psi(B^+)$. Under the mapping ψ the equation $Lu=f$ in B^+ is transformed to an equation of the same form in D^+ (see page 97). The constants λ, K for the transformed equation can be estimated in terms of the mapping ψ and their values for the original equation. Furthermore, since $u\in W_0^{1,2}(\Omega)$, the transformed solution $v=u\circ\psi^{-1}\in W^{1,2}(D^+)$ and satisfies $\eta v\in W_0^{1,2}(D^+)$ for all $\eta\in C_0^1(D')$. Accordingly, let us now suppose that $u\in W^{1,2}(D^+)$ satisfies $Lu=f$ in D^+ and $\eta u\in W_0^{1,2}(D^+)$ for any $\eta\in C_0^1(D')$. Then for $|h|<\text{dist}(\text{supp}.\ \eta,\partial D')$ and $1\leqslant k\leqslant n-1$, we have $\eta^2\Delta_k^h u\in W_0^{1,2}(D^+)$. Consequently the proof of Theorem 8.8 will apply and we can conclude that $D_{ij}u\in L^2(\psi(B_\rho\cap\Omega))$ for any $\rho<R$, provided i or $j\neq n$. The remaining second derivative $D_{nn}u$ can be estimated directly from the equation (8.18). Hence, returning to the original domain Ω with the mapping $\psi^{-1}\in C^2$ we obtain that $u\in W^{2,2}(B_\rho\cap\Omega)$. Since x_0 is an arbitrary point of $\partial\Omega$ and $u\in W^{2,2}_{\text{loc}}(\Omega)$ by Theorem 8.8, we infer that $u\in W^{2,2}(\Omega)$. Finally by choosing a finite number of points $x^{(i)}\in\partial\Omega$ such that the balls $B_\rho(x^{(i)})$ cover $\partial\Omega$, we obtain the estimate (8.25) from (8.17) and (8.26). □

Note that the conditions, $u\in W^{2,2}(D^+)$, $\eta u\in W_0^{1,2}(D^+)$ for $\eta\in C_0^1(D')$, imply that also $\eta D_k u\in W_0^{1,2}(D^+)$ provided $1\leqslant k\leqslant n-1$. Namely, by Lemma 7.23 we have $\eta\,\Delta_k^h u\in W_0^{1,2}(D^+)$ and

$$\|\eta\,\Delta_k^h u\|_{W^{1,2}(D^+)}\leqslant\|\eta\|_{C^1(D^+)}\,\|u\|_{W^{2,2}(D^+)}$$

for sufficiently small h. It follows, by Theorem 5.12, that there exists a sequence $\{\eta\,\Delta_k^{h_j}u\}$ converging weakly in the Hilbert space $W_0^{1,2}(D^+)$. The limit of this sequence is clearly the function $\eta D_k u$. Further global regularity of solutions of the equation $Lu=f$ then follows in the same manner as Theorem 8.10 from Theorem 8.8. Accordingly we have the following extension of Theorems 8.10 and 8.11.

Theorem 8.13. *Let us assume in addition to the hypotheses of Theorem* 8.10, *that* $\partial\Omega\in C^{k+2}$ *and that there exists a function* $\varphi\in W^{k+2,2}(\Omega)$ *for which* $u-\varphi\in W_0^{1,2}(\Omega)$. *Then we have also* $u\in W^{k+2,2}(\Omega)$ *and*

$$(8.27)\qquad \|u\|_{W^{k+2,2}(\Omega)}\leqslant C(\|u\|_{L^2(\Omega)}+\|f\|_{W^{k,2}(\Omega)}+\|\varphi\|_{W^{k+2,2}(\Omega)})$$

where $C=C(n,\lambda,K,k,\partial\Omega)$. *If the functions* a^{ij}, b^i, c^i, d, f *and* φ *belong to* $C^\infty(\bar\Omega)$ *and* $\partial\Omega$ *is of class* C^∞, *then the solution* u *is also in* $C^\infty(\bar\Omega)$.

Combining Theorems 8.3 and 8.13, we have an existence theorem for the classical Dirichlet problem for equation (8.21) that was previously obtained in Chapter 6 (see Theorems 6.14 and 6.19).

Theorem 8.14. *Let the operator L (given by (8.21)) be strictly elliptic in Ω and have $C^\infty(\bar{\Omega})$ coefficients with $c\leqslant 0$ in Ω. Then if $\partial\Omega\in C^\infty$, there exists a unique solution $u\in C^\infty(\bar{\Omega})$ of the Dirichlet problem, $Lu=f$, $u=\varphi$ on $\partial\Omega$ for arbitrary f, $\varphi\in C^\infty(\bar{\Omega})$.*

The existence theorems of Chapter 6 can now be obtained from Theorem 8.14 by approximation arguments. Of course we still require the apriori estimates of Chapter 6 to guarantee the convergence of the approximating solutions.

8.5. Global Boundedness of Weak Solutions

We derive here results asserting the global boundedness of $W^{1,2}(\Omega)$ solutions of equation (8.3) that are bounded on $\partial\Omega$. An interesting feature of the test function techniques to be used is that they depend not so much on the linearity of the operator L but rather on a *nonlinear structure* satisfied by L. To be more explicit, let us write (8.3) in the form

$$D_iA^i(x,u,Du)+B(x,u,Du)=0 \tag{8.28}$$

where

$$\begin{aligned} A^i(x,z,p)&=a^{ij}(x)p_j+b^i(x)z-f^i(x),\\ B(x,z,p)&=c^i(x)p_i+d(x)z-g(x), \end{aligned} \tag{8.29}$$

for $(x,z,p)\in\Omega\times\mathbb{R}\times\mathbb{R}^n$.

A weakly differentiable function u is then called a weak subsolution (supersolution, solution) of equation (8.28) in Ω if the functions $A^i(x,u,Du)$ and $B(x,u,Du)$ are locally integrable and

$$\int_\Omega (D_iv A^i(x,u,Du)-vB(x,u,Du))\,dx\leqslant(\geqslant,=)0 \tag{8.30}$$

for all $v\geqslant 0$, $\in C_0^1(\Omega)$.

Writing $\mathbf{b}=(b^1,\dots,b^n)$, $\mathbf{c}=(c^1,\dots,c^n)$, $\mathbf{f}=(f^1,\dots,f^n)$ and using condition (8.5) and the Schwarz inequality, we have the estimates

$$\begin{aligned} p_iA^i(x,z,p)&\geqslant\frac{\lambda}{2}|p|^2-\frac{1}{\lambda}(|\mathbf{b}z|^2+|\mathbf{f}|^2)\\ |B(x,z,p)|&\leqslant|\mathbf{c}|\,|p|+|dz|+|g|. \end{aligned} \tag{8.31}$$

Equation (8.3) is accordingly said to satisfy the structural inequalities (8.31). For our purposes below, we may even simplify the form of these inequalities by writing

$$\bar{z} = |z| + k, \qquad \bar{b} = \lambda^{-2}(|\mathbf{b}|^2 + |\mathbf{c}|^2 + k^{-2}|\mathbf{f}|^2) + \lambda^{-1}(|d| + k^{-1}|g|) \tag{8.32}$$

for some $k > 0$. We obtain then, for any $0 < \varepsilon < 1$,

$$\begin{aligned} p_i A^i(x, z, p) &\geqslant \frac{\lambda}{2}(|p|^2 - 2\bar{b}\bar{z}^2), \\ |\bar{z}B(x, z, p)| &\leqslant \frac{\lambda}{2}\left(\varepsilon|p|^2 + \frac{\bar{b}}{\varepsilon}\bar{z}^2\right). \end{aligned} \tag{8.33}$$

We now prove:

Theorem 8.15. *Let the operator L satisfy conditions* (8.5), (8.6) *and suppose that $f^i \in L^q(\Omega)$, $i = 1, \ldots, n$, $g \in L^{q/2}(\Omega)$ for some $q > n$. Then if u is a $W^{1,2}(\Omega)$ subsolution (supersolution) of equation* (8.3) *in Ω satisfying $u \leqslant 0$ ($\geqslant 0$) on $\partial\Omega$, we have*

$$\sup_{\Omega} u(-u) \leqslant C(\|u^+(u^-)\|_2 + k) \tag{8.34}$$

where $k = \lambda^{-1}(\|\mathbf{f}\|_q + \|g\|_{q/2})$ and $C = C(n, \nu, q, |\Omega|)$.

Proof. We assume that u is a subsolution of (8.3). For $\beta \geqslant 1$ and $N > k$, let us define a function $H \in C^1[k, \infty)$ by setting $H(z) = z^\beta - k^\beta$ for $z \in [k, N]$ and taking H to be linear for $z \geqslant N$. Let us next set $w = \bar{u}^+ = u^+ + k$ and take

$$v = G(w) = \int_k^w |H'(s)|^2\, ds \tag{8.35}$$

in the integral inequality (8.30). By the chain rule, Theorem 7.8, v is a legitimate test function in (8.30) and on substitution we obtain, using the structure (8.33),

$$\begin{aligned} \int_\Omega |Dw|^2 G'(w)\, dx &\leqslant \int_\Omega \left(\bar{b}G'(w)w^2 + \frac{2}{\lambda} G(w)|B(x, u, Du)|\right) dx \\ &\leqslant \varepsilon \int_\Omega G'(w)|Dw|^2\, dx + \left(1 + \frac{1}{\varepsilon}\right) \int_\Omega \bar{b}G'(w)w^2\, dx \end{aligned}$$

since $G(s) \leqslant sG'(s)$ and $Du = Dw$ when $v = G(w) > 0$. Hence, taking $\varepsilon = \frac{1}{2}$, we obtain

$$\int_\Omega G'(w)|Dw|^2\, dx \leqslant 6 \int_\Omega \bar{b}G'(w)w^2\, dx,$$

that is, by (8.35),

$$\int_\Omega |DH(w)|^2\,dx \leqslant 6\int_\Omega \bar{b}|H'(w)w|^2\,dx.$$

Since $H(w)\in W_0^{1,2}(\Omega)$, we may apply the Sobolev inequality (7.26) and the Hölder inequality to obtain

$$\|H(w)\|_{2\hat{n}/(\hat{n}-2)} \leqslant C\left(\int_\Omega \bar{b}(H'(w)w)^2\,dx\right)^{1/2}$$

$$\leqslant C\|\bar{b}\|_{q/2}^{1/2}\|H'(w)w\|_{2q/(q-2)}$$

where $\hat{n}=n$ for $n>2$, $2<\hat{2}<q$, $C=C(n)$ for $n>2$, and $C=C(\hat{2},|\Omega|)$ for $n=2$. It is clear that the structure (8.33) and consequently the above estimate continue to hold for $k=0$ provided in (8.32) the terms involving f and g are set equal to zero. Choosing k as in the statement of the theorem, we thus have

$$\|H(w)\|_{2\hat{n}/(\hat{n}-2)} \leqslant C\|wH'(w)\|_{2q/(q-2)} \tag{8.36}$$

where $C=C(n,\nu,|\Omega|)$. To proceed further, we recall the definition of H and let $N\to\infty$ in the estimate (8.36). It follows then, for any $\beta\geqslant 1$, that the inclusion $w\in L^{2\beta q/(q-2)}(\Omega)$ implies the stronger inclusion, $w\in L^{2\beta\hat{n}/(\hat{n}-2)}(\Omega)$, and moreover, setting $q^*=2q/(q-2)$, $\chi=\hat{n}(q-2)/q(\hat{n}-2)>1$, we obtain

$$\|w\|_{\beta\chi q^*} \leqslant (C\beta)^{1/\beta}\|w\|_{\beta q^*}. \tag{8.37}$$

The result is now obtained by iteration of the estimate (8.37). Namely, by induction, we may assume $w\in \bigcap_{1\leqslant p<\infty} L^p(\Omega)$. Let us take $\beta=\chi^m$, $m=0,1,2,\ldots$, so that by (8.37)

$$\|w\|_{\chi^N q^*} \leqslant \prod_0^{N-1}(C\chi^m)^{\chi^{-m}}\|w\|_{q^*}$$

$$\leqslant C^\sigma\chi^\tau\|w\|_{q^*},\quad \sigma=\sum_0^{N-1}\chi^{-m},\quad \tau=\sum_0^{N-1}m\chi^{-m}$$

$$\leqslant C\|w\|_{q^*}$$

where $C=C(n,\nu,q,|\Omega|)$. Letting $N\to\infty$, we therefore obtain

$$\sup_\Omega w \leqslant C\|w\|_{q^*},$$

whence by the interpolation inequality (7.10) we have

$$\sup_\Omega w \leqslant C\|w\|_2.$$

The desired estimate (8.34) follows from the definition $w=u^+ +k$. The result for supersolutions is obtained by replacing u with $-u$. □

The above technique of iteration of L^p norms was introduced by Moser [MJ 1]. The proof of Theorem 8.15 may also be effected by other choices of test functions; (see [LU 4] or [ST 4]).

Let us now suppose that in the statement of Theorem 8.5 the hypothesis $u\leqslant 0$ on $\partial\Omega$ is generalized to $u\leqslant l$ on $\partial\Omega$ for some constant l. Then, since $L(u-l)=Lu-Ll=Lu-l(D_ib^i+d)$, the conclusion of the theorem will hold for the function $u-l$ with k replaced by $\bar{k}=k+\lambda^{-1}|l|(\|\mathbf{b}\|_q+\|d\|_{q/2})$. That is, a subsolution (supersolution) u of (8.3) will satisfy an estimate

$$\sup_{\Omega} u(-u)\leqslant C(\|u\|_2+\bar{k}+|l|) \tag{8.38}$$

where as before $k=\lambda^{-1}(\|\mathbf{f}\|_q+\|g\|_{q/2})$ and $C=C(n, \nu, q, |\Omega|)$. In particular, if u is a solution then (8.38) holds for $|u|$.

We propose next to derive an estimate for $\sup_{\Omega} u$ independent of $\|u\|_2$, that is, an *apriori bound* which extends the weak maximum principle, Theorem 8.1. From the estimate (8.16), $\|u\|_2$ can be bounded independently of u for solutions of (8.3) provided L is one-to-one. This is the case, for example, if (8.8) holds. This bound may be extended to subsolutions through the weak maximum principle and the existence theorem, Theorem 8.3. For, if u is a subsolution of (8.3), and (8.8) holds, we may define a function v to be the solution of the generalized Dirichlet problem $Lv=g+D_if^i$, $v=u$ on $\partial\Omega$. By Theorem 8.1, $u\leqslant v$ in Ω and hence $\|u^+\|_2\leqslant\|v\|_2$. We therefore have an estimate

$$\sup_{\Omega} u\leqslant\sup_{\partial\Omega} u^+ +Ck$$

for subsolutions of (8.3), where C is a constant independent of u. However, we now show that this result can be derived from the *nonlinear* structure (8.31) without use of the linear existence theory and, moreover, that the constant C is determined by the same quantities as in the estimate (8.34).

Theorem 8.16. *Let the operator L satisfy conditions* (8.5), (8.6) *and* (8.8), *and suppose that $f^i\in L^q(\Omega)$, $i=1,\ldots,n$, $g\in L^{q/2}(\Omega)$ for some $q>n$. Then if u is a $W^{1,2}(\Omega)$ subsolution (supersolution) of equation* (8.3) *we have*

$$\sup_{\Omega} u(-u)\leqslant\sup_{\partial\Omega} u^+(u^-)+Ck \tag{8.39}$$

where $k=\lambda^{-1}(\|\mathbf{f}\|_q+\|g\|_{q/2})$ and $C=C(n, \nu, q, |\Omega|)$.

Proof. Let us suppose that u is a subsolution of (8.3). By the assumption (8.8), $l=\sup_{\partial\Omega} u^+$ is a supersolution and hence there is no loss of generality in assuming $l=0$. Proceeding as in the proof of Theorem 8.1, we have

$$\int_\Omega (a^{ij}D_j u D_i v-(b^i+c^i)vD_i u)\,dx\leqslant\int_\Omega (f^i D_i v-gv)\,dx \tag{8.40}$$

for all non-negative v in $W_0^{1,2}(\Omega)$ such that $uv\leqslant 0$. The weak inequality (8.40) clearly satisfies a structure condition (8.31) with $b^i=d=0$ and with $\mathbf{c}$ replaced by $\mathbf{b}+\mathbf{c}$. Let us assume $k>0$ and put $M=\sup_\Omega u^+$. In (8.40) we then choose the test function

$$v=\frac{u^+}{M+k-u^+}\in W_0^{1,2}(\Omega)$$

and obtain, using (8.31),

$$\frac{\lambda}{2}\int_\Omega\frac{|Du^+|^2\,dx}{(M+k-u^+)^2}\leqslant\frac{1}{M+k}\int_\Omega\left(\frac{|\mathbf{b}+\mathbf{c}|u^+|Du^+|}{(M+k-u^+)}+\frac{u^+|g|}{(M+k-u^+)}+\frac{(M+k)|\mathbf{f}|^2}{2\lambda(M+k-u^+)^2}\right)dx.$$

Consequently, by the definition of k, we have

$$\int_\Omega\frac{|Du^+|^2\,dx}{(M+k-u^+)^2}\leqslant C+\frac{2}{\lambda}\int_\Omega\frac{|\mathbf{b}+\mathbf{c}|\,|Du^+|}{(M+k-u^+)}\,dx,\quad C=C(|\Omega|).$$

Let us now define

$$w=\log\frac{M+k}{M+k-u^+},$$

so that from the Schwarz inequality we obtain

$$\int_\Omega|Dw|^2\,dx\leqslant C\Big(1+\lambda^{-2}\int_\Omega|\mathbf{b}+\mathbf{c}|^2\,dx\Big)$$

$$\leqslant C(\nu,|\Omega|),$$

and hence, by the Sobolev inequality (7.26),

$$\|w\|_2\leqslant C(n,\nu,|\Omega|). \tag{8.41}$$

The proof is completed by showing that w is also a subsolution of an equation of the form (8.3). Letting $\eta \in C_0^1(\Omega)$ satisfy $\eta \geqslant 0$, $\eta u \geqslant 0$ in Ω, we substitute in (8.40) the test function

$$v = \frac{\eta}{(M+k-u^+)}.$$

Then we obtain

$$\int_\Omega (a^{ij}D_jwD_i\eta + \eta a^{ij}D_iwD_jw - (b^i+c^i)\eta D_iw)\,dx \leqslant \int_\Omega \left(\frac{-\eta g}{(M+k-u^+)} + \frac{(D_i\eta + \eta D_iw)f^i}{(M+k-u^+)}\right) dx.$$

Therefore

$$\int_\Omega (a^{ij}D_jwD_i\eta - (b^i+c^i)\eta D_iw)\,dx + \lambda \int_\Omega \eta |Dw|^2\,dx \leqslant \int_\Omega \left\{\left(\frac{|g|}{k} + \frac{|\mathbf{f}|^2}{2\lambda k^2}\right)\eta + \frac{f^iD_i\eta}{(M+k-u^+)}\right\} dx + \frac{\lambda}{2}\int_\Omega \eta|Dw|^2\,dx$$

and consequently

$$\int_\Omega (a^{ij}D_jwD_i\eta - (b^i+c^i)\eta D_iw)\,dx \leqslant \int_\Omega (\hat{g}\eta + \hat{f}^iD_i\eta)\,dx \tag{8.42}$$

where $\hat{g} = |g|/k + |\mathbf{f}|^2/2\lambda k^2$, $\hat{f}^i = f^i/(M+k-u^+)$, and evidently $\|\hat{g}\|_{q/2} \leqslant 2\lambda$, $\|\hat{\mathbf{f}}\|_q \leqslant \lambda$. Hence we can apply Theorem 8.15 to obtain

$$\begin{aligned} \sup_\Omega w &\leqslant C(1 + \|w\|_2), \quad C = C(n, \nu, q, |\Omega|) \\ &\leqslant C \quad \text{by (8.41)}. \end{aligned}$$

Therefore $(M+k)/k \leqslant C$ and from this the desired estimate (8.39) follows. The result for supersolutions is obtained by replacing u with $-u$. □

Theorem 8.16 can be viewed as the generalized version of the classical apriori estimate, Theorem 3.7. We remark that the result is still valid if b^i is replaced by $-c^i$ in condition (8.8); (see Theorem 9.7). Furthermore, it is clear from the above proof that the boundedness of the coefficients b^i, c^i and d can be replaced by the condition $\bar{b} \in L^{q/2}(\Omega)$, $q > n$.

8.6. Local Properties of Weak Solutions

We shift our attention now from global to local behavior. Denoting the matrix $[a^{ij}(x)]$ by $\mathbf{a}(x)$, $x \in \Omega$, we add an additional structural inequality to (8.31) and (8.33), namely

$$|\mathbf{A}(x, z, p)| \leqslant |\mathbf{a}|\,|p| + |\mathbf{b}z| + |\mathbf{f}|. \tag{8.43}$$

By dividing equation (8.3) by the constant $\lambda/2$, we can assume that $\lambda = 2$ in the structural inequalities. Collecting these inequalities together under this assumption, we thus have,

$$\begin{aligned} &|\mathbf{A}(x, z, p)| \leqslant |\mathbf{a}|\,|p| + 2(\bar{b})^{1/2}\bar{z}, \\ &p \cdot \mathbf{A}(x, z, p) \geqslant |p|^2 - 2\bar{b}\bar{z}^2, \\ &|\bar{z}B(x, z, p)| \leqslant \varepsilon |p|^2 + \frac{1}{\varepsilon}\bar{b}\bar{z}^2, \end{aligned} \tag{8.44}$$

for any $0 < \varepsilon \leqslant 1$, where $\bar{z}$ and $\bar{b}$ are defined by (8.32) with $\lambda = 2$. For the development of the local results, we define the quantity k by

$$k = k(R) = \lambda^{-1}(R^{\delta}\|\mathbf{f}\|_q + R^{2\delta}\|g\|_{q/2}) \tag{8.45}$$

where $R > 0$ and $\delta = 1 - n/q$. We shall establish a local analogue of Theorem 8.15, namely:

Theorem 8.17. *Let the operator L satisfy conditions* (8.5), (8.6) *and suppose that $f^i \in L^q(\Omega)$, $i = 1, \ldots, n$, $g \in L^{q/2}(\Omega)$ for some $q > n$. Then if u is a $W^{1,2}(\Omega)$ subsolution (supersolution) of equation* (8.3) *in Ω, we have, for any ball $B_{2R}(y) \subset \Omega$ and $p > 1$,*

$$\sup_{B_R(y)} u(-u) \leqslant C(R^{-n/p}\|u^+(u^-)\|_{L^p(B_{2R}(y))} + k(R)) \tag{8.46}$$

where $C = C(n, \Lambda/\lambda, \nu R, q, p)$.

The crucial result in our development of local properties of weak solutions and the subsequent nonlinear theory will be the following *weak Harnack inequality* for supersolutions.

Theorem 8.18. *Let the operator L satisfy conditions* (8.5), (8.6) *and suppose that $f^i \in L^q(\Omega)$, $g \in L^{q/2}(\Omega)$ for some $q > n$. Then if u is a $W^{1,2}(\Omega)$ supersolution of equation* (8.3) *in Ω, non-negative in a ball $B_{4R}(y) \subset \Omega$ and $1 \leqslant p < n/(n-2)$, we have*

$$R^{-n/p}\|u\|_{L^p(B_{2R}(y))} \leqslant C(\inf_{B_R(y)} u + k(R)) \tag{8.47}$$

where $C = C(n, \Lambda/\lambda, \nu R, q, p)$.

In the following we shall abbreviate $B_R(y)=B_R$ for any R, the center y to be understood. For the case in Theorem 8.17 when u is a bounded, non-negative subsolution, it is convenient to prove Theorems 8.17 and 8.18 jointly. The full strength of Theorem 8.17 can then be obtained by varying the test functions used. The essence of this idea has already been demonstrated in the proof of Theorem 8.15. Accordingly, it is left to the reader to make the necessary extension of our proof below. Broadly speaking, the scheme of the joint proof follows the Moser iteration method (see [MJ 2]) introduced in the previous section combined with the John-Nirenberg result (Theorem 7.21), which is employed to bridge a vital gap in the iteration scheme. The test functions are again constructed from power functions but in order to establish Theorem 8.18 the exponents of these powers must be unrestricted real numbers. The detailed proof now follows.

We assume initially that $R=1$ and $k>0$. The general case is later recovered through a simple coordinate transformation: $x \to x/R$, and by letting k tend to zero. Let us define, for $\beta \neq 0$ and non-negative $\eta \in C_0^1(B_4)$, the test function

$$v=\eta^2\bar{u}^\beta \quad (\bar{u}=u+k). \tag{8.48}$$

By the chain and product rules, v is a valid test function in (8.30) and also

$$Dv=2\eta D\eta\bar{u}^\beta+\beta\eta^2\bar{u}^{\beta-1}Du, \tag{8.49}$$

so that by substitution into (8.30) we obtain

$$\begin{aligned}\beta\int_\Omega \eta^2\bar{u}^{\beta-1}Du\cdot \mathbf{A}(x,u,Du)\,dx+2\int_\Omega \eta D\eta\cdot \mathbf{A}(x,u,Du)\bar{u}^\beta\,dx\\ -\int_\Omega \eta^2\bar{u}^\beta B(x,u,Du)\,dx\\ \leqslant 0 \quad \text{if } u \text{ is a subsolution,}\\ \geqslant 0 \quad \text{if } u \text{ is a supersolution.}\end{aligned} \tag{8.50}$$

Using the structural inequalities (8.44), we can estimate, for any $0<\varepsilon\leqslant 1$,

$$\begin{aligned}\eta^2\bar{u}^{\beta-1}Du\cdot \mathbf{A}(x,u,Du) &\geqslant \eta^2\bar{u}^{\beta-1}|Du|^2-2\bar{b}\eta^2\bar{u}^{\beta+1}\\ |\eta D\eta\cdot\mathbf{A}(x,u,Du)\bar{u}^\beta| &\leqslant |\mathbf{a}|\eta|D\eta|\bar{u}^\beta|Du|+2\bar{b}^{1/2}\eta|D\eta|\bar{u}^{\beta+1}\\ &\leqslant \frac{\varepsilon}{2}\eta^2\bar{u}^{\beta-1}|Du|^2+\left(1+\frac{|\mathbf{a}|^2}{2\varepsilon}\right)|D\eta|^2\bar{u}^{\beta+1}\\ &\quad+\bar{b}\eta^2\bar{u}^{\beta+1}\\ |\eta^2\bar{u}^\beta B(x,u,Du)| &\leqslant \varepsilon\eta^2\bar{u}^{\beta-1}|Du|^2+\frac{1}{\varepsilon}\bar{b}\eta^2\bar{u}^{\beta+1}.\end{aligned} \tag{8.51}$$

We assume henceforth that $\beta > 0$ if u is a subsolution and $\beta < 0$ if u is a supersolution. By choosing $\varepsilon = \min\{1, |\beta|/4\}$, we then obtain from (8.50) and (8.51)

$$\int_\Omega \eta^2 \bar{u}^{\beta-1}|Du|^2\,dx \leqslant C(|\beta|)\int_\Omega (\bar{b}\eta^2 + (1+|\mathbf{a}|^2)|D\eta|^2)\bar{u}^{\beta+1}\,dx, \tag{8.52}$$

where $C(|\beta|)$ is bounded if $|\beta|$ is bounded away from zero. It is now convenient to introduce a function w defined by

$$w = \begin{cases} \bar{u}^{(\beta+1)/2} & \text{if } \beta \neq -1 \\ \log \bar{u} & \text{if } \beta = -1. \end{cases}$$

Letting $\gamma = \beta + 1$, we may rewrite (8.52)

$$\int_\Omega |\eta D w|^2\,dx \leqslant \begin{cases} C(|\beta|)\gamma^2 \displaystyle\int_\Omega (\bar{b}\eta^2 + (1+|\mathbf{a}|^2)|D\eta|^2)w^2\,dx & \text{if } \beta \neq -1 \\ C \displaystyle\int_\Omega (\bar{b}\eta^2 + (1+|\mathbf{a}|^2)|D\eta|^2)\,dx & \text{if } \beta = -1. \end{cases} \tag{8.53}$$

The desired iteration process can now be developed from the first part of (8.53). For from the Sobolev inequality (7.26) we have

$$\|\eta w\|^2_{2\hat{n}/(\hat{n}-2)} \leqslant C\int_\Omega (|\eta D w|^2 + |w D\eta|^2)\,dx$$

where $\hat{n} = n$ for $n > 2$, $2 < \hat{2} < q$ and $C = C(\hat{n})$. Using the Hölder inequality (7.7) followed by the interpolation inequality (7.10), we obtain, for any $\varepsilon > 0$,

$$\begin{aligned} \int_\Omega \bar{b}(\eta w)^2\,dx &\leqslant \|\bar{b}\|_{q/2}\|\eta w\|^2_{2q/(q-2)} \\ &\leqslant \|\bar{b}\|_{q/2}(\varepsilon\|\eta w\|_{2\hat{n}/(\hat{n}-2)} + \varepsilon^{-\sigma}\|\eta w\|_2)^2 \end{aligned}$$

where $\sigma = \hat{n}/(q-\hat{n})$. Hence, by substitution into (8.53) and appropriate choice of ε, we obtain

$$\|\eta w\|_{2\hat{n}/(\hat{n}-2)} \leqslant C(1+|\gamma|)^{\sigma+1}\|(\eta + |D\eta|)w\|_2 \tag{8.54}$$

where $C = C(\hat{n}, \Lambda, \nu, q, |\beta|)$ is bounded when $|\beta|$ is bounded away from zero. It is now desirable to specify the cut-off function η more precisely. Let r_1, r_2 be such that $1 \leqslant r_1 < r_2 \leqslant 3$ and set $\eta \equiv 1$ in B_{r_1}, $\eta \equiv 0$ in $\Omega - B_{r_2}$ with $|D\eta| \leqslant 2/(r_2 - r_1)$. Writing $\chi = \hat{n}/(\hat{n}-2)$ we then have from (8.54)

$$\|w\|_{L^{2\chi}(B_{r_1})} \leqslant \frac{C(1+|\gamma|)^{\sigma+1}}{r_2 - r_1}\|w\|_{L^2(B_{r_2})}. \tag{8.55}$$

For $r<4$ and $p\neq 0$, let us now introduce the quantities

$$\Phi(p, r)=\left(\int_{B_r} |\bar{u}|^p\,dx\right)^{1/p}. \tag{8.56}$$

By Problem 7.1, we have

$$\Phi(\infty, r)=\lim_{p\to\infty} \Phi(p, r)=\sup_{B_r} \bar{u},$$

and

$$\Phi(-\infty, r)=\lim_{p\to-\infty} \Phi(p, r)=\inf_{B_r} \bar{u}.$$

From inequality (8.55), we now obtain

$$\begin{aligned}\Phi(\chi\gamma, r_1) &\leqslant \left(\frac{C(1+|\gamma|)^{\sigma+1}}{r_2-r_1}\right)^{2/|\gamma|}\Phi(\gamma, r_2) \quad \text{if } \gamma>0\\ \Phi(\gamma, r_2) &\leqslant \left(\frac{C(1+|\gamma|)^{\sigma+1}}{r_2-r_1}\right)^{2/|\gamma|}\Phi(\chi\gamma, r_1) \quad \text{if } \gamma<0.\end{aligned} \tag{8.57}$$

These inequalities can now be iterated to yield the desired estimates. For example, when u is a subsolution we have $\beta>0$ and $\gamma>1$. Hence, taking $p>1$, we set $\gamma=\gamma_m=\chi^m p$ and $r_m=1+2^{-m}$, $m=0, 1, \ldots$, so that, by inequality (8.57),

$$\begin{aligned}\Phi(\chi^m p, 1) &\leqslant (C\chi)^{2(1+\sigma)\Sigma m\chi^{-m}}\Phi(p, 2)\\ &= C\Phi(p, 2), \quad C=C(\hat{n}, \Lambda, \nu, q, p).\end{aligned}$$

Consequently, letting m tend to infinity, we have

$$\sup_{B_1} \bar{u} \leqslant C\|\bar{u}\|_{L^p(B_2)}, \tag{8.58}$$

and, by means of the transformation: $x\to x/R$, the estimate (8.46) is established. For the case when u is a supersolution, that is when $\beta<0$ and $\gamma<1$, we may prove in a similar manner, for any p, p_0 such that $0<p_0<p<\chi$,

$$\begin{aligned}\Phi(p, 2) &\leqslant C\Phi(p_0, 3)\\ \Phi(-p_0, 3) &\leqslant C\Phi(-\infty, 1), \quad C=C(\hat{n}, \Lambda, q, p, p_0).\end{aligned} \tag{8.59}$$

The conclusion of Theorem 8.18 will thus follow if we can show that, for some $p_0>0$,

$$\Phi(p_0, 3) \leqslant C\Phi(-p_0, 3). \tag{8.60}$$

In order to establish (8.60) we turn to the second of the estimates (8.53). Let B_{2r} be any ball of radius $2r$, lying in $B_4(=B_4(y))$, and choose the cut-off function η so that $\eta\equiv 1$ in B_r, $\eta\equiv 0$ in $\Omega - B_4$ and $|D\eta|\leqslant 2/r$. From (8.53), with the aid of the Hölder inequality (7.7), we then obtain

$$(8.61)\qquad \int_{B_r} |Dw|\,dx \leqslant Cr^{n/2}\left(\int_{B_r} |Dw|^2\,dx\right)^{1/2} \leqslant Cr^{n-1},\quad C=C(n,\Lambda,\nu).$$

Hence, by Theorem 7.21, there exists a constant $p_0>0$ depending on n, Λ and ν such that, for

$$w_0=\frac{1}{|B_3|}\int_{B_3} w\,dx,$$

we have

$$\int_{B_3} e^{p_0|w-w_0|}\,dx\leqslant C(n,\Lambda,\nu),$$

and thus

$$\int_{B_3} e^{p_0w}\,dx\int_{B_3} e^{-p_0w}\,dx\leqslant C\,e^{p_0w_0}\,e^{-p_0w_0}=C.$$

Recalling the definition of w, we obtain the estimate (8.60) and consequently Theorem 8.18 with $R=1$ and $k>0$. The full result then follows by means of the transformation: $x\mapsto x/R$ and by letting k tend to zero. □

The strong maximum principle for subsolutions of the equation $Lu=0$, the Harnack inequality for solutions of $Lu=0$ and the local Hölder continuity of solutions of equation (8.3) may all be derived as consequences of the weak Harnack inequality. We treat these interesting local results in turn.

8.7. The Strong Maximum Principle

Theorem 8.19. *Let the operator L satisfy conditions* (8.5), (8.6) *and* (8.8) *and let $u\in W^{1,2}(\Omega)$ satisfy $Lu\geqslant 0$ in Ω. Then, if for some ball $B\subset\subset\Omega$ we have*

$$(8.62)\qquad \sup_B u=\sup_\Omega u\geqslant 0,$$

the function u must be constant in Ω and equality holds in (8.8) *when $u\not\equiv 0$.*

Proof. Writing $B = B_R(y)$, there is no loss of generality in assuming $B_{4R}(y) \subset \Omega$. Let $M = \sup_{\Omega} u$ and apply the weak Harnack inequality (8.47) with $p=1$ to the supersolution $v = M - u$. We obtain thus

$$R^{-n} \int_{B_{2R}} (M-u)\, dx \leqslant C \inf_{B} (M-u) = 0.$$

Consequently $u \equiv M$ in B_{2R} and by an argument similar to that of Theorem 2.2, we obtain $u \equiv M$ in Ω. □

Theorem 8.19 shows that in an appropriately generalized sense, a subsolution of $Lu = 0$ cannot possess an interior positive maximum. For continuous subsolutions the statement reduces to the usual classical one. Note that the strong minimum principle for supersolutions of $Lu = 0$ will follow immediately by replacement of u with $-u$, and that the weak maximum principle, Theorem 8.1, for $C^0(\Omega)$ subsolutions is a direct consequence.

8.8. The Harnack Inequality

By combining Theorems 8.17 and 8.18, we obtain the full Harnack inequality.

Theorem 8.20. *Let the operator L satisfy conditions* (8.5) *and* (8.6), *and let $u \in W^{1,2}(\Omega)$ satisfy $u \geqslant 0$ in Ω and $Lu = 0$ in Ω. Then for any ball $B_{4R}(y) \subset \Omega$, we have*

$$\sup_{B_R(y)} u \leqslant C \inf_{B_R(y)} u \tag{8.63}$$

where $C = C(n, \Lambda/\lambda, \nu R)$.

Examination of the dependence of the constants C on Λ in the estimates (8.54) and (8.61) shows that the constant C in (8.63) can be estimated by

$$C \leqslant C_0^{(\Lambda/\lambda + \nu R)}, \quad C_0 = C_0(n).$$

When the matrix $\mathbf{a}$ is symmetric this estimate may be refined even further; (see Problem 8.3). By an argument similar to that of Theorem 2.5, we can deduce from Theorem 8.20 the following form of the Harnack inequality.

Corollary 8.21. *Let L and u satisfy the hypotheses of Theorem* 8.20. *Then for any $\Omega' \subset\subset \Omega$, we have*

$$\sup_{\Omega'} u \leqslant C \inf_{\Omega'} u \tag{8.64}$$

where $C = C(n, \Lambda/\lambda, \nu, \Omega', \Omega)$.

8.9. Hölder Continuity

The following result is basic to the theory of second order quasilinear equations. Indeed, its discovery by De Giorgi [DG 1] and Nash [NA] for operators of the form $Lu = D_i(a^{ij}(x)D_ju)$ essentially opened up the theory of quasilinear equations in more than two variables.

Theorem 8.22. *Let the operator L satisfy conditions (8.5), (8.6), and suppose that $f^i \in L^q(\Omega)$, $i=1,\ldots,n$, $g \in L^{q/2}(\Omega)$ for some $q>n$. Then if u is a $W^{1,2}(\Omega)$ solution of equation (8.3) in Ω, it follows that u is locally Hölder continuous in Ω, and for any ball $B_0 = B_{R_0}(y) \subset \Omega$ and $R \leqslant R_0$ we have*

$$\operatorname*{osc}_{B_R(y)} u \leqslant CR^\alpha(R_0^{-\alpha} \sup_{B_0} |u| + k) \tag{8.65}$$

where $C = C(n, \Lambda/\lambda, \nu, q, R_0)$ and $\alpha = \alpha(n, \Lambda/\lambda, \nu R_0, q)$ are positive constants, and $k = \lambda^{-1}(\|\mathbf{f}\|_q + \|g\|_{q/2})$.

Proof. We may assume without loss of generality that $R \leqslant R_0/4$. Let us write $M_0 = \sup_{B_0}|u|$, $M_4 = \sup_{B_{4R}} u$, $m_4 = \inf_{B_{4R}} u$, $M_1 = \sup_{B_R} u$, $m_1 = \inf_{B_R} u$. Then we have

$$L(M_4 - u) = M_4(D_ib^i + d) - D_if^i - g$$
$$L(u - m_4) = -m_4(D_ib^i + d) + D_if^i + g.$$

Hence, if we set

$$\bar{k}(R) = \lambda^{-1}R^\delta(\|\mathbf{f}\|_q + M_0\|\mathbf{b}\|_q) + \lambda^{-1}R^{2\delta}(\|g\|_{q/2} + M_0\|d\|_{q/2}),$$
$$\delta = 1 - n/q$$

and apply the weak Harnack inequality (8.47) with $p=1$ to the functions $M_4 - u$, $u - m_4$ in B_{4R}, we obtain

$$R^{-n}\int_{B_{2R}} (M_4 - u)\,dx \leqslant C(M_4 - M_1 + \bar{k}(R)),$$

$$R^{-n}\int_{B_{2R}} (u - m_4)\,dx \leqslant C(m_1 - m_4 + \bar{k}(R)).$$

Hence by addition,

$$M_4 - m_4 \leqslant C(M_4 - m_4 + m_1 - M_1 + \bar{k}(R))$$

so that, writing

$$\omega(R) = \operatorname*{osc}_{B_R} u = M_1 - m_1,$$

we have

$$\omega(R) \leqslant \gamma\omega(4R) + \bar{k}(R)$$

where $\gamma = 1 - C^{-1}$, $C = C(n, \Lambda/\lambda, \nu R_0, q)$. The following simple lemma then implies the desired result. □

Lemma 8.23. *Let ω be a non-decreasing function on an interval $(0, R_0]$ satisfying, for all $R \leqslant R_0$, the inequality*

$$\omega(\tau R) \leqslant \gamma\omega(R) + \sigma(R) \tag{8.66}$$

where σ is also non-decreasing and $0 < \gamma, \tau < 1$. Then, for any $\mu \in (0, 1)$ and $R \leqslant R_0$, we have

$$\omega(R) \leqslant C\left(\left(\frac{R}{R_0}\right)^{\alpha} \omega(R_0) + \sigma(R^{\mu} R_0^{1-\mu})\right) \tag{8.67}$$

where $C = C(\gamma, \tau)$ and $\alpha = \alpha(\gamma, \tau, \mu)$ are positive constants.

Proof. Let us fix initially some number $R_1 \leqslant R_0$. Then for any $R \leqslant R_1$ we have

$$\omega(\tau R) \leqslant \gamma\omega(R) + \sigma(R_1)$$

since σ is non-decreasing. We now iterate this inequality to get, for any positive integer m,

$$\begin{aligned}\omega(\tau^m R_1) &\leqslant \gamma^m \omega(R_1) + \sigma(R_1) \sum_{i=0}^{m-1} \gamma^i \\ &\leqslant \gamma^m \omega(R_0) + \frac{\sigma(R_1)}{1-\gamma}.\end{aligned}$$

For any $R \leqslant R_1$, we can choose m such that

$$\tau^m R_1 < R \leqslant \tau^{m-1} R_1.$$

Hence

$$\begin{aligned}\omega(R) &\leqslant \omega(\tau^{m-1} R_1) \\ &\leqslant \gamma^{m-1}\omega(R_0) + \frac{\sigma(R_1)}{1-\gamma} \\ &\leqslant \frac{1}{\gamma}\left(\frac{R}{R_1}\right)^{\log\gamma/\log\tau} \omega(R_0) + \frac{\sigma(R_1)}{1-\gamma}.\end{aligned}$$

Now let $R_1 = R_0^{1-\mu} R^\mu$ so that we have from the preceding

$$\omega(R) \leqslant \frac{1}{\gamma}\left(\frac{R}{R_0}\right)^{(1-\mu)(\log\gamma/\log\tau)} \omega(R_0) + \frac{\sigma(R_0^{1-\mu}R^\mu)}{1-\gamma}. \quad \square$$

Theorem 8.22 follows by choosing μ such that $(1-\mu)\log\gamma/\log\tau < \mu\delta$. An alternative proof based on Theorem 8.17 rather than Theorem 8.18 is outlined in Problem 8.6.

By combining Theorems 8.17 and 8.22 we have the following interior Hölder estimate for weak solutions of equation (8.3).

Theorem 8.24. *Let the operator L satisfy conditions* (8.5) *and* (8.6), *and suppose that $f^i \in L^q(\Omega)$, $i=1,\ldots,n$, $g \in L^{q/2}(\Omega)$ for some $q>n$. Then, if $u \in W^{1,2}(\Omega)$ satisfies equation* (8.3) *in Ω, we have for any $\Omega' \subset\subset \Omega$ the estimate*

$$\|u\|_{C^\alpha(\bar{\Omega}')} \leqslant C(\|u\|_{L^2(\Omega)} + k), \tag{8.68}$$

where $C = C(n, \Lambda/\lambda, \nu, q, d')$, $d' = \text{dist}\ (\Omega', \partial\Omega)$, $\alpha = \alpha(n, \Lambda/\lambda, \nu d') > 0$ and $k = \lambda^{-1}(\|\mathbf{f}\|_q + \|g\|_{q/2})$.

Proof. The estimate (8.68) follows by taking $R_0 = d'$ in Theorem 8.22 and using Theorem 8.17 to estimate $\sup|u|$. $\square$

Remark. It is clear from the above proofs that the constants C in estimates (8.46), (8.47) and (8.63) are non-decreasing with respect to the argument νR, the constant C in (8.65) is non-decreasing with respect to R_0, the constants α in (8.65) and (8.68) are non-increasing with respect to the arguments νR_0 and $\nu d'$ respectively. When $\nu = 0$, the constants C in (8.46), (8.47) and (8.63) and α in (8.65) and (8.68) will be independent of R, R_0 and d' and hence independent of the domains involved in the assertions of Theorems 8.17, 8.18, 8.20, 8.22 and 8.24.

8.10. Local Estimates at the Boundary

Our previous definition of inequality of $W^{1,2}(\Omega)$ functions on the boundary $\partial\Omega$ can be generalized in the following way. Let T be any subset of $\bar{\Omega}$ and u be a $W^{1,2}(\Omega)$ function. Then we shall say $u \leqslant 0$ on T in the sense of $W^{1,2}(\Omega)$ if u^+ is the limit in $W^{1,2}(\Omega)$ of a sequence of functions in $C_0^1(\bar{\Omega} - T)$. One sees that if u is continuous on T, this definition is satisfied if $u \leqslant 0$ on T in the usual sense. When $T = \partial\Omega$, this definition coincides with our earlier one in Section 8.1. Other definitions of inequality on T will follow as previously indicated there. We shall establish the following extensions of Theorems 8.17 and 8.18.

Theorem 8.25. *Let the operator L satisfy* (8.5), (8.6) *and suppose that $f^i \in L^q(\Omega)$, $i=1,\ldots,n$, $g \in L^{q/2}(\Omega)$ for some $q>n$. Then if u is a $W^{1,2}(\Omega)$ subsolution of equation* (8.3) *in Ω, we have for any $y \in \mathbb{R}^n$, $R>0$* and $p>1$,

$$\sup_{B_R(y)} u_M^+ \leqslant C(R^{-n/p}\|u_M^+\|_{L^p(B_{2R}(y))} + k(R)) \tag{8.69}$$

where

$$M = \sup_{\partial\Omega \cap B_{2R}} u^+,$$

$$u_M^+(x) = \begin{cases} \sup\{u(x), M\}, & x \in \Omega, \\ M, & x \notin \Omega, \end{cases}$$

and k is given by (8.45), $C = C(n, \Lambda/\lambda, \nu R, q, p)$.

Theorem 8.26. *Let the operator L satisfy conditions* (8.5), (8.6) *and suppose that* $f^i \in L^q(\Omega)$, $g \in L^{q/2}(\Omega)$ *for some* $q > n$. *Then if u is a* $W^{1,2}(\Omega)$ *supersolution of equation* (8.3) *in* Ω *and is non-negative in* $\Omega \cap B_{4R}(y)$ *for some ball* $B_{4R}(y) \subset \mathbb{R}^n$, *we have, for any p such that* $1 \leqslant p < n/(n-2)$,

$$R^{-n/p} \|u_m^-\|_{L^p(B_{2R}(y))} \leqslant C(\inf_{B_R(y)} u_m^- + k(R)) \tag{8.70}$$

where

$$m = \inf_{\partial\Omega \cap B_{4R}} u,$$

$$u_m^-(x) = \begin{cases} \inf\{u(x), m\}, & x \in \Omega, \\ m, & x \notin \Omega, \end{cases}$$

and $C = C(n, \Lambda/\lambda, \nu R, q, p)$.

Proof. A reduction to the proof of Theorems 8.17 and 8.18 is made as follows. We set $\bar{u} = u_m^+ + k$ if u is a subsolution and $\bar{u} = u_m^- + k$ if u is a supersolution. Then as test functions in the integral inequalities (8.30) we choose

$$v = \eta^2 \begin{cases} \bar{u}^\beta - (M+k)^\beta & \text{if } \beta > 0 \\ \bar{u}^\beta - (m+k)^\beta & \text{if } \beta < 0, \end{cases} \tag{8.71}$$

where $\eta \in C_0^1(B_{4R})$ is to be further specified. Since the structure (8.44) holds in the support of v, for $\bar{z} = \bar{u}$ and $p = Du$, and since $v \leqslant \eta^2 \bar{u}^\beta$, we arrive again at the estimate (8.52) for $\bar{u}$. The desired estimates (8.69) and (8.70) are then obtained as in the proof of Theorems 8.17 and 8.18. □

A global continuity result cannot be derived from Theorem 8.26 unless some restriction is placed on the domain Ω. We shall say that Ω satisfies an *exterior cone condition* at a point $x_0 \in \partial\Omega$ if there exists a finite right circular cone $V = V_{x_0}$ with vertex x_0 such that $\bar{\Omega} \cap V_{x_0} = x_0$. An exterior cone condition is clearly satisfied wherever an exterior sphere condition holds. We now have the following extension of the Hölder estimate (8.65).

Theorem 8.27. *Let the operator L satisfy conditions* (8.5), (8.6) *and suppose that* $f^i \in L^q(\Omega)$, $i = 1, \ldots, n$, $g \in L^{q/2}(\Omega)$ *for some* $q > n$. *Then if u is a* $W^{1,2}(\Omega)$ *solution of* *equation* (8.3) *in* Ω *and* Ω *satisfies an exterior cone condition at a point* $x_0 \in \partial\Omega$, *we have for any* $0 < R \leqslant R_0$, *and* $B_0 = B_{R_0}(x_0)$,

$$\text{(8.72)} \qquad \operatorname*{osc}_{\Omega \cap B_R} u \leqslant C\{R^\alpha(R_0^{-\alpha} \sup_{\Omega \cap B_0}|u| + k) + \sigma(\sqrt{RR_0})\}$$

where $\sigma(R) = \operatorname*{osc}_{\partial\Omega \cap B_R(x_0)} u$, *and* $C = C(n, \Lambda/\lambda, \nu, q, R_0, V_{x_0})$,

$\alpha = \alpha(n, \Lambda/\lambda, \nu R_0, q, V_{x_0})$ *are positive constants.*

In the following we shall abbreviate $\Omega \cap B_R(x_0) = \Omega_R$ for any R, $\partial\Omega \cap B_R(x_0) = (\partial\Omega)_R$, the point $x_0 \in \partial\Omega$ to be understood.

Proof. We follow the proof of Theorem 8.22. Assume initially that $R \leqslant \inf\{R_0/4, \text{height } V_{x_0}\}$ and write $M_0 = \sup_{\Omega_{R_0}} |u|$, $M_4 = \sup_{\Omega_{4R}} u$, $m_4 = \inf_{\Omega_{4R}} u$, $M_1 = \sup_{\Omega_R} u$, $m_1 = \inf_{\Omega_R} u$. Then, applying the estimate (8.70) to each of the functions $M_4 - u$, $u - m_4$ in $B_{4R}(x_0)$, we obtain

$$(M_4 - M)\frac{|B_{2R}(x_0) - \Omega|}{R^n} \leqslant R^{-n} \int_{B_{2R}(x_0)} (M_4 - u)^-_{M_4 - M}\, dx$$
$$\leqslant C(M_4 - M_1 + \bar{k}(R))$$
$$(m - m_4)\frac{|B_{2R}(x_0) - \Omega|}{R^n} \leqslant R^{-n} \int_{B_{2R}(x_0)} (u - m_4)^-_{m - m_4}\, dx$$
$$\leqslant C(m_1 - m_4 + \bar{k}(R))$$

where $M = \sup_{(\partial\Omega)_{4R}} u$, $m = \inf_{(\partial\Omega)_{4R}} u$. Using the exterior cone condition we thus have

$$M_4 - M \leqslant C(M_4 - M_1 + \bar{k}(R))$$
$$m - m_4 \leqslant C(m_1 - m_4 + \bar{k}(R))$$

so that by addition we get

$$\operatorname*{osc}_{\Omega_R} u \leqslant \gamma \operatorname*{osc}_{\Omega_{4R}} u + \bar{k}(R) + \operatorname*{osc}_{(\partial\Omega)_{4R}} u,$$

where $\gamma = 1 - 1/C$, $C = C(n, \Lambda/\lambda, \nu R_0, q, V_{x_0})$. The estimate (8.72) then follows from Lemma 8.23. □

If the hypotheses of Theorem 8.27 are satisfied and $\sigma(R) \to 0$ as $R \to 0$, then the estimate (8.72) implies that $u(x_0) = \lim_{x \to x_0} u(x)$ is well defined. The following global continuity result then follows immediately from Theorems 8.22 and 8.27.

Corollary 8.28. *In addition to the hypotheses of Theorem 8.27, let us assume that Ω satisfies an exterior cone condition at every point $x_0 \in \partial\Omega$ and that $\operatorname{osc}_{\partial\Omega \cap B_R(x_0)} u \to 0$ as $R \to 0$ for all $x_0 \in \partial\Omega$. Then the function u is uniformly continuous in Ω.*

A uniform Hölder estimate may also be obtained from Theorem 8.27 if the domain Ω is further restricted. Namely, let us say that Ω satisfies a *uniform exterior cone condition* on $T \subset \partial\Omega$ if Ω satisfies an exterior cone condition at every $x_0 \in T$ and the cones V_{x_0} are all congruent to some fixed cone V. We can then assert the following extension of Theorem 8.24.

Theorem 8.29. *Let the operator L satisfy conditions (8.5), (8.6), let $f^i \in L^q(\Omega)$, $i = 1, \ldots, n$, $g \in L^{q/2}(\Omega)$ for some $q > n$, and suppose that Ω satisfies a uniform exterior cone condition on a boundary portion T. Then if $u \in W^{1,2}(\Omega)$ satisfies equation (8.3) in Ω and there exist constants $K, \alpha_0 > 0$ such that*

$$\operatorname*{osc}_{\partial\Omega \cap B_R(x_0)} u \leqslant KR^{\alpha_0} \quad \forall x_0 \in T, R > 0,$$

it follows that $u \in C^\alpha(\Omega \cup T)$ for some $\alpha > 0$ and, for any $\Omega' \subset\subset \Omega \cup T$,

$$\|u\|_{C^\alpha(\bar{\Omega}')} \leqslant C(\sup_\Omega |u| + K + k) \tag{8.73}$$

where $\alpha = \alpha(n, \Lambda/\lambda, \nu d', V, q, \alpha_0)$, $C = C(n, \Lambda/\lambda, \nu, V, q, \alpha_0, d')$,

$$d' = \operatorname{dist}(\Omega', \partial\Omega - T) \quad \textit{and} \quad k = \lambda^{-1}(\|\mathbf{f}\|_q + \|g\|_{q/2}).$$

If $\Omega' = \Omega$, d' is to be replaced by diam Ω.

Proof. Let $y \in \Omega'$, $\delta = \operatorname{dist}(y, \partial\Omega) < d'$. By Theorem 8.22 with $R_0 = \delta$, we have for any $x \in B_\delta$

$$\frac{|u(x) - u(y)|}{|x - y|^\alpha} \leqslant C(\delta^{-\alpha} \sup_{B_\delta} |u| + k).$$

Now choose $x_0 \in \partial\Omega$ such that $|x_0 - y| = \delta$. By the estimate (8.72) with $R = 2\delta$, $R_0 = 2d'$, we then obtain

$$\delta^{-\alpha} \operatorname*{osc}_{B_\delta} u \leqslant \delta^{-\alpha} \operatorname*{osc}_{\Omega_{2\delta}} u \leqslant C(\sup_\Omega |u| + k + K)$$

provided $2\alpha \leqslant \alpha_0$. Hence for any $x \in B_\delta(y)$, we have (taking $u(x_0) = 0$)

$$\frac{|u(x) - u(y)|}{|x - y|^\alpha} \leqslant C(\sup_\Omega |u| + k + K). \tag{8.74}$$

By applying the estimate (8.72) again, with $R=2|x-y|$, $R_0=2d'$, we see that (8.74) will also hold for $d' \geqslant |x-y| \geqslant \delta$. □

In effect, the preceding theorem combines separate Hölder estimates in the interior and on the boundary into a partially interior or a global Hölder estimate. Note that if $u, v \in W^{1,2}(\Omega)$ and $u-v \in W_0^{1,2}(\Omega)$, then $\underset{\partial\Omega \cap B_R(y)}{\operatorname{osc}} u \to 0$ as $R \to 0$ for all $y \in \partial\Omega$ provided $v \in C^0(\bar{\Omega})$, and $\underset{\partial\Omega \cap B_R(y)}{\operatorname{osc}} u \leqslant KR^{\alpha_0}$ for all $y \in \partial\Omega$, $R>0$ provided $v \in C^{\alpha_0}(\bar{\Omega})$. The remark following the proof of Theorem 8.24 is of course also pertinent with regard to the constants C in estimates (8.69), (8.70) and α in estimates (8.72) and (8.73).

An existence theorem for equation (8.3) for continuous boundary values follows from Theorem 8.3 and Corollary 8.28.

Theorem 8.30. *Let the operator L satisfy conditions (8.5), (8.6), (8.8) and $f^i \in L^q(\Omega)$, $g \in L^{q/2}(\Omega)$ for some $q>n$, and suppose that Ω satisfies an exterior cone condition at each point of $\partial\Omega$. Then for $\varphi \in C^0(\partial\Omega)$, there exists a unique function $u \in W_{\mathrm{loc}}^{1,2}(\Omega) \cap C^0(\bar{\Omega})$ satisfying $Lu=g+D_i f^i$ in Ω, $u=\varphi$ on $\partial\Omega$.*

Proof. Let $\{\varphi_m\}$ be a sequence in $C^1(\bar{\Omega})$ converging uniformly to φ on $\partial\Omega$. By Theorem 8.3 and Corollary 8.28 there exists a sequence $\{u_m\}$ in $W^{1,2}(\Omega) \cap C^0(\bar{\Omega})$ such that $Lu_m=g+D_i f^i$ in Ω and $u_m=\varphi_m$ on $\partial\Omega$. By Theorem 8.1, we have

$$\sup_{\Omega} |u_{m_1}-u_{m_2}| \leqslant \sup_{\partial\Omega} |\varphi_{m_1}-\varphi_{m_2}| \to 0 \quad \text{as } m_1, m_2 \to \infty,$$

so that $\{u_m\}$ converges uniformly to a function $u \in C^0(\bar{\Omega})$ satisfying $u=\varphi$ on $\partial\Omega$. Furthermore by the estimate (8.52) we then have, for any $\Omega' \subset\subset \Omega$,

$$\int_{\Omega'} |D(u_{m_1}-u_{m_2})|^2 \, dx \to 0 \quad \text{as } m_1, m_2 \to \infty.$$

Consequently $u \in W_{\mathrm{loc}}^{1,2}(\Omega)$ and satisfies equation (8.3) in Ω. The uniqueness of the solution u follows by applying Theorem 8.1 in domains $\Omega' \subset\subset \Omega$. □

The Wiener Criterion. If we make no restriction on the domain Ω in the hypotheses of Theorem 8.30, then we would obtain from the above procedure a bounded function $u \in C^0(\Omega) \cap W_{\mathrm{loc}}^{1,2}(\Omega)$ such that $Lu = g + D_i f^i$ in Ω; in addition $u(x) \to \varphi(x_0)$ as $x \to x_0 \in \partial\Omega$ if Ω satisfies an exterior cone condition at x_0. Any point $x_0 \in \partial\Omega$ where $u(x) \to \varphi(x_0)$ for arbitrary choice of φ, g, f^i is called a *regular point* for the operator L. Using the methods in [HR] or [LSW], one can show that the regular points for L coincide with those for the Laplacian as defined in Chapter 2. The barrier considerations of Chapter 6 are also applicable here. Note also, that it follows from the proof of Theorem 8.27 that the exterior cone condition can be relaxed to the condition

$$\liminf_{R\to 0} \frac{|B_R(x_0)-\Omega|}{R^n} > 0. \tag{8.75}$$

More generally, by further development of the techniques of this section, we can establish the sufficiency of the Wiener criterion (2.37) for regular points. To accomplish this, we first prove estimates similar to the weak Harnack inequalities, Theorems 8.18 and 8.26, but involving the gradient of the supersolution u. Let us consider first the interior case and suppose accordingly that the hypotheses of Theorem 8.18 are satisfied. It is then evident from the proof of Theorem 8.18 (in particular (8.52)) that, along with (8.47), we may also obtain the estimate

$$\left(R^{2-n}\int_{B_{2R}} (\bar{u})^{-p}|Du|^2\,dx\right)^{1/(2-p)} \leqslant C(\inf_{B_R} u + k(R))$$

for $1 < p < n/(n-2)$ and $C = C(n, \Lambda/\lambda, \nu R, q, p)$. But then by Hölder's inequality,

$$\begin{aligned} R^{1-n}\int_{B_{2R}} |Du| &\leqslant \left(R^{-n}\int_{B_{2R}} (\bar{u})^{p}\right)^{1/2}\left(R^{2-n}\int_{B_{2R}} (\bar{u})^{-p}|Du|^2\right)^{1/2} \\ &\leqslant C(\inf_{B_R} u + k(R)) \end{aligned} \tag{8.76}$$

where $C = C(n, \Lambda/\lambda, \nu R, q)$ if p is fixed, say $p = n/(n-1)$.

If we assume in addition the hypotheses of Theorem 8.26 and consider its proof, we see that u may be replaced by u_m^- in (8.76), and hence

$$R^{1-n}\int_{\Omega\cap B_{2R}} |Du_m^-| \leqslant C\left(\inf_{\Omega\cap B_R} u_m^- + k(R)\right) \tag{8.77}$$

where $C = C(n, \Lambda/\lambda, \nu R, q)$. To proceed further, let us fix a cutoff function $\eta \in C_0^1(B_{2R})$ such that $0 \leqslant \eta \leqslant 1$, $\eta = 1$ on B_R, $|D\eta| \leqslant 2/R$, and then insert $v = \eta^2(m - u_m^-)$ as a test function in (8.30), noting that

$$m = \sup_{B_{2R}} u_m^- \quad \text{if } B_{4R} \cap \partial\Omega \neq \phi,$$

which will be assumed in the following. Normalizing $R = 1$ and using the conditions (8.5), (8.6), we obtain

$$\int_{B_2} \eta^2|Du_m^-|^2\,dx \leqslant C(m + k)\int_{B_2} (\bar{u} + |D\bar{u}|)\,dx.$$

Hence, writing

$$w = \eta u_m^-$$

and using (8.70) and (8.77), we have

$$\int_{B_2} |Dw|^2 \leqslant C(m + k) \int_{B_2} (\bar{u} + |D\bar{u}|)\, dx$$
$$\leqslant C(m + k)\left(\inf_{B_1} u_m^- + k\right).$$

Consequently, for general R we have the estimate

$$R^{2-n} \int_{B_{2R}} |Dw|^2 \leqslant C(m + k)\left(\inf_{B_R} u_m^- + k(R)\right). \tag{8.78}$$

We now recall from (2.36) that the capacity of the set $B_R - \Omega$ is given by

$$\operatorname{cap}(B_R - \Omega) = \inf_{v \in K} \int |Dv|^2, \tag{8.79}$$

where

$$K = \{v \in C_0^1(\mathbb{R}^n) \,|\, v = 1 \quad \text{on } B_R - \Omega\}.$$

Since $u_m^- = m$ on $B_R - \Omega$ and $C_0^1(B_{2R})$ is dense in $W_0^{1,2}(B_{2R})$, and using the fact that $\operatorname{cap}(B_R - \Omega) \leqslant CR^{n-2}$, we obtain from (8.78),

$$mR^{2-n} \operatorname{cap}(B_R - \Omega) \leqslant C \inf_{B_R} (u_m^- + k(R)). \tag{8.80}$$

Hence, if u is a solution of equation (8.3) in Ω and $y = x_0 \in \partial\Omega$, we obtain, with the notation in the proof of Theorem 8.27,

$$(M_4 - M)\chi(R) \leqslant C(M_4 - M_1 + \bar{k}(R))$$
$$(m - m_4)\chi(R) \leqslant C(m_1 - m_4 + \bar{k}(R)),$$

where

$$\chi(R) = R^{2-n} \operatorname{cap}(B_R - \Omega).$$

Thus, by addition, we obtain the oscillation estimate

$$\operatorname*{osc}_{\Omega_R} u \leqslant \left(1 - \frac{\chi(R)}{C}\right) \operatorname*{osc}_{\Omega_{4R}} u + \frac{\chi(R)}{C} \operatorname*{osc}_{(\partial\Omega)_{4R}} u + \bar{k}(R). \tag{8.81}$$

We leave it to the reader to check that if (2.37) holds for $\lambda = 1/4$, then an estimate for the modulus of continuity of u at x_0 is determined by iteration of (8.81), in an argument similar to that in Lemma 8.23 (Problem 8.8). We can thus state the following extension of Theorem 8.30.

Theorem 8.31. *Let the operator L satisfy conditions (8.5), (8.6), (8.8) and $f^i \in L^q(\Omega)$, $g \in L^{q/2}(\Omega)$ for some $q > n$, and suppose that the Wiener condition (2.37) holds at each point of $\partial\Omega$. Then for $\varphi \in C^0(\partial\Omega)$, there exists a unique function $u \in W^{1,2}_{loc}(\Omega) \cap C^0(\bar{\Omega})$ satisfying $Lu = g + D_i f^i$ in Ω, $u = \varphi$ on $\partial\Omega$.*

Finally, to conclude this section we remark that the results of Sections 8.6 to 8.10 are still valid when the condition (8.6) on the coefficients $\mathbf{b}$, $\mathbf{c}$ and d is replaced by $\mathbf{b}, \mathbf{c} \in L^q(\Omega)$, $d \in L^{q/2}(\Omega)$, $q>n$. Setting

$$\bar{b} = \lambda^{-2}(|\mathbf{b}|^2 + |\mathbf{c}|^2) + \lambda^{-1} d,$$
$$\nu^2 = \|\bar{b}\|_{q/2}$$

we then need to replace the quantity νR by νR^δ in the estimates of Theorems 8.17 to 8.29. It is also possible in certain of the preceding local results to weaken both the uniform and the strict ellipticity of the operator L; (see [TR 4, 7], [FKS]).

8.11. Hölder Estimates for the First Derivatives

When the principal coefficients in (8.3) are Hölder continuous, the existence and regularity theory can be patterned after the Schauder theory of Chapter 6, and yields similar results. The starting point is again Poisson's equation, in the form

$$\Delta u = g + D_i f^i. \tag{8.82}$$

If $g, f \in L^\infty(\Omega)$ and $f \in C^\alpha(\Omega)$ for some $\alpha \in (0, 1)$, it is easily shown that the Newtonian potential of the right-hand side, given by

$$w(x) = \int_\Omega \Gamma(x-y) g(y)\, dy + \int_\Omega D_i \Gamma(x-y) f^i(y)\, dy,$$

is a weak solution of (8.82) and consequently the estimates of Section 4.5 will apply to weak solutions of (8.82); (see Problem 7.20). In particular, we recall the interior and boundary estimates (4.45) and (4.46). The considerations of Lemma 6.1 show that the same estimates hold for weak solutions of

$$L_0 u \equiv A^{ij} D_{ij} u = g + D_i f^i,$$

where L_0 is a constant coefficient elliptic operator.

We can now apply the perturbation technique of Section 6.1 to the equation

$$Lu = g + D_i f^i, \tag{8.83}$$

where L is given by (8.1). For any point $x_0 \in \Omega$, we "freeze" the coefficients a^{ij} at x_0 and rewrite the equation as

$$\begin{aligned} a^{ij}(x_0)D_{ij}u &= D_i\{(a^{ij}(x_0) - a^{ij}(x))D_j u - b^i(x)u\} \\ &\quad - c^i(x)D_i u - d(x)u + g + D_i f^i \\ &= G(x) + D_i F^i(x), \end{aligned} \tag{8.84}$$

where

$$F^i(x) = (a^{ij}(x_0) - a^{ij}(x))D_j u - b^i(x)u + f^i(x)$$

$$G(x) = -c^i(x)D_i u - d(x)u + g(x).$$

With x_0 fixed, this equation is of the form (8.82) with $f^i = F^i$, $g = G$.

Let us assume in the following that L is strictly elliptic, satisfying (8.5), the coefficients $a^{ij}, b^i \in C^\alpha(\overline{\Omega})$, $c^i, d, g \in L^\infty(\Omega)$, and $f \in C^\alpha(\overline{\Omega})$. Suppose

$$\max_{i,j=1,\dots,n} \{|a^{ij}, b^i|_{0,\alpha;\Omega}, |c^i, d|_{0;\Omega}\} \leqslant K. \tag{8.85}$$

Then we assert the following interior and global estimates:

Theorem 8.32. *Let $u \in C^{1,\alpha}(\Omega)$ be a weak solution of (8.83) in a bounded domain Ω. Then for any subdomain $\Omega' \subset\subset \Omega$ we have*

$$|u|_{1,\alpha;\Omega'} \leqslant C(|u|_{0;\Omega} + |g|_{0;\Omega} + |f|_{0,\alpha;\Omega}), \tag{8.86}$$

for $C = C(n, \lambda, K, d')$, where λ is given by (8.5), K by (8.85) and $d' = \operatorname{dist}(\Omega', \partial\Omega)$.

Theorem 8.33. *Let $u \in C^{1,\alpha}(\overline{\Omega})$ be a weak solution of (8.83) in a $C^{1,\alpha}$ domain Ω, satisfying $u = \varphi$ on $\partial\Omega$, where $\varphi \in C^{1,\alpha}(\overline{\Omega})$. Then we have*

$$|u|_{1,\alpha} \leqslant C(|u|_0 + |\varphi|_{1,\alpha} + |g|_0 + |f|_{0,\alpha}). \tag{8.87}$$

for $C = C(n, \lambda, K, \partial\Omega)$, where λ and K are as above.

The proof of these results is essentially the same as that of Theorems 6.1 and 6.6, but is based now on the estimates (4.45) and (4.46), applied to (8.84). We note that the proof of (8.87), which is reduced to the boundary estimate (4.46), requires a preliminary flattening of the boundary. Since the hypotheses concerning (8.83) are invariant under $C^{1,\alpha}$ mappings, it suffices that the domain Ω be of class $C^{1,\alpha}$, and thus Theorem 8.33 is stated for domains and boundary values in this class. The dependence on $\partial\Omega$ of the constant C in (8.87) is through the $C^{1,\alpha}$ norms of the mappings that flatten the boundary.

The global estimate in Theorem 8.33 leads directly to the basic existence theorem for (8.83).

Theorem 8.34. *Let Ω be a $C^{1,\alpha}$ domain and L an operator satisfying* (8.5), (8.8) *and* (8.85) *with $K < \infty$. Let $g \in L^\infty(\Omega)$, $f^i \in C^\alpha(\bar{\Omega})$ and $\varphi \in C^{1,\alpha}(\bar{\Omega})$. Then the generalized Dirichlet problem*

$$(8.88) \qquad Lu = g + D_i f^i \quad \text{in } \Omega, \quad u = \varphi \quad \text{on } \partial\Omega$$

is uniquely solvable in $C^{1,\alpha}(\bar{\Omega})$.

Proof. Argument by approximation. Let L_k be a sequence of operators with sufficiently smooth (say $C^2(\bar{\Omega})$) coefficients a_k^{ij}, b_k^i, c_k^i, d_k, such that $a_k^{ij} \to a^{ij}$, $b_k^i \to b^i$ uniformly in Ω and $c_k^i \to c^i$, $d_k \to d$ in L^1 as $k \to \infty$; it can be assumed that the approximating coefficients also satisfy (8.5), (8.8) and (8.85). In addition, let $f_k^i, g_k, \varphi_k \in C^3(\bar{\Omega})$, and as $k \to \infty$ let $f_k^i \to f^i$ with $|f_k^i|_{0,\alpha} \leqslant c|f^i|_{0,\alpha}$, $\varphi_k \to \varphi$ with $|\varphi_k|_{1,\alpha} \leqslant c|\varphi|_{1,\alpha}$, and $g_k \to g$ in $L^1(\Omega)$ with $|g_k|_{0,\Omega} \leqslant c\|g\|_{\infty;\Omega}$. Finally, let $\{\Omega_k\}$ be a sequence of $C^{2,\alpha}$ domains exhausting Ω, such that $\partial\Omega_k \to \partial\Omega$ and the surfaces $\partial\Omega_k$ are uniformly in $C^{1,\alpha}$, (see Problem 6.9).

Under these assumptions, the smooth approximating Dirichlet problems

$$(8.89) \qquad L_k u = g_k + D_i f_k^i \quad \text{in } \Omega_k,\ u = \varphi_k \quad \text{on } \partial\Omega_k$$

have unique $C^{2,\alpha}(\bar{\Omega}_k)$ solutions satisfying the $C^{1,\alpha}$ estimate (8.87). Since Theorem 8.16 implies

$$|u_k|_0 \leqslant \sup_{\partial\Omega} |u_k| + C(|g_k|_0 + |f_k|_0),$$

we infer the uniform $C^{1,\alpha}$ estimate

$$(8.90) \qquad \begin{aligned} |u_k|_{1,\alpha;\Omega_k} &\leqslant C(|\varphi_k|_{1,\alpha;\Omega_k} + |g_k|_{0;\Omega_k} + |f_k|_{0,\alpha;\Omega_k}) \\ &\leqslant C(|\varphi|_{1,\alpha;\Omega} + |g|_{0,\Omega} + |f|_{0,\alpha;\Omega}) \end{aligned}$$

where the constant C is independent of k. Letting $k \to \infty$ in the weak form of (8.83), we obtain in the limit a (unique) $C^{1,\alpha}(\bar{\Omega})$ weak solution u of (8.88), which also satisfies (8.90). This solution is also unique within the larger class of $W^{1,2}(\Omega)$ functions for which $u - \varphi \in W_0^{1,2}(\Omega)$, by virtue of Theorem 8.1. □

The estimate (8.87), which was stated for weak $C^{1,\alpha}$ solutions, can now be seen to hold for $W^{1,2}(\Omega)$ solutions under the same hypotheses. Let $u \in W^{1,2}(\Omega)$ be a solution of (8.83) satisfying the hypotheses of Theorem 8.33, with $u - \varphi \in W_0^{1,2}(\Omega)$; then u is bounded (by Theorem 8.16) and for a sufficiently large positive constant σ, u is also the unique $C^{1,\alpha}(\bar{\Omega})$ solution v of the generalized Dirichlet problem

$$\begin{aligned} (L - \sigma)v &= g + D_i f^i - \sigma u \quad \text{in } \Omega \\ v &= \varphi \quad \text{on } \partial\Omega. \end{aligned}$$

Thus, we have:

Corollary 8.35. *Under the hypotheses of Theorem* 8.33, *if* $u \in W^{1,2}(\Omega)$ *and* $u - \varphi \in W_0^{1,2}(\Omega)$, *the same conclusion* (8.87) *remains valid.*

Local $C^{1,\alpha}$ regularity can be derived similarly by first approximating the solution u by smooth functions. In fact, we have the following extension of Corollary 8.35, the details of which are left to the reader.

Corollary 8.36. *Let T be a (possibly empty) $C^{1,\alpha}$ boundary portion of a domain Ω, and suppose $u \in W^{1,2}(\Omega)$ is a weak solution of* (8.83) *such that $u = 0$ on T (in the sense of $W^{1,2}(\Omega)$). Then $u \in C^{1,\alpha}(\Omega \cup T)$, and for any $\Omega' \subset\subset \Omega \cup T$ we have*

$$|u|_{1,\alpha;\Omega'} \leqslant C(|u|_{0;\Omega} + |g|_{0;\Omega} + |f|_{0,\alpha;\Omega}) \tag{8.91}$$

for $C = C(n, \lambda, K, d', T)$ where λ and K are as in Theorem 8.32 *and now $d' = \operatorname{dist}(\Omega', \partial\Omega - T)$.*

In this result, if $\varphi \in C^{1,\alpha}(\bar{\Omega})$ and $u = \varphi$ on T (in the sense of $W^{1,2}(\Omega)$), then $|\varphi|_{1,\alpha;\Omega}$ appears in the right member. This is seen by simply replacing u by $u - \varphi$ and applying (8.91).

Remark. In all the results of this section, if $g \in L^p(\Omega)$, $p = n/(1-\alpha)$, the same conclusions are valid provided $\|g\|_{p;\Omega}$ replaces $|g|_{0;\Omega}$ throughout. The corresponding proofs are essentially unchanged; (see Remark at the end of Section 4.5).

8.12. The Eigenvalue Problem

The Fredholm theory, as expressed by Theorem 8.6, guarantees that an elliptic operator of the form (8.1) will have at most a countable set of eigenvalues. We prove directly in this section that a self-adjoint operator has eigenvalues and consider some of their basic properties. Although the existence of eigenvalues follows from standard functional analysis, we consider it worthwhile to go through the demonstration of existence for the special case under consideration.

Let us now suppose that the operator L is self-adjoint so that it can be written as

$$Lu = D_i(a^{ij}D_ju + b^iu) - b^iD_iu + cu$$

where $[a^{ij}]$ is symmetric. The associated quadratic form on $H = W_0^{1,2}(\Omega)$ is then given by

$$\mathscr{L}(u, u) = \int_\Omega (a^{ij}D_iuD_ju + 2b^iuD_iu + cu^2)\,dx.$$

The ratio

$$J(u) = \frac{\mathscr{L}(u, u)}{(u, u)}, \qquad u \not\equiv 0, \quad u \in H,$$

is called the *Rayleigh quotient* of L. We commence by studying the variational problem of minimizing J. First, it is clear by Lemma 8.4 that J is bounded from below, so that we may define

$$\sigma = \inf_H J. \tag{8.92}$$

We claim now that σ is the minimum eigenvalue of L on H, that is, there exists a nontrivial function $u \in H$ such that

$$Lu + \sigma u = 0, \tag{8.93}$$

and σ is the smallest number for which this is possible. To show this we choose a *minimizing sequence* $\{u_m\} \subset H$ such that $\|u_m\|_2 = 1$ and $J(u_m) \to \sigma$. By (8.5) and (8.6), we have that $\{u_m\}$ is bounded in H, and hence by the compactness of the imbedding $H \to L^2(\Omega)$ (Theorem 7.22), a subsequence, which we take as $\{u_m\}$ itself, converges in $L^2(\Omega)$ to a function u with $\|u\|_2 = 1$. Since $Q(u) = \mathscr{L}(u, u)$ is quadratic we also have for any l, m

$$Q\left(\frac{u_l - u_m}{2}\right) + Q\left(\frac{u_l + u_m}{2}\right) = \tfrac{1}{2}(Q(u_m) + Q(u_l))$$

so that

$$Q\left(\frac{u_l - u_m}{2}\right) \leqslant \tfrac{1}{2}(Q(u_m) + Q(u_l)) - \sigma\left\|\frac{u_l + u_m}{2}\right\|_2^2 \to 0 \quad \text{as } m, l \to \infty.$$

Again using Lemma 8.4, we see that $\{u_m\}$ is a Cauchy sequence in H. Hence $u_m \to u$ in H, and moreover $Q(u) = \sigma$. The verification of the Euler equation (8.93) is standard in the calculus of variations: It follows by setting

$$f(t) = J(u + tv)$$

for $v \in H$ and calculating

$$f'(0) = 2(\mathscr{L}(u, v) - \sigma(u, v)) = 0.$$

The number σ is easily seen to be the minimum eigenvalue since any smaller eigenvalue would contradict the formula (8.92). If we arrange the eigenvalues of L in increasing order $\sigma_1, \sigma_2, \ldots$, and designate their corresponding eigenspaces by $V_1, V_2, \ldots$, we may characterize the higher eigenvalues of L through the formula

$$\sigma_m = \inf\{J(u) \mid u \not\equiv 0, \quad (u, v) = 0 \; \forall v \in \{V_1, \ldots, V_{m-1}\}\}. \tag{8.94}$$

The solvability of these variational problems is established essentially as in the case $m = 1$ and, furthermore, the process (8.94) is seen to exhaust all possible eigenvalues of L, the resulting eigenfunctions forming a complete set in $L^2(\Omega)$. We can therefore assert:

Theorem 8.37. *Let L be a self-adjoint operator satisfying* (8.5) *and* (8.6). *Then L has a countably infinite discrete set of eigenvalues, $\Sigma = \{\sigma_m\}$, given by* (8.94), *whose eigenfunctions span H.*

Solutions of the Dirichlet problem for L can now be represented by eigenfunction expansions according to standard procedures; (e.g., see [CH]). We may also apply the preceding regularity considerations of this chapter to eigenfunctions. In particular, they belong to $L^\infty(\Omega) \cap C^\alpha(\Omega)$ for some $\alpha > 0$, by virtue of Theorems 8.15 and 8.24; and to $C^\alpha(\overline{\Omega})$ if Ω is sufficiently smooth (Theorem 8.29). If the coefficients of L belong to $C^\infty(\Omega)$, then so also will the eigenfunctions (Corollary 8.11).

To complete this section we observe a special property of the minimum eigenvalue σ_1.

Theorem 8.38. *Let L be a self-adjoint operator satisfying* (8.5) *and* (8.6). *Then the minimum eigenvalue is simple and has a positive eigenfunction.*

Proof. If u is an eigenfunction of σ_1, then it follows from the formula (8.92) that $|u|$ is one also. But then by the Harnack inequality, Theorem 8.21, we must have $|u|$ *positive* (a.e.) in Ω and hence σ_1 has a positive eigenfunction. This argument also shows that the eigenfunctions of σ_1 are either positive or negative and hence it is impossible that two of them are orthogonal, whence V_1 must be one-dimensional and σ_1 simple. □

Notes

The Hilbert space or variational approach to the Dirichlet problem for linear, elliptic equations can be traced back as far as the works of Hilbert [HI] and Lebesgue [LE] for Laplace's equation. During this century it has been developed by many authors including, in particular, Friedrichs [FD 1, 2] and Gårding [GA]. The reader is referred to the books [AG], [BS] and [FR] for further discussion. The generalized Dirichlet problem, which we treat in Section 8.2, was also considered by Ladyzhenskaya and Ural'tseva [LU 4] and Stampacchia [ST 4, 5]. These authors derived the Fredholm alternative, Theorem 8.6, but their existence and uniqueness results were restricted by smallness or coercivety conditions. The weak maximum principle, Theorem 8.1, although a simple consequence of the weak Harnack inequality in [TR 1], appears to have been first noted in the literature by Chicco [CI 1]; (see also [HH]). We have followed the proof of Trudinger [TR 7] which has the advantage of being readily extended to non-uniformly elliptic equations. Given the Fredholm alternative, the existence result, Theorem 8.3, is an immediate consequence of the weak maximum principle.

Higher order differentiability theorems for weak solutions, as in Sections 8.3 and 8.4, were proved by various authors including Friedrichs [FD 2], Browder [BW 1], Lax [LX] and Nirenberg [NI 1, 2]; see also [AG], [BS] and [FR].

The global bound, Theorem 8.15, appears in the works [LU 4] and [ST 4, 5] and is an extension of an earlier version by Stampacchia [ST 1, 2]. Our proof, through the Moser iteration technique, follows that of Serrin [SE 2]. The apriori bound, Theorem 8.16, is due to Trudinger [TR 7].

The local pointwise estimates, which comprise the rest of Chapter 8, all stem from the pioneering work of De Giorgi [DG 1], where the special cases of Theorems 8.17 and 8.22 for equations of the form

$$Lu = D_i(a^{ij}(x)\, D_j u) = 0 \tag{8.95}$$

were established; (see also Nash [NA]). De Giorgi's work was extended to linear equations, of the form treated here, by Morrey [MY 4], Stampacchia [ST 3] and to quasilinear equations in divergence form by Ladyzhenskaya and Ural'tseva [LU 2]. An interesting new proof of De Giorgi's result was proposed by Moser [MJ 1]. This proof can also be extended to more general classes of equations (see [LU 4]), and indeed could have been employed by us to derive Theorems 8.22, 8.24, as well as the boundary estimate, Theorem 8.29, (see Problem 8.6). A Harnack inequality for weak solutions of equation (8.95) was established by Moser [MJ 2] and extended to quasilinear equations in divergence form by Serrin [SE 2] and Trudinger [TR 1]. We have based our treatment of local estimates in Theorems 8.18, 8.26 on the weak Harnack inequality derived in [TR 1]. We note here that, in the case of equations in two variables, the Hölder estimate and Harnack inequality can be deduced by simpler methods; see [MY 3], [BN] and Problem 8.5. For sharp results in the case of two variables see [PS] and [WI 3]. In Section 8.10, the treatment of Theorems 8.25 to 8.30 follows [TR 1] while the proof of the sufficiency of the Wiener criterion is adapted from Gariepy and Ziemer [GZ 2].

The methods and results of Sections 8.1 and 8.2 can be extended to treat other boundary value problems. In particular, we can consider a generalized version of the *mixed boundary value problem*

$$\begin{aligned} Lu &= D_i f^i + g \quad \text{in } \Omega \\ u &= \varphi_1 \quad \text{on } \partial\Omega - \Gamma \\ Nu &\equiv a^{ij}(x)\nu_i D_j u + b^i(x)\nu_i u + \sigma(x)u = \varphi_2 \quad \text{on } \Gamma \end{aligned} \tag{8.96}$$

where Γ is a relatively open C^1 portion of $\partial\Omega$, and $\nu = (\nu_1, \ldots \nu_n)$ is the outer normal to $\partial\Omega$ on Γ. For $\varphi_1 \in W^{1,2}(\Omega)$ and $\sigma, \varphi_2 \in L^2(\Gamma)$, a function $u \in W^{1,2}(\Omega)$ is called a *generalized solution* of the boundary value problem (8.96) if $u - \varphi_1 \in W_0^{1,2}(\Omega \cup \Gamma)$ and

$$\mathfrak{L}(u, v) = \int_\Omega (f^i D_i v - gv)\, dx + \int_\Gamma (\varphi_2 - f^i \nu_i - \sigma u)\, v\, ds, \tag{8.97}$$

for all $v \in W_0^{1,2}(\Omega \cup \Gamma)$. Here $W_0^{1,2}(\Omega \cup \Gamma)$ denotes the closure of $C_0^1(\Omega \cup \Gamma)$ in $W^{1,2}(\Omega)$. We again obtain a weak maximum principle; namely, if conditions (8.5), (8.6) hold, together with the inequality (cf. (8.8))

$$\int_\Omega (dv - b^i D_i v)\, dx - \int_\Gamma \sigma v\, ds \leqslant 0 \quad \forall v \geqslant 0,\ v \in C_0^1(\Omega \cup \Gamma), \tag{8.98}$$

then any function $u \in W^{1,2}(\Omega)$, satisfying $\mathfrak{L}(u, v) + \int_\Gamma \sigma uv\, ds \leqslant 0$ for all non-negative $v \in W_0^{1,2}(\Omega \cap \Gamma)$, must either satisfy $\sup_\Omega u \leqslant \sup_\Gamma u^+$ or be a positive constant. It follows then that generalized solutions of (8.96) are unique provided either $\Gamma \neq \partial\Omega, \sigma + v_i b^i \not\equiv 0$ on Γ, or $L1 \neq 0$. If all of these last three conditions are fulfilled, then generalized solutions of (8.96) must only differ by a constant. An analogue of the existence theorem, Theorem 8.2, can again be concluded from a Fredholm alternative. Maximum principles for mixed boundary value problems are treated in the papers [CI 3] and [TR 11]; in the latter work the above assertions are derived for a general class of non-uniformly elliptic equations.

Finally, we remark that through an approach of Campanato, depending on certain integral characterizations of Hölder spaces, the Schauder theory may be derived directly from the Hilbert space theory and thereby independently of the potential theory of Chapter 4; (see [CM 2], [GT 4]).

Problems

8.1. Show that in the weak maximum principle, Theorem 8.1, provided $u \leqslant 0$ on $\partial\Omega$, condition (8.8) can be replaced by either the condition

$$\int_\Omega (dv + c^i D_i v)\, dx \leqslant 0 \quad \forall v \geqslant 0, \in C_0^1(\Omega), \tag{8.99}$$

or the condition

$$\begin{bmatrix} \mathbf{a} & \mathbf{b} \\ -\mathbf{c} & -\mathrm{d} \end{bmatrix} \geqslant 0 \quad \text{a.e. } (\Omega); \text{ (see Theorem 10.7)} \tag{8.100}$$

8.2. Let $u \in W^{1,2}(\Omega)$ be a weak solution of the equation $Lu = g + D_i f^i$ in Ω, where L satisfies the conditions (8.5), (8.6) and $g, f^i \in L^2(\Omega)$, $i = 1, \ldots, n$. Show that for any subdomain $\Omega' \subset\subset \Omega$, we have

$$\|u\|_{W^{1,2}(\Omega')} \leqslant C(\|u\|_2 + \|\mathbf{f}\|_2 + \|g\|_2) \tag{8.101}$$

where $C = C(n, \Lambda/\lambda, \nu, d')$, $d' = \operatorname{dist}(\Omega', \partial\Omega)$.

8.3. Show that if the matrix $\mathbf{a}=[a^{ij}]$ is symmetric then the constant C in the Harnack inequality, Theorem 8.20, can be estimated by

$$C \leqslant C_0^{(\sqrt{\Lambda/\lambda}+\nu R)}, \quad C_0 = C_0(n).$$

8.4. Using Theorem 8.8 and regularization, (as in Section 7.2), show that Theorem 3.9 is valid for functions $u \in W^{1,2}(\Omega)$. Hence, by considering the function

$$u(x, y) = |xy| \log(|x| + |y|),$$

in the domain

$$\Omega = \{(x, y) \in \mathbb{R}^2 \mid |x| + |y| < 1\} \subset \mathbb{R}^2,$$

demonstrate the sharpness of Theorem 3.9.

8.5. (a) Let u be a function in $C^1(B_R(0))$, $B_R(0) \subset \mathbb{R}^2$, and write for $0 < r < R$,

$$\omega(r) = \operatorname*{osc}_{\partial B_r} u,$$

$$D(r) = \int_{B_r} |Du|^2 \, dx$$

where $B_r = B_r(0)$. If ω is non-decreasing, show that for $0 < r < R$,

$$\omega(r) \leqslant \sqrt{\pi D(R)/\log(R/r)}.$$

(b) Prove the Harnack inequality for divergence structure equations in two variables:

$$Lu = D_i(a^{ij}D_j u) + b^i D_i u = 0, \quad i, j = 1, 2$$

satisfying conditions (8.5) and (8.6), as follows. If the solution u is positive in the disc $B_R(0) \subset \mathbb{R}^2$, show that the Dirichlet integral of the function $v = \log u$ is bounded in every disc $B_r(0)$, $0 < r < R$, in terms of Λ/λ, ν, r and R. (See the case $\beta = -1$ in the proof of Theorem 8.18). Apply the weak maximum principle, Theorem 8.1, and part (a) to obtain the result

$$C^{-1}u(0) \leqslant u(x) \leqslant Cu(0)$$

for $|x| \leqslant R/2$, where $C = C(\lambda, \Lambda, R.)$ (Cf. [BN].)

8.6. (a) Using the hypotheses and notation of Theorem 8.22, show that the functions

$$w_1 = \log \frac{M_4 - m_4 + \bar{k}(R)}{2(M_4 - u) + \bar{k}(R)},$$

$$w_2 = \log \frac{M_4 - m_4 + \bar{k}(R)}{2(u - m_4) + \bar{k}(R)}$$

are subsolutions in B_{4R} of equations with structures similar to equation (8.3). (Cf. the proof of Theorem 8.16).

(b) By applying Theorem 8.17 to the functions w_1, w_2 in part (a), using the Poincaré inequality (7.45) and the case $\beta = -1$ in the proof of Theorem 8.18, give an alternative proof of the Hölder estimate (Theorem 8.22). Note that one of the functions w_1, w_2 is nonpositive on a set S such that $|S| \geqslant \frac{1}{2}|B_{4R}|$.

8.7. Let Ω be a bounded measurable set in $\mathbb{R}^n$. Prove that

$$|\Omega|^{1-2/n} \leqslant \gamma(n) \operatorname{cap} \Omega,$$

where $\gamma(n)$ is the constant in the Sobolev inequality (7.26) when $p = 2$.

8.8. Prove that (8.81) implies a modulus of continuity at the boundary point x_0.

8.9. Using Theorem 8.16, show that the existence and uniqueness theorem, Theorem 8.3, holds for unbounded domains Ω with finite measure. (Note that the compactness results of Theorem 7.22 do not necessarily hold for such domains [AD].)

Chapter 9

Strong Solutions

Until now in this work we have concentrated on either weak or classical solutions of second-order elliptic equations; a weak solution need only be once weakly differentiable while a classical solution must be at least twice continuously differentiable. The formulation of the weak solution concept depended on the operator L under consideration having a "divergence form" while the concept of classical solution made sense for operators with completely arbitrary coefficients. In this chapter our concern is with the intermediate situation of *strong* solutions. For operators in the general form

$$Lu = a^{ij}(x)D_{ij}u + b^i(x)D_iu + c(x)u \tag{9.1}$$

with coefficients a^{ij}, b^i, c, where $i, j = 1, \ldots, n$, defined on a domain $\Omega \subset \mathbb{R}^n$ and a function f on Ω, a strong solution of the equation

$$Lu = f \tag{9.2}$$

is a twice weakly differentiable function on Ω satisfying the equation (9.2) almost everywhere in Ω. The attachment of such a solution to prescribed boundary values on $\partial\Omega$ in the Dirichlet problem may be considered in a generalized sense analogous to that in Chapter 8 or in the classical sense of Chapter 6, where they are taken on continuously. With the aid of a regularity argument, we have already derived in Chapter 8 an existence theorem (Theorem 8.9) for strong solutions that are not necessarily classical. In this case, the boundary values are assumed in the generalized sense. However, the results of Section 8.10, in particular, Theorem 8.30, provide conditions for the continuous assumption of boundary values.

This chapter can be viewed as consisting of two strands. The first is the development of a theory of solutions in Sobolev spaces $W^{2,p}(\Omega)$, $p > 1$ analogous to the Schauder theory in the Hölder spaces $C^{2,\alpha}(\overline{\Omega})$; this theory is generally known as the "L^p theory" and the relevant L^p estimates are themselves of great importance in elliptic theory. The other strand is analogous to our work in Chapters 3 and 8 on maximum principles and local properties of solutions, and the pointwise estimates established in Section 9.7 will also be important in Part II, in particular for the treatment of fully nonlinear equations in Chapter 17. The natural solution space for these considerations is the Sobolev space $W^{2,n}(\Omega)$ and, indeed, the combination of the two strands in this chapter facilitates an attractive theory in this space.

9.1. Maximum Principles for Strong Solutions

In this section, we treat the extension of the classical maximum principle in Chapter 3 to strong solutions, in particular to solutions in the Sobolev space $W^{2,n}_{\text{loc}}(\Omega)$. Recall that an operator L of the form (9.1) is *elliptic* in the domain Ω if the coefficient matrix $\mathscr{A} = [a^{ij}]$ is positive everywhere in Ω. For such operators we will let $\mathscr{D}$ denote the determinant of $\mathscr{A}$ and set $\mathscr{D}^* = \mathscr{D}^{1/n}$ so that $\mathscr{D}^*$ is the geometric mean of the eigenvalues of $\mathscr{A}$ and

$$0 < \lambda \leqslant \mathscr{D}^* \leqslant \Lambda$$

where as before λ, Λ denote, respectively, the minimum and maximum eigenvalues of $\mathscr{A}$. Our conditions on the coefficients of L and inhomogeneous term f in the equation (9.2) will now take the form

$$(9.3) \qquad |b|/\mathscr{D}^*, f/\mathscr{D}^* \in L^n(\Omega), \qquad c \leqslant 0 \quad \text{in } \Omega.$$

The following weak maximum principle, of A. D. Aleksandrov, can now be formulated as an extension of the apriori bound, Theorem 3.7.

Theorem 9.1. *Let $Lu \geqslant f$ in a bounded domain Ω and $u \in C^0(\bar{\Omega}) \cap W^{2,n}_{\text{loc}}(\Omega)$. Then*

$$(9.4) \qquad \sup_{\Omega} u \leqslant \sup_{\partial\Omega} u^+ + C\|f/\mathscr{D}^*\|_{L^n(\Omega)}$$

where C is a constant depending only on n, diam Ω *and $\|b/\mathscr{D}^*\|_{L^n(\Omega)}$.*

The Sobolev embedding theorem, in particular Corollary 7.11, guarantees that functions in $W^{2,n}_{\text{loc}}(\Omega)$ are at least continuous in Ω. If u is not also assumed continuous on $\bar{\Omega}$ in the hypotheses of Theorem 9.1, the conclusion (9.4) can be modified by replacing $\sup_{\partial\Omega} u^+$ by $\limsup_{x\to\partial\Omega} u^+$.

The proof of Theorem 9.1 depends upon the notions of contact set and normal mapping and certain aspects of it will be important for later considerations. If u is an arbitrary continuous function on Ω, we define the *upper contact set* of u, denoted Γ^+ or Γ^+_u, to be the subset of Ω where the graph of u lies below a support hyperplane in $\mathbb{R}^{n+1}$, that is,

$$(9.5) \qquad \Gamma^+ = \{y \in \Omega \mid u(x) \leqslant u(y) + p\cdot(x - y) \quad \text{for all } x \in \Omega, \text{ for some } p = p(y) \in \mathbb{R}^n\}.$$

Clearly, u is a concave function in Ω if and only if $\Gamma^+ = \Omega$. When $u \in C^1(\Omega)$ we must have $p = Du(y)$ in (9.5) as any support hyperplane must then be a tangent hyperplane to the graph of u. Furthermore, when $u \in C^2(\Omega)$, the Hessian matrix $D^2u = [D_{ij}u]$ is nonpositive on Γ^+. In general, the set Γ^+ is closed relative to Ω.

For an arbitrary function $u \in C^0(\Omega)$ we define the *normal mapping*, $\chi(y) = \chi_u(y)$ of a point $y \in \Omega$ to be the set of "slopes" of support hyperplanes at y lying above the graph of u, that is,

$$\chi(y) = \{p \in \mathbb{R}^n \mid u(x) \leqslant u(y) + p \cdot (x - y) \quad \text{for all } x \in \Omega\}. \tag{9.6}$$

Clearly $\chi(y)$ is nonempty if and only if $y \in \Gamma^+$. Furthermore, when $u \in C^1(\Omega)$, then $\chi(y) = Du(y)$ on Γ^+; that is, χ is the gradient vector field of u on Γ^+. As a useful example of a nondifferentiable function u, let us take Ω to be a ball $B = B_R(z)$ and u to be the function whose graph is a cone with base Ω and vertex (z, a) for some positive $a \in \mathbb{R}$, that is,

$$u(x) = a\left(1 - \frac{|x - z|}{R}\right).$$

Then, we have

$$\chi(y) = \begin{cases} \dfrac{-a(y - z)}{R|y - z|} & \text{for } y \neq z \\ B_{a/R}(0) & \text{for } y = z. \end{cases} \tag{9.7}$$

First, we prove:

Lemma 9.2. *For $u \in C^2(\Omega) \cap C^0(\bar{\Omega})$ we have*

$$\sup_{\Omega} u \leqslant \sup_{\partial\Omega} u + \frac{d}{\omega_n^{1/n}} \left(\int_{\Gamma^+} |\det D^2 u| \right)^{1/n} \tag{9.8}$$

where $d = \operatorname{diam} \Omega$.

Proof. By replacing u with $u - \sup_{\partial\Omega} u$, it suffices to assume $u \leqslant 0$ on $\partial\Omega$. The n-dimensional Lebesgue measure of the normal image of Ω is given by

$$\begin{aligned} |\chi(\Omega)| &= |\chi(\Gamma^+)| \\ &= |Du(\Gamma^+)| \\ &\leqslant \int_{\Gamma^+} |\det D^2 u| \end{aligned} \tag{9.9}$$

since $D^2 u \leqslant 0$ on Γ^+. Formula (9.9) can be realised as a consequence of the classical change of variables formula by considering, for positive ε, the mapping $\chi_\varepsilon = \chi - \varepsilon I$, whose Jacobian matrix $D^2 u - \varepsilon I$ is then strictly negative in a neighbourhood of Γ^+, and by subsequently letting $\varepsilon \to 0$. Furthermore, the mapping χ_ε is readily shown to be one-to-one on Γ^+, so that the equality holds in (9.9).

Let us now show that u can be estimated in terms of $|\chi(\Omega)|$. Suppose that u takes a positive maximum at a point $y \in \Omega$, and let k be the function whose graph is the cone K with vertex $(y, u(y))$ and base $\partial\Omega$. Then $\chi_k(\Omega) \subset \chi_u(\Omega)$ since for each supporting hyperplane to K, there exists a parallel hyperplane tangent to the graph of u. Now let $\tilde{k}$ be the function whose graph is the cone $\tilde{K}$ with vertex $(y, u(y))$ and base $B_d(y)$. Clearly, $\chi_{\tilde{k}}(\Omega) \subset \chi_k(\Omega)$; and, consequently,

$$|\chi_{\tilde{k}}(\Omega)| \leqslant |\chi_u(\Omega)|.$$

But then using (9.7) and (9.9), we have

$$\omega_n\left(\frac{u(y)}{d}\right)^n \leqslant \int_{\Gamma^+} |\det D^2u|$$

and hence

$$u(y) \leqslant \frac{d}{\omega_n^{1/n}}\left(\int_{\Gamma^+} |\det D^2u|\right)^{1/n}$$

as required. □

The special case of the estimate (9.4), when $b = 0$, follows from Lemma 9.2 via the matrix inequality

$$\det A \det B \leqslant \left(\frac{\text{trace } AB}{n}\right)^n, \qquad A, B \text{ symmetric} \geqslant 0. \tag{9.10}$$

Taking $A = -D^2u$, $B = [a^{ij}]$, we therefore have on Γ^+

$$\begin{aligned} |\det D^2u| &= \det(-D^2u) \\ &\leqslant \frac{1}{\mathscr{D}}\left(\frac{-a^{ij}D_{ij}u}{n}\right)^n. \end{aligned}$$

We formulate the resulting estimate as follows for later reference.

Lemma 9.3. *For $u \in C^2(\Omega) \cap C^0(\overline{\Omega})$, we have*

$$\sup_\Omega u \leqslant \sup_{\partial\Omega} u + \frac{d}{n\omega_n^{1/n}}\left\|\frac{a^{ij}D_{ij}u}{\mathscr{D}^*}\right\|_{L^n(\Gamma^+)}. \tag{9.11}$$

The full estimate (9.4) is derived from the following generalization of Lemmas 9.2 and 9.3.

Lemma 9.4. *Let g be a nonnegative, locally integrable function on $\mathbb{R}^n$. Then for any $u \in C^2(\Omega) \cap C^0(\bar{\Omega})$, we have*

$$(9.12)\qquad \int_{B_{\tilde{M}}(0)} g \leqslant \int_{\Gamma^+} g(Du)|\det D^2u| \leqslant \int_{\Gamma^+} g(Du)\left(-\frac{a^{ij}D_{ij}u}{n\mathscr{D}^*}\right)^n$$

where

$$\tilde{M} = (\sup_{\Omega} u - \sup_{\partial\Omega} u)/d, \qquad d = \operatorname{diam} \Omega.$$

Proof. The proof of Lemma 9.4 follows those of Lemmas 9.2 and 9.3, which correspond to the special case $g \equiv 1$. Instead of (9.9), we have the more general formula

$$(9.13)\qquad \int_{\chi_u(\Omega)} g \leqslant \int_{\Gamma^+} g(Du)|\det D^2u|$$

and since $\chi_{\tilde{k}}(\Omega) \subset \chi_u(\Omega)$ the estimate (9.12) follows by virtue of (9.7) and (9.10). □

Let us now suppose that $u \in C^0(\bar{\Omega}) \cap C^2(\Omega)$ and satisfies $Lu \geqslant f$ in Ω with condition (9.3). For the weight function g we take

$$g(p) = (|p|^{n/n-1} + \mu^{n/n-1})^{1-n}$$

for some $\mu > 0$, to be fixed later. Then using Hölder's inequality, we have, in $\Omega^+ = \{x \in \Omega \mid u(x) > 0\}$,

$$-\frac{a^{ij}D_{ij}u}{n\mathscr{D}^*} \leqslant \frac{b^iD_iu - f}{n\mathscr{D}^*} \leqslant \frac{|b||Du| + |f|}{n\mathscr{D}^*} \leqslant \frac{(|b|^n + \mu^{-n}|f|^n)^{1/n}}{ng^{1/n}\mathscr{D}^*}.$$

Hence, by (9.12)

$$\int_{B_{\tilde{M}}} g \leqslant \frac{1}{n^n} \int_{\Gamma^+} (|b|^n + \mu^{-n}|f|^n)/\mathscr{D}.$$

The integral on the left-hand side can be estimated using a further consequence of Hölder's inequality,

$$g(p) \geqslant 2^{2-n}(|p|^n + \mu^n)^{-1}.$$

Accordingly, we obtain

$$\omega_n \log\left(\frac{\tilde{M}^n}{\mu^n} + 1\right) \leqslant \frac{2^{n-2}}{n^n} \int_{\Gamma^+} (|b|^n + \mu^{-n}|f|^n)/\mathscr{D}.$$

If $f \not\equiv 0$, we choose $\mu = \|f/\mathscr{D}^*\|_{L^n(\Gamma^+)}$ thus to obtain

$$\tilde{M} \leqslant \left\{\exp\left[\frac{2^{n-2}}{n^n\omega_n} \int_{\Gamma^+} \left(1 + \frac{|b|^n}{\mathscr{D}}\right)\right] - 1\right\}^{1/n} \|f/\mathscr{D}^*\|_{L^n(\Gamma^+)}; \tag{9.14}$$

while for $f \equiv 0$, we let $\mu \to 0$ so that (9.14) is again satisfied.

This establishes the estimate (9.4) for functions $u \in C^0(\bar{\Omega}) \cap C^2(\Omega)$. The extension to functions $u \in C^0(\bar{\Omega}) \cap W^{2,n}_{\text{loc}}(\Omega)$ can be carried out by an approximation argument. Suppose first that L is uniformly elliptic in Ω with the ratio $|b|/\lambda$ also bounded. Let $\{u_m\}$ be a sequence of functions in $C^2(\Omega)$ converging in the sense of $W^{2,n}_{\text{loc}}(\Omega)$ to u. For arbitrary $\varepsilon > 0$, we can then assume that u_m converges to u in $W^{2,n}(\Omega_\varepsilon)$ and $u_m \leqslant \varepsilon + \sup_{\partial\Omega} u$ on $\partial\Omega_\varepsilon$ for some domain $\Omega_\varepsilon \subset\subset \Omega$. Consequently, by applying (9.4) to the functions u_m (with Ω replaced by Ω^+), we obtain

$$\sup_{\Omega_\varepsilon} u_m \leqslant \varepsilon + \sup_{\partial\Omega} u^+ + \frac{C}{\lambda} \|a^{ij}D_{ij}(u_m - u) + b^iD_i(u_m - u)\|_{L^n(\Omega_\varepsilon)} + C\|f/\mathscr{D}^*\|_{L^n(\Omega_\varepsilon)},$$

and hence, letting $m \to \infty$ and using the fact that $\{u_m\}$ converges uniformly to u on Ω_ε, we have

$$\sup_{\Omega_\varepsilon} u \leqslant \varepsilon + \sup_{\partial\Omega} u^+ + C\|f/\mathscr{D}^*\|_{L^n(\Omega_\varepsilon)} \tag{9.15}$$

from which (9.4) follows as $\varepsilon \to 0$.

To remove the above restrictions on L, we consider for $\eta > 0$, the operators

$$L_\eta = \eta(\Lambda + |b|)\Delta + L.$$

We obtain accordingly from (9.15),

$$\sup_{\Omega_\varepsilon} u \leqslant \varepsilon + \sup_{\partial\Omega} u^+ + C\left\{\left\|\frac{\eta(\Lambda + |b|)\Delta u}{\mathscr{D}^*_\eta}\right\|_{L^n(\Omega_\varepsilon)} + \left\|\frac{f}{\mathscr{D}^*}\right\|_{L^n(\Omega_\varepsilon)}\right\}.$$

so that letting $\eta \to 0$ and using the dominated convergence theorem, we get inequality (9.15) again. Theorem 9.1 finally follows by letting $\varepsilon \to 0$. □

Note that the estimate (9.14) does not yield (9.11) when $b = 0$. In fact, the constant in (9.11) can be improved, and a corresponding sharp improvement of (9.14) can be obtained by explicitly integrating the function g and optimizing the choice of μ; (see Problems 9.1, 9.2). Also the dependence on diam Ω can be replaced by a dependence on $|\hat{\Omega}|$, where $\hat{\Omega}$ is the convex hull of Ω; (see Problem 9.3).

The case $f \equiv 0$ in Theorem 9.1 yields an extension of the weak maximum principles, Theorem 3.1 and Corollary 3.2. The following uniqueness result for the Dirichlet problem for strong solutions which extends Theorem 3.3, also follows automatically.

Theorem 9.5. *Let L be elliptic in the bounded domain Ω and satisfy (9.3). Suppose that u and v are functions in $W^{2,n}_{\text{loc}}(\Omega) \cap C^0(\overline{\Omega})$ satisfying $Lu = Lv$ in Ω, $u = v$ on $\partial\Omega$. Then $u = v$ in Ω.*

We can also deduce from the weak maximum principle a generalization of the strong maximum principle, Theorem 3.5. As in the hypotheses of Theorem 3.5, we assume that the operator L is uniformly elliptic in Ω and that $|b|/\lambda$, c/λ are bounded.

Theorem 9.6. *If $u \in W^{2,n}_{\text{loc}}(\Omega)$ satisfies $Lu \geqslant 0$ in Ω and $c = 0$ ($c \leqslant 0$), then u cannot achieve a maximum (nonnegative maximum) in Ω unless it is a constant.*

Proof. If u is differentiable we can follow the proof of Theorem 3.5 with Theorem 9.1 being used in place of Corollary 3.2. In general a slight modification of the proof of Theorem 3.5 suffices. If we assume, contrary to the theorem, that u is nonconstant in Ω and assumes its maximum M in Ω, there must exist concentric balls $B_\rho(y) \subset B_R(y) \subset \Omega$ such that $u < M$ in $\overline{B}_\rho(y)$ and $u(x_0) = M$ for some $x_0 \in B_R(y)$. But then using Theorem 9.1 and the auxiliary function v defined in the proof of Lemma 3.4 with $v(x_0) = 0$, we have $M - u - \varepsilon v > 0$, for some $\varepsilon > 0$, in the annular region $A = B_R(y) - B_\rho(y)$, which is a contradiction at $x = x_0$. □

9.2. L^p Estimates: Preliminary Analysis

Our route to the basic L^p estimates of this chapter is via interpolation. In this section, we develop some preliminary analysis: namely, a cube decomposition procedure, also necessary for the Hölder estimates in Section 9.7; and the Marcinkiewicz interpolation theorem that is applied in the next section.

Cube Decomposition

Let K_0 be a cube in $\mathbb{R}^n$, f a nonnegative integrable function defined in K_0 and t a positive number satisfying

$$\int_{K_0} f \leqslant t|K_0|.$$

By bisection of the edges of K_0, we subdivide K_0 into 2^n congruent subcubes with disjoint interiors. Those subcubes K, which satisfy

$$\int_K f \leqslant t|K|, \tag{9.16}$$

are similarly subdivided and the process is repeated indefinitely. Let $\mathscr{S}$ denote the set of subcubes K thus obtained that satisfy

$$\int_K f > t|K|,$$

and for each $K \in \mathscr{S}$ denote by $\tilde{K}$ the subcube whose subdivision gives K. Since $|\tilde{K}|/|K| = 2^n$, we have for any $K \in \mathscr{S}$.

$$t < \frac{1}{|K|}\int_K f \leqslant 2^n t. \tag{9.17}$$

Furthermore, setting $F = \bigcup_{K \in \mathscr{S}} K$ and $G = K_0 - F$, we have

$$f \leqslant t \quad \text{a.e. in } G. \tag{9.18}$$

The last inequality (9.18) is a consequence of Lebesgue's differentiation theorem [SN], as each point of G lies in a nested sequence of cubes satisfying (9.16) with diameters tending to zero.

For the pointwise estimates in Section 9.7, we also need to consider the set

$$\tilde{F} = \bigcup_{K \in \mathscr{S}} \tilde{K}$$

which satisfies, by (9.16),

$$\int_{\tilde{F}} f \leqslant t|\tilde{F}|. \tag{9.19}$$

In particular, when f is the characteristic function χ_Γ of a measurable subset Γ of K_0, we obtain from (9.18) and (9.19) that

$$|\Gamma| = |\Gamma \cap \tilde{F}| \leqslant t|\tilde{F}|. \tag{9.20}$$

9.3. The Marcinkiewicz Interpolation Theorem

Let f be a measurable function on a domain Ω (bounded or unbounded) in $\mathbb{R}^n$. The *distribution function* $\mu = \mu_f$ of f is defined by

$$\mu(t) = \mu_f(t) = |\{x \in \Omega \,|\, f(x) > t\}| \tag{9.21}$$

for $t > 0$, and measures the relative size of f. Note that μ is a decreasing function on $(0, \infty)$. The basic properties of the distribution function are embodied in the following lemma.

Lemma 9.7. *For any $p > 0$ and $|f|^p \in L^1(\Omega)$, we have*

$$\mu(t) \leqslant t^{-p} \int_\Omega |f|^p, \tag{9.22}$$

$$\int_\Omega |f|^p = p \int_0^\infty t^{p-1}\mu(t)\, dt. \tag{9.23}$$

Proof. Clearly,

$$\mu(t)t^p \leqslant \int_{|f| \geqslant t} |f|^p \leqslant \int_\Omega |f|^p$$

for all $t > 0$ and hence (9.22) follows. Next, suppose that $f \in L^1(\Omega)$. Then, by Fubini's theorem,

$$\int_\Omega |f| = \int_\Omega \int_0^{|f(x)|} dt\, dx = \int_0^\infty \mu(t)\, dt,$$

and the result (9.23) for general p follows by a change of variables. □

We prove the Marcinkiewicz interpolation theorem in the following restricted form:

Theorem 9.8. *Let T be a linear mapping from $L^q(\Omega) \cap L^r(\Omega)$ into itself, $1 \leqslant q < r < \infty$ and suppose there are constants T_1 and T_2 such that*

$$\mu_{Tf}(t) \leqslant \left(\frac{T_1 \|f\|_q}{t}\right)^q, \qquad \mu_{Tf}(t) \leqslant \left(\frac{T_2 \|f\|_r}{t}\right)^r \tag{9.24}$$

for all $f \in L^q(\Omega) \cap L^r(\Omega)$ and $t > 0$. Then T extends as a bounded linear mapping from $L^p(\Omega)$ into itself for any p such that $q < p < r$, and

$$\|Tf\|_p \leqslant C T_1^\alpha T_2^{1-\alpha} \|f\|_p \tag{9.25}$$

for all $f \in L^q(\Omega) \cap L^r(\Omega)$, where

$$\frac{1}{p} = \frac{\alpha}{q} + \frac{1-\alpha}{r}$$

and C depends only on p, q and r.

Proof. For $f \in L^q(\Omega) \cap L^r(\Omega)$ and $s > 0$, we write

$$f = f_1 + f_2,$$

where

$$f_1(x) = \begin{cases} f(x) & \text{if } |f(x)| > s \\ 0 & \text{if } |f(x)| \leqslant s, \end{cases}$$

$$f_2(x) = \begin{cases} 0 & \text{if } |f(x)| > s \\ f(x) & \text{if } |f(x)| \leqslant s. \end{cases}$$

Then

$$|Tf| \leqslant |Tf_1| + |Tf_2|,$$

and hence

$$\begin{aligned} \mu(t) = \mu_{Tf}(t) &\leqslant \mu_{Tf_1}(t/2) + \mu_{Tf_2}(t/2). \\ &\leqslant \left(\frac{2T_1}{t}\right)^q \int_\Omega |f_1|^q + \left(\frac{2T_2}{t}\right)^r \int_\Omega |f_2|^r. \end{aligned}$$

Therefore, by Lemma 9.7 we have

$$\int_\Omega |Tf|^p = p\int_0^\infty t^{p-1}\mu(t)\,dt$$

$$\leqslant p(2T_1)^q \int_0^\infty t^{p-1-q}\left(\int_{|f|>s} |f|^q\right) dt$$

$$+ p(2T_2)^r \int_0^\infty t^{p-1-r}\left(\int_{|f|\leqslant s} |f|^r\right) dt$$

Let us now choose s as a function of t; in particular, we take $t = As$ for some positive number A to be fixed later. Then, we have

$$\int_\Omega |Tf|^p \leqslant p(2T_1)^q A^{p-q}\int_0^\infty s^{p-1-q}\left(\int_{|f|>s} |f|^q\right) ds$$

$$+ p(2T_2)^r A^{p-r}\int_0^\infty s^{p-1-r}\left(\int_{|f|\leqslant s} |f|^r\right) ds.$$

But

$$\int_0^\infty s^{p-1-q}\left(\int_{|f|>s} |f|^q\right) ds = \int_\Omega |f|^q \int_0^{|f|} s^{p-1-q}\,ds$$

$$= \frac{1}{p-q}\int_\Omega |f|^p,$$

and, similarly,

$$\int_0^\infty s^{p-1-r}\left(\int_{|f|\leqslant s} |f|^r\right) ds = \int_\Omega |f|^r\left(\int_{|f|}^\infty s^{p-1-r}\,ds\right)$$

$$= \frac{1}{r-p}\int_\Omega |f|^p.$$

Consequently,

$$\int_\Omega |Tf|^p \leqslant \left\{\frac{p}{p-q}(2T_1)^q A^{p-q} + \frac{p}{r-p}(2T_2)^r A^{p-r}\right\} \int_\Omega |f|^p$$

for any positive number A. By taking the value of A for which the expression in braces is a minimum, namely,

$$A = 2T_1^{q/(r-q)} T_2^{r/(r-q)},$$

we thus obtain

$$\|Tf\|_p \leqslant 2\left(\frac{p}{p-q} + \frac{p}{r-p}\right)^{1/p} T_1^\alpha T_2^{1-\alpha} \|f\|_p$$

as required, with $C = 2\{p(r-q)/[(p-q)(r-p)]\}^{1/p}$ in the statement of the theorem. □

9.4. The Calderon-Zygmund Inequality

In this section we establish the basic L^p estimates for Poisson's equation through a further consideration of the Newtonian potential, previously treated in Chapter 4. Let Ω be a bounded domain in $\mathbb{R}^n$ and f a function in $L^p(\Omega)$ for some $p \geqslant 1$. Recall that the Newtonian potential of f is the function $w = Nf$ defined by the convolution.

$$w(x) = \int_\Omega \Gamma(x-y) f(y)\, dy, \tag{9.26}$$

where Γ is the fundamental solution of Laplace's equation given by (4.1). The following result, which embraces a special case of the Calderon-Zygmund inequality, is the L^p analogue of the Hölder estimate of Lemma 4.4.

Theorem 9.9. *Let $f \in L^p(\Omega)$, $1 < p < \infty$, and let w be the Newtonian potential of f. Then $w \in W^{2,p}(\Omega)$, $\Delta w = f$ a.e. and*

$$\|D^2 w\|_p \leqslant C\|f\|_p \tag{9.27}$$

where C depends only on n and p. Furthermore, when $p = 2$ we have

$$\int_{R^n} |D^2 w|^2 = \int_\Omega f^2 \tag{9.28}$$

Proof. (i) Let us deal first with the case $p = 2$. If $f \in C_0^\infty(\mathbb{R}^n)$, we have $w \in C^\infty(\mathbb{R}^n)$ and, by Lemma 4.3, $\Delta w = f$. Consequently, for any ball B_R containing the support of f,

$$\int_{B_R} (\Delta w)^2 = \int_{B_R} f^2.$$

Applying Green's first identity (2.10) twice, we then obtain

$$\begin{aligned}\int_{B_R} |D^2 w|^2 &= \int_{B_R} \sum (D_{ij} w)^2 \\ &= \int_{B_R} f^2 + \int_{\partial B_R} Dw \cdot \frac{\partial}{\partial \nu} Dw.\end{aligned}$$

Using (2.14) we have

$$Dw = O(R^{1-n}),\ D^2 w = O(R^{-n})$$

uniformly on ∂B_R as $R \to \infty$, whence the identity (9.28) follows. To extend (9.28) to arbitrary $f \in L^2(\Omega)$, we observe first that, by Lemma 7.12, N is a bounded mapping from L^p into itself for $1 \leqslant p < \infty$. The full strength of Theorem 9.9 in the case $p = 2$ then follows by approximation. Indeed, if the sequence $\{f_m\} \subset C_0^\infty(\Omega)$ converges to f in $L^2(\Omega)$, the sequence of Newtonian potentials $\{Nf_m\}$ converges to w in $W^{2,2}(\Omega)$.

(ii) For fixed i, j, we now define the linear operator $T: L^2(\Omega) \to L^2(\Omega)$ by

$$Tf = D_{ij} w.$$

From Lemma 9.7 and (9.28), we have

$$\mu(t) = \mu_{Tf}(t) \leqslant \left(\frac{\|f\|_2}{t}\right)^2 \tag{9.29}$$

for all $t > 0$ and $f \in L^2(\Omega)$. We now show that, in addition,

$$\mu(t) \leqslant \frac{C\|f\|_1}{t} \tag{9.30}$$

for all $t > 0$ and $f \in L^2(\Omega)$, thereby making possible the application of the Marcinkiewicz interpolation theorem. To accomplish this we first extend f to vanish outside Ω and fix a cube $K_0 \supset \Omega$, so that for fixed $t > 0$ we have

$$\int_{K_0} f \leqslant t|K_0|.$$

The cube K_0 is now decomposed according to the procedure described in Section 9.2 yielding a sequence of parallel subcubes $\{K_l\}_{l=1}^{\infty}$ such that

$$t < \frac{1}{|K_l|} \int_{K_l} |f| < 2^n t \tag{9.31}$$

and

$$|f| \leqslant t \quad \text{a.e. on } G = K_0 - \cup K_l.$$

The function f is now split into a "good part" g defined by

$$g(x) = \begin{cases} f(x) & \text{for } x \in G \\ \dfrac{1}{|K_l|} \displaystyle\int_{K_l} f & \text{for } x \in K_l, \quad l = 1, 2, \ldots, \end{cases}$$

and a "bad part" $b = f - g$. Clearly,

$$|g| \leqslant 2^n t \quad \text{a.e.,}$$

$$b(x) = 0 \qquad \text{for } x \in G,$$

$$\int_{K_l} b = 0 \qquad \text{for } l = 1, 2, \ldots.$$

Since T is linear, $Tf = Tg + Tb$; and, hence,

$$\mu_{Tf}(t) \leqslant \mu_{Tg}(t/2) + \mu_{Tb}(t/2).$$

(iii) *Estimation of* Tg: By (9.29)

$$\begin{aligned} \mu_{Tg}(t/2) &\leqslant \frac{4}{t^2} \int g^2 \\ &\leqslant \frac{2^{n+2}}{t} \int |g| \\ &\leqslant \frac{2^{n+2}}{t} \int |f| \end{aligned}$$

(iv) *Estimation of Tb*: Writing

$$b_l = b\chi_{K_l} = \begin{cases} b & \text{on } K_l \\ 0 & \text{elsewhere,} \end{cases}$$

we have

$$Tb = \sum_{l=1}^{\infty} Tb_l.$$

Let us now fix some l and a sequence $\{b_{lm}\} \subset C_0^\infty(K_l)$ converging to b_l in $L^2(\Omega)$ and satisfying

$$\int_{K_l} b_{lm} = \int_{K_l} b_l = 0.$$

Then for $x \notin K_l$, we have the formula

$$\begin{aligned} Tb_{lm}(x) &= \int_{K_l} D_{ij}\Gamma(x - y) b_{lm}(y)\, dy \\ &= \int_{K_l} \{D_{ij}\Gamma(x - y) - D_{ij}\Gamma(x - \bar{y})\} b_{lm}(y)\, dy, \end{aligned}$$

where $\bar{y} = \bar{y}_l$ denotes the center of K_l. Letting $\delta = \delta_l$ denote the diameter of K_l, we then obtain (with an estimation similar to that of the integral I_6 in the proof of Lemma 4.4),

$$|Tb_{lm}(x)| \leqslant C(n)\delta[\operatorname{dist}(x, K_l)]^{-n-1} \int_{K_l} |b_{lm}(y)|\, dy.$$

Letting $B_l = B_\delta(\bar{y})$ denote the concentric ball of radius δ, we obtain by integration

$$\begin{aligned} \int_{K_0 - B_l} |Tb_{lm}| &\leqslant C(n)\delta \int_{|x| \geqslant \delta/2} \frac{dx}{|x|^{n+1}} \int_{K_l} |b_{lm}| \\ &\leqslant C(n) \int_{K_l} |b_{lm}|. \end{aligned}$$

Consequently, letting $m \to \infty$, writing $F^* = \cup B_l, G^* = K_0 - F^*$ and summing over l, we get

$$\int_{G^*} |Tb| \leqslant C(n) \int |b|$$

$$\leqslant C(n) \int |f|,$$

so that by Lemma 9.7

$$|\{x \in G^* \,|\, |Tb| > t/2\}| \leqslant \frac{C\|f\|_1}{t}.$$

However, by (9.31),

$$|F^*| \leqslant \omega_n n^{n/2} |F|$$

$$\leqslant \frac{C\|f\|_1}{t},$$

and thus (9.30) holds.

(v) To conclude the proof of Theorem 9.9, we note that (9.29) and (9.30) fulfill the hypotheses of the Marcinkiewicz interpolation theorem (Theorem 9.8) with $q = 1, r = 2$. Consequently,

$$\|Tf\|_p \leqslant C(n, p)\|f\|_p \tag{9.32}$$

for all $1 < p \leqslant 2$ and $f \in L^2(\Omega)$. The inequality (9.32) is extended to $p > 2$ by duality. For if $f, g \in C_0^\infty(\Omega)$, then

$$\begin{aligned}\int_\Omega (Tf)g &= \int_\Omega w D_{ij} g \\ &= \int_\Omega \int_\Omega \Gamma(x - y) f(y) D_{ij} g(x) \, dx \, dy \\ &= \int_\Omega f Tg \\ &\leqslant \|f\|_p \|Tg\|_{p'}.\end{aligned}$$

Thus if $p > 2$, we have from (9.32),

$$\|Tf\|_p = \sup\left\{\int_\Omega (Tf)g \mid \|g\|_{p'} = 1\right\}$$

$$\leqslant C(n, p')\|f\|_p,$$

so that (9.32) holds for all $1 < p < \infty$. As in the case $p = 2$, we can then infer the full conclusion of Theorem 9.9 by approximation. □

Note that T can be defined as a bounded operator on $L^p(\Omega)$ even if Ω is unbounded, in which case the conclusion of Theorem 9.9 still holds provided $n \geqslant 3$. Other approaches to the derivation of inequality (9.27) are discussed in the Notes to this chapter.

The L^p estimates for solutions of Poisson's equation follow immediately from Theorem 9.9.

Corollary 9.10. *Let Ω be a domain in $\mathbb{R}^n$, $u \in W_0^{2,p}(\Omega)$, $1 < p < \infty$. Then*

$$\|D^2u\|_p \leqslant C\|\Delta u\|_p \tag{9.33}$$

where $C = C(n, p)$. If $p = 2$,

$$\|D^2u\|_2 = \|\Delta u\|^2. \tag{9.34}$$

9.5. L^p Estimates

In this section, we derive interior and global L^p estimates for the second derivatives of elliptic equations of the forms (9.1) and (9.2). The technique of perturbation from the constant coefficient case is similar to that used for the derivation of the Schauder estimates in Sections 6.1 and 6.2. We first deal with interior estimates; the following theorem is analogous to Theorem 6.1.

Theorem 9.11. *Let Ω be an open set in $\mathbb{R}^n$ and $u \in W_{\mathrm{loc}}^{2,p}(\Omega) \cap L^p(\Omega)$, $1 < p < \infty$, a strong solution of the equation $Lu = f$ in Ω where the coefficients of L satisfy, for positive constants λ, Λ,*

$$a^{ij} \in C^0(\Omega), \quad b^i, c \in L^\infty(\Omega), \quad f \in L^p(\Omega);$$

$$a^{ij}\xi_i\xi_j \geqslant \lambda|\xi|^2 \quad \forall \xi \in \mathbb{R}^n; \tag{9.35}$$

$$|a^{ij}|, |b^i|, |c| \leqslant \Lambda,$$

where $i, j = 1, \ldots, n$. Then for any domain $\Omega' \subset\subset \Omega$,

$$\|u\|_{2,p;\Omega'} \leqslant C(\|u\|_{p;\Omega} + \|f\|_{p;\Omega}), \tag{9.36}$$

where C depends on n, p, λ, Λ, Ω', Ω and the moduli of continuity of the coefficients a^{ij} on Ω'.

Proof. For a fixed point $x_0 \in \Omega'$, we let L_0 denote the constant coefficient operator given by

$$L_0 u = a^{ij}(x_0) D_{ij} u.$$

By means of the linear transformation Q used in the proof of Lemma 6.1, we obtain from Corollary 9.10, the estimate

$$\|D^2 v\|_{p;\Omega} \leqslant \frac{C}{\lambda} \|L_0 v\|_{p;\Omega} \tag{9.37}$$

for any $v \in W_0^{2,p}(\Omega)$, where $C = C(n, p)$ as in (9.33). Consequently, if v has support in a ball $B_R = B_R(x_0) \subset\subset \Omega$, we have

$$L_0 v = (a^{ij}(x_0) - a^{ij}) D_{ij} v + a^{ij} D_{ij} v,$$

and by (9.37)

$$\|D^2 v\|_p \leqslant \frac{C}{\lambda} (\sup_{B_R} |a - a(x_0)| \, \|D^2 v\|_p + \|a^{ij} D_{ij} v\|_p),$$

where $a = [a^{ij}]$. Since a is uniformly continuous on Ω', there exists a positive number δ such that

$$|a - a(x_0)| \leqslant \lambda / 2C$$

if $|x - x_0| < \delta$, and hence

$$\|D^2 v\|_p \leqslant C \|a^{ij} D_{ij} v\|_p$$

provided $R \leqslant \delta$, where $C = C(n, p, \lambda)$.

For $\sigma \in (0, 1)$, we now introduce a cutoff function $\eta \in C_0^2(B_R)$ satisfying $0 \leqslant \eta \leqslant 1$, $\eta = 1$ in $B_{\sigma R}$, $\eta = 0$ for $|x| \geqslant \sigma' R$, $\sigma' = (1 + \sigma)/2$, $|D\eta| \leqslant 4/(1 - \sigma)R$, $|D^2 \eta| \leqslant 16/(1 - \sigma)^2 R^2$. Then, if $u \in W_{\mathrm{loc}}^{2,p}(\Omega)$ satisfies $Lu = f$ in Ω and $v = \eta u$, we obtain

$$\begin{aligned} \|D^2 u\|_{p;B_{\sigma R}} &\leqslant C \|\eta a^{ij} D_{ij} u + 2 a^{ij} D_i \eta D_j u + u a^{ij} D_{ij} \eta\|_{p;B_R} \\ &\leqslant C \left(\|f\|_{p;B_R} + \frac{1}{(1 - \sigma)R} \|Du\|_{p;B_{\sigma' R}} \right. \\ &\qquad \left. + \frac{1}{(1 - \sigma)^2 R^2} \|u\|_{p;B_R} \right) \end{aligned}$$

provided $R \leqslant \delta \leqslant 1$, where $C = C(n, p, \lambda, \Lambda)$.

Introducing the weighted seminorms

$$\Phi_k = \sup_{0<\sigma<1} (1-\sigma)^k R^k \|D^k u\|_{p;B_{\sigma R}}, \qquad k = 0, 1, 2,$$

we, therefore, have

$$\Phi_2 \leqslant C(R^2\|f\|_{p;B_R} + \Phi_1 + \Phi_0). \tag{9.38}$$

We claim now that Φ_k satisfy an interpolation inequality

$$\Phi_1 \leqslant \varepsilon\Phi_2 + \frac{C}{\varepsilon}\Phi_0 \tag{9.39}$$

for any $\varepsilon > 0$, where $C = C(n)$. By its invariance under coordinate stretching it suffices to prove (9.39) for the case $R = 1$.

For $\gamma > 0$, we fix $\sigma = \sigma_\gamma$ so that

$$\begin{aligned}\Phi_1 &\leqslant (1-\sigma_\gamma)\|Du\|_{p;B_\sigma} + \gamma \\ &\leqslant \varepsilon(1-\sigma)^2\|D^2u\|_{p;B_\sigma} + \frac{C}{\varepsilon}\|u\|_{p;B_\sigma} + \gamma\end{aligned}$$

by Theorem 7.28, so that letting $\gamma \to 0$, we obtain (9.39). Using (9.39) in (9.38), we then get

$$\Phi_2 \leqslant C(R^2\|f\|_{p;B_R} + \Phi_0),$$

that is,

$$\|D^2u\|_{p;B_{\sigma R}} \leqslant \frac{C}{(1-\sigma)^2R^2}(R^2\|f\|_{p;B_R} + \|u\|_{p;B_R}), \tag{9.40}$$

where $C = C(n, p, \lambda, \Lambda)$ and $0 < \sigma < 1$.

The desired estimate (9.36) follows by taking $\sigma = \frac{1}{2}$ and covering Ω' with a finite number of balls of radius $R/2$ for $R \leqslant \min\{\delta, \operatorname{dist}(\Omega', \partial\Omega)\}$. □

In order to extend Theorem 9.11 to the boundary $\partial\Omega$, we first consider the case of a flat boundary portion. Letting

$$\begin{aligned}\Omega^+ &= \Omega \cap \mathbb{R}^n_+ = \{x \in \Omega \mid x_n > 0\}, \\ (\partial\Omega)^+ &= (\partial\Omega) \cap \mathbb{R}^n_+ = \{x \in \partial\Omega \mid x_n > 0\},\end{aligned}$$

we have the following extension of Corollary 9.10.

Lemma 9.12. *Let $u \in W_0^{1,1}(\Omega^+)$, $f \in L^p(\Omega^+)$, $1 < p < \infty$, saitisfy $\Delta u = f$ weakly in Ω^+ with $u = 0$ near $(\partial\Omega)^+$. Then $u \in W^{2,p}(\Omega^+) \cap W_0^{1,p}(\Omega^+)$ and*

$$(9.41) \qquad \|D^2 u\|_{p;\Omega^+} \leqslant C\|f\|_{p;\Omega^+},$$

where $C = C(n, p)$.

Proof. We extend u and f to all of $\mathbb{R}^n_+$ by setting $u = f = 0$ in $\mathbb{R}^n_+ - \Omega$, and then to all of $\mathbb{R}^n$ by odd reflection, that is, by setting

$$u(x', x_n) = -u(x', -x_n), \qquad f(x', x_n) = -f(x', -x_n)$$

for $x_n < 0$, where $x' = (x_1, \ldots, x_{n-1})$. It follows that the extended functions satisfy $\Delta u = f$ weakly in $\mathbb{R}^n$. To show this we take an arbitrary test function $\varphi \in C_0^1(\mathbb{R}^n)$, and for $\varepsilon > 0$ let η be an even function in $C^1(\mathbb{R})$ such that $\eta(t) = 0$ for $|t| \leqslant \varepsilon$, $\eta(t) = 1$ for $|t| \geqslant 2\varepsilon$ and $|\eta'| \leqslant 2/\varepsilon$. Then

$$\begin{aligned} -\int \eta f \varphi &= \int Du \cdot D(\eta\varphi) \\ &= \int \eta Du \cdot D\varphi + \int \varphi \eta' D_n u. \end{aligned}$$

Now

$$\begin{aligned} \left| \int \varphi \eta' D_n u \right| &= \left| \int_{0 < x_n < 2\varepsilon} (\varphi(x', x_n) - \varphi(x', -x_n))\eta' D_n u \right| \\ &\leqslant 8 \max |D\varphi| \int_{0 < x_n < 2\varepsilon} |D_n u| \\ &\to 0 \qquad \text{as } \varepsilon \to 0. \end{aligned}$$

Consequently, letting $\varepsilon \to 0$, we obtain

$$-\int f\varphi = \int Du \cdot D\varphi,$$

so that $u \in W^{1,1}(\mathbb{R}^n)$ is a weak solution of $\Delta u = f$.

Since u also has compact support in $\mathbb{R}^n$, the regularization $u_h \in C_0^\infty(\mathbb{R}^n)$, and satisfies $\Delta u_h = f_h$ in $\mathbb{R}^n$. Hence, by Lemma 7.2 and Corollary 9.10, $u_h \to u$ in $W^{2,p}(\mathbb{R}^n)$ as $h \to 0$ and, moreover, u satisfies the estimate (9.33). However, then the estimate (9.41) follows with constant C twice that in (9.33). Since $u_h(x', 0) = 0$, we also obtain $u \in W_0^{1,p}(\Omega^+)$. □

For the global estimate, we require that boundary values are taken on in the sense of $W^{1,p}(\Omega)$. If T is a subset of $\partial\Omega$ and $u \in W^{1,p}(\Omega)$, we say that $u = 0$ on T in the sense of $W^{1,p}(\Omega)$ if u is the limit in $W^{1,p}(\Omega)$ of a sequence of functions in $C^1(\Omega)$ vanishing near T. For the case $p = 2$, this coincides with the definition used in Section 8.10 and, when u is continuous on T, is implied by u vanishing on T in the usual pointwise sense. With the aid of Lemma 9.12 we now derive a local boundary estimate.

Theorem 9.13. *Let Ω be a domain in $\mathbb{R}^n$ with a $C^{1,1}$ boundary portion $T \subset \partial\Omega$. Let $u \in W^{2,p}(\Omega)$, $1 < p < \infty$, be a strong solution of $Lu = f$ in Ω with $u = 0$ on T, in the sense of $W^{1,p}(\Omega)$, where L satisfies (9.35) with $a^{ij} \in C^0(\Omega \cup T)$. Then, for any domain $\Omega' \subset\subset \Omega \cup T$,*

$$\|u\|_{2,p;\Omega'} \leqslant C(\|u\|_{p;\Omega} + \|f\|_{p;\Omega}) \tag{9.42}$$

where C depends on n, p, λ, Λ, T, Ω', Ω and the moduli of continuity of the coefficients a^{ij} on Ω'.

Proof. Since $T \in C^{1,1}$, for each point $x_0 \in T$ there is a neighbourhood $\mathcal{N} = \mathcal{N}_{x_0}$ and a diffeomorphism $\psi = \psi_{x_0}$ from N onto the unit ball $B = B_1(0)$ in $\mathbb{R}^n$ such that $\psi(\mathcal{N} \cap \Omega) \subset \mathbb{R}^n_+$, $\psi(\mathcal{N} \cap \partial\Omega) \subset \partial\mathbb{R}^n_+$, $\psi \in C^{1,1}(\mathcal{N})$, $\psi^{-1} \in C^{1,1}(B)$. As in Lemma 6.5, writing $y = \psi(x) = (\psi_1(x), \ldots, \psi_n(x))$, $\tilde{u}(y) = u(x)$, $x \in \mathcal{N}$, $y \in B$, we have

$$\tilde{L}\tilde{u} = \tilde{a}^{ij}D_{ij}\tilde{u} + \tilde{b}^iD_i\tilde{u} + \tilde{c}\tilde{u} = \tilde{f}$$

in B^+, where

$$\tilde{a}^{ij}(y) = \frac{\partial\psi_i}{\partial x_r}\frac{\partial\psi_j}{\partial x_s}a^{rs}(x), \qquad \tilde{b}^i(y) = \frac{\partial^2\psi_i}{\partial x_r\,\partial x_s}a^{rs}(x) + \frac{\partial\psi_i}{\partial x_r}b^r(x),$$

$$\tilde{c}(y) = c(x), \qquad \tilde{f}(y) = f(x)$$

so that $\tilde{L}$ satisfies conditions similar to (9.35) with constants $\tilde{\lambda}$, $\tilde{\Lambda}$ depending on λ, Λ and ψ. Furthermore, $\tilde{u} \in W^{2,p}(B^+)$, and $\tilde{u} = 0$ on $B \cap \partial\mathbb{R}^n_+$ in the sense of $W^{1,p}(B^+)$.

We now proceed as in the proof of Theorem 9.11 with the ball $B_R(x_0)$ replaced by the half ball $B^+_R(0) \subset B$ and with Lemma 9.12 used in place of Corollary 9.10. We obtain thus, instead of (9.40), the estimate

$$\|D^2\tilde{u}\|_{p;B^+_{\sigma R}} \leqslant \frac{C}{(1-\sigma)^2R^2}\{R^2\|\tilde{f}\|_{p;B^+_R} + \|\tilde{u}\|_{p;B^+_R}\}$$

provided $R \leqslant \delta \leqslant 1$, where C depends on n, p, λ, Λ and ψ; and δ depends on the moduli of continuity of a^{ij} at x_0 and also on ψ. Taking $\sigma = \frac{1}{2}$ and $\tilde{\mathcal{N}} = \tilde{\mathcal{N}}_{x_0} = \psi^{-1}(B_{\delta/2})$ we therefore have, on returning to our original coordinates.

$$\|D^2u\|_{p;\tilde{\mathcal{N}}} \leqslant C(\|u\|_{p;\mathcal{N}} + \|f\|_{p;\mathcal{N}})$$

where $C = C(n, p, \lambda, \Lambda, \delta, \psi)$. Finally, by covering $\Omega' \cap T$ with a finite number of such neighbourhoods $\tilde{\mathcal{N}}$ and using also the interior estimate (9.36), we obtain the desired estimate (9.42). $\square$

When $T = \partial\Omega$ in Theorem 9.13 we may take $\Omega' = \Omega$ to obtain a global $W^{2,p}(\Omega)$ estimate. This estimate can, in fact, be refined as follows.

Theorem 9.14. *Let Ω be a $C^{1,1}$ domain in $\mathbb{R}^n$ and suppose the operator L satisfies the conditions (9.35) with $a^{ij} \in C^0(\overline{\Omega})$, $i, j = 1, \ldots, n$. Then if $u \in W^{2,p}(\Omega) \cap W_0^{1,p}(\Omega)$, $1 < p < \infty$, we have*

$$\|u\|_{2,p;\Omega} \leqslant C\|Lu - \sigma u\|_{p;\Omega} \tag{9.43}$$

for all $\sigma \geqslant \sigma_0$, where C and σ_0 are positive constants depending only on n, p, λ, Λ, Ω and the moduli of continuity of the coefficients a^{ij}.

Proof. We define a domain Ω_0 in $\mathbb{R}^{n+1}(x, t)$ by

$$\Omega_0 = \Omega \times (-1, 1)$$

together with the operator L_0, given by

$$L_0 v = Lv + D_{tt} v,$$

for $v \in W^{2,p}(\Omega_0)$. Then, if $u \in W^{2,p}(\Omega) \cap W_0^{1,p}(\Omega)$, the function v, given by

$$v(x, t) = u(x) \cos \sigma^{1/2} t,$$

belongs to $W^{2,p}(\Omega_0)$ and vanishes on $\partial\Omega \times (-1, 1)$ in the sense of $W^{1,p}(\Omega_0)$. Furthermore,

$$L_0 v = \cos \sigma^{1/2} t(Lu - \sigma u),$$

so that by Theorem 9.13 with $\Omega' = \Omega \times (-\varepsilon, \varepsilon)$, $0 < \varepsilon \leqslant \frac{1}{2}$, we get

$$\|D_{tt} v\|_{p;\Omega'} \leqslant C(\|Lu - \sigma u\|_{p;\Omega} + \|u\|_{p;\Omega}),$$

where C depends on the quantities listed in the statement of the Theorem. But now, taking $\varepsilon = \pi/3\sigma^{1/2}$, we have

$$\begin{aligned}
\|D_{tt} v\|_{p;\Omega'} &= \sigma\|v\|_{p;\Omega'} \\
&\geqslant \sigma \cos(\sigma^{1/2}\varepsilon)(2\varepsilon)^{1/p}\|u\|_{p;\Omega} \\
&\geqslant \frac{1}{2}\left(\frac{2\pi}{3}\right)^{1/p} \sigma^{1-1/2p}\|u\|_{p;\Omega}
\end{aligned}$$

so that if σ is sufficiently large

$$\|u\|_{p;\Omega} \leqslant C\|Lu - \sigma u\|_{p;\Omega}, \tag{9.44}$$

and hence (9.43) follows from Theorem 9.13. □

Note that when $p \geqslant n$, Theorem 9.14 follows directly from Theorems 9.1 and 9.13. In the Hilbert space approach in Chapter 8, the analogue of Theorem 9.14 is Lemma 8.4 and, indeed, it is possible to deduce the case $p = 2$ from Lemma 8.4. We remark also that by invoking the Sobolev imbedding theorem (see, in particular, Corollary 7.11) in the proofs of Theorems 9.12, 9.13 and 9.14, we may weaken the conditions on the lower order coefficients of L to $b_i \in L^q(\Omega)$, $c \in L^r(\Omega)$, where $q > n$ if $p \leqslant n$, $q = p$ if $p > n$, $r > n/2$ if $p \leqslant n/2$, $r = p$ if $p > n/2$.

9.6. The Dirichlet Problem

The main objective of this section is the following existence and uniqueness theorem for the Dirichlet problem for strong solutions.

Theorem 9.15. *Let Ω be a $C^{1,1}$ domain in $\mathbb{R}^n$, and let the operator L be strictly elliptic in Ω with coefficients $a^{ij} \in C^0(\overline{\Omega})$, $b^i, c \in L^\infty$, with $i, j = 1, \ldots, n$ and $c \leqslant 0$. Then, if $f \in L^p(\Omega)$ and $\varphi \in W^{2,p}(\Omega)$, with $1 < p < \infty$, the Dirichlet problem $Lu = f$ in Ω, $u - \varphi \in W_0^{1,p}(\Omega)$ has a unique solution $u \in W^{2,p}(\Omega)$.*

Proof. There are various methods for deducing Theorem 9.15 from previous existence theorems in Chapters 4, 6 and 8. For example, the case $p \geqslant n$ may be derived from either Theorems 6.14 or 8.14 by appropriate approximation (Problem 9.7), or alternatively from the special case of Poisson's equation by the method of continuity (Problem 9.8). Our treatment is based on the case $p = 2$ already covered by Theorems 8.9 and 8.12 under stronger coefficient hypotheses. We shall need the following regularity result which in fact is a refinement of Theorems 9.11 and 9.13.

Lemma 9.16. *In addition to the hypotheses of Theorem 9.13, suppose that $f \in L^q(\Omega)$ for some $q \in (p, \infty)$. Then, $u \in W^{2,q}_{\mathrm{loc}}(\Omega \cup T)$, $u = 0$ on T in the sense of $W^{1,q}(\Omega)$, and consequently, u satisfies the estimate (9.42) with p replaced by q.*

Proof. We first treat the interior case when T is empty. Returning to the proof of Theorem 9.11, we fix a ball $B_R = B_R(x_0)$ and a cutoff function η, and set $v = \eta u$, $g = a^{ij}D_{ij}v$, so that

$$L_0 v = (a^{ij}(x_0) - a^{ij}(x))D_{ij}v + g.$$

Since $Lu = f$, it follows from the Sobolev imbedding theorem that $g \in L^r(\Omega)$ where $1/r = \max\{(1/q), (1/p) - (1/n)\}$. By means of the linear transformation Q, we can diagonalize the matrix $[a^{ij}(x_0)]$ so that the operator L_0 becomes the Laplacian, and hence

$$\Delta\tilde{v} = (\delta^{ij} - \tilde{a}^{ij}(x))D_{ij}\tilde{v} + \tilde{g},$$

where $\tilde{v}, \tilde{a}^{ij}, \tilde{g}$ correspond to v, a^{ij}, g, respectively. By taking the Newtonian potential, we then obtain the equation

$$\tilde{v} = N[(\delta^{ij} - a^{ij}(x))D_{ij}\tilde{v}] + N\tilde{g}.$$

Consequently, the function v satisfies an equation of the form

$$v = Tv + h, \tag{9.45}$$

where by virtue of the Calderon-Zygmund estimate (Theorem 9.9) T is a bounded linear mapping from $W^{2,p}(B_R)$ into itself for any $p \in (1, \infty)$, $h \in L^r(B_R)$ and if, as in the proof of Theorem 9.11, $R \leqslant \delta$ we must have $\|T\| \leqslant \frac{1}{2}$. Therefore, by the contraction mapping principle (Theorem 5.1), (9.45) has a unique solution $v \in W^{2,p}(B_R)$ for any $p \in [1, r]$. Hence, $\eta u \in W^{2,r}(\Omega)$, and, since $x_0 \in \Omega$ is arbitrary, we obtain $u \in W^{2,r}_{\text{loc}}(\Omega)$. If now $r = q$, we are done.

Otherwise, the desired interior regularity follows by using the Sobolev imbedding theorem and repeating the above argument. The case of local boundary regularity is handled similarly with $x_0 \in T$ and the ball $B_R(x_0)$ replaced by the half-ball $B_R^+(0)$ as in the proof of Theorem 9.13. □

The uniqueness assertion of Theorem 9.15 follows from Lemma 9.16, for if the operator L satisfies the hypotheses of Theorem 9.15 and the functions $u, v \in W^{2,p}(\Omega)$ satisfy $Lu = Lv$ in Ω, $u - v \in W_0^{1,p}(\Omega)$, we have, by Lemma 9.16, $u - v \in W^{2,q}(\Omega) \cap W_0^{1,q}(\Omega)$ for all $1 < q < \infty$. Now using the uniqueness result, Theorem 9.5, and the Sobolev imbedding (Theorem 7.10), we conclude $u = v$. From the uniqueness, we can derive an apriori bound which extends Corollary 9.14.

Lemma 9.17. *Let the operator L satisfy the hypotheses of Theorem 9.15. Then there exists a constant C (independent of u) such that*

$$\|u\|_{2,p;\Omega} \leqslant C\|Lu\|_{p;\Omega} \tag{9.46}$$

for all $u \in W^{2,p}(\Omega) \cap W_0^{1,p}(\Omega)$, $1 < p < \infty$.

Proof. We argue by contradiction. If (9.46) is not true, there must exist a sequence $\{v_m\} \subset W^{2,p}(\Omega) \cap W_0^{1,p}(\Omega)$ satisfying

$$\|v_m\|_{p;\Omega} = 1; \qquad \|Lv_m\|_{p;\Omega} \to 0.$$

By virtue of the apriori estimate (Theorem 9.13), the compactness of the imbedding $W_0^{1,p}(\Omega) \to L^p(\Omega)$, and the weak compactness of bounded sets in $W^{2,p}(\Omega)$ (Problem 5.5), there exists a subsequence, which we relabel as $\{v_m\}$, converging weakly to a function $v \in W^{2,p}(\Omega) \cap W_0^{1,p}(\Omega)$ satisfying $\|v\|_{p;\Omega} = 1$. Since

$$\int_\Omega gD^\alpha v_m \to \int_\Omega gD^\alpha v$$

for all $|\alpha| \leqslant 2$ and $g \in L^{p/(p-1)}(\Omega)$, we must have

$$\int_\Omega gLv = 0$$

for all $g \in L^{p/(p-1)}(\Omega)$; hence $Lv = 0$ and $v = 0$ by the uniqueness assertion, which contradicts the condition $\|v\|_p = 1$. □

We are now in a position to prove Theorem 9.15. First we observe that if the principal coefficients $a^{ij} \in C^{0,1}(\bar{\Omega})$ and $p \geqslant 2$, the result follows directly from Theorems 8.9 and 8.12, and Lemma 9.16. In the general case, we replace u by $u - \varphi$, to reduce to zero boundary values, and approximate the coefficients a^{ij} uniformly by a sequence $\{a^{ij}_m\} \subset C^{0,1}(\bar{\Omega})$ with, in the case $p < 2$, the function f being approximated in $L^p(\Omega)$ by a sequence $\{f_m\} \subset L^2(\Omega)$. If $\{u_m\}$ denotes the sequence of solutions of the corresponding Dirichlet problems, we infer from Lemma 9.17 that the sequence $\{u_m\}$ is bounded in $W^{2,p}(\Omega)$. Consequently, by Problem 5.5 again, a subsequence converges weakly in $W^{2,p}(\Omega) \cap W^{1,p}_0(\Omega)$ to a function u that (by an argument similar to that in Lemma 9.17) satisfies $Lu = f$ in Ω. □

Theorem 9.15 may also be derived from a Fredholm alternative, which follows from Corollary 9.14; (Problem 9.9). When $p > n/2$, we obtain an existence theorem for continuous boundary values.

Corollary 9.18. *Let Ω be a $C^{1,1}$ domain in $\mathbb{R}^n$, and let the operator L be strictly elliptic in Ω with coefficients $a^{ij} \in C^0(\bar{\Omega})$, $b^i, c \in L^\infty$, $i, j = 1, \ldots, n$ and $c \leqslant 0$. Then, if $f \in L^p(\Omega)$, $p > n/2$, $\varphi \in C^0(\partial\Omega)$, the Dirichlet problem $Lu = f$ in Ω, $u = \varphi$ on $\partial\Omega$, has a unique solution $u \in W^{2,p}_{\text{loc}}(\Omega) \cap C^0(\bar{\Omega})$.*

Proof. The uniqueness assertion follows from Theorem 9.5 and Lemma 9.16; (in fact, it is clearly valid for arbitrary Ω and $p > 1$). To get the existence, we let $\{\varphi_m\} \subset W^{2,p}(\Omega)$ converge uniformly to φ on $\partial\Omega$, and let $u_m \in W^{2,p}(\Omega)$ be the solution of the Dirichlet problem $Lu_m = f$ in Ω, $u_m = \varphi_m$ on $\partial\Omega$, known to exist by Theorem 9.15. The differences, $u_l - u_m$, clearly satisfy

$$L(u_l - u_m) = 0 \quad \text{in } \Omega, \qquad u_l - u_m = \varphi_l - \varphi_m \quad \text{on } \partial\Omega.$$

Hence by Theorems 9.1 and 9.11, and Lemma 9.16, we obtain that $\{u_m\}$ converges in $C^0(\bar{\Omega}) \cap W^{2,p}_{\text{loc}}(\Omega)$ to a solution of the Dirichlet problem, $Lu = f$ in Ω, $u = \varphi$ on $\partial\Omega$. □

By invoking barrier considerations similar to those in Chapter 6, we may extend Corollary 9.18 to allow for more general domains Ω. We consider this type of result in Section 9.7 in conjunction with continuity estimates at the boundary.

To conclude this section we state a theorem concerning higher-order regularity that improves Theorems 6.17 and 6.19 for classical solutions. The proof may be effected by using difference quotients as in these theorems, or by an argument similar to that of Lemma 9.16. The details are left to the reader; (Problem 9.10).

Theorem 9.19. *Let u be a $W^{2,p}_{\text{loc}}(\Omega)$ solution of the elliptic equation $Lu = f$ in a domain Ω, where the coefficients of L belong to $C^{k-1,1}(\Omega)$, $(C^{k-1,\alpha}(\Omega))$, $f \in W^{k,q}_{\text{loc}}(\Omega)$, $(C^{k-1,\alpha}(\Omega))$, with $1 < p, q < \infty$, $k \geqslant 1$, $0 < \alpha < 1$. Then $u \in W^{k+2,q}_{\text{loc}}(\Omega)$, $(C^{k+1,\alpha}(\Omega))$.*

Furthermore, if $\Omega \in C^{k+1,1}$, $(C^{k+1,\alpha})$, *L is strictly elliptic in* Ω *with coefficients in* $C^{k-1,1}(\bar{\Omega})$, $(C^{k-1,\alpha}(\bar{\Omega}))$, *and* $f \in W^{k,q}(\Omega)$, $(C^{k-1,\alpha}(\bar{\Omega}))$, *then* $u \in W^{k+2,q}(\Omega)$, $(C^{k+1,\alpha}(\bar{\Omega}))$.

9.7. A Local Maximum Principle

In this and the following sections, we focus attention on local pointwise estimates for operators in the general form (9.1) and derive results corresponding to those for divergence structure operators in Sections 8.6 through 8.10. We shall assume throughout the rest of this chapter that the operator L, as given by (9.1), is strictly elliptic with bounded coefficients in the domain Ω, and accordingly, we fix constants γ and ν such that

$$\frac{\Lambda}{\lambda} \leqslant \gamma, \qquad \left(\frac{|b|}{\lambda}\right)^2, \frac{|c|}{\lambda} \leqslant \nu \tag{9.47}$$

in Ω. We prove in this section the following analogue of the subsolution estimate, Theorem 8.17.

Theorem 9.20. *Let* $u \in W^{2,n}(\Omega)$ *and suppose* $Lu \geqslant f$, *where* $f \in L^n(\Omega)$. *Then for any ball* $B = B_{2R}(y) \subset \Omega$ *and* $p > 0$, *we have*

$$\sup_{B_R(y)} u \leqslant C\left\{\left(\frac{1}{|B|}\int_B (u^+)^p\right)^{1/p} + \frac{R}{\lambda}\|f\|_{L^n(B)}\right\} \tag{9.48}$$

where $C = C(n, \gamma, \nu R^2, p)$.

Proof. Without loss of generality, we can assume that $B = B_1(0)$, the general case being recovered by means of the coordinate transformation $x \to (x - y)/2R$. We also assume initially that $u \in C^2(\Omega) \cap W^{2,n}(\Omega)$. For $\beta \geqslant 1$, a cutoff function η is defined by

$$\eta(x) = (1 - |x|^2)^\beta. \tag{9.49}$$

By differentiation, we obtain

$$D_i\eta = -2\beta x_i(1 - |x|^2)^{\beta-1},$$
$$D_{ij}\eta = -2\beta\delta_{ij}(1 - |x|^2)^{\beta-1} + 4\beta(\beta - 1)x_i x_j(1 - |x|^2)^{\beta-2}.$$

Setting $v = \eta u$, we then have

$$\begin{aligned} a^{ij}D_{ij}v &= \eta a^{ij}D_{ij}u + 2a^{ij}D_i\eta D_j u + ua^{ij}D_{ij}\eta \\ &\geqslant \eta(f - b^iD_iu - cu) + 2a^{ij}D_i\eta D_ju + ua^{ij}D_{ij}\eta. \end{aligned}$$

Letting $\Gamma^+ = \Gamma_v^+$ denote the upper contact set of v in the ball B, we clearly have $u > 0$ on Γ^+; and furthermore, using the concavity of v on Γ^+, we can estimate there

$$\begin{aligned} |Du| &= \frac{1}{\eta}|Dv - uD\eta| \\ &\leqslant \frac{1}{\eta}(|Dv| + u|D\eta|) \\ &\leqslant \frac{1}{\eta}\left(\frac{v}{1-|x|} + u|D\eta|\right) \\ &\leqslant 2(1+\beta)\eta^{-1/\beta}u. \end{aligned}$$

Thus, on Γ^+ we have the inequality

$$\begin{aligned} -a^{ij}D_{ij}v &\leqslant \{(16\beta^2 + 2n\beta)\Lambda\eta^{-2/\beta} + 2\beta|b|\eta^{-1/\beta} + c\}v + \eta f \\ &\leqslant C\lambda\eta^{-2/\beta}v + f, \end{aligned}$$

where $C = C(n, \beta, \gamma, \nu)$. Therefore, applying Lemma 9.3, we obtain for $\beta \geqslant 2$,

$$\begin{aligned} \sup_B v &\leqslant C\left(\|\eta^{-2/\beta}v^+\|_{n;B} + \frac{1}{\lambda}\|f\|_{n;B}\right) \\ &\leqslant C\left\{(\sup v^+)^{1-2/\beta}\|(u^+)^{2/\beta}\|_{n;B} + \frac{1}{\lambda}\|f\|_{n;B}\right\}. \end{aligned}$$

Choosing $\beta = 2n/p$ (provided $p \leqslant n$) and using Young's inequality (7.6) in the form

$$(\sup v^+)^{1-2/\beta} \leqslant \varepsilon \sup v^+ + \varepsilon^{1-\beta/2}$$

for $\varepsilon > 0$, we then get

$$\sup_B v \leqslant C\left\{\left(\int_B (u^+)^p\right)^{1/p} + \frac{1}{\lambda}\|f\|_{n;B}\right\}$$

and the estimate (9.48) follows. The extension to $u \in W^{2,n}(\Omega)$ follows directly by approximation and is left to the reader. □

By replacement of u with $-u$, Theorem 9.20 extends automatically to supersolutions and solutions of the equation $Lu = f$.

Corollary 9.21. *Let $u \in W^{2,n}(\Omega)$ and suppose $Lu \leqslant f, (=f)$, where $f \in L^n(\Omega)$. Then, for any ball $B = B_{2R}(y) \subset \Omega$ and $p > 0$, we have*

$$(9.50) \qquad \sup_{B_R(y)} -u, (|u|) \leqslant C\left\{\left(\frac{1}{|B|}\int_B (u^-)^p, (|u|^p)\right)^{1/p} + \frac{R}{\lambda}\|f\|_{L^n(B)}\right\}$$

where $C = C(n, \gamma, \nu R^2, p)$.

Note that when we take $p = 1$ in (9.48), we obtain an extension of the mean value inequality for non-negative subharmonic functions, namely,

$$(9.51) \qquad u(y) \leqslant \frac{C}{R^n}\int_{B_R(y)} u$$

provided $Lu, u \geqslant 0$ in $B_R(y)$ and $C = C(n, \gamma, \nu R^2)$. Theorem 9.20 also continues to hold under more general coefficient conditions; (see Notes).

9.8. Hölder and Harnack Estimates

We present in this section a treatment of the Hölder and Harnack estimates, of Krylov and Safonov [KS 1, 2], that are the analogues for uniformly elliptic operators in the general form of the De Giorgi, Nash and Moser estimates for divergence form operators. Also important for our treatment of fully nonlinear elliptic operators in Chapter 17 is the weak Harnack inequality for non-negative supersolutions from which the Hölder and Harnack estimates are readily derived.

Theorem 9.22. *Let $u \in W^{2,n}(\Omega)$ satisfy $Lu \leqslant f$ in Ω, where $f \in L^n(\Omega)$, and suppose that u is non-negative in a ball $B = B_{2R}(y) \subset \Omega$. Then*

$$(9.52) \qquad \left(\frac{1}{|B_R|}\int_{B_R} u^p\right)^{1/p} \leqslant C\left(\inf_{B_R} u + \frac{R}{\lambda}\|f\|_{L^n(B)}\right)$$

where p and C are positive constants depending only on n, γ and νR^2.

Proof. Again let us assume initially that $B = B_1(0)$ and also that $\lambda \equiv 1$, (by replacing L, f by L/λ, f/λ). Setting

$$(9.53) \qquad \begin{aligned} &\bar{u} = u + \varepsilon + \|f\|_{n;B}, \\ &w = -\log \bar{u}, \qquad v = \eta w, \qquad g = f/\bar{u}, \end{aligned}$$

where $\varepsilon > 0$, and η is given by (9.49), we then obtain, using Schwarz's inequality,

$$\begin{aligned}-a^{ij}D_{ij}v &= -\eta a^{ij}D_{ij}w - 2a^{ij}D_i\eta D_jw - wa^{ij}D_{ij}\eta \\ &\leqslant \eta(-a^{ij}D_iwD_jw + b^iD_iw + |c| + g) \\ &\quad - 2a^{ij}D_i\eta D_jw - wa^{ij}D_{ij}\eta \\ &\leqslant \frac{2}{\eta}a^{ij}D_i\eta D_j\eta - wa^{ij}D_{ij}\eta + (|b|^2 + |c| + g).\end{aligned}$$

Next we calculate

$$a^{ij}D_{ij}\eta = -2\beta a^{ii}(1-|x|^2)^{\beta-1} + 4\beta(\beta-1)a^{ij}x_ix_j(1-|x|^2)^{\beta-2}$$

so that $a^{ij}D_{ij}\eta \geqslant 0$ if

$$2(\beta-1)a^{ij}x_ix_j + a^{ii}|x|^2 \geqslant a^{ii}$$

in particular if

$$2\beta|x|^2 \geqslant n\Lambda.$$

Consequently if $0 < \alpha < 1$ and β is chosen so that

$$\beta \geqslant \frac{n\gamma}{2\alpha}$$

we have $a^{ij}D_{ij}\eta \geqslant 0$ for all $|x| \geqslant \alpha$. Hence we obtain, on $B^+ = \{x \in B \mid w(x) > 0\}$, the inequality

$$\begin{aligned}-a^{ij}D_{ij}v &\leqslant 4\beta^2(1-|x|^2)^{\beta-2}|x|^2 \\ &\quad + v\chi(B_\alpha)\sup_{B_\alpha}\left(-\frac{a^{ij}D_{ij}\eta}{\eta}\right) + (|b|^2 + |c| + g)\eta \\ &\leqslant 4\beta^2\Lambda + |b|^2 + |c| + g + \frac{2n\beta\Lambda}{1-\alpha^2}v\chi(B_\alpha).\end{aligned}$$

Therefore, applying Lemma 9.3, and noting that $\|g\|_{n;B} \leqslant 1$, we get a bound for v, namely,

$$\sup_B v \leqslant C(1 + \|v^+\|_{n;B_\alpha}), \tag{9.54}$$

where $C = C(n, \alpha, \gamma, \nu)$.

To facilitate eventual application of the cube decomposition procedure in Section 9.2 it is convenient at this point to switch from balls to cubes. For any point $y \in \mathbb{R}^n$ and $R > 0$ we shall denote by $K_R(y)$ the open cube, parallel to the coordinate axes, with center y and side length $2R$. If $\alpha < \dfrac{1}{\sqrt{n}}$ we have $K_\alpha = K_\alpha(0) \subset\subset B$ and hence from (9.54),

$$\begin{aligned}\sup_B v &\leqslant C(1 + \|v^+\|_{n; K_\alpha}) \\ &\leqslant C(1 + |K_\alpha^+|^{1/n} \sup_B v^+),\end{aligned}$$

where $K_\alpha^+ = \{x \in K_\alpha \,|\, v > 0\}$. Hence, if

$$|K_\alpha^+|/|K_\alpha| \leqslant \theta = [2(2\alpha)^n C]^{-1},$$

then

$$\sup_B v \leqslant 2C,$$

where $C = C(n, \alpha, \gamma, \nu)$ is the constant in (9.54). Let us now choose $\alpha = 1/3n$ and fix θ accordingly. Using the transformation $x \to \alpha(x - z)/r$, we therefore obtain for any cube $K = K_r(z)$, such that $B_{3nr}(z) \subset B$ and

$$(9.55) \qquad |K^+| \leqslant \theta |K|,$$

the estimate

$$(9.56) \qquad \sup_{K_{3r}(z)} w \leqslant C(n, \gamma, \nu).$$

The proof of Theorem 9.22 is now completed with the aid of the following measure theoretic lemma.

Lemma 9.23. *Let K_0 be a cube in $\mathbb{R}^n$, $w \in L^1(K_0)$; and, for $k \in \mathbb{R}$, set*

$$\Gamma_k = \{x \in K_0 \,|\, w(x) \leqslant k\}.$$

Suppose there exist positive constants $\delta < 1$ and C such that

$$(9.57) \qquad \sup_{K_0 \cap K_{3r}(z)} (w - k) \leqslant C,$$

whenever k and $K = K_r(z) \subset K_0$ satisfy

$$(9.58) \qquad |\Gamma_k \cap K| \geqslant \delta |K|.$$

Then it follows that, for all k,

$$(9.59) \qquad \sup_{K_0} (w - k) \leqslant C\left(1 + \frac{\log(|\Gamma_k|/|K_0|)}{\log \delta}\right).$$

Proof. We first show by induction that

$$\sup_{K_0} (w - k) \leqslant mC,$$

for any natural number m and $k \in \mathbb{R}$ satisfying $|\Gamma_k| \geqslant \delta^m |K_0|$. This assertion is clearly true by hypothesis for $m = 1$. Suppose now it holds for some $m \in \mathbb{N}$, and that $|\Gamma_k| \geqslant \delta^{m+1}|K_0|$. Let $\tilde{\Gamma}_k$ be defined by

$$\tilde{\Gamma}_k = \bigcup \{K_{3r}(z) \cap K_0 \,|\, |K_r(z) \cap \Gamma_k| \geqslant \delta |K_r(z)|\}.$$

By our cube decomposition procedure in Section 9.2, in particular inequality (9.20) with $t = \delta$, we obtain that either $\tilde{\Gamma}_k = K_0$ or that

$$|\tilde{\Gamma}_k| \geqslant \frac{1}{\delta} |\Gamma_k|$$
$$\geqslant \delta^m |K_0|$$

and hence, replacing k by $k + C$, we obtain

$$\sup_{K_0} (w - k) \leqslant (m + 1)C,$$

which guarantees the validity of the above assertion for $m + 1$. The estimate (9.59) now follows by the appropriate choice of m. □

To apply Lemma 9.23, we take $\delta = 1 - \theta$, $K_0 = K_\alpha(0)$, $\alpha = 1/3n$ and note that the estimate (9.56) still holds when w is replaced by $w - k$. Let

$$\mu_t = |\{x \in K_0 \,|\, \bar{u}(x) > t\}|$$

denote the distribution function of $\bar{u}$ in K_0 to obtain from (9.53) and (9.59) with $t = e^{-k}$, the estimate

$$\mu_t \leqslant C(\inf_{K_0} \bar{u}/t)^\kappa, \qquad t > 0, \tag{9.60}$$

where C and κ are positive constants depending only on n, γ and ν. Replacing the cube K_0 by the inscribed ball $B_\alpha(0)$, $\alpha = 1/3n$ and using Lemma 9.7, we then obtain

$$\int_{B_\alpha} (\bar{u})^p \leqslant C(\inf_{B_\alpha} \bar{u})^p, \tag{9.61}$$

for $p < \kappa$, say $p = \kappa/2$. The weak Harnack inequality in the form (9.52) then follows by letting $\varepsilon \to 0$, using a covering argument to extend (9.61) to arbitrary $\alpha < 1$, (in particular $\alpha = \frac{1}{2}$), and finally invoking the coordinate transformation $x \to (x - y)/2R$. □

By adapting the proof of the Hölder estimate for divergence structure operators (Theorem 8.22), (with only a minor modification to offset not having $p = 1$ in the weak Harnack inequality (9.52)), we conclude from Theorem 9.22 the following Hölder estimate for operators in general form.

Corollary 9.24. *Let $u \in W^{2,n}(\Omega)$ satisfy the equation $Lu = f$ in Ω. Then, for any ball $B_0 = B_{R_0}(y) \subset \Omega$ and $R \leqslant R_0$, we have*

$$\operatorname*{osc}_{B_R(y)} u \leqslant C\left(\frac{R}{R_0}\right)^{\alpha}(\operatorname*{osc}_{B_0} u + kR_0), \tag{9.62}$$

where $C = C(n, \gamma, \nu R_0^2)$, $\alpha = \alpha(n, \gamma, \nu R_0^2)$ are positive constants and $k = \|f - cu\|_{n;B_0}$.

Also by combining Theorem 9.22 with the subsolution estimate (Theorem 9.20), we obtain the full Harnack inequality.

Corollary 9.25. *Let $u \in W^{2,n}(\Omega)$ satisfy $Lu = 0$, $u \geqslant 0$ in Ω. Then for any ball $B_{2R}(y) \subset \Omega$, we have*

$$\sup_{B_R(y)} u \leqslant C \inf_{B_R(y)} u, \tag{9.63}$$

where $C = C(n, \gamma, \nu R^2)$.

9.9. Local Estimates at the Boundary

The local maximum principle (Theorem 9.20) may be extended to balls intersecting the boundary $\partial\Omega$ as follows.

Theorem 9.26. *Let $u \in W^{2,n}(\Omega) \cap C^0(\bar{\Omega})$ satisfy $Lu \geqslant f$ in Ω, $u \leqslant 0$ on $B \cap \partial\Omega$ where $f \in L^n(\Omega)$ and $B = B_{2R}(y)$ is a ball in $\mathbb{R}^n$. Then, for any $p > 0$, we have*

$$\sup_{\Omega \cap B_R(y)} u \leqslant C\left\{\left(\frac{1}{|B|}\int_{B \cap \Omega} (u^+)^p\right)^{1/p} + \frac{R}{\lambda}\|f\|_{L^n(B \cap \Omega)}\right\}, \tag{9.64}$$

where $C = C(n, \gamma, \nu R^2, p)$.

Proof. It suffices to establish the estimate (9.64) for $u \in C^2(\Omega) \cap C^0(\bar{\Omega})$ satisfying $u \leqslant 0$ on $B \cap \partial\Omega$. We extend u to the whole of the ball B by setting $u = 0$ in $B - \Omega$. Although u then does not necessarily belong to $C^2(B)$, the argument of Theorem 9.20 may still be applied since Γ^+, the upper contact set of the function v, will lie in $B \cap \Omega$, and $v \in C^2(\Gamma^+)$ is sufficient for the application of Lemma 9.3. □

It is worth recording here the general observation, used in the proof of Theorem 9.26, that if u fulfills the hypotheses of Lemma 9.3, the estimate (9.11) continues to hold when Γ^+ is replaced by the upper contact set of u with respect to any larger domain $\tilde{\Omega}$, (to which u is extended by setting $u = 0$ in $\tilde{\Omega} - \Omega$), and with $d = \operatorname{diam} \tilde{\Omega}$.

The weak Harnack inequality (Theorem 9.22) admits the following extension to the boundary.

Theorem 9.27. *Let $u \in W^{2,n}(\Omega)$ satisfy $Lu \leqslant f$ in Ω, $u \geqslant 0$ in $B \cap \Omega$, where $B = B_{2R}(y)$ is a ball in $\mathbb{R}^n$. Set $m = \inf_{B \cap \partial\Omega} u$ and*

$$u_m^-(x) = \begin{cases} \inf\{u(x), m\} & \text{for } x \in B \cap \Omega \\ m & \text{for } x \in B - \Omega. \end{cases}$$

Then

$$\left(\frac{1}{|B_R|}\int_{B_R} (u_m^-)^p\right)^{1/p} \leqslant C\left(\inf_{\Omega \cap B_R} u + \frac{R}{\lambda}\|f\|_{L^n(B \cap \Omega)}\right), \tag{9.65}$$

where p and C are positive constants depending only on n, γ and νR^2. If we assume only $u \in W^{2,n}_{\text{loc}}(\Omega)$, then (9.65) holds with $m = \liminf_{x \to B\,\partial\Omega} u$.

Proof. We adapt the proof of Theorem 9.22 by replacing u with u_m^-. The estimate (9.56) then follows when the function w is replaced by $w - k$ for $k \geqslant -\log m$, and we thus infer the estimate (9.60) for $0 < t \leqslant m$. But $\mu_t = 0$, if $t > m$, hence (9.65) follows as before. The final assertion of Theorem 9.27 is a consequence of the remark following Theorem 9.1. □

Global and boundary modulus of continuity estimates follow as consequences of Theorem 9.27. Corresponding to the divergence structure results (Theorems 8.27 and 8.29), we obtain the following estimates.

Corollary 9.28. *Let $u \in W^{2,n}_{\text{loc}}(\Omega)$ satisfy $Lu = f$ in Ω where $f \in L^n(\Omega)$, and suppose that Ω satisfies an exterior cone condition at a point $y \in \partial\Omega$. Then, for any $0 < R < R_0$ and ball $B_0 = B_{R_0}(y)$, we have*

$$\operatorname*{osc}_{\Omega \cap B_R} u \leqslant C\left\{\left(\frac{R}{R_0}\right)^{\alpha}\left(\operatorname*{osc}_{\Omega \cap B_0} u + \bar{k}R_0\right) + \sigma\sqrt{RR_0}\right\}, \tag{9.66}$$

where $C = C(n, \gamma, \nu R_0^2, V_y)$, $\alpha = \alpha(n, \gamma, \nu R_0^2, V_y)$ are positive constants, V_y is the exterior cone at y and

$$\sigma(r) = \operatorname*{osc}_{\partial\Omega \cap B_r} u = \limsup_{x \to \partial\Omega \cap B_r} u - \liminf_{x \to \partial\Omega \cap B_r} u$$

for $0 < r \leqslant R_0$.

Corollary 9.29. *Let $u \in W^{2,n}(\Omega) \cap C^0(\bar{\Omega})$ satisfy $Lu = f$ in Ω, $u = \varphi$ on $\partial\Omega$ where $f \in L^n(\Omega)$, $\varphi \in C^\beta(\bar{\Omega})$ for some $\beta > 0$, and suppose that $\partial\Omega$ satisfies a uniform exterior cone condition. Then $u \in C^\alpha(\bar{\Omega})$ and*

$$|u|_{\alpha;\Omega} \leqslant C, \tag{9.67}$$

where α and C are positive constants depending on n, γ, ν, β, Ω, $|\varphi|_{\delta;\Omega}$ and $|u|_{0;\Omega}$.

We note here that modulus of continuity estimates at the boundary also arise from barrier constructions as in Section 6.3. However, Theorem 9.28 may be used in place of barrier arguments to solve the Dirichlet problem by means of the Perron process. To see this, let us suppose that the operator L satisfies the hypotheses of Theorem 6.11; and let $u \in C^2(\Omega)$ be the Perron solution of the Dirichlet problem, $Lu = f$, $u = \varphi$ on $\partial\Omega$, whose existence is asserted by Theorem 6.11. Corollary 9.28 then provides an estimate for the modulus of continuity of u at a point $y \in \partial\Omega$, where Ω satisfies an exterior cone condition, in terms of the modulus of continuity of the function φ at y. If Ω satisfies an exterior cone condition everywhere on $\partial\Omega$, we then conclude $u \in C^0(\bar{\Omega})$ and $u = \varphi$ on $\partial\Omega$ thus solving the above Dirichlet problem.

Consequently, in the existence theorem (Theorem 6.13), we may replace the exterior sphere condition by an exterior cone condition that would be satisfied, for example, by Lipschitz domains; (see also Problem 6.3). Utilizing our results in Section 9.5, in particular Corollary 9.18, we may further extend Theorem 6.13 to cover continuous coefficients.

Theorem 9.30. *Let L be strictly elliptic in a bounded domain Ω with coefficients $a^{ij} \in C^0(\Omega) \cap L^\infty(\Omega)$, $b^i, c \in L^\infty(\Omega)$ and $c \leqslant 0$, and suppose that Ω satisfies an exterior cone condition at every boundary point. Then if $f \in L^p(\Omega)$, $p \geqslant n$, the Dirichlet problem, $Lu = f$ in Ω, $u = \varphi$ on $\partial\Omega$, has a unique solution $u \in W^{2,p}_{loc}(\Omega) \cap C^0(\bar{\Omega})$.*

Proof. To complete the proof of Theorem 9.30, it suffices, in view of the remarks preceding the theorem, to establish the existence of an analogue in $W^{2,p}_{loc}(\Omega)$ of a Perron solution under the hypotheses of the theorem. Again, we may imitate the Perron method for subharmonic functions in Chapter 2 making crucial use of the strong maximum principle (Theorem 9.6), the solvability of the Dirichlet problem in balls with continuous boundary data (Corollary 9.18), and the interior estimates (Theorem 9.11) coupled with the weak relative compactness of bounded sets in $W^{2,p}(\Omega')$, $\Omega' \subset\subset \Omega$. The details are left to the reader. □

Boundary Hölder Estimates for the Gradient

An interesting and useful Hölder estimate for the traces of the gradients of solutions on the boundary may also be deduced from the interior Harnack (or weak Harnack) inequalities. This result was established by Krylov [KV 5] in connection with its application to the theory of fully nonlinear equations, which we describe in Section 17.8. For these purposes it suffices to restrict to a flat boundary portion where the solution is assumed to vanish and to consider operators of the form

$$Lu = a^{ij} D_{ij} u$$

satisfying the uniform ellipticity condition (9.47). More general results are readily formulated.

Theorem 9.31. *Let $u \in W^{2,n}_{loc}(B^+) \cap C^0(\bar{B}^+)$ satisfy the equation $Lu = f$ in the half ball $B^+ = B_{R_0}(0) \cap \mathbb{R}^n_+$ with $f \in L^\infty(B^+)$ and $u = 0$ on $T = B_{R_0} \cap \partial R^n_+$. Then for any $R \leqq R_0$, we have*

$$\text{(9.68)} \qquad \operatorname*{osc}_{B_R^+} \frac{u}{x_n} \leqq C\left(\frac{R}{R_0}\right)^\alpha \left(\operatorname*{osc}_{B^+} \frac{u}{x_n} + R_0 \sup_{B^+} \frac{|f|}{\lambda}\right)$$

where α and C are positive constants depending only on n and γ.

Proof. We proceed under the assumption that the function $v = u/x_n$ is bounded in B^+; (local boundedness at least is guaranteed by the barrier considerations in Section 6.3). Assuming initially that $u \geqq 0$ in B^+, we first prove the following assertion: There exists $\delta = \delta(n, \gamma) > 0$ such that

$$\text{(9.69)} \qquad \inf_{\substack{|x'| < R, \\ x_n = \delta R}} v \leqq 2\left(\inf_{B_{R/2,\delta}} v + \frac{R}{\lambda} \sup_{B^+} |f|\right)$$

for any $R \leqq R_0$, where

$$B_{R,\delta} = \{x \mid |x'| < R, \quad 0 < x_n < \delta R\}.$$

To prove (9.69) it is convenient to normalize so that $\lambda = R = 1$ and $\inf_{|x'| < R} v(x', \delta R) = 1$. We consider in $B_{1,\delta}$ the barrier function

$$w(x) = \left(1 - |x'|^2 + (1 + \sup|f|) \frac{(x_n - \delta)}{\sqrt{\delta}}\right) x_n .$$

By straightforward computation, we obtain $Lw \geqq f$ for sufficiently small $\delta = \delta(n, \gamma)$ and $w \leqq u$ on $\partial B_{1,\delta}$, so by the maximum principle, Theorem 9.1, we have $w \leqq u$ in $B_{1,\delta}$. Thus on $B_{1/2,\delta}$,

$$v \geqq 1 - |x'|^2 + (1 + \sup|f|) \frac{(x_n - \delta)}{\sqrt{\delta}} \geqq \frac{1}{2} - \sup|f|,$$

again for sufficiently small δ. Removal of the normalization yields (9.69) as required. We now define

$$B^*_{R/2,\delta} = \{x \mid |x'| < R, \quad \delta R/2 < x_n < 3\delta R/2\},$$

noting that

$$\frac{2u}{3\delta R} \leqq v \leqq \frac{2u}{\delta R}.$$

in $B^*_{R/2,\delta}$. Consequently by the Harnack inequality, Corollary 9.25, we obtain

$$\begin{aligned}(9.70)\qquad \sup_{B^*_{R/2,\delta}} v &\leq C\Big(\inf_{B^*_{R/2,\delta}} v + R\sup|f/\lambda|\Big)\\ &\leq C\Big(\inf_{\substack{|x'|<R,\\ x_n=\delta R}} v + R\sup|f/\lambda|\Big)\\ &\leq C\Big(\inf_{B_{R/2,\delta}} v + R\sup|f/\lambda|\Big), \qquad \text{by (9.69).}\end{aligned}$$

We now drop the assumption $u \geq 0$ and set $M = \inf\limits_{B_{2R,\delta}} v$, $m = \inf\limits_{B_{2R,\delta}} v$. Applying (9.70) to the functions $M - v$, $v - m$ and adding the resulting inequalities, we get the standard oscillation estimate

$$\operatorname*{osc}_{B_{R/2,\delta}} v \leq \sigma\Big(\operatorname*{osc}_{B_{2R,\delta}} v + C\,R\sup|f/\lambda|\Big)$$

where $C > 0$ and $\sigma < 1$ depend only on n and γ. From this (9.68) follows by virtue of Lemma 8.23. □

Theorem 9.31 shows in fact that the gradient Du exists on T and is Hölder continuous there, satisfying the estimate

$$(9.71)\qquad \operatorname*{osc}_{|x'|<R} Du(x',0) \leq C\left(\frac{R}{R_0}\right)^{\alpha}\left(\operatorname*{osc}_{B^+}\frac{u}{x_n} + \frac{R}{\lambda}\sup_{B^+}|f|\right)$$

Moreover the term $\operatorname*{osc}\limits_{B^+}\dfrac{u}{x_n}$ on the right hand side of (9.68) or (9.71) can be replaced by either $\operatorname*{osc}\limits_{B^+} u/R_0$ or $\sup\limits_{B^+}|Du|$. Global $C^{1,\alpha}$ estimates can be deduced from Theorem 9.31 on combination with appropriate interior estimates which typically hold for nonlinear equations; (see Problem 13.1 and Section 17.8).

Notes

The maximum and uniqueness principles for strong solutions, as formulated in Theorems 9.1, 9.5 and 9.6, are due to Aleksandrov [AL 2], [AL 3], although the essential case covered by Lemma 9.3 appears in Bakelman [BA 3]. A detailed analysis of the form of the constant C in the estimate (9.4) is carried out in the papers [AL 4, 5]. In all these results it is impossible to replace the L^n norms by L^p norms for $p < n$ [AL 6]. Indeed the example (8.22) shows that the uniqueness result (Theorem 9.5) no longer holds if we only assume the functions $u, v \in W^{2,p}_{\text{loc}}(\Omega) \cap C^0(\bar{\Omega})$, $p < n$. Different versions of the maximum principle (Theorem 9.1) were discovered by Bony [BY] and Pucci [PU 3].

The Calderon-Zygmund inequality was discovered by Calderon and Zygmund [CZ], and we have largely followed their original proof, as expounded by Stein [SN], making use of the cube decomposition procedure (which extends the one dimensional case due to Riesz [RZ]) and the Marcinkiewicz interpolation theorem [MZ]. Our proof differs from those in [CZ] and [SN] in that we do not use the

Fourier transform to get the L^2 estimate. The operator T in the proof of Theorem 9.9 is a special case of a *singular integral operator*, and it was this class of operators that was the main object of study in [CZ] and [SN]. An alternative proof of the Calderon-Zygmund inequality is presented in the monographs [BS], [MY 5]. A further proof, also based on interpolation, uses the space of functions of bounded mean oscillation, BMO; see [CS], [FS].

The L^p estimates for second-order elliptic equations, presented in Section 9.4, were derived by Koselev [KO] and Greco [GC] and extended to higher-order equations and systems by various authors, including Slobodeckii [SL] Browder [BW 3] and Agmon, Douglis and Nirenberg [ADN 1, 2]. The existence theorem (Theorem 9.15) appears in Chicco [CI 4, 5], although derived differently there; and a further approach to the Dirichlet problem is given by P. L. Lions [LP 1]. The regularity argument in Lemma 9.16 is taken from Morrey [MY 5], where the L^p theory is also treated.

The pointwise estimates of Sections 9.6, 9.7 and 9.8 stem from the fundamental work of Krylov and Safonov, (see [KS 1, 2], [SF]) where the Hölder and Harnack estimates of Corollaries 9.24, 9.25 are established, for $c \leqslant 0$. In fact, the more general situation of *parabolic* equations is treated in [KS 1, 2]. Our presentation in Section 9.7, adapted from [TR 12], carries over their basic ideas. The local maximum principle (Theorem 9.20) was also proved in [TR 12], under more general coefficient conditions, namely $\Lambda/\mathscr{D}^*, b/\mathscr{D}^* \in L^q(\Omega), q > n, c/\mathscr{D}^*, f/\mathscr{D}^* \in L^n(\Omega)$. The estimates of Section 9.7 may similarly be extended to allow $b/\lambda \in L^{2n}(\Omega)$, c/λ, $f/\lambda \in L^n(\Omega)$, although the condition of uniform ellipticity seems essential for the proofs in this case. Extensions to quasilinear equations are treated in [TR 12], [LU 7] and [MV 1]; (see also Chapter 15).

We have also included in the present edition a proof of the boundary Hölder gradient estimate of Krylov [KV 5], utilizing a simplification of Caffarelli. Krylov's original proof was presented in the form of a problem in the English second edition.

Problems

9.1. Prove that in the estimates (9.8) and (9.11) we can replace d by $d/2$ and also by considering a spherical cone show that the resulting estimates are sharp; (see [AL 4, 5]).

9.2. By explicitly integrating the function g and optimizing the choice of μ in the proof of Theorem 9.1, deduce an improvement of the estimate (9.14); (see [AL 4, 5]).

9.3. Derive the estimate (9.11) in the form

$$(9.68)\qquad \sup_{\Omega} u \leqslant \sup_{\partial\Omega} u^+ + C(n)\,|\hat{\Omega}|^{1/n} \left\| \frac{a^{ij} D_{ij} u}{\mathscr{D}^*} \right\|_{L^n(\Gamma^+)},$$

where $\hat{\Omega}$ denotes the convex hull of Ω; (see [TA 5] for a sharp version of (9.68) when $n = 2$).

9.4. Prove the following more general version of the Marcinkiewicz interpolation theorem:

Theorem 9.31. *Let T be a linear mapping from $L^q(\Omega) \cap L^r(\Omega)$ into $L^{\bar{q}}(\Omega) \cap L^{\bar{r}}(\Omega)$, $1 \leqslant q < r < \infty$, $1 \leqslant \bar{q} \leqslant \bar{r} < \infty$, and suppose there are constants T_1 and T_2 such that*

$$(9.69) \qquad \mu_{Tf}(t) \leqslant \left(\frac{T_1 \|f\|_q}{t}\right)^{\bar{q}}, \qquad \mu_{Tf}(t) \leqslant \left(\frac{T_2 \|f\|_r}{t}\right)^{\bar{r}}$$

for all $f \in L^q(\Omega) \cap L^r(\Omega)$ and $t > 0$. Then T extends as a bounded mapping from $L^p(\Omega)$ to $L^{\bar{p}}(\Omega)$ for any $p, \bar{p}$ satisfying

$$1/p = \sigma/q + (1-\sigma)/r, \qquad 1/\bar{p} = \sigma/\bar{q} + (1-\sigma)/\bar{r}$$

for some $\sigma \in (0, 1)$.

Examine also the case when r or $\bar{r} = \infty$.

9.5. With the notation of Section 7.8, use the general Marcinkiewicz interpolation theorem to show that the potential operator V_μ maps $L^p(\Omega)$ continuously into $L^q(\Omega)$ for $p > 1$ and $\delta = \mu$.

9.6. Using Lemma 9.12, show that for a $C^{1,1}$ domain Ω the subspace

$$\{u \in C^2(\bar{\Omega}) \,|\, u = 0 \quad \text{on } \partial\Omega\}$$

is dense in $W^{2,p}(\Omega) \cap W_0^{1,p}(\Omega)$, $1 < p < \infty$.

9.7. Deduce Theorem 9.15 directly by approximation based on either Theorems 6.14 or 8.14.

9.8. Deduce Theorem 9.15 for the special case of Poisson's equation from the Riesz representation theorem, and use the method of continuity (Theorem 5.2) to get the full result.

9.9. Starting from Corollary 9.14, establish a Fredholm alternative for operators of the form (9.1) in the Sobolev spaces $W^{2,p}(\Omega)$, $1 < p < \infty$; and again deduce Theorem 9.15.

9.10. Prove Theorem 9.19.

9.11. Suppose the operator L satisfies the hypotheses of Theorem 9.22 in an annular region, $A = B_R(y) - B_\rho(y) \subset \mathbb{R}^n$. If $u \in W^{2,n}(A)$ satisfies $Lu \leqslant f$, $u \geqslant 0$ in A, where $f \in L^n(A)$, prove that for any $\rho < r < R$

$$(9.70) \qquad \inf_{B_r - B_\rho} u \geqslant \kappa(\inf_{\partial B_\rho} u - R\|f\|_{n;A}),$$

where κ is a positive constant depending only on n, ρ/R, r/R, γ and νR^2. Use this result to deduce Theorem 9.22 from Theorem 9.20.

Problem 9.12. By considering the operator

$$Lu = \Lambda \, \Delta u + (\lambda - \Lambda) \frac{x_i x_j}{|x|^2} D_{ij} u$$

and the function

$$u(x) = |x|^{1-(n-1)\gamma}, \quad \gamma = \Lambda/\lambda$$

suitably redefined near 0, show that the exponent p in the weak Harnack inequality Theorem 9.22 must satisfy

$$p < \frac{n}{(n-1)\gamma - 1}.$$

Part II

Quasilinear Equations

Chapter 10

Maximum and Comparison Principles

The purpose of this chapter is to provide various maximum and comparison principles for quasilinear equations which extend corresponding results in Chapter 3. We consider second order, quasilinear operators Q of the form

$$Qu = a^{ij}(x, u, Du)D_{ij}u + b(x, u, Du), \quad a^{ij} = a^{ji}, \tag{10.1}$$

where $x = (x_1, \ldots, x_n)$ is contained in a domain Ω of $\mathbb{R}^n$, $n \geqslant 2$, and, unless otherwise stated, the function u belongs to $C^2(\Omega)$. The coefficients of Q, namely the functions $a^{ij}(x, z, p)$, $i, j = 1, \ldots, n$, $b(x, z, p)$ are assumed to be defined for all values of (x, z, p) in the set $\Omega \times \mathbb{R} \times \mathbb{R}^n$. Two operators of the form (10.1) will be called *equivalent* if one is a multiple of the other by a fixed positive function in $\Omega \times \mathbb{R} \times \mathbb{R}^n$. Equations $Qu = 0$ corresponding to equivalent operators Q will also be called equivalent.

We adopt the following definitions:

Let $\mathscr{U}$ be a subset of $\Omega \times \mathbb{R} \times \mathbb{R}^n$. Then Q is *elliptic in* $\mathscr{U}$ if the coefficient matrix $[a^{ij}(x, z, p)]$ is positive for all $(x, z, p) \in \mathscr{U}$. If $\lambda(x, z, p)$, $\Lambda(x, z, p)$ denote respectively the minimum and maximum eigenvalues of $[a^{ij}(x, z, p)]$, this means that

$$0 < \lambda(x, z, p)\,|\xi|^2 \leqslant a^{ij}(x, z, p)\xi_i\xi_j \leqslant \Lambda(x, z, p)\,|\xi|^2 \tag{10.2}$$

for all $\xi = (\xi_1, \ldots, \xi_n) \in \mathbb{R}^n - \{0\}$ and for all $(x, z, p) \in \mathscr{U}$. If, further, Λ/λ is bounded in $\mathscr{U}$, we shall call Q *uniformly elliptic in* $\mathscr{U}$. If Q is elliptic (uniformly elliptic) in the whole set $\Omega \times \mathbb{R} \times \mathbb{R}^n$, then we shall simply say that Q is *elliptic (uniformly elliptic) in* Ω. If $u \in C^1(\Omega)$ and the matrix $[a^{ij}(x, u(x), Du(x))]$ is positive for all $x \in \Omega$, we shall say that Q is *elliptic with respect to* u. We also define a scalar function $\mathscr{E}$, which will prove to be quite important, by

$$\mathscr{E}(x, z, p) = a^{ij}(x, z, p)p_ip_j. \tag{10.3}$$

If Q is elliptic in $\mathscr{U}$, we have by (10.2)

$$0 < \lambda(x, z, p)|p|^2 \leqslant \mathscr{E}(x, z, p) \leqslant \Lambda(x, z, p)|p|^2 \tag{10.4}$$

for all $(x, z, p) \in \mathscr{U}$.

The operator Q is of *divergence form* if there exists a differentiable vector function $\mathbf{A}(x, z, p)=(A^1(x, z, p), \ldots, A^n(x, z, p))$ and a scalar function $B(x, z, p)$ such that

$$(10.5) \qquad Qu=\operatorname{div}\mathbf{A}(x, u, Du)+B(x, u, Du), \quad u\in C^2(\Omega);$$

that is, in (10.1),

$$a^{ij}(x, z, p)=\frac{1}{2}(D_{p_i}A^j(x, z, p)+D_{p_j}A^i(x, z, p)).$$

Unlike the case of linear operators, a quasilinear operator with smooth coefficients is not necessarily expressible in divergence form.

The operator Q is *variational* if it is the Euler-Lagrange operator corresponding to a multiple integral

$$\int_\Omega F(x, u, Du)\, dx$$

where F is a differentiable scalar function; that is, Q is of divergence form (10.5) and

$$(10.6) \qquad A^i(x, z, p)=D_{p_i}F(x, z, p), \quad B(x, z, p)=-D_zF(x, z, p).$$

The ellipticity of Q is equivalent to the strict convexity of the function F with respect to the p variables.

Examples

(i) $Qu=\Delta u+(\alpha-2)\dfrac{D_iu\, D_ju}{(1+|Du|^2)}D_{ij}u, \quad \alpha\geqslant 1.$

Here

$$\lambda(x, z, p)=\begin{cases} 1 & \text{if } \alpha\geqslant 2,\\ \dfrac{1+(\alpha-1)|p|^2}{1+|p|^2} & \text{if } \alpha\leqslant 2,\end{cases}$$

$$\Lambda(x, z, p)=\begin{cases} \dfrac{1+(\alpha-1)|p|^2}{1+|p|^2} & \text{if } \alpha\geqslant 2,\\ 1 & \text{if } \alpha\leqslant 2,\end{cases}$$

and

$$\mathscr{E}(x, z, p)=|p|^2(1+(\alpha-1)|p|^2)/(1+|p|^2),$$

so that Q is elliptic for all $\alpha \geqslant 1$ and uniformly elliptic only if $\alpha > 1$. Writing

$$Qu = (1 + |Du|^2)^{1-\alpha/2} \operatorname{div} (1 + |Du|^2)^{\alpha/2-1} Du,$$

we see that Q is equivalent to a divergence form operator, and moreover that Q is equivalent to the variational operator associated with the integral

$$\int_\Omega (1 + |Du|^2)^{\alpha/2}\, dx.$$

The equation $Qu = 0$ coincides with Laplace's equation when $\alpha = 2$, and with the minimal surface equation when $\alpha = 1$. For other values of α, this equation arises in the cracking of plates and the modelling of blast furnaces.

(ii) $Qu = \Delta u + \beta D_i u D_j u D_{ij} u, \quad \beta \geqslant 0.$
Here

$$\lambda(x, z, p) = 1$$
$$\Lambda(x, z, p) = 1 + \beta |p|^2$$
$$\mathscr{E}(x, z, p) = |p|^2 (1 + \beta |p|^2)$$

so that Q is elliptic for all $\beta \geqslant 0$ and uniformly elliptic only when $\beta = 0$ (that is when Q is the Laplacian). For $\beta > 0$, Q is equivalent to the variational operator associated with the integral

$$\int_\Omega \exp\left(\frac{\beta}{2}|Du|^2\right) dx.$$

Note that when $\beta \geqslant 1$, the minimum and maximum eigenvalues of Q are proportional to those in the case $\alpha = 1$ in the previous example. However, the existence results for these operators will turn out to differ substantially due to the different growth properties of their $\mathscr{E}$ functions.

(iii) *The Equation of Prescribed Mean Curvature*
Let $u \in C^2(\Omega)$ and suppose the graph of u in $\mathbb{R}^{n+1}$ has mean curvature $H(x)$ at the point $(x, u(x))$, $x \in \Omega$. (The mean curvature is understood with respect to the normal direction along which x_{n+1} is increasing). It follows (see Appendix to Chapter 14) that u satisfies the equation

$$\mathfrak{M}u = (1 + |Du|^2)\Delta u - D_i u D_j u D_{ij} u = nH(1 + |Du|^2)^{3/2}. \tag{10.7}$$

Here

$$\lambda(x, z, p) = 1,$$
$$\Lambda(x, z, p) = 1 + |p|^2,$$

and

$$\mathscr{E}(x, z, p) = |p|^2.$$

The operator $\mathfrak{M}$ in (10.7) is equivalent to the operator Q in example (i) when $\alpha = 1$.

(iv) *The Equation of Gas Dynamics*

The stationary irrotational flow of an ideal compressible fluid is described by the equation of continuity, $\operatorname{div}(\rho Du) = 0$, where u is the velocity potential of the flow and the fluid density p satisfies a density–speed relation $\rho = \rho(|Du|)$. In the case of a perfect gas this relation takes the form

$$\rho = \left(1 - \frac{\gamma - 1}{2}|Du|^2\right)^{1/(\gamma - 1)},$$

where the constant γ is the ratio of specific heats of the gas and $\gamma > 1$. The equation satisfied by the velocity potential u is then

$$\Delta u - \frac{D_i u D_j u}{1 - \dfrac{\gamma - 1}{2}|Du|^2} D_{ij} u = 0, \tag{10.8}$$

which has the eigenvalues

$$\lambda = \frac{1 - \dfrac{\gamma + 1}{2}|Du|^2}{1 - \dfrac{\gamma - 1}{2}|Du|^2}, \quad \Lambda = 1.$$

Equation (10.8) is elliptic—and the flow is *subsonic*—when $|Du| < [2/(\gamma + 1)]^{1/2}$, but is hyperbolic when $[2/(\gamma + 1)]^{1/2} < |Du| < [2/(\gamma - 1)]^{1/2}$. We note that (10.8) becomes the minimal surface equation when $\gamma = -1$.

(v) *The Equation of Capillarity*

The equilibrium shape of a liquid surface with constant surface tension in a uniform gravity field is governed by the equation of *capillarity*,

$$\operatorname{div}\left(\frac{Du}{\sqrt{1 + |Du|^2}}\right) = \kappa u, \tag{10.9}$$

or equivalently,

$$\mathfrak{M}u = \kappa u(1 + |Du|^2)^{3/2},$$

where $\mathfrak{M}$ is the operator defined in (10.7), u is the height of the liquid above an undisturbed reference surface, and κ is a constant that is positive or negative according to whether the gravitational field is acting downward or upward. In the absence of gravity, the equation is replaced by the constant mean curvature equation, with H equal to a constant in (10.7). The function $\mathscr{E}$ and the eigenvalues λ, Λ are the same as for (10.7). The natural physical boundary condition for the height u in (10.9), when the fluid is constrained by a fixed rigid boundary, is

$$\frac{\partial u/\partial \nu}{\sqrt{1+|Du|^2}} = \cos \gamma,$$

where the *contact angle* γ is the angle between the liquid surface and the fixed boundary, measured within the fluid; ν is the corresponding normal to the fixed boundary.

10.1. The Comparison Principle

If L is a linear operator satisfying the hypotheses of the weak maximum principle, Corollary 3.2, and if $u, v \in C^0(\bar{\Omega}) \cap C^2(\Omega)$ satisfy the inequalities $Lu \geqslant Lv$ in Ω, $u \leqslant v$ on $\partial\Omega$, we have immediately from Corollary 3.2 that $u \leqslant v$ in Ω. This comparison principle has the following extension to quasilinear operators.

Theorem 10.1. *Let $u, v \in C^0(\bar{\Omega}) \cap C^2(\Omega)$ satisfy $Qu \geqslant Qv$ in Ω, $u \leqslant v$ on $\partial\Omega$, where*

(i) *the operator Q is locally uniformly elliptic with respect to either u or v;*
(ii) *the coefficients a^{ij} are independent of z;*
(iii) *the coefficient b is non-increasing in z for each $(x, p) \in \Omega \times \mathbb{R}^n$;*
(iv) *the coefficients a^{ij}, b are continuously differentiable with respect to the p variables in $\Omega \times \mathbb{R} \times \mathbb{R}^n$.*

It then follows that $u \leqslant v$ in Ω. Furthermore, if $Qu > Qv$ in Ω, $u \leqslant v$ on $\partial\Omega$ and conditions (i), (ii) *and* (iii) *hold, (but not necessarily* (iv)), *we have the strict inequality $u < v$ in Ω.*

Proof. Let us assume that Q is elliptic with respect to u. Then we have

$$\begin{aligned} Qu - Qv = {} & a^{ij}(x, Du)D_{ij}(u-v) + (a^{ij}(x, Du) - a^{ij}(x, Dv))D_{ij}v \\ & + b(x, u, Du) - b(x, u, Dv) + b(x, u, Dv) - b(x, v, Dv) \geqslant 0 \end{aligned}$$

so that by writing

$$w = u - v$$

$$a^{ij}(x) = a^{ij}(x, Du)$$

$$[a^{ij}(x, Du) - a^{ij}(x, Dv)]D_{ij}v + b(x, u, Du) - b(x, u, Dv) = b^i(x)D_i w$$

we see that

$$Lw = a^{ij}(x)D_{ij}w + b^i D_i w \geqslant 0$$

on $\Omega^+ = \{x \in \Omega \mid w(x) > 0\}$ and $w \leqslant 0$ on $\partial\Omega$. Note that the existence of the locally bounded functions b^i is guaranteed by condition (iv) and the theorem of the mean. Consequently, using conditions (i) and (iv) we have from Theorem 3.1 that $w \leqslant 0$ in Ω. If $Qu > Qv$ in Ω, the function w cannot assume a non-negative maximum in Ω; (see the proof of Theorem 3.1). Hence $w < 0$ in Ω. If Q is elliptic at v the result follows from the minimum principle for supersolutions. □

A uniqueness theorem for the Dirichlet problem for quasilinear elliptic operators follows immediately from Theorem 10.1.

Theorem 10.2. *Let $u, v \in C^0(\bar{\Omega}) \cap C^2(\Omega)$, satisfy $Qu = Qv$ in Ω, $u = v$ on $\partial\Omega$, and suppose that conditions* (i) *to* (iv) *in Theorem* 10.1 *hold. Then $u \equiv v$ in Ω.*

Condition (ii) in the hypotheses of Theorems 10.1 and 10.2 might appear unnecessarily restrictive. However, we shall show below that the conclusions of Theorems 10.1 and 10.2 are not generally valid when the principal coefficients depend on z. The comparison principle, Theorem 10.1, will be useful in the establishment of boundary gradient estimates in Chapter 13.

By using the maximum principle for strong solutions (Theorem 9.1) in place of Corollary 3.2, we see that Theorems 10.1 and 10.2 remain valid when the functions $u, v \in C^0(\bar{\Omega}) \cap C^1(\Omega) \cap W^{2,n}_{loc}(\Omega)$.

10.2. Maximum Principles

Using Theorem 10.1 we can derive the following quasilinear extension of the apriori bound (Theorem 3.7), which also illustrates the significance of the $\mathscr{E}$ function.

Theorem 10.3. *Let Q be elliptic in Ω and suppose that there exist non-negative constants μ_1 and μ_2 such that*

$$\frac{b(x, z, p)\,\text{sign}\, z}{\mathscr{E}(x, z, p)} \leqslant \frac{\mu_1|p| + \mu_2}{|p|^2} \quad \forall (x, z, p) \in \Omega \times \mathbb{R} \times \mathbb{R}^n. \tag{10.10}$$

Then, if $u \in C^0(\bar{\Omega}) \cap C^2(\Omega)$ satisfies $Qu \geqslant 0 (=0)$ in Ω, we have

$$\sup_{\Omega} u(|u|) \leqslant \sup_{\partial\Omega} u^+(|u|) + C\mu_2 \tag{10.11}$$

where $C = C(\mu_1, \text{diam}\, \Omega)$.

Proof. Let $u \in C^0(\bar{\Omega}) \cap C^2(\Omega)$ and satisfy $Qu \geqslant 0$ in Ω, and define the operator $\bar{Q}$ by

$$\bar{Q}v = a^{ij}(x, u, Dv)D_{ij}v + b(x, u, Dv).$$

We choose a comparison function v as in the proof of Theorem 3.7; namely, for $\mu_2 > 0$ we set

$$v(x) = \sup_{\partial\Omega} u^+ + \mu_2(e^{\alpha d} - e^{\alpha x_1}),$$

where Ω is assumed to lie in the slab, $0 < x_1 < d$, and $\alpha \geqslant \mu_1 + 1$. Then we have in $\Omega^+ = \{x \in \Omega \mid u(x) > 0\}$,

$$\begin{aligned}\bar{Q}v &= -\mu_2\alpha^2 a^{11}(x, u, Dv)e^{\alpha x_1} + b(x, u, Dv)\\ &\leqslant -\frac{e^{-\alpha x_1}}{\mu_2}\mathscr{E}(x, u, Dv)\left(1 - \frac{\mu_1}{\alpha} - \frac{e^{-\alpha x_1}}{\alpha^2}\right) \quad \text{by (10.10)}\\ &< 0 \leqslant \bar{Q}u.\end{aligned}$$

Hence, by Theorem 10.1, we have $u \leqslant v$ in Ω. The result for $\mu_2 = 0$ follows by letting μ_2 tend to zero. □

For uniformly elliptic operators, condition (10.10) is equivalent to a condition of the form

$$\frac{b(x, z, p) \operatorname{sign} z}{\lambda(x, z, p)} \leqslant \mu_1|p| + \mu_2, \qquad \forall (x, z, p) \in \Omega \times \mathbb{R} \times \mathbb{R}^n. \tag{10.12}$$

An example of a nonuniformly elliptic operator that satisfies (10.10), but not (10.12), is given by

$$Qu = \Delta u + D_i u D_j u D_{ij} u + (1 + |Du|^2).$$

It is clear from the proof of Theorem 10.3 that in the hypotheses we need only have assumed that (i) $\mathscr{E} > 0$ in $\Omega \times \mathbb{R} \times \mathbb{R}^n$; (ii) Q is elliptic with respect to u; and (iii) there exists a fixed vector $p_0 \in \mathbb{R}^n$ such that (10.10) hold for all (x, z, tp_0) with $(x, z, t) \in \Omega \times \mathbb{R} \times \mathbb{R}$. Further maximum principles are treated in Problems 10.1, 10.2.

The condition (10.12) may alternatively be generalized to nonuniformly elliptic operators by using the Aleksandrov maximum principle (Theorem 9.1). Following the notation in Chapter 9, we set

$$\mathscr{D} = \det [a^{ij}(x, z, p)], \qquad \mathscr{D}^* = \mathscr{D}^{1/n}.$$

Theorem 10.4. *Let Q be elliptic in Ω and suppose there exist non-negative constants μ_1 and μ_2 such that*

$$\frac{b(x, z, p) \operatorname{sign} z}{\mathscr{D}^*} \leqslant \mu_1|p| + \mu_2 \qquad \forall (x, z, p) \in \Omega \times \mathbb{R} \times \mathbb{R}^n. \tag{10.13}$$

Then, if $u \in C^0(\bar{\Omega}) \cap C^2(\Omega)$ *satisfies* $Qu \geqslant 0\ (=0)$ *in* Ω, *we have*

$$\sup_{\Omega} u(|u|) \leqslant \sup_{\partial\Omega} u^+(|u|) + C\mu_2 \tag{10.14}$$

where $C = C(\mu_1, \text{diam}\,\Omega)$.

Proof. In the subdomain $\Omega^+ = \{x \in \Omega \mid u(x) > 0\}$, we have

$$\begin{aligned} 0 \leqslant Qu &= a^{ij}D_{ij}u + b \,\text{sign}\, u \\ &\leqslant a^{ij}D_{ij}u + [\mu_1(\text{sign}\, D_iu)D_iu + \mu_2]\mathscr{D}^*, \end{aligned}$$

and therefore the estimate (10.14) for $\sup_{\Omega} u$ follows by Theorem 9.1. The full estimate (10.14) is obtained by replacing u with $-u$ in the proof. □

Theorem 10.4 is in fact implicit in the proof of Theorem 9.1. More generally, we have from Lemma 9.4, the following result.

Theorem 10.5. *Let* Q *be elliptic in the bounded domain* Ω *and suppose there exist non-negative functions* $g \in L^n_{\text{loc}}(\mathbb{R}^n)$, $h \in L^n(\Omega)$ *such that*

$$\frac{b(x, z, p)\,\text{sign}\, z}{n\mathscr{D}^*} \leqslant \frac{h(x)}{g(p)} \qquad \forall (x, z, p) \in \Omega \times \mathbb{R} \times \mathbb{R}^n, \tag{10.15}$$

$$\int_{\Omega} h^n\, dx < \int_{\mathbb{R}^n} g^n\, dp = g_\infty. \tag{10.16}$$

Then if $u \in C^0(\bar{\Omega}) \cap C^2(\Omega)$ *satisfies* $Qu \geqslant 0\ (=0)$ *in* Ω, *we have*

$$\sup_{\Omega} u(|u|) \leqslant \sup_{\partial\Omega} u^+(|u|) + C\,\text{diam}\,\Omega. \tag{10.17}$$

where C *depends on* g *and* h.

The quantity g_∞, as is the case in Theorem 9.1, may be infinite so that (10.16) becomes superfluous. If the function g is positive and G is defined by

$$G^{-1}(t) = \int_{B_t(0)} g^n\, dp$$

so that $G\colon (0, g_\infty) \to (0, \infty)$, then the constant C in (10.17) is given by

$$C = G\left(\int_{\Omega} h^n\right).$$

To complete this section we consider the application of Theorem 10.5 to the equation of prescribed mean curvature (10.7). Here

$$\mathscr{D} = (1 + |p|^2)^{n-1},$$

so that we may take

$$g(p) = (1 + |p|^2)^{-(n+2)/2n}.$$

By a calculation, we obtain

$$g_\infty = \int_{\mathbb{R}^n} \frac{dp}{(1 + |p|^2)^{n/2+1}} = \omega_n,$$

and hence we have the following estimate:

Corollary 10.6. *Let $u \in C^0(\bar{\Omega}) \cap C^2(\Omega)$ be a solution of the prescribed mean curvature equation* (10.7) *in the bounded domain Ω. Then, if*

$$H_0 = \int |H(x)|^n \, dx < \omega_n, \tag{10.18}$$

we have

$$\sup_\Omega |u| \leq \sup_{\partial\Omega} |u| + C \operatorname{diam} \Omega, \tag{10.19}$$

where $C = C(n, H_0)$.

Finally we remark that the estimates of this section continue to be valid for subsolutions or solutions u assumed only in $C^0(\bar{\Omega}) \cap W^{2,n}_{\text{loc}}(\Omega)$.

10.3. A Counterexample

The following example shows that Theorems 10.1 and 10.2 cannot in general be extended to allow the principal coefficients a^{ij} to depend on u. We consider an operator Q of the form

$$Qu = \Delta u + g(r, u) \frac{x_i x_j}{r^2} D_{ij} u, \quad r = |x|, \tag{10.20}$$

in the spherical shell $\Omega = \{x \in \mathbb{R}^n \mid 1 < |x| < 2\}$. If $u = u(r)$, the equation $Qu = 0$ is equivalent to the ordinary differential equation

$$u'' + u' \left(\frac{n-1}{r(1+g)} \right) = 0.$$

Let v and w be polynomials satisfying the conditions

(i) $v(1)=w(1), \quad v(2)=w(2)$;

(ii) $v', w'>0 \quad \text{in } [1, 2]$;

(iii) $v'(1)<w'(1), \quad v'(2)>w'(2)$;

(iv) $v'', w''<0 \quad \text{in } [1, 2]$;

(v) $\dfrac{w''(1)}{w'(1)}=\dfrac{v''(1)}{v'(1)}, \quad \dfrac{w''(2)}{w'(2)}=\dfrac{v''(2)}{v'(2)}$;

and define for $1\leqslant r\leqslant 2$, $v\leqslant u\leqslant w$,

$$f(r, u)=\frac{u-v}{w-v}\left(\frac{v''}{v'}-\frac{w''}{w'}\right)-\frac{v''}{v'},$$

$$g(r, u)=-1+\frac{n-1}{rf(r, u)}.$$

Then, writing $v(x)=v(|x|)$, $w(x)=w(|x|)$, we see that $Qv=Qw=0$ in Ω and $v=w$ on $\partial\Omega$. Also Q is elliptic with respect to both v and w. Furthermore, by extending f in an appropriate way to the strip $[1, 2]\times\mathbb{R}$, we can obtain an operator Q that is uniformly elliptic in Ω and whose coefficients belong to $C^\infty(\bar{\Omega}\times\mathbb{R})$.

10.4. Comparison Principles for Divergence Form Operators

Interesting variants of Theorem 10.1 can be obtained when the operator Q is of divergence form (10.5). We recall from Chapter 8 that a function u, weakly differentiable in Ω, satisfies $Qu\geqslant 0$ $(=0, \leqslant 0)$ in Ω if the functions $A^i(x, u, Du)$, $B(x, u, Du)$ are locally integrable in Ω and

$$(10.21) \qquad Q(u, \varphi)=\int_\Omega (\mathbf{A}(x, u, Du)\cdot D\varphi-B(x, u, Du)\varphi)\,dx\leqslant 0 \;(=0, \geqslant 0)$$

for all non-negative $\varphi\in C_0^1(\Omega)$. The following theorem provides three alternative criteria for a comparison principle.

Theorem 10.7. *Let $u, v\in C^1(\bar{\Omega})$ satisfy $Qu\geqslant 0$ in Ω, $Qv\leqslant 0$ in Ω and $u\leqslant v$ on $\partial\Omega$, where the functions $\mathbf{A}$, B are continuously differentiable with respect to the z, p variables in $\bar{\Omega}\times\mathbb{R}\times\mathbb{R}^n$, the operator Q is elliptic in Ω, and the function B is non-increasing in z for fixed $(x, p)\in\Omega\times\mathbb{R}^n$. Then, if either*

(i) *the vector function $\mathbf{A}$ is independent of z; or*

(ii) *the function B is independent of p; or*

(iii) *the $(n+1)\times(n+1)$ matrix,*

$$\begin{bmatrix} D_{p_j}A^i(x, z, p) & -D_{p_j}B(x, z, p) \\ D_zA^i(x, z, p) & -D_zB(x, z, p)\end{bmatrix}\geqslant 0 \text{ in } \Omega\times\mathbb{R}\times\mathbb{R}^n;$$

it follows that $u\leqslant v$ in Ω.

Proof. Let us define

$$w = u - v,$$

$$u_t = tu + (1-t)v, \quad 0 \leqslant t \leqslant 1,$$

$$a^{ij}(x) = \int_0^1 D_{p_j} A^i(x, u_t, Du_t)\, dt,$$

$$b^i(x) = \int_0^1 D_z A^i(x, u_t, Du_t)\, dt,$$

$$c^i(x) = \int_0^1 D_{p_i} B(x, u_t, Du_t)\, dt,$$

$$d(x) = \int_0^1 D_z B(x, u_t, Du_t)\, dt.$$

Then we have

$$\begin{aligned} (10.22) \qquad 0 &\geqslant Q(u, \varphi) - Q(v, \varphi) \\ &= \int_\Omega \{(\mathbf{A}(x, u, Du) - \mathbf{A}(x, v, Dv)) \cdot D\varphi \\ &\quad - (B(x, u, Du) - B(x, v, Dv))\varphi\}\, dx \\ &= \int_\Omega \{(a^{ij}(x) D_j w + b^i(x) w) D_i \varphi - (c^i(x) D_i w + d(x) w)\varphi\}\, dx \end{aligned}$$

for all non-negative $\varphi \in C_0^1(\Omega)$. Hence $Lw \geqslant 0$ where L is the linear operator given by

$$Lw = D_i(a^{ij} D_j w + b^i w) + c^i D_i w + dw.$$

Since $u, v \in C^1(\bar{\Omega})$, there exist by the hypotheses positive constants λ, Λ such that

$$a^{ij}(x)\xi_i \xi_j \geqslant \lambda |\xi|^2 \quad \forall \xi \in \mathbb{R}^n,\ x \in \Omega.$$
$$|a^{ij}|, |b^i|, |c^i|, |d| \leqslant \Lambda \text{ in } \Omega, \quad d \leqslant 0 \text{ in } \Omega,$$

and therefore L is strictly elliptic in Ω with bounded coefficients. The conclusion of Theorem 10.7 can now be obtained directly from the theory of Chapter 8. In particular if condition (i) holds, then $b^i = 0$ in Ω so that by the weak maximum principle, Theorem 8.1, we have $w \leqslant 0$ in Ω. Although the remainder of Theorem 10.7 follows immediately from Problem 8.1, we include the full proof here. If condition (ii) holds, then $c^i=0$ in Ω. Note that this condition on L is equivalent to the preceding condition on the adjoint operator L^*. For $\varepsilon>0$, we define

$$\varphi=\frac{w^+}{w^++\varepsilon}\in W_0^{1,2}(\Omega),$$

and obtain from substitution in (10.22),

$$\begin{aligned}\lambda\int_\Omega\left|D\log\left(1+\frac{w^+}{\varepsilon}\right)\right|^2dx&\leqslant\int_\Omega\frac{a^{ij}(x)D_iw^+D_jw^+}{(w^++\varepsilon)^2}dx\\&\leqslant\Lambda\int_\Omega\frac{w^+}{w^++\varepsilon}\left|D\log\left(1+\frac{w^+}{\varepsilon}\right)\right|dx\\&\leqslant\Lambda\int_\Omega\left|D\log\left(1+\frac{w^+}{\varepsilon}\right)\right|dx.\end{aligned}$$

Hence using Young's inequality (7.6), we have

$$\int_\Omega\left|D\log\left(1+\frac{w^+}{\varepsilon}\right)\right|^2dx\leqslant\left(\frac{\Lambda}{\lambda}\right)^2|\Omega|;$$

it follows from Poincaré's inequality (7.44) that

$$\int_\Omega\left|\log\left(1+\frac{w^+}{\varepsilon}\right)\right|^2dx\leqslant C(n,\lambda,\Lambda,|\Omega|).$$

Letting $\varepsilon\to0$, we see that w^+ must vanish in Ω, that is $w\leqslant0$ in Ω.

Finally, if condition (iii) holds, we choose $\varphi=w^+$ in Ω and obtain on substitution in (10.22) that

$$a^{ij}D_iw^+D_jw^++(b^i-c^i)w^+D_iw^+-d(w^+)^2=0\quad\text{in }\Omega,$$

so that, by Young's inequality (7.6),

$$|Dw^+|^2\leqslant n\left(\frac{2\Lambda}{\lambda}\right)^2|w^+|^2\quad\text{in }\Omega.$$

Hence, for any $\varepsilon>0$, we have

$$\left|D\log\left(1+\frac{w^+}{\varepsilon}\right)\right|\leqslant\frac{2\sqrt{n}\,\Lambda}{\lambda}\frac{w^+}{w^++\varepsilon}\leqslant\frac{2\sqrt{n}\,\Lambda}{\lambda}$$

and, since $w^+=0$ on $\partial\Omega$, it follows that

$$\left|\log\left(1+\frac{w^+}{\varepsilon}\right)\right|\leqslant\frac{2\sqrt{n}\,\Lambda}{\lambda}\operatorname{diam}\Omega.$$

Letting $\varepsilon\to 0$ we have, as before, $w^+=0$ in Ω, and therefore $w\leqslant 0$ in Ω. □

Note that when condition (i) holds in Theorem 10.7, we need only assume that $u,\ v\in C^0(\bar\Omega)\cap C^1(\Omega)$ and that the derivatives of the coefficients belong to $C^0(\Omega\times\mathbb{R}\times\mathbb{R}^n)$. This is easily seen by applying the result of Theorem 10.7 to a subdomain $\Omega'\subset\subset\Omega$. A similar extension is valid in the other cases provided the coefficients satisfy an appropriate uniform structure condition.

10.5. Maximum Principles for Divergence Form Operators

When the operator Q is of divergence form, we can derive maximum principles under different hypotheses from those of Theorems 10.3 and 10.4. We shall assume that the functions $\mathbf{A}$ and B in (10.5) satisfy the following structure conditions:

For all $(x, z, p)\in\Omega\times\mathbb{R}\times\mathbb{R}^n$ and some $\alpha\geqslant 1$,

$$p\cdot\mathbf{A}(x,z,p)\geqslant|p|^\alpha-|a_1 z|^\alpha-a_2^\alpha,$$

(10.23)

$$B(x,z,p)\operatorname{sign}z\leqslant\begin{cases}b_0|p|^{\alpha-1}+|b_1 z|^{\alpha-1}+b_2^{\alpha-1} & \text{if } \alpha>1,\\ b_0 & \text{if } \alpha=1;\end{cases}$$

here a_1, a_2, b_0, b_1, b_2 are non-negative constants. The first inequality in (10.23) can be viewed as a weak ellipticity condition (see Problem 10.3). The development below is similar to the derivation of global estimates for weak solutions of linear elliptic equations in Chapter 8.

Lemma 10.8. *Let $u\in C^0(\bar\Omega)\cap C^1(\Omega)$ satisfy $Qu\geqslant 0$ in Ω, and suppose that Q satisfies the structure conditions* (10.23). *Then we have*

$$\sup_\Omega u\leqslant C\{\|u^+\|_\alpha+(a_1+b_1)\sup_{\partial\Omega}u^+ +a_2+b_2\}+\sup_{\partial\Omega}u^+ \tag{10.24}$$

where $C=C(n,\alpha,a_1,b_0,b_1,|\Omega|)$.

Proof. Let us assume initially that $u\in C^1(\bar\Omega)$ and $u\leqslant 0$ on $\partial\Omega$ so that $\sup_{\partial\Omega}u^+=0$. The proof follows that of Theorem 8.15 with this difference: in the present case, u is assumed bounded at the outset and hence it is unnecessary to truncate the power functions used as test functions. By writing

$$k=a_2+b_2, \qquad \bar{z}=|z|+k, \qquad \bar{b}=a_1^\alpha+b_0^\alpha+b_1^{\alpha-1}+1,$$

we obtain from inequalities (10.23), with the help of Young's inequality,

$$\begin{aligned} &p\cdot\mathbf{A}(x,z,p)\geqslant|p|^\alpha-\bar{b}|\bar{z}|^\alpha \\ &\bar{z}B(x,z,p)\ \mathrm{sign}\ z\leqslant\begin{cases}\mu|p|^\alpha+(\mu^{1-\alpha}+1)\bar{b}\bar{z}^\alpha & \text{if } \alpha>1,\\ \bar{b}\bar{z} & \text{if } \alpha=1,\end{cases}\end{aligned} \tag{10.25}$$

where $\mu>0$. Hence substituting in the integral inequality (10.21) the function

$$\varphi=w^\beta-k^\beta$$

where $w=\bar{u}^+=u^++k$, $\beta\geqslant1$, and choosing $\mu=\beta/2$, we obtain

$$\int_\Omega w^{\beta-1}|Dw|^\alpha\,dx\leqslant C\bar{b}\int_\Omega w^{\alpha+\beta-1}\,dx,$$

where $C=C(\beta)$. By the Sobolev inequality (7.26), there exists a number $s>\alpha$ such that

$$\|w^r-k^r\|_s\leqslant Cr\left(\int_\Omega w^{\beta-1}|Dw|^\alpha\,dx\right)^{1/\alpha}$$

where $r=(\alpha+\beta-1)/\alpha$ and $C=C(n,s,|\Omega|)$. Consequently we have

$$\|w\|_{rs}\leqslant(Cr)^{1/r}(\bar{b})^{1/\alpha r}\|w\|_{r\alpha}$$

for all $r\geqslant1$, and the estimate (10.24) with $\sup_{\partial\Omega}u^+=0$ now follows by the iteration argument of Theorem 8.15. To dispense with the initial assumptions concerning u, we replace u by $u-L$ where $L=\sup_{\partial\Omega}u^+$, and approximate Ω by domains $\Omega'\subset\subset\Omega$. □

Using Lemma 10.8, we can now derive the following apriori estimates for subsolutions and solutions of the equation $Qu=0$.

Theorem 10.9. *Let $u\in C^0(\bar{\Omega})\cap C^1(\Omega)$ satisfy $Qu\geqslant0$ $(=0)$ in Ω and suppose that Q satisfies the structure conditions (10.23) with $\alpha>1$, $b_1=0$ and either b_0 or $a_1=0$. Then we have the estimate*

$$\sup_\Omega u(|u|)\leqslant C(a_2+b_2+a_1\sup_{\partial\Omega}u^+(|u|))+\sup_{\partial\Omega}u^+(|u|) \tag{10.26}$$

where $C=C(n,\alpha,a_1,b_0,|\Omega|)$.

Proof. As in the proof of Lemma 10.8 we can assume initially that $u \in C^1(\bar{\Omega})$ and $u \leqslant 0$ on $\partial\Omega$. We also assume that $k>0$. The two cases $b_0=0$, $a_1=0$ will be considered separately.

(i) Suppose $b_0=0$. We then substitute

$$\varphi=\frac{1}{k^{\alpha-1}}-\frac{1}{w^{\alpha-1}},\quad w=\bar{u}^+,$$

into the integral inequality (10.21) to obtain

$$(\alpha-1)\int_\Omega\left|\frac{Dw}{w}\right|^\alpha dx \leqslant \alpha\bar{b}|\Omega|,$$

so that

$$\int_\Omega\left|D\log\frac{w}{k}\right|^\alpha dx \leqslant \frac{\alpha}{\alpha-1}\bar{b}|\Omega|.$$

Hence, by Poincaré's inequality (7.44),

$$\int_\Omega\left|\log\frac{w}{k}\right|^\alpha dx \leqslant C\bar{b}$$

where $C = C(n, \alpha, |\Omega|)$. Now writing $M = \sup_\Omega w$, we have by the proof of Lemma 10.8

$$\left(\frac{M}{k}\right)^\alpha \leqslant C\int_\Omega\left(\frac{w}{k}\right)^\alpha dx$$
$$\leqslant C\left(\frac{M}{k}\right)^\alpha\left(\log\frac{M}{k}\right)^{-\alpha}\int_\Omega\left(1+\left|\log\frac{w}{k}\right|^\alpha\right)dx,$$

so that

$$\left|\log\frac{M}{k}\right|^\alpha \leqslant C\int_\Omega\left(1+\left|\log\frac{w}{k}\right|^\alpha\right)dx \leqslant C.$$

Hence $M \leqslant Ck$ where $C=C(n, \alpha, a_1, |\Omega|)$.

(ii) Suppose $a_1=0$. The proof is then similar to that of Theorem 8.16. Writing again $M=\sup_\Omega w$, we substitute

$$\varphi=\frac{1}{|M-w+k|^{\alpha-1}}-\frac{1}{M^{\alpha-1}}$$

into (10.21) to obtain

$$(\alpha-1)\int_\Omega\left|\frac{Dw}{M-w+k}\right|^\alpha dx\leqslant b_0\int_\Omega\left|\frac{Dw}{M-w+k}\right|^{\alpha-1}dx$$
$$+\left\{\left(\frac{a_2}{k}\right)^\alpha+\left(\frac{b_2}{k}\right)^{\alpha-1}\right\}|\Omega|.$$

Using Young's inequality (7.5), we then have

$$\int_\Omega\left|D\log\frac{M}{M-w+k}\right|^\alpha dx\leqslant C\bar{b}|\Omega|$$

where $C=C(\alpha)$, and hence, by Poincaré's inequality (7.44),

$$\int_\Omega\left|\log\frac{M}{M-w+k}\right|^\alpha dx\leqslant C\bar{b}, \tag{10.27}$$

where $C=C(n,\alpha,|\Omega|)$. To proceed further, we take

$$\varphi=\frac{\eta}{(M-w+k)^{\alpha-1}}$$

in (10.21), where $\eta\geqslant 0$, $\operatorname{supp}\eta\subset\operatorname{supp}u^+$ and $\eta\in C_0^1(\Omega)$. We then obtain from the structure conditions (10.23) the inequality

$$\int_\Omega\frac{\mathbf{A}\cdot D\eta}{(M-w+k)^{\alpha-1}}dx\leqslant\int_\Omega\left\{b_0\left|\frac{Dw}{M-w+k}\right|^{\alpha-1}+\alpha\left(\frac{a_2}{k}\right)^\alpha+\left(\frac{b_2}{k}\right)^{\alpha-1}\right\}\eta\,dx.$$
$$\leqslant\int_\Omega\left\{b_0\left|D\log\frac{M}{M-w+k}\right|^{\alpha-1}+\alpha\right\}\eta\,dx.$$

The function $\bar{w}=\log[M/(M-w+k)]$ accordingly satisfies $\bar{Q}\bar{w}\geqslant 0$ in $\Omega^+=\{x\in\Omega\,|\,u(x)>0\}$, where the operator $\bar{Q}$ fulfills the structure conditions (10.23) with $a_1=b_1=0$ and $a_2, b_2\leqslant\alpha$. Hence, by Lemma 10.8,

$$\sup_\Omega\bar{w}\leqslant C(\|\bar{w}\|_\alpha+1)$$
$$\leqslant C(n,\alpha,b_0,|\Omega|)\quad\text{by (10.27)}$$

Consequently $M\leqslant Ck$. The case $k=0$ is obtained by letting k tend to zero. By removing the conditions, $u\in C^1(\bar{\Omega})$, $u\leqslant 0$ on $\partial\Omega$, as in the proof of Lemma 10.8, we obtain the estimate (10.26) in each case. □

As a byproduct of the existence theory for the equation of prescribed mean curvature in Chapter 16, we shall see that Theorem 10.9 cannot be extended to allow $\alpha=1$ in its hypotheses. The following estimate, which includes the case $\alpha=1$, requires that the structure constants a_1, b_0 and b_1 be sufficiently small.

Theorem 10.10. *Let $u\in C^0(\bar{\Omega})\cap C^1(\Omega)$ satisfy $Qu\geqslant 0(=0)$ in Ω and suppose that Q satisfies the structure conditions* (10.23). *Then there exists a positive constant $C_0=C_0(\alpha,n)$ such that if*

$$(10.28)\qquad (a_1^\alpha+b_0^\alpha+b_1^{\alpha-1})|\Omega|^{\alpha/n}<C_0,$$

we have the estimate

$$(10.29)\qquad \sup_\Omega u(|u|)\leqslant C\{(a_1+b_1)\sup_{\partial\Omega}u^+(|u|)+a_2+b_2\}+\sup_{\partial\Omega}u^+(|u|),$$

where $C=C(n,\alpha,a_1,b_0,b_1,|\Omega|)$.

Proof. By virtue of Lemma 10.8 we need only estimate $\|u^+\|_\alpha$. As in the preceding proofs we assume initially that $u\in C^1(\bar{\Omega})$ and $u\leqslant 0$ on $\partial\Omega$. Substituting $\varphi=u^+=v$ into the integral inequality (10.21), we obtain by (10.23)

$$\int_\Omega |Dv|^\alpha dx\leqslant\int_\Omega\{(a_1^\alpha+b_1^{\alpha-1})v^\alpha+b_0v|Dv|^{\alpha-1}+a_2^\alpha+b_2^{\alpha-1}v\}\,dx$$

$$\leqslant\int_\Omega\left\{\left(a_1^\alpha+b_1^{\alpha-1}+\frac{b_0^\alpha}{\alpha\varepsilon^{\alpha-1}}\right)v^\alpha+(1-1/\alpha)\varepsilon|Dv|^\alpha+a_2^\alpha+b_2^{\alpha-1}v\right\}dx$$

for arbitrary $\varepsilon>0$, by inequality (7.6). Taking, in particular, $\varepsilon=\alpha^{1/(1-\alpha)}$ for $\alpha\neq 1$, and using the Poincaré inequality (7.44) we obtain for $\alpha\geqslant 1$

$$\int_\Omega v^\alpha\,dx\leqslant C(n,\alpha)|\Omega|^{\alpha/n}\int_\Omega\{(a_1^\alpha+b_1^{\alpha-1}+b_0^\alpha)v^\alpha+a_2^\alpha+b_2^{\alpha-1}v\}\,dx.$$

Hence if

$$C(n,\alpha)|\Omega|^{\alpha/n}(a_1^\alpha+b_1^{\alpha-1}+b_0^\alpha)<1,$$

we have

$$\int_\Omega v^\alpha\,dx\leqslant C(a_2^\alpha+b_2^\alpha)$$

and the desired estimate (10.29) follows. □

Note that when $\alpha = 1$ in Theorem 10.10, the constants b_1 and b_2 are not present in inequalities (10.28) and (10.29). By invoking the Poincaré inequality (7.44) in the sharp form

$$\int_\Omega |v|\,dx \leqslant \frac{1}{n}(|\Omega|/\omega_n)^{1/n}\int_\Omega |Dv|\,dx, \quad v \in W_0^{1,1}(\Omega), \tag{10.30}$$

we can in this case take

$$C_0 = C_0(1, n) = n\omega_n^{1/n}.$$

By writing the equation of prescribed mean curvature (10.7) in its divergence form,

$$\operatorname{div}\frac{Du}{\sqrt{1+|Du|^2}} = nH, \tag{10.31}$$

we see that it satisfies the structure conditions (10.23) with $\alpha = 1$, $a_1 = 0$, $a_2 = 1$, $b_2 = n\sup_\Omega |H|$. Hence, if the function H satisfies

$$H_0 = \sup_\Omega |H| < (\omega_n/|\Omega|)^{1/n}, \tag{10.32}$$

we have for any $C^2(\Omega) \cap C^0(\bar{\Omega})$ subsolution (solution) of equation (10.31) the estimate

$$\sup_\Omega u(|u|) \leqslant \sup_{\partial\Omega} u(|u|) + C(n, |\Omega|, H_0). \tag{10.33}$$

We conclude this section by noting that the structure conditions (10.23) can be generalized to allow the quantities a_1, a_2, b_0, b_1, b_2 to be nonnegative measurable functions. In particular if we assume $a_1, a_2, b_0, \bar{b}_1, \bar{b}_2, \in L^q(\Omega)$ where q is such that $q \geqslant \alpha$ and $q > n$ and $\bar{b}_1 = b_1^{1-1/\alpha}$, $\bar{b}_2 = b_2^{1-1/\alpha}$, then Lemma 10.8 and Theorem 10.9 continue to hold provided in inequalities (10.24) and (10.26), a_1, a_2, b_0, b_1, b_2 are replaced respectively by $\|a_1\|_q$, $\|a_2\|_q$, $\|b_0\|_q$, $\|\bar{b}_1\|^{\alpha/(\alpha-1)}$, $\|\bar{b}_2\|_q^{\alpha/(\alpha-1)}$ and the constants C depend additionally on q. Theorem 10.10 can be similarly extended, the condition (10.28) being replaced by

$$\|a_1^\alpha + b_0^\alpha + b_1^{\alpha-1}\|_\beta < C_0 \tag{10.34}$$

where $\beta = \max(1, n/\alpha)$. For the example of the equation of prescribed mean curvature (10.31) mentioned above, we obtain the more general result (already demonstrated in Corollary 10.6 by different means) that if the function H satisfies

$$\int_\Omega |H|^n\,dx < \omega_n, \tag{10.35}$$

then the maximum principle (10.33) holds with $H_0 = \|H\|_n$. The proofs of these assertions are basically the same as the case when a_1, a_2, b_0, b_1, b_2 are constant; (see Problem 10.4). Finally we remark that all the results of this section and their proofs are still applicable if the function u belongs to the Sobolev space $W^{1,\infty}(\Omega)$ instead of the space $C^2(\Omega) \cap C^0(\bar{\Omega})$.

Notes

The early results of this chapter, Theorems 10.1, 10.2, and 10.3 are basically variants of the Hopf maximum principle, Theorem 3.1. The counterexample in Section 10.3 is due to Meyers [ME 2]. Parts (i) and (ii) of the comparison principle, Theorem 10.7, were proved in Trudinger [TR 10]. Part (ii) extends an earlier result of Douglas, Dupont and Serrin [DDS]. Part (iii) was essentially proved in Serrin [SE 3]. The maximum principle, Theorem 10.9, is a new result in this work although the technique of its proof has already been demonstrated in [TR 7]. For further maximum principles of quasilinear equations the reader is referred to the works [SE 3], [SE 5].

We also note here that the form (10.30) of the Poincaré inequality is a consequence of the isoperimetric inequality; (see for example [FE]). Theorem 10.5 and Corollary 10.6 appear in Bakelman [BA 5].

Problems

Use the comparison principle, Theorem 10.1, to establish the following maximum principles in Problems 10.1, 10.2.

10.1. Let Q be elliptic in $\Omega \times \mathbb{R} \times \{0\}$ with coefficients $a^{ij}, b, i, j = 1, \ldots, n$, differentiable with respect to the p variables in $\Omega \times \mathbb{R} \times \mathbb{R}^n$. Suppose that there exists a constant M such that

$$(10.36) \qquad zb(x, z, 0) \leqslant 0 \quad \text{for } x \in \Omega, |z| \geqslant M.$$

Then, if $u \in C^0(\bar{\Omega}) \cap C^2(\Omega)$ satisfies $Qu \geqslant 0 (=0)$ in Ω, we have

$$(10.37) \qquad \max_{\Omega} u(|u|) \leqslant \max \{M, \max_{\partial\Omega} u^+(|u|)\}.$$

10.2. Let Ω lie in a ball B_R of radius R and suppose that Q is elliptic in Ω and that

$$(10.38) \qquad (\text{sign } z)\, b(x, z, p) \leqslant \frac{|p|}{R} \mathscr{T}(x, z, p), \quad \mathscr{T} = \text{trace } [a^{ij}].$$

for all $x \in \Omega$, $|z| \geqslant M$, $|p| \geqslant L$ for constants M and L. Then, if $u \in C^0(\bar{\Omega}) \cap C^2(\Omega)$ satisfies $Qu \geqslant 0 (=0)$ in Ω, we have ([SE 3])

$$(10.39) \qquad \max_{\Omega} u(|u|) \leqslant \max \{M, \max_{\partial\Omega} u^+(|u|)\} + 2LR.$$

(Hint: Follow the proof of Theorem 10.3 with the half-space, $x_1 > 0$, replaced by B_R.)

10.3. Let Q be an operator in the divergence form (10.5). Show that

$$p \cdot \mathbf{A}(x, z, p) = \int_0^1 s^{-2} \mathscr{E}(x, z, sp)\, ds + p \cdot \mathbf{A}(x, z, 0). \tag{10.40}$$

Hence show that if $\mathscr{E} \geqslant c|p|^{\alpha}$, where $c > 0$ and $\alpha > 1$, then

$$p \cdot \mathbf{A}(x, z, p) \geqslant \frac{c}{\alpha - 1} |p|^{\alpha} + p \cdot \mathbf{A}(x, z, 0). \tag{10.41}$$

10.4. Verify the assertions made at the end of Section 10.5.

10.5. Let $\mathscr{A}(p) = [a^{ij}(p)]$ be the coefficient matrix of the minimal surface operator given by

$$a^{ij}(p) = (1 + |p|^2)\delta_{ij} - p_i p_j, \qquad p \in \mathbb{R}^n.$$

Verify that 1 is an eigenvalue of $\mathscr{A}$ with eigenvector p and that $1 + |p|^2$ is the only other eigenvalue with eigenspace consisting of vectors orthogonal to p.

10.6. Apply Theorem 10.5 to the equations

$$(1 + |Du|^2)\Delta u - D_i u D_j u D_{ij} u = nH(x)(1 + |Du|)^s,$$

where $0 \leqslant s < \infty$.

Chapter 11

Topological Fixed Point Theorems and Their Application

In this chapter the solvability of the classical Dirichlet problem for quasilinear equations is reduced to the establishment of certain apriori estimates for solutions. This reduction is achieved through the application of topological fixed point theorems in appropriate function spaces. We shall first formulate a general criterion for solvability and illustrate its application in a situation where the required apriori estimates are readily derived from our previous results. The derivation of these apriori estimates under more general hypotheses will be the major concern of the ensuing chapters.

The fixed point theorems required for the treatment presented here are obtained as infinite dimensional extensions of the Brouwer fixed point theorem, which asserts that a continuous mapping of a closed ball in $\mathbb{R}^n$ into itself has at least one fixed point.

11.1. The Schauder Fixed Point Theorem

The Brouwer fixed point theorem can be extended to infinite dimensional spaces in various ways. We require first the following extension to Banach spaces.

Theorem 11.1. *Let $\mathfrak{S}$ be a compact convex set in a Banach space $\mathfrak{B}$ and let T be a continuous mapping of $\mathfrak{S}$ into itself. Then T has a fixed point, that is, $Tx=x$ for some $x \in \mathfrak{S}$.*

Proof. Let k be any positive integer. Since $\mathfrak{S}$ is compact, there exists a finite number of points $x_1 \ldots, x_N \in \mathfrak{S}$, where $N=N(k)$, such that the balls $B^i=B_{1/k}(x_i)$, $i=1, \ldots, N$, cover $\mathfrak{S}$. Let $\mathfrak{S}_k \subset \mathfrak{S}$ be the convex hull of $\{x_1, \ldots, x_N\}$, and define the mapping $J_k\colon \mathfrak{S} \to \mathfrak{S}_k$ by

$$J_k x = \frac{\sum \operatorname{dist}(x, \mathfrak{S}-B^i)x_i}{\sum \operatorname{dist}(x, \mathfrak{S}-B^i)}.$$

Clearly J_k is continuous and for any $x \in \mathfrak{S}$

$$(11.1) \qquad \|J_k x - x\| \leqslant \frac{\sum \operatorname{dist}(x, \mathfrak{S} - B^i)\|x_i - x\|}{\sum \operatorname{dist}(x, \mathfrak{S} - B^i)} < \frac{1}{k}.$$

The mapping $J_k \circ T$ when restricted to $\mathfrak{S}_k$ is accordingly a continuous mapping of $\mathfrak{S}_k$ into itself and hence, by virtue of the Brouwer fixed point theorem, possesses a fixed point $x^{(k)}$. (Note that $\mathfrak{S}_k$ is homeomorphic to a closed ball in some Euclidean space.) Since $\mathfrak{S}$ is compact, a subsequence of the sequence $x^{(k)}$, $k = 1, 2, \ldots$, converges to some $x \in \mathfrak{S}$. We claim that x is a fixed point of T. For, applying (11.1) to $Tx^{(k)}$, we have

$$\|x^{(k)} - Tx^{(k)}\| = \|J_k \circ Tx^{(k)} - Tx^k\| < \frac{1}{k},$$

and, since T is continuous, we infer $Tx = x$. □

As will be demonstrated in the next chapter, Theorem 11.1 is applicable to broad classes of equations in two variables. For later purposes we note the following extension of Theorem 11.1.

Corollary 11.2. *Let $\mathfrak{S}$ be a closed convex set in a Banach space $\mathfrak{B}$ and let T be a continuous mapping of $\mathfrak{S}$ into itself such that the image $T\mathfrak{S}$ is precompact. Then T has a fixed point.*

In the above theorems we note an essential difference from the contraction mapping principle, Theorem 5.1, in that the fixed points whose existence is asserted are not necessarily unique.

11.2. The Leray-Schauder Theorem: a Special Case

A continuous mapping between two Banach spaces is called *compact* (or *completely continuous*) if the images of bounded sets are precompact (that is, their closures are compact). The following consequence of Corollary 11.2 is the fixed point result most often applied in our approach to the Dirichlet problem for quasilinear equations.

Theorem 11.3. *Let T be a compact mapping of a Banach space $\mathfrak{B}$ into itself, and suppose there exists a constant M such that*

$$(11.2) \qquad \|x\|_{\mathfrak{B}} < M$$

for all $x \in \mathfrak{B}$ and $\sigma \in [0, 1]$ satisfying $x = \sigma Tx$. Then T has a fixed point.

Proof. We can assume without loss of generality that $M = 1$. Let us define a mapping T^* by

$$T^*x=\begin{cases} Tx & \text{if } \|Tx\|\leqslant 1,\\ \dfrac{Tx}{\|Tx\|} & \text{if } \|Tx\|\geqslant 1.\end{cases}$$

Then T^* is a continuous mapping of the closed unit ball $\bar{B}$ in $\mathfrak{B}$ into itself. Since $T\bar{B}$ is precompact the same is true of $T^*\bar{B}$. Hence by Corollary 11.2 the mapping T^* has a fixed point x. We claim that x is also a fixed point of T. For, suppose that $\|Tx\|\geqslant 1$. Then $x=T^*x=\sigma Tx$ if $\sigma=1/\|Tx\|$, and $\|x\|=\|T^*x\|=1$, which contradicts (11.2) with $M=1$. Hence $\|Tx\|<1$ and consequently $x=T^*x=Tx$. □

Remark. Theorem 11.3 implies that if T is *any* compact mapping of a Banach space into itself (whether or not (11.2) holds), then for *some* $\sigma\in(0,1]$ the mapping σT possesses a fixed point. Furthermore, if the estimate (11.2) holds then σT has a fixed point for all $\sigma\in[0,1]$.

In order to apply Theorem 11.3 to the Dirichlet problem for quasilinear equations, we fix a number $\beta\in(0,1)$ and take the Banach space $\mathfrak{B}$ to be the Hölder space $C^{1,\beta}(\bar{\Omega})$, where Ω is a bounded domain in $\mathbb{R}^n$. Let Q be the operator given by

$$Qu=a^{ij}(x,u,Du)D_{ij}u+b(x,u,Du) \tag{11.3}$$

and assume that Q is elliptic in $\bar{\Omega}$, that is, the coefficient matrix $[a^{ij}(x,z,p)]$ is positive for all $(x,z,p)\in\bar{\Omega}\times\mathbb{R}\times\mathbb{R}^n$. We also assume, for some $\alpha\in(0,1)$, that the coefficients a^{ij}, $b\in C^{\alpha}(\bar{\Omega}\times\mathbb{R}\times\mathbb{R}^n)$, that the boundary $\partial\Omega\in C^{2,\alpha}$ and that φ is a given function in $C^{2,\alpha}(\bar{\Omega})$. For all $v\in C^{1,\beta}(\bar{\Omega})$, the operator T is defined by letting $u=Tv$ be the unique solution in $C^{2,\alpha\beta}(\bar{\Omega})$ of the *linear* Dirichlet problem,

$$a^{ij}(x,v,Dv)D_{ij}u+b(x,v,Dv)=0 \text{ in } \Omega,\quad u=\varphi \text{ on } \partial\Omega. \tag{11.4}$$

The unique solvability of the problem (11.4) is guaranteed by the linear existence result, Theorem 6.14. The solvability of the Dirichlet problem, $Qu=0$ in Ω, $u=\varphi$ on $\partial\Omega$, in the space $C^{2,\alpha}(\bar{\Omega})$ is thus equivalent to the solvability of the equation $u=Tu$ in the Banach space $\mathfrak{B}=C^{1,\beta}(\bar{\Omega})$. The equation $u=\sigma Tu$ in $\mathfrak{B}$ is equivalent to the Dirichlet problem

$$Q_\sigma u=a^{ij}(x,u,Du)D_{ij}u+\sigma b(x,u,Du)=0 \text{ in } \Omega,\quad u=\sigma\varphi \text{ on } \partial\Omega. \tag{11.5}$$

By applying Theorem 11.3, we can then prove the following criterion for existence.

Theorem 11.4. *Let Ω be a bounded domain in $\mathbb{R}^n$ and suppose that Q is elliptic in $\bar{\Omega}$ with coefficients a^{ij}, $b\in C^{\alpha}(\bar{\Omega}\times\mathbb{R}\times\mathbb{R}^n)$, $0<\alpha<1$. Let $\partial\Omega\in C^{2,\alpha}$ and $\varphi\in C^{2,\alpha}(\bar{\Omega})$. Then, if for some $\beta>0$ there exists a constant M, independent of u and σ, such that every $C^{2,\alpha}(\bar{\Omega})$ solution of the Dirichlet problems, $Q_\sigma u=0$ in Ω, $u=\sigma\varphi$ on $\partial\Omega$, $0\leqslant\sigma\leqslant 1$, satisfies*

$$\|u\|_{C^{1,\beta}(\bar{\Omega})}<M, \tag{11.6}$$

it follows that the Dirichlet problem, $Qu=0$ in Ω, $u=\varphi$ on $\partial\Omega$, is solvable in $C^{2,\alpha}(\bar{\Omega})$.

Proof. In view of the remarks preceding the statement of the theorem, it only remains to show that the operator T is continuous and compact. By virtue of the global Schauder estimate, Theorem 6.6, T maps bounded sets in $C^{1,\beta}(\bar{\Omega})$ into bounded sets in $C^{2,\alpha\beta}(\bar{\Omega})$ which (by Arzela's theorem) are precompact in $C^2(\bar{\Omega})$ and $C^{1,\beta}(\bar{\Omega})$. In order to show the continuity of T, we let v_m, $m=1,2,\ldots$, converge to v in $C^{1,\beta}(\bar{\Omega})$. Then, since the sequence $\{Tv_m\}$ is precompact in $C^2(\bar{\Omega})$, every subsequence in turn has a convergent subsequence. Let $\{T\bar{v}_m\}$ be such a convergent subsequence with limit $u\in C^2(\bar{\Omega})$. Then since

$$\begin{aligned} &a^{ij}(x,v,Dv)D_{ij}u+b(x,v,Dv)\\ &\qquad=\lim_{m\to\infty}\{a^{ij}(x,\bar{v}_m,D\bar{v}_m)D_{ij}T\bar{v}_m+b(x,\bar{v}_m,D\bar{v}_m)\}=0,\end{aligned}$$

we must have $u=Tv$, and hence the sequence $\{Tv_m\}$ itself converges to u. □

11.3. An Application

Theorem 11.4 reduces the solvability of the Dirichlet problem $Qu=0$ in Ω, $u=\varphi$ on $\partial\Omega$ to the apriori estimation in the space $C^{1,\beta}(\bar{\Omega})$, for some $\beta>0$, of the solutions of a related family of problems. In practice it is desirable to break the derivation of the apriori estimates into four stages:

I. Estimation of $\sup_\Omega |u|$;

II. Estimation of $\sup_{\partial\Omega} |Du|$ in terms of $\sup_\Omega |u|$;

III. Estimation of $\sup_\Omega |Du|$ in terms of $\sup_{\partial\Omega} |Du|$ and $\sup_\Omega |u|$;

IV. Estimation of $[Du]_{\beta;\Omega}$, for some $\beta>0$, in terms of $\sup_\Omega |Du|$, $\sup_\Omega |u|$.

Step I has already been treated in Chapter 10; (see Theorems 10.3, 10.4 and 10.9). Steps II and III are treated in Chapters 14 and 15. In Chapter 13 it will be shown that Step IV can be carried out under very general hypotheses on Q. We shall illustrate the overall procedure here by considering a problem where the required estimates are readily obtained from some of the results in earlier chapters. Namely, let us suppose that either Q has the special divergence form,

$$Qu=\operatorname{div}\mathbf{A}(Du), \tag{11.7}$$

or that $n=2$ and Q has the form

$$Qu=a^{ij}(x,u,Du)D_{ij}u,\quad i,j=1,2. \tag{11.8}$$

We shall demonstrate in Chapter 14 that geometric conditions on the boundary $\partial\Omega$ play an important role in the solvability of the Dirichlet problem for quasilinear equations. For our present purposes, we shall require that the *boundary manifold*

$$\Gamma=(\partial\Omega, \varphi)=\{(x, z) \in \partial\Omega \times \mathbb{R} \mid z=\varphi(x)\}$$

satisfies a *bounded slope condition*, that is, for every point $P \in \Gamma$ there exist two planes in $\mathbb{R}^{n+1}$, $z=\pi_P^+(x)$ and $z=\pi_P^-(x)$, passing through P such that:

(i) $\pi_P^-(x) \leqslant \varphi(x) \leqslant \pi_P^+(x) \quad \forall x \in \partial\Omega$;
(ii) the slopes of these planes are uniformly bounded, independently of P, by a constant K; that is $|D\pi_P^\pm| \leqslant K$ for all $P \in \Gamma$.

If $\partial\Omega \in C^2$, $\varphi \in C^2(\bar{\Omega})$ and $\partial\Omega$ is *uniformly convex* (that is, its principal curvatures are bounded away from zero), then Γ satisfies a bounded slope condition; (see [HA]). We now can assert the following existence result.

Theorem 11.5. *Let Q have either of the forms* (11.7) *or* (11.8), *and suppose that Q, Ω and φ satisfy the hypotheses of Theorem* 11.4. *Then if, in addition, the boundary manifold $(\partial\Omega, \varphi)$ satisfies the bounded slope condition, it follows that the Dirichlet problem, $Qu=0$ in Ω, $u=\varphi$ on $\partial\Omega$, is solvable in $C^{2,\alpha}(\bar{\Omega})$.*

Proof. Since $Q_\sigma=Q$, we must estimate the solutions of the Dirichlet problems, $Qu=0$ in Ω, $u=\sigma\varphi$ on $\partial\Omega$, $0 \leqslant \sigma \leqslant 1$. Let us take in turn the different stages described above.

I. From the weak maximum principle, (Theorems 3.1 or 10.3), we have

$$\sup_\Omega |u|=\sigma \sup_{\partial\Omega} |\varphi| \leqslant \sup_{\partial\Omega} |\varphi|. \tag{11.9}$$

II. The bounded slope condition provides linear barriers which serve to estimate Du on $\partial\Omega$. For clearly

$$a^{ij}(x, u, Du)D_{ij}\pi_P^\pm=0,$$

and hence, by the weak maximum principle,

$$\sigma\pi_P^-(x) \leqslant u(x) \leqslant \sigma\pi_P^+(x),$$

for all $x \in \Omega$. Consequently we have

$$\sup_{\partial\Omega} |Du| \leqslant \sigma K \leqslant K, \tag{11.10}$$

where K is the assumed bound for the slopes of $\pi_P^\pm$.

III. Steps III and IV will follow from the fact that the derivatives $D_k u$, $k=1, \ldots, n$, are weak solutions of simple linear divergence structure equations of the type treated in Chapter 8. Let us first suppose that Q has the form (11.7) and write the equation $Qu=0$ in its integral form,

$$\int_\Omega \mathbf{A}(Du)\cdot D\eta\, dx=0 \quad \forall \eta \in C_0^1(\Omega). \tag{11.11}$$

Fixing k, replacing η by $D_k\eta$ and then integrating by parts, we obtain

$$\int_\Omega D_{p_j}A^i(Du)D_{kj}uD_i\eta\, dx=0 \quad \forall \eta \in C_0^1(\Omega),$$

or, if $w=D_k u$,

$$\int_\Omega a^{ij}(Du)D_jwD_i\eta\, dx=0 \quad \forall \eta \in C_0^1(\Omega).$$

The function $w\in C^1(\bar{\Omega})$ is thus a weak solution of the linear elliptic equation

$$D_i(\bar{a}^{ij}(x)D_jw)=0, \tag{11.12}$$

where $\bar{a}^{ij}(x)=a^{ij}(Du(x))$, and hence, by the weak maximum principle in Section 3.6 (see also Theorem 8.1), we have

$$\sup_\Omega |Du|=\sup_{\partial\Omega}|Du|\leqslant K. \tag{11.13}$$

Next, if Q has the form (10.8), the equation $Qu=0$ is equivalent to

$$\frac{a^{11}}{a^{22}}D_{11}u+\frac{2a^{12}}{a^{22}}D_{12}u+D_{22}u=0,$$

so that

$$\int_\Omega\left(\frac{a^{11}}{a^{22}}D_{11}u+\frac{2a^{12}}{a^{22}}D_{12}u+D_{22}u\right)\eta\, dx=0 \quad \forall \eta \in C_0^1(\Omega).$$

Replacing η by $D_1\eta$ and integrating by parts, we obtain for $w=D_1u$,

$$\int_\Omega\left\{\left(\frac{a^{11}}{a^{22}}D_1w+\frac{2a^{12}}{a^{22}}D_2w\right)D_1\eta+D_2wD_2\eta\right\}dx=0,$$

and hence w is a weak solution of the linear elliptic equation

$$D_i(a_1^{ij}(x)D_jw)=0, \quad i,j=1,2, \tag{11.14}$$

with coefficient matrix

$$[a_1^{ij}(x)] = \begin{bmatrix} \frac{a^{11}}{a^{22}}(x, u(x), Du(x)) & \frac{2a^{12}}{a^{22}}(x, u(x), Du(x)) \\ 0 & 1 \end{bmatrix}.$$

Similarly, it follows that $D_2 u$ is a weak solution of a corresponding linear elliptic equation and hence again, by the weak maximum principle, the estimate (11.13) holds.

IV. The equations (11.12) and (11.14) for the derivatives $D_k u$ will satisfy the hypotheses of Theorem 8.24 with constants λ and Λ depending on $\sup_\Omega |u|$, $\sup_\Omega |Du|$ and the coefficients a^{ij}. Consequently we obtain an interior Hölder estimate for Du, that is, for any subdomain $\Omega' \subset\subset \Omega$, we have

$$(11.15) \qquad [Du]_{\beta;\Omega'} \leqslant Cd^{-\beta}$$

where the positive constants C and β are independent of u and σ, and $d = \text{dist}(\Omega', \partial\Omega)$. We cannot however infer a global Hölder estimate for Du directly from the results of Chapter 8. Instead we proceed as follows. We first use the smoothness of $\partial\Omega$ to map portions of $\partial\Omega$ into the hyperplane $x_n = 0$. The derivatives $D_{y_k} u$, $k = 1, \ldots, n-1$, with respect to the transformed coordinates $y_1, \ldots, y_n$ can then be estimated by means of Theorem 8.29. The remaining derivative $D_{y_n} u$ is finally estimated by using the equation itself together with Morrey's estimate, Theorem 7.19. The details of this procedure are carried out in Chapter 13 for the general divergence structure equation. The resulting estimate

$$(11.16) \qquad [Du]_{\beta;\Omega} \leqslant C$$

with positive constants β and C independent of u and σ completes the proof of Theorem 11.5. □

We note here that from the results of Chapter 14 it will follow that the bounded slope condition in the hypothesis of Theorem 11.5 can be replaced by the boundedness of the quantity

$$\frac{\Lambda(x, z, p)|p|}{\mathscr{E}(x, z, p)}$$

for $x \in \bar{\Omega}$, $|z| \leqslant \sup_{\partial\Omega} |\varphi|$, $|p| \geqslant 1$; (see Theorem 14.1). Furthermore, if Ω is convex, the bounded slope condition can be replaced by the boundedness of the quantity

$$\frac{\Lambda(x, z, p)}{a^{ij}(x, z, p)(p_i - D_i\varphi)(p_j - D_j\varphi)}$$

for $x \in \bar{\Omega}$, $|z| \leqslant \sup_{\partial\Omega} |\varphi|$, $|p| \geqslant 1$ (see Theorem 14.2).

11.4. The Leray-Schauder Fixed Point Theorem

For certain applications it is desirable to replace the family of Dirichlet problems, $Q_\sigma u=0$ in Ω, $u=\sigma\varphi$ on $\partial\Omega$, $0\leqslant\sigma\leqslant 1$, used in Theorem 11.4 by other families which depend differently on the parameter σ. Accordingly, we shall require the following generalization of Theorem 11.3.

Theorem 11.6. *Let $\mathfrak{B}$ be a Banach space and let T be a compact mapping of $\mathfrak{B}\times[0,1]$ into $\mathfrak{B}$ such that $T(x,0)=0$ for all $x\in\mathfrak{B}$. Suppose there exists a constant M such that*

$$(11.17)\qquad \|x\|_{\mathfrak{B}}<M$$

for all $(x,\sigma)\in\mathfrak{B}\times[0,1]$ satisfying $x=T(x,\sigma)$. Then the mapping T_1 of $\mathfrak{B}$ into itself given by $T_1x=T(x,1)$ has a fixed point.

Theorem 11.6 will be derived from the following consequence of Corollary 11.2.

Lemma 11.7. *Let $B=B_1(0)$ denote the unit ball in $\mathfrak{B}$ and let T be a continuous mapping of $\bar{B}$ into $\mathfrak{B}$ such that $T\bar{B}$ is precompact and $T\,\partial B\subset B$. Then T has a fixed point.*

Proof. We define a mapping T^* by

$$T^*x=\begin{cases} Tx & \text{if } \|Tx\|\leqslant 1\\ \dfrac{Tx}{\|Tx\|} & \text{if } \|Tx\|\geqslant 1.\end{cases}$$

Clearly T^* is a continuous mapping of $\bar{B}$ into itself and since $T\bar{B}$ is precompact, the same is true of $T^*\bar{B}$. Hence, by Corollary 11.2, T^* has a fixed point x and since $T\,\partial B\subset B$ we must have $\|x\|<1$ and therefore $x=Tx$.

Proof of Theorem 11.6. We can assume without loss of generality that $M=1$. For $0<\varepsilon\leqslant 1$, let us define a mapping T^* from $\bar{B}$ into $\mathfrak{B}$ by

$$T^*x=T^*_\varepsilon x=\begin{cases} T\left(\dfrac{x}{\|x\|},\dfrac{1-\|x\|}{\varepsilon}\right) & \text{if } 1-\varepsilon\leqslant\|x\|\leqslant 1,\\ T\left(\dfrac{x}{1-\varepsilon},1\right) & \text{if } \|x\|<1-\varepsilon.\end{cases}$$

The mapping T^* is clearly continuous, $T^*\bar{B}$ is precompact by the compactness of T and $T^*\partial B=0$. Hence, by Lemma 11.7, the mapping T^* has a fixed point $x(\varepsilon)$. We now set

$$\varepsilon=\frac{1}{k},\qquad x_k=x\left(\frac{1}{k}\right),\qquad \sigma_k=\begin{cases} k(1-\|x_k\|) & \text{if } 1-\dfrac{1}{k}\leqslant\|x_k\|\leqslant 1,\\ 1 & \text{if } \|x_k\|<1-\dfrac{1}{k},\end{cases}$$

where $k=1, 2, \ldots$. By the compactness of T we can assume, by passing to a subsequence if necessary, that the sequence $\{(x_k, \sigma_k)\}$ converges in $\mathfrak{B}\times[0,1]$ to (x, σ). It then follows that $\sigma=1$. For if $\sigma<1$, we must have $\|x_k\|\geqslant 1-1/k$ for sufficiently large k and hence $\|x\|=1$, $x=T(x,\sigma)$, which contradicts (11.17). Since $\sigma=1$, we then have by the continuity of T that $T^*_{1/k}x_k\to T(x,1)$, and therefore x is also a fixed point of T_1. □

We note that Theorem 11.3 corresponds to the special case of Theorem 11.6 where $T(x,\sigma)=\sigma T_1 x$. Now let Q be an operator of the form (11.3) and suppose that Q, Ω and φ satisfy the hypotheses of Theorem 11.4. In order to apply Theorem 11.6 to the Dirichlet problem, $Qu=0$ in Ω, $u=\varphi$ on $\partial\Omega$, we imbed this problem in a family of problems,

$$Q_\sigma u=a^{ij}(x, u, Du; \sigma)D_{ij}u+b(x, u, Du; \sigma)=0 \text{ in } \Omega,$$
$$u=\sigma\varphi \text{ on } \partial\Omega, \quad 0\leqslant\sigma\leqslant 1,$$

such that:

(i) $Q_1=Q$, $b(x, z, p; 0)=0$;
(ii) the operators Q_σ are elliptic in $\bar\Omega$ for all $\sigma\in[0,1]$;
(iii) the coefficients $a^{ij}, b\in C^0(C^\alpha(\bar\Omega\times\mathbb{R}\times\mathbb{R}^n); [0,1])$, that is, $a^{ij}, b\in C^\alpha(\bar\Omega\times\mathbb{R}\times\mathbb{R}^n)$ for each $\sigma\in[0,1]$ and considered as mappings from $[0,1]$ into $C^\alpha(\bar\Omega\times\mathbb{R}\times\mathbb{R}^n)$, the functions a^{ij}, b are continuous.

For all $v\in C^{1,\beta}(\bar\Omega)$, $\sigma\in[0,1]$ the operator T is defined by letting $u=T(v,\sigma)$ be the unique solution in $C^{2,\alpha\beta}(\bar\Omega)$ of the *linear* Dirichlet problem,

$$a^{ij}(x, v, Dv; \sigma)D_{ij}u+b(x, v, Dv; \sigma)=0 \text{ in } \Omega, \quad u=\sigma\varphi \text{ on } \partial\Omega.$$

From condition (i) above we see that the solvability of the Dirichlet problem, $Qu=0$ in Ω, $u=\varphi$ on $\partial\Omega$, in the space $C^{2,\alpha}(\bar\Omega)$ is equivalent to the solvability of the equation $u=T(u,1)$ in the Banach space $C^{1,\beta}(\bar\Omega)$, and that $T(u,0)=0$ for all $v\in C^{1,\beta}(\bar\Omega)$. The continuity and compactness of the mapping T are assured by conditions (ii) and (iii); the details of this argument are similar to the proof of Theorem 11.4 and are left to the reader. Hence we can conclude from Theorem 11.6 the following generalization of Theorem 11.4.

Theorem 11.8. *Let Ω be a bounded domain in $\mathbb{R}^n$ with boundary $\partial\Omega\in C^{2,\alpha}$ and let $\varphi\in C^{2,\alpha}(\bar\Omega)$. Let $\{Q_\sigma, 0\leqslant\sigma\leqslant 1\}$ be a family of operators satisfying conditions* (i), (ii), (iii) *above and suppose that for some $\beta>0$ there exists a constant M, independent of u and σ, such that every $C^{2,\alpha}(\bar\Omega)$ solution of the Dirichlet problems $Q_\sigma u=0$ in Ω, $u=\sigma\varphi$ on $\partial\Omega$, $0\leqslant\sigma\leqslant 1$, satisfies*

$$\|u\|_{C^{1,\beta}(\bar\Omega)}<M.$$

Then the Dirichlet problem, $Qu=0$ in Ω, $u=\varphi$ on $\partial\Omega$, is solvable in $C^{2,\alpha}(\bar\Omega)$.

We note here that by means of the theory of topological degree (see [LS]), the hypotheses of Theorems 11.6 and 11.8 can be weakened slightly. However, since the improvement so obtained is not relevant to the particular applications in this book, we have chosen to avoid the theory of degree altogether.

11.5. Variational Problems

In this section we consider variational problems and, in particular, their relationship with elliptic partial differential equations. Let Ω be a bounded domain in $\mathbb{R}^n$ and F a given function in $C^1(\Omega \times \mathbb{R} \times \mathbb{R}^n)$. We consider the functional I defined on $C^{0,1}(\bar{\Omega})$ by

$$I(u) = \int_\Omega F(x, u, Du)\, dx. \tag{11.18}$$

Note that, since $u \in C^{0,1}(\bar{\Omega})$, the gradient Du exists almost everywhere and is measurable and bounded; (see Section 7.3). Now let φ be a given $C^{0,1}(\bar{\Omega})$ function and consider $I(u)$ for all functions u in the set

$$\mathscr{C} = \{u \in C^{0,1}(\bar{\Omega}) | u = \varphi \text{ on } \partial\Omega\}.$$

The problem we now consider is the following:

$\mathscr{P}$: Find $u \in \mathscr{C}$ such that $I(u) \leqslant I(v)$ for all $v \in \mathscr{C}$.

Let us suppose that u is a solution of $\mathscr{P}$ and let η belong to the space

$$\mathscr{C}_0 = \{\eta \in C^{0,1}(\bar{\Omega}) \mid \eta = 0 \text{ on } \partial\Omega\};$$

then the function $v = u + t\eta$ must belong to $\mathscr{C}$ for every $t \in \mathbb{R}$. Thus $I(u) \leqslant I(u + t\eta)$ for all $t \in \mathbb{R}$, or defining $\mathscr{I}(t) = I(u + t\eta)$, we have $\mathscr{I}(0) \leqslant \mathscr{I}(t)$ for all $t \in \mathbb{R}$, and so $\mathscr{I}$ has a minimum at 0, whence $\mathscr{I}'(0) = 0$. By differentiation, we therefore obtain the equation

$$\int_\Omega \{D_{p_i} F(x, u, Du) D_i \eta + D_z F(x, u, Du)\eta\}\, dx = 0 \tag{11.19}$$

for all $\eta \in \mathscr{C}_0$, that is the function u is a weak solution of the *Euler-Lagrange equation*

$$Qu = \operatorname{div} D_p F(x, u, Du) - D_z F(x, u, Du) = 0. \tag{11.20}$$

Moreover if $F \in C^2(\Omega \times \mathbb{R} \times \mathbb{R}^n)$ and $u \in C^2(\Omega) \cap C^{0,1}(\bar{\Omega})$, then u is a solution of the classical Dirichlet problem $Qu=0$ in Ω in Ω, $u=\varphi$ on $\partial\Omega$. Hence the solvability of problem $\mathscr{P}$ implies the solvability of the Dirichlet problem for equation (11.20).

Let us call the functional I *regular* if the integrand F is strictly convex with respect to the p variables. Clearly if $F \in C^2(\Omega \times \mathbb{R} \times \mathbb{R}^n)$, then the regularity of I is equivalent to the ellipticity of the Euler-Lagrange operator Q. We suppose now that the function $u \in C^{0,1}(\Omega)$ satisfies (11.20) and $u=\varphi$ on $\partial\Omega$. Then we have

$$\begin{aligned}\mathscr{I}(t) &= \mathscr{I}(0)+t\mathscr{I}'(0)+\frac{t^2}{2}\mathscr{I}''(\zeta)\\ &= \mathscr{I}(0)+\frac{t^2}{2}\mathscr{I}''(\zeta)\end{aligned}$$

for some ζ such that $|\zeta| \leqslant |t|$. If we now assume that the function F is jointly convex in z and p so that the matrix

$$\begin{bmatrix} D_{p_ip_j}F & D_{p_iz}F \\ D_{p_jz}F & D_{zz}F \end{bmatrix}$$

is non-negative in $\Omega \times \mathbb{R} \times \mathbb{R}^n$, we have

$$\begin{aligned}\mathscr{I}''(\zeta) = \int_\Omega \{ & D_{p_ip_j}F(x, u+\zeta\eta, Du+\zeta D\eta)D_i\eta D_j\eta \\ & +2D_{p_iz}F(x, u+\zeta\eta, Du+\zeta D\eta)\eta D_i\eta \\ & +D_{zz}F(x, u+\zeta\eta, Du+\zeta D\eta)\eta^2\}\, dx \geqslant 0,\end{aligned}$$

and hence $\mathscr{I}(0) \leqslant \mathscr{I}(t)$ for all $t \in \mathbb{R}$. Consequently the function u is a solution of the variational problem $\mathscr{P}$ and moreover if I is regular, we see from Theorem 10.1 that u is uniquely determined. Therefore we have proved

Theorem 11.9. *Let I be regular with F jointly convex in z and p. Then the variational problem $\mathscr{P}$ can have at most one solution. Furthermore, the solvability of $\mathscr{P}$ is equivalent to the solvability of the Dirichlet problem for the Euler-Lagrange equation, $Qu=0$, $u=\varphi$ on $\partial\Omega$, in the space $C^{0,1}(\bar{\Omega})$.*

Alternative Approaches

By utilizing direct procedures in the calculus of variations, we can develop alternative approaches to the Dirichlet problem for variational operators Q. Direct methods which involve the enlargement of the set $\mathscr{C}$ to a subset of an appropriate space of weakly differentiable functions are treated in [LU 4] and [MY 5]. We describe briefly a further approach which has the advantages that the integrand F need not be C^2 and that the solutions obtained are automatically in $C^{0,1}(\bar{\Omega})$. Let us define, for $K \in \mathbb{R}$,

$$\mathscr{C}_K = \{u \in \mathscr{C} \mid \|u\|_{C^{0,1}(\bar{\Omega})} \leqslant K\}$$

and consider the problem

$\mathscr{P}_K$: Find $u \in \mathscr{C}_K$ such that $I(u) \leqslant I(v)$ for all $v \in \mathscr{C}_K$.

For $\mathscr{P}_K$ we have the following existence result.

Theorem 11.10. *Let $F \in C^1(\Omega \times \mathbb{R} \times \mathbb{R}^n)$ and suppose that $F, D_z F, D_{p_i} F \in C^0(\bar{\Omega} \times \mathbb{R} \times \mathbb{R}^n)$, $i = 1, \ldots, n$. Then, if F is convex with respect to p, the problem $\mathscr{P}_K$ is solvable for any K such that $\mathscr{C}_K$ is non-empty.*

Proof. We show that the functional I is lower semicontinuous on $\mathscr{C}_K$ with respect to uniform convergence in Ω. Since I is also bounded below on $\mathscr{C}_K$ and $\mathscr{C}_K$ is precompact in $C^0(\bar{\Omega})$, the result follows. Hence let $\{u_m\} \subset \mathscr{C}_K$ converge uniformly to a function $u \in \mathscr{C}_K$. We then have

$$\begin{aligned}
I(u_m) - I(u) &= \int_\Omega [F(x, u_m, Du_m) - F(x, u, Du)]\, dx \\
&= \int_\Omega [F(x, u_m, Du_m) - F(x, u, Du_m)]\, dx \\
&\quad + \int_\Omega [F(x, u, Du_m) - F(x, u, Du)]\, dx \\
&\geqslant - \sup_{\Omega \times \mathscr{C}_K} |D_z F| \int_\Omega |u_m - u|\, dx \\
&\quad + \int_\Omega D_{p_i} F(x, u, Du) D_i(u_m - u)\, dx
\end{aligned} \tag{11.21}$$

by the convexity of F with respect to p. For fixed i, let us set $\varphi = D_{p_i} F(x, u, Du)$ and suppose first that $\varphi \in C_0^1(\Omega)$. Integrating by parts, we then have

$$\int_\Omega \varphi D_i(u_m - u)\, dx = -\int_\Omega (u_m - u) D_i \varphi\, dx \to 0 \quad \text{as } m \to \infty.$$

If $\varphi \notin C_0^1(\Omega)$, then since $\varphi \in L^\infty(\Omega)$ there exists, for $\varepsilon > 0$, a function $\varphi_\varepsilon \in C_0^1(\Omega)$ such that

$$\int_\Omega |\varphi_\varepsilon - \varphi|\, dx < \frac{\varepsilon}{2K}.$$

Then we have

$$\left|\int_\Omega \varphi D_i(u_m-u)\,dx\right| \leqslant \left|\int_\Omega \varphi_\varepsilon D_i(u_m-u)\,dx\right| + \int_\Omega |\varphi_\varepsilon-\varphi|\,|D_i(u_m-u)|\,dx.$$

But $\int_\Omega \varphi_\varepsilon D_i(u_m-u)\,dx \to 0$ as $m\to\infty$, and as $u_m, u \in \mathscr{C}_K$, $|D_i(u_m-u)| \leqslant 2K$, so that

$$\int_\Omega |\varphi_\varepsilon-\varphi|\,|D_i(u_m-u)|\,dx < \varepsilon.$$

Hence

$$\limsup_{m\to\infty}\left|\int_\Omega \varphi D_i(u_m-u)\,dx\right| \leqslant \varepsilon$$

and, since ε can be chosen arbitrarily, we conclude from (11.21) that

$$\liminf_{m\to\infty} I(u_m) \geqslant I(u),$$

that is, I is lower semicontinuous on $\mathscr{C}_K$ with respect to uniform convergence. □

Let us call a solution of problem $\mathscr{P}_K$, a K-quasisolution of problem $\mathscr{P}$. If there exists some space in which the family of all K-quasisolutions for $K \in \mathbb{R}$ is relatively compact, we can obtain a generalized solution of problem $\mathscr{P}$ as the limit of a sequence of quasisolutions $\{u_m\}$ corresponding to constants $\{K_m\}$, $K_m \to \infty$. The following theorem shows that the solvability of problem $\mathscr{P}$, as formulated, is a consequence of an apriori bound in $C^{0,1}(\bar{\Omega})$ for quasisolutions.

Theorem 11.11. *Let u be a K-quasisolution of problem $\mathscr{P}$ such that*

$$|u|_{C^{0,1}(\bar{\Omega})} < K. \tag{11.22}$$

Then, if $F \in C^1(\Omega \times \mathbb{R} \times \mathbb{R}^n)$ is jointly convex with respect to z and p, the function u is also a solution of problem $\mathscr{P}$.

Proof. Let $v \in \mathscr{C}$. Then by (11.22) we have

$$w = u + \varepsilon(v-u) \in \mathscr{C}_K$$

for some $\varepsilon>0$. Since u solves $\mathscr{P}_K$, we have

$$\int_\Omega F(x, u, Du)\, dx \leqslant \int_\Omega F(x, w, Dw)\, dx,$$

but, as $w=(1-\varepsilon)u+\varepsilon v$ and F is convex in (z, p),

$$\int_\Omega F(x, w, Dw)\, dx \leqslant (1-\varepsilon) \int_\Omega F(x, u, Du)\, dx + \varepsilon \int_\Omega F(x, v, Dv)\, dx.$$

Consequently

$$\int_\Omega F(x, u, Du)\, dx \leqslant \int_\Omega F(x, v, Dv)\, dx. \quad \square$$

The combination of Theorems 11.10 and 11.11 can be viewed as analogous to Theorems 11.4 and 11.8. It is practicable to carry out the required estimation of quasisolutions in three steps corresponding to steps (i), (ii) and (iii) of the existence procedure described in Section 11.3. Namely:

(i)′ Estimation of $\sup_\Omega |u|$;

(ii)′ Using (i)′, estimation of

$$l'(u)=\sup_{x\in\Omega,\, y\in\partial\Omega} \frac{|u(x)-u(y)|}{|x-y|};$$

(iii)′ Using (ii)′, estimation of

$$l(u)=\sup_{x, y\in\Omega} \frac{|u(x)-u(y)|}{|x-y|}.$$

It turns out that many of our estimates in Chapters 10, 15 and 16, (in particular the comparison principle, Theorem 10.7), can be adapted to hold for quasisolutions of variational problems, thereby facilitating the above steps. Furthermore, under appropriate hypotheses on Q and $\partial\Omega$, we can obtain by regularity considerations classical solutions of the Dirichlet problem $Qu=0$, $u=\varphi$ on $\partial\Omega$.

We note here also that the methods described above can be extended to the class of divergence structure operators using the theory of *monotone operators* and also to encompass *obstacle problems*. In these situations the solvability of problem $\mathscr{P}_K$ is generalized to that of a variational inequality. The reader is referred to the works [BW 3], [HS], [LL], [LST], [PA], [WL], [KST] for further information.

Notes

The Schauder fixed point theorem, Theorem 11.1, was established in [SC 1] and applied by Schauder to nonlinear equations in [SC 3]. Theorems 11.3 and 11.6 are special cases of the Leray-Schauder theorem [LS]. Our proofs of these results follow respectively those of Schaefer [SH] and Browder [BW 2]. The application to the Dirichlet problem, Theorem 11.5, is adapted from Gilbarg [GL 2]. For Equation (11.7) Morrey ([MY 5] p. 98) proves Theorem 11.5 assuming only the bounded slope condition, without the regularity hypotheses on Ω and φ in Theorem 11.4.

In the first edition of this work we proved the Brouwer fixed point theorem following [DS]. In recent years many elegant and simple proofs have appeared in the literature.

Chapter 12

Equations in Two Variables

The theory of quasilinear elliptic equations in two dimensions is in many respects simpler and in some respects more general than that in higher dimensions. This chapter is concerned with aspects of the theory that are specifically two-dimensional in character, although the basic results on quasilinear equations can be extended to higher dimensions by other methods. As will be seen, the special features of this theory are founded on strong apriori estimates that are valid for general *linear* equations in two variables.

12.1. Quasiconformal Mappings

Various function theoretic concepts and methods play a special role in the theory of elliptic equations in two variables; (see [CH], for example). Here we shall be concerned mainly with apriori estimates arising from the theory of quasiconformal mappings. A continuously differentiable mapping $p=p(x, y)$, $q=q(x, y)$ from a domain Ω in the $z=(x, y)$ plane to the $w=(p, q)$ plane is *quasiconformal*, or *K-quasiconformal*, in Ω if for some constant $K>0$ we have

$$p_x^2+p_y^2+q_x^2+q_y^2\leqslant 2K(p_xq_y-p_yq_x) \tag{12.1}$$

for all $(x, y)\in\Omega$. Although it suffices for the present purposes that p and q are in $C^1(\Omega)$, the results developed in this section will apply as well to continuous p, q in $W^{1,2}_{\text{loc}}$, that is, to continuous p, q that have locally square integrable weak derivatives.

When $K<1$, (12.1) is seen to imply that p and q are constant and we shall therefore assume $K\geqslant 1$. For $K=1$, the mapping $w(z)=p(z)+iq(z)$ defines an analytic function of z. When $K\geqslant 1$, the inequality (12.1) has the geometric meaning that at points of non-vanishing Jacobian the mapping between the z and w planes preserves orientation and takes infinitesimal circles into infinitesimal ellipses of uniformly bounded eccentricity, in which the ratio of minor to major axis is bounded below by $\alpha=K-(K^2-1)^{1/2}>0$. This remark can be verified by direct calculation.

It will also be of interest to consider the more general class of mappings $(x, y)\rightarrow(p, q)$ defined by the inequality

$$(12.2)\qquad p_x^2+p_y^2+q_x^2+q_y^2\leqslant 2K(p_xq_y-p_yq_x)+K',$$

where K, K' are constants, with $K\geqslant 1$, $K'\geqslant 0$. Although the geometric meaning is no longer the same, we shall refer to the mappings obeying (12.2) as (K, K')-quasiconformal. In the subsequent development it will be seen that mappings satisfying (12.1) and (12.2) arise naturally from elliptic equations in two variables, with p and $-q$ representing the first derivatives of the solution.

The object of this section is the derivation of apriori interior Hölder estimates for (K, K')-quasiconformal mappings. The main result will be the consequence of lemmas concerning the Dirichlet integral

$$(12.3)\qquad \mathfrak{D}(r; z)=\iint\limits_{B_r(z)}|Dw|^2\,dx\,dy=\iint\limits_{B_r(z)}(|w_x|^2+|w_y|^2)\,dx\,dy$$

of a (K, K')-quasiconformal mapping w taken over disks $B_r(z)$. When there is no ambiguity, we shall write $\mathfrak{D}(r)$ for $\mathfrak{D}(r; z)$ and B_r for $B_r(z)$.

Lemma 12.1. *Let $w=p+iq$ be (K, K')-quasiconformal in a disk $B_R=B_R(z_0)$, satisfying* (12.2) *with $K>1$, $K'\geqslant 0$, and suppose $|p|\leqslant M$ in B_R. Then for all $r\leqslant R/2$,*

$$(12.4)\qquad \mathfrak{D}(r)=\iint\limits_{B_r}|Dw|^2\,dx\,dy\leqslant C\left(\frac{r}{R}\right)^{2\alpha},\quad \alpha=K-(K^2-1)^{1/2},$$

where $C=C_1(K)(M^2+K'R^2)$. If $K'=0$, the conclusion remains valid for $K=1$.

Proof. We first establish an estimate for the Dirichlet integral in the disk of radius $R/2$. From (12.2) we have in any concentric disk $B_r\subset B_R$

$$(12.5)\qquad \mathfrak{D}(r)=\iint\limits_{B_r}|Dw|^2\,dx\,dy\leqslant 2K\iint\limits_{B_r}\frac{\partial(p, q)}{\partial(x, y)}\,dx\,dy+K'\pi r^2$$

$$=2K\int\limits_{C_r}p\frac{\partial q}{\partial s}\,ds+K'\pi r^2,$$

where s denotes arc length along the circle $C_r=\partial B_r$ described in the counterclockwise direction. Using the fact that $\mathfrak{D}'(r)=\int\limits_{C_r}|Dw|^2\,ds$, we observe

$$(12.6)\qquad \int\limits_{C_r}p\frac{\partial q}{\partial s}\,ds\leqslant\left(\int\limits_{C_r}p^2\,ds\int\limits_{C_r}|Dq|^2\,ds\right)^{1/2}$$

$$\leqslant\left(\int\limits_{C_r}p^2\,ds\int\limits_{C_r}|Dw|^2\,ds\right)^{1/2}\leqslant M(2\pi r\mathfrak{D}'(r))^{1/2}.$$

Inserting this into (12.5), and replacing r by R in the second member on the right, we obtain

$$[\mathfrak{D}(r)-k_1]^2\leqslant k_2 r\mathfrak{D}'(r), \tag{12.7}$$

where $k_1=\pi R^2K'$, $k_2=8\pi M^2K^2$. Now either $\mathfrak{D}(R/2)\leqslant k_1$, in which case we have the desired estimate; or if not, then $\mathfrak{D}(r)>k_1$ for some $r=r_0<R/2$ and hence for all larger r. The differential inequality (12.7) can then be integrated in $r_0<r_1\leqslant r\leqslant r_2<R$ to yield

$$\frac{1}{\mathfrak{D}(r_1)-k_1}\geqslant\int_{r_1}^{r_2}\frac{\mathfrak{D}'(r)\,dr}{[\mathfrak{D}(r)-k_1]^2}\geqslant\frac{1}{k_2}\log\frac{r_2}{r_1}.$$

Taking $r_1=R/2$, $r_2=R$, we obtain

$$\mathfrak{D}(R/2)\leqslant\frac{8\pi}{\log 2}M^2K^2+\pi R^2K'. \tag{12.8}$$

We note that the derivation of this estimate involved no restrictions on K, K' other than the non-negativity of K. We note also that it is not possible in general to obtain such an estimate in the full disk B_R, as is shown by the set of analytic function $w_n=z^n$, $n=1, 2, \ldots$, all of which satisfy $|w_n|\leqslant 1$ in $|z|\leqslant 1$, but

$$\iint\limits_{|z|<1}|Dw_n|^2\,dx\,dy\to\infty\quad\text{as } n\to\infty;$$

on the other hand, $\iint\limits_{|z|<1-\delta}|Dw_n|^2\,dx\,dy\leqslant C(\delta)<\infty$ for any fixed $\delta>0$, where $C(\delta)$ is independent of n.

We proceed now from the bound (12.8) on the Dirichlet integral in $B_{R/2}$ to a growth estimate for $\mathfrak{D}(r)$. From the inequalities,

$$|p_xq_y|\leqslant\frac{\alpha}{2}p_x^2+\frac{1}{2\alpha}q_y^2,\qquad |p_yq_x|\leqslant\frac{\alpha}{2}q_x^2+\frac{1}{2\alpha}p_y^2,\quad(\alpha>0),$$

we obtain

$$J=p_xq_y-p_yq_x\leqslant\frac{\alpha}{2}|w_x|^2+\frac{1}{2\alpha}|w_y|^2.$$

Hence, writing (12.2) in the form,

$$|w_x|^2+|w_y|^2\leqslant 2KJ+K',$$

and substituting $\alpha=K-(K^2-1)^{1/2}$ (or, equivalently, $K=(1+\alpha^2)/2\alpha$), we find

$$|w_x|^2 \leqslant \frac{1}{\alpha^2}|w_y|^2+\frac{2K'}{1-\alpha^2}.$$

Thus

$$\begin{aligned} (12.9) \qquad |w_x|^2 &= \frac{1}{1+\alpha^2}(|w_x|^2+\alpha^2|w_x|^2) \\ &\leqslant \frac{1}{1+\alpha^2}\left(|Dw|^2+\frac{2\alpha^2 K'}{1-\alpha^2}\right). \end{aligned}$$

Since (12.2) is invariant under rotation, this inequality remains valid if any directional derivative w_s replaces w_x.

We shall apply (12.9) to obtain a more precise estimation of $\int_{C_r} pq_s\,ds$ in (12.5). Let $\bar{p}=\bar{p}(r)$ denote the mean value of p over the circle C_r. Then

$$(12.10) \qquad \int_{C_r} pq_s\,ds=\int_{C_r}(p-\bar{p})q_s\,ds \leqslant \frac{1}{2}\int_{C_r}\left[\frac{(p-\bar{p})^2}{r}+rq_s^2\right]ds.$$

We now make use of the Wirtinger inequality [HLP], which states

$$\int_0^{2\pi}[p(r,\theta)-\bar{p}]^2\,d\theta \leqslant \int_0^{2\pi} p_\theta^2\,d\theta,$$

that is,

$$(12.11) \qquad \int_{C_r}(p-\bar{p})^2\,ds \leqslant r^2\int_{C_r} p_s^2\,ds.$$

(This result is easily proved by expanding $p=p(r,\theta)$ in a Fourier series in θ and applying Parseval's equality.) Inserting (12.11) into (12.10), we see that

$$\int_{C_r} pq_s\,ds \leqslant \frac{r}{2}\int_{C_r}(p_s^2+q_s^2)\,ds=\frac{r}{2}\int_{C_r}|w_s|^2\,ds$$

and hence by (12.9),

$$\int_{C_r} pq_s\,ds \leqslant \frac{r}{2(1+\alpha^2)}\int_{C_r}|Dw|^2\,ds+\frac{2\pi\alpha^2 K'}{1-\alpha^4}r^2.$$

Now substituting this inequality into (12.5), and again using the relation

$$\int_{C_r} |Dw|^2 \, ds = \mathfrak{D}'(r),$$

we arrive at the differential inequality,

$$\mathfrak{D}(r) \leqslant \frac{r}{2\alpha} \mathfrak{D}'(r) + kr^2, \quad k = \pi K' \left(1 + \frac{2\alpha}{1-\alpha^2}\right). \tag{12.12}$$

This implies

$$-\frac{d}{dr}(r^{-2\alpha} \mathfrak{D}(r)) \leqslant 2\alpha k r^{1-2\alpha},$$

from which it follows by integration between r and r_0

$$\mathfrak{D}(r) \leqslant \left[\mathfrak{D}(r_0) + \frac{\alpha}{1-\alpha} k r_0^2\right] \left(\frac{r}{r_0}\right)^{2\alpha}. \tag{12.13}$$

Letting $r_0 = R/2$ and inserting the bound (12.8), we obtain the desired estimate (12.4), with $C = \max\{C_2(K), C_3(K)\}(M^2 + K'R^2)$, where

$$C_2 = \frac{32\pi}{\log 2} K^2, \qquad C_3 = \pi\left[4 + \frac{\alpha}{1-\alpha}\left(1 + \frac{2\alpha}{1-\alpha^2}\right)\right].$$

We observe finally that when $K' = 0$ the arguments are unaffected by allowing $K = 1$, and C reduces to $C_2 M^2$. □

The following calculus lemma of Morrey provides the essential step from an estimate of the growth of the Dirichlet integral to a Hölder estimate on the function itself.

Lemma 12.2. *Let $w \in C^1(\Omega)$, and let $\tilde{\Omega} \subset\subset \Omega$ with* dist $(\tilde{\Omega}, \partial\Omega) > R$. *Suppose there are positive constants C, α and R' such that*

$$\mathfrak{D}(r; z) = \iint_{B_r(z)} |Dw|^2 \, dx \, dy \leqslant C r^{2\alpha}$$

for all disks $B_r(z)$ with center $z \in \tilde{\Omega}$ and radius $r \leqslant R' \leqslant R$. Then for all $z_1, z_2 \in \tilde{\Omega}$ such that $|z_2 - z_1| \leqslant R'$, we have

$$|w(z_2) - w(z_1)| \leqslant 2\sqrt{\frac{C}{\alpha}} |z_2 - z_1|^\alpha.$$

This lemma is an immediate corollary of Theorem 7.19 for the case $n=2$ after applying the Schwarz inequality. A proof of the lemma in the form stated above is given in [FS].

The preceding lemmas are collected in the following apriori Hölder estimate for (K, K')-quasiconformal mappings.

Theorem 12.3. *Let w be (K, K')-quasiconformal in a domain Ω, with $K>1$, $K'\geqslant 0$, and suppose $|w|\leqslant M$. Let $\tilde{\Omega}\subset\subset\Omega$ with $\operatorname{dist}(\tilde{\Omega}, \partial\Omega)>d$. Then for all $z_1, z_2\in\tilde{\Omega}$, we have*

$$|w(z_2)-w(z_1)|\leqslant C\left|\frac{z_2-z_1}{d}\right|^\alpha, \quad \alpha=K-(K^2-1)^{1/2}, \tag{12.14}$$

where $C=C_1(K)(M+d\sqrt{K'})$. If $K'=0$, then $C=C_1(K)M$ and the conclusion is valid also for $K=1$.

Proof. Suppose first that $|z_2-z_1|\leqslant d/2$. The conditions of Lemmas 12.1 and 12.2 then apply with $R=d$ and $R'=d/2$, so that we have

$$|w(z_2)-w(z_1)|\leqslant L\left|\frac{z_2-z_1}{d}\right|^\alpha,$$

where

$$L=C(K)(M^2+K'd^2)^{1/2}\leqslant C(K)(M+d\sqrt{K'}).$$

If $|z_2-z_1|>d/2$, then

$$|w(z_2)-w(z_1)|\leqslant 2M\leqslant 2M\left|\frac{z_2-z_1}{\frac{1}{2}d}\right|^\alpha\leqslant 4M\left|\frac{z_2-z_1}{d}\right|^\alpha$$

The theorem is therefore proved with $C_1(K)=\max(4, C(K))$. □

Remarks. (1) The exponent $\alpha=K-(K^2-1)^{1/2}$ is the best (i.e., the largest) for which Lemma 12.1 and Theorem 12.3 are true. This can be seen from the example of the K-quasiconformal mapping $w(z)=r^\alpha e^{i\theta}$, $\alpha=K-(K^2-1)^{1/2}$, which has precisely the Hölder exponent α at $z=0$. The same results (for $K\geqslant 1$, $K'\geqslant 0$) with a smaller exponent α can be obtained from a slightly more direct proof of Lemma 12.1 starting with (12.5) and dispensing with (12.9); in this case the sharp form of the Wirtinger inequality (12.11) is not required (cf. [NI 1]).

(2) Counterexamples show that Lemma 12.1 and Theorem 12.3 are not true for the exponent $\alpha=K-(K^2-1)^{1/2}$ when $K=1$ and $K'>0$, that is, for $\alpha=1$ (see Problem 12.1). However, if a mapping satisfies (12.2) with $K=1$, it satisfies such an inequality with any larger value of K and the corresponding results in Lemma 12.1 and Theorem 12.3 then apply with exponent α arbitrarily close to 1.

(3) If $\tilde{\Omega}$ is bounded and can be covered by N disks of diameter $d/2$, one sees from the proof that Theorem 12.3 remains valid under the weaker hypothesis $|p| \leqslant M$, with the constant C in (12.14) now depending also on N and hence on the diameter of $\tilde{\Omega}$.

(4) *Global estimates.* If $w=p+iq$ is (K, K')-quasiconformal in a C^1 domain Ω and $w \in C^1(\bar{\Omega})$, then Theorem 12.3 can be strengthened to an apriori global Hölder estimate for w. In particular, if $|w| \leqslant M$ and $p=0$ on $\partial\Omega$, then w satisfies a global Hölder condition with Hölder coefficient and exponent depending only on K, K', M and Ω. To outline the proof, let $\partial\Omega$ be the union of a finite number of overlapping arcs each of which can be straightened by a suitable C^1 diffeomorphism $(x, y) \to (\xi, \eta)$ defined in the neighbourhood of the arc. The function w is quasiconformal in the (ξ, η) variables with constants κ, κ' depending on K, K' and Ω. By reflection across $\eta=0$, so that $p(\xi, -\eta)=-p(\xi, \eta)$ and $q(\xi, -\eta)=q(\xi, \eta)$ in the extended (ξ, η) plane, the function $p+iq$ is seen to define a (κ, κ')-quasiconformal mapping to which the preceding interior estimates apply. Returning to the (x, y) plane, we thus obtain a Hölder estimate for w valid in $\bar{\Omega}$; that is

$$|w(z_1)-w(z_2)| \leqslant C|z_1-z_2|^{\alpha}, \quad z_1, z_2 \in \bar{\Omega},$$

where $\alpha=\alpha(K, K', \Omega)$ and $C=C(K, K', \Omega, M)$. If $p=\tilde{p}$ on $\partial\Omega$, where $\tilde{p} \in C^1(\bar{\Omega})$ and $|\tilde{p}|_{1;\Omega} \leqslant M'$, then by considering $p-\tilde{p}$ in place of p, we see that w satisfies the same kind of global estimate, with α and C now depending on K, K', M, M', and Ω.

12.2. Hölder Gradient Estimates for Linear Equations

The results of the preceding section will now be applied to obtain interior Hölder estimates for the first derivatives of solutions of uniformly elliptic equations,

$$(12.15) \qquad Lu=au_{xx}+2bu_{xy}+cu_{yy}=f,$$

where a, b, c, f are defined in a domain Ω of the $z=(x, y)$ plane. Let $\lambda=\lambda(z)$, $\Lambda=\Lambda(z)$ denote the eigenvalues of the coefficient matrix, so that

$$(12.16) \qquad \lambda(\xi^2+\eta^2) \leqslant a\xi^2+2b\xi\eta+c\eta^2 \leqslant \Lambda(\xi^2+\eta^2) \quad \forall(\xi, \eta) \in \mathbb{R}^2;$$

and assuming L is uniformly elliptic in Ω, we have

$$(12.17) \qquad \frac{\Lambda}{\lambda} \leqslant \gamma$$

for some constant $\gamma \geqslant 1$. We suppose also that $\sup_{\Omega}(|f|/\lambda) \leqslant \mu < \infty$. Dividing (12.15) by the minimum eigenvalue λ, we may assume that $\lambda=1$ and that (12.16) holds with $\lambda=1$ and $\Lambda=\gamma$, while $|f|\leqslant\mu$. Let us make this assumption in the following. Setting

$$p=u_x, \qquad q=u_y,$$

we can write (12.15) as the system

$$\frac{a}{c}p_x+\frac{2b}{c}p_y+q_y=\frac{f}{c}, \quad p_y=q_x. \tag{12.18}$$

By formal differentiation, p is seen to be a solution (in the weak sense) of the uniformly elliptic equation of divergence form,

$$\left(\frac{a}{c}p_x+\frac{2b}{c}p_y-\frac{f}{c}\right)_x+(p_y)_y=0,$$

and a similar equation holds for q; (see the proof of Theorem 11.5). The Hölder estimates on p and q that are derived in this section can also be obtained from the methods developed in Chapter 8 for equations of divergence form. For $n=2$ the details are not fundamentally different from those based on quasiconformal mappings presented here.

Multiplying the left member of (12.18) by cp_x, we obtain

$$p_x^2+p_y^2\leqslant ap_x^2+2bp_xp_y+cp_y^2=cJ+fp_x, \quad J=q_xp_y-q_yp_x,$$

and similarly

$$q_x^2+q_y^2\leqslant aJ+fq_y.$$

Adding these inequalities and noting that $2\leqslant a+c=1+\Lambda\leqslant 1+\gamma$, we have

$$\begin{aligned}|Dp|^2+|Dq|^2&\leqslant(a+c)J+f(p_x+q_y)\\&\leqslant(1+\gamma)J+\tfrac{1}{2}(1+\gamma)\mu(|p_x|+|q_y|).\end{aligned} \tag{12.19}$$

Inserting the inequality,

$$(1+\gamma)\mu(|p_x|+|q_y|)\leqslant\varepsilon(p_x^2+q_y^2)+\frac{1}{2\varepsilon}(1+\gamma)^2\mu^2, \quad \varepsilon>0,$$

and fixing $\varepsilon=1$ (the particular choice of $\varepsilon<2$ is inessential for our purposes), we obtain from (12.19).

$$|Dp|^2+|Dq|^2\leqslant 2(1+\gamma)J+(1+\gamma)^2\mu^2/2. \tag{12.20}$$

Hence $w=p-iq$ (or $q+ip$) defines a (K, K')-quasiconformal mapping, satisfying (12.2) with

$$(12.21)\qquad K=1+\gamma, \qquad K'=(1+\gamma)^2\mu^2/2.$$

By taking ε sufficiently small, the constant K can be made arbitrarily close to $(1+\gamma)/2$.

If $f=0$, one obtains directly from (12.16) and (12.17) the inequality

$$|Dw|^2=|Dp|^2+|Dp|^2\leqslant(a+c)J\leqslant(1+\gamma)J.$$

In this case the mapping $w=p-iq$ is K-quasiconformal with the constant $K=(1+\gamma)/2$. An elementary but more careful calculation shows that the *smallest* quasiconformality constant does not exceed $K=(\gamma+1/\gamma)/2$ (see Problem 12.3, also [TA 1]).

We now establish the basic estimate for solutions of (12.15) required for the ensuing nonlinear theory. We use the notation $d_z=\text{dist}\,(z, \partial\Omega), d_{1,2}=\min\,(d_{z_1}, d_{z_2})$ and the interior norms and seminorms defined in (4.17) and (6.10); in particular,

$$[u]^*_{1,\alpha}=\sup_{z_1, z_2\in\Omega} d^{1+\alpha}_{1,2}\frac{|Du(z_2)-Du(z_1)|}{|z_2-z_1|^\alpha}, \qquad |f/\lambda|^{(2)}_0=\sup_{z\in\Omega} d^2_z|f/\lambda|.$$

Theorem 12.4. *Let u be a bounded $C^2(\Omega)$ solution of*

$$Lu=au_{xx}+2bu_{xy}+cu_{yy}=f,$$

where L is uniformly elliptic, satisfying (12.16), (12.17) *in a domain Ω of $\mathbb{R}^2$. Then for some $\alpha=\alpha(\gamma)>0$, we have*

$$(12.22)\qquad [u]^*_{1,\alpha}\leqslant C(|u|_0+|f/\lambda|^{(2)}_0), \quad C=C(\gamma).$$

The significant feature of this result is that the estimate (12.22) depends only on bounds on the coefficients and not on any regularity properties. This is in contrast with the Schauder estimates (Theorem 6.2) which depend as well on the Hölder constants of the coefficients. The Hölder estimates of Chapter 8 for divergence form equations in n variables (Theorem 8.24) are also independent of regularity properties of the coefficients, but those estimates concern the solution itself and not its derivatives. The validity of the analogue of Theorem 12.4 for $n>2$ remains in doubt.

Proof of Theorem 12.4. Let z_1, z_2 be any pair of points in Ω, set $2d=d_{1,2}$ and define $\Omega'=\{z\in\Omega \mid d_z>d\}$, $\Omega''=\{z\in\Omega' \mid \text{dist}\,(z, \partial\Omega')>d\}$. We note that z_1, $z_2\in\bar{\Omega}''$. We now apply Theorem 12.3 with Ω', Ω'' in place of Ω, $\tilde{\Omega}$ respectively and $K=1+\gamma$, $K'=[(1+\gamma)\sup_{\Omega'}|f/\lambda|]^2/2$. The inequality (12.14) for $w=p-iq$, stated in terms of the gradient Du, becomes for $\alpha=(1+\gamma)-(\gamma^2+2\gamma)^{1/2}$

$$d^\alpha \frac{|Du(z_2)-Du(z_1)|}{|z_2-z_1|^\alpha} \leqslant C(\sup_{\Omega'}|Du|+d\sup_{\Omega'}|f/\lambda|), \quad (C=C(\gamma)),$$

$$\leqslant \frac{C}{d}(\sup_{\Omega'} d_z|Du(z)|+\sup_{\Omega'} d_z|f/\lambda|);$$

hence

$$d_{1,2}^{1+\alpha} \frac{|Du(z_2)-Du(z_1)|}{|z_2-z_1|^\alpha} \leqslant C(\sup_{\Omega} d_z|Du(z)|+\sup_{\Omega} d_z^2|f/\lambda|),$$

which implies

$$[u]^*_{1,\alpha;\Omega'} \leqslant C([u]^*_{1;\Omega'}+|f/\lambda|^{(2)}_{0;\Omega'}, \quad \text{for any } \Omega' \subset\subset \Omega.$$

The interpolation inequality (6.8) for $j=k=1$, $\beta=0$, namely,

$$(12.23) \qquad [u]^*_1 \leqslant \varepsilon[u]^*_{1,\alpha}+C_1|u|_0, \quad (C_1=C_1(\varepsilon)),$$

gives

$$[u]^*_{1,\alpha} \leqslant C(\varepsilon[u]^*_{1,\alpha}+C_1|u|_0+|f/\lambda|^{(2)}_0).$$

Choosing ε so that $C\varepsilon=\frac{1}{2}$, we obtain for an appropriate constant $C=C(\gamma)$

$$[u]^*_{1,\alpha} \leqslant C(|u|_0+|f/\lambda|^{(2)}_0),$$

which gives the required result (12.22). □

From (12.22) and the interpolation inequality (12.23) follows the norm estimate

$$(12.24) \qquad |u|^*_{1,\alpha} \leqslant C(|u|_0+|f/\lambda|^{(2)}_0), \quad C=C(\gamma).$$

Global Estimates

Theorem 12.4 can be extended to a $C^{1,\alpha}(\bar{\Omega})$ estimate under suitable smoothness hypotheses on the boundary data and the solution itself. For suppose, in addition to the hypotheses of Theorem 12.4, that $u \in C^2(\bar{\Omega})$, where Ω is a C^2 domain, and that $u=0$ on $\partial\Omega$. Then we can assert a global bound: $|u|_{1,\alpha;\Omega} \leqslant C$, where $\alpha=\alpha(\gamma, \Omega)$ and $C=C(\gamma, \Omega, |u|_0, |f/\lambda|_0)$. We outline the proof. As a normalization, we set $v=u/(1+|Du|_0)$, so that v satisfies $Lv=f/(1+|Du|_0)$ and $|Dv| \leqslant 1$. Let the boundary curve $\partial\Omega$ be covered by a finite number of overlapping arcs, each of which can be straightened into a segment of $\eta=0$ by a suitable C^2 diffeomorphism $(x, y) \to (\xi, \eta)$ defined in the neighborhood of the arc. As in the derivation of

(12.20), the mapping $p=p(\xi,\eta)$, $q=q(\xi,\eta)$, where $p=v_\xi$, $q=-v_\eta$, is (κ,κ')-quasiconformal in (ξ,η) with constants $\kappa=\kappa(\gamma,\Omega)$, $\kappa'=\kappa'(\gamma,\Omega,|f/\lambda|_0)$ (we recall $|Dv|\leqslant 1$). Also, $p=0$ on $\eta=0$. The same argument as in Remark 4 at the end of Section 12.1 shows that p and q, and hence Dv, satisfy a global Hölder condition in Ω, in which the Hölder exponent α depends only on γ and Ω, and the Hölder coefficient depends also on $|f/\lambda|_0$. Thus,

$$[u]_{1,\alpha}\leqslant C(1+|Du|_0),\quad C=C(\gamma,\Omega,|f/\lambda|_0),$$

and by interpolation (see Lemma 6.35), we obtain a bound

$$|u|_{1,\alpha;\Omega}\leqslant C(\gamma,\Omega,|u|_0,|f/\lambda|_0),\quad \alpha=\alpha(\gamma,\Omega).$$

If $\varphi\in C^2(\bar\Omega)$ and $u=\varphi$ on $\partial\Omega$, then by considering $u-\varphi$ in place of u in the preceding, and recalling that $|u|_0$ can be estimated in terms of $\sup\limits_{\partial\Omega}|\varphi|$ and $|f/\lambda|_0$ (Theorem 3.7), we infer the apriori global bound

$$|u|_{1,\alpha;\Omega}\leqslant C=C(\gamma,\Omega,|\varphi|_2,|f/\lambda|_0),\quad \alpha=(\gamma,\Omega).$$

It should be emphasized that this estimate is independent of any regularity properties of f and the coefficients of L. It is clear from the details that the dependence on Ω can be stated in terms of its dimensions and the bounds on the first and second derivatives of the mapping $\xi=\xi(x,y)$, $\eta=\eta(x,y)$, that is, in terms of the C^2 properties of $\partial\Omega$.

Later in this chapter we make the following application of the above result. Let Ω be a $C^{2,\beta}$ domain for some $\beta>0$ and let f and the coefficients of L lie in $C^\beta(\Omega)$. *Suppose* $u\in C^2(\Omega)\cap C^0(\bar\Omega)$, $\varphi\in C^{2,\beta}(\bar\Omega)$, *satisfy* $Lu=f$ *in* Ω, $u=\varphi$ *on* $\partial\Omega$. *Then* $u\in C^{1,\alpha}(\bar\Omega)$ *and* $|u|_{1,\alpha;\Omega}\leqslant C$, *where* $\alpha=\alpha(\gamma,\Omega)$ *and* $C=C(\gamma,\Omega,|\varphi|_2,|f/\lambda|_0)$. We note that u is assumed only continuous on $\partial\Omega$. The proof follows from the preceding paragraph by an approximation argument. Namely, if a_m,b_m,c_m,f_m, $m=1,2,\ldots$, are suitably chosen functions in $C^\beta(\bar\Omega)$ converging to a, b, c, f, uniformly in compact subdomains of Ω, the corresponding solutions u_m of the Dirichlet problems, $L_mu_m=f_m$ in Ω, $u_m=\varphi$ on $\partial\Omega$, are in $C^2(\bar\Omega)$ and (by the preceding) satisfy a uniform $C^{1,\alpha}(\bar\Omega)$ bound $|u_m|_{1,\alpha}\leqslant C$ for some α and C independent of m. By the Schauder interior estimates and uniqueness, the sequence $\{u_m\}$ converges to the given solution u of $Lu=f$. It follows that u also satisfies the same $C^{1,\alpha}(\bar\Omega)$ bound $|u|_{1,\alpha}\leqslant C$, which was our assertion.

12.3. The Dirichlet Problem for Uniformly Elliptic Equations

In this section we prove existence by a procedure that is a variant of the one outlined in Chapter 11. In the Dirichlet problem treated here the details are generally simpler than in later problems and some steps of the program in Chapter 11 can be omitted.

We consider the Dirichlet problem for quasilinear elliptic equations of the general form,

$$(12.25)\qquad Qu = a(x, y, u, u_x, u_y)u_{xx} + 2b(x, y, u, u_x, u_y)u_{xy} + c(x, y, u, u_x, u_y)u_{yy} + f(x, y, u, u_x, u_y) = 0,$$

defined in a bounded domain Ω in the (x, y) plane. Concerning the operator Ω we shall assume:

(i) The functions $a = a(x, y, u, p, q), \ldots, f = f(x, y, u, p, q)$ are defined for all (x, y, u, p, q) in $\Omega \times \mathbb{R} \times \mathbb{R}^2$ and, in addition, $a, b, c, f, \in C^{\beta}(\Omega \times \mathbb{R} \times \mathbb{R}^2)$ for some $\beta \in (0, 1)$.

(ii) The operator Q is uniformly elliptic in Ω for bounded u; that is, the eigenvalues $\lambda = \lambda(x, y, u, p, q)$, $\Lambda = \Lambda(x, y, u, p, q)$ of the coefficient matrix satisfy

$$(12.26)\qquad 1 \leqslant \frac{\Lambda}{\lambda} \leqslant \gamma(|u|) \quad \forall (x, y, u, p, q) \in \Omega \times \mathbb{R} \times \mathbb{R}^2,$$

where γ is non-decreasing.

(iii) The function f satisfies the structure conditions,

$$(12.27)\qquad \frac{|f|}{\lambda} \leqslant \mu(|u|)(1 + |p| + |q|)$$

$$(12.28)\qquad \frac{f}{\lambda} \operatorname{sign} u \leqslant \nu(1 + |p| + |q|) \quad \forall (x, y, u, p, q) \in \Omega \times \mathbb{R} \times \mathbb{R}^2,$$

where μ is non-decreasing and ν is a non-negative constant. These correspond to the conditions on lower order terms in linear equations. Equations satisfying (12.28) are discussed in Chapter 10.

We now establish the following existence theorem.

Theorem 12.5. *Let Ω be a domain in $\mathbb{R}^2$ satisfying an exterior sphere condition, and let φ be a continuous function on $\partial\Omega$. Then if Q is an elliptic quasilinear operator satisfying conditions* (i)–(iii), *the Dirichlet problem*

$$(12.29)\qquad Qu = 0 \text{ in } \Omega, \quad u = \varphi \text{ on } \partial\Omega,$$

has a solution $u \in C^{2,\beta}(\Omega) \cap C^0(\bar{\Omega})$.

Proof. We first prove the theorem under the more restrictive hypothesis,

$$(12.30)\qquad \frac{|f|}{\lambda} \leqslant \mu < \infty \quad (\mu = \text{const}),$$

in place of (12.27). The argument is based on reduction to the Schauder fixed point theorem (Theorem 11.1). To define the mapping T appearing in the statement of that theorem, we make the following observation. Let v be any bounded function with locally Hölder continuous first derivatives in Ω, and let $\bar{a}=\bar{a}(x, y)=a(x, y, v, v_x, v_y), \ldots$, be the locally Hölder continuous functions in Ω obtained by inserting v for u in the coefficients of Q. Since $|f|/\lambda$ is bounded, it follows from Theorem 6.13 that the linear Dirichlet problem,

$$(12.31) \qquad \bar{a}u_{xx}+2\bar{b}u_{xy}+\bar{c}u_{yy}+\bar{f}=0 \text{ in } \Omega, \quad u=\varphi \text{ on } \partial\Omega,$$

has a unique solution $u \in C^2(\Omega) \cap C^0(\bar{\Omega})$. We observe from Theorem 3.7 that

$$|u|_0=\sup_{\Omega}|u| \leqslant \sup_{\partial\Omega}|\varphi|+C_1\mu=M_0, \quad C_1=C_1 \text{ (diam } \Omega).$$

Furthermore, if $\sup_{\Omega}|v| \leqslant M_0$ and if we set $\gamma_0=\gamma(M_0)$, we have from Theorem 12.4

$$(12.32) \qquad \begin{aligned} |u|^*_{1,\alpha} &\leqslant C(|u|_0+\mu(\text{diam } \Omega)^2), \quad \alpha=\alpha(\gamma_0), \ C=C(\gamma_0), \\ &\leqslant C(M_0+\mu(\text{diam } \Omega)^2)=K. \end{aligned}$$

We note especially that this estimate depends only on the bound M_0 of the function v used in defining the coefficients of equation (12.31).

Let us introduce the Banach space

$$C^{1,\alpha}_*=C^{1,\alpha}_*(\Omega)=\{u \in C^{1,\alpha}(\Omega) \mid |u|^*_{1,\alpha;\Omega}<\infty\},$$

where α is the Hölder exponent in (12.32). We can define a mapping T of the set,

$$\mathfrak{S}=\{v \in C^{1,\alpha}_* \mid |v|^*_{1,\alpha} \leqslant K, |v|_0 \leqslant M_0\},$$

by letting $u=Tv$ be the unique solution of the linear Dirichlet problem (12.31) for $v \in \mathfrak{S}$. By virtue of (12.32) and the bound $|u|_0 \leqslant M_0$, we have $u \in \mathfrak{S}$ and hence T maps $\mathfrak{S}$ into itself. Since $\mathfrak{S}$ is convex and is closed in the Banach space

$$C^1_*=\{u \in C^1(\Omega) \mid |u|^*_{1;\Omega}<\infty\},$$

we may conclude from the Schauder fixed point theorem (Corollary 12.2) that T has a fixed point, $u=Tu$, in $\mathfrak{S}$ provided the mapping T is continuous in C^1_* and the image $T\mathfrak{S}$ is precompact. This will provide a solution of the problem (12.29) under the hypothesis (12.30).

To prove $T\mathfrak{S}$ is precompact in C^1_*, we observe first that the set $\mathfrak{S}$, and hence $T\mathfrak{S}$, is equicontinuous at each point of Ω. We claim that the functions of $T\mathfrak{S}$ are also equicontinuous at each point $z_0 \in \partial\Omega$. For let w be the barrier function in the argument preceding Theorem 6.13. The function w, which depends only on the ellipticity modulus γ_0 (in equation (12.31)) and the radius of the exterior disk at z_0, has the property that for any $\varepsilon>0$ and a suitable constant k_ε independent of $v \in \mathfrak{S}$, the solution $u = Tv$ of (12.31) satisfies the inequality

$$|u(z) - \varphi(z_0)| \leqslant \varepsilon + k_\varepsilon w(z) \quad \text{in } \Omega. \tag{12.33}$$

Since $w(z) \to 0$ as $z \to z_0$, this implies the equicontinuity of the set $T\mathfrak{S}$ at z_0. The functions of $T\mathfrak{S}$ are therefore equicontinuous on $\bar{\Omega}$. Since $T\mathfrak{S}$ is a bounded equicontinuous set in $C_*^{1,\alpha}$, it is precompact in C_*^1 (see Lemma 6.33).

The continuity of T in C_*^1 is proved in a similar way: Let $v, v_n \in \mathfrak{S}$, $n = 1, 2, \ldots$, and assume $|v_n - v|_1^* \to 0$ as $n \to \infty$. Consider $u = Tv$ and the sequence $u_n = Tv_n$, $n = 1, 2, \ldots$; we wish to show $|u_n - u|_1^* \to 0$. By the Schauder interior estimates and the Remark following Corollary 6.3, it follows that a suitable (renumbered) subsequence $\{u_m\} \subset \{u_n\}$ converges uniformly with its first and second derivatives in compact subsets of Ω to a solution $\tilde{u}$ in Ω of the limit equation (12.31) obtained by inserting v into the coefficients of Q. We claim that $\tilde{u}(z) \to \varphi(z_0)$ for all $z_0 \in \partial\Omega$ and hence (by uniqueness) $\tilde{u} = u = Tv$. For by the same barrier argument as above we may assert (12.33) with u_m replacing u, from which we obtain in the limit $\tilde{u}(z) \to \varphi(z_0)$ as $z \to z_0$. Thus $\tilde{u} = u$ and we have $Tv_m \to Tv$ on $\bar{\Omega}$ for the subsequence $\{v_m\}$.

Since the sequence $\{Tv_m\}$ is contained in $T\mathfrak{S}$, which is precompact in C_*^1, a suitable subsequence of $\{Tv_m\}$ converges in the C_*^1 norm to Tv. The same argument, repeated on arbitrary subsequences of $\{v_n\}$, shows that $|Tv_n - Tv|_1^* \to 0$ for the entire sequence. This establishes the continuity of T on C_*^1 and, as already observed, we may conclude the existence of a fixed point, $u = Tu$, in $\mathfrak{S}$.

The theorem is thus proved in the special case that $|f|/\lambda$ is bounded on $\Omega \times \mathbb{R} \times \mathbb{R}^2$, in particular when $f \equiv 0$. We return now to the original hypothesis (iii). It will be convenient in the following to assume $\lambda(x, y, u, p, q) \equiv 1$ in $\Omega \times \mathbb{R} \times \mathbb{R}^2$, which can always be achieved by dividing the functions a, b, c f by λ. In this case (12.27), (12.28) become

$$(f| \leqslant \mu(|u|)(1 + |p| + |q|). \tag{12.34}$$

$$f \text{ sign } u \leqslant v(1 + |p| + |q|), \quad v = \text{const}. \tag{12.35}$$

We now proceed by *truncation* of f to reduce the given problem (12.29) to the above case of bounded f. Namely, let ψ_N denote the function given by

$$\psi_N(t) = \begin{cases} t, & |t| \leqslant N \\ N \text{ sign } t, & |t| > N \end{cases}$$

and define the truncation of f by

$$f_N(x, y, u, p, q) = f(x, y, \psi_N(u), \psi_N(p), \psi_N(q)).$$

From (12.34) we have $|f_N| \leqslant \mu N(1 + 2N)$. Consider now the family of problems,

$$\begin{aligned} Q_N u = a(x, y, y, Du)u_{xx} + 2b(x, y, u, Du)u_{xy} + c(x, y, u, Du)u_{yy} \\ + f_N(x, y, u, Du) = 0, \\ u = \varphi \quad \text{on } \partial\Omega. \end{aligned} \tag{12.36}$$

By virtue of (12.35) and Theorem 10.3, any solution u in this family is subject to the bound, independent of N,

$$\sup_{\Omega}|u| \leqslant \sup_{\partial\Omega}|\varphi| + C_1(\nu, \operatorname{diam}\Omega) = M. \tag{12.37}$$

From the preceding treatment of problem (12.29) with bounded f the problem (12.36)—in which f_N is bounded—is seen to have a solution $u_N \in C^{1,\alpha}_*(\Omega) \cap C^{2,\beta}(\Omega) \cap C^0(\bar{\Omega})$, $\alpha = \alpha(\gamma)$, $\gamma = \gamma(M)$. Furthermore, from Theorem 12.4 we infer the estimate,

$$[u_N]^*_{1,\alpha} \leqslant C(|u_N|_0 + |f_N|^{(2)}_0), \tag{12.38}$$

where $C = C(\gamma)$, $f_N = f_N(x, y, u_N, Du_N)$. By (12.34) and (12.37), this becomes

$$[u_N]^*_{1,\alpha} \leqslant C(1 + [u_N]^*_1), \quad C = C(M, \gamma, \mu, \operatorname{diam}\Omega), \; \mu = \mu(M).$$

The interpolation inequality (12.23), with $C\varepsilon = \frac{1}{2}$, now yields the uniform bound, independent of N,

$$|u_N|^*_{1,\alpha} \leqslant C = C(M, \gamma, \mu, \operatorname{diam}\Omega). \tag{12.39}$$

Applying the preceding estimate and the Schauder interior estimates (Corollary 6.3) to the family of equations $Q_N u_N = 0$ on compact subsets, we obtain a (renumbered) subsequence $\{u_n\} \subset \{u_N\}$ that converges to a solution u of $Qu = 0$ in Ω satisfying the estimate (12.39).

It remains to show that u also satisfies the boundary condition $u = \varphi$ and for this purpose we apply a barrier argument very similar to that given above. By virtue of (12.34) and (12.37), each u_n is the solution of a linear equation,

$$Q_n v = a^{ij}_n D_{ij} v + b^i_n D_i v + f_n = 0, \quad i, j = 1, 2,$$

where $a^{11}_n(x, y) = a(x, y, u_n, Du_n), \ldots$, and where $b^i_n(x, y)$, $f_n(x, y)$ are bounded independently of n. At any point $z_0 \in \partial\Omega$ the barrier argument preceding Theorem 6.13 applies to this family of equations with the barrier function w in (6.45), which depends only on γ, μ and the radius of the exterior disk at z_0. We therefore obtain for any $\varepsilon > 0$ and a suitable constant k_ε independent of n the inequality

$$|u_n(z) - \varphi(z_0)| \leqslant \varepsilon + k_\varepsilon w(z) \quad \text{in } \Omega.$$

Letting $n \to \infty$, we infer the same inequality with u in place of u_n, and hence $u(z) \to \varphi(z_0)$ as $z \to z_0$. This completes the proof of the theorem. □

Remarks. (1) The proof of the preceding theorem is based only on interior derivative estimates, thus allowing more general conditions on the coefficients and boundary data. With the use of global estimates, a simple modification of the proof yields a $C^{2,\beta}(\bar{\Omega})$ solution when $\partial\Omega$ and φ are in $C^{2,\beta}$ and $a, b, c, f \in C^{\beta}(\bar{\Omega} \times \mathbb{R} \times \mathbb{R}^2)$; (cf. Problem 12.5). Under these same hypotheses the solution provided by Theorem 12.5 must also lie in $C^{2,\beta}(\bar{\Omega})$. To prove this assertion, we observe first that the equation $Qu=0$, after inserting u into the coefficients, has coefficients in $C^{\beta}(\Omega)$, while $u \in C^{2,\beta}(\Omega) \cap C^0(\bar{\Omega})$ and $u=\varphi$ on $\partial\Omega$, where $\varphi \in C^{2,\beta}(\bar{\Omega})$. According to the results on global estimates for linear equations at the end of Section 12.2, the solution u lies in $C^{1,\alpha}(\bar{\Omega})$ for some α. The coefficients of Qu are therefore in $C^{\alpha\beta}(\bar{\Omega})$. It follows from Theorem 6.14 and uniqueness that $u \in C^{2,\alpha\beta}(\bar{\Omega})$ and hence the coefficients of Qu are in $C^{\beta}(\bar{\Omega})$. We infer in the same way that $u \in C^{2,\beta}(\bar{\Omega})$, as asserted.

(2) Condition (12.28) was imposed to insure a uniform bound on the magnitude of all possible solutions of the Dirichlet problem for $Q_N u=0$. Whenever such a bound is known apriori, condition (12.28) may be omitted.

(3) The linear growth condition (12.27) on $|f|/\lambda$ was required to obtain the bound (12.39) by interpolation for $[u]_1^*$ in terms of $[u]_{1,\alpha}^*$ and $|u|_0$. Whenever an apriori bound on the gradient of solutions is known for the family of equations $Q_N u=0$, this growth condition is superfluous; (such examples will be discussed in Chapter 15).

(4) The hypothesis that Ω satisfy an exterior sphere condition can be replaced by any other condition that guarantees the existence of a barrier for strictly elliptic linear equations $Lu=f$, where f and the coefficients of L are bounded. For example, it suffices that Ω satisfy an exterior cone condition (see Problem 6.3).

12.4. Non-Uniformly Elliptic Equations

In our study of non-uniformly elliptic equations we shall see that, unlike the preceding section in which the domain is essentially arbitrary, the solvability of the Dirichlet problem is in general closely connected with the geometry of the domain. This feature of non-uniformly elliptic problems has already been observed in the linear theory (Sect. 6.6). The results of the present section emphasize the important role of convexity of the domain in insuring solvability of the Dirichlet problem for general quasilinear elliptic equations of the form,

$$(12.40) \qquad Qu=a(x, y, u, u_x, u_y)u_{xx}+2b(x, y, u, u_x, u_y)u_{xy}+c(x, u, y, u, u_x, u_y)u_{yy}=0.$$

Let Ω be a bounded domain in $\mathbb{R}^2$ and φ be a function defined on $\partial\Omega$. The Dirichlet problem for equation (12.40) will be formulated in terms of the boundary curve

$$\Gamma=(\partial\Omega, \varphi)=\{(z, \varphi(z)) \in \mathbb{R}^3 \mid z \in \partial\Omega\}.$$

We recall (as in Sect. 11.3) that Γ and φ satisfy a *bounded slope condition* (with constant K) if for every point $P=(z_0, \varphi(z_0)) \in \Gamma$ there are planes in $\mathbb{R}^3$,

$$u=\pi_p^{\pm}(z)=\mathbf{a}^{\pm}\cdot(z-z_0)+\varphi(z_0), \quad \mathbf{a}^{\pm}=\mathbf{a}^{\pm}(z_0),$$

passing through P such that:

$$\text{(12.41)} \quad \begin{array}{ll} \text{(i)} & \pi_p^-(z)\leqslant\varphi(z)\leqslant\pi_p^+(z) \quad \forall z \in \partial\Omega; \\ \text{(ii)} & |D\pi_p^{\pm}|=|\mathbf{a}^{\pm}(z_0)|\leqslant K \quad \forall z_0 \in \partial\Omega. \end{array}$$

Condition (i) states that for each P the curve Γ is bounded above and below on the cylinder $\partial\Omega\times\mathbb{R}$ by the planes $u=\pi_p^+(z)$ and $u=\pi_p^-(z)$, and coincides with them at P. Condition (ii) states that the slopes of the planes are uniformly bounded, independently of P, by the constant K. It is evident that the bounded slope condition implies the continuity of φ.

We make the following remarks concerning the bounded slope condition.

(1) Whatever the domain Ω, Γ lies in a plane (and satisfies a bounded slope condition) if and only if φ is the restriction of a linear function on $\partial\Omega$. However, if Γ does not lie in a plane and satisfies a bounded slope condition, then Ω must be convex. For we have from (12.41)

$$0\not\equiv\pi_p^+(z)-\pi_p^-(z)=(\mathbf{a}^+-\mathbf{a}^-)\cdot(z-z_0)\geqslant 0 \quad \forall z \in \partial\Omega,$$

and thus there is a supporting line $(\mathbf{a}^+-\mathbf{a}^-)\cdot(z-z_0)=0$ at each $z_0 \in \partial\Omega$, which implies the convexity of Ω.

(2) Suppose $\partial\Omega$ is convex and $\Gamma=(\partial\Omega, \varphi)$ satisfies a bounded slope condition. Let $P_i=(z_i, \varphi(z_i))$, $i=1, 2, 3$, be three distinct points on Γ. If z_1, z_2, z_3 are collinear then P_1, P_2, P_3 are also collinear on Γ, for otherwise these points would determine a vertical plane, contradicting (12.41). Thus φ is linear on the straight segments of Ω.

(3) Closely related to and, in fact, equivalent to the bounded slope condition is the following *three-point condition*, which appears often in the literature on minimal surfaces and non-parametric variational problems in two independent variables. Let Ω be bounded and convex; then the curve $\Gamma=(\partial\Omega, \varphi)$ is said to satisfy a three-point condition with constant K if every set of three distinct points on Γ lies in a plane of slope $\leqslant K$. At the end of this section we prove the equivalence of the bounded slope and three-point conditions, with the same constant K.

It follows from the three-point condition that the plane determined by any three non-collinear points of Γ must have slope $\leqslant K$. Thus, if Ω is strictly convex (that is, the open straight segment joining any two points of $\partial\Omega$ lies entirely in Ω), then the slope of *every* plane intersecting Γ in at least three points does not exceed K and, conversely, this stronger form of the three-point condition obviously implies that $\partial\Omega$ is strictly convex.

(4) It is not difficult to show that if $\varphi \in C^2$, $\partial\Omega \in C^2$ and the curvature of $\partial\Omega$ is everywhere positive, then $\Gamma=(\partial\Omega, \varphi)$ satisfies a bounded slope condition, with a constant depending on the minimum curvature of $\partial\Omega$ and bounds on the first and second derivatives of φ; (see [SC 3], [HA]).

The solution of the Dirichlet problem for (12.40) will require an apriori bound for the gradient which is provided by the following lemma.

Lemma 12.6. *Let Ω be a bounded domain in $\mathbb{R}^2$, and let φ be a function defined on $\partial\Omega$ satisfying a bounded slope condition with constant K. Suppose $u \in C^2(\Omega) \cap C^0(\bar{\Omega})$ satisfies a linear elliptic equation,*

$$Lu = au_{xx} + 2bu_{xy} + cu_{yy} = 0 \quad \text{in } \Omega, \tag{12.42}$$

with $u = \varphi$ on $\partial\Omega$. Then

$$\sup_{\Omega} |Du| \leqslant K \tag{12.43}$$

We emphasize that L is only required to be elliptic and that no other conditions are placed on the coefficients. It is of interest that the result is valid even more generally, for arbitrary saddle surfaces $u = u(x, y)$ (see [RA], [NU]).

Proof of Lemma 12.6. We observe first that if φ is the restriction of a linear function on $\partial\Omega$, then by uniqueness (Theorem 3.3) the solution u coincides with this function in Ω and the conclusion (12.43) holds trivially. It follows from the remark (1) above that Ω may be assumed convex.

From (12.42) and the ellipticity of L we have

$$0 \leqslant au_{xx}^2 + 2bu_{xx}u_{xy} + cu_{xy}^2 = c(u_{xy}^2 - u_{xx}u_{yy}),$$
$$0 \leqslant au_{xy}^2 + 2bu_{xy}u_{yy} + cu_{yy}^2 = a(u_{xy}^2 - u_{xx}u_{yy}).$$

Accordingly, $u_{xx}u_{yy} - u_{xy}^2 \leqslant 0$ and the equality holds only at points where $D^2u = 0$. (We note that $u = u(x, y)$ is a saddle surface.) Consider now any point $z_0 = (x_0, y_0) \in \Omega$ where $u_{xx}u_{yy} - u_{xy}^2 < 0$ and let $u_0(x, y) = Ax + By + C$ define the tangent plane Π to the surface $u = u(x, y)$ at (x_0, y_0). The function $w = u - u_0$ is a solution of (12.42) in Ω, and the set on which $w = 0$ divides a small disk about z_0 into precisely four domains $D_1, \ldots, D_4$ in which w is alternately positive and negative, say $w > 0$ in D_1, D_3 and $w < 0$ in D_2, D_4. Let $D_1' \supset D_1$, $D_3' \supset D_3$ be components of the set in Ω where $w > 0$, and let $D_2' \supset D_2$, $D_4' \supset D_4$ be similarly defined. Then each of the domains $D_1', \ldots, D_4'$ has at least two boundary points on $\partial\Omega$ where $w = 0$, for otherwise the weak maximum principle (Theorem 3.1) would imply $w \equiv 0$ in the domain. It follows that Π intersects the boundary curve $\Gamma = (\partial\Omega, \varphi)$ in at least four points. From remark (3) above we infer that if any three of these points are not collinear, then Π has slope not exceeding K. On the other hand, if the points on $\partial\Omega$ where $w = 0$ form a collinear set Σ, then (by remark (2)) φ and u are linear and identical with u_0 on the straight segment of $\partial\Omega$ containing Σ. This would imply $w \equiv 0$ in one of the domains D_i', which is a contradiction. Thus the slope of the tangent plane to the solution surface cannot exceed K at points where $D^2u \neq 0$.

Now consider the set S on which $D^2u = 0$. We may assume $S \neq \Omega$, for otherwise $u(x, y)$ would be linear and the conclusion is trivial. If $z_0 \in S$ and z_0 is the limit of points where $D^2u \neq 0$, then by continuity the tangent plane at z_0 must have inclination not in excess of K. The remaining possibility is that z_0 is an interior point of S, in which case z_0 is contained in an open component G of the set in which $D^2u=0$. On G, we have that u is linear and that the tangent plane coincides with the surface $u=u(x, y)$. Since any boundary point of G that is interior to Ω is the limit of points where $D^2u \neq 0$, we conclude $|Du| \leqslant K$ everywhere on G and in particular at z_0. This completes the proof. □

We are now in the position to establish the following existence theorem for equation (12.40), which extends Theorem 11.5 for the case $n=2$.

Theorem 12.7. *Let Ω be a bounded domain in $\mathbb{R}^2$ and assume that the equation,*

$$Qu = a(x, y, u, u_x, u_y)u_{xx} + 2b(x, y, u, u_x, u_y)u_{xy} + c(x, y, u, u_x, u_y)u_{yy} = 0,$$

is elliptic in Ω with coefficients $a, b, c \in C^\beta(\Omega \times \mathbb{R} \times \mathbb{R}^2)$ for some $\beta \in (0, 1)$. Let φ be a function defined on $\partial\Omega$ satisfying a bounded slope condition with constant K. Then the Dirichlet problem

$$Qu = 0 \text{ in } \Omega, \qquad u = \varphi \text{ on } \partial\Omega$$

has a solution $u \in C^{2,\beta}(\Omega) \cap C^0(\bar{\Omega})$, with $\sup_\Omega |Du| \leqslant K$.

Proof. It suffices to assume Ω is convex, since otherwise φ is the restriction of a linear function and the result is trivial. We divide the coefficients a, b, c by the maximum eigenvalue $\Lambda = \Lambda(x, y, u, p, q)$ and denote again by $Qu=0$ the equation thus obtained. The operator Q now has maximum eigenvalue 1, but the minimum eigenvalue λ may approach zero in $\Omega \times \mathbb{R} \times \mathbb{R}^2$. We consider the family of equations

$$Q_\varepsilon u = Qu + \varepsilon \Delta u = 0, \quad \varepsilon > 0, \tag{12.44}$$

with the boundary condition $u=\varphi$ on $\partial\Omega$. For each ε this equation is uniformly elliptic and Theorem 12.5 asserts the existence of a solution $u_\varepsilon \in C^{2,\beta}(\Omega) \cap C^0(\bar{\Omega})$ such that $u_\varepsilon = \varphi$ on $\partial\Omega$. (Only the first part of the proof of Theorem 12.5 is needed here.) By Lemma 12.6, the solution u_ε satisfies the uniform gradient estimate,

$$\sup_\Omega |Du_\varepsilon| \leqslant K, \tag{12.45}$$

independently of ε. From the maximum principle we also have $|u_\varepsilon| \leqslant \sup_{\partial\Omega} |\varphi|$. As a consequence, in every subdomain $\Omega' \subset\subset \Omega$, the linear equations,

$$Q_\varepsilon u = a_\varepsilon u_{xx} + 2b_\varepsilon u_{xy} + c_\varepsilon u_{yy} + \varepsilon \Delta u = 0 \tag{12.46}$$

obtained by setting $a_\varepsilon(x, y) = a(x, y, u_\varepsilon, Du_\varepsilon), \ldots$, have minimum eigenvalues, $\lambda_\varepsilon = \lambda(x, y, y_\varepsilon, Du_\varepsilon)$ for $(x, y) \in \Omega'$, that are uniformly bounded below by a constant $\lambda(\Omega') > 0$ depending only on Ω'; the upper eigenvalues are of course bounded by 2 for all $\varepsilon < 1$. Theorem 12.4 now asserts that in subsets $\Omega'' \subset\subset \Omega'$ the solutions of (12.46) with $u = \varphi$ on $\partial\Omega$, in particular the solutions u_ε, satisfy a uniform Hölder bound for the gradient (independent of ε):

$$[Du]_{\alpha;\,\Omega''} \leqslant C, \qquad \alpha = \alpha(\lambda(\Omega')),$$

where $C = C(\lambda(\Omega'), \sup_{\partial\Omega} |\varphi|, \text{dist}(\Omega'', \partial\Omega'))$. Accordingly, the coefficients $a_\varepsilon, b_\varepsilon, c_\varepsilon$ are locally Hölder continuous with exponent $\alpha\beta$ in Ω' and have uniform bounds in $C^{\alpha\beta}(\bar{\Omega}'')$. Since Ω' and Ω'' are arbitrary it follows from Corollary 6.3 and the accompanying Remark that the family of solutions u_ε of (12.46) are equicontinuous with their first and second derivatives on compact subsets of Ω and hence, by the usual diagonalization process, there is a sequence $\{u_{\varepsilon_n}\}$ of the family $\{u_\varepsilon\}$ converging in Ω to a solution u_0 of $Qu = 0$ as $\varepsilon_n \to 0$. The uniform gradient bound (12.45) guarantees that the convergence is uniform on $\bar{\Omega}$ and hence $u_0 = \varphi$ on $\partial\Omega$. This completes the proof of the theorem. □

Remarks. (1) That some geometric condition on Ω, such as convexity, must be imposed in general is indicated by classical counterexamples such as that of the minimal surface equation,

$$(1 + u_y^2)u_{xx} - 2u_x u_y u_{xy} + (1 + u_x^2)u_{yy} = 0, \tag{12.47}$$

in the annulus, $a < r < b$, $r = (x^2 + y^2)^{1/2}$. If the boundary condition is $\varphi = h$ ($=$const>0) on $r = a$, $\varphi = 0$ on $r = b$, and h is sufficiently small, the boundary value problem has the well-known catenoid solution. However, if h is sufficiently large, there is no solution taking on the prescribed boundary values. It will be seen in Chapter 14 that the Dirichlet problem for the minimal surface equation (12.47) is solvable for arbitrary C^2 boundary values if and only if Ω is convex.

(2) The smoothness hypotheses implicit in the boundary slope condition cannot be relaxed in general to allow continuous boundary values φ. Counterexamples show that the Dirichlet problem need not have a solution for continuous boundary values even when the boundary curve is a circle and the coefficients of the equation are arbitrarily smooth; (see [FN 2]). Solvability for continuous boundary values will be discussed in Chapters 15 and 16.

(3) The essential step in the proof of Theorem 12.7 is the reduction to the uniformly elliptic case, which is made possible by the existence of an apriori global gradient bound (Lemma 12.6). Such gradient bounds can also be established under suitable structure conditions on the operator Q and geometric assumptions concerning the domain Ω. They are discussed in Chapters 14 and 15. In the case of convex Ω it is sometimes possible to replace the planes of the bounded slope condition by suitable super- and subfunctions with respect to Q and φ and to proceed with the argument in essentially the same way. Thus, if Ω is convex and

the operator Q is such that $\Lambda/\lambda \leqslant \mu(|u|)(1+p^2+q^2)$, the Dirichlet problem for equation (12.40) is solvable for arbitrary $\varphi \in C^2$ whether or not the bounded slope condition is satisfied (see Section 14.2). This includes, for example, the minimal surface equation (12.47).

(4) If Q, $\partial\Omega$ and φ satisfy the additional smoothness hypotheses of Theorem 11.4, that is, $a, b, c \in C^\beta(\bar{\Omega} \times \mathbb{R} \times \mathbb{R}^2)$, $\partial\Omega \in C^{2,\beta}$, $\varphi \in C^{2,\beta}(\partial\Omega)$, then the solution provided by Theorem 12.7 lies in $C^{2,\beta}(\bar{\Omega})$. The argument proceeds essentially as in Remark 1 following Theorem 12.5, after the observation that the gradient bound $|Du| \leqslant K$ makes the equation $Qu=0$ uniformly elliptic.

Equivalence of the Bounded Slope and Three-Point Conditions

Assume first that $\Gamma=(\partial\Omega, \varphi)$ satisfies a bounded slope condition with constant K, Ω being convex. Let z_1, z_2, z_3 be three non-collinear points on $\partial\Omega$, and let

$$u=\pi_i^\pm(z)=\mathbf{a}_i^\pm \cdot (z-z_i)+\varphi(z_i), \quad i=1, 2, 3,$$

be the corresponding upper and lower planes at the points $P_i=(z_i, \varphi(z_i)) \in \Gamma$ satisfying (12.41). Also let

$$u=\pi(z)=\mathbf{a}\cdot z+b=\mathbf{a}\cdot(z-z_i)+\varphi(z_i), \quad i=1, 2, 3,$$

be the plane passing through P_1, P_2, P_3. We wish to show $|\mathbf{a}| \leqslant K$. (If z_1, z_2, z_3, and hence P_1, P_2, P_3, are collinear, the plane $u=\pi(z)$ with slope $\leqslant K$ can be determined by continuity.) From (12.41) it follows

$$\mathbf{a}_i^- \cdot (z_j-z_i) \leqslant \mathbf{a}\cdot(z_j-z_i) \leqslant \mathbf{a}_i^+ \cdot (z_j-z_i), \quad i, j=1, 2, 3. \tag{12.48}$$

Since z_1, z_2, z_3 are the vertices of a non-degenerate triangle and $\mathbf{a}$ is a vector in $\mathbb{R}^2$, we have for some $i=1, 2$, or 3, either

$$\mathbf{a}=\pm \sum_j c_j(z_j-z_i), \quad c_j \geqslant 0.$$

From (12.48) we infer either

$$|\mathbf{a}|^2 \leqslant \mathbf{a}_i^+ \cdot \sum_j c_j(z_j-z_i)=\mathbf{a}_i^+ \cdot \mathbf{a} \leqslant K|\mathbf{a}|$$

or

$$|\mathbf{a}|^2 \leqslant \mathbf{a}_i^- \cdot \sum_j c_j(z_i-z_j)=\mathbf{a}_i^- \cdot \mathbf{a} \leqslant K|\mathbf{a}|,$$

and hence $|\mathbf{a}| \leqslant K$. Thus *the bounded slope condition implies the three-point condition with the same constant K.*

Conversely, let $\Gamma=(\partial\Omega, \varphi)$ satisfy the three-point condition with constant K over the convex domain Ω. Let $A=(z_A, \varphi(z_A))$ be any point of Γ such that z_A is not an interior point of a straight segment of $\partial\Omega$. There exists a sequence of triangles with vertices at non-collinear points $A, B_i, C_i \in \Gamma$, $i=1, 2, \ldots$, such that $B_i, C_i \to A$ as $i \to \infty$ and the planes Δ_i determined by A, B_i, C_i converge to a limiting plane Δ. It can be assumed that the segments AB_i have a limiting direction, which determines a straight line L through A lying in Δ whose projection L_z on the z plane is a supporting line of Ω. Obviously L coincides with the tangent to Γ at A whenever the tangent exists. Let $Q \in \Gamma$, $Q \notin L$, and let Π denote the plane determined by Q and L. The slope of Π does not exceed K since the sequence of planes determined by A, Q and B_i have the same limit slope as Π. Consider now the set of planes containing L and the points $Q \in \Gamma$, $Q \notin L$, and let these planes be defined by the linear functions $u = \pi_Q(z)$. In the z plane let H denote the half-plane on the side of L containing Ω. If $\pi_Q(z) \geqslant_{Q'}(z)$ for some $z \in H$, then the same inequality holds for all $z \in H$ and, in particular, for all $z \in \partial\Omega$. Thus, there are upper and lower planes at A defined by

$$\pi^+(z) = \sup_{\substack{Q\in\Gamma\\ Q\notin L}} \pi_Q(z), \qquad \pi^-(z) = \inf_{\substack{Q\in\Gamma\\ Q\notin L}} \pi_Q(z), \quad z \in H.$$

The planes $u=\pi^+(z)$ and $u=\pi^-(z)$ by the preceding discussion have slopes not exceeding K and lie respectively above and below Γ (over $\partial\Omega$). Hence they satisfy the bounded slope condition at A with constant K. It is clear that if A lies on a straight segment of Γ, then the line containing this segment can replace L in the above argument. Thus *the three-point condition implies the bounded slope condition with the same constant K.*

Notes

Hölder estimates for quasiconformal mappings (Section 12.1) have been derived in various ways starting with Morrey [MY 1]; (for references see [FS]). The development here is due to Finn and Serrin [FS], who derived the estimate (12.14) with its optimal Hölder exponent and, when $K'=0$ in (12.2), also obtained this result with a Hölder coefficient CM in which C is an absolute constant.

The basic $C^{1,\alpha}$ Hölder estimates for solutions of linear equations in Section 12.2 are due to Morrey [MY 1] and, in simpler form, to Nirenberg [NI 1]. The presentation here is a variant of the latter, but follows Morrey in using growth estimates for the Dirichlet integral. The idea of applying these apriori $C^{1,\alpha}$ estimates for *linear* equations to the Dirichlet problem for the quasilinear equation (12.40) appears in [MY 1], which contains a flaw, however. The idea was carried to completion and simplified in the details by Nirenberg who applied the Schauder fixed point theorem in arriving at the general existence theorem described below. This approach is the basis for the development in Sections 12.3 and 12.4.

Until the late 1950's the theory of the Dirichlet problem for nonlinear elliptic equations was confined largely to the case of two independent variables. The pioneering work of Bernstein [BE 1–4], which was somewhat restrictive in its hypotheses, was extended by Leray and Schauder [SC 3], [LS], who obtained solutions of the Dirichlet problem for (12.40) under the assumption of $C^{2,\beta}$ coefficients and sufficiently smooth boundary data by using methods based (as in Bernstein's work) on apriori estimates of the second derivatives of solutions. These results were improved and simplified by the above contributions of Morrey and Nirenberg, the latter proving that if (12.40) is elliptic in $\bar{\Omega}$ with coefficients in $C^\beta(\bar{\Omega}\times\mathbb{R}\times\mathbb{R}^2)$, where $\partial\Omega$ is a $C^{2,\beta}$ uniformly convex curve, and $\varphi\in C^{2,\beta}(\partial\Omega)$, then the Dirichlet problem for (12.40) has a solution in $C^{2,\beta}(\bar{\Omega})$; (if $\varphi\in C^{1,1}(\partial\Omega)$ then a solution exists in $C^{2,\beta}(\Omega)\cap C^0(\bar{\Omega})$). Theorem 12.7 extends this result by weakening the hypotheses on the coefficients and on the boundary data by requiring the latter to satisfy only a bounded slope condition, without additional regularity assumptions. The solution thus obtained also lies in $C^{2,\beta}(\bar{\Omega})$ when the hypotheses are the same as in Nirenberg's theorem (see Remark 4 after Theorem 12.7). The proof of Theorem 12.7 depends only on apriori *interior* $C^{1,\alpha}$ estimates for linear equations (Theorem 12.4). That such estimates suffice for an existence proof was observed already by Nirenberg ([NI 1], p. 146).

Concerning the Dirichlet problem for the uniformly elliptic equation (12.25), we mention the contributions of Bers and Nirenberg [BN], Ladyzhenskaya and Ural'tseva [LU 4] and von Wahl [WA]. In [BN], existence results—for both the Dirichlet and Neumann problems—are obtained in the context of $W^{2,2}$ solutions (which are also in $C^{2,\alpha}$ if the coefficients in (12.25) are in C^α). It assumes a linear growth condition for $f(x,y,u,p,q)$ analogous to (12.27), and the methods of proof in [BN] and Theorem 12.5 are similar in being based on $C^{1,\alpha}$ estimates for linear equations and the Schauder fixed point theorem. Both [BN] and Theorem 12.5 make no apriori assumptions concerning the solutions. However, if a uniform bound on $\sup_\Omega |u|$ is assumed for all solutions u of (12.29) under the following hypotheses: $a, b, c, f\in C^\beta(\bar{\Omega}\times\mathbb{R}\times\mathbb{R}^2)$ in (12.25), $\partial\Omega\in C^{2,\beta}$, $\varphi\in C^{2,\beta}(\bar{\Omega})$, and $|f/\lambda|\leqslant C(1+p^2+q^2)$ (weakening the condition (12.27)), then [WA] establishes an apriori $C^{1,\alpha}(\bar{\Omega})$ bound on solutions of the Dirichlet problem, thereby extending a similar result in [LU 4]. (This estimate does not follow from the methods of the present chapter without an apriori gradient bound.) Existence of a $C^{2,\beta}(\bar{\Omega})$ solution of the Dirichlet problem (12.29) can now be obtained by application of the Leray–Schauder fixed point theorem.

The proof of the gradient bound in Lemma 12.6 is suggested by the introductory remarks of Finn in [FN 2]. It is usually inferred from a theorem of Radó [RA], which asserts that if $u=u(x,y)$ is a saddle surface over a convex domain Ω and u is continuous on $\bar{\Omega}$ with boundary values satisfying a three-point condition with constant K, then u satisfies a Lipschitz condition with constant K in Ω. In this theorem $u(x,y)$ is considered to define a saddle surface if it is continuous and the function $u(x,y)-(ax+by+c)$ satisfies the weak maximum–minimum principle for all constants a, b, c. In particular, this will be the case if $u(x,y)$ satisfies (12.42) or represents a surface having non-positive Gaussian curvature. An elementary (but still not simple) proof is due to von Neumann [NU]. Hartman and Nirenberg [HN], [NI 4] have extended the result to higher dimensions.

With regard to the bounded slope condition, Hartman [HA] analyzes the relation between the regularity properties of φ and $\partial\Omega$ when the function φ satisfies a bounded slope condition over the boundary $\partial\Omega$ of a bounded convex domain Ω in $\mathbb{R}^n$. In the process he establishes the equivalence of the bounded slope and $(n+1)$-point conditions for $n\geqslant 2$; however, the relation between the constants in the two conditions remains unclear for $n>2$. Among the results we mention the following. (i) If $\partial\Omega\in C^{1,\alpha}, 0\leqslant\alpha\leqslant 1$, and φ satisfies a bounded slope condition over $\partial\Omega$, then $\varphi\in C^{1,\alpha}(\partial\Omega)$. (ii) If $\partial\Omega\in C^{1,1}$ and is uniformly convex, then φ satisfies a bounded slope condition over $\partial\Omega$ if and only if $\varphi\in C^{1,1}$. The latter result follows from (i) and the fact that if Ω is *any* uniformly convex domain and φ is the restriction to $\partial\Omega$ of a $C^{1,1}$ function, then φ satisfies a bounded slope condition over $\partial\Omega$.

Problems

12.1. Show that the function $w(z)=z\log|z|$ is (K, K')-quasiconformal with $K=1$, $K'>0$ and does not satisfy (12.14) with the exponent $\alpha=1$.

12.2. Prove Theorem 12.3 for quasiconformal mappings satisfying (12.2) in which p, q are continuous and lie in $W^{1,2}_{loc}$ (cf. [FS]).

12.3. Let $Lu=au_{xx}+2bu_{xy}+cu_{yy}$ be uniformly elliptic in a domain Ω with $\gamma=\sup_\Omega(\Lambda/\lambda)$, where $\lambda=\lambda(x,y)$ and $\Lambda=\Lambda(x,y)$ are the minimum and maximum eigenvalues of the coefficient matrix. If $Lu=0$, show that $p-iq=u_x-iu_y$ is K-quasiconformal in Ω for all $K\geqslant\frac{1}{2}(\gamma+1/\gamma)$. [It suffices to show that

$$\sup\frac{r^2+2s^2+t^2}{s^2-rt}=\frac{\lambda}{\Lambda}+\frac{\Lambda}{\lambda},$$

where the supremum is taken over all r, s, t satisfying $ar+2bs+ct=0$ for fixed $(x,y)\in\Omega$.] Hence infer that when $f=0$ Theorem 12.4 holds for any Hölder exponent $\alpha\leqslant 1/\gamma$ (cf. [TA 1] for $f\neq 0$).

12.4. In Theorem 12.4, assume $f/\lambda\in L^p(\Omega)$, $p>2$. If $\Omega'\subset\subset\Omega$, prove that for some $\alpha\in(0,1)$ we have

$$|u|_{1,\alpha;\Omega'}\leqslant C(|u|_{0;\Omega}+\|f/\lambda\|_{L^p(\Omega)}),$$

where $C=C(\gamma, p, \operatorname{dist}(\Omega',\partial\Omega))$ and $\alpha=\alpha(\gamma,p)$. (Either $u\in C^2(\Omega)$ or $u\in W^{2,2}(\Omega)$ may be assumed.)

12.5. In Theorem 12.5, assume $\Omega\in C^{2,\beta}$, $\varphi\in C^{2,\beta}(\bar\Omega)$ and $a,b,c,f\in C^\beta(\bar\Omega\times\mathbb{R}\times\mathbb{R}^2)$. By applying the $C^{1,\alpha}(\bar\Omega)$ estimates discussed after Theorem 12.4, modify the proof of Theorem 12.5 to show that problem (12.29) has a solution in $C^{2,\beta}(\bar\Omega)$. [Replace

C_*^1 and $C_*^{1,\alpha}$ by $C^{1,\alpha}(\bar{\Omega})$ and define the set $\mathfrak{S}$ and the operator T accordingly. Use Theorem 6.14 in place of Theorem 6.13, and global in place of interior Schauder estimates. A simpler proof is obtained by considering the family of equations, $u=\sigma Tu$, $\sigma\in[0,1]$, corresponding to the Dirichlet problems

$$Q_\sigma u = au_{xx}+2bu_{xy}+cu_{yy}+\sigma f=0 \text{ in } \Omega, \quad u=\sigma\varphi \text{ on } \partial\Omega,$$

where $a=a(x,y,u,u_x,u_y)$, etc. Show that the solutions are uniformly bounded in $C^{1,\alpha}(\bar{\Omega})$ and apply Theorem 11.3].

12.6. Assume the operator Q in equation (12.40) is uniformly elliptic on bounded subsets of $\Omega\times\mathbb{R}\times\mathbb{R}^2$. Prove Theorem 12.7 without the use of Theorem 12.5 by replacing the set $\mathfrak{S}$ in Theorem 12.5 with

$$\mathfrak{S}'=\{v\in C_*^{1,\alpha} \mid |v|_{1,\alpha}^*\leqslant K', \quad |Dv|\leqslant K\}$$

where K is the constant of the bounded slope condition and K' is a constant determined as in (12.32).

Chapter 13

Hölder Estimates for the Gradient

In this chapter we derive interior and global Hölder estimates for the derivatives of solutions of quasilinear elliptic equations of the form

$$(13.1)\qquad Qu = a^{ij}(x, u, Du)D_{ij}u + b(x, u, Du) = 0$$

in a bounded domain Ω. From the global results we shall see that Step IV of the existence procedure described in Chapter 11 can be carried out if, in addition to the hypotheses of Theorem 11.4, we assume that either the coefficients a^{ij} are in $C^1(\bar{\Omega} \times \mathbb{R} \times \mathbb{R}^n)$ or that Q is of divergence form or that $n=2$. The estimates of this chapter will be established through a reduction to the results of Chapter 8, in particular to Theorems 8.18, 8.24, 8.26 and 8.29.

13.1. Equations of Divergence Form

Let us suppose now that Q is equivalent to an elliptic operator of the form,

$$(13.2)\qquad Qu = \operatorname{div} \mathbf{A}(x, u, Du) + B(x, u, Du),$$

where the vector function $\mathbf{A} \in C^1(\Omega \times \mathbb{R} \times \mathbb{R}^n)$ and $B \in C^0(\Omega \times \mathbb{R} \times \mathbb{R}^n)$. Then if $u \in C^1(\Omega)$ satisfies $Qu=0$ in Ω, we have

$$\int_\Omega \{\mathbf{A}(x, u, Du) \cdot D\zeta - B(x, u, Du)\zeta\}\, dx = 0 \quad \forall \zeta \in C_0^1(\Omega).$$

Fixing k, $1 \leqslant k \leqslant n$, replacing ζ by $D_k\zeta$, and integrating by parts, we then obtain

$$\int_\Omega \{(D_{p_j}A^i D_{jk}u + \delta_k A^i)D_i\zeta + BD_k\zeta\}\, dx = 0 \quad \forall \zeta \in C_0^1(\Omega),$$

where δ_k is the differential operator defined by

$$\delta_k A^i(x, z, p) = p_k D_z A^i(x, z, p) + D_{x_k} A^i(x, z, p)$$

and the arguments of $D_{p_j}A^i$, $\delta_k A^i$ and B are x, $u(x)$, $Du(x)$. Hence, writing

$$\bar{a}^{ij}(x) = D_{p_j}A^i(x, u(x), Du(x))$$
$$f^i_k(x) = \delta_k A^i(x, u(x), Du(x)) + \delta^i_k B(x, u(x), Du(x))$$

where $[\delta^i_k]$ is the identity matrix, we see that the derivative $w = D_k u$ satisfies

$$\int_\Omega (\bar{a}^{ij}(x) D_j w + f^i_k(x)) D_i \zeta \, dx = 0 \qquad \forall \zeta \in C^1_0(\Omega); \tag{13.3}$$

that is, w is a generalized solution of the linear elliptic equation

$$Lw = D_i(\bar{a}^{ij} D_j w) = -D_i f^i_k.$$

By replacing Ω if necessary by a strictly contained subdomain we can assume that L is strictly elliptic in Ω and that the coefficients $\bar{a}^{ij}$, f^i_k are bounded, that is the hypotheses of Theorems 8.22 and 8.24 are satisfied. Accordingly, choosing λ_K, Λ_K, μ_K such that

$$\begin{aligned} 0 < \lambda_K &\leqslant \lambda(x, z, p), \\ \Lambda_K &\geqslant |D_{p_j} A^i(x, z, p)|, \\ \mu_K &\geqslant |\delta_j A^i(x, z, p)| + |B(x, z, p)|, \end{aligned} \tag{13.4}$$

for all $x \in \Omega$, $|z| + |p| \leqslant K$, $i, j = 1, \ldots, n$, we obtain the following interior estimate.

Theorem 13.1. *Let $u \in C^2(\Omega)$ satisfy $Qu = 0$ in Ω where Q is elliptic in Ω and is of divergence form (13.2) with $\mathbf{A} \in C^1(\Omega \times \mathbb{R} \times \mathbb{R}^n)$, $B \in C^0(\Omega \times \mathbb{R} \times \mathbb{R}^n)$. Then for any $\Omega' \subset\subset \Omega$ we have the estimate*

$$[Du]_{\alpha;\Omega'} \leqslant C d^{-\alpha} \tag{13.5}$$

where

$$C = C(n, K, \Lambda_K/\lambda_K, \mu_K/\lambda_K, \operatorname{diam} \Omega),$$
$$K = |u|_{1;\Omega} = \sup_\Omega (|u| + |Du|),$$
$$d = \operatorname{dist}(\Omega', \partial\Omega) \text{ and } \alpha = \alpha(n, \Lambda_K/\lambda_K) > 0.$$

In order to extend Theorem 13.1 to a global Hölder estimate in Ω we assume that Q is elliptic in $\bar{\Omega}$ with $\mathbf{A} \in C^1(\bar{\Omega} \times \mathbb{R} \times \mathbb{R}^n)$, $B \in C^0(\bar{\Omega} \times \mathbb{R} \times \mathbb{R}^n)$, that $\partial\Omega \in C^2$ and that $u = \varphi$ on $\partial\Omega$ where $\varphi \in C^2(\bar{\Omega})$. By replacing u with $u - \varphi$ we can assume without loss of generality that $u = 0$ on $\partial\Omega$. Since $\partial\Omega \in C^2$, there exists for each $x_0 \in \partial\Omega$ a ball $B = B(x_0)$ and a one-to-one mapping ψ from B onto an open set

$D \subset \mathbb{R}^n$ such that

$$\psi(B \cap \Omega) \subset \mathbb{R}^n_+ = \{x \in \mathbb{R}^n \mid x_n > 0\},$$
$$\psi(B \cap \partial\Omega) \subset \partial\mathbb{R}^n_+ \quad \text{and} \quad \psi \in C^2(B), \quad \psi^{-1} \in C^2(D).$$

Writing $y = \psi(x)$, $v(y) = u \circ \psi^{-1}(y)$, $B^+ = B \cap \Omega$, $D^+ = \psi(B^+)$, we have $D_{y_k} v = 0$ on $\partial D^+ \cap \partial\mathbb{R}^n_+$, $k = 1, \ldots, n-1$, and that the equation $Qu = 0$ in B^+ is equivalent to the equation

$$(13.6) \qquad \bar{Q}v = D_{y_i}\bar{A}^i(x, u, Du) + \bar{B}(x, u, Du) = 0, \quad x = \psi^{-1}(y),$$

in D^+ where the functions $\bar{\mathbf{A}}$ and $\bar{B}$ are given by

$$\bar{A}^i = \frac{\partial y_i}{\partial x_r} A^r, \quad \bar{B} = -\frac{\partial}{\partial y_i}\left(\frac{\partial y_i}{\partial x_r}\right) A^r + B.$$

The derivatives $w = D_{y_k} v$, $k = 1, \ldots, n$, will consequently be generalized solutions in D^+ of the linear elliptic equations

$$Lw = D_i(\bar{a}^{ij} D_j w) = -D_i f^i_k,$$

where

$$\bar{a}^{ij}(y) = \frac{\partial y_i}{\partial x_r} \frac{\partial y_j}{\partial x_s} D_{p_s} A^r,$$

$$f^i_k(y) = \frac{\partial y_i}{\partial x_r} \frac{\partial x_s}{\partial y_k}\left(\frac{\partial^2 y_l}{\partial x_j \, \partial x_s} D_{p_j} A^r D_{y_l} u + \delta_s A^r\right) + \delta^i_k B$$
$$+ \left[\frac{\partial}{\partial y_k}\left(\frac{\partial y_i}{\partial x_r}\right) - \delta^i_k \frac{\partial}{\partial y_j}\left(\frac{\partial y_j}{\partial x_r}\right)\right] A^r,$$

the arguments of $D_{p_s} A^r$, $\delta_s A^r$, A^r and B being $x = \psi^{-1}(y)$, $u(x) = v(y)$ and $Du(x) = Du(\psi^{-1}(y))$. By replacing $B(x_0)$ if necessary by a smaller concentric ball, we can then assume that the Jacobian matrix $[D\psi] = [\partial y_i / \partial x_j]$ is bounded from above and below in B by positive constants, and consequently the operator L is strictly elliptic in D^+ with bounded coefficients $\bar{a}^{ij}, f^i_k$. By applying Theorem 8.29, we thus have for any $D' \subset\subset D$

$$(13.7) \qquad [D_{y_k} v]_{\alpha; D' \cap D^+} \leqslant C, \quad k = 1, \ldots, n-1,$$

where $C = C(n, K, \Lambda_K/\lambda_K, \mu_K/\lambda_K, \Omega, d)$, $K = |u|_{1;\Omega}$, $d = \text{dist}\,(D' \cap D^+, \partial D)$ and $\alpha = \alpha(n, \Lambda_K/\lambda_K, \Omega) > 0$.

The remaining derivative $D_{y_n} v$ can be estimated as follows. Let $y_0 \in D^+ \cap D'$, $R \leqslant d/3$, $B_{2R} = B_{2R}(y_0)$, $\eta \in C^1_0(B_{2R})$, and let c be a constant such that $c = w(y_0)$ if $B_{2R} \subset D^+$, $c = 0$ if $B_{2R} \cap \partial\mathbb{R}^n_+ \neq \phi$. The function $\zeta = \eta^2(w - c)$ then belongs to $W^{1,2}_0(D^+)$ for $w = D_{y_k} v$, $k = 1, \ldots, n-1$. By substitution into the integral identity

(13.3) with $\Omega = D^+$, we then have

$$\int_{D^+} \eta^2 \bar{a}^{ij} D_i w D_j w \, dy \leqslant \int_{D^+} \{|2\eta(w-c)\bar{a}^{ij} D_i \eta D_j w|$$

$$+|\eta^2 f_k^i D_i w| + |2\eta(w-c) f_k^i D_i \eta|\} \, dy,$$

so that by the Schwarz inequality (7.6) and the ellipticity of L, we obtain

$$\int_{D^+} \eta^2 |Dw|^2 \, dy \leqslant C \int_{D^+} (\eta^2 + |D\eta|^2 (w-c)^2) \, dy$$

where $C = C(n, K, \Lambda_K/\lambda_K, \mu_K/\lambda_K, \Omega)$. Now let us further require η to be such that $0 \leqslant \eta \leqslant 1$, $\eta = 1$ in $B_R = B_R(y_0)$ and $|D\eta| \leqslant 2/R$. We thus obtain

$$\int_{B_R} |Dw|^2 \, dy \leqslant C\, R^{n-2} (R^2 + \sup_{B_{2R}} (w-c)^2)$$

$$\leqslant C\, R^{n-2+2\alpha}$$

by (13.7), where C and α depend on the same quantities as in (13.7). Therefore, since $k = 1, \ldots, n-1$, we have

$$\int_{B_R} |D_{ij} v|^2 \, dy \leqslant C\, R^{n-2+2\alpha} \tag{13.8}$$

provided $j \neq n$. To proceed further we solve equation (13.6) for $D_{nn} v$, so that we can write

$$D_{nn} v = b^{ij} D_{ij} v + b, \quad i = 1, \ldots, n, \quad j = 1, \ldots, n-1,$$

for certain functions b^{ij}, b bounded in terms of $D\psi$, K, Λ_K/λ_K and μ_K/λ_K. Hence by (13.8) we have

$$\int_{B_R} |D_{ni} v|^2 \, dy \leqslant C\, R^{n-2+2\alpha}, \quad i = 1, \ldots, n,$$

so that using inequality (7.8) and Morrey's estimate, Theorem 7.19, we can conclude that the estimate (13.7) is also valid for $k = n$. Returning to the domain Ω by means of the mapping ψ^{-1}, we thus have

$$[Du]_{\alpha; B' \cap \Omega} \leqslant C \tag{13.9}$$

for any concentric ball $B'\subset\subset B$, where $C=C(n, K, \Lambda_K/\lambda_K, \mu_K/\lambda_K, \Omega, B')$. Finally, by choosing a finite number of points $x_0\in\partial\Omega$ and balls B' covering $\partial\Omega$, we obtain the following global Hölder estimate from (13.5) and (13.9).

Theorem 13.2. *Let $u\in C^2(\bar\Omega)$ satisfy $Qu=0$ in Ω where Q is elliptic in $\bar\Omega$ and is of divergence form* (13.2) *with $\mathbf{A}\in C^1(\bar\Omega\times\mathbb{R}\times\mathbb{R}^n)$, $B\in C^0(\bar\Omega\times\mathbb{R}\times\mathbb{R}^n)$. Then if $\partial\Omega\in C^2$ and $u=\varphi$ on $\partial\Omega$, where $\varphi\in C^2(\bar\Omega)$, we have the estimate*

$$(13.10)\qquad [Du]_{\alpha;\Omega}\leqslant C$$

where

$$C=C(n, K, \Lambda_K/\lambda_K, \mu_K/\lambda_K, \Omega, \Phi),$$
$$K=|u|_{1;\Omega},\qquad \Phi=|\varphi|_{2;\Omega}\qquad \textit{and}\quad \alpha=\alpha(n, \Lambda_K/\lambda_K, \Omega)>0.$$

An inspection of the proofs of Theorems 13.1 and 13.2 shows that the estimates (13.5) and (13.10) continue to hold if we only assume $u\in C^{0,1}(\bar\Omega)\cap W^{2,2}(\Omega)$, and $\varphi\in W^{2,q}(\Omega)$ for some $q>n$. In this generality we now must take $\Phi=\|\varphi\|_{W^{2,q}(\Omega)}$, and α will depend as well on q.

13.2. Equations in Two Variables

If $u\in C^2(\Omega)$ satisfies the elliptic equation (13.1) in $\Omega\subset\mathbb{R}^2$, then the derivatives $w_1=D_1u$, $w_2=D_2u$ are generalized solutions in Ω of the linear elliptic equations

$$(13.11)\qquad \begin{aligned} L_1w_1&=D_1\left(\frac{a^{11}}{a^{22}}D_1w_1+\frac{2a^{12}}{a^{22}}D_2w_2\right)+D_{22}w_1=-D_1\frac{b}{a^{22}},\\ L_2w_2&=D_{11}w_2+D_2\left(\frac{2a^{12}}{a^{11}}D_1w_2+\frac{a^{22}}{a^{11}}D_2w_2\right)=-D_2\frac{b}{a^{11}}. \end{aligned}$$

Consequently the methods for equations of divergence form also apply here. Accordingly, if λ_K, Λ_K and μ_K satisfy

$$(13.12)\qquad \begin{aligned} &0<\lambda_K<\lambda(x,z,p),\\ &\Lambda_K\geqslant|a^{ij}(x,z,p)|,\\ &\mu_K\geqslant|b(x,z,p)|, \end{aligned}$$

for all $x\in\Omega$, $|z|+|p|\leqslant K$, $i,j=1,2$, we have the following estimates.

Theorem 13.3. *Let $u\in C^2(\Omega)$ satisfy $Qu=0$ in $\Omega\subset\mathbb{R}^2$, where Q is elliptic in Ω and the coefficients a^{ij}, $b\in C^0(\Omega\times\mathbb{R}\times\mathbb{R}^2)$. Then for any $\Omega'\subset\subset\Omega$, we have*

$$(13.13)\qquad [Du]_{\alpha;\Omega'}\leqslant Cd^{-\alpha},$$

where

$$C=C(K,\Lambda_K/\lambda_K,\mu_K/\lambda_K,\operatorname{diam}\Omega),$$
$$K=|u|_{1;\Omega},\quad d=\operatorname{dist}(\Omega',\partial\Omega)\quad\textit{and}\quad \alpha=\alpha(\Lambda_K/\lambda_K)>0.$$

Theorem 13.4. *Let $u\in C^2(\bar\Omega)$ satisfy $Qu=0$ in $\Omega\subset\mathbb{R}^2$ where Q is elliptic in $\bar\Omega$ and the coefficients a^{ij}, $b\in C^0(\bar\Omega\times\mathbb{R}\times\mathbb{R}^2)$. Then if $\partial\Omega\in C^2$, $Q\in C^2(\bar\Omega)$ and $u=\varphi$ on $\partial\Omega$, we have the estimate*

$$[Du]_{\alpha;\Omega}\leqslant C \tag{13.14}$$

where

$$C=C(K,\Lambda_K/\lambda_K,\mu_K/\lambda_K,\Omega,\Phi),$$
$$K=|u|_{1;\Omega},\ \Phi=|\varphi|_{2;\Omega}\quad\textit{and}\quad \alpha=\alpha(\Lambda_K/\lambda_K,\Omega)>0.$$

Note that we have given an alternative derivation of Theorem 13.3 in Chapter 12 using the method of quasiconformal mappings. We note also that in the case of two variables the proof of the Hölder estimate for linear divergence form equations is simpler than for more variables (see Problem 8.5). The remark following Theorem 13.2 of course applies as well to Theorems 13.3 and 13.4.

13.3. Equations of General Form; the Interior Estimate

We shall treat elliptic equations of the general form (13.1) by showing that certain combinations of the derivatives of solutions are generalized *subsolutions* of linear elliptic equations of divergence form. The desired Hölder estimates are then obtained through application of the weak Harnack inequalities, Theorems 8.18 and 8.26.

Let us suppose that the coefficients a^{ij} and b of Q, are respectively in $C^1(\Omega\times\mathbb{R}\times\mathbb{R}^n)$ and $C^0(\Omega\times\mathbb{R}\times\mathbb{R}^n)$. Let $Qu=0$ in Ω and assume initially that $u\in C^3(\Omega)$. By differentiation with respect to x_k, $k=1,\ldots,n$, we obtain

$$\begin{aligned}a^{ij}(x,u,Du)D_{kij}u+D_{p_l}a^{ij}(x,u,Du)D_{lk}u\,D_{ij}u\\ +\delta_k a^{ij}(x,u,Du)D_{ij}u+D_kb(x,u,Du)=0\end{aligned}\tag{13.15}$$

where δ_k is the differential operator defined by

$$\delta_k g(x,z,p)=D_{x_k}g(x,z,p)+p_kD_zg(x,z,p).$$

The equation (13.15) can be written in the following divergence form

$$\begin{aligned}D_i(a^{ij}D_{kj}u)+(D_{p_l}a^{ij}-D_{p_j}a^{il})D_{lk}uD_{ij}u\\ +\delta_k a^{ij}D_{ij}u-\delta_i a^{ij}D_{kj}u+D_kb=0.\end{aligned}\tag{13.16}$$

Let us now write $v=|Du|^2$, multiply equation (13.16) by $D_k u$ and sum the resulting equations from $k=1$ to n. We then obtain

$$\text{(13.17)} \quad -a^{ij}D_{ki}uD_{kj}u+\tfrac{1}{2}D_i(a^{ij}D_j v)+\tfrac{1}{2}(D_{p_l}a^{ij}-D_{p_j}a^{il})D_{ij}uD_l v \\ +D_k u\,\delta_k a^{ij}D_{ij}u-\tfrac{1}{2}\,\delta_i a^{ij}D_j v+D_k(bD_k u)-b\,\Delta u=0.$$

For $\gamma\in\mathbb{R}$ and $r=1,\dots,n$, we next define functions

$$\text{(13.18)} \quad w=w_r=\gamma D_r u+v,$$

and by combining (13.16) and (13.17) obtain the equation,

$$\text{(13.19)} \quad -2a^{ij}D_{ki}uD_{kj}u+D_i(a^{ij}D_j w+(2D_i u+\gamma\delta^{ir})b) \\ +(D_{p_l}a^{ij}-D_{p_j}a^{il})D_{ij}uD_l w+[(2D_k u+\gamma\delta^{kr})\delta_k a^{ij}-2b\delta^{ij})]D_{ij}u-\delta_i a^{ij}D_j w \\ =0.$$

Setting
$$\bar{a}^{ij}(x)=a^{ij}(x,u(x),Du(x)),$$
$$a_l^{ij}(x)=(D_{p_l}a^{ij}-D_{p_j}a^{il})(x,u(x),Du(x)),$$
$$f_r^i(x)=(2D_i u(x)+\gamma\delta^{ir})b(x,u(x),Du(x)),$$
$$b^{ij}(x)=[(2D_k u(x)+\gamma\delta^{kr})\delta_k a^{ij}-2\,\delta^{ij}b](x,u(x),Du(x)),$$
$$c^j(x)=\delta_i a^{ij}(x,u(x),Du(x)),$$

we write the equation (13.19) in its integral form

$$\text{(13.20)} \quad \int_\Omega \{(\bar{a}^{ij}D_j w+f_r^i)D_i\zeta+(2\bar{a}^{ij}D_{ki}uD_{kj}u-a_l^{ij}D_{ij}uD_l w+c^iD_i w-b^{ij}D_{ij}u)\zeta\}\,dx \\ =0$$

for all $\zeta\in C_0^1(\Omega)$. We claim now that the integral identity (13.20) continues to be valid if we only assume that $u\in C^2(\Omega)$. To see this, we let $\{u_m\}\subset C^3(\Omega)$ approach u in the sense of $C^2(\Omega)$, that is $\{D^\beta u_m\}$, converges uniformly to $D^\beta u$ on compact subsets of Ω for all $|\beta|\leqslant 2$. Since $Qu=0$ in Ω, we have $Qu_m\to 0$ uniformly on compact subsets of Ω and hence letting $m\to\infty$ in the integral identity for u_m corresponding to (13.20), we obtain (13.20). We note that a similar approximation argument shows that in fact we need only assume $u\in C^{0,1}(\Omega)\cap W^{2,2}(\Omega)$.

To proceed further, we need to remove the terms involving $D_{ij}u$ from (13.20). Since Q is elliptic we have

$$\bar{a}^{ij}D_{ki}uD_{kj}u\geqslant\lambda|D^2u|^2\geqslant 0$$

where $\lambda=\lambda(x,u(x),Du(x))$. Using the Schwarz inequality, we then have from (13.20)

(13.21) $$\int_\Omega (\bar{a}^{ij}D_j w + f_r^i)D_i\zeta\,dx \leqslant \int_\Omega \left\{(\lambda + \frac{1}{\lambda}\sum|a_i^{ij}|^2)\,|Dw|^2 + \frac{1}{\lambda}\sum(|c^i|^2 + |b^{ij}|^2)\right\}\zeta\,dx$$

for all non-negative $\zeta \in C_0^1(\Omega)$. The reduction to the linear theory is finally achieved by replacing ζ in (13.21) by $\zeta\,e^{2\chi w}$ where $\chi = \sup_\Omega (1 + \lambda^{-2}\sum|a_i^{ij}|^2)$. Setting

$$\tilde{a}^{ij}(x) = e^{2\chi w(x)}\,\bar{a}^{ij}(x)$$
$$\tilde{f}^i(x) = e^{2\chi w(x)}\,f_r^i(x)$$
$$\tilde{g}(x) = \frac{1}{\lambda}\,e^{2\chi w(x)}\,(\chi\sum|f_r^i|^2 + \sum|c^i|^2 + \sum|b^{ij}|^2)$$

we thus obtain

(13.22) $$\int_\Omega \{(\tilde{a}^{ij}D_i w + \tilde{f}^i)D_i\zeta - \tilde{g}\zeta\}\,dx \leqslant 0$$

for all non-negative $\zeta \in C_0^1(\Omega)$; that is, the function w satisfies the inequality

(13.23) $$Lw = D_i(\tilde{a}^{ij}D_j w) \geqslant -(\tilde{g} + D_i\tilde{f}^i)$$

in the generalized sense. By replacing Ω if necessary by a strictly contained subdomain we can assume that the operator L is strictly elliptic in Ω and that the coefficients $\tilde{a}^{ij}$, $\tilde{f}^i$ and $\tilde{g}$ are bounded.

Let us take $\gamma > 0$ and write $w_r^\pm = \pm\gamma D_r u + v$. We now show that the validity of the inequality (13.23) for all sufficiently large $\gamma \in \mathbb{R}$ and $r = 1, \ldots, n$ is sufficient for the derivation of Hölder estimates for the derivatives $D_r u$, $r = 1, \ldots, n$. For, by choosing γ sufficiently large, we can ensure that the functions $w_r^\pm$ behave in a certain sense the same as $\pm D_r u$. It turns out that a convenient choice of γ is $\gamma = 10nM$ where $M = \sup|Du|$. For with this choice if $\mathfrak{S}$ is an arbitrary subset of Ω and r is chosen so that

$$\operatorname*{osc}_{\mathfrak{S}} D_r u \geqslant \operatorname*{osc}_{\mathfrak{S}} D_i u, \quad i = 1, \ldots, n,$$

it is readily seen that

(13.24) $$8n\,M \operatorname*{osc}_{\mathfrak{S}} D_r u \leqslant \operatorname*{osc}_{\mathfrak{S}} w_r^\pm \leqslant 12n\,M \operatorname*{osc}_{\mathfrak{S}} D_r u.$$

Furthermore, writing $w^\pm = w_r^\pm$ for brevity and setting $W^\pm = \sup_{\mathfrak{S}} w^\pm$, we have

$$\inf_{\mathfrak{S}} \sum_{+,-} (W^{\pm} - w^{\pm}) \geqslant 10n\, M(\sup_{\mathfrak{S}} D_r u - \inf_{\mathfrak{S}} D_r u) + 2 \inf_{\mathfrak{S}} v - 2 \sup_{\mathfrak{S}} v \tag{13.25}$$
$$\geqslant 6n\, M \operatorname{osc}_{\mathfrak{S}} D_r u$$
$$\geqslant \tfrac{1}{2} \operatorname{osc}_{\mathfrak{S}} w^{\pm}$$

We are now in a convenient position to apply the weak Harnack inequality, Theorem 8.18. Let us take $\mathfrak{S} = B_{4R}(y) \subset \Omega$. The functions $u = W^{\pm} - w^{\pm}$ will be non-negative supersolutions in $\mathfrak{S}$ of the equation

$$Lu = -(\tilde{g} + D_i \tilde{f}^i).$$

Accordingly, choosing λ_K and μ_K such that

$$0 < \lambda_K < \lambda(x, z, p),$$
$$\mu_K \geqslant |a^{ij}(x, z, p)| + |D_{p_k} a^{ij}(x, z, p)| + |D_z a^{ij}(x, z, p)| + |D_{x_k} a^{ij}(x, z, p)| + |b(x, z, p)|, \tag{13.26}$$

for all $x \in \Omega$, $|z| + |p| \leqslant K$, $i, j, k = 1, \ldots, n$, and setting $p = 1$ in Theorem 8.18, we obtain the estimates

$$R^{-n} \int_{B_{2R}} (W^{\pm} - w^{\pm})\, dx \leqslant C(W^{\pm} - \sup_{B_R} w^{\pm} + \sigma(R)), \tag{13.27}$$

where $C = C(n, K, \mu_K/\lambda_K)$, $K = |u|_{1;\Omega}$ and $\sigma(R) = R + R^2$. Using (13.25), we see that the inequality

$$\frac{1}{\omega_n (2R)^n} \int_{B_{2R}} (W^{\pm} - w^{\pm})\, dx \geqslant \frac{1}{4} \operatorname{osc}_{B_{4R}} w^{\pm},$$

holds for either w^+ or w^-. Let us suppose it is true for w^+. Then from (13.27) we have

$$\operatorname{osc}_{B_{4R}} w^+ \leqslant C(W^+ - \sup_{B_R} w^+ + \sigma(R))$$
$$\leqslant C(\operatorname{osc}_{B_{4R}} w^+ - \operatorname{osc}_{B_R} w^+ + \sigma(R)),$$

and consequently, writing $\omega(R) = \operatorname{osc}_{B_R} w^+$, we have

$$\omega(R) \leqslant \gamma \omega(4R) + \sigma(R),$$

where $\gamma = 1 - C^{-1}$, $C = C(n, K, \mu_K/\lambda_K)$.

We need now the following extension of Lemma 8.23.

Lemma 13.5. *Let $\omega_1, \ldots, \omega_N, \bar{\omega}_1, \ldots \bar{\omega}_N \geqslant 0$ be non-decreasing functions on an interval $(0, R_0)$ such that for each $R \leqslant R_0$ a function $\bar{\omega}_r$ can be found satisfying the inequalities*

$$\begin{aligned} &\bar{\omega}_r(R) \geqslant \delta_0 \omega_i(R), \quad i=1, \ldots, N, \quad \text{and} \\ (13.28) \qquad &\bar{\omega}_r(\delta R) \leqslant \gamma \bar{\omega}_r(R) + \sigma(R) \end{aligned}$$

where σ is also non-decreasing, $\delta_0 > 0$ and $0 < \gamma, \delta < 1$. Then for any $\mu \in (0, 1)$ and $R \leqslant R_0$, we have

$$(13.29) \qquad \omega_i(R) \leqslant C\left\{\left(\frac{R}{R_0}\right)^\alpha \omega_0 + \sigma(R^\mu R_0^{1-\mu})\right\}$$

where $C = C(N, \delta_0, \delta, \gamma)$ and $\alpha = \alpha(N, \delta_0, \delta, \gamma, \mu)$ are positive constants and $\omega_0 = \max_{i=1,\ldots,N} \bar{\omega}_i(R_0)$.

The proof of Lemma 13.5 follows by a simple modification of Lemma 8.23 and is therefore omitted. If we now let $B_0 = B_{R_0}(y)$ be any ball contained in Ω, it then follows by Lemma 13.5 with $N = 2n$, $\delta_0 = 8nM$, $\delta = \frac{1}{4}$, and $\mu = \frac{1}{2}$, that for any $R \leqslant R_0$, $i = 1, \ldots, n$

$$(13.30) \qquad \operatorname*{osc}_{B_R(y)} D_i u \leqslant CR^\alpha$$

where $C = C(n, K, \mu_K/\lambda_K, R_0)$, $\alpha = \alpha(n, K, \mu_K/\lambda_K)$. Consequently we obtain the desired interior estimate asserted in the following theorem.

Theorem 13.6. *Let $u \in C^2(\Omega)$ satisfy $Qu = 0$ in Ω where Q is elliptic in Ω and the coefficients $a^{ij} \in C^1(\Omega \times \mathbb{R} \times \mathbb{R}^n)$, $b \in C^0(\Omega \times \mathbb{R} \times \mathbb{R}^n)$. Then for any $\Omega' \subset\subset \Omega$ we have the estimate*

$$(13.31) \qquad [Du]_{\alpha;\Omega'} \leqslant Cd^{-\alpha}$$

where $C = C(n, K, \mu_K/\lambda_K, \operatorname{diam}\Omega)$, $K = |u|_{1;\Omega}$, $d = \operatorname{dist}(\Omega', \partial\Omega)$ *and* $\alpha = \alpha(n, K, \mu_K/\lambda_K)$.

13.4. Equations of General Form; the Boundary Estimate

Let us suppose now that the operator Q is elliptic in $\bar{\Omega}$ and that the coefficients a^{ij} and b are respectively in $C^1(\bar{\Omega} \times \mathbb{R} \times \mathbb{R}^n)$ and $C^0(\bar{\Omega} \times \mathbb{R} \times \mathbb{R}^n)$. Let $\partial\Omega \in C^2$ and assume $u = \varphi$ on $\partial\Omega$ where $\varphi \in C^2(\bar{\Omega})$. By replacing u by $u - \varphi$, we can assume without loss of generality that $u = 0$ on $\partial\Omega$. Furthermore, if $\mathscr{U}$ is an open subset of Ω and $x \to y = \psi(x)$ defines a $C^2(\mathscr{U})$ coordinate transformation, we have for $x \in \mathscr{U}$

$$(13.32)\qquad Qu=a^{kl}(x,u,Du)\frac{\partial y_i}{\partial x_k}\frac{\partial y_j}{\partial x_l}D_{y_iy_j}u+a^{ij}(x,u,Du)\frac{\partial^2 y_k}{\partial x_i\partial x_j}D_{y_k}u+b(x,u,Du).$$

Consequently the equation $Qu=0$ in $\mathscr{U}$ is transformed into an equation $\bar{Q}v=0$ in $\psi(\mathscr{U})$ where $v=u\circ\psi^{-1}$ and $\bar{Q}$ satisfies the same hypotheses as Q. It therefore suffices to consider equation (13.1) in the neighborhood of a flat boundary portion. Accordingly, let D be an open set in $\mathbb{R}^n$ such that $D^+=D\cap\Omega\subset\mathbb{R}^n_+=\{x\in\mathbb{R}^n|x_n>0\}$ and $D\cap\partial\Omega\subset\partial\mathbb{R}^n_+$. We define functions v' and w by

$$(13.33)\qquad v'=\sum_{i=1}^{n-1}|D_iu|^2,\qquad w=w_r=\gamma D_ru+v',\quad r=1,\ldots,n-1,\quad \gamma\in\mathbb{R}.$$

It is evident that $w=0$ on $D\cap\partial\Omega$ and that w satisfies equation (13.20) provided the summation over the index k is taken from 1 to $n-1$. Under this restriction we then have

$$\bar{a}^{ij}D_{ki}uD_{kj}u\geq\lambda\sum_{j\neq n}|D_{ij}u|^2.$$

The missing derivative $D_{nn}u$ is estimated by writing equation (13.1) in the form

$$(13.34)\qquad D_{nn}u=-\frac{1}{a^{nn}}\Big(\sum_{(i,j)\neq(n,n)}a^{ij}D_{ij}u+b\Big).$$

Inserting (13.34) into (13.20) and proceeding as in the preceding section, we arrive at the integral inequality (13.22) with Ω replaced by D^+ and χ and $\tilde{g}$ replaced respectively by

$$\chi=\sup_{\Omega}\,(1+\lambda^{-2}\sum|a_l^{ij}|^2)(1+\lambda^{-2}\sum|a^{ij}|^2)$$

and

$$\tilde{g}=\lambda^{-1}e^{2\chi w}\Big(\chi\sum|f_r^i|^2+\sum|c^i|^2+b^2(1+\lambda^{-2}\sum|a_l^{ij}|^2)+\sum|b^{ij}|^2(1+\lambda^{-2}\sum|a^{ij}|^2)\Big).$$

By considering balls centered at $y\in D\cap\partial\Omega$, applying the boundary weak Harnack inequality (Theorem 8.26) instead of the interior Harnack inequality (Theorem 8.18) and following the proof of the preceding section, we obtain a boundary Hölder estimate for the tangential derivatives of u. That is, for any ball $B_0=B_{R_0}(y)\subset D$ with center $y\in D\cap\partial\Omega$, we have for any $R\leq R_0$

$$(13.35)\qquad \operatorname*{osc}_{D^+\cap B_R(y)}D_iu\leq CR^\alpha,\quad i=1,\ldots,n-1,$$

where $C=C(n,K,\mu_K/\lambda_K,R_0)$, $K=|u|_{1;\Omega}$ and $\alpha=\alpha(n,K,\mu_K/\lambda_K)>0$.

The passage from (13.35) to an estimate for the remaining derivative $D_n u$ is similar to the corresponding step for divergence form equations in Section 13.1. Let $D' \subset\subset D$, $d = \text{dist}(D' \cap D^+, \partial D)$, $y \in D^+ \cap D'$, $R \leqslant d/3$, $B_{2R} = B_{2R}(y)$, $\eta \in C_0^1(B_{2R})$, and let c be a constant such that $c = \inf_{B_{2R}} w$ if $B_{2R} \subset D^+$, $c = 0$ if $B_{2R} \cap \partial\Omega \neq \phi$. The function $\zeta = \eta^2(w-c)^+ = \eta^2 \sup(w-c, 0)$ is non-negative and belongs to $W_0^{1,2}(D^+)$ for w given by (13.33). By substitution into the integral inequality (13.22) with $\Omega = D^+$, we then have

$$\int_{w \geqslant c} \eta^2 |Dw|^2 \, dx \leqslant C \int_{D^+} (\eta^2 + |D\eta|^2 (w-c)^2) \, dx$$

where $C = C(n, K, \mu_K/\lambda_K)$. Now let us further require η to be such that $0 \leqslant \eta \leqslant 1$, $\eta = 1$ in $B_R = B_R(y)$ and $|D\eta| \leqslant 2/R$. We then obtain

$$\begin{aligned} \int_{B_R^+} |Dw|^2 \, dx &\leqslant CR^{n-2}(R^2 + \sup_{B_{2R}} (w-c)^2), \quad B_R^+ = \{x \in B_R \mid w(x) \geqslant c\}, \\ &\leqslant CR^{n-2}(R^2 + (\operatorname{osc}_{B_{2R}} w)^2) \\ &\leqslant CR^{n-2+2\alpha} \quad \text{by (13.30) and (13.35)}, \end{aligned} \tag{13.36}$$

where $C = C(n, K, \mu_K/\lambda_K, d)$ and $\alpha = \alpha(n, K, \mu_K/\lambda_K)$. By taking $\gamma = 0$ and $\gamma = 1$ in (13.36), we then have for $B_{2R} \subset D^+$ (in which case $B_R^+ = B_R = D^+ \cap B_R$)

$$\begin{aligned} \int_{D^+ \cap B_R} |DD_r u|^2 \, dx &\leqslant 2 \int_{D^+ \cap B_R} (|Dv'|^2 + |Dw|^2) \, dx, \\ &\leqslant CR^{n-2+2\alpha}, \quad r = 1, \ldots, n-1. \end{aligned} \tag{13.37}$$

If $B_{2R} \cap \partial\Omega \neq \phi$, we require $\gamma = 0, \pm 1$ in (13.36). The estimate (13.37) then follows again since at each point of $D' \cap B_R$ at least one of the functions $w^{\pm} = \pm D_r u + v'$ is non-negative. Therefore we have the estimate

$$\int_{D^+ \cap B_R} |D_{ij} u|^2 \, dx \leqslant CR^{n-2+2\alpha} \tag{13.38}$$

for any $y \in D' \cap D^+$, $R \leqslant d/3$, provided $j \neq n$. By virtue of (13.34), the estimate (13.38) is also valid for $i = j = n$. Hence, by Theorem 7.19, we have

$$[Du]_{\alpha; D' \cap D^+} \leqslant C \tag{13.39}$$

where $C = C(n, K, \mu_k/\lambda_k, d)$ and $\alpha = \alpha(n, K, \mu_k/\lambda_k) > 0$. Finally, by returning to the original domain Ω and boundary values φ we obtain the fundamental global Hölder estimate of Ladyzhenskaya and Ural'tseva.

Theorem 13.7. *Let $u \in C^2(\bar{\Omega})$ satisfy $Qu=0$ in Ω where Q is elliptic in $\bar{\Omega}$ and the coefficients $a^{ij} \in C^1(\bar{\Omega} \times \mathbb{R} \times \mathbb{R}^n)$, $b \in C^0(\bar{\Omega} \times \mathbb{R} \times \mathbb{R}^n)$. Then if $\partial\Omega \in C^2$, $\varphi \in C^2(\bar{\Omega})$ with $u=\varphi$ on $\partial\Omega$, we have the estimate*

$$[Du]_{\alpha;\Omega} \leqslant C \tag{13.40}$$

where

$$C=C(n, K, \mu_K/\lambda_K, \Omega, \Phi),$$
$$K=|u|_{1;\Omega}, \Phi=|\varphi|_{2;\Omega} \quad \text{and} \quad \alpha=\alpha(n, K, \mu_K/\lambda_K, \Omega)>0.$$

An inspection of the proofs of Theorems 13.6 and 13.7 shows that the estimates (13.31) and (13.40) continue to hold if we only assume $u \in C^{0,1}(\bar{\Omega}) \cap W^{2,2}(\Omega)$ and $\varphi \in W^{2,q}(\Omega)$ for some $q>n$. In this generality we now must take $\Phi=\|\varphi\|_{W^{2,q}(\Omega)}$ and the constants C and α will depend on q.

Furthermore, it is evident from the development of this chapter that the global and interior Hölder estimates can be realized as special cases of a partially interior estimate. Namely, let us suppose that the operator Q satisfies the hypotheses of Theorem 13.7 and let T be a C^2 boundary portion of $\partial\Omega$ such that $u, \varphi \in C^2(\Omega \cup T)$ with $u=\varphi$ on T. Then, if $Qu=0$ in Ω, we have for any $\Omega' \subset\subset \Omega \cup T$ the estimate

$$[Du]_{\alpha;\Omega'} \leqslant C, \tag{13.41}$$

where

$$C=C(n, K, \mu_K/\lambda_K, T, \Phi, d),$$
$$d=\text{dist}(\Omega', \partial\Omega-T) \quad \text{and} \quad \alpha=\alpha(n, K, \mu_K/\lambda_K, T)>0.$$

13.5. Application to the Dirichlet Problem

By combining Theorems 11.4, 13.2, 13.4, 13.7, we obtain the following fundamental existence theorem.

Theorem 13.8. *Let Ω be a bounded domain in $\mathbb{R}^n$ and suppose that the operator Q is elliptic in $\bar{\Omega}$ with coefficients $a^{ij} \in C^1(\bar{\Omega} \times \mathbb{R} \times \mathbb{R}^n)$, $b \in C^\alpha(\bar{\Omega} \times \mathbb{R} \times \mathbb{R}^n)$, $0<\alpha<1$. Let $\partial\Omega \in C^{2,\alpha}$ and $\varphi \in C^{2,\alpha}(\bar{\Omega})$. Then, if there exists a constant M, independent of u and σ, such that every $C^{2,\alpha}(\bar{\Omega})$ solution of the Dirichlet problems,*

$$\begin{aligned} Q_\sigma u &= a^{ij}(x, u, Du)\, D_{ij}u+\sigma b(x, u, Du)=0 \text{ in } \Omega, \\ u &= \sigma\varphi \text{ on } \partial\Omega, \quad 0 \leqslant \sigma \leqslant 1, \end{aligned} \tag{13.42}$$

satisfies

$$(13.43) \qquad \|u\|_{C^1(\bar\Omega)} = \sup_\Omega |u| + \sup_\Omega |Du| < M,$$

it follows that the Dirichlet problem, $Qu=0$ *in* Ω, $u=\varphi$ *on* $\partial\Omega$ *is solvable in* $C^{2,\alpha}(\bar\Omega)$. *If* Q *is of divergence form or if* $n=2$ *we need only assume* $a^{ij} \in C^\alpha(\bar\Omega \times \mathbb{R} \times \mathbb{R}^n)$.

The solvability of the Dirichlet problem is reduced by Theorem 13.8 to the apriori estimation in the space $C^1(\bar\Omega)$ of a related family of problems. Provided the hypotheses of Theorem 13.8 hold, we therefore need only carry out the first three steps of the existence procedure described in Chapter 11, Section 3. Furthermore by invoking the more general Theorem 11.8 instead of Theorem 11.4, we see that the family of problems (13.42) can be replaced in the statement of Theorem 13.8 by any family of the form

$$(13.44) \qquad \begin{aligned} Q_\sigma u = a^{ij}(x, u, Du; \sigma)D_{ij}u + b(x, u, Du; \sigma) = 0 \text{ in } \Omega, \\ u = \sigma\varphi \text{ on } \partial\Omega, \quad 0 \leqslant \sigma \leqslant 1 \end{aligned}$$

where

(i) $Q_1 = Q$, $b(x, z, p; 0) = 0$.
(ii) the operators Q_σ are elliptic in $\bar\Omega$ for all $\sigma \in [0, 1]$,
(iii) the coefficients a^{ij}, b are sufficiently smooth; for example, $a^{ij}, b \in C^1(\bar\Omega \times \mathbb{R} \times \mathbb{R}^n \times [0, 1])$.

Notes

The Hölder estimates in Theorems 13.1, 13.2, 13.6 and 13.7 are due to Ladyzhenskaya and Ural'tseva, [LU 2, 3, 4]. Our derivation of Theorems 13.6 and 13.7, adapted from [TR 1], differs somewhat from theirs, although we retain their key idea of a reduction to a divergence structure inequality for the functions w in (13.33). Note that we may use the weak Harnack inequality, Theorem 9.22, in place of divergence structure results in the derivation of Theorem 13.6 (see [TR 13]). Divergence structure theory can also be avoided in Theorem 13.7 by means of Krylov's boundary gradient Hölder estimate, Theorem 9.31; (see Problem 13.1).

Problem 13.1. Using the interpolation inequality, Lemma 6.32, show that the interior estimates (13.5), (13.13), (13.31) can be cast more explicitly in the form

$$(13.45) \qquad [Du]_{\alpha;\Omega'} \leqq C(d^{-1-\alpha} \sup_{\Omega''} |u| + 1)$$

for any $\Omega' \subset\subset \Omega'' \subset \Omega$ where C and α depend on the same quantities as before and $d = \operatorname{dist}(\Omega', \partial\Omega'')$. By combining (13.45) with the boundary estimate (9.68) (analogously to the proof of Theorem 8.29) deduce the *global* estimates of (13.10), (13.14), (13.40) for solutions $u \in C^{0,1}(\bar\Omega) \cap C^2(\Omega)$. We remark that these global estimates may also be derived directly from the present forms of (13.5), (13.13), (13.31); (see [TR 14]).

Chapter 14

Boundary Gradient Estimates

An examination of the proof of Theorem 11.5 shows that for elliptic operators of the forms (11.7) or (11.8) the solvability of the classical Dirichlet problem with smooth data depends only upon the fulfillment of Step II of the existence procedure, that is, upon the existence of a boundary gradient estimate. In this chapter we provide a variety of hypotheses for the general equation,

$$(14.1)\qquad Qu=a^{ij}(x, u, Du)D_{ij}u+b(x, u, Du)=0$$

in $\Omega\subset\mathbb{R}^n$, that guarantee a boundary gradient estimate for solutions. These hypotheses are combinations of structural conditions on Q and geometric conditions on the domain Ω. It will be seen that the gradient bound aspect of the theory of quasilinear elliptic equations is not as profound as other aspects such as the Hölder estimates of Chapters 6 and 13. The boundary gradient estimates are tied through the classical maximum principle to judicious and generally natural choices of barrier functions. Nevertheless these estimates are of considerable importance since they seem to be the principal factor in determining the solvability character of the Dirichlet problem. This will be evidenced by the non-existence results at the end of the chapter.

A description of the *barrier method* to be employed below is appropriate at this juncture. This is a modification of ideas already met in Chapters 2 and 6. Let Q be an elliptic operator of the form,

$$(14.2)\qquad Qu=a^{ij}(x, Du)D_{ij}u+b(x, u, Du),$$

where $b(x, z, p)$ is non-increasing in z, and suppose that $u\in C^2(\Omega)\cap C^0(\bar{\Omega})$ satisfies $Qu=0$ in Ω. Suppose that, in some neighborhood $\mathcal{N}=\mathcal{N}_{x_0}$ of a point $x_0\in\partial\Omega$, there exist two functions $w^\pm=w^\pm_{x_0}\in C^2(\mathcal{N}\cap\Omega)\cap C^1(\mathcal{N}\cap\bar{\Omega})$ such that

(i) $\pm Qw^\pm<0$ in $\mathcal{N}\cap\Omega$
(ii) $w^\pm(x_0)=u(x_0)$
(iii) $w^-(x)\leqslant u(x)\leqslant w^+(x),\quad x\in\partial(\mathcal{N}\cap\Omega)$.

It then follows from the comparison principle, Theorem 10.1, applied to the domain $\mathcal{N}\cap\Omega$, that

$$w^-(x)\leqslant u(x)\leqslant w^+(x)\quad\text{for all } x\in\mathcal{N}\cap\Omega,$$

and hence by (ii)

$$\frac{w^-(x)-w^-(x_0)}{|x-x_0|}\leqslant\frac{u(x)-u(x_0)}{|x-x_0|}\leqslant\frac{w^+(x)-w^+(x_0)}{|x-x_0|}.$$

Consequently, provided they exist at x_0, the normal derivatives of $w^\pm$ and u satisfy

$$\frac{\partial w^-}{\partial \nu}(x_0)\leqslant\frac{\partial u}{\partial \nu}(x_0)\leqslant\frac{\partial w^+}{\partial \nu}(x_0). \tag{14.3}$$

We call the functions $w^\pm$ respectively *upper* and *lower barriers* at x_0 for the operator Q and function u. Their existence at all points $x_0 \in \partial\Omega$, with uniformly bounded gradients, implies the desired boundary gradient estimate for u satisfying $Qu=0$ in Ω.

Before setting about the construction of barriers, it is convenient to have certain transformation formulae at our disposal. First, let I be some interval in $\mathbb{R}$ and set $u=\psi(v)$ where $\psi \in C^2(I)$, $\psi'\neq 0$ on I. We then have, for $v(x)\in I$,

$$Qv=\psi' a^{ij}D_{ij}v+\frac{\psi''}{(\psi')^2}\mathscr{E}+b \tag{14.4}$$

where the arguments of a^{ij}, $\mathscr{E}$ and b are x, $u=\psi(v)$ and $Du=\psi' Dv$. Next, let us set $u=v+\varphi$ for some $\varphi\in C^2(\bar{\Omega})$. Then we have for $x\in\Omega$,

$$\tilde{Q}v=Qu=a^{ij}D_{ij}v+a^{ij}D_{ij}\varphi+b, \tag{14.5}$$

where the arguments of a^{ij} and b are x, $u=v+\varphi$ and $Du=Dv+D\varphi$. Defining the function $\mathscr{F}$ by

$$\mathscr{F}(x,z,p,q)=a^{ij}(x,z,p)(p_i-q_i)(p_j-q_j), \qquad (x,z,p,q)\in\Omega\times\mathbb{R}\times\mathbb{R}^n\times\mathbb{R}^n, \tag{14.6}$$

we see that for the transformed operator $\tilde{Q}$ in (14.5) we have

$$\tilde{\mathscr{E}}(x,v,Dv)=\mathscr{F}(x,u,Du,D\varphi). \tag{14.7}$$

Although the formula (14.5) has been implicitly used in Chapter 13 to effect a subtraction of boundary values, the explicit relation (14.7) is important for our purposes here. As will be evident from the considerations of this chapter, the formulae (14.4) and (14.5) foreshadow to some extent the nature of the structure conditions required for our barrier constructions. We shall assume throughout this chapter that the operator Q is elliptic in the domain Ω.

14.1. General Domains

We commence with a barrier construction that is applicable to arbitrary smooth domains. Suppose that Ω satisfies an exterior sphere condition at a point $x_0 \in \partial\Omega$ so that there exists a ball $B = B_R(y)$ with $x_0 \in \bar{B} \cap \bar{\Omega} = \bar{B} \cap \partial\Omega$. Let us define the distance function $d(x) = \text{dist}(x, \partial B)$ and set $w = \psi(d)$ where $\psi \in C^2[0, \infty)$ and $\psi' > 0$. By virtue of formula (14.4), we have for any $u \in C^1(\bar{\Omega}) \cap C^2(\Omega)$

$$\begin{aligned} \bar{Q}w &= a^{ij}(x, u(x), Dw)D_{ij}w + b(x, u(x), Dw) \\ &= \psi' a^{ij} D_{ij} d + b + \frac{\psi''}{(\psi')^2}\mathscr{E} \qquad (14.8) \\ &\leqslant \frac{(n-1)}{R}\psi'\Lambda + b + \frac{\psi''}{(\psi')^2}\mathscr{E}; \end{aligned}$$

the last inequality follows since

$$D_{ij}d = |x-y|^{-3}(|x-y|^2\,\delta_{ij} - (x_i - y_i)(x_j - y_j)).$$

We now suppose that Q satisfies a structure condition. Namely, we assume the existence of a non-decreasing function μ such that

$$|p|\Lambda + |b| \leqslant \mu(|z|)\mathscr{E} \qquad (14.9)$$

for all $(x, z, p) \in \Omega \times \mathbb{R} \times \mathbb{R}^n$ with $|p| \geqslant \mu(|z|)$. Using the condition (14.9) in (14.8), we obtain

$$\bar{Q}w \leqslant \left(\frac{\psi''}{(\psi')^2} + \nu\right)\mathscr{E} \qquad (14.10)$$

provided $\psi' \geqslant \mu = \mu(M)$, where $\nu = (1 + (n-1)/R)\mu$, $M = \sup_{\Omega} |u|$. Consider now the function ψ given by

$$\psi(d) = \frac{1}{\nu}\log(1 + kd), \quad k > 0, \qquad (14.11)$$

and the neighborhood $\mathscr{N} = \mathscr{N}_{x_0} = \{x \in \bar{\Omega} \mid d(x) < a\}$, $a > 0$. Clearly $\psi'' = -\nu(\psi')^2$ in $\mathscr{N}$. Furthermore,

$$\psi(a) = \frac{1}{\nu}\log(1 + ka) = M \quad \text{if } ka = e^{\nu M} - 1 \qquad (14.12)$$

and

$$\psi'(d)=\frac{k}{\nu(1+kd)}\geqslant\frac{k}{\nu(1+ka)} \quad \text{in } \mathcal{N}\cap\Omega \tag{14.13}$$
$$=\frac{k}{\nu\, e^{\nu M}}$$
$$\geqslant\mu \quad \text{if } k\geqslant\mu\nu\, e^{\nu M}.$$

Consequently, if k and a are chosen to satisfy the relations

$$k\geqslant\mu\nu\, e^{\nu M}, \qquad ka=e^{\nu M}-1, \tag{14.14}$$

the function $w^+=\psi(d)$ is an upper barrier at x_0 for the operator $\bar{Q}$ and the function u provided $u=0$ on $\mathcal{N}\cap\partial\Omega$. Similarly the function $w^-=-\psi(d)$ is a corresponding lower barrier. Hence if also $Qu=0$ in Ω, we obtain from (14.3) the estimate

$$|Du(x_0)|\leqslant\psi'(0)=\mu\, e^{\nu M} \quad \text{if equality holds in (14.14).} \tag{14.15}$$

Since the estimates to follow throughout this chapter are derived by essentially the same argument as above but with the surface ∂B replaced by other surfaces, it is worth illustrating the situation by the accompanying Figure 2.

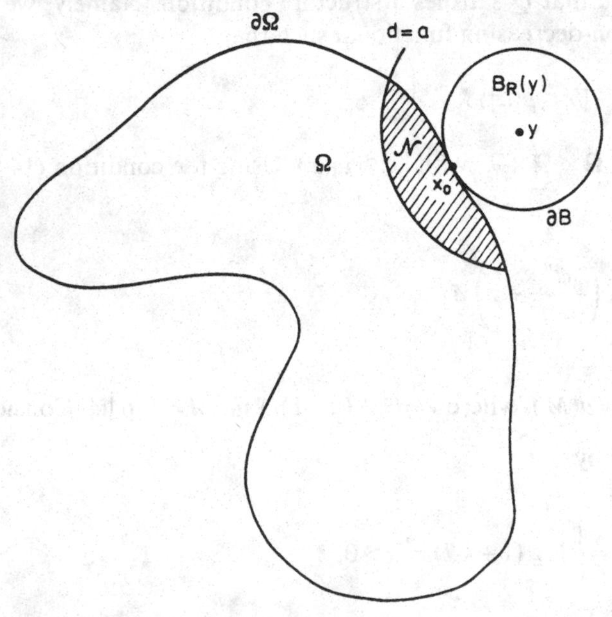

Figure 2

We now extend the estimate (14.15) to non-zero boundary values. Let $\varphi\in C^2(\bar{\Omega})$ and suppose that $u=\varphi$ on $\partial\Omega$. Then we require the transformed operator given by (14.5) to satisfy the structure condition (14.9). It therefore suffices that

$$(|p-D\varphi|+|D^2\varphi|)\Lambda+|b|\leqslant\bar{\mu}(|z|)\mathscr{F}(x, z, p, D\varphi) \tag{14.16}$$

for all $(x, z, p) \in \Omega \times \mathbb{R} \times \mathbb{R}^n$ such that $|p-D\varphi| \geqslant \bar{\mu}(|z|)$ for some non-decreasing function $\bar{\mu}$. Since

$$\begin{aligned} \mathscr{F}(x, z, p, q) &= a^{ij}(x, z, p)(p_i - q_i)(p_j - q_j) \\ &\geqslant \tfrac{1}{2}\mathscr{E} - a^{ij}q_iq_j \quad \text{by the Schwarz inequality}, \\ &\geqslant \tfrac{1}{2}\mathscr{E} - \Lambda|q|^2, \end{aligned} \tag{14.17}$$

we see that the structure condition (14.9) will imply (14.16) provided we choose

$$\bar{\mu} = 4(\mu(1+|\varphi|_1^2) + |\varphi|_2).$$

Consequently, the estimate (14.15) will hold with u replaced by $u-\varphi$ and μ replaced by $\bar{\mu}$. We can therefore assert the following boundary gradient estimate.

Theorem 14.1. *Let $u \in C^2(\Omega) \cap C^1(\bar{\Omega})$ satisfy $Qu=0$ in Ω and $u=\varphi$ on $\partial\Omega$. Suppose that Ω satisfies a uniform exterior sphere condition and $\varphi \in C^2(\bar{\Omega})$. Then if the structure condition* (14.9) *holds, we have*

$$|Du| \leqslant C \quad \text{on } \partial\Omega \tag{14.18}$$

where $C = C(n, M, \mu(M), \Phi, \delta)$, $M = \sup_\Omega |u|$, $\Phi = |\varphi|_{2;\Omega}$ and δ is the radius of the assumed exterior spheres.

It is often convenient to write condition (14.9) in the form

$$p\Lambda, b = O(\mathscr{E}) \quad \text{as } |p| \to \infty \tag{14.19}$$

where the limit behavior with respect to $|p|$ is understood to be uniform in $\Omega \times (-N, N)$ for any $N>0$. In particular, if Q is uniformly elliptic in $\Omega \times (-N, N) \times \mathbb{R}^n$ for any $N > 0$, that is $\Lambda = O(\lambda)$, and if also $b = O(\lambda|p|^2)$, then the structure condition (14.9) is fulfilled.

14.2 Convex Domains

We consider in this section barrier constructions that are applicable to convex and uniformly convex domains. Suppose that Ω satisfies an exterior plane condition at a point $x_0 \in \partial\Omega$, so that there exists a hyperplane $\mathscr{P}$ with $x_0 \in \mathscr{P} \cap \bar{\Omega} = \mathscr{P} \cap \partial\Omega$. Setting $d(x) = \text{dist}\,(x, \mathscr{P})$ and $w = \psi(d)$ we then obtain, for any $u \in C^1(\bar{\Omega}) \cap C^2(\Omega)$,

$$\begin{aligned} \bar{Q}w &= \psi' a^{ij} D_{ij} d + b + \frac{\psi''}{(\psi')^2}\mathscr{E} \\ &= b + \frac{\psi''}{(\psi')^2}\mathscr{E} \end{aligned} \tag{14.20}$$

Hence if $b = O(\mathscr{E})$, so that for some non-decreasing function μ we have

(14.21) $\quad |b| \leqslant \mu(|z|)\mathscr{E} \quad \text{for } |p| \geqslant \mu(|z|),$

the barrier argument of the preceding section is applicable with $v = \mu(M)$, $M = \sup_\Omega |u|$. Consequently we obtain an estimate for $Du(x_0)$ provided $Qu = 0$ in Ω and $u = 0$ on $\partial\Omega$. In order to extend this result to non-zero boundary values φ, we require that $\Lambda D^2\varphi$ and $b = O(\mathscr{F})$, that is,

(14.22) $\quad \Lambda|D^2\varphi| + |b| \leqslant \bar{\mu}(|z|)\mathscr{F} \quad \text{for } |p - D\varphi| \geqslant \bar{\mu}(|z|)$

for some non-decreasing function $\bar{\mu}$. We therefore have the following estimate.

Theorem 14.2. *Let $u, \varphi \in C^2(\Omega) \cap C^1(\bar{\Omega})$ satisfy $Qu = 0$ in Ω and $u = \varphi$ on $\partial\Omega$ and suppose that Ω is convex. Then if the structure condition* (14.22) *holds, we have*

(14.23) $\quad |Du| \leqslant C \quad \text{on } \partial\Omega$

where $C = C(n, M, \bar{\mu}(M), |\varphi|_{1;\Omega})$, $M = \sup_\Omega |u|$.

As in the preceding section, the structure condition (14.22) can be replaced in the hypotheses of Theorem 14.2 by conditions that are independent of the boundary values φ. In particular, either the condition

(14.24) $\quad \Lambda = o(\mathscr{E}),\ b = O(\mathscr{E}) \quad \text{as } |p| \to \infty,$

or the condition

(14.25) $\quad \Lambda, b = O(\lambda|p|^2) \quad \text{as } |p| \to \infty$

implies the validity of (14.22) for some function $\bar{\mu}$ depending on $|\varphi|_{2;\Omega}$. The first implication is a consequence of inequality (14.17); the second follows from the inequality

(14.26) $\quad \mathscr{F}(x, z, p, q) \geqslant \lambda(x, z, p)|p - q|^2, \quad (x, z, p, q) \in \Omega \times \mathbb{R} \times \mathbb{R}^n \times \mathbb{R}^n.$

We can therefore assert the following consequence of Theorem 14.2.

Corollary 14.3. *Let $u \in C^2(\Omega) \cap C^1(\bar{\Omega})$ satisfy $Qu = 0$ in Ω and $u = \varphi$ on $\partial\Omega$. Suppose that Ω is convex and $\varphi \in C^2(\bar{\Omega})$. Then, if either of the structure conditions* (14.24), (14.25) *hold, we have*

(14.27) $\quad |Du| \leqslant C \quad \text{on } \partial\Omega,$

where $C = C(n, M, \bar{\mu}, |\varphi|_{2;\Omega})$.

Corollary 14.3 is in particular applicable to the minimal surface operator $\mathfrak{M}$ given by

$$\mathfrak{M}u = (1+|Du|^2)\,\Delta u - D_iuD_juD_{ij}u. \tag{14.28}$$

Here $\lambda = 1$, $\Lambda = 1 + |p|^2$ (see Chapter 10, Examples (i) and (iii)), so that the boundary gradient estimate is valid in convex domains for the equation $\mathfrak{M}u = 0$. This result which is already sharp in two dimensions, will be improved for higher dimensions in the next section.

Let us next suppose that the domain Ω satisfies an *enclosing sphere condition* at a point $x_0 \in \partial\Omega$, that is there exists a ball $B = B_R(y) \supset \Omega$ with $x_0 \in \partial B$. Setting $d(x) = \text{dist}\,(x, \partial B)$ and $w = \psi(d)$, we thus obtain, for any $u \in C^1(\bar{\Omega}) \cap C^2(\Omega)$,

$$\begin{aligned} \bar{Q}w &= \psi' a^{ij}D_{ij}d + b + \frac{\psi''}{(\psi')^2}\mathscr{E} \\ &\leqslant -\frac{\psi'}{R}(\mathscr{T} - \mathscr{E}^*) + b + \frac{\psi''}{(\psi')^2}\mathscr{E}, \end{aligned} \tag{14.29}$$

where $\mathscr{T}(x, z, p) = \text{trace}\,[a^{ij}(x, z, p)] = a^{ii}(x, z, p)$ and $\mathscr{E}^* = \mathscr{E}/|p|^2$. The last relation follows since

$$D_{ij}d = -|x-y|^{-3}(|x-y|^2\delta_{ij} - (x_i - y_i)(x_j - y_j)).$$

A boundary gradient estimate can now be derived from (14.29) if b is bounded in terms of either $\mathscr{T}$ or $\mathscr{E}$. Indeed, let $\varphi \in C^2(\bar{\Omega})$ and assume there exists a non-decreasing function $\bar{\mu}$ such that

$$|a^{ij}D_{ij}\varphi + b| \leqslant \frac{1}{R}|p - D\varphi|\mathscr{T} + \bar{\mu}\mathscr{F} \quad \text{for } |p - D\varphi| \geqslant \bar{\mu}. \tag{14.30}$$

The domain Ω is said to be *uniformly convex* if it satisfies an enclosing sphere condition at each boundary point with a ball of fixed radius R. Our previous barrier constructions then yield the following estimate.

Theorem 14.4. *Let $u, \varphi \in C^2(\Omega) \cap C^1(\bar{\Omega})$ satisfy $Qu = 0$ in Ω and $u = \varphi$ on $\partial\Omega$. Suppose that Ω is uniformly convex. Then if the structure condition* (14.30) *holds, we have*

$$|Du| \leqslant C \quad \text{on } \partial\Omega \tag{14.31}$$

where $C = C(n, M, \bar{\mu}(M), R, |\varphi|_{1;\Omega})$.

It is clear that the structure condition (14.30) with $\bar{\mu}$ depending on $|\varphi|_{2;\Omega}$ is implied by either the condition

$$b = o(\Lambda|p|) + O(\lambda|p|^2) \quad \text{as } |p| \to \infty \tag{14.32}$$

or the conditions

$$\Lambda = O(\lambda|p|^2),\ |b| \leqslant \frac{|p|}{R}\mathscr{T} + O(\lambda|p|^2) \quad \text{as } |p| \to \infty. \tag{14.33}$$

Therefore we have the following consequence of Theorem 14.4.

Corollary 14.5. *Let $u \in C^2(\Omega) \cap C^1(\bar{\Omega})$ satisfy $Qu=0$ in Ω and $u=\varphi$ on $\partial\Omega$. Suppose that Ω is uniformly convex and $\varphi \in C^2(\bar{\Omega})$. Then, if either of the structure conditions* (14.32) *or* (14.33) *hold, we have*

$$|Du| \leqslant C \quad \text{on } \partial\Omega \tag{14.34}$$

where $C=C(n, M, \bar{\mu}(M), |\varphi|_2, R)$.

Corollary 14.5 is in particular applicable to the prescribed mean curvature equation

$$\mathfrak{M}u = nH(x, u, Du)(1+|Du|^2)^{3/2}. \tag{14.35}$$

Here $\mathscr{T} = 1+(n-1)(1+|p|^2)$ so that a boundary gradient estimate will hold for solutions of (14.35) in uniformly convex domains provided the function H satisfies

$$|H| \leqslant \frac{(n-1)}{nR} \quad \text{for } |p| \geqslant \mu(|z|). \tag{14.36}$$

This result will be improved for dimensions higher than two in the next section.

The structure condition (14.32) is obviously satisfied when $b=0$. In this case Corollary 14.5 can be derived by means of linear barrier functions, since the boundary manifold $(\partial\Omega, \varphi)$ will satisfy a bounded slope condition. Moreover when $b = o(\Lambda|p|)$ in (14.32) in Corollary 14.5, a barrier of the form $w = kd, d = \text{dist}(x, \partial B)$ for sufficient large k will suffice for the proof. It is also clear from the above proofs that the results of this section will continue to be valid if the structure conditions assumed in their hypotheses only hold for x in some neighborhood of $\partial\Omega$.

Remark. With slight modifications the considerations of this and the preceding section can be combined, so that a boundary gradient estimate is determined only by the *local* behavior of the boundary $\partial\Omega$. It is convenient here to formulate such a result for C^2 domains. Let $\kappa=\kappa(x_0)$ be the minimum of the principal curvatures of $\partial\Omega$ at x_0 and let $\nu=\nu(x_0)$ denote the inner normal to $\partial\Omega$ at x_0. Suppose there exists a non-decreasing function $\bar{\mu}$ such that for each $x_0 \in \partial\Omega$ there is an $\varepsilon>0$ for which

$$|a^{ij}D_{ij}\varphi + b| < \kappa\mathscr{T}|p| + \bar{\mu}(|z|)\mathscr{F} \quad \forall (x, z, p) \in \Omega \times \mathbb{R} \times \mathbb{R}^n \tag{14.37}$$

with

$$|x-x_0|<\varepsilon, \quad \left|\frac{p-D\varphi}{|p-D\varphi|}\pm\nu\right|<\varepsilon \quad \text{and} \quad |p-D\varphi|\geqslant\bar{\mu}.$$

Then we have the estimate

$$(14.38) \qquad |Du|\leqslant C \quad \text{on } \partial\Omega$$

where $C=C(n, M, \bar{\mu}(M), x)$. If $\kappa(x_0)\geqslant\kappa_0=\text{constant}$, for all $x_0\in\partial\Omega$ we can take the non-strict inequality in (14.37) provided κ is replaced by κ_0. As an illustration of the possible type of behavior encompassed here, let us consider the equation

$$(14.39) \qquad Qu=D_{11}u+(1+|D_2u|^N)D_{22}u=0, \quad N\geqslant 0.$$

A boundary gradient estimate will then hold for a convex domain Ω and arbitrary $\varphi\in C^2(\bar{\Omega})$, provided $\kappa(x_0)>0$ whenever $\nu(x_0)\neq(\pm 1, 0)$, that is provided the curvature of $\partial\Omega$ is positive except possibly when the tangent line is parallel to the x_1 axis.

14.3. Boundary Curvature Conditions

So far in this chapter we have constructed barriers in terms of the distance function of an exterior surface of constant curvature (plane or sphere), the curvature of the latter being the significant factor in determining the structure condition imposed on the operator Q. By allowing more general exterior surfaces, our previous convexity conditions can be considerably relaxed in more than two dimensions. In the following we shall assume that $\partial\Omega\in C^2$ and use the boundary $\partial\Omega$ itself as an appropriate exterior surface. Setting $d(x) = \text{dist}(x, \partial\Omega)$ we see from Lemma 14.16 that $d\in C^2(\Gamma)$ where $\Gamma=\{x\in\bar{\Omega} \mid d(x)<d_0\}$ for some $d_0>0$. Therefore if $w=\psi(d)$ where $\psi\in C^2[0, \infty)$ and $\psi'>0$ we have by formula (14.4), for any $u\in C^1(\bar{\Omega})\cap C^2(\Omega)$,

$$(14.40) \qquad \begin{aligned} \bar{Q}w &= a^{ij}(x, u(x), Dw)D_{ij}w+b(x, u(x), Dw) \\ &= \psi' a^{ij}D_{ij}d+b+\frac{\psi''}{(\psi')^2}\,\mathscr{E}. \end{aligned}$$

As a preliminary illustration of the general theory of this section, let us consider the special case of the minimal surface operator $\mathfrak{M}$; that is, we take

$$(14.41) \qquad a^{ij}(x, z, p)=(1+|p|^2)\delta_{ij}-p_ip_j.$$

We then have

$$(14.42) \qquad \begin{aligned} a^{ij}D_{ij}d &= (1+|\psi'|^2)\Delta d-|\psi'|^2 D_idD_jdD_{ij}d \\ &= (1+|\psi'|^2)\Delta d \quad \text{since } |Dd|=1,\ D_idD_{ij}d=0 \\ &\leqslant -(n-1)(1+|\psi'|^2)H' \quad \text{by Lemma 14.17,} \end{aligned}$$

where H' is the mean curvature of $\partial\Omega$ at the point $y=y(x)$ on $\partial\Omega$ closest to x. Hence, if $\partial\Omega$ has non-negative *mean curvature* everywhere, we obtain

$$\bar{Q}w \leqslant b + \frac{\psi''}{|\psi'|^2}\mathscr{E}$$

as in the convex case of the previous section. A boundary gradient estimate then follows as in the preceding section for arbitrary $C^2(\bar{\Omega})$ boundary values if $b=O(|p|^2)$. We shall show in the next section that this result is sharp for the minimal surface equation $\mathfrak{M}u=0$. Using relations (14.40) and (14.42) we can also conclude a corresponding sharp result for the equation of prescribed mean curvature (14.35). However, let us first return to the general situation. We assume that the coefficients of Q are decomposed in such a way that for $p\neq 0$ we have

$$\begin{aligned} a^{ij} &= \Lambda a^{ij}_\infty + a^{ij}_0, \quad i,j=1,\ldots,n, \\ b &= |p|\Lambda b_\infty + b_0 \end{aligned} \tag{14.43}$$

where

$$a^{ij}_\infty(x,z,p)=a^{ij}_\infty(x,p/|p|), \qquad b_\infty(x,z,p)=b_\infty(x,z,p/|p|),$$
$$a^{ij}_\infty(x,\sigma)\xi_i\xi_j \geqslant 0 \quad \text{for all } x\in\Omega,\ |\sigma|=1,\ \xi\in\mathbb{R}^n$$

and b_∞ is non-increasing in z. For example, in the case of the minimal surface operator $\mathfrak{M}$ we can take

$$a^{ij}_\infty = \delta_{ij} - p_ip_j/|p|^2, \qquad a^{ij}_0 = p_ip_j/|p|^2.$$

Using the matrix $[a^{ij}_\infty]$, we introduce a generalized notion of mean curvature as follows. Namely, let y be a point of $\partial\Omega$ and let ν denote the unit inner normal to $\partial\Omega$ at y, $\kappa_1,\ldots,\kappa_{n-1}$ the principal curvatures of $\partial\Omega$ at y and $a_1,\ldots,a_n$ the diagonal elements of the matrix $[a^{ij}_\infty]$ with respect to a corresponding principal coordinate system at y. We then define

$$\mathscr{H}^\pm(y) = \sum_{i=1}^{n-1} a_i(y,\pm\nu)\kappa_i. \tag{14.44}$$

Since $a_i \geqslant 0$, $i=1,\ldots,n$, the quantities $\mathscr{H}^\pm$ are a weighted average of the curvatures of $\partial\Omega$ at y. Furthermore, in the special case of the minimal surface operator $\mathfrak{M}$, we have $a_i=1$, $i=1,\ldots,n-1$, $a_n=0$ and hence

$$\mathscr{H}^+(y)=\mathscr{H}^-(y)=\sum_{i=1}^{n-1}\kappa_i=(n-1)H'(y)$$

where $H'(y)$ denotes the mean curvature of $\partial\Omega$ at y. Through Lemma 14.17 we see that the curvatures $\mathscr{K}^{\pm}$ are connected with the distance function d by the formula

$$\mathscr{K}^{\pm}(y) = -a_{\infty}^{ij}(y, \pm Dd(y))D_{ij}d(y). \tag{14.45}$$

In order to generalize our earlier result for the minimal surface equation let us suppose that the inequality

$$\mathscr{K}^{+} \geqslant b_{\infty}(y, u, v) \tag{14.46}$$

holds at each point $y \in \partial\Omega$. In addition we assume that the functions a_{∞}^{ij}, $b_{\infty} \in C^1(\Gamma \times \mathbb{R} \times \mathbb{R}^n)$ and that Q satisfies the structure condition

$$\Lambda, |p| a_0^{ij}, b_0 = O(\mathscr{E}) \quad \text{as } |p| \to \infty, i, j = 1, \ldots, n, \tag{14.47}$$

so that for some non-decreasing function μ we have

$$\Lambda + |p| \sum |a_0^{ij}| + |b_0| \leqslant \mu(|z|)\mathscr{E} \quad \text{for } |p| \geqslant \mu(|z|). \tag{14.48}$$

As previously we shall assume initially that the function u vanishes on $\partial\Omega$. Then taking w as before we have

$$\begin{aligned}\bar{\bar{Q}}w &= a^{ij}(x, u(x), Dw)D_{ij}w + |Dw|\Lambda(x, u(x), Dw)b_{\infty}(x, w, Dw) \\ &\quad + b_0(x, u(x), Dw) \\ &= \psi'\Lambda(a_{\infty}^{ij}D_{ij}d + b_{\infty}) + \psi' a_0^{ij}D_{ij}d + b_0 + \frac{\psi''}{|\psi'|^2}\mathscr{E}\end{aligned}$$

by (14.4) and (14.43), where

$$\begin{aligned}a_{\infty}^{ij}D_{ij}d + b_{\infty} &= a_{\infty}^{ij}(x, v)D_{ij}d(x) + b_{\infty}(x, w, v) \\ &\leqslant a_{\infty}^{ij}(x, v)D_{ij}d(y) + b_{\infty}(x, 0, v) \quad \text{by Lemma 14.17} \\ &\leqslant (a_{\infty}^{ij}(x, v) - a_{\infty}^{ij}(y, v))D_{ij}d(y) + b_{\infty}(x, 0, v) - b_{\infty}(y, 0, v) \\ &\quad \text{by (14.45) and (14.46)} \\ &\leqslant Kd.\end{aligned}$$

Here $y = y(x)$ is the point on $\partial\Omega$ closest to x, $v = Dd(y) = Dd(x)$ is the inner unit normal to $\partial\Omega$ at y and the constant K is given by

$$K = \sup_{x \in \Gamma} \frac{|(a_{\infty}^{ij}(x, v) - a_{\infty}^{ij}(y, v))D_{ij}d(y) + b_{\infty}(x, u(y), v) - b_{\infty}(y, u(y), v)|}{|x - y|}. \tag{14.49}$$

Hence by (14.48) we obtain

$$\bar{\bar{Q}}w \leqslant K\psi' d\mu\mathscr{E} + (1+\sup_{\Gamma}|D^2 d|)\mu\mathscr{E} + \frac{\psi''}{|\psi'|^2}\mathscr{E}$$
$$\leqslant \left(\nu + \frac{\psi''}{|\psi'|^2}\right)\mathscr{E}$$

provided $\psi' d \leqslant 1$ and $\psi' \geqslant \mu$ where $\nu = (K+1+\sup_{\Gamma}|D^2 d|)\mu$, $\mu = \mu(M)$ and $M = \sup_{\Gamma}|u|$. Consequently by choosing ψ by formula (14.11) and taking k large enough to guarantee $a \leqslant d_0$, the function $w^+ = w$ will be an upper barrier at every point of $\partial\Omega$ for the operator $\bar{\bar{Q}}$ and the function u. A lower barrier is similarly constructed provided instead of (14.46) the inequality

$$\mathscr{K}^- \geqslant -b_\infty(y, u(y), -\nu) \tag{14.46)'$$

holds at each point $y \in \partial\Omega$. Hence, if both (14.46) and (14.46)′ hold and $Qu = 0$ in Ω, then u will satisfy the estimate (14.3) at each $x_0 \in \partial\Omega$. In order to extend these results to non-zero boundary values φ, we require that Λ, $|p|a_0^{ij}$ and $b_0 = O(\mathscr{F})$, that is,

$$\Lambda + |p|\sum|a_0^{ij}| + |b_0| \leqslant \bar{\mu}(|z|)\mathscr{F} \quad \text{for } |p - D\varphi| \geqslant \bar{\mu}(|z|) \tag{14.50}$$

for some non-decreasing function $\bar{\mu}$. For the transformed operator $\tilde{Q}$ given by (14.5) we can take $\tilde{a}_\infty^{ij} = a_\infty^{ij}$, $\tilde{b}_\infty(x, z+\varphi, p) = b_\infty(x, z, p)$ so that conditions (14.46) and (14.46)′ are unchanged. We therefore have the following estimate.

Theorem 14.6. *Let $u \in C^2(\Omega) \cap C^1(\bar{\Omega})$, $\varphi \in C^2(\bar{\Omega})$ satisfy $Qu = 0$ in Ω and $u = \varphi$ on $\partial\Omega$. Suppose that Q satisfies the structure conditions* (14.43), (14.50) *and that at each point $y \in \partial\Omega$ the inequalities*

$$\mathscr{K}^+ \geqslant b_\infty(y, \varphi(y), \nu), \qquad \mathscr{K}^- \geqslant -b_\infty(y, \varphi(y), -\nu) \tag{14.51}$$

hold. Then we have the estimate

$$|Du| \leqslant C \quad \text{on } \partial\Omega \tag{14.52}$$

where $C = C(n, M, \bar{\mu}(M), \Omega, K, |\varphi|_{2;\Omega})$, $M = \sup_{\Omega}|u|$ and K is given by (14.49) *with u replaced by φ.*

As in the case of convex Ω treated in Section 14.2, the structure condition (14.50) can be replaced in the hypotheses of the above theorem by conditions that are independent of the boundary values φ. In particular either the condition

$$\Lambda = o(\mathscr{E}), \quad |p|a_0^{ij}, b_0 = O(\mathscr{E}) \quad \text{as } |p| \to \infty, \tag{14.53}$$

or the condition

$$(14.54)\qquad \Lambda, |p|a_0^{ij}, b_0 = O(\lambda|p|^2) \quad \text{as } |p| \to \infty,$$

implies the validity of (14.50) for some function $\bar{\mu}$ depending on $|\varphi|_{2;\Omega}$. We can therefore assert the following consequence of Theorem 14.6.

Corollary 14.7. *Let $u \in C^2(\Omega) \cap C^1(\bar{\Omega})$, $\varphi \in C^2(\bar{\Omega})$ satisfy $Qu=0$ in Ω and $u=\varphi$ on $\partial\Omega$. Suppose that, in addition to (14.43), either of the structure conditions (14.53), (14.54) hold and that the inequalities (14.51) are satisfied on $\partial\Omega$. Then we have the estimate*

$$(14.55)\qquad |Du| \leqslant C \quad \text{on } \partial\Omega,$$

where $C=C(n, M, \bar{\mu}, \Omega, K, |\varphi|_{2;\Omega})$.

The application of Corollary 14.7 to the equation of prescribed mean curvature,

$$(14.56)\qquad \mathfrak{M}u = nH(x,u)(1+|Du|^2)^{3/2},$$

where $H \in C^1(\bar{\Omega} \times \mathbb{R})$ and $D_zH \geqslant 0$, now yields the following result.

Corollary 14.8. *Let u be a $C^2(\Omega) \cap C^1(\bar{\Omega})$ solution of equation (14.35) in Ω with $u=\varphi$ on $\partial\Omega$ where $\varphi \in C^2(\bar{\Omega})$. Suppose that the mean curvature H' of $\partial\Omega$ is such that*

$$(14.57)\qquad H'(y) \geqslant \frac{n}{n-1}|H(y, \varphi(y))| \quad \forall y \in \partial\Omega.$$

Then we have the estimate

$$(14.58)\qquad |Du| \leqslant C \quad \text{on } \partial\Omega,$$

where $C=C(n, M, \Omega, H_1, |\varphi|_{2;\Omega})$, $M=\sup_\Omega |u|$, $H_1 = \sup_{\Omega\times(-M,M)} (|H|+|DH|)$.

The sharpness of Corollary 14.8 will be demonstrated in the next section. Note that the result could have been derived more directly by proceeding along the lines of our earlier treatment of the minimal surface equation, $\mathfrak{M}u=0$.

So far the results in this section have necessitated some control over the behavior of the maximum eigenvalue Λ with respect to $\mathscr{F}$, $\mathscr{E}$ or λ. We move on now to consider a situation where such controls are not imposed but where, to compensate, the inequalities (14.51) must hold in the strict sense everywhere on the boundary $\partial\Omega$. The forerunner of this situation is the case in Corollary 14.5 where the structure condition (14.32) holds. To proceed further we assume that in the decomposition (14.43) the coefficients a_∞^{ij}, b_∞ are continuous on $\partial\Omega \times \mathbb{R} \times \mathbb{R}^n$ and that the coefficients

$$(14.59)\qquad a_0^{ij} = o(\Lambda), b_0 = o(|p|\Lambda) \quad \text{as } |p| \to \infty.$$

Let $u \in C^0(\bar{\Omega}) \cap C^2(\Omega)$ satisfy $Qu=0$ in Ω and $u=\varphi$ on $\partial\Omega$ where $\varphi \in C^2(\bar{\Omega})$ and suppose that the strict inequalities

$$\text{(14.60)} \qquad \mathscr{K}^+ > b_\infty(y, \varphi(y), \nu), \qquad \mathscr{K}^- > -b_\infty(y, \varphi(y), -\nu)$$

hold everywhere on $\partial\Omega$. Assume initially $\varphi \equiv 0$ so that setting $w=kd$ for some positive constant k, we have

$$\begin{aligned}\bar{Q}w &= a^{ij}(x, u(x), Dw)D_{ij}w + |Dw|\Lambda(x, u(x), Dw)b_\infty(x, w, Dw)\\ &\quad + b_0(x, u(x), Dw)\\ &= k\Lambda(a^{ij}_\infty(x, Dd)D_{ij}d + b_\infty(x, w, Dd)) + ka^{ij}_0 D_{ij}d + b_0\\ &= \Lambda\{k(a^{ij}_\infty(x, Dd)D_{ij}d + b_\infty(x, w, Dd)) + o(k)\}\end{aligned}$$

by (14.59). Now by Lemma 14.16, the relation (14.45) and the fact that b_∞ is non-increasing in z, there exist positive constants χ and $a \leqslant d_0$ such that

$$a^{ij}_\infty(x, Dd)D_{ij}d + b_\infty(x, w, Dd) \leqslant -\chi$$

in the neighborhood $\mathscr{N} = \{x \in \bar{\Omega} \mid d(x) < a\}$. Hence we obtain

$$\bar{Q}w \leqslant \Lambda(-k\chi + o(k)) < 0 \quad \text{for sufficiently large } k.$$

Consequently, by choosing k large enough so that also $ka \geqslant \sup_\Omega |u|$, the function $w^+ = w$ will be an upper barrier at every point of $\partial\Omega$ for the operator $\bar{Q}$ and the function u. Similarly the function $w^- = -w$ will be a corresponding lower barrier and we obtain $|Du| \leqslant k$ on $\partial\Omega$ for $u \in C^1(\bar{\Omega})$. This result extends automatically to non-zero boundary values φ by replacement of u by $u-\varphi$ and use of (14.5). Thus we have proved the following estimate.

Theorem 14.9. *Let $u \in C^2(\Omega) \cap C^1(\bar{\Omega})$, $\varphi \in C^2(\bar{\Omega})$ satisfy $Qu=0$ in Ω and $u=\varphi$ on $\partial\Omega$. Suppose that Q satisfies the structure conditions* (14.43), (14.59) *and that the inequalities* (14.60) *hold at each point $y \in \partial\Omega$. Then we have the estimate*

$$\text{(14.61)} \qquad |Du| \leqslant C \quad \text{on } \partial\Omega$$

where $C = C(n, M, a^{ij}_\infty, a^{ij}_0, b_\infty, b_0, \Omega, |\varphi|_{2;\Omega})$ and $M = \sup_\Omega |u|$.

The dependence of the constant C in Theorem 14.9 on the coefficients a^{ij}_0, b_0 arises through the structure conditions (14.59), and the dependence on the coefficients a^{ij}_∞, b_∞ through their moduli of continuity on $\partial\Omega \times \mathbb{R} \times \mathbb{R}^n$.

To conclude this section, let us note the relationship of the results here to those of the preceding sections. If we choose $a^{ij}_\infty = b_\infty = 0$ in the decomposition (14.43), then Theorem 14.6 reduces to Theorem 14.1 for $\partial\Omega \in C^2$. Next although the results in Section 14.2 on convex domains are not special cases of Theorems 14.6 and 14.9 they are encompassed by slight variants of these theorems which are treated in Problems 14.2 and 14.3.

14.4. Non-Existence Results

We present here some non-existence results which show that many of the conditions in the theorems of the preceding sections cannot be significantly relaxed. Since the class of equations considered in this section include equations for which Steps I and III of the existence procedure, described in Chapter 11, are readily established, it follows that the non-solvability of the Dirichlet problem for these cases is due to the lack of a boundary gradient estimate. Indeed the non-existence of such an estimate for these equations can be demonstrated directly by techniques analogous to those used below.

The main tool for our treatment of non-existence results is the following variant of the comparison principle, Theorem 10.1.

Theorem 14.10. *Let Ω be a bounded domain in $\mathbb{R}^n$ and Γ a relatively open C^1 portion of $\partial\Omega$. Then if Q is an elliptic operator of the form (14.2) and $u \in C^0(\bar{\Omega}) \cap C^2(\Omega \cup \Gamma)$, $v \in C^0(\bar{\Omega}) \cap C^2(\Omega)$ satisfy $Qu > Qv$ in Ω, $u \leqslant v$ on $\partial\Omega - \Gamma$, $\partial v/\partial \nu = -\infty$ on Γ, it follows that $u \leqslant v$ in Ω.*

Proof. By Theorem 10.1, we have

$$\sup_{\Omega}(u-v) \leqslant \sup_{\Gamma}(u-v)^+.$$

Since

$$\frac{\partial}{\partial \nu}(u-v) = \frac{\partial u}{\partial \nu} - \frac{\partial v}{\partial \nu} = \infty$$

on Γ, the function $u-v$ cannot achieve a maximum value on Γ. Hence $u \leqslant v$ in Ω. □

In order to apply Theorem 14.10, we let $y \in \partial\Omega$ be fixed, $\delta = \operatorname{diam} \Omega, 0 < a < \delta/2$, and consider the function w defined by

$$w(x) = m + \psi(r), \quad r = |x-y|,$$

where $m \in \mathbb{R}$ and $\psi \in C^2(a, \delta)$ is such that $\psi(\delta) = 0$, $\psi' \leqslant 0$, $\psi'(a) = -\infty$. Using (14.8), we obtain for $u \in C^2(\Omega)$, $r > a$,

$$\bar{Q}w = a^{ij}(x, u(x), Dw)D_{ij}w + b(x, u(x), Dw)$$

$$= \frac{\psi'}{r}(\mathcal{T} - \mathcal{E}^*) + \frac{\psi''}{(\psi')^2}\mathcal{E} + b, \tag{14.62}$$

the arguments of $\mathcal{T} = \operatorname{trace}[a^{ij}]$, $\mathcal{E}^* = \mathcal{E}/|p|^2$, $\mathcal{E}$ and b being x, $u(x)$ and Dw. We now wish to choose the function ψ in such a way that $\bar{Q}w < 0$ in the domain $\tilde{\Omega} = \{x \in \Omega | r > a, |u(x)| > M\}$ for some constant M. If this is done and $Qu = 0$ in Ω we then have by Theorem 14.10

$$\sup_{\Omega - B_a(y)} u \leqslant M + m + \psi(a) \tag{14.63}$$

where

$$m = \sup_{\partial\Omega - B_a(y)} u^+.$$

The estimate (14.63) can be regarded as a preliminary stage in the establishment of non-existence results.

We shall consider two different cases.

(i) First, let us suppose that

$$b \leqslant -|p|^\theta \mathscr{E} \tag{14.64}$$

for $x \in \Omega$, $|z| \geqslant M$, $|p| \geqslant L$ where M, L and θ are positive constants. Then setting

$$\psi(r) = K[(\delta - a)^\beta - (r-a)^\beta] \tag{14.65}$$

where $\beta = \theta/(1+\theta)$ and $K \in \mathbb{R}$ we obtain, for sufficiently large K, $\bar{Q}w < 0$ in $\tilde{\Omega}$. Hence the estimate (14.63) holds in this case.

(ii) Next, let us suppose that

$$\begin{aligned} &b \leqslant 0, \\ &\mathscr{E} \leqslant \mu(\mathscr{T} - \mathscr{E}^*)|p|^{1-\theta} \end{aligned} \tag{14.66}$$

for $x \in \Omega$, $|z| \geqslant M$, $|p| \geqslant 0$, where μ, θ and M are positive constants. Then we have, in $\tilde{\Omega}$,

$$\bar{Q}w \leqslant \left(-\frac{|\psi'|^\theta}{\mu r} + \frac{\psi''}{(\psi')^2}\right)\mathscr{E} < 0$$

for the specific choice

$$\psi(r) = \left(\frac{\mu}{\beta}\right)^\beta \int_r^\delta \left(\log\frac{t}{a}\right)^{-\beta} dt \tag{14.67}$$

where $\beta = 1/(1+\theta)$. Hence again the estimate (14.63) follows. Note that the function ψ given by (14.67) satisfies

$$\lim_{a \to 0} \psi(a) = 0.$$

Now let us assume that the domain Ω satisfies an interior sphere condition at the point y so that there exists a ball $B = B_R(x_0) \subset \Omega$ with $y \in \bar{B} \cap \partial\Omega$. We then consider the function w^* defined by

$$w^*(x)=m^*+\chi(r), \quad r=|x-x_0| \tag{14.68}$$

where $m^*\in\mathbb{R}$ and $\chi\in C^2(0,\ R-\varepsilon)$, $0<\varepsilon<R$, is such that $\chi(0)=0$, $\chi'\geqslant 0$ and $\chi'(R-\varepsilon)=\infty$. By (14.62) we have, for $r>0$,

$$\bar{Q}w^*=\frac{\chi'}{r}(\mathscr{T}-\mathscr{E}^*)+\frac{\chi''}{(\chi')^2}\mathscr{E}+b. \tag{14.69}$$

Let us impose now a stronger condition than (14.64), namely that

$$b+\frac{|p|}{R'}\mathscr{T}\leqslant -|p|^\theta\,\mathscr{E}, \quad 0<R'<R, \tag{14.70}$$

for $x\in\Omega$, $|z|\geqslant M$, $|p|\geqslant L$. Then setting

$$\chi(r)=K\{(R-\varepsilon)^\beta-(R-\varepsilon-r)^\beta\}, \quad \beta=\frac{\theta}{1+\theta}. \tag{14.71}$$

we obtain, for sufficiently large K, $\bar{Q}w^*<0$ in the domain $\mathring{\Omega}=\{x\in\Omega\,|\,|x-y|<R,\ R'<|x-x_0|<R-\varepsilon, |u(x)|>M\}$. Hence by Theorem 14.10 we have

$$\sup_{\mathring{\Omega}} u\leqslant M+m^*+\chi(R-\varepsilon) \tag{14.72}$$

where

$$m^*=\sup_{|x-y|=R} u.$$

Combining the estimates (14.72) and (14.63) with $a=R-R'$, and letting ε tend to zero, we therefore obtain the estimate

$$u(y)\leqslant 2M+m+K[(\delta-R)^\beta+R^\beta] \tag{14.73}$$

where K depends on θ and L and

$$m=\sup_{\partial\Omega-B_R(y)} u.$$

The estimate (14.73) shows that the boundary values of the function u cannot be specified arbitrarily on $\partial\Omega$. Since the above argument can be repeated with u replaced by $-u$, the structure condition (14.70) can be relaxed to

$$|b|\geqslant\frac{|p|}{R'}\mathscr{T}+|p|^\theta\,\mathscr{E} \tag{14.74}$$

for $x \in \Omega$, $|z| \geqslant M$, $|p| \geqslant L$. We have thus proved the following non-existence result.

Theorem 14.11. *Let Ω be a bounded domain in $\mathbb{R}^n$ and suppose that the operator Q satisfies the structure condition (14.74) where R is the radius of the largest ball contained in Ω. Then there exists a function $\varphi \in C^\infty(\bar{\Omega})$ such that the Dirichlet problem $Qu = 0$ in Ω, $u = \varphi$ on $\partial\Omega$ is not solvable.*

Theorem 14.11 implies that both Theorems 14.1 and 14.4 are sharp in the sense that the quantities $\mathscr{E}$, $\mathscr{F}$ in the structure conditions (14.9), (14.30) cannot be replaced by $|p|^\theta \mathscr{E}$, $|p|^\theta \mathscr{F}$ for some $\theta > 0$, (even if the operator Q is the Laplacian!). Furthermore, in Corollary 14.5, we can neither replace (14.32) by the condition, $b = O(\Lambda |p|)$ as $|p| \to \infty$, nor in the second inequality in (14.33), replace $\lambda |p|^2$ by $\lambda |p|^{2+\theta}$ for some $\theta > 0$.

We shall use case (ii) above to demonstrate the need for the geometric restrictions in Section 14.3. Let us assume that the decomposition (14.43) is valid with b_∞ independent of z and a^{ij}_∞, b_∞ continuous on $\partial\Omega \times \mathbb{R} \times \mathbb{R}^n$ and satisfying the structure conditions (14.59). In addition we impose on the operator Q the structure conditions

$$\text{(14.75)} \qquad \begin{aligned} &b \leqslant 0 \\ &\mathscr{E} \leqslant \mu \Lambda |p|^{1-\theta} \end{aligned}$$

for $x \in \Omega$, $|z| \geqslant M$, $|p| \geqslant 0$, where μ, θ and M are positive constants. It is easy to show that the conditions (14.43), (14.59), (14.75) imply that condition (14.66) holds for $x \in \Omega$, $M \leqslant |z| \leqslant \bar{M}$, $p \in \mathbb{R}^n$ for some constant $\bar{M}$ and a possibly different constant μ than that in (14.75). Moreover, since (14.75) also implies, by the classical maximum principle, Theorem 3.1, that

$$\text{(14.76)} \qquad \sup_\Omega u \leqslant M + \sup_{\partial\Omega} u,$$

we can assume below that (14.66) is applicable in $\Omega^+ = \{x \in \Omega \mid u(x) > 0\}$ and that also in Ω^+ the quantities in (14.59) are bounded in terms of $\sup_{\partial\Omega} u$. Let us now suppose that $\partial\Omega \in C^2$ and that

$$\text{(14.77)} \qquad \mathscr{K}^-(y) < -b_\infty(y, -\nu(y))$$

where $\mathscr{K}^-$ is given by (14.44) and ν denotes the unit inner normal to $\partial\Omega$ at y. The argument which follows is analogous to the proof of Theorem 14.11, the interior sphere at y being replaced by another quadric surface. In fact, condition (14.77) implies that for sufficiently small a, there exists a quadric surface $\mathscr{S}$ tangent to $\partial\Omega$ at y such that (i) $\mathscr{S}$ has a unique parallel projection onto the tangent plane at y, (ii) $\mathscr{S} \cap B_a(y) \subset \bar{\Omega}$ and (iii) the curvature $\mathscr{K}^-$ corresponding to $\mathscr{S}$, $\mathscr{K}^-_{\mathscr{S}}$, satisfies

$$\text{(14.78)} \qquad \mathscr{K}^-_{\mathscr{S}}(y) \leqslant \mathscr{K}^-(y) + \eta$$

for some $\eta>0$. We now consider the function w^* defined by

$$w^*(x)=m^*+\chi(d), \quad d=\operatorname{dist}(x, \mathscr{S}), \tag{14.79}$$

where $m^* \in \mathbb{R}$ and $\chi \in C^2(\varepsilon, a)$, $0<\varepsilon<a$, is such that $\chi(2a)=0$, $\chi' \leqslant 0$, $\chi'(\varepsilon)=-\infty$. By (14.40) we have, in the domain $\tilde{\Omega}=\{x \in \Omega \,|\, |x-y|<a, \varepsilon<d<a, u(x)>M\}$,

$$\begin{aligned}
\bar{Q}w^* &= \chi' \Lambda (a^{ij}_{\infty} D_{ij} d + b_{\infty}) + \chi' a^{ij}_0 D_{ij} d + b_0 + \frac{\chi''}{(\chi')^2} \mathscr{E} \\
&\leqslant \chi' \Lambda (\eta + o(1)) + \frac{\chi''}{(\chi')^2} \mathscr{E} \quad \text{by (14.59) and (14.78)}, \\
&\leqslant \chi' \Lambda (\eta + o(1) + \mu \chi'' |\chi'|^{-2-\theta}) \quad \text{by (14.75)}.
\end{aligned}$$

Setting

$$\chi(d)=K[(2a-\varepsilon)^{\beta}-(d-\varepsilon)^{\beta}], \tag{14.80}$$

where $\beta=\theta/(1+\theta)$, we then obtain, for sufficiently large K,

$$\bar{Q}w^*<0$$

in $\tilde{\Omega}$. Hence by Theorem 14.10, we have

$$\begin{aligned}
\sup_{\tilde{\Omega}} u &\leqslant M+m^*+\chi(\varepsilon) \\
&\leqslant M+m^*+K(2a)^{\beta}
\end{aligned} \tag{14.81}$$

where

$$m^*=\sup_{|x-y|=a} u.$$

Combining the estimates (14.81), (14.63) and letting ε tend to zero, we therefore obtain the estimate

$$u(y) \leqslant 2M+m+\psi(a)+K(2a)^{\beta} \tag{14.82}$$

where ψ is given by (14.67) and

$$m=\sup_{\partial\Omega - B_a(y)} u.$$

Again, the estimate (14.82) shows that u cannot be prescribed arbitrarily on $\partial\Omega$. The sharpness of Theorem 14.6 is thus evidenced by the following non-existence result.

Theorem 14.12. *Let Ω be a bounded C^2 domain in $\mathbb{R}^n$ and suppose that the operator Q satisfies the structure conditions* (14.43), (14.59) *and* (14.75). *Then if the inequality*

$$\mathscr{K}^-(y) < -b_\infty(y, -\nu(y)) \tag{14.83}$$

holds at some point $y \in \partial\Omega$, there exists a function $\varphi \in C^\infty(\bar{\Omega})$ such that the Dirichlet problem, $Qu = 0$ in Ω, $u = \varphi$ on $\partial\Omega$, is not solvable.

If we replace u by $-u$ in the above considerations, we obtain a similar conclusion with the inequality (14.83) replaced by

$$\mathscr{K}^+(y) < b_\infty(y, \nu(y)), \tag{14.84}$$

provided b is replaced by $-b$ in the structure conditions (14.75). Also if $M = 0$ in (14.76) we can choose $\sup_{\partial\Omega} |\varphi|$ to be arbitrarily small. Specializing to the equation of prescribed mean curvature (14.56) with $H(x, z) \equiv H(x)$, we therefore have

Corollary 14.13. *Let Ω be a bounded C^2 domain in $\mathbb{R}^n$ and suppose that at some point $y \in \partial\Omega$ the mean curvature H' of $\partial\Omega$ is such that*

$$H'(y) < \frac{n}{n-1} |H(y)| \tag{14.85}$$

where $H \in C^0(\bar{\Omega})$ is either non-positive or non-negative in Ω. Then, for any $\varepsilon > 0$, there exists a function $\varphi \in C^\infty(\bar{\Omega})$ with $\sup |\varphi| \leqslant \varepsilon$, such that the Dirichlet problem, $Qu = 0$ in Ω, $u = \varphi$ on $\partial\Omega$, is not solvable.

Combining Corollaries 14.8 and 14.13 with our treatment of the existence procedure for the special case (11.7) in Section 11.3, we obtain the following sharp criterion of Jenkins and Serrin for solvability of the minimal surface equation [JS 2].

Theorem 14.14. *Let Ω be a bounded $C^{2,\gamma}$ domain in $\mathbb{R}^n$, $0 < \gamma < 1$. Then the Dirichlet problem $\mathfrak{M}u = 0$ in Ω, $u = \varphi$ on $\partial\Omega$, is solvable for arbitrary $\varphi \in C^{2,\gamma}(\bar{\Omega})$ if and only if the mean curvature H' of $\partial\Omega$ is non-negative at every point of $\partial\Omega$.*

The prescribed mean curvature equation will be studied further in Chapter 16. We note here that the restriction that b_∞ is independent of z in Theorem 14.12 can be removed and consequently that condition (14.85) in Corollary 14.13 can be replaced by

$$H'(y) < \sup_{z \in \mathbb{R}} \frac{n}{n-1} |H(y, z)|; \tag{14.86}$$

(see Problem 14.5). It is also worth comparing the results of this section with the existence theorems of Section 15.5 in the next chapter.

14.5. Continuity Estimates

The barrier constructions of Sections 14.1, 14.2 and 14.3 can be adapted to provide boundary modulus of continuity estimates for $C^0(\bar{\Omega}) \cap C^2(\Omega)$ solutions of equation (14.1). In particular, we observe that if in the hypotheses of any of the gradient estimates of these sections, we only assume that $u \in C^0(\bar{\Omega}) \cap C^2(\Omega)$ then we obtain, in place of an estimate for $\sup_{\partial\Omega} |Du|$, a bound for the quantity

$$\sup_{\substack{x\in\Omega \\ y\in\partial\Omega}} \frac{|u(x)-u(y)|}{|x-y|}.$$

Furthermore a boundary modulus of continuity estimate still results when we only assume that the boundary values $\varphi \in C^0(\partial\Omega)$. To show this we fix a point y on $\partial\Omega$ and for arbitrary $\varepsilon>0$, choose $\delta>0$ such that $|\varphi(x)-\varphi(y)|<\varepsilon$ whenever $|x-y|<\delta$. We then define functions $\varphi^\pm \in C^2(\bar{\Omega})$ by

$$\varphi^\pm(x)=\varphi(y)\pm\left(\varepsilon+\frac{2\sup|\varphi|}{\delta^2}|x-y|^2\right). \tag{14.87}$$

Clearly, on $\partial\Omega$ we have

$$\varphi^- \leqslant \varphi \leqslant \varphi^+$$

Hence if the operator Q and the functions $\varphi^\pm$ satisfy any of the conditions of the estimates derived in Sections 14.1, 14.2 and 14.3, we obtain an estimate

$$\varphi^- + w^- \leqslant u \leqslant \varphi^+ + w^+ \tag{14.88}$$

in $\mathcal{N} \cap \Omega$ where $w^\pm$ are the relevant barrier functions and $\mathcal{N}$ is some neighborhood of y. Since in all our previous barrier constructions we had $w^+=-w^-=w$, it follows from (14.88) that

$$|u(x)-\varphi(y)| \leqslant \varepsilon + w(x) + \frac{2\sup|\varphi|}{\delta^2}|x-y|^2 \tag{14.89}$$

in $\mathcal{N} \cap \Omega$. We can therefore assert the following continuity estimates.

Theorem 14.15. *Let $u, \varphi \in C^0(\bar{\Omega}) \cap C^2(\Omega)$ satisfy $Qu=0$ in Ω and $u=\varphi$ on $\partial\Omega$. Suppose that the operator Q and domain Ω satisfy the structural and geometric conditions of either Theorem* 14.1, *Corollary* 14.3, *Corollary* 14.5, *Corollary* 14.7 *or Theorem* 14.9. *Then the modulus of continuity of u on $\partial\Omega$ can be estimated in terms of the modulus of continuity of φ on $\partial\Omega$, $\sup_{\partial\Omega}|\varphi|$, $\sup_{\Omega}|u|$, Ω and the coefficients of Q.*

14.6. Appendix: Boundary Curvatures and the Distance Function

Let Ω be a domain in $\mathbb{R}^n$ having non-empty boundary $\partial\Omega$. The *distance function* d is defined by

$$d(x) = \text{dist}\,(x, \partial\Omega). \tag{14.90}$$

It is readily shown that d is uniformly Lipschitz continuous. For let $x, y \in \mathbb{R}^n$ and choose $z \in \partial\Omega$ such that $|y-z| = d(y)$. Then

$$d(x) \leqslant |x-z| \leqslant |x-y| + d(y)$$

so that by interchanging x and y we have

$$|d(x) - d(y)| \leqslant |x-y|. \tag{14.91}$$

Now let $\partial\Omega \in C^2$. For $y \in \partial\Omega$, let $\nu(y)$, and $T(y)$ denote respectively the unit *inner* normal to $\partial\Omega$ at y and the tangent hyperplane to $\partial\Omega$ at y. The curvatures of $\partial\Omega$ at a fixed point $y_0 \in \partial\Omega$ are determined as follows. By a rotation of coordinates we can assume that the x_n coordinate axis lies in the direction $\nu(y_0)$. In some neighborhood $\mathcal{N} = \mathcal{N}(y_0)$ of y_0, $\partial\Omega$ is then given by $x_n = \varphi(x')$ where $x' = (x_1, \ldots, x_{n-1})$, $\varphi \in C^2(T(y_0) \cap \mathcal{N})$ and $D\varphi(y_0') = 0$. The curvature of $\partial\Omega$ at y_0 is then described by the orthogonal invariants of the Hessian matrix $[D^2\varphi]$ evaluated at y_0'. The eigenvalues of $[D^2\varphi(y_0')]$, $\kappa_1, \ldots, \kappa_{n-1}$, are called the *principal curvatures* of $\partial\Omega$ at y_0 and the corresponding eigenvectors are called the *principal directions* of $\partial\Omega$ at y_0. The *mean curvature of* $\partial\Omega$ at y_0 is given by

$$H(y_0) = \frac{1}{n-1} \sum \kappa_i = \frac{1}{n-1}\, \Delta\varphi(y_0'). \tag{14.92}$$

By a further rotation of coordinates we can assume that the $x_1, \ldots, x_{n-1}$ axes lie along principal directions corresponding to $\kappa_1, \ldots, \kappa_{n-1}$ at y_0. Let us call such a coordinate system a *principal coordinate system* at y_0. The Hessian matrix $[D^2\varphi(y_0')]$ with respect to the principal coordinate system at y_0 described above is given by

$$[D^2\varphi(y_0')] = \text{diag}\,[\kappa_1, \ldots, \kappa_{n-1}]. \tag{14.93}$$

The unit inner normal vector $\bar{\nu}(y') = \nu(y)$ at the point $y = (y', \varphi(y')) \in \mathcal{N} \cap \partial\Omega$ is given by

$$\nu_i(y) = \frac{-D_i\varphi(y')}{\sqrt{1+|D\varphi(y')|^2}}, \quad i = 1, \ldots, n-1, \quad \nu_n(y) = \frac{1}{\sqrt{1+|D\varphi(y')|^2}}. \tag{14.94}$$

Hence with respect to the principal coordinate system at y_0, we have

$$D_j\bar{\nu}_i(y_0') = -\kappa_i\delta_{ij}, \quad i, j = 1, \ldots, n-1. \tag{14.95}$$

For $\mu>0$, let us set $\Gamma_\mu=\{x\in\bar{\Omega} \mid d(x)<\mu\}$. The following lemma relates the smoothness of the distance function d in Γ_μ to that of the boundary $\partial\Omega$.

Lemma 14.16. *Let Ω be bounded and $\partial\Omega\in C^k$ for $k\geqslant 2$. Then there exists a positive constant μ depending on Ω such that $d\in C^k(\Gamma_\mu)$.*

Proof. The conditions on Ω imply that $\partial\Omega$ satisfies a *uniform* interior sphere condition; that is, at each point $y_0\in\partial\Omega$ there exists a ball B depending on y_0 such that $\bar{B}\cap(\mathbb{R}^n-\Omega)=y_0$ and the radii of the balls B are bounded from below by a positive constant, which we take to be μ. It is easy to show that μ^{-1} bounds the principal curvatures of $\partial\Omega$. Also, for each point $x\in\Gamma_\mu$, there will exist a unique point $y=y(x)\in\partial\Omega$ such that $|x-y|=d(x)$. The points x and y are related by

$$x=y+\nu(y)\,d. \tag{14.96}$$

We show that this equation determines y and d as C^{k-1} functions of x. For a fixed point $x_0\in\Gamma_\mu$, let $y_0=y(x_0)$ and choose a principal coordinate system at y_0. We define a mapping $\mathbf{g}=(g^1,\dots,g^n)$ from $\mathscr{U}=(T(y_0)\cap\mathscr{N}(y_0))\times\mathbb{R}$ into $\mathbb{R}^n$ by

$$\mathbf{g}(y',d)=y+\nu(y)d,\quad y=(y',\varphi(y')).$$

Clearly $\mathbf{g}\in C^{k-1}(\mathscr{U})$, and the Jacobian matrix of $\mathbf{g}$ at $(y_0',d(x))$ is given by

$$[D\mathbf{g}]=\operatorname{diag}[1-\kappa_1 d,\dots,1-\kappa_{n-1}d,1]. \tag{14.97}$$

Since the Jacobian of $\mathbf{g}$ at $(y_0',d(x_0))$ is given by

$$\det[D\mathbf{g}]=(1-\kappa_1 d(x_0))\cdots(1-\kappa_{n-1}d(x_0))>0\quad\text{since } d(x_0)<\mu, \tag{14.98}$$

it follows from the inverse mapping theorem that for some neighborhood $\mathscr{M}=\mathscr{M}(x_0)$, the mapping y' is contained in $C^{k-1}(\mathscr{M})$. From (14.96) we have $Dd(x)=\nu(y(x))=\nu(y'(x))\in C^{k-1}(\mathscr{M})$ for $x\in\mathscr{M}$. Hence $d\in C^k(\mathscr{M})$, and thus $d\in C^k(\Gamma_\mu)$. □

An expression for the Hessian matrix of d at points close to $\partial\Omega$ is an immediate consequence of the proof of Lemma 14.16.

Lemma 14.17. *Let Ω and μ satisfy the conditions of Lemma 14.16 and let $x_0\in\Gamma_\mu$, $y_0\in\partial\Omega$ be such that $|x_0-y_0|=d(x_0)$. Then, in terms of a principal coordinate system at y_0, we have*

$$[D^2d(x_0)]=\operatorname{diag}\left[\frac{-\kappa_1}{1-\kappa_1 d},\dots,\frac{-\kappa_{n-1}}{1-\kappa_{n-1}d},0\right]. \tag{14.99}$$

Proof. Since

$$Dd(x_0)=\nu(y_0)=(0,0,\dots,1)$$

we have $D_{in}d(x_0) = 0$, $i = 1, \ldots, n$. To obtain the other derivatives we write, for $i, j = 1, \ldots, n-1$,

$$\begin{aligned} D_{ij}d(x_0) &= D_j \nu_i \circ y(x_0) \\ &= D_k \bar{\nu}_i(y_0) D_j y_k(x_0) \\ &= \begin{cases} \dfrac{-\kappa_i}{1-\kappa_i d} & \text{if } i=j \\ 0 & \text{if } i \neq j \end{cases} \end{aligned}$$

by (14.95) and (14.97). □

Note that the result of Lemma 14.17 is equivalent to the geometrically evident statement that the circles of principal curvature to $\partial\Omega$ at y_0 and to the parallel surface through x_0 at x_0 are concentric.

Mean Curvature

Let us derive now a formula for the mean curvature of a C^2 hypersurface $\mathfrak{S}$ in terms of its given representation. Let $y_0 \in \mathfrak{S}$ and suppose that in a neighborhood $\mathcal{N}$ of y_0, $\mathfrak{S}$ is given by $\psi(x) = 0$ where $\psi \in C^2(\mathcal{N})$ and $|D\psi| > 0$ in $\mathcal{N}$. The unit normal to $\mathfrak{S}$ at a point $y \in \mathfrak{S} \cap \mathcal{N}$ (directed towards positive ψ) is given by

$$\nu = \frac{D\psi}{|D\psi|}. \tag{14.100}$$

Let $\kappa_1, \ldots, \kappa_{n-1}$ be the principal curvatures of $\mathfrak{S}$ at y_0. Then with respect to a corresponding principal coordinate system at y_0, one can show that

$$\begin{aligned} D_i\left(\frac{D_j\psi}{|D\psi|}\right) &= -\kappa_i \delta_{ij}, \quad i, j = 1, \ldots, n-1, \\ D_i\left(\frac{D_n\psi}{|D\psi|}\right) &= 0, \qquad i = 1, \ldots, n. \end{aligned} \tag{14.101}$$

Consequently the eigenvalues of the matrix $[D_j(D_i\psi/|D\psi|)]$ at y_0 with respect to the original coordinates are $-\kappa_1, \ldots, -\kappa_{n-1}, 0$ and hence the mean curvature of $\mathfrak{S}$ at y_0 is given by

$$H(y_0) = -\frac{1}{n-1}\left(D_i\left[\frac{D_i\psi}{|D\psi|}\right]\right)_{x=y_0}. \tag{14.102}$$

In particular if $\mathfrak{S}$ is the graph in $\mathbb{R}^{n+1}$ of a function of n variables $u \in C^2(\Omega)$, that is, $\mathfrak{S}$ is defined by $x_{n+1} = u(x_1, \ldots, x_n)$, the mean curvature of $\mathfrak{S}$ at $x_0 \in \Omega$ is given by

$$(14.103)\qquad H(x_0)=\frac{1}{n}\left(D_i\left[\frac{D_iu}{\sqrt{1+|Du|^2}}\right]\right)_{x=x_0}.$$

$\mathfrak{S}$ is called a *minimal surface* if $H(x_0)=0$ for all $x_0 \in \Omega$.

Note that we also obtain from (14.101) the following formula for the sum of the squares of the principal curvatures $\kappa_1, \ldots, \kappa_n$ at x_0,

$$(14.104)\qquad \mathscr{C}^2=\sum_{i=1}^{n}\kappa_i^2(x_0)=\sum_{i,j=1}^{n}D_iv_jD_jv_i(x_0),$$

where

$$v_i=\frac{D_iu}{\sqrt{1+|Du|^2}},\quad i=1,\ldots,n.$$

Finally, the Gauss curvature of $\mathfrak{S}$ at x_0 is given by

$$\begin{aligned}(14.105)\qquad K(x_0)&=\prod_{i=1}^{n}\kappa_i\\&=\det\,[D_iv_j]\\&=\frac{\det D^2u}{(1+|Du|^2)^{(n+2)/2}}.\end{aligned}$$

Notes

Many of the basic features of our treatment of boundary gradient estimates are already present in the early work of Bernstein for equations in two variables, [BE 1, 2, 3, 4, 5, 6]. Indeed, Bernstein employed auxiliary functions such as the function ψ in (14.4) for similar purposes and also considered the question of non-existence. Bernstein's work in two variables was continued by Leray [LR], who also considered the relationship between the solvability of the Dirichlet problem in a domain Ω and the geometric nature of the boundary $\partial\Omega$. Finn showed that convexity was a necessary and sufficient condition for the solvability of the Dirichlet problem for the minimal surface equation in two variables [FN 2, 4].

The first striking results for equations in more than two variables were proved by Jenkins and Serrin [JS 2], (see Theorem 14.14), and Bakel'man [BA 5, 6]. A general theory embracing many interesting examples was finally developed by Serrin [SE 3] to whom the results of Sections 14.3 and 14.4 are due. We have adopted some minor simplifications from [TR 5] in the exposition of Section 14.3.

Problems

14.1. Suppose that in the hypotheses of Theorem 14.6 we have a_∞^{ij}, b_∞ independent of x and $\varphi \equiv 0$. Show that the estimate (14.52) continues to hold when the restriction on Λ in (14.50) is removed.

14.2. Suppose that only the coefficient b of Q is decomposed according to (14.43). Show that the result of Theorem 14.6 continues to hold provided we take $a_0^{ij}=0$ in (14.50) and replace the inequalities (14.51) by

$$\kappa(y) \geqslant \pm b_\infty(y, \varphi(y), \pm \nu) \tag{14.106}$$

where $\kappa(y)$ is the minimum principal curvature of $\partial\Omega$ at y.

14.3. Again, suppose that only the coefficient b of Q is decomposed according to (14.43). Show that the result of Theorem 14.9 continues to hold provided we take $a_0^{ij}=0$ in (14.59) and replace (14.60) by

$$\kappa(y) > \pm b_\infty(y, \varphi(y), \pm \nu). \tag{14.107}$$

14.4. Show that the results of Problems 14.2 and 14.3 can be sharpened by replacing the maximum eigenvalue Λ by the trace $\mathcal{T}$ and compare these results with Theorem 14.5.

14.5. Derive the assertion at the end of Section 14.4 (see [SE 3]).

Chapter 15

Global and Interior Gradient Bounds

In this chapter we are mainly concerned with the derivation of apriori estimates for the gradients of $C^2(\Omega)$ solutions of quasilinear elliptic equations of the form

$$Qu = a^{ij}(x, u, Du)D_{ij}u + b(x, u, Du) = 0 \tag{15.1}$$

in terms of the gradients on the boundary $\partial\Omega$ and the magnitudes of the solutions. The resulting estimates facilitate the establishment of Step III of the existence procedure described in Section 11.3. On combination with the estimates of Chapters 10, 13 and 14, they yield existence theorems for large classes of quasilinear elliptic equations including both uniformly elliptic equations and equations of form similar to the prescribed mean curvature equation (10.7). Since the methods of this chapter involve the differentiation of equation (15.1), our hypotheses will generally require structural conditions to be satisfied by the derivatives of the coefficients a^{ij}, b. In Section 15.4 we shall see that these derivative conditions can be relaxed somewhat for equations in divergence form, where different types of arguments are appropriate.

We shall also consider in this chapter the derivation of apriori interior gradient estimates. Such estimates lead to existence theorems for Dirichlet problems where only continuous boundary values are assigned. Interior gradient bounds for equations of mean curvature type will be treated in Chapter 16.

15.1. A Maximum Principle for the Gradient

We commence with a gradient bound under relatively simple hypotheses which will also serve as an illustration of the general technique to be applied in the following section. Let us suppose that the principal coefficients of the operator Q can be written as follows:

$$a^{ij}(x, z, p) = a^{ij}_*(p) + \tfrac{1}{2}(p_i c_j(x, z, p) + c_i(x, z, p)p_j) \tag{15.2}$$

where $a^{ij}_* \in C^1(\mathbb{R}^n)$, $c_i \in C^1(\Omega \times \mathbb{R} \times \mathbb{R}^n)$, $i, j = 1, \ldots, n$, and the matrix $[a^{ij}_*]$ is non-negative. The following are examples of such decompositions:

(i) If Q is elliptic with principal coefficients a^{ij} depending only on p, then clearly we can take $a_*^{ij}=a^{ij}$ and $c_i=0$;

(ii) The Euler–Lagrange equation corresponding to a multiple integral of the form

$$\int_\Omega F(x, u, |Du|)\, dx, \tag{15.3}$$

where $F\in C^2(\Omega\times\mathbb{R}\times\mathbb{R})$, $D_tF\neq 0$, $t=|p|$, can be written as

$$\begin{aligned}\Delta u+\{(|Du|D_{tt}F/DF)-1\}|Du|^{-2}D_iuD_juD_{ij}u\\ +(|Du|^2D_{tz}F+D_iuD_{tx_i}F-|Du|D_zF)/D_tF=0,\end{aligned} \tag{15.4}$$

so that the decomposition (15.2) is valid with

$$a_*^{ij}=\delta^{ij},\quad c_i=\{(|p|D_{tt}F/D_tF)-1\}|p|^{-2}p_i$$

provided $c_i\in C^1(\Omega\times\mathbb{R}\times\mathbb{R}^n)$, $i=1,\ldots,n$.

(iii) In the special case of two variables we can write

$$a^{ij}=(\mathscr{T}-\mathscr{E}^*)\delta^{ij}+\tfrac{1}{2}(p_id_j+d_ip_j) \tag{15.5}$$

where

$$\begin{aligned}&\mathscr{T}=\text{trace}\,[a^{ij}]=a^{11}+a^{22}\\ &\mathscr{E}^*=\mathscr{E}/|p|^2=a^{ij}p_ip_j/|p|^2,\\ &d_1=[(a^{11}-a^{22})p_1+2a^{12}p_2]/|p|^2,\\ &d_2=[2a^{12}p_1+(a^{22}-a^{11})p_2]/|p|^2.\end{aligned}$$

Hence for $n=2$, the equation (15.1) is equivalent to the equation

$$\Delta u+c_iD_juD_{ij}u+b^*=0 \tag{15.6}$$

where $c_i=d_i/(\mathscr{T}-\mathscr{E}^*)$, $b^*=b/(\mathscr{T}-\mathscr{E}^*)$; the decomposition (15.2) is clearly valid for equation (15.6) with $a_*^{ij}=\delta^{ij}$.

The basic idea in our treatment of gradient bounds, which goes back as far as Bernstein's work [BE 1], involves differentiation of equation (15.1) with respect to x_k, $k=1,\ldots,n$, followed by multiplication by D_ku and summation over k. The maximum principle is then applied to the resulting equation in the function $v=|Du|^2$. By (15.2) we can write equation (15.1) in the form

$$a_*^{ij}(Du)D_{ij}u+\tfrac{1}{2}c_i(x,u,Du)D_iv+b(x,u,Du)=0.$$

Assuming that the solution $u \in C^3(\Omega)$ we then obtain, by differentiation with respect to x_k,

$$a_*^{ij}D_{ijk}u + D_{p_l}a_*^{ij}D_{lk}uD_{ij}u + \tfrac{1}{2}(c_iD_{ik}v + D_kc_iD_iv) + D_{p_l}bD_{lk}u + D_kuD_zb + D_{x_k}b = 0.$$

Multiplying by D_ku and summing over k, we thus have

$$-2\,a_*^{ij}D_{ik}uD_{jk}u + a^{ij}D_{ij}v + (D_{p_i}a_*^{lj}D_{lj}u + D_{p_i}b + D_kuD_kc_i)D_iv + 2\,\delta bv = 0 \tag{15.7}$$

where δ is the differential operator, acting on $C^1(\Omega \times \mathbb{R} \times \mathbb{R}^n)$, defined by

$$\delta = D_z + |p|^{-2}p_iD_{x_i}. \tag{15.8}$$

We next apply the following consequence of Schwarz's inequality

$$a_*^{ij}D_{ik}uD_{jk}u \geqslant (a_*^{ij}D_{ij}u)^2/\text{trace }[a_*^{ij}] = (b + \tfrac{1}{2}c_iD_iv)^2/\text{trace }[a_*^{ij}], \tag{15.9}$$

and hence obtain

$$a^{ij}D_{ij}v + B_iD_iv \geqslant 2\left(\frac{b^2}{\mathcal{T}_*} - v\,\delta b\right) \tag{15.10}$$

where $\mathcal{T}_* = \text{trace }[a_*^{ij}]$ and

$$B_i = D_{p_i}a_*^{lj}D_{lj}u + D_{p_i}b + D_kuD_kc_i - 2\mathcal{T}_*^{-1}bc_i - \tfrac{1}{2}\mathcal{T}_*^{-1}c_ic_jD_jv.$$

Consequently, if the inequality

$$|p|^2\delta b \leqslant \frac{b^2}{\mathcal{T}_*} \tag{15.11}$$

holds in $\Omega \times \mathbb{R} \times \mathbb{R}^n$, we obtain immediately from the classical maximum principle, Theorem 3.1, that $\sup_\Omega v = \sup_{\partial\Omega} v$. In order to extend this result to $C^2(\Omega)$ solutions we write equation (15.7) in the divergence form

$$-2a_*^{ij}D_{ik}uD_{jk}u + D_i(a^{ij}D_jv) + (D_{p_i}a_*^{lj}D_{lj}u - D_ja^{ij} + D_kuD_kc_i)D_iv + 2D_k(bD_ku) - 2\,\Delta ub = 0,$$

which has the corresponding integral form

$$\begin{aligned} 2\int_\Omega \eta a_*^{ij}D_{ik}uD_{jk}u\,dx + \int_\Omega (a^{ij}D_jv + 2bD_iu)D_i\eta\,dx \\ - \int_\Omega \{(D_{p_i}a_*^{lj}D_{lj}u - D_ja^{ij} + D_kuD_kc_i)D_iv\}\eta\,dx + \int_\Omega 2b\eta\,\Delta u\,dx = 0 \end{aligned} \tag{15.12}$$

for all $\eta \in C_0^1(\Omega)$ (see equation 13.17). An approximation argument as in Section 13.3 shows that equation (15.12) continues to be valid for $u \in C^2(\Omega)$. Integrating the term $\int_\Omega 2b\eta\, \Delta u\, dx$ by parts, and then proceeding as above for the case of $C^3(\Omega)$ solutions we obtain the weak inequality

$$D_i(a^{ij}D_jv)+(B_i-D_ja^{ij})D_iv \geqslant 2(b^2/\mathcal{T}_*-v\,\delta b) \tag{15.13}$$

for $v \in C^1(\bar\Omega)$ and the estimate $\sup_\Omega v=\sup_{\partial\Omega} v$ follows from the maximum principle, Theorem 8.1.

We have therefore proved the following gradient maximum principle.

Theorem 15.1. *Let $u \in C^2(\Omega) \cap C^1(\bar\Omega)$ satisfy equation* (15.1) *in the bounded domain Ω and suppose that Q is elliptic in Ω with coefficients satisfying* (15.2) *and* (15.11). *Then we have the estimate*

$$\sup_\Omega |Du| = \sup_{\partial\Omega} |Du| \tag{15.14}$$

It is clear from the above proof that conditions (15.2) and (15.11) in the hypotheses of Theorem 15.1 need only hold for $|z| \leqslant \sup_\Omega |u|$. Furthermore, if they also hold for $|p| \geqslant L$ for some constant L, we obtain in place of (15.14) the estimate

$$\sup_\Omega |Du| \leqslant \max\,\{\sup_{\partial\Omega} |Du|,\, L\}. \tag{15.15}$$

The estimate (15.15) follows by applying Theorem 15.1 in the domain $\Omega_L=\{x \in \Omega \,|\, |Du|>L\}$.

Note that when Theorem 15.1 is applied to examples (ii) and (iii) above where $a_*^{ij}=\delta^{ij}$ the condition (15.11) reduces to

$$|p|^2\,\delta b \leqslant \frac{b^2}{n} \tag{15.16}$$

In particular if $b=0$, we obtain a gradient maximum principle.

15.2. The General Case

The technique to be employed in this section is basically a modification of that in the preceding section through the use of an auxiliary function. Assuming initially that u is a $C^2(\Omega)$ solution of equation (15.1), we set $m=\inf_\Omega u$, $M=\sup_\Omega u$ and let ψ be a strictly increasing function in $C^3[\bar m, \bar M]$ with $m=\psi(\bar m)$, $M=\psi(\bar M)$. Defining the function $\bar{u}$ by $u = \psi(\bar{u})$, so that

$$D_i u = \psi' D_i \bar{u}$$
$$D_{ij} u = \psi'' D_i \bar{u} D_j \bar{u} + \psi' D_{ij} \bar{u},$$

we have the equation (see also (14.4)),

$$\psi' a^{ij}(x, u, Du) D_{ij}\bar{u} + b(x, u, Du) + \frac{\psi''}{(\psi')^2}\mathscr{E}(x, u, Du) = 0. \tag{15.17}$$

Let us now write $v = |Du|^2$, $\bar{v} = |D\bar{u}|^2$ and apply the operator $D_k u D_k$ to the equation (15.17). We obtain thus

$$\begin{aligned} -a^{ij} D_{ik}\bar{u} D_{jk}\bar{u} + \tfrac{1}{2} a^{ij} D_{ij}\bar{v} + \frac{1}{2(\psi')^2}\bigg(\psi' D_{p_l} a^{ij} D_{ij}\bar{u} + D_{p_l} b \\ + \frac{\psi''}{(\psi')^2} D_{p_l}\mathscr{E}\bigg) D_l v + \bigg(\psi' \delta a^{ij} D_{ij}\bar{u} + \delta b + \frac{\psi''}{(\psi')^2}\delta\mathscr{E}\bigg)\bar{v} \\ - \frac{\psi''}{(\psi')^2}\bigg(b + \frac{\psi''}{(\psi')^2}\mathscr{E}\bigg)\bar{v} + \frac{1}{\psi'}\bigg[\frac{\psi''}{(\psi')^2}\bigg]'\mathscr{E}\bar{v} = 0. \end{aligned} \tag{15.18}$$

Next, letting δ be the operator acting on $C^1(\Omega \times \mathbb{R} \times \mathbb{R}^n)$ defined by

$$\delta = p_i D_{p_i} \tag{15.19}$$

and introducing the function

$$\omega = \frac{\psi''}{(\psi')^2} \in C^1[\bar{m}, \bar{M}], \tag{15.20}$$

we obtain, with the aid of the relation

$$D_l v = 2\psi'' \bar{v} D_l u + (\psi')^2 D_l \bar{v},$$

the following equation:

$$\begin{aligned} -a^{ij} D_{ik}\bar{u} D_{jk}\bar{u} + \tfrac{1}{2} a^{ij} D_{ij}\bar{v} + \tfrac{1}{2}(\psi' D_{p_i} a^{jk} D_{jk}\bar{u} + D_{p_i} b + \omega D_{p_i}\mathscr{E}) D_i \bar{v} \\ + \psi' \bar{v}(\omega\bar{\delta} a^{ij} + \delta a^{ij}) D_{ij}\bar{u} \\ + \left\{\frac{\omega'}{\psi'}\mathscr{E} + \omega^2(\bar{\delta} - 1)\mathscr{E} + \omega(\delta\mathscr{E} + (\bar{\delta} - 1) b) + \delta b\right\}\bar{v} = 0. \end{aligned} \tag{15.21}$$

Equation (15.21) can be further generalized by combining it with equation (15.17). Letting r and s be arbitrary scalar functions on $\Omega \times \mathbb{R} \times \mathbb{R}^n$ we obtain, by adding

$\bar{v}(\omega(r+1)+s)$ times equation (15.17) to equation (15.21), the equation

$$\begin{aligned}(15.22)\qquad &-a^{ij}D_{ik}\bar{u}D_{jk}\bar{u}+\tfrac{1}{2}a^{ij}D_{ij}\bar{v}+\tfrac{1}{2}(\psi' D_{p_i}a^{jk}D_{jk}\bar{u}+D_{p_i}b+\omega D_{p_i}\mathscr{E})D_i\bar{v}\\ &+\psi'\bar{v}(\omega(\delta+r+1)a^{ij}+(\delta+s)a^{ij})D_{ij}\bar{u}\\ &+\left\{\frac{\omega'}{\psi'}\mathscr{E}+\omega^2(\delta+r)\mathscr{E}+\omega((\delta+s)\mathscr{E}+(\delta+r)b)+(\delta+s)b\right\}\bar{v}=0.\end{aligned}$$

At this point it is convenient to write the principal coefficients a^{ij} in the form

$$(15.23)\qquad a^{ij}(x,z,p)=a^{ij}_*(x,z,p)+\tfrac{1}{2}[p_ic_j(x,z,p)+c_i(x,z,p)p_j]$$

where $a^{ij}_*, c_i\in C^1(\Omega\times\mathbb{R}\times(\mathbb{R}^n-\{0\}))$ and $[a^{ij}_*]$ is a positive, symmetric matrix. Clearly the decomposition (15.2) with positive $[a^{ij}_*]$ is a special case of (15.23). Furthermore the principal coefficients of any elliptic operator Q can be written in the form (15.23) by simply choosing $a^{ij}_*=a^{ij}$ and $c_i=0$. Indeed for our intended application to uniformly elliptic equations there is nothing to be gained by taking non-trivial c_i. However for the minimal surface operator, a decomposition of the form (15.23) with the matrix $[a^{ij}_*]$ proportional to the identity matrix is crucial for our derivation of global gradient bounds.

Returning to equation (15.22) we now substitute the relation (15.23) to obtain

$$\begin{aligned}(15.24)\qquad &-a^{ij}_*D_{ik}\bar{u}D_{jk}\bar{u}+\tfrac{1}{2}a^{ij}D_{ij}\bar{v}+\tfrac{1}{2}\{\psi' D_{p_i}a^{jk}D_{jk}\bar{u}-\psi' c_jD_{ij}\bar{u}\\ &+v[\omega(\delta+r+1)+\delta+s+1]c_i+D_{p_i}b+\omega D_{p_i}\mathscr{E}\}D_i\bar{v}\\ &+\psi'\bar{v}[\omega(\delta+r+1)a^{ij}_*+(\delta+s)a^{ij}_*]D_{ij}\bar{u}\\ &+\left\{\frac{\omega'}{\psi'}\mathscr{E}+\omega^2(\delta+r)\mathscr{E}+\omega[(\delta+s)\mathscr{E}+(\delta+r)b]+(\delta+s)b\right\}\bar{v}=0.\end{aligned}$$

By Cauchy's inequality (7.6) we can estimate

$$\begin{aligned}&\psi'\bar{v}[\omega(\delta+r+1)a^{ij}_*+(\delta+s)a^{ij}_*]D_{ij}\bar{u}\\ &\leqslant\lambda_*\sum|D_{ij}\bar{u}|^2+\frac{v\bar{v}}{4\lambda_*}\sum|[\omega(\delta+r+1)+\delta+s]a^{ij}_*|^2\\ &\leqslant a^{ij}_*D_{ik}\bar{u}D_{jk}\bar{u}+\frac{v\bar{v}}{4\lambda_*}\sum|[\omega(\delta+r+1)+\delta+s]a^{ij}_*|^2\end{aligned}$$

where λ_* denotes the minimum eigenvalue of the matrix $[a^{ij}_*]$. Hence by substitution into (15.24) we finally obtain the inequality

$$(15.25)\qquad a^{ij}D_{ij}\bar{v}+B_iD_i\bar{v}+2G\mathscr{E}\bar{v}\geqslant 0$$

where the coefficients B_i and G are given by

$$\begin{aligned}(15.26)\qquad &B_i=\psi'(D_{p_i}a^{jk}D_{jk}\bar{u}-c_jD_{ij}\bar{u})+v[\omega(\delta+r+1)+\delta+s+1]c_i\\ &\qquad+D_{p_i}b+\omega D_{p_i}\mathscr{E}\\ &G=\frac{\omega'}{\psi'}+\alpha\omega^2+\beta\omega+\gamma,\end{aligned}$$

and

$$\alpha=\frac{1}{\mathscr{E}}\Big((\delta+r)\mathscr{E}+\frac{|p|^2}{4\lambda_*}\sum|(\delta+r+1)a_*^{ij}|^2\Big),$$

(15.27) $$\beta=\frac{1}{\mathscr{E}}\Big((\delta+s)\mathscr{E}+(\delta+r)b+\frac{|p|^2}{2\lambda_*}[(\delta+r+1)a_*^{ij}][(\delta+s)a_*^{ij}]\Big),$$

$$\gamma=\frac{1}{\mathscr{E}}\left[\frac{|p|^2}{4\lambda_*}\sum|(\delta+s)a_*^{ij}|^2+(\delta+s)b\right].$$

In order to get a global gradient bound for the solution u, we need to choose the functions ψ, r and s in such a way that

(15.28) $$G\leqslant 0 \quad \text{for } x\in\Omega, \quad \text{and} \quad |Du(x)|\geqslant L$$

for sufficiently large L. For, if (15.28) is satisfied, it follows from the weak maximum principle, Theorem 3.1, that

$$\sup_{\Omega_L}\bar{v}=\sup_{\partial\Omega_L}\bar{v}, \quad (\Omega_L=\{x\in\Omega \mid |Du(x)|\geqslant L\}),$$

and hence

(15.29) $$\sup_{\Omega}|Du|\leqslant\max\left\{\left[\frac{\max\psi'}{\min\psi'}\right]\sup_{\partial\Omega}|Du|,\, L\right\}.$$

We digress here momentarily to remark that the estimate (15.29) would remain valid if the solution u is only assumed to lie in $C^2(\Omega)$. For, in this case the equation (15.18) would continue to hold in the weak form

$$\int_\Omega \eta a^{ij}D_{ik}\bar{u}D_{jk}\bar{u}\,dx+\tfrac{1}{2}\int_\Omega a^{ij}D_i\bar{v}D_j\eta\,dx$$

(15.30) $$+\tfrac{1}{2}\int_\Omega\left\{D_ja^{ij}D_i\bar{v}-\frac{1}{(\psi')^2}(\psi'D_{p_i}a^{jk}D_{jk}\bar{u}+D_{p_i}b+\omega D_{p_i}\mathscr{E})D_i v\right\}\eta\,dx$$

$$-\int_\Omega\left\{\psi'\,\delta a^{ij}D_{ij}\bar{u}+\delta b-\omega(b+\omega\mathscr{E})+\frac{\omega'}{\psi'}\mathscr{E}\right\}\bar{v}\eta\,dx=0$$

$$\text{for all } \eta\geqslant 0,\in C_0^1(\Omega)$$

(see Sections 13.3 or 14.1). Applying the same argument as above for the case $u\in C^3(\Omega)$ we deduce from (15.30) the weak form of inequality (15.25), namely

(15.31) $$\int_\Omega\{a^{ij}D_i\bar{v}D_j\eta+(D_ja^{ij}-B_i)D_i\bar{v}\eta-2G\mathscr{E}\bar{v}\eta\}\,dx\leqslant 0$$

for all non-negative $\eta \in C_0^1(\Omega)$. The estimate (15.29) is then a consequence of the weak maximum principle, Theorem 8.1.

Let us now consider conditions on the coefficients of Q which will ensure that inequality (15.28) is satisfied for some ψ, r and s. In order to guarantee that α, β and γ are bounded from above, we impose the structure conditions

$$(15.32)\qquad (\bar{\delta}+r+1)a_*^{ij},\ (\delta+s)a_*^{ij} = O\left[\frac{\sqrt{\lambda_*\mathscr{E}}}{|p|}\right] \quad \text{as } |p|\to\infty$$

$$\bar{\delta}\mathscr{E},\ \delta\mathscr{E},\ (\bar{\delta}+r)b,\ (\delta+s)b \leqslant O(\mathscr{E}) \quad \text{as } |p|\to\infty.$$

Here the limit behavior is understood to be uniform for $(x, z)\in\Omega\times[m, M]$. We now define the numbers

$$(15.33)\qquad a, b, c = \varlimsup_{|p|\to\infty}\ \sup_{\Omega\times[m,M]}\ \alpha, \beta, \gamma$$

so that inequality (15.28) is implied by the Riccati differential inequality

$$(15.34)\qquad \chi'(z)+a\chi^2(z)+b\chi(z)+c+\varepsilon\leqslant 0$$

holding on the interval $[m, M]$ for some positive number ε. To get from the solution χ to our auxiliary function ψ we use the relation $\chi=\omega\circ\psi^{-1}$. It turns out that inequality (15.34) cannot be solved for arbitrary a, b and c unless $\operatorname{osc}_{\Omega} u=M-m$ is sufficiently small (see Problem 15.1). The inequality can be solved however when either $a\leqslant 0$, $c\leqslant 0$ or $b\leqslant -2\sqrt{|ac|}$. Let us consider now the two important cases $a\leqslant 0$, $c\leqslant 0$; (the case $b\leqslant -2\sqrt{|ac|}$ is left to the reader in Problem 15.2). We simplify our calculations somewhat by setting $\varphi(z)=\psi'\circ\psi^{-1}(z)$. Then if χ satisfies inequality (15.34) we can determine φ through the relation

$$\frac{\varphi'}{\varphi}=\chi.$$

(i) $a\leqslant 0$. If the strict inequality $a<0$ holds, the quadratic equation

$$a\chi^2+b\chi+c+\varepsilon=0$$

has a positive real root $\chi=\kappa$, if ε is chosen so that $c+\varepsilon>0$, in which case inequality (15.34) is solved by taking $\chi=\kappa$. Consequently we obtain $(\log\varphi)'=\kappa$ and hence we can choose

$$\varphi(z)=e^{\kappa z}.$$

If, on the other hand, $a=0$, the inequality (15.34) can be solved by taking

$$\chi(z)=\frac{|c+\varepsilon|}{|b|+\varepsilon}\,e^{2(|b|+\varepsilon)(M-z)}$$

in which case we can choose

$$\varphi(z)=\exp -\left\{\frac{|c+\varepsilon|}{2(|b|+\varepsilon)^2}\, e^{2(|b|+\varepsilon)(M-z)}\right\}.$$

(ii) $c\leqslant 0$. If $c<0$ we can take $\varphi(z)=e^{\kappa z}$ for some $\kappa>0$ as in (i). If $c=0$, inequality (15.34) can be solved, for sufficiently small ε, by taking

$$\chi(z)=e^{A(m-z-1)} \quad \text{where } A=|a|+|b|+1.$$

In this case we can choose

$$\varphi(z)=\exp\left\{-\frac{1}{A}\, e^{A(m-z-1)}\right\}.$$

Note that in the cases considered above φ is monotone increasing and hence in the estimate (15.29) we have $\max \psi'=\varphi(M)$, $\min \psi'=\varphi(m)$.

The development of this section has accordingly led us to the establishment of the following theorem.

Theorem 15.2. *Let $u\in C^2(\Omega)\cap C^1(\bar{\Omega})$ satisfy equation* (15.1) *in the bounded domain Ω. Suppose that the operator Q is elliptic in Ω and that there exist scalar multipliers r and s such that the structure conditions* (15.32) *are fulfilled together with either of the conditions $a\leqslant 0$ or $c\leqslant 0$ (the numbers a and c being defined by* (15.27) *and* (15.33)*). Then we have the estimate*

$$\sup_{\Omega} |Du|\leqslant C \tag{15.35}$$

where C depends on the quantities in (15.32), $\operatorname{osc}_{\Omega} u$ *and* $\sup_{\partial\Omega} |Du|$.

We illustrate the application of Theorem 15.2 by considering some important special cases.

(i) *Uniformly elliptic equations.* If Q is uniformly elliptic in Ω, that is $a^{ij}=O(\lambda)$ as $|p|\to\infty$, and if also the function $b=O(\mathscr{E})=O(\lambda|p|^2)$ as $|p|\to\infty$, then conditions (15.32) with $a_*^{ij}=a^{ij}$, $c_i=0$, $r=s=0$ are often referred to as *natural* conditions (see [LU 4]). If we restrict these conditions slightly by requiring that

$$\delta a^{ij}=o(\lambda),\quad \delta b=o(\lambda|p|^2)\quad \text{as } |p|\to\infty \tag{15.36}$$

then $c\leqslant 0$, and hence $C^2(\Omega)$ solutions of the equation $Qu=0$ satisfy an apriori global gradient bound. We shall show in the next section that the restriction (15.36) is unnecessary.

(ii) *Prescribed mean curvature equation.* By writing equation (10.7) in the form

$$\Delta u-\frac{D_iuD_ju}{(1+|Du|^2)}\, D_{ij}u-nH(x)\sqrt{1+|Du|^2}=0 \tag{15.37}$$

where $H \in C^1(\bar{\Omega})$ we can choose

$$a_*^{ij} = \delta^{ij}, \qquad c_i = -\frac{p_i}{1+|p|^2}, \qquad r = -1 \quad \text{and} \quad s = 0$$

so that by calculation we have

$$\alpha = -1 + \frac{2}{1+|p|^2}, \qquad \beta = \frac{n\,H(x)\sqrt{1+|p|^2}}{|p|^2},$$

$$\gamma = -\frac{np_i D_i H(1+|p|^2)^{3/2}}{|p|^4}$$

and hence

$$a = -1, \qquad b = 0, \qquad c = n \sup_{\Omega} |DH|.$$

Consequently, a global gradient bound holds for $C^2(\Omega)$ solutions by virtue of the case $a<0$ of Theorem 15.2. In particular, we can deduce using (15.29) the estimate

$$\sup_{\Omega} |Du| \leqslant c_1 + c_2 n \sup_{\Omega} |H| + c_3 \sup_{\partial\Omega} |Du| \exp\left(c_4 n \sup_{\Omega} |DH| \operatorname{osc}_{\Omega} u\right) \tag{15.38}$$

where c_1, c_2, c_3 and c_4 are constants.

(iii) *Equations where* $a_*^{ij} = \delta^{ij}$. As shown in the preceding section, if equation (15.1) is the Euler–Lagrange equation of a multiple integral of the form (15.3) or if the dimension $n=2$, then we can take $a_*^{ij} = \delta^{ij}$. Using (15.27) we obtain in these cases

$$\alpha = \frac{1}{\mathscr{E}}(\delta + r)\mathscr{E} + \frac{(r+1)^2}{4}\frac{|p|^2}{\mathscr{E}},$$

$$\beta = \frac{1}{\mathscr{E}}[(\delta+s)\mathscr{E} + (\delta+r)b] + \frac{(r+1)s}{2}\frac{|p|^2}{\mathscr{E}}, \tag{15.39}$$

$$\gamma = \frac{1}{\mathscr{E}}(\delta+s)b + \frac{s^2}{4}\frac{|p|^2}{\mathscr{E}}.$$

Choosing the particular values $r=-1$, $s=0$ we have

$$\alpha = \frac{1}{\mathscr{E}}(\delta - 1)\mathscr{E},$$

$$\beta = \frac{1}{\mathscr{E}}[\delta\mathscr{E} + (\delta - 1)b],$$

$$\gamma = \frac{1}{\mathscr{E}}\delta b.$$

Note that the same formulae arise whenever

$$a_*^{ij}(p)=a_*^{ij}(p/|p|)$$

since in these cases we have $\delta a_*^{ij}=0$.

15.3. Interior Gradient Bounds

An interior gradient bound for $C^2(\Omega)$ solutions of equation (15.1) can also be deduced from equation (15.24). Let $B=B_R(y)$ be a ball strictly lying in Ω and let η be a function in $C^2(\bar{B})$ such that $0\leqslant\eta\leqslant 1$ in $\bar{B}$, $\eta=0$ on ∂B, $\eta(y)=1$ and $\eta>0$ in B. A typical example of such a function, which we shall use below, is given by

$$\eta(x)=\left(1-\frac{|x-y|^2}{R^2}\right)^\beta,\ \beta\geqslant 1. \tag{15.40}$$

We now consider the function w defined by

$$w=\eta\bar{v}$$

Clearly $w\in C^1(\bar{B})$, $w(y)=\bar{v}(y)$, $w=0$ on ∂B, and the derivatives of w are given by

$$\begin{aligned}D_iw&=\eta D_i\bar{v}+\bar{v}D_i\eta\\ D_{ij}w&=\eta D_{ij}\bar{v}+D_i\bar{v}D_j\eta+D_i\eta D_j\bar{v}+\bar{v}D_{ij}\eta.\end{aligned}$$

Note that to guarantee the existence of the second derivatives $D_{ij}w$ we require that $u\in C^3(\Omega)$; this restriction can however be avoided by utilizing the weak form of equation (15.24). Multiplying equation (15.24) by η and substituting for $\bar{v}$, we now obtain the equation

$$\begin{aligned}&-\eta a_*^{ij}D_{ik}\bar{u}D_{jk}\bar{u}+\frac{1}{2}a^{ij}D_{ij}w+\frac{1}{2}\left(B_i-\frac{2}{\eta}a^{ij}D_j\eta\right)D_iw\\ &\quad+\psi'w[\omega(\bar{\delta}+r+1)+(\delta+s)]a_*^{ij}D_{ij}\bar{u}\\ &\quad+\left\{\frac{\omega'}{\psi'}\mathscr{E}+\omega^2(\bar{\delta}+r)\mathscr{E}+\omega[(\delta+s)\mathscr{E}+(\bar{\delta}+r)b]+(\delta+s)b\right.\\ &\quad\left.+\frac{1}{\eta^2}a^{ij}D_i\eta D_j\eta-\frac{1}{2\eta}a^{ij}D_{ij}\eta-\frac{1}{2\eta}B_iD_i\eta\right\}w=0,\end{aligned} \tag{15.41}$$

where B_i is given by (15.26). At this point we can introduce further scalar multipliers t_i, $i=1,\ldots,n$. For, if t_i, $i=1,\ldots,n$, are arbitrary scalar functions on $\Omega\times\mathbb{R}\times\mathbb{R}^n$ we can write, using equation (15.17) and setting $\partial_i=D_{p_i}$,

$$\begin{aligned}B_i &= \psi'[(\partial_i+t_i)a^{jk}D_{jk}\bar{u}-c_jD_{ij}\bar{u}]+v[\omega(\delta+r+1)+(\delta+s+1)]c_i\\&\quad+(\partial_i+t_i)b+\omega(\partial_i+t_i)\mathscr{E}\\&=\psi'(\partial_i+t_i)a_*^{jk}D_{jk}\bar{u}-\frac{v}{2\eta}D_j\eta(\partial_i+t_i)c_j+\frac{(\psi')^2}{2\eta}(\partial_i+t_i)c_jD_jw\\&\quad+v[\omega(\bar{\delta}+r+1)+(\delta+s+1)]c_i+(\partial_i+t_i)b+\omega(\partial_i+t_i)\mathscr{E}.\end{aligned}$$

Hence by substitution into equation (15.41), we have

$$\begin{aligned}&-\eta a_*^{ij}D_{ik}\bar{u}D_{jk}\bar{u}+\tfrac{1}{2}a^{ij}D_{ij}w+\tfrac{1}{2}\tilde{B}_iD_iw\\&\quad+\psi'w[\omega(\delta+r+1)+(\delta+s)+(\tilde{\delta}+t)]a_*^{ij}D_{ij}\bar{u}\\&\quad+\left\{\frac{\omega'}{\psi'}\mathscr{E}+\omega^2(\delta+r)\mathscr{E}+\omega[(\delta+\tilde{\delta}+s+t)\mathscr{E}+(\tilde{\delta}+r)b\right.\\&\quad-\frac{v}{2\eta}D_i\eta(\tilde{\delta}+r+1)c_i]+(\delta+\tilde{\delta}+s+t)b\\&\quad-\frac{v}{2\eta}D_i\eta(\delta+\tilde{\delta}+s+t+1)c_i\\&\quad\left.+\frac{1}{\eta^2}a^{ij}D_i\eta D_j\eta-\frac{1}{2\eta}a^{ij}D_{ij}\eta\right\}w=0\end{aligned}\tag{15.42}$$

where

$$\tilde{\delta}=-\frac{1}{2\eta}D_i\eta\,\partial_i,\qquad t=-\frac{t_i}{2\eta}D_i\eta\tag{15.43}$$

and

$$\tilde{B}_i=B_i-\frac{2}{\eta}a^{ij}D_j\eta-\frac{v}{2\eta}D_j\eta(\partial_j+t_j)c_i.$$

We then obtain, instead of inequality (15.25), the inequality

$$a^{ij}D_{ij}w+\tilde{B}_iD_iw+2\tilde{G}\mathscr{E}w\geqslant 0\tag{15.44}$$

where $\tilde{G}$ is given by

$$\tilde{G}=\frac{\omega'}{\psi'}+\tilde{\alpha}\omega^2+\tilde{\beta}\omega+\tilde{\gamma},\tag{15.45}$$

and

$$\tilde{\alpha}=\alpha,$$

$$\tilde{\beta}=\beta+\frac{1}{\mathscr{E}}\left\{-\frac{|p|^2}{2\eta}D_i\eta(\delta+r+1)c_i+(\tilde{\delta}+t)\mathscr{E}\right.$$
$$\left.+\frac{|p|^2}{2\lambda_*}[(\delta+r+1)a_*^{ij}][(\tilde{\delta}+t)a_*^{ij}]\right\},$$

(15.46)
$$\tilde{\gamma}=\gamma+\frac{1}{\mathscr{E}}\left\{\frac{|p|^2}{4\lambda_*}\sum|(\tilde{\delta}+t)a_*^{ij}|^2+(\tilde{\delta}+t)b\right.$$
$$-\frac{v}{2\eta}D_i\eta(\delta+\tilde{\delta}+s+t+1)c_i+\frac{1}{\eta^2}a^{ij}D_i\eta D_j\eta$$
$$\left.-\frac{1}{2\eta}a^{ij}D_{ij}\eta+\frac{|p|^2}{2\lambda_*}[(\delta+s)a_*^{ij}][(\tilde{\delta}+t)a_*^{ij}]\right\}.$$

Unless there exist special relationships between the function η and the coefficients of Q, we need to add more structure conditions to ensure that the coefficients $\tilde{\beta}$ and $\tilde{\gamma}$ behave similarly to β and γ. Let us therefore assume, in addition to the structure conditions (15.32), the conditions

$$|p|^\theta(\partial_k+t_k)a_*^{ij}=O(\sqrt{\lambda_*\mathscr{E}/|p|})\qquad\text{as }|p|\to\infty,$$

(15.47)
$$|p|^{2\theta}\Lambda,\ |p|^\theta(\partial_i+t_i)\mathscr{E},\ |p|^\theta(\partial_i+t_i)b=O(\mathscr{E})\qquad\text{as }|p|\to\infty,$$
$$|p|^\theta(\delta+r+1)c_i,\ |p|^\theta(\delta+s+1)c_i,\ |p|^{2\theta}(\partial_i+t_i)c_i=O(\mathscr{E}/|p|^2)$$
$$\text{as }|p|\to\infty,$$

with $i,j,k=1,\ldots,n$, for some $\theta>0$. As in (15.32), the limit behavior is understood to be uniform for $(x,z)\in(m,M)$.

Using (15.47) and the function η given by (15.40), with $\beta=2/\theta$, we can then estimate (for sufficiently large $|Du|$)

$$|\beta-\tilde{\beta}|\leqslant\frac{C|D\eta|}{\eta|Du|^\theta}\leqslant\frac{C\beta}{Rw^{1/\beta}},$$

$$|\gamma-\tilde{\gamma}|\leqslant C\left\{\frac{|D\eta|^2}{\eta^2|Du|^{2\theta}}+\frac{|D\eta|}{\eta|Du|^\theta}\right\}\leqslant\left\{\frac{\beta^2}{R^2w^{2/\beta}}+\frac{\beta}{Rw^{1/\beta}}\right\},$$

where C depends on n and the quantities in (15.32) and (15.47). Consequently, applying the considerations of the preceding section we obtain, instead of (15.29), the estimate

(15.48) $$|Du(y)|\leqslant C(1+R^{-1/\theta}),$$

provided either $a \leqslant 0$, $c \leqslant 0$, $b \leqslant -2\sqrt{ac}$, or $\operatorname*{osc}_{B_R(y)} u$ is less than some constant depending on a, b, c. Moreover, we can clearly replace the ball $B = B_R(y)$ by any ball intersecting Ω, and therefore obtain, for any $y \in \Omega$, the same estimate with C depending in addition on $\sup_{\partial\Omega \cap B} |Du|$. Note also that by invoking (15.30), rather than (15.18), the above considerations are also applicable to $C^2(\Omega)$ solutions. Therefore, we have the following estimate.

Theorem 15.3. *Let $u \in C^2(\Omega)$ satisfy (15.1) in the domain Ω. Suppose that the operator Q is elliptic in Ω and that there exist scalar multipliers r, s, t_i, $i = 1, \ldots, n$ such that the structure conditions (15.32) and (15.47) are fulfilled. Then, if any of the conditions $a \leqslant 0$, $c \leqslant 0$, $b \leqslant -2\sqrt{ac}$ hold, (the numbers a, b, c being defined by (15.27) and (15.33)), we have the estimate*

$$(15.49) \qquad |Du(y)| \leqslant C(1 + R^{-1/\theta})$$

for any $y \in \Omega$ and $B = B_R(y)$, where C depends on the quantities in (15.32), (15.47), $\operatorname*{osc}_{B \cap \Omega} u$ and $\sup_{B \cap \partial\Omega} |Du|$. If a, b and c are arbitrary, the estimate (15.49) continues to hold for sufficiently small R depending on a, b, c and the modulus of continuity of u at y.

For uniformly elliptic equations, under the *natural* conditions, estimates for the moduli of continuity of solutions (and hence gradient estimates) follow from the considerations of Section 9.7. In fact, we have the following estimate which, in some sense, extends Corollary 9.25.

Lemma 15.4. *Let $u \in W^{2,n}(\Omega)$ satisfy the inequality*

$$(15.50) \qquad |Lu| \leqslant \lambda(\mu_0 |Du|^2 + f)$$

in Ω, where the operator L, given by $Lu = a^{ij}D_{ij}u$, satisfies (9.47), $\mu_0 \in \mathbb{R}$ and $f \in L^n(\Omega)$. Then, for any ball $B_0 = B_{R_0}(y) \subset \Omega$ and $R \leqslant R_0$, we have

$$(15.51) \qquad \operatorname*{osc}_{B_R(y)} u \leqslant C\left(\frac{R}{R_0}\right)^{\alpha} (\operatorname*{osc}_{B_0} u + R\|f\|_{n;\Omega})$$

where C and α depend only on n, Λ/λ and $\mu_0 M$; $M = |u|_{0;\Omega}$.

Proof. Let us suppose first that $u \geqslant 0$ satisfies

$$(15.52) \qquad Lu \leqslant \lambda(\mu_0 |Du|^2 + f)$$

in Ω, where $\mu_0 > 0$, $f \geqslant 0$, and set

$$w = \frac{1}{\mu_0}(1 - e^{-\mu_0 u})$$

It follows then that

$$Lw \leqslant f$$

in Ω, so that the weak Harnack inequality (Theorem 9.22) is applicable to w. But since

$$w \leqslant u \leqslant e^{\mu_0 M} w,$$

the weak Harnack inequality (9.48) is satisfied also by u, with constant C depending in addition on $\mu_0 M$. The Hölder estimate (15.51) then follows.

Combining Theorem 15.3 and Lemma 15.4, we conclude both interior and global gradient bounds under the conditions

$$\begin{aligned} &\Lambda, (\bar{\delta} + r + 1)\, a^{ij}, (\delta + s)\, a^{ij}, |p|^{\theta} (\partial_k + t_k)\, a^{ij} = O(\lambda), \\ &b, |p|^{\theta} (\partial_i + t_i)\, b = O(\lambda |p|^2), (\bar{\delta} + r)\, b, (\delta + s)\, b \leqslant O(\lambda |p|^2), \end{aligned} \tag{15.53}$$

as $|p| \to \infty$, for some $\theta > 0$.

Theorem 15.5. *Let $u \in C^2(\Omega)$ satisfy (15.1) in the domain Ω and suppose that the operator Q is elliptic in Ω and satisfies the structure conditions (15.53), with multipliers $r, s, t_i, i = 1, \ldots, n$. Then, for any point $y \in \Omega$, we have*

$$|Du(y)| \leqslant C(1 + d_y^{-1/\theta}) \tag{15.54}$$

where C depends on n, the quantities in (15.53), $|u|_{0;\Omega}$, and $d_y = \text{dist}\,(y, \partial\Omega)$. If $u \in C^2(\Omega) \cap C^1(\bar{\Omega})$, we also have

$$|Du(y)| \leqslant C \tag{15.55}$$

where C depends in addition on $\sup\limits_{\partial\Omega} |Du|$.

15.4. Equations in Divergence Form

Let $u \in C^2(\Omega)$ satisfy the divergence form equation

$$Qu = \text{div}\, \mathbf{A}(x, u, Du) + B(x, u, Du) = 0 \tag{15.56}$$

in the domain Ω. Then it was shown in Section 13.1 that the derivatives $D_k u$, $k = 1, \ldots, n$ satisfy the linear divergence form equations

$$\int_{\Omega} (\bar{a}^{ij} D_j D_k u + f_k^i) D_i \zeta \, dx = 0 \quad \forall \zeta \in C_0^1(\Omega), \tag{15.57}$$

where

$$\begin{aligned}\bar{a}^{ij}(x) &= D_{p_j}A^i(x, u(x), Du(x)),\\ f^i_k(x) &= \delta_k A^i(x, u(x), Du(x)) + \delta^{ik}B(x, u(x), Du(x)), \quad \delta_k = p_k D_z + D_{x_k}.\end{aligned} \tag{15.58}$$

In order to proceed further we shall assume that Q is elliptic in the sense that

$$\bar{a}^{ij}(x, z, p)\xi_i\xi_j = D_{p_j}A^i(x, z, p)\xi_i\xi_j \geqslant \nu(|z|)(1+|p|)^\tau|\xi|^2 \tag{15.59}$$

for all $\xi \in \mathbb{R}^n$, $(x, z, p) \in \Omega \times \mathbb{R} \times \mathbb{R}^n$, where τ is some real number and ν is a positive, non-increasing function on $\mathbb{R}$. A global gradient bound for non-uniformly elliptic Q can then be derived under the additional structure conditions

$$|p|D_z\mathbf{A}, D_x\mathbf{A}, B = o(|p|)(1+|p|)^\tau \quad \text{as } |p| \to \infty. \tag{15.60}$$

To show this we put $M = \sup_\Omega |u|$, $M_1 = \sup_\Omega |Du|$ and apply the maximum principle, Theorem 8.16, to equation (15.57) in the domain $\tilde{\Omega} = \{x \in \Omega \mid |Du(x)| > M_1/2\sqrt{n}\}$. We obtain thus

$$\sup_{\tilde{\Omega}} |D_k u| \leqslant \sup_{\partial\tilde{\Omega}} |D_k u| + C(2\sqrt{n})^{|\tau|}\|(1+|Du|)^{-\tau} f^i_k\|_q \tag{15.61}$$

for $q > n$, $k = 1, \ldots, n$, where $C = C(n, \nu(M), q, |\Omega|)$. Taking $q = \infty$ and using conditions (15.60), we therefore have

$$M_1 = \sup_\Omega |Du| \leqslant C\,(\sup_{\partial\Omega} |Du| + \sigma(M_1)) \tag{15.62}$$

where $\sigma(t) = o(t)$ as $t \to \infty$. Consequently, an apriori estimate for M_1 follows.

Theorem 15.6. *Let $u \in C^2(\Omega)$ satisfy equation* (15.56) *in the bounded domain Ω and suppose that the structure conditions* (15.59), (15.60) *are fulfilled. Then we have the estimate*

$$\sup_\Omega |Du| \leqslant C(1 + \sup_{\partial\Omega} |Du|) \tag{15.63}$$

where C depends on n, τ, $\nu(M)$ and the quantities in (15.60).

Condition (15.60) requires that the coefficient b in equation (15.1) satisfies $b = o(\lambda|p|)$ as $|p| \to \infty$. In order to relax this condition we must either impose additional conditions on the coefficient matrix $D_p\mathbf{A}$ such as uniform ellipticity or assume that the solution u vanishes on $\partial\Omega$; (see Problems 15.4, 15.5). Note that the existence of a global gradient bound is reduced, through the estimate (15.61), to the existence of suitable integral estimates for the functions $(1+|Du|)^{-\tau} f^i_k$. Instead of pursuing global bounds any further, at this stage, we shall now turn to a consideration of interior gradient estimates for uniformly elliptic equations. Let us assume that the operator Q is uniformly elliptic in Ω in the sense that

$$(15.64)\qquad |D_p\mathbf{A}(x, z, p)| \leqslant \mu(|z|)(1+|p|)^\tau$$

for all $(x, z, p) \in \Omega \times \mathbb{R} \times \mathbb{R}^n$, where μ is a positive, non-decreasing function on $\mathbb{R}$. Henceforth we shall also assume that $\tau > -1$, in which case conditions (15.59), (15.64) imply respectively the inequalities

$$(15.65)\qquad \begin{aligned} &p\cdot\mathbf{A}(x, z, p) - p\cdot\mathbf{A}(x, z, 0) \geqslant \frac{\nu}{\tau+1}|p|^{\tau+2},\\ &|A^i(x, z, p) - A^i(x, z, 0)| \leqslant \frac{\mu}{\tau+1}(1+|p|)^{\tau+1}, \quad i=1, \ldots, n. \end{aligned}$$

Finally we shall take, in place of (15.60), the more general condition

$$(15.66)\qquad g(x, z, p) = (1+|p|)|D_z\mathbf{A}| + |D_x\mathbf{A}| + |B| \leqslant \mu(|z|)(1+|p|)^{\tau+2}$$

for all $(x, z, p) \in \Omega \times \mathbb{R} \times \mathbb{R}^n$. Conditions (15.59), (15.64), (15.66) can be regarded as *natural* for divergence structure operators. The derivation of an apriori interior gradient bound under these conditions is accomplished in three stages:

(i) *Reduction to an L^p estimate.* We replace the test function ζ in (15.57) by $\zeta D_k u$ and sum the resulting equations over k. Setting $v = |Du|^2$ we obtain thus

$$\int_\Omega \zeta \bar{a}^{ij} D_{ik}u D_{jk}u\, dx + \int_\Omega (\tfrac{1}{2}\bar{a}^{ij} D_j v + D_k u f^i_k) D_i\zeta\, dx + \int_\Omega \zeta f^i_k D_{ik}u\, dx = 0.$$

Hence by Young's inequality (7.6) we have

$$(15.67)\qquad \int_\Omega (\bar{a}^{ij} D_j v + 2D_k u f^i_k) D_i\zeta\, dx \leqslant \frac{1}{2}\int_\Omega \zeta\lambda^{-1} \sum (f^i_k)^2\, dx$$

for all non-negative $\zeta \in C^1_0(\Omega)$. Therefore, setting

$$\bar{v} = \int_0^v (1+\sqrt{t})^\tau\, dt,$$

we may write (15.67) as

$$(15.68)\qquad \int_\Omega (\tilde{a}^{ij} D_j\bar{v} + 2D_k u f^i_k) D_i\zeta \leqslant \frac{1}{2}\int_\Omega \zeta\lambda^{-1} \sum (f^i_k)^2\, dx,$$

where

$$\tilde{a}^{ij} = \bar{a}^{ij}(1+|Du|)^{-\tau}.$$

The interior estimate for weak subsolutions of linear equations, Theorem 8.17, is thus applicable to inequality (15.68). Hence, using (15.66), we have for any ball $B_{2R}=B_{2R}(y)\subset\Omega$ and $q>n$, the estimate

$$\sup_{B_R(y)} \bar{v}\leqslant C\{R^{-n/2}\|\bar{v}\|_{L^2(B_{2R})}+\|(1+|Du|)^{\tau+4}\|_{L^{q'}(B_{2R})}\}$$

where $C=C(n, \nu(M), \mu(M), \tau, q, \operatorname{diam}\Omega)$, $M=\sup_{B_{2R}(y)}|u|$. Consequently for sufficiently large p we have

$$\sup_{B_R(y)} v\leqslant C(n, \nu(M), \mu(M), \tau, \operatorname{diam}\Omega, R^{-n}\int_{B_{2R}(y)} v^p\,dx). \tag{15.69}$$

(ii) *Reduction to a Hölder estimate.* We now need to utilize the weak form of equation (15.57), viz.

$$Q(u,\varphi)=\int_\Omega (A^i(x,u,Du)D_i\varphi-B(x,u,Du)\varphi)\,dx=0\quad\forall\varphi\in C_0^1(\Omega). \tag{15.70}$$

From (15.65) we see that the function $\mathbf{A}$ satisfies inequalities

$$\begin{aligned}|\mathbf{A}(x,z,p)|&\leqslant\mu_1(|z|)(1+|p|)^{\tau+1}\\ p\cdot\mathbf{A}(x,z,p)&\geqslant\nu_1(|z|)|p|^{\tau+2}-\mu_1(|z|)\end{aligned} \tag{15.71}$$

for $(x,z,p)\in\Omega\times\mathbb{R}\times\mathbb{R}^n$, where μ_1 and ν_1 are respectively positive non-decreasing and positive non-increasing functions depending on τ, μ, ν and $\sup_\Omega|\mathbf{A}(x,u,0)|$. Let us substitute into (15.70) the test function

$$\varphi=\eta^2[u-u(y)]$$

where $\eta\in C_0^1(B_{2R})$, $B_{2R}=B_{2R}(y)\subset\Omega$. Using (15.66) and (15.71) we obtain thus

$$\begin{aligned}\nu_1\int_\Omega\eta^2|Du|^{\tau+2}\,dx&\leqslant\mu_1\int_\Omega\eta^2\,dx\\ &\quad+\mu\int_\Omega\eta^2|u(x)-u(y)|(1+|Du|)^{\tau+2}\,dx\\ &\quad+2\mu_1\int_\Omega|\eta D\eta|\,|u(x)-u(y)|(1+|Du|)^{\tau+1}\,dx\\ &\leqslant\mu_1\int_\Omega\eta^2\,dx+4(\mu+\mu_1)2^\tau\omega(R)\int_\Omega\eta^2(1+|Du|^{\tau+2})\,dx\\ &\quad+\mu_1\omega(R)\int_\Omega(1+|Du|)^\tau|D\eta|^2\,dx\end{aligned}$$

where

$$\omega(R)=\sup_{B_{2R}}|u(x)-u(y)|.$$

Hence, if R is chosen small enough to ensure that

$$\omega(R)\leqslant\frac{\nu_1 2^{-\tau}}{8(\mu+\mu_1)},$$

we have

$$\int_\Omega \eta^2|Du|^{\tau+2}\,dx\leqslant C\int_\Omega(\eta^2+\omega(R)(1+|Du|)^\tau|D\eta|^2)\,dx \tag{15.72}$$

where $C=C(\mu,\mu_1,\nu_1,\tau)$. Let us now replace the function η in (15.72) by

$$\eta v^{(\beta+1)/2},\quad \beta>0,$$

to obtain the estimate

$$\begin{aligned}\int_\Omega \eta^2|Du|^{2\beta+\tau+4}\,dx\leqslant C\int_\Omega\{&|Du|^{2(\beta+1)}(\eta^2+\omega(R)(1+|Du|)^\tau|D\eta|^2)\\ &+(\beta+1)^2\omega(R)(1+|Du|)^\tau\eta^2v^{\beta-1}|Dv|^2\}\,dx.\end{aligned} \tag{15.73}$$

To estimate the last term in the above inequality we choose

$$\zeta=\eta^2v^\beta$$

in inequality (15.67). We obtain thus, using conditions (15.59), (15.64) and (15.66),

$$\begin{aligned}\beta\nu\int_\Omega(1+|Du|)^\tau\eta^2v^{\beta-1}|Dv|^2\,dx&\\ \leqslant C\int_\Omega\{&(1+|Du|)^\tau\eta v^\beta|D\eta|\,|Dv|+(1+|Du|)^{\tau+3}\\ &\times(\eta v^\beta|D\eta|+\beta\eta^2v^{\beta-1}|Dv|)+\eta^2(1+|Du|)^{\tau+4}v^\beta\}\,dx\end{aligned}$$

where $C=C(\mu,\nu)$. Hence, by Young's inequality (7.6),

$$\begin{aligned}\int_\Omega(1+|Du|)^\tau\eta^2v^{\beta-1}|Dv|^2\,dx&\\ \leqslant C\int_\Omega&(1+|Du|)^{2\beta+\tau+2}(\eta^2(1+|Du|)^2+|D\eta|^2)\,dx\end{aligned}$$

where $C=C(\mu, \nu, \beta)$. Consequently if $\omega(R)$ is sufficiently small we obtain, by substitution into (15.73),

$$\int_\Omega \eta^2(1+|Du|)^{2\beta+\tau+4}\,dx \leqslant C\int_\Omega (\eta^2(1+|Du|)^{2(\beta+1)}+|D\eta|^2(1+|Du|)^{2\beta+\tau+2})\,dx$$

where $C=C(\mu, \nu, \beta, \tau)$. Replacing η by $\eta^{(2\beta+\tau+4)/2}$ and using Young's inequality (7.6), we then obtain

$$\int_\Omega [\eta(1+|Du|)]^{2\beta+\tau+4}\,dx \leqslant C$$

where $C=C(\mu, \nu, \mu_1, \nu_1, \beta, \tau, \sup|D\eta|)$. In particular, if $\eta\equiv 1$ on $B_R(y)$ and $|D\eta|\leqslant 2/R$, it follows that, for any $p\geqslant 1$,

$$\int_{B_R(y)} (1+|Du|)^p\,dx \leqslant C \tag{15.74}$$

where $C=C(\mu, \nu, \mu_1, \nu_1, p, \tau, R^{-1})$. Combining the estimates (15.69) and (15.74) we obtain for any ball $B_0=B_{R_0}(y)\subset\Omega$ and $0<\alpha<1$ the estimate

$$|Du(y)|\leqslant C \tag{15.75}$$

where $C=C(n, \tau, \nu, \mu, \nu_1, \mu_1, \alpha, [u]_{\alpha,y})$ and

$$[u]_{\alpha,y}=\sup_{B_0}\frac{|u(x)-u(y)|}{|x-y|^\alpha};$$

that is, the derivation of an interior gradient bound is reduced to the existence of an interior Hölder estimate for u. We note here that an estimate of the modulus of continuity of u would in fact suffice to complete the proof. Moreover, if the structure conditions (15.66) were strengthened so that $g = o(|p|^{\tau+2})$ as $|p|\to\infty$, the above considerations yield an interior gradient bound independent of the modulus of continuity of u. In this case we can also weaken the uniform ellipticity of Q to allow $D_{p_j}A^i = O(|p|^{\tau+\sigma})$ as $|p|\to\infty, i,j = 1,\ldots, n$, where $\sigma < 1$ (see Problem 15.6).

(iii) *A Hölder estimate for weak solutions of equation* (15.66). Let us write inequalities (15.71) together with the condition on B in (15.66) in the form

$$\begin{aligned} &|\mathbf{A}(x, z, p)|\leqslant a_0|p|^{m-1}+\chi^{m-1}\\ &p\cdot\mathbf{A}(x, z, p)\geqslant \nu_0|p|^m-\chi^m\\ &|B(x, z, p)|\leqslant b_0|p|^m+\chi^m\end{aligned} \tag{15.76}$$

where $m=\tau+2>1$ and a_0, b_0, v_0 and χ are positive constants depending possibly on $M=\sup_\Omega |u|$. Then we have the following Hölder estimate.

Theorem 15.7. *Let $u \in C^1(\Omega)$ be a weak solution of equation (15.56) in the domain Ω and suppose that Q satisfies the structure conditions (15.76). Then for any ball $B_0=B_{R_0}(y)\subset\Omega$ and $R\leqslant R_0$ we have the estimate*

$$\operatorname*{osc}_{B_R(y)} u \leqslant C(1+R_0^{-\alpha}M_0)R^\alpha \tag{15.77}$$

where $C=C(n, a_0, b_0, v_0, \chi, R_0, m, M_0)$ and $\alpha=\alpha(n, a_0, b_0, v_0, m, M_0)$ are positive constants and $M_0=\sup_{B_0} |u|$.

Proof. Theorem 15.7 can be derived by essentially the same method as used for Theorem 8.22 or alternatively by following the method described in Problem 8.6. We shall not present all the details here. The essential ingredient in our proof of Theorem 8.22 was the weak Harnack inequality Theorem 8.18. Setting

$$k=\chi(R+R^{m/(m-1)})$$

we obtain for non-negative weak supersolutions of equation (15.56) the analogous weak Harnack inequality

$$R^{-n/p}\|u\|_{L^p(B_{2R}(y))}\leqslant C(\inf_{B_R(y)} u+k) \tag{15.78}$$

where $C=C(n, a_0, b_0, v_0, m, M_0, p)$ and $1\leqslant p<n/(n-m)^+$. The proof of (15.78) can be modelled on that of Theorem 8.18 provided that in place of (8.48), we take as test functions, in the inequality $Q(u, \varphi)\geqslant 0$,

$$\varphi=\eta^m\bar{u}^\beta \exp(-b_0\bar{u}), \quad \beta<0,$$

and use the Sobolev inequality (7.26) with $p=m$ instead of $p=2$. The passage from the weak Harnack inequality to the Hölder estimate (15.77) can be accomplished according to the proof of Theorem 8.22. Note that the case where $m>n$ can be handled directly from the Sobolev inequality, Theorem 7.17. □

Combining Theorem 15.7 with our previous estimate (15.75) we finally arrive at the following interior gradient estimate.

Theorem 15.8. *Let $u \in C^2(\Omega)$ satisfy equation (15.57) in the domain Ω and suppose that the structure conditions (15.60), (15.64) and (15.66) are fulfilled with $\tau>-1$. Then we have the estimate*

$$|Du(y)|\leqslant C \tag{15.79}$$

for any $y \in \Omega$, where C depends on n, τ, $\nu(M_0)$, $\mu(M_0)$, $\sup_\Omega |\mathbf{A}(x, u, 0)|$, M_0, M_0/d and $M_0 = \sup_{B_d(y)} |u|$, $d = \text{dist}(y, \partial\Omega)$.

By applying an argument similar to the proof of Theorem 8.29, we can conclude from the interior estimate (15.79) (and the boundary Lipschitz estimate, Theorem 14.1) the following global estimate.

Theorem 15.9. *Let $u \in C^2(\Omega) \cap C^0(\bar{\Omega})$ satisfy equation (15.57) in the bounded domain Ω and suppose that the structure conditions (15.60), (15.64) and (15.66) are fulfilled with $\tau > -1$. Assume also that Ω satisfies a uniform exterior sphere condition and that $u = \varphi$ on $\partial\Omega$ for $\varphi \in C^2(\bar{\Omega})$. Then we have the estimate*

$$\sup_\Omega |Du| \leqslant C \tag{15.80}$$

where C depends on n, τ, $\nu(M)$, $\mu(M)$, $\sup_\Omega |\mathbf{A}(x, u, 0)|$, $\partial\Omega$, $|\varphi|_2$ and $M = \sup_\Omega |u|$.

15.5. Selected Existence Theorems

It is not feasible to present here a comprehensive account of existence theorems for the classical Dirichlet problem that follow by combination of the results of Chapters 10, 13, 14 and 15. Instead we shall present a selection of results which hopefully will serve as an illustration of the scope of the theory.

(i) *Uniformly elliptic equations in the general form* (15.1). We assume the structure conditions

$$\begin{aligned} a^{ij}, \bar{\delta}a^{ij} &= O(\lambda), \\ \delta a^{ij} &= o(\lambda), \\ b &= O(\lambda|p|^2) \\ \bar{\delta}b &\leqslant O(\lambda|p|^2) \\ \delta b &\leqslant o(\lambda|p|^2), \end{aligned} \tag{15.81}$$

as $|p| \to \infty$, uniformly for $x \in \Omega$ and bounded z. Then we have, by Theorems 10.3, 13.8, 14.1, 15.2 and 15.5.

Theorem 15.10. *Let Ω be a bounded domain in $\mathbb{R}^n$ and suppose that the operator Q is elliptic in $\bar{\Omega}$ with coefficients a^{ij}, $b \in C^1(\bar{\Omega} \times \mathbb{R} \times \mathbb{R}^n)$ satisfying the structure conditions (15.81) or (15.53) together with condition (10.10) (or (10.36)). Then, if $\partial\Omega \in C^{2,\gamma}$ and $\varphi \in C^{2,\gamma}(\bar{\Omega})$, $0 < \gamma < 1$, there exists a solution $u \in C^{2,\gamma}(\bar{\Omega})$ of the Dirichlet problem $Qu = 0$ in Ω, $u = \varphi$ on $\partial\Omega$.*

In general, if condition (10.10) (or (10.36)) is not assumed in the hypotheses of Theorem 15.10, then the Dirichlet problem $Qu = 0$ in Ω, $u = \varphi$ on $\partial\Omega$, is solvable provided the family of solutions of the problems (13.42) is uniformly bounded in Ω. Note that by Theorem 10.1 the solution is unique if the coefficients a^{ij} are independent of z and the coefficient b satisfies $D_z b \leqslant 0$. In this case, the conditions $\delta a^{ij} = o(\lambda), \delta b \leqslant o(\lambda |p|^2)$ in (15.81) would be implied by the conditions $D_x a^{ij} = O(\lambda)$, $D_x b = O(\lambda |p|^2)$.

(ii) *Uniformly-elliptic equations in the divergence form* (15.56). Here we assume the structure equations

$$
\begin{aligned}
|p|^\tau &\leqslant O(\lambda)\\
D_p \mathbf{A} &= O(|p|^\tau),\\
|p| D_z \mathbf{A}, D_x \mathbf{A}, B &= O(|p|^{\tau+2})
\end{aligned}
\tag{15.82}
$$

as $|p| \to \infty$, uniformly for $x \in \Omega$ and bounded z, with $\tau > -1$. We then have, by Theorems 10.9, 13.8, 14.1 and 15.9.

Theorem 15.11. *Let Ω be a bounded domain in $\mathbb{R}^n$ and suppose that the operator Q is elliptic in $\bar{\Omega}$ with coefficients $A^i \in C^{1,\gamma}(\bar{\Omega} \times \mathbb{R} \times \mathbb{R}^n)$, $i = 1, \ldots, n$, $B \in C^\gamma(\bar{\Omega} \times \mathbb{R} \times \mathbb{R}^n)$, $0 < \gamma < 1$, satisfying the structure conditions* (15.82) *together with the hypotheses of Theorem* 10.9 *for $\alpha = \tau + 2$. Then, if $\partial\Omega \in C^{2,\gamma}$ and $\varphi \in C^{2,\gamma}(\bar{\Omega})$, there exists a solution $u \in C^{2,\gamma}(\bar{\Omega})$ of the Dirichlet problem $Qu = 0$ in Ω, $u = \varphi$ on $\partial\Omega$.*

In order to conclude Theorem 15.11 from the estimates in Theorems 10.9, 14.1 and 15.9, we take in Theorem 13.8 the family of Dirichlet problems given by

$$
\begin{aligned}
Q_\sigma u &= \operatorname{div}\{\sigma \mathbf{A} + (1-\sigma)(1+|Du|^2)^{\tau/2} Du\} + \sigma B = 0,\\
u &= \sigma\varphi \text{ on } \partial\Omega, \quad 0 \leqslant \sigma \leqslant 1.
\end{aligned}
\tag{15.83}
$$

Note that the family given by (13.42) does not necessarily satisfy the hypotheses of Theorem 10.9. As in the previous case, if we do not assume the hypotheses of a specific maximum principle such as Theorem 10.9, then the Dirichlet problem $Qu = 0$ in Ω, $u = \varphi$ on $\partial\Omega$ is solvable provided the solutions of a related family of problems such as (15.83) are uniformly bounded on Ω. Note that, by Theorem 10.7, the solution of the problem $Qu = 0$, $u = \varphi$ on $\partial\Omega$, is unique provided either the coefficients $\mathbf{A}$ are independent of z or the coefficient B is independent of p and also provided B is non-increasing in z.

Equations of the form

$$
Qu = a^{ij}(x, u) D_{ij} u + b(x, u, Du)
\tag{15.84}
$$

with $a^{ij} \in C^1(\Omega \times \mathbb{R})$ can be written in the divergence form (15.57) with

$$
\begin{aligned}
A^i(x, z, p) &= a^{ij}(x, z) p_j,\\
B(x, z, p) &= b(x, z, p) - D_z a^{ij}(x, z) p_i p_j - D_{x_i} a^{ij}(x, z) p_j.
\end{aligned}
$$

Therefore, using Theorems 10.3, 13.8, 14.1 and 15.9, we have

Theorem 15.12. *Let Ω be a bounded domain in $\mathbb{R}^n$ and suppose that the operator Q given by (15.84) is elliptic in $\bar{\Omega}$ with coefficients*

$$a^{ij} \in C^1(\bar{\Omega} \times \mathbb{R}), \qquad b \in C^\gamma(\bar{\Omega} \times \mathbb{R} \times \mathbb{R}^n), 0<\gamma<1,$$

satisfying $b = O(|p|^2)$ as $|p| \to \infty$, uniformly for $x \in \Omega$ and bounded z, together with condition (10.10) (or 10.36)). Then, if $\partial\Omega \in C^{2,\gamma}$ and $\varphi \in C^{2,\gamma}(\bar{\Omega})$, there exists a solution $u \in C^{2,\gamma}(\bar{\Omega})$ of the Dirichlet problem $Qu = 0$ in Ω, $u = \varphi$ on $\partial\Omega$.

(iii) *Non-uniformly elliptic equations in general domains.* Let us now assume that the coefficients of equation (15.1) can be decomposed according to (15.23) such that the following structure conditions are fulfilled

$$\begin{aligned} |p|a^{ij}, b &= O(\mathscr{E}), \\ \bar{\delta}a^{ij}_* &= O(\sqrt{\lambda_* \mathscr{E}}/|p|), \\ \delta a^{ij}_* &= o(\sqrt{\lambda_* \mathscr{E}}/|p|), \\ \bar{\delta}b &\leqslant O(\mathscr{E}), \\ \delta b &\leqslant o(\mathscr{E}), \end{aligned} \tag{15.85}$$

as $|p| \to \infty$, uniformly for $x \in \Omega$ and bounded z. Then, by combining Theorems 10.3, 13.8, 14.1 and 15.2, we have

Theorem 15.13. *Let Ω be a bounded domain in $\mathbb{R}^n$ with coefficients a^{ij}, $b \in C^1(\bar{\Omega} \times \mathbb{R} \times \mathbb{R}^n)$ satisfying the structure conditions (15.85) together with condition (10.10) (or (10.36)). Then, if $\partial\Omega \in C^{2,\gamma}$ and $\varphi \in C^{2,\gamma}(\bar{\Omega})$, $0 < \gamma < 1$, there exists a solution $u \in C^{2,\gamma}(\bar{\Omega})$ of the Dirichlet problem $Qu = 0$ in Ω, $u = \varphi$ on $\partial\Omega$.*

Theorem 15.13 is clearly an extension of Theorem 15.10 and the remarks following the latter are also pertinent here. Note that when a decomposition of the form (15.2) is valid we have $\delta a^{ij}_* = 0$, and moreover if $a^{ij}_*(p) = a^{ij}_*(p/|p|)$ we also have $\bar{\delta}a^{ij}_* = 0$.

(iv) *Non-uniformly elliptic equations in convex domains.* Let us now assume that the decomposition (15.2) is valid and

$$\begin{aligned} b &= o(|p|\Lambda), \\ \delta b &\leqslant O(b^2/|p|^2\mathscr{T}_*), \quad (\mathscr{T}_* = \operatorname{trace}[a^{ij}_*]), \end{aligned} \tag{15.86}$$

as $|p| \to \infty$, uniformly in $x \in \Omega$ and bounded z. Then from Theorems 13.8, 15.1 and Corollary 14.5 we have

Theorem 15.14. *Let Ω be a uniformly convex domain in $\mathbb{R}^n$ and suppose that the operator Q is elliptic in $\bar{\Omega}$ with coefficients a^{ij}, $b \in C^1(\bar{\Omega} \times \mathbb{R} \times \mathbb{R}^n)$ satisfying the*

structure conditions (15.86). *Then, if* $\partial\Omega \in C^{2,\gamma}$ *and* $\varphi \in C^{2,\gamma}(\bar{\Omega})$, $0<\gamma<1$, *there exists a solution* $u \in C^{2,\gamma}(\bar{\Omega})$ *of the Dirichlet problem* $Qu=0$ *in* Ω, $u=\varphi$ *on* $\partial\Omega$, *provided the family of* $C^{2,\gamma}(\bar{\Omega})$ *solutions of the problems*

$$\begin{aligned} Q_\sigma u = a^{ij}(x, u, Du)D_{ij}u + \sigma b(\sigma x + (1-\sigma)x_0, \sigma u, Du) = 0, \\ u = \sigma\varphi \text{ on } \partial\Omega, \quad 0 \leqslant \sigma \leqslant 1, \end{aligned} \tag{15.87}$$

is uniformly bounded in Ω, *for some fixed* $x_0 \in \Omega$.

An existence theorem for non-uniformly elliptic equations in the divergence form (15.57) follows in a similar fashion when Theorem 15.6 is used in place of Theorem 15.1 above. We assume that the coefficients A^i and B in (15.56) satisfy for some $\tau \in \mathbb{R}$, the conditions

$$\begin{aligned} |p|^\tau \leqslant O(\lambda), \\ |p| D_z \mathbf{A}, D_x \mathbf{A}, B = o(|p|^{\tau+1}) \end{aligned} \tag{15.88}$$

as $|p| \to \infty$, uniformly for $x \in \Omega$ and bounded z. Then we have

Theorem 15.15. *Let* Ω *be a uniformly convex domain in* $\mathbb{R}^n$ *and suppose that the operator* Q *is elliptic in* $\bar{\Omega}$ *with coefficients* $A^i \in C^{1,\gamma}(\bar{\Omega} \times \mathbb{R} \times \mathbb{R}^n)$, $i=1, \ldots, n$, $B \in C^\gamma(\bar{\Omega} \times \mathbb{R} \times \mathbb{R}^n)$, $0<\gamma<1$, *satisfying the structure conditions* (15.88). *Then, if* $\partial\Omega \in C^{2,\gamma}$ *and* $\varphi \in C^{2,\gamma}(\bar{\Omega})$, *there exists a solution* $u \in C^{2,\gamma}(\bar{\Omega})$ *of the Dirichlet problem* $Qu=0$ *in* Ω, $u=\varphi$ *on* $\partial\Omega$, *provided the family of* $C^{2,\gamma}(\bar{\Omega})$ *solutions of the problems* (13.42) *is uniformly bounded in* Ω.

(v) *Problems with boundary curvature conditions.* Let us now consider operators that are decomposed according to both (14.43) and (15.23). In particular, we shall assume that

$$\begin{aligned} a^{ij}(x, z, p) &= \Lambda a^{ij}_\infty(x, p/|p|) + a^{ij}_0(x, z, p) \\ &= a^{ij}_*(x, z, p/|p|) + \tfrac{1}{2}[p_i c_j(x, z, p) + c_i(x, z, p)p_j], \\ b(x, z, p) &= |p| \Lambda b_\infty(x, z, p/|p|) + b_0(x, z, p) \end{aligned} \tag{15.89}$$

where $a^{ij}_\infty \in C^1(\bar{\Omega} \times B_1(0))$, $a^{ij}_*, b_\infty \in C^1(\bar{\Omega} \times \mathbb{R} \times B_1(0))$, $i, j=1, \ldots, n$, the matrices $[a^{ij}_\infty]$, $[a^{ij}_*]$ are non-negative and symmetric, and b_∞ is non-increasing with respect to z. We shall impose the following structure conditions

$$\begin{aligned} \delta\mathscr{E} &\leqslant \mathscr{E} + o(\mathscr{E}), \\ \delta\mathscr{E}, \delta b, (\delta - 1)b_0 &\leqslant O(\mathscr{E}), \\ \delta a^{ij}_* &= O(\sqrt{\mathscr{E}}/|p|), \\ a^{ij}_0 &= o(\Lambda) \\ b_0 &= o(|p|) \end{aligned} \tag{15.90}$$

as $|p| \to \infty$, uniformly for $x \in \Omega$ and bounded z. Then we have, by Theorems 13.8, 14.9 and 15.2,

Theorem 15.16. *Let Ω be a bounded domain in $\mathbb{R}^n$ and suppose that the operator Q is elliptic in $\bar{\Omega}$ with coefficients a^{ij}, $b \in C^1(\bar{\Omega} \times \mathbb{R} \times \mathbb{R}^n)$ satisfying the structure conditions* (15.89), (15.90). *Then, if $\partial\Omega \in C^{2,\gamma}$, $\varphi \in C^{2,\gamma}(\bar{\Omega})$, $0<\gamma<1$, and the inequalities*

$$(15.91) \qquad \mathscr{K}^+ > b_\infty(y, \varphi(y), \mathbf{v}), \qquad \mathscr{K}^- > -b_\infty(y, \varphi(y), -\mathbf{v})$$

hold at each point $y \in \partial\Omega$, where $\mathbf{v}$ is the unit inner normal at y and $\mathscr{K}^+, \mathscr{K}^-$, given by (14.44), *are non-negative, it follows that the Dirichlet problem $Qu=0$ in Ω, $u=\varphi$ on $\partial\Omega$, is solvable provided the family of $C^{2,\gamma}(\bar{\Omega})$ solutions of the problems* (13.42) *is uniformly bounded in Ω.*

In order to permit the non-strict inequalities in (15.91) we need to strengthen the structure conditions (15.90) so that

$$(15.92) \qquad \begin{aligned} \delta\mathscr{E} &\leqslant \mathscr{E} + o(\mathscr{E}), \\ 1, \delta\mathscr{E}, \delta b, \bar{\delta} b_0 &\leqslant O(\mathscr{E}), \\ \delta a_*^{ij} &= O(\sqrt{\mathscr{E}}/|p|), \\ a_0^{ij} &= O(\mathscr{E}/|p|), \\ b_0 &= O(\mathscr{E}), \end{aligned}$$

as $|p| \to \infty$, uniformly for $x \in \Omega$ and bounded z. Then we have, by Theorems 13.8, 14.6, 15.2,

Theorem 15.17. *Let Ω be a bounded domain in $\mathbb{R}^n$ and suppose that the operator Q is elliptic in $\bar{\Omega}$ with coefficients a^{ij}, $b \in C^1(\bar{\Omega} \times \mathbb{R} \times \mathbb{R}^n)$ satisfying the structure conditions* (15.89), (15.92). *Then, if $\partial\Omega \in C^{2,\gamma}$, $\varphi \in C^{2,\gamma}(\bar{\Omega})$, $0<\gamma<1$, and the inequalities*

$$(15.93) \qquad \mathscr{K}^+ \geqslant b_\infty(y, \varphi(y), \mathbf{v}), \quad \mathscr{K}^- \geqslant -b_\infty(y, \varphi(y), -\mathbf{v}), \quad \mathscr{K}^\pm \geqslant 0$$

hold at each point $y \in \partial\Omega$, it follows that the Dirichlet problem $Qu=0$ in Ω, $u=\varphi$ on $\partial\Omega$, is solvable provided the family of $C^{2,\gamma}(\bar{\Omega})$ solutions of the problems (13.42) *is uniformly bounded in Ω.*

15.6. Existence Theorems for Continuous Boundary Values

By means of the interior estimates, Theorems 15.3 and 15.6, certain of the existence theorems of the preceding section can be extended to hold when the function φ is only assumed to be continuous on $\partial\Omega$. The basic procedure is to approximate the function φ uniformly on $\partial\Omega$ by functions $\varphi_m \in C^{2,\gamma}(\Omega)$ and solve the Dirichlet problems, $Qu_m = 0$ in Ω, $u = \varphi_m$ on $\partial\Omega$, for functions $u_m \in C^{2,\gamma}(\bar{\Omega})$. The interior estimates, Theorems 15.3 and 15.8, on combination with the interior Hölder derivative estimates, Theorems 13.1 or 13.3 and the Schauder interior estimate, Theorem 6.2, guarantee that a subsequence of the sequence $\{u_m\}$ converges uniformly on compact subsets of Ω, together with its first and second derivatives, to a function $u \in C^{2,\gamma}(\Omega)$ satisfying $Qu = 0$ in Ω. The modulus of continuity estimates, Theorem 14.15, ensure that $u \in C^0(\bar{\Omega}) \cap C^2(\Omega)$ and moreover that $u = \varphi$ on $\partial\Omega$. For this procedure to work we also require that the sequence $\{u_m\}$ be uniformly bounded, that is we require a maximum principle as in Chapter 10. By approximation of the domain Ω by $C^{2,\gamma}$ domains we can also, through Theorem 14.15, remove the restriction that $\partial\Omega \in C^{2,\gamma}$. Let us now state two existence theorems for general domains which can be obtained through this procedure. In Section 16.3, we shall consider analogous results for the minimal surface and prescribed mean curvature equations.

Theorem 15.18. *Let Ω be a bounded domain in $\mathbb{R}^n$ satisfying an exterior sphere condition at each point of the boundary $\partial\Omega$. Let Q be an elliptic operator Ω with coefficients $a^{ij}, b \in C^1(\Omega \times \mathbb{R} \times \mathbb{R}^n)$ satisfying the hypotheses of Theorems* 15.3 *(or* 15.5*) and* 14.1 *together with condition* (10.10) *(or* (10.36)*). Then for any function $\varphi \in C^0(\partial\Omega)$, there exists a solution $u \in C^0(\bar{\Omega}) \cap C^2(\Omega)$ of the Dirichlet problem $Qu = 0$ in Ω, $u = \varphi$ on $\partial\Omega$.*

For equations in the divergence form (15.57), we have

Theorem 15.19. *Let Ω be a bounded domain in $\mathbb{R}^n$ satisfying an exterior sphere condition at each point of the boundary $\partial\Omega$. Let Q be a divergence structure operator with coefficients $A^i \in C^{1,\gamma}(\Omega \times \mathbb{R} \times \mathbb{R}^n)$, $i = 1, \ldots, n$, $B \in C^{\gamma}(\Omega \times \mathbb{R} \times \mathbb{R}^n)$, $0 < \gamma < 1$, satisfying the hypotheses of Theorem* 15.8, *together with the hypotheses of Theorem* 10.9 *for $\alpha = \tau + 2$. Then, for any function $\varphi \in C^0(\partial\Omega)$, there exists a solution $u \in C^0(\bar{\Omega}) \cap C^{2,\gamma}(\Omega)$ of the Dirichlet problem $Qu = 0$ in Ω, $u = \varphi$ on $\partial\Omega$.*

Notes

The essential ideas in the maximum principle approach to gradient estimates, presented in Sections 15.1 and 15.2, go back to Bernstein [BE 3, 6]. Bernstein's method was substantially developed by Ladyzhenskaya [LA] and Ladyzhenskaya and Ural'tseva [LU 2, 4] to yield both global and interior gradient estimates for uniformly elliptic equations. Later Serrin [SE 3], by exploiting representations of the form (15.2), extended these results to encompass equations with similar features to the prescribed mean curvature equation (10.7). Theorems 15.2 and 15.3 are very close to results in [LU 5, 6] (see also [IV 1, 2]), although they, as well as Theorem 15.1, are formulated similarly to the statements in [SE 4]. Rather than use multipliers r, s, t, the authors in [LU 5, 6] point out that their hypotheses need only be satisfied by an operator equivalent to the given one. Our treatment differs from those in [LU 5, 6] and [SE 4], in that we consider $C^2(\Omega)$ solutions instead of $C^3(\Omega)$ solutions. Also, our proofs and results will continue to be valid for solutions in the space $C^{0,1}(\bar{\Omega}) \cap W^{2,2}(\Omega)$.

The global gradient bound for solutions of divergence structure equations, Theorem 15.6, is due to Trudinger [TR 3] while the estimates of Theorems 15.7, 15.8 and 15.9 for uniformly elliptic divergence structure equations are due to Ladyzhenskaya and Ural'tseva [LU 1, 2]. Our proof of Theorem 15.8 differs in some aspects from that in [LU 2]. For further gradient estimates for non-uniformly elliptic, divergence structure equations, see [IO], [OS 1, 2]. Some of the existence theorems in Sections 15.5 and 15.6 are already formulated in the literature, see for example [LU 4], [SE 3]. The survey paper [ED] provides a clear account of various aspects of the execution of the existence procedure. We note also that many of the above existence theorems extend to $C^{1,\alpha}$ boundary data, $0 < \alpha < 1$; (see [LB 1]).

Finally, we remark that Lemma 15.4 resolved positively the long standing problem as to whether the natural conditions (that is, the case $\theta = 1, r = s = t_i = 0$ in (15.53)) alone suffice for interior and global gradient bounds; (see [LU 7], [TR 12]).

Problems

15.1. Show that Theorem 15.2 continues to hold for positive a and c provided $\operatorname{osc}_{\Omega} u$ is sufficiently small.

15.2. Show that Theorems 15.2 and 15.3 continue to hold for positive a and c provided $b < -2\sqrt{ac}$.

15.3. Show that Theorem 15.6 continues to hold if the structure condition (15.60) is replaced by

$$(15.94)\qquad |p|D_z\mathbf{A},\ D_x\mathbf{A},\ B = o(|p|^\gamma) \quad \text{as } |p| \to \infty,$$

where $\gamma = \tau + 1 + (\tau + 2)/n$, provided $\tau > -1$ and $u = 0$ on $\partial\Omega$.

15.4. Show that Theorem 15.6 continues to hold if the structure condition (15.60) is replaced by

$$(15.95)\qquad |p|D_z\mathbf{A},\ D_x\mathbf{A},\ B = o(|p|^\gamma) \quad \text{as } |p| \to \infty,$$

where $\gamma = \tau + 1 + 1/n$, provided $\tau \leqslant -1$ and $u = 0$ on $\partial\Omega$.

15.5. Show that an interior gradient bound is valid for $C^2(\Omega)$ solutions of equation (15.56) provided that the structure conditions (15.60),

$$(15.64)' \qquad D_p\mathbf{A} = O(|p|^{\tau+\sigma}),$$

$$(15.66)' \qquad g = o(|p|^{\tau+2})$$

hold for $\tau > -1$, $\sigma < 1$.

15.6. Consider the Dirichlet problem

$$(15.96) \qquad \varepsilon\Delta u + g(Du) = f(x) \quad \text{in } \Omega, \qquad u = 0 \quad \text{on } \partial\Omega,$$

in a bounded convex domain $\Omega \subset \mathbb{R}^n$, where $\varepsilon > 0$, $g \in C^1(\mathbb{R}^n)$, $f \in C^1(\bar{\Omega})$, $\sqrt{|p|}/g = O(1)$ as $|p| \to \infty$, and $g(0) \leqslant f(x)$ for all $x \in \Omega$. Using Theorem 15.1, together with linear barrier functions, deduce the unique solvability of (15.96), for sufficiently small ε, in the space $C^{0,1}(\bar{\Omega}) \cap C^2(\Omega)$. By letting $\varepsilon \to 0$, show that there exists a $C^{0,1}(\bar{\Omega})$ solution of the *first order* Dirichlet problem

$$(15.97) \qquad g(Du) = f(x) \quad \text{a.e. } (\Omega), \quad u = 0 \quad \text{on } \partial\Omega,$$

and using suitable approximation show that the above conditions on f and g can be replaced by $g \in C^0(\mathbb{R}^n)$, $g \to \infty$ as $|p| \to \infty$, $f \in L^\infty(\Omega)$, $g(0) \leqslant f(x)$ for all $x \in \Omega$. Find a similar result for arbitrary domains satisfying exterior sphere conditions. In a certain sense the solution of (15.97) thus obtained is unique. (Cf. [LP 3, 5]).

Chapter 16

Equations of Mean Curvature Type

In this chapter we focus attention on both the prescribed mean curvature equation,

$$\mathfrak{M}u=(1+|Du|^2)\Delta u-D_iuD_juD_{ij}u=nH(1+|Du|^2)^{3/2}, \tag{16.1}$$

and a related family of equations in two variables. Our main concern is with interior derivative estimates for solutions. We shall see that not only can interior gradient bounds be established for solutions of these equations but that also their non-linearity leads to strong second derivative estimates which distinguish them from uniformly elliptic equations such as Laplace's equation. In particular we shall derive an extension of the classical result of Bernstein that a $C^2(\mathbb{R}^2)$ solution of the minimal surface equation in $\mathbb{R}^2$ must be a linear function (Theorem 16.12).

The approach to interior gradient bounds, in this chapter, differs considerably from those in Chapter 15 (although it does have some features in common with the divergence structure case of Section 15.4). An interior gradient bound for a solution of equation (16.1), Theorem 16.5, is derived through a consideration of the tangential gradient and Laplace operators on the hypersurface in $\mathbb{R}^{n+1}$ given by the graph of the solution u. The basic estimates on hypersurfaces are supplied in Section 16.1.

The study of the general class of equations of prescribed mean curvature type in two variables is taken up in Section 16.4. Interior first and second derivative estimates (Theorems 16.20 and 16.21) arise through a treatment in Section 16.5 of quasiconformal mappings between surfaces in $\mathbb{R}^3$, which extends that of Section 12.1.

Sections 16.4 to 16.8 and a portion of Section 16.1 of this chapter were written in collaboration with L. M. Simon and they are essentially his contribution.

16.1. Hypersurfaces in $\mathbb{R}^{n+1}$

A subset $\mathfrak{S}$ of $\mathbb{R}^{n+1}$ is called a C^k hypersurface in $\mathbb{R}^{n+1}$ if locally $\mathfrak{S}$ can be represented as the graph of a C^k function over an open subset of $\mathbb{R}^n$. Our concern here is with C^2 hypersurfaces and for convenience we shall assume that $\mathfrak{S}$ can be globally represented as the level surface of a C^2 function, that is, there exists an open subset

$\mathscr{U}$ of $\mathbb{R}^{n+1}$ and a function $\Phi \in C^2(\mathscr{U})$, with $D\Phi \neq 0$ on $\mathscr{U}$, such that

$$\mathfrak{S} = \{x \in \mathscr{U} \mid \Phi(x) = 0\}. \tag{16.2}$$

For the applications of this chapter, $\mathfrak{S}$ will always be the graph of a function $u \in C^2(\Omega)$ where Ω is a domain in $\mathbb{R}^n$. In this case we can take $\mathscr{U} = \Omega \times \mathbb{R}$ and

$$\Phi = x_{n+1} - u(x'), \quad x' = (x_1, \ldots, x_n) \in \Omega. \tag{16.3}$$

When $\mathfrak{S}$ is given by (16.2) the normal $\boldsymbol{\nu}$ to $\mathfrak{S}$ (in the direction of increasing Φ), is given by

$$\boldsymbol{\nu} = \frac{D\Phi}{|D\Phi|}.$$

For $g \in C^1(\mathscr{U})$, the *tangential gradient* δg of g on $\mathfrak{S}$ is defined by

$$\delta g = Dg - (\boldsymbol{\nu} \cdot Dg)\boldsymbol{\nu}. \tag{16.4}$$

For any point $y \in \mathfrak{S}$, the tangential gradient $\delta g(y)$ is the projection of the gradient $Dg(y)$ onto the tangent plane to $\mathfrak{S}$ at y. Clearly we have

$$\boldsymbol{\nu} \cdot \delta g = 0 \tag{16.5}$$

and

$$|\delta g|^2 = |Dg|^2 - |\boldsymbol{\nu} \cdot Dg|^2, \tag{16.6}$$

so that

$$|\delta g| \leqslant |Dg| \tag{16.7}$$

and

$$|\nu_{n+1}|\,|Dg| \leqslant |\delta g| \quad \text{if } D_{n+1} g \equiv 0. \tag{16.8}$$

Furthermore, it is clear that δg only depends on the values of g on $\mathfrak{S}$. For, suppose $\bar{g} \in C^1(\mathscr{U})$ satisfies $g = \bar{g}$ on $\mathfrak{S}$. Then $D(g - \bar{g}) = k\boldsymbol{\nu}$ for some constant k and hence

$$\delta(g - \bar{g}) = k\boldsymbol{\nu} - k\boldsymbol{\nu} = 0.$$

Next, by formula (14.102) in Section 14.6, we have

$$\begin{aligned} \delta_i \nu_i &= D_i\left[\frac{D_i\Phi}{|D\Phi|}\right] - \nu_i \nu_j D_j \nu_i \\ &= -nH - \tfrac{1}{2}\nu_j D_j |\boldsymbol{\nu}|^2 \\ &= -nH \end{aligned}$$

where H denotes the mean curvature of $\mathfrak{S}$ with respect to $\boldsymbol{\nu}$. Hence we have the formula

$$H = -\frac{1}{n}\delta_i\nu_i. \tag{16.9}$$

The following lemma provides an integration by parts formula for the differential operator δ.

Lemma 16.1. *Letting dA denote the area element in $\mathfrak{S}$, we have*

$$\int_{\mathfrak{S}} \delta g\, dA = -n\int_{\mathfrak{S}} gH\boldsymbol{\nu}\, dA \tag{16.10}$$

for all $g \in C_0^1(\mathscr{U})$.

Proof. We shall establish formula (16.10) for the case where $\mathfrak{S}$ is the graph of a C^2 function. The general case then follows by means of a partition of unity. Accordingly, let us assume that Φ is given by (16.3) so that at points $(x, u(x)) \in \mathfrak{S}$, we have

$$\nu_i = -\frac{D_i u}{v}, \quad i = 1, \ldots, n,$$

$$\nu_{n+1} = \frac{1}{v}, \tag{16.11}$$

$$H = \frac{1}{n}D_i\left[\frac{D_i u}{v}\right],$$

and

$$dA = v\, dx,$$

where

$$v = \sqrt{1 + |Du|^2}.$$

Defining $\tilde{g} \in C^1(\mathscr{U})$ by

$$\tilde{g}(x, x_{n+1}) = g(x, u(x)),$$

we obviously have $\tilde{g} = g$ on $\mathfrak{S}$ and hence $\delta\tilde{g} = \delta g$. Therefore, for $i \leqslant n$,

$$\begin{aligned}\int_{\mathfrak{S}} \delta_i g\, dA &= \int_{\mathfrak{S}} \delta_i \tilde{g}\, dA \\ &= \int_{\Omega} (D_i g - \nu_i\nu_j D_j g) v\, dx\end{aligned}$$

$$= -\int_\Omega \tilde{g}\left\{D_i v - \sum_{j=1}^n D_j(\nu_i \nu_j v)\right\} dx \quad \text{by integration by parts,}$$

$$= -\int_\Omega \tilde{g}\left\{D_i v - \sum_{j=1}^n \nu_j D_j(\nu_i v) + nvH\nu_i\right\} dx \quad \text{by (16.11),}$$

$$= -n\int_\Omega \tilde{g}H\nu_i v\, dx - \int_\Omega \tilde{g}\left(\sum_{j=1}^n \nu_j D_{ij}u + D_i v\right) dx$$

$$= -n\int_\Omega \tilde{g}H\nu_i v\, dx \quad \text{by (16.11).}$$

For $i=n+1$, we have

$$\int_\mathfrak{S} \delta_{n+1}g\, dA = \int_\mathfrak{S} \delta_{n+1}\tilde{g}\, dA$$

$$= -\int_\Omega \nu_{n+1}\sum_{j=1}^n \nu_j D_j\tilde{g}v\, dx$$

$$= \int_\Omega \tilde{g}\sum_{j=1}^n D_j\nu_j\, dx$$

$$= -n\int_\Omega \tilde{g}H\, dx$$

$$= -n\int_\mathfrak{S} gH\nu_{n+1}\, dA. \quad \square$$

Note that the case $i=n+1$ in Lemma 16.1 is equivalent to the integral form of the prescribed mean curvature equation.

For $g\in C^2(\mathscr{U})$, the *Laplacian* (or *Laplace-Beltrami operator*) of g on $\mathfrak{S}$ is defined by

$$(16.12)\qquad \Delta g = \delta_i\delta_i g.$$

From the integration by parts formula (16.10) and (16.5) we have

$$(16.13)\qquad \int_\mathfrak{S} \varphi\,\Delta g\, dA = \int_\mathfrak{S} g\,\Delta\varphi\, dA = \int_\mathfrak{S} -\delta g\cdot\delta\varphi\, dA$$

for all $\varphi\in C_0^2(\mathscr{U})$.

We proceed now to derive some important inequalities involving the operators δ and Δ on $\mathfrak{S}$. These inequalities will include useful extensions of the mean value inequalities, Theorem 2.1, and the potential representation, Lemma 7.14, to hypersurfaces in $\mathbb{R}^{n+1}$. If y is a point on $\mathfrak{S}$, $r=|x-y|$ and $\psi\in C^2(\mathbb{R})$ we have, by calculation,

$$\delta_i\psi(r)=(\delta_{ij}-\nu_i\nu_j)\frac{x_j-y_j}{r}\psi'(r),$$

$$\Delta\psi(r)=\frac{n\psi'(r)}{r}+\left[\frac{\psi''(r)}{r^2}-\frac{\psi'(r)}{r^3}\right](r^2-|\nu_i(x_i-y_i)|^2)+\frac{n\psi'(r)}{r}H\nu_i(x_i-y_i) \tag{16.14}$$

$$=\frac{n\psi'(r)}{r}+\left[\psi''-\frac{\psi'(r)}{r}\right]|\delta r|^2+\frac{n\psi'(r)}{r}H\boldsymbol{\nu}\cdot(x-y)$$

since, by (16.6),

$$|\delta r|^2=1-\frac{|\boldsymbol{\nu}\cdot(x-y)|^2}{r^2}.$$

In particular, let χ be a non-negative, non-increasing function in $C^1(\mathbb{R})$ with support in the interval $(-\infty, 1)$ and set

$$\psi(r)=\int_r^\infty \tau\chi(\tau/\rho)\,d\tau$$

where $0<\rho<R$ and the ball $B_R(y)\subset\mathscr{U}$. We then have by (16.14)

$$\begin{aligned}\Delta\psi(r)&=-\{n\chi(r/\rho)+r\chi'(r/\rho)|\delta r|^2/\rho+n\chi(r/\rho)H\boldsymbol{\nu}\cdot(x-y)\}\\ &=\rho^{n+1}D_\rho[\rho^{-n}\chi(r/\rho)]+r\chi'(r/\rho)(1-|\delta r|^2)/\rho-n\chi(r/\rho)H\boldsymbol{\nu}\cdot(x-y)\\ &\leqslant\rho^{n+1}D_\rho[\rho^{-n}\chi(r/\rho)]-n\chi(r/\rho)H\boldsymbol{\nu}\cdot(x-y).\end{aligned} \tag{16.15}$$

In the special case where $\mathfrak{S}$ is a minimal surface, that is $H\equiv 0$, inequality (16.15) reduces to

$$\Delta\psi(r)\leqslant\rho^{n+1}D_\rho[\rho^{-n}\chi(r/\rho)]. \tag{16.16}$$

The relations (16.15), (16.16) are fundamental to our treatment of interior estimates. We illustrate their application by first considering the minimal surface case, $H\equiv 0$. Let g be a non-negative function in $L^1(\mathfrak{S})$ and suppose that

$$\int_{\mathfrak{S}} g\,\Delta\varphi\,dA\geqslant 0 \tag{16.17}$$

for all non-negative $\varphi \in C_0^2(\mathscr{U})$. By choosing $\varphi = \psi$ in (16.17), we obtain immediately from (16.16) the inequality

$$D_\rho\left[\frac{1}{\rho^n}\int_{\mathfrak{S}} g\chi(r/\rho)\,dA\right] \geqslant 0,$$

that is, the function I_χ given by

$$(16.18) \qquad I_\chi(\rho) = \frac{1}{\omega_n \rho^n}\int_{\mathfrak{S}} g\chi(r/\rho)\,dA$$

is non-decreasing in ρ. Letting χ approximate the characteristic function of the interval $(-\infty, 1)$, in an appropriate fashion, we obtain that the function I given by

$$(16.19) \qquad I(\rho) = \frac{1}{\omega_n \rho^n}\int_{\mathfrak{S} \cap B_\rho(y)} g\,dA$$

is also non-decreasing. Since

$$(16.20) \qquad \lim_{\rho \to 0} I(\rho) = g(y)$$

for almost all (with respect to dA) $y \in \mathfrak{S}$, we conclude the *mean value inequality*

$$(16.21) \qquad g(y) \leqslant \frac{1}{\omega_n R^n}\int_{\mathfrak{S} \cap B_R(y)} g\,dA$$

for almost all $y \in \mathfrak{S}$ with $B_R(y) \subset \mathscr{U}$. Let us call a function $g \in C^2(\mathscr{U})$ *subharmonic* (*harmonic*) on $\mathfrak{S}$ if $\Delta g \geqslant 0 (=0)$ on $\mathfrak{S}$. Using (16.13), we thus have

Lemma 16.2. *Let g be a non-negative, subharmonic function on a C^2 minimal hypersurface $\mathfrak{S}$. Then for any point $y \in \mathfrak{S}$ and ball $B_R(y) \subset \mathscr{U}$, the inequality* (16.21) *is valid.*

When $\mathfrak{S}$ is a hyperplane, inequality (16.21) reduces to the mean value inequality for non-negative subharmonic functions in Euclidean space $\mathbb{R}^n$. Note however that in this case we do not need to assume that g is non-negative. When g is a positive constant, we obtain from (16.21) the estimate

$$(16.22) \qquad A(\mathfrak{S} \cap B_R(y)) \geqslant \omega_n R^n$$

where $A(\mathfrak{S} \cap B_R(y))$ denotes the area of $\mathfrak{S} \cap B_R(y)$. Henceforth we shall write

$\mathfrak{S}_R(y)=\mathfrak{S}\cap B_R(y)$ and generally abbreviate $\mathfrak{S}_R(y)=\mathfrak{S}_R$ when there is no ambiguity.

We turn our attention now to the case of a general hypersurface $\mathfrak{S}$ and $C^1(\mathscr{U})$ function g. Although the procedure adopted above in the minimal surface case can be generalized, we shall instead proceed somewhat differently, thereby giving an alternative proof of Lemma 16.2. Let y be a point on $\mathfrak{S}$ and suppose that the ball $B_R(y)\subset\mathscr{U}$. Let χ be a non-negative, non-increasing function in $C^1[0,\infty)$ with support in the interval $[0, R]$. Defining $\psi\in C^2(\mathscr{U})$ by

$$\psi(r)=\int_r^\infty \tau\chi(\tau)\,d\tau,$$

we have, by (16.15),

$$\begin{aligned}\Delta\psi(r)&=-\{n\chi(r)+r\chi'(r)|\delta r|^2+n\chi(r)H\nu\cdot(x-y)\}\\&=-(n\chi(r)+r\chi'(r))+r\chi'(r)(1-|\delta r|^2)-n\chi(r)H\nu\cdot(x-y).\end{aligned}\tag{16.23}$$

Therefore, substituting into (16.13), we obtain

$$\begin{aligned}&\int_{\mathfrak{S}}(n\chi(r)+r\chi'(r))g\,dA-\int_{\mathfrak{S}}r\chi'(r)(1-|\delta r|^2)g\,dA\\&\qquad=-n\int_{\mathfrak{S}}\chi(r)gH\nu\cdot(x-y)\,dA+\int_{\mathfrak{S}}\delta\psi\cdot\delta g\,dA.\end{aligned}\tag{16.24}$$

If we further restrict χ so that

$$\chi(r)=\varepsilon^{-n}-R^{-n}\quad\text{for } r<\varepsilon,$$

where $0<\varepsilon<R$, we can write the above relation as

$$\begin{aligned}&n(\varepsilon^{-n}-R^{-n})\int_{\mathfrak{S}_\varepsilon}g\,dA+\int_{\mathfrak{S}_R-\mathfrak{S}_\varepsilon}(n\chi(r)+r\chi'(r))g\,dA\\&\qquad-\int_{\mathfrak{S}_R-\mathfrak{S}_\varepsilon}r\chi'(r)(1-|\delta r|^2)g\,dA\\&\qquad=-n\int_{\mathfrak{S}_R}\chi(r)gH\nu\cdot(x-y)\,dA+\int_{\mathfrak{S}_R}\delta\psi\cdot\delta g\,dA.\end{aligned}\tag{16.25}$$

The form of (16.25) suggests that we should choose χ so that $n\chi(r)+r\chi'(r)=$ constant in the interval $[\varepsilon, R)$. Accordingly, let us define a function χ_ε by

$$\chi_\varepsilon(r)=\begin{cases}\varepsilon^{-n}-R^{-n} & \text{for } 0\leqslant r<\varepsilon,\\ r^{-n}-R^{-n} & \text{for } \varepsilon\leqslant r<R,\\ 0 & \text{for } r\geqslant R.\end{cases}$$

We cannot immediately replace χ by χ_ε in (16.25). However, this can be accomplished by replacing χ by a sequence $\{\chi_m\}$ of non-negative, non-increasing functions in $C^1[0, \infty)$, with support in $[0, R)$ and uniformly bounded derivatives. By further requiring that $\{\chi_m\}$ converges uniformly to χ_ε and that the sequence of derivatives $\{\chi_m'\}$ converges pointwise to the function

$$\chi_\varepsilon'(r)=\begin{cases}0 & \text{for } 0\leqslant r<\varepsilon,\\ -nr^{-(n+1)} & \text{for } \varepsilon\leqslant r<R,\\ 0 & \text{for } r\geqslant R,\end{cases}$$

we can conclude, from (16.25),

$$(\varepsilon^{-n}-R^{-n})\int_{\mathfrak{S}_\varepsilon} g\,dA+\int_{\mathfrak{S}_R-\mathfrak{S}_\varepsilon} gr^{-n}(1-|\delta r|^2)\,dA$$

$$=R^{-n}\int_{\mathfrak{S}_R-\mathfrak{S}_\varepsilon} g\,dA-(\varepsilon^{-n}-R^{-n})\int_{\mathfrak{S}_\varepsilon} gHv\cdot(x-y)\,dA \tag{16.26}$$

$$-\int_{\mathfrak{S}_R-\mathfrak{S}_\varepsilon} g(r^{-n}-R^{-n})Hv\cdot(x-y)\,dA+\frac{1}{n}\int_{\mathfrak{S}_R}\delta\psi_\varepsilon(r)\cdot\delta g\,dA,$$

where

$$\psi_\varepsilon(r)=\int_r^R \tau\chi_\varepsilon(\tau)\,d\tau.$$

Letting ε tend to zero, we thus have

$$g(y)+\frac{1}{\omega_n}\int_{\mathfrak{S}_R}\frac{(1-|\delta r|^2)}{r^n}g\,dA$$

$$=\frac{1}{\omega_n R^n}\int_{\mathfrak{S}_R} g\,dA-\frac{1}{\omega_n}\int_{\mathfrak{S}_R} g(r^{-n}-R^{-n})Hv\cdot(x-y)\,dA \tag{16.27}$$

$$+\frac{1}{n\omega_n}\int_{\mathfrak{S}_R}\delta\psi(r)\cdot\delta g\,dA,$$

where

$$\psi(r)=\int_r^R \tau(\tau^{-n}-R^{-n})\,d\tau.$$

Lemma 16.2 is now an immediate consequence of the identities (16.27) and (16.13). Using the inequality

$$\text{(16.28)}\qquad \begin{aligned}|Hv\cdot(x-y)| &\leqslant r^{-2}|v\cdot(x-y)|^2+\tfrac{1}{4}H^2r^2\\ &=1-|\delta r|^2+\tfrac{1}{4}H^2r^2\end{aligned}$$

we can deduce, from (16.27), the estimate

$$\text{(16.29)}\qquad \begin{aligned}g(y)\leqslant &\frac{1}{\omega_nR^n}\int_{\mathfrak{S}_R} g|\delta r|^2\,dA+\frac{1}{4\omega_n}\int_{\mathfrak{S}_R} gH^2r^2(r^{-n}-R^{-n})\,dA\\ &-\frac{1}{n\omega_n}\int_{\mathfrak{S}_R}(r^{-n}-R^{-n})(x-y)\cdot\delta g\,dA.\end{aligned}$$

Thus we have proved the following generalization of Lemma 16.2.

Lemma 16.3. *Let g be a non-negative function in $C^1(\mathscr{U})$. Then for any point $y\in\mathfrak{S}$ and ball $B_R(y)\subset\mathscr{U}$, the inequality (16.29) is valid.*

The derivation of the interior gradient bound in the following section will be based upon Lemma 16.3. Note that for $g\in C^2(\mathscr{U})$ we can express the last term in (16.29) as

$$-\frac{1}{n\omega_n}\int_{\mathfrak{S}_R}\psi(r)\,\Delta g\,dA.$$

Further consequences of inequalities (16.26) and (16.29) will be required for the treatment of equations in two variables later in this chapter. In particular, if we set $n=2$ in inequality (16.29), we obtain

$$\text{(16.30)}\qquad g(y)\leqslant\frac{1}{\pi R^2}\int_{\mathfrak{S}_R} g(1+\tfrac{1}{4}H^2R^2)\,dA+\frac{1}{2\pi}\int_{\mathfrak{S}_R} r(r^{-2}-R^{-2})|\delta g|\,dA.$$

Setting $g\equiv1$ in (16.30), we have the estimate

$$\text{(16.31)}\qquad 1\leqslant\frac{A(\mathfrak{S}_R)}{\pi R^2}+\frac{1}{4\pi}\int_{\mathfrak{S}_R} H^2\,dA,$$

where $A(\mathfrak{S}_R)$ denotes the area of $\mathfrak{S}_R$. Consequently, if $\mathfrak{S}$ is a compact hypersurface in $\mathbb{R}^3$ (or more generally if $\mathscr{U} = \mathbb{R}^3$ and $A(\mathfrak{S}_R) = o(R^2)$ as $R \to \infty$), we have

$$\int_{\mathfrak{S}} H^2 \, dA \geqslant 4\pi. \tag{16.32}$$

Furthermore, one can show that equality holds in (16.32) if and only if $\mathfrak{S}$ is a sphere, [TR 9].

Next, defining I by (16.19), we obtain from (16.26) and (16.28)

$$\begin{aligned}\omega_n I(\varepsilon) &= \varepsilon^{-n} \int_{\mathfrak{S}_\varepsilon} g \, dA \\ &\leqslant R^{-n} \int_{\mathfrak{S}_R} g|\delta r|^2 \, dA + \varepsilon^{1-n} \int_{\mathfrak{S}_\varepsilon} g|H| \, dA \\ &\quad + \frac{1}{4} \int_{\mathfrak{S}_R} gH^2 r^2 (r^{-n} - R^{-n}) \, dA + \frac{1}{n} \int_{\mathfrak{S}_R} \delta\psi_\varepsilon(r) \cdot \delta g \, dA.\end{aligned}$$

Hence, using Young's inequality (7.6), we obtain the estimate

$$\begin{aligned}\sup_{(0,R)} I(\rho) &\leqslant \frac{1}{\omega_n R^n} \int_{\mathfrak{S}_R} g[n|\delta r|^2 + |HR|^n + nH^2 r^2 R^n (r^{-n} - R^{-n})] \, dA \\ &\quad + \frac{1}{\omega_n} \int_{\mathfrak{S}_R} r(r^{-n} - R^{-n})|\delta g| \, dA.\end{aligned} \tag{16.33}$$

Note that the last term in (16.33) can be replaced by

$$\frac{1}{\omega_n} \int_{\mathfrak{S}_R} \psi(r)(-\Delta g)^+ \, dA,$$

when $g \in C^2(\mathscr{U})$. Specializing inequality (16.33) to the case $n=2$, $g \equiv 1$, we obtain the estimate

$$\sup_{(0,R)} \frac{A(\mathfrak{S}_\rho)}{\rho^2} \leqslant \frac{2A(\mathfrak{S}_R)}{R^2} + 3 \int_{\mathfrak{S}_R} H^2 \, dA. \tag{16.34}$$

We next determine an estimate for the quantity

$$J(\rho) = D_\rho \left[\int_{\mathfrak{S}_\rho} g|\delta r|^2 \, dA \right].$$

Choosing χ as in the proof of Lemma 16.2 and using (16.13) and (16.5) we have

$$D_\rho\left[\int_{\mathfrak{S}} \chi(r/\rho) g|\delta r|^2\, dA\right] = \frac{n}{\rho}\int_{\mathfrak{S}} \chi(r/\rho)[1 + H\nu\cdot(x-y)]g\, dA$$

$$-\frac{1}{\rho}\int_{\mathfrak{S}} \delta\psi(r)\cdot\delta g\, dA.$$

Hence as χ approaches the characteristic function of the interval $(-\infty, 1)$, we see that $J(\rho)$ is well defined for all $\rho\in(0, R)$ and moreover that

$$\begin{aligned} J(\rho) &\leqslant \frac{1}{\rho}\int_{\mathfrak{S}_\rho} [n|g|(1+|H|r) + r|\delta g|]\, dA \\ &\leqslant \frac{n}{\rho}\int_{\mathfrak{S}_\rho} |g|\, dA + \int_{\mathfrak{S}_\rho} (n|Hg| + |\delta g|)\, dA. \end{aligned} \tag{16.35}$$

In particular, for $g\equiv 1$, we have

$$\begin{aligned} D_\rho\left[\int_{\mathfrak{S}_\rho} |\delta r|^2\, dA\right] &\leqslant n\left[\frac{A(\mathfrak{S}_\rho)}{\rho} + \int_{\mathfrak{S}_\rho} |H|\, dA\right] \\ &\leqslant 10\rho\left[\frac{A(\mathfrak{S}_R)}{R^2} + \int_{\mathfrak{S}_R} H^2\, dA\right] \quad \text{by (16.34)}, \end{aligned} \tag{16.36}$$

if $n=2$.

The *potential* type relations (16.27), (16.29) and (16.30) are somewhat analogous to Lemmas 7.14 and 7.16 in Chapter 7. Indeed, we shall now use (16.30) to derive an analogue of the Morrey estimate, Theorem 7.19, for two dimensional surfaces. The following lemma will be applied in Section 16.5 to conclude a Hölder estimate for generalized quasiconformal mappings.

Lemma 16.4. *Let $g\in C^1(\mathscr{U})$, $n=2$ and suppose there exist constants $K>0$ and $\beta\in(0, 1)$ such that*

$$\int_{\mathfrak{S}_\rho(\bar{y})} |\delta g|\, dA \leqslant K\rho(\rho/R)^\beta \tag{16.37}$$

for all $\bar{y}\in\mathfrak{S}_{R/4}(y)$ and all $\rho\leqslant R/4$. Then

$$\sup_{x\in\mathfrak{S}^*_\rho(y)} |g(x)-g(y)| \leqslant CK(\rho/R)^\beta \left\{\frac{A(\mathfrak{S}_R)}{R^2} + \int_{\mathfrak{S}_R} H^2\, dA\right\} \tag{16.38}$$

where C is a constant, and where $\mathfrak{S}_\rho^*(y)$ denotes the component of $\mathfrak{S}_\rho(y)$ which contains y.

Proof. We commence by writing (16.30) in the form

$$g(y) \leqslant \frac{1}{\pi}\left\{\int_{\mathfrak{S}_R} g(R^{-2}+H^2/4)\,dA + \int_0^R \rho^{-2}\int_{\mathfrak{S}_\rho} |\delta g|\,dA\,d\rho\right\}, \tag{16.39}$$

and defining, for $\rho \leqslant R/4$,

$$g_1 = \sup_{\mathfrak{S}_\rho^*(y)} g, \quad g_0 = \inf_{\mathfrak{S}_\rho^*(y)} g.$$

If

$$g_1 - g_0 \leqslant 6\beta^{-1}K(\rho/R)^\beta,$$

then Lemma 16.4 is established with $C = 6\beta^{-1}$ by (16.34). If

$$g_1 - g_0 > 6\beta^{-1}K(\rho/R)^\beta,$$

then we let N be the largest integer such that

$$N \leqslant \beta(g_1 - g_0)/6K(\rho/R)^\beta,$$

and we subdivide the interval $[g_0, g_1]$ into N pairwise disjoint intervals $I_1, I_2, \ldots, I_N$ of length $\geqslant 6\beta^{-1}K(\rho/R)^\beta$. For each $j=1,\ldots N$, we then let ψ_j be a non-negative $C^1(\mathbb{R})$ function with support contained in I_j, $\max\psi_j = 1$ and $\max\psi_j' \leqslant \beta/2K(\rho/R)^\beta$. (It is clear that such a function exists, because length $I_j \geqslant 6\beta^{-1}K(\rho/R)^\beta$.) Since $\mathfrak{S}_\rho^*(y)$ is connected, we know that for each $j=1,\ldots N$, there is a point $x^{(j)} \in \mathfrak{S}_\rho^*(y)$ such that $\psi_j[g(x^{(j)})] = 1$. Then using (16.39) with $x^{(j)}$ in place of y, ρ in place of R and $\psi_j \circ g$ in place of g, we obtain

$$1 \leqslant \frac{1}{\pi}\int_{\mathfrak{S}_\sigma(x^{(j)})} \psi_j(g)(\rho^{-2}+H^2/4)\,dA + \int_0^\rho \sigma^{-2}\int_{\mathfrak{S}_\sigma(x^{(j)})} |\delta\psi_j \circ g|\,dA\,d\sigma,$$

for all $\rho \leqslant R/4$. Consequently, using (16.37), we obtain

$$\begin{aligned}\int_0^\rho \sigma^{-2}\int_{\mathfrak{S}_\sigma(x^{(j)})} |\delta\psi_j \circ g|\,dA\,d\sigma &\leqslant \beta R^\beta(2K\rho^\beta)^{-1}\int_0^\rho \sigma^{-2}\int_{\mathfrak{S}_\sigma(x^{(j)})} |\delta g|\,dA\,d\sigma \\ &\leqslant \beta R^\beta(2K\rho^\beta)^{-1}KR^{-\beta}\int_0^\rho \sigma^{\beta-1}\,d\sigma \\ &= \tfrac{1}{2}.\end{aligned}$$

Combining the last two inequalities then gives

$$1 \leqslant \frac{1}{\pi} \int_{\mathfrak{S}_\rho(x^{(j)})} \psi_j(g)(\rho^{-2} + H^2/4)\, dA + \tfrac{1}{2},$$

so that

$$1 \leqslant \frac{2}{\pi} \int_{\mathfrak{S}_\rho(x^{(j)})} \psi_j(g)(\rho^{-2} + H^2/4)\, dA$$

$$\leqslant \frac{2}{\pi} \int_{\mathfrak{S}_{2\rho}(y)} \psi_j(g)(\rho^{-2} + H^2/4)\, dA.$$

Summing over $j = 1, \ldots, N$, and noting that $\sum \psi_j \leqslant 1$, we then deduce

$$N \leqslant \frac{2}{\pi} \int_{\mathfrak{S}_{2\rho}(y)} (\rho^{-2} + H^2)\, dA$$

$$\leqslant C\left\{\frac{A(\mathfrak{S}_R)}{R^2} + \int_{\mathfrak{S}_R} H^2\, dA\right\} \quad \text{by (16.34)}$$

and hence the estimate (16.38) follows. □

Remark. If the function g has compact support, we obtain, by letting R approach infinity in (16.27) and (16.29), the relations

$$g(y) + \frac{1}{\omega_n} \int_{\mathfrak{S}} \frac{(1 - |\delta r|^2)}{r^n} g\, dA$$

$$= \frac{1}{\omega_n} \int_{\mathfrak{S}} \frac{H\nu \cdot (x - y)}{r^n} g\, dA - \frac{1}{n\omega_n} \int_{\mathfrak{S}} \frac{(x - y) \cdot \delta g}{r^n}\, dA, \tag{16.40}$$

$$g(y) \leqslant \frac{1}{4\omega_n} \int_{\mathfrak{S}} g H^2 r^{2-n}\, dA - \frac{1}{n\omega_n} \int_{\mathfrak{S}} \frac{(x - y) \cdot \delta g}{r^n}\, dA.$$

These inequalities can be used to establish imbedding theorems of the Sobolev type (Problems 16.1, 16.2). We remark that the Sobolev inequality, Theorem 7.10, has been extended to minimal hypersurfaces in $\mathbb{R}^{n+1}$ by Miranda [MD 1] and to arbitrary submanifolds by Allard [AA], Michael and Simon [MSI].

16.2. Interior Gradient Bounds

Let Ω be a domain in $\mathbb{R}^n$ and u a function in $C^2(\Omega)$. If $\mathfrak{S}$ denotes the graph of u in $\mathbb{R}^{n+1}$, then the mean curvature (with respect to the upper normal) of $\mathfrak{S}$, at the point $x=(x', u(x'))\in\mathfrak{S}$, is given by

$$(16.41)\qquad \begin{aligned} H(x') &= -\frac{1}{n} D_i\nu_i(x') \\ &= \frac{1}{n} D_i\left[\frac{D_i u}{v}\right](x'), \end{aligned}$$

where $v=\sqrt{1+|Du|^2}$.

The object of this section is to establish a bound for $Du(x')$ in terms of H and dist $(x', \partial\Omega)$. We commence by writing the relation (16.41) in the integral form

$$(16.42)\qquad \int_\Omega (\boldsymbol{\nu}\cdot D\varphi - nH\varphi)\,dx'=0$$

for all $\varphi\in C_0^1(\Omega)$. Replacing φ by $D_k\varphi$, $k=1,\ldots,n$, and integrating by parts, we obtain

$$(16.43)\qquad \int_\Omega (D_k\boldsymbol{\nu}\cdot D\varphi - nD_kH\varphi)\,dx'=0$$

for all $\varphi\in C_0^1(\Omega)$ (cf. equation (13.3)). In (16.43) and henceforth we assume that $H\in C^1(\Omega)$. Now let us replace φ in (16.43) by $\nu_k\varphi$ where $\varphi\in C_0^1(\Omega)$. Then we obtain the equation

$$(16.44)\qquad \int_\Omega D_k\nu_i D_i\nu_k\varphi\,dx' + \int_\Omega (\nu_k D_k\nu_i D_i\varphi - n\nu_k D_kH\varphi)\,dx'=0$$

for all $\varphi\in C_0^1(\Omega)$. By calculation we have

$$\begin{aligned} \nu_k D_k\nu_i &= -\frac{\nu_k}{v}(\delta_{ij}-\nu_i\nu_j)D_{jk}u \\ &= -v(\delta_{ij}-\nu_i\nu_j)D_j\left(\frac{1}{v}\right) \\ &= -v\,\delta_i\nu_{n+1} \end{aligned}$$

by (15.4), $i=1,\ldots,n$. Hence, by means of formula (14.104) we can write the above relation as

$$\int_\Omega \mathscr{C}^2\varphi\,dx' - \int_\Omega v(\delta_i \nu_{n+1} D_i\varphi - n\delta_{n+1} H\varphi)\,dx' = 0, \tag{16.45}$$

where

$$\mathscr{C}^2 = \sum_{i,j=1}^{n} D_i\nu_j D_j\nu_i = \sum_{i,j=1}^{n+1} (\delta_i\nu_j)^2$$

Next, let us set $\mathscr{U} = \Omega \times \mathbb{R}$ and suppose that $\varphi \in C_0^1(\mathscr{U})$. By replacing φ in (16.45) by the function

$$\tilde{\varphi}(x', x_{n+1}) = \varphi(x', u(x')),$$

and using the relations

$$\sum_{i=1}^{n+1} \delta_i\nu_{n+1}\delta_i\varphi = \sum_{i=1}^{n+1} \delta_i\nu_{n+1}\delta_i\tilde{\varphi} = \sum_{i=1}^{n} \delta_i\nu_{n+1} D_i\tilde{\varphi},$$

we can therefore conclude from (16.45) the identity

$$\int_{\mathfrak{S}} (\mathscr{C}^2\nu_{n+1}\varphi - \delta_i\nu_{n+1}\delta_i\varphi + n\delta_{n+1}H\varphi)\,dA = 0 \tag{16.46}$$

for all $\varphi \in C_0^1(\mathscr{U})$. Note that the functions ν and H in (16.46) are independent of x_{n+1}. Also, in (16.46) and in the remainder of this section, we follow the summation convention that repeated indices indicate summation from 1 to $n+1$. Defining the function w by

$$w = \log v = -\log \nu_{n+1}, \tag{16.47}$$

we obtain, by replacing φ by φv in (16.46), the inequality

$$\int_{\mathfrak{S}} (\delta w \cdot \delta\varphi + |\delta w|^2\,\varphi - n\nu \cdot DH\,\varphi)\,dA \leqslant 0 \tag{16.48}$$

for all non-negative $\varphi \in C_0^1(\mathscr{U})$. Inequality (16.48) is the weak form of the inequality

$$\Delta w \geqslant |\delta w|^2 - n\nu \cdot DH. \tag{16.49}$$

In particular, if $\mathfrak{S}$ is a minimal surface, that is, the function u satisfies the minimal surface equation in Ω, we see that w is weakly *subharmonic* on $\mathfrak{S}$ and hence Lemma 16.2 is applicable to w. In the general case, we can apply Lemma 16.3 to obtain, for any point $y' \in \Omega$,

$$w(y') \leqslant \frac{1}{\omega_n R^n} \int\limits_{\mathfrak{S}_R(y)} w\, dA + \frac{1}{4\omega_n} \int\limits_{\mathfrak{S}_R(y)} w H^2 r^2 (r^{-n} - R^{-n})\, dA \tag{16.50}$$
$$+ \frac{1}{\omega_n} \int\limits_{\mathfrak{S}_R(y)} \psi(r)\, \nu \cdot DH\, dA$$

provided $R < \operatorname{dist}(y', \partial\Omega)$, $y = (y', u(y'))$ and ψ is given by (16.27). Setting

$$H_0 = \sup_\Omega |H|, \tag{16.51}$$
$$H_1 = \sup_\Omega (\nu \cdot DH)^+ \leqslant \sup_\Omega |DH|,$$

we have from (16.50)

$$w(y') \leqslant \frac{1}{\omega_n R^n} \int\limits_{\mathfrak{S}_R} w\, dA + \frac{H_0^2}{4\omega_n} \int\limits_{\mathfrak{S}_R} w r^2 (r^{-n} - R^{-n})\, dA$$
$$+ \frac{H_1}{\omega_n} \int\limits_{\mathfrak{S}_R} \psi(r)\, dA$$
$$= \frac{1}{\omega_n R^n} \int\limits_{\mathfrak{S}_R} w\, dA + \frac{nH_0^2}{4\omega_n} \int_0^R \rho^{-n-1} \int\limits_{\mathfrak{S}_\rho} w r^2\, dA\, d\rho \tag{16.52}$$
$$+ \frac{H_1}{\omega_n} \int_0^R \rho(\rho^{-n} - R^{-n}) A(\mathfrak{S}_\rho)\, d\rho \quad \text{(by Fubini's theorem)}$$
$$\leqslant \frac{1}{\omega_n R^n} \int\limits_{\mathfrak{S}_R} w\, dA + \frac{nH_0^2}{4\omega_n} \int_0^R \rho^{1-n} \int\limits_{\mathfrak{S}_\rho} w\, dA\, d\rho$$
$$+ \frac{H_1}{\omega_n} \int_0^R \rho^{1-n} A(\mathfrak{S}_\rho)\, d\rho.$$

The estimation of w, and consequently Du, is thus reduced to the estimation of $A(\mathfrak{S}_\rho)$ and $\int_{\mathfrak{S}_\rho} w\, dA$ for $0 < \rho \leqslant R$.

Estimation of $A(\mathfrak{S}_\rho)$.

Let us assume that $3R<\text{dist}\,(y', \partial\Omega)$ and also, without loss of generality, that $y'=0$ and $u(y')=0$. For $\rho\leqslant R$, we define the function u_ρ by

$$u_\rho=\begin{cases}\rho & \text{for } u\geqslant\rho\\ u & \text{for } -\rho\leqslant u\leqslant\rho\\ -\rho & \text{for } u\leqslant-\rho\end{cases}$$

and substitute the test function

$$\varphi=\eta u_\rho$$

into the integral identity (16.42), where η is a uniformly Lipschitz continuous function satisfying $\eta\equiv 1$ for $|x'|<\rho$, $\eta\equiv 0$ for $|x'|>2\rho$ and $|D\eta|\leqslant 1/\rho$. Note that the identity (16.42) clearly holds for all $\varphi\in W_0^{1,1}(\Omega)$ and hence for all uniformly Lipschitz continuous φ with support in Ω. We obtain thus

$$\int\limits_{|x'|,\,|u|<\rho}\frac{|Du|^2}{v}\,dx'\leqslant\rho\int\limits_{|x'|<2\rho}(|D\eta|+n|H|\eta)\,dx'.$$

Consequently

$$\begin{aligned}A(\mathfrak{S}_\rho)&=\int\limits_{|x'|^2+u^2<\rho^2}v\,dx'\\ &\leqslant\int\limits_{|x'|,\,|u|<\rho}v\,dx'\\ &\leqslant C(n)\Big\{\rho^n+\rho\int\limits_{|x'|<2\rho}|H|\eta\,dx'\Big\}\\ &\leqslant C(n)\rho^n(1+H_0R).\end{aligned}\tag{16.53}$$

Estimation of $\int_{\mathfrak{S}_\rho} w\,dA$.

Let us now substitute

$$\varphi=\eta w(u_\rho+\rho)$$

into (16.42) where η is as above. We obtain thus

$$\int_{|x'|,|u|<\varrho} \frac{w|Du|^2}{v}\,dx' \leqslant 2\varrho \int_{|x'|<2\varrho,u>-\varrho} (w|D\eta|+\eta|Dw|+n|Hw\eta|)\,dx'.$$

In order to estimate $\int \eta|Dw|\,dx'$, we replace φ by φ^2 in inequality (16.48) so that

$$\int_{\mathfrak{S}} \varphi^2|\delta w|^2\,dA \leqslant -2\int_{\mathfrak{S}} \varphi\,\delta w\cdot\delta\varphi\,dA - n\int_{\mathfrak{S}} \varphi^2 \boldsymbol{v}\cdot DH\,dA$$

for all $\varphi \in C_0^1(\Omega\times\mathbb{R})$. Using Cauchy's inequality (7.6), we obtain

$$\int_{\mathfrak{S}} \varphi^2|\delta w|^2\,dA \leqslant 4\int_{\mathfrak{S}} |\delta\varphi|^2\,dA + nH_1\int_{\mathfrak{S}} \varphi^2\,dA.$$

In particular, let us choose φ such that

$$\varphi = \eta\tau(x_{n+1}),$$

where $0\leqslant\tau\leqslant 1$, $\tau\equiv 1$ in $(-\rho, \sup_{|x'|<2\rho} u)$, $\tau\equiv 0$ outside $(-2\rho, \rho+\sup_{|x'|<2\rho} u)$, $|\tau'|<2/\rho$ and η is as above. We then have

$$\int_{\mathfrak{S}} \varphi^2|\delta w|^2\,dA \leqslant (8\rho^{-2}+nH_1)A(\mathfrak{S}\cap \operatorname{supp}\varphi).$$

Using (16.8), we can then conclude that

$$\begin{aligned}\int_{|x'|<2\rho,u>-\rho} \eta|Dw|\,dx' &\leqslant \int_{\mathfrak{S}} \varphi|\delta w|\,dA\\ &\leqslant (8\rho^{-2}+nH_1)^{1/2}A(\mathfrak{S}\cap\operatorname{supp}\varphi) \quad \text{by Schwarz's inequality,}\\ &\leqslant (8+nH_1R^2)^{1/2}\rho^{-1}\int_{|x'|<2\rho,u>-2\rho} v\,dx'.\end{aligned}$$

Since $w\leqslant v$, we also have

$$\int_{|x'|<2\rho,u>-\rho} w\,dx' \leqslant \int_{|x'|<2\rho,u>-\rho} v\,dx'.$$

It therefore only remains to estimate $\int v\,dx'$, and this we accomplish by taking

$$\varphi = \eta\max\{u+2\rho, 0\}$$

in (16.42), where $\eta \equiv 1$ for $|x'| < 2\rho$, $\eta \equiv 0$ for $|x'| > 3\rho$ and $|D\eta| \leqslant 1/\rho$. We obtain accordingly

$$\int_{|x'| < 2\rho,\, u > -2\rho} v\, dx' \leqslant C(n)\rho^n(1 + H_0 R)(1 + \rho^{-1} \sup_{|x'| < 3\rho} u).$$

Combining the above estimates, we therefore have

$$\begin{aligned}\int_{\mathfrak{S}_\rho} w\, dA &\leqslant \int_{|x'|,\, |u| < \rho} wv\, dx' \\ &\leqslant C(n)\rho^n(1 + H_0 R)(1 + H_1 R^2)^{1/2}(1 + \rho^{-1} \sup_\Omega u)\end{aligned} \tag{16.54}$$

Our desired interior gradient bound now follows by combining the estimates (16.52) (16.53), (16.54) and exponentiating.

Theorem 16.5. *Let Ω be a domain in $\mathbb{R}^n$ and u a function in $C^2(\Omega)$. Then, for any point $y' \in \Omega$, we have the estimate*

$$|Du(y')| \leqslant C_1 \exp\{C_2 \sup_\Omega (u - u(y'))/d\}, \tag{16.55}$$

where $d = \operatorname{dist}(y', \partial\Omega)$ and where $C_1 = C_1(n, dH_0, d^2 H_1)$, $C_2 = C_2(n, dH_0, d^2 H_1)$, ($H_0$ and H_1 being given by (16.51)).

As an immediate consequence of Theorem 16.5, we have the following interior estimate for non-negative functions.

Corollary 16.6. *Let Ω be a domain in $\mathbb{R}^n$ and u a non-negative function in $C^2(\Omega)$. Then for any point $y \in \Omega$, we have the estimate*

$$|Du(y')| \leqslant C_1 \exp\{C_2 u(y')/d\} \tag{16.56}$$

where C_1, C_2 and d are as in Theorem 16.5.

The exponential form of the estimates (16.55) and (16.56) is interesting in that it cannot be improved. This is evidenced in the case of two dimensional minimal surfaces by an example in [FN 4]. In the following section, we shall apply Theorem 16.5 to the Dirichlet problem, with continuous boundary values, for the minimal surface and prescribed mean curvature equations. A further application to the minimal surface equation is treated by Problem 16.4. Let us conclude this section by noting the interior estimates for higher order derivatives which now follow from Theorem 16.5, the Hölder estimate Theorem 12.1 and the Schauder interior estimates Theorem 6.2 and Problem 6.1.

Corollary 16.7. *Let Ω be a domain in $\mathbb{R}^n$ and u a function in $C^2(\Omega)$ whose graph has mean curvature $H \in C^k(\Omega)$, $k \geqslant 1$. Then $u \in C^{k+1}(\Omega)$ and for any point $y \in \Omega$, and multi-index β, $|\beta| = k+1$,*

$$|D^\beta u(y)| \leqslant C \tag{16.57}$$

where $C = C(n, k, |H|_{k;\Omega}, d, \sup|u|)$, $d = \operatorname{dist}(y, \partial\Omega)$.

16.3. Application to the Dirichlet Problem

In this section we study the solvability of the Dirichlet problem with continuous boundary data for both the minimal surface and prescribed mean curvature equations. For the minimal surface equation we have the following extension of Theorem 14.14.

Theorem 16.8. *Let Ω be a bounded C^2 domain in $\mathbb{R}^n$. Then the Dirichlet problem $\mathfrak{M}u = 0$ in Ω, $u = \varphi$ on $\partial\Omega$, is solvable for arbitrary $\varphi \in C^0(\partial\Omega)$ if and only if the mean curvature of the boundary $\partial\Omega$ is everywhere non-negative.*

Proof. Let us assume initially that $\partial\Omega \in C^{2,\alpha}$ for some $\alpha > 0$ and that the mean curvature of $\partial\Omega$ is everywhere non-negative. Let $\{\varphi_m\}$ be a sequence of functions in $C^{2,\alpha}(\bar{\Omega})$ which converges uniformly on $\partial\Omega$ to φ. By Theorem 14.14, the Dirichlet problems $\mathfrak{M}u_m = 0$ in Ω, $u_m = \varphi_m$ on $\partial\Omega$, are uniquely solvable in $C^{2,\alpha}(\bar{\Omega})$ and from the comparison principle, Theorem 10.1, we have

$$\sup_\Omega |u_{m_1} - u_{m_2}| \leqslant \sup_{\partial\Omega} |\varphi_{m_1} - \varphi_{m_2}| \to 0 \quad \text{as } m_1, m_2 \to \infty.$$

Consequently the sequence $\{u_m\}$ converges uniformly on Ω to some function $u \in C^0(\bar{\Omega})$ with $u = \varphi$ on $\partial\Omega$. Applying Corollary 16.7, together with Arzela's theorem we then obtain $u \in C^2(\Omega)$ and $\mathfrak{M}u = 0$ in Ω. The result for C^2 domains Ω follows by approximation of Ω by $C^{2,\alpha}$ domains. The non-existence part of Theorem 16.8 is an immediate consequence of Theorem 14.14. □

Existence results for the inhomogeneous equation (16.1) depend on the establishment of an appropriate maximum principle for solutions. By combining Theorem 13.8, Corollary 14.8 and Theorem 15.2, we first have the following basic result for smooth boundary data (see also Theorem 15.15).

Theorem 16.9. *Let Ω be a bounded domain in $\mathbb{R}^n$ with $\partial\Omega \in C^{2,\alpha}$ for some α, $0 < \alpha < 1$, and let $\varphi \in C^{2,\alpha}(\bar{\Omega})$. Let $H \in C^1(\bar{\Omega})$ and suppose that the mean curvature of $\partial\Omega$, H', satisfies*

$$H'(y) \geqslant \frac{n}{n-1} |H(y)| \tag{16.58}$$

at each point $y \in \partial\Omega$. Then there exists a unique function $u \in C^{2,\alpha}(\bar{\Omega})$ satisfying equation (16.1) in Ω and $u = \varphi$ on $\partial\Omega$ provided the family of solutions of the Dirichlet problems,

$$\mathfrak{M}u = \sigma nH(x)(1+|Du|^2)^{3/2} \text{ in } \Omega, \quad u = \sigma\varphi \text{ on } \partial\Omega, \tag{16.59}$$

is uniformly bounded in Ω.

The necessity of the condition (16.58) is demonstrated by Corollary 14.13. Let us now determine a further necessary condition for solvability. Namely, by the integral form (16.42) of equation (16.1), we have for any $\eta \in C_0^1(\Omega)$

$$\left|\int_\Omega H\eta\, dx\right| \leqslant \frac{1}{n}\int_\Omega |\nu \cdot D\eta|\, dx$$

$$\leqslant \frac{1}{n}\sup_\Omega \frac{|Du|}{\sqrt{1+|Du|^2}}\int_\Omega |D\eta|\, dx$$

and hence, writing

$$1-\varepsilon_0 = \sup_\Omega \frac{|Du|}{\sqrt{1+|Du|^2}},$$

we obtain

$$\left|\int_\Omega H\eta\, dx\right| \leqslant \frac{(1-\varepsilon_0)}{n}\int_\Omega |D\eta|\, dx, \tag{16.60}$$

for all $\eta \in C_0^1(\Omega)$, and some $\varepsilon_0 > 0$. However, it is clear, from the proof of Theorem 10.10, that condition (16.60) is also sufficient to ensure that an apriori estimate for $\sup_\Omega |u|$ holds for $C^1(\bar{\Omega})$ solutions u of equation (16.1). Accordingly we have, from Theorem 16.9, the following sharp existence theorem.

Theorem 16.10. *Let Ω be a bounded domain in $\mathbb{R}^n$ with $\partial\Omega \in C^{2,\alpha}$ for some α, $0<\alpha<1$, and let $\varphi \in C^{2,\alpha}(\bar{\Omega})$. Let $H \in C^1(\bar{\Omega})$ satisfy (16.58) and (16.60). Then the Dirichlet problem $\mathfrak{M}u = nH(1+|Du|^2)^{3/2}$, $u = \varphi$ on $\partial\Omega$, is uniquely solvable for $u \in C^{2,\alpha}(\bar{\Omega})$. Furthermore, if we only assume that $\varphi \in C^0(\partial\Omega)$, the problem is uniquely solvable for $u \in C^0(\bar{\Omega}) \cap C^2(\Omega)$.*

Note that condition (16.60) is implied by the condition

$$\int_\Omega |H^\pm|^n\, dx < \omega_n, \tag{16.61}$$

(cf. (10.35)). A more general condition than (16.61) is given in [GT 2]. The second assertion in Theorem 16.10 follows from the first assertion there in the same way that Theorem 16.8 follows from Theorem 14.14. Note that, if we only have $u \in C^0(\bar{\Omega}) \cap C^2(\Omega)$, then the function H need only satisfy condition (16.60) with $\varepsilon_0 = 0$.

When the function H is constant in Ω it turns out that the condition (16.60) in Theorem 16.10 is redundant. To show this we assume that (16.58) holds and let Ω_1 be the subset of Ω consisting of points having a unique closest point on $\partial\Omega$. An examination of the proofs of Lemmas 14.16 and 14.17 then shows that

$$\Delta d \leqslant -n|H|$$

in Ω_1, where $d(x) = \operatorname{dist}(x, \partial\Omega)$. Let us now set, for $u \in C^0(\bar{\Omega}) \cap C^2(\Omega)$, $x \in \Omega_1$,

$$v(x) = \sup_{\partial\Omega} |u| + \frac{e^{\mu\delta}}{\mu}(1 - e^{-\mu d(x)}),$$

where $\delta = \operatorname{diam} \Omega$ and $\mu = 1 + n|H|$. It then follows that

$$\begin{aligned}
\mathfrak{M}v &= [-\mu + (1 + |Dv|^2)\,\Delta d]\, e^{\mu(\delta - d)} \\
&\leqslant -[\mu + n|H|(1 + |Dv|^2)]\, e^{\mu(\delta - d)} \\
&\leqslant -n|H|(1 + |Dv|^2)^{3/2},
\end{aligned}$$

so that the function v is a supersolution of equation (16.1) in the open set Ω_1. Consequently if the function u satisfies (16.1) in Ω, then, by the comparison principle, Theorem 10.1, $w = u - v$ assumes a maximum value in $\bar{\Omega}$ either on $\partial\Omega$ or in $\Omega - \Omega_1$. Now let y be a point in $\Omega - \Omega_1$ and γ be a straight line segment from y to $\partial\Omega$, normal to $\partial\Omega$. If the maximum value of w on γ is taken on at y we must have $Du(y) \neq 0$ and also that the maximum value of u on γ occurs at y. This shows that w cannot take on a maximum value on $\Omega - \Omega_1$. Therefore we obtain the estimate

$$\sup_{\Omega} |u| \leqslant \sup_{\partial\Omega} |u| + (e^{\mu\delta} - 1)/\mu. \tag{16.62}$$

Combining (16.62) with Theorem 16.9 and Corollary 14.13, we thus have the following existence theorem for the equation of constant mean curvature.

Theorem 16.11. *Let Ω be a bounded C^2 domain in $\mathbb{R}^n$. Then the Dirichlet problem $\mathfrak{M}u = nH(1 + |Du|^2)^{3/2}$, $u = \varphi$ on $\partial\Omega$, is solvable for constant H and arbitrary $\varphi \in C^0(\partial\Omega)$ if and only if the mean curvature of $\partial\Omega$, H', satisfies $H' \geqslant n|H|/(n-1)$ everywhere on $\partial\Omega$.*

16.4. Equations in Two Independent Variables

So far this chapter has been concerned with the prescribed mean curvature equation and, in particular, the minimal surface equation. Now, in the case of two independent variables, a somewhat more general class of equations will be considered. We shall consider equations of the form

$$Qu = a^{ij}(x, u, Du)D_{ij}u + b(x, u, Du) = 0, \tag{16.63}$$

where $x = (x_1, x_2) \in \Omega$, Ω is a domain in $\mathbb{R}^2$ and where a^{ij}, b, $i, j = 1, 2$, denote given real-valued functions on $\Omega \times \mathbb{R} \times \mathbb{R}^2$ with

$$|\xi|^2 - \frac{(p \cdot \xi)^2}{1+|p|^2} \leqslant a^{ij}(x, z, p)\xi_i\xi_j \leqslant \gamma\left[|\xi|^2 - \frac{(p \cdot \xi)^2}{1+|p|^2}\right] \tag{16.64}$$

for all $(x, z, p) \in \Omega \times \mathbb{R} \times \mathbb{R}^2$ and all $\xi = (\xi_1, \xi_2) \in \mathbb{R}^2$; and

$$|b(x, z, p)| \leqslant \mu\sqrt{1+|p|^2} \tag{16.65}$$

for all $(x, z, p) \in \Omega \times \mathbb{R} \times \mathbb{R}^2$. Here γ and μ denote fixed constants.

Note that the minimal surface equation, $\mathfrak{M}u = 0$, can be written in the form (16.63) with

$$a^{ij}(x, z, p) = \delta^{ij} - \frac{p_i p_j}{1+|p|^2}, \quad b = 0;$$

in this case (16.64) and (16.65) hold with $\gamma = 1$ and $\mu = 0$. More generally, any equation which arises as the non-parametric Euler equation of an elliptic parametric functional (see Appendix 16.8), is of the form (16.63), (16.64), (16.65). But quite apart from these examples, the class of equations (16.63), (16.64), (16.65), which we call the class of equations of *mean curvature type*, is both natural and interesting in that it is completely characterized as follows:

Suppose u is a $C^2(\Omega)$ function with graph

$$\mathfrak{S} = \{(x, z) \in \mathbb{R}^3 \mid x \in \Omega, z = u(x)\}.$$

Then there exist real-valued functions a^{ij}, b such that (16.63), (16.64), (16.65) hold if and only if the *principal curvatures* κ_1, κ_2 of $\mathfrak{S}$ (see Section 14.6) are related at each point of $\mathfrak{S}$ by an equation of the form

$$\alpha_1\kappa_1 + \alpha_2\kappa_2 + \beta = 0, \tag{16.63$'$}$$

with $\alpha_1, \alpha_2, \beta$ satisfying

$$1 \leqslant \alpha_i \leqslant \gamma, \quad i = 1, 2, \tag{16.64$'$}$$

$$|\beta| \leqslant \mu. \tag{16.65$'$}$$

To demonstrate that this characterization is valid, we let d denote the distance function of $\mathfrak{S}$ defined for $X=(x, z)\in\Omega\times\mathbb{R}$ by setting $d(X)=\text{dist}\,(X, \mathfrak{S})$ if $z>u(x)$ and $d(X) = -\text{dist}\,(X, \mathfrak{S})$ if $z<u(x)$. Since d is C^2 (see Lemma 14.16) and $d(x, u(x))\equiv 0$, $x\in\Omega$, we then have, by the chain rule, the identities

$$D_i d(X)+D_i u(x)D_3 d(X)=0$$

and

$$D_{ij}d(X)+D_i u(x)D_{3j}d(X)+D_j u(x)D_{3i}d(X)+D_i u(x)D_j u(x)D_{33}d(X)+D_3 d(X)D_{ij}u(x)=0, \tag{16.66}$$

$i, j=1, 2$, where $X=(x, u(x))$. Since $D_3 d(X)=v^{-1}$, $v=\sqrt{1+|Du(x)|^2}$, (16.63) then implies

$$\sum_{i,j=1}^{3} a_*^{ij}(x)D_{ij}d(X)+v^{-1}b_*(x)=0, \tag{16.67}$$

where $b_*(x)=b(x, u(x), Du(x))$ and where the 3×3 matrix $[a_*^{ij}(x)]$ is defined by setting $a_*^{ij}(x)=a^{ij}(x, u(x), Du(x))$ for $i, j=1, 2$, and

$$a_*^{i3}(x)=a_*^{3i}(x)=\sum_{j=1}^{2} D_j u(x)a_*^{ij}(x), \quad i=1, 2,$$

$$a_*^{33}(x)=\sum_{i,j=1}^{2} D_i u(x)D_j u(x)a_*^{ij}(x).$$

Note that these last relations are equivalent to

$$\sum_{j=1}^{3} a_*^{ji}(x)\nu_j=\sum_{j=1}^{3} a_*^{ij}(x)\nu_j=0, \quad i=1, 2, 3,$$

where $\nu=v^{-1}(-Du(x), 1)\,(=Dd(X))$ is the upward unit normal of $\mathfrak{S}$. Next we let $\mathbf{q}$ be the matrix of an isometry which transforms the coordinate system to a *principal* coordinate system at X (see Section 14.6), so that $\mathbf{q}^t[D_{ij}d(X)]\mathbf{q} = \text{diag}\,[\kappa_1, \kappa_2, 0]$, where κ_1, κ_2 are principal curvatures of $\mathfrak{S}$ at X. Thus (16.67) can be written in the form (16.63)′, with α_1, α_2 the first two elements or the leading diagonal of $\mathbf{q}^t[a_*^{ij}(x)]\mathbf{q}$ and with $\beta=v^{-1}b_*(x)$. (16.65)′ is now true by (16.65). To check (16.64)′, we first note that

$$\sum_{i,j=1}^{3} a_*^{ij}(x)\xi_i\xi_j=\sum_{i,j=1}^{2} a_*^{ij}(x)(\xi_i+\xi_3 D_i u(x))(\xi_j+\xi_3 D_j u(x)), \quad \xi\in\mathbb{R}^3,$$

and it then follows from (16.64) that

$$|\xi'|^2 \leqslant \sum_{i,j=1}^{3} a_*^{ij}(x)\xi_i\xi_j \leqslant \gamma|\xi'|^2, \quad \xi' = \xi - (\mathbf{v}\cdot\xi)\mathbf{v}, \ \mathbf{v} = v^{-1}(-Du(x), 1).$$

(16.64)′ now easily follows from this.

To prove the converse implication we suppose that (16.63)′, (16.64)′, and (16.65)′ hold at $X = (x, u(x)) \in \mathfrak{S}$, we let $[a_*^{ij}(x)] = \mathbf{q}\,\mathrm{diag}\,[\alpha_1, \alpha_2, 0]\mathbf{q}'$, where $\mathbf{q}$ is as above, and we let $b_*(x) = v\beta$. Then (16.67) holds and consequently, since we still have the relations

$$\sum_{j=1}^{3} a_*^{ij}(x)\nu_j = 0, \quad i = 1, 2, 3,$$

an application of (16.66) yields

$$\sum_{i,j=1}^{2} a_*^{ij}(x)D_{ij}u + b_*(x) = 0.$$

We then define, for $i, j = 1, 2$,

$$a^{ij}(x, z, p) = \begin{cases} a_*^{ij}(x) & \text{if } z = u(x) \text{ and } p = Du(x) \\ \delta_{ij} - \dfrac{p_i p_j}{1+|p|^2} & \text{otherwise,} \end{cases}$$

and

$$b(x, z, p) = \begin{cases} b_*(x) & \text{if } z = u(x) \text{ and } p = Du(x) \\ 0 & \text{otherwise.} \end{cases}$$

(16.63), (16.64), (16.65) are now easily checked.

The treatment of equations of the form (16.63), (16.64), (16.65), to be given here, is in many respects analogous to the treatment given in Chapter 12 for *uniformly* elliptic equations in two independent variables; as in Chapter 12 we shall begin by considering quasiconformal maps, although here it will be necessary to consider mappings between surfaces in $\mathbb{R}^3$ rather than mappings in the plane. Also, as in Chapter 12, the principal result is a Hölder estimate for quasiconformal maps. A special consequence of this general estimate is a Hölder estimate for the unit normal of the graph of a solution u of (16.63), (16.64), (16.65). Using this estimate, apriori bounds for the principal curvatures of the graph of u and for the gradient of u will be established. One of the most striking results of the theory of equations of the form (16.63), (16.64), (16.65) is the following generalization of a classical theorem of Bernstein:

Theorem 16.12. *Let $u \in C^2(\mathbb{R}^2)$ satisfy equation* (16.63) *on the whole of $\mathbb{R}^2$, with $b \equiv 0$, and suppose that* (16.64) *holds with $\Omega = \mathbb{R}^2$. Then u is a linear function.*

We shall see below that this theorem is also a consequence of the Hölder estimate for the unit normal of the graph of u.

16.5. Quasiconformal Mappings

In this section, we shall consider mappings between C^2 hypersurfaces $\mathfrak{S}, \mathfrak{T} \subset \mathbb{R}^3$. The surfaces $\mathfrak{S}$ and $\mathfrak{T}$ are assumed to be *oriented*, that is, it is assumed that there exist unit normal vectors $\mathbf{v}$, $\boldsymbol{\mu}$ defined and continuous over all of $\mathfrak{S}, \mathfrak{T}$ respectively. For our application in the next section $\mathfrak{S}$ and $\mathfrak{T}$ will be graphs and hence given by representations of the form (16.2), (16.3).

Points of $\mathbb{R}^3$ will be denoted $X = (x_1, x_2, x_3)$. $Y = (y_1, y_2, y_3)$ will always denote a fixed point of $\mathfrak{S}$ and, for $\rho > 0$, we set $\mathfrak{S}_\rho(Y) = \mathfrak{S} \cap B_\rho(Y)$. R will denote a fixed positive constant such that $\mathfrak{S}_R(Y) \subset\subset \mathfrak{S}$. We shall often use $\mathfrak{S}_\rho$ as an abbreviation for $\mathfrak{S}_\rho(Y)$.

We shall need the classical version of Stokes' theorem: if $\mathbf{v} = (v_1, v_2, v_3)$ is a C^1 vector function defined in a neighborhood of $\mathfrak{S}$ and if $\mathscr{G} \subset\subset \mathfrak{S}$ is such that $\partial\mathscr{G}$ consists of a finite union of simple closed C^1 curves, then

$$\int_{\mathscr{G}} \mathbf{v} \cdot \operatorname{curl} \mathbf{v}\, dA \equiv \int_{\mathscr{G}} \mathbf{v} \cdot D \times \mathbf{v}\, dA = \int_{\partial\mathscr{G}} v_i\, dx_i \equiv \int_{\partial\mathscr{G}} \mathbf{t} \cdot \mathbf{v}\, ds, \tag{16.68}$$

where A denotes surface area on $\mathfrak{S}$, $\partial\mathscr{G}$ is appropriately oriented, s denotes arc length on $\partial\mathscr{G}$ and $\mathbf{t}$ is the unit tangent vector of $\partial\mathscr{G}$. In this section we follow the convention that repeated indices imply summation from 1 to 3. If we take $\mathbf{v} = fDg$ in (16.68), where f, g are respectively C^1 and C^2 functions defined in a neighborhood of $\mathfrak{S}$, then, by virtue of the operator identity $D \times D = 0$, we get from (16.68)

$$\int_{\mathscr{G}} \mathbf{v} \cdot Df \times Dg\, dA = \int_{\partial\mathscr{G}} f\, dg \equiv \int_{\partial\mathscr{G}} f \frac{dg}{ds}\, ds,$$

where $dg/ds \equiv \mathbf{t} \cdot Dg$ is the directional derivative of g in the direction of $\mathbf{t}$. Since only first derivatives of f, g appear here, it is easy to see that the above identity is valid if both f and g are merely C^1. Note that, because of the vector identities $a \times a = 0$, $a \cdot a \times b = 0$ for $a, b \in \mathbb{R}^3$, we can write

$$\mathbf{v} \cdot Df \times Dg = \mathbf{v} \cdot \delta f \times \delta g$$

on $\mathfrak{S}$, where δ is the tangential gradient operator on $\mathfrak{S}$ defined by (16.4). Hence for arbitrary C^1 functions f, g on $\mathfrak{S}$ we have the identity

$$\int_{\mathscr{G}} \mathbf{v} \cdot \delta f \times \delta g\, dA = \int_{\partial\mathscr{G}} f\, dg \equiv \int_{\partial\mathscr{G}} f \frac{dg}{ds}\, ds. \tag{16.69}$$

Our basic assumption concerning $\mathfrak{T}$ will be that there exists a vector function $\omega=(\omega_1, \omega_2, \omega_3)$ which is C^1 in some neighborhood of $\mathfrak{T}$ and such that

$$(16.70) \qquad \sup_{\mathfrak{T}} |\omega| + \sup_{\mathfrak{T}} |D\omega| \leqslant \Lambda_0 \quad \text{and} \quad \boldsymbol{\mu} \cdot D \times \omega \equiv 1 \text{ on } \mathfrak{T},$$

where Λ_0 is a constant. By applying Stokes' formula (16.68) on a subset $\mathscr{G} \subset \mathfrak{T}$ (with $\boldsymbol{\mu}$ in place of $\boldsymbol{\nu}$ and ω in place of $\mathbf{v}$), we then have

$$(16.71) \qquad A(\mathscr{G}) = \int_{\partial\mathscr{G}} \omega_i \, dx_i,$$

that is, the area of a subset $\mathscr{G} \subset \mathfrak{T}$ can be expressed as a boundary integral taken over $\partial\mathscr{G}$.

An example of special interest for us here will be the case when $\mathfrak{T}$ is the upper hemisphere $\{X=(x_1, x_2, x_3) \mid |X|=1, \ x_3>0\}$ of the unit sphere. In this case, taking $\boldsymbol{\mu}(X)=X$, we have (16.70) with

$$\omega(X) = (-x_2/(1+x_3), \ x_1/(1+x_3), \ 0).$$

It is worth pointing out (although we shall have no need of it here) that if $\mathfrak{L}$ is an arbitrary, connected, oriented, compact C^2 surface in $\mathbb{R}^3$, if $\mathfrak{K} \subset \mathfrak{L}$ is compact with non-empty interior and if we take

$$\mathfrak{T} = \mathfrak{L} - \mathfrak{K},$$

then there always exists a vector field ω satisfying (16.70). A proof of this assertion involves a straightforward application of the theory of differential forms and the de Rham cohomology groups; the reader is referred to the discussion in [SI 6].

Let us now consider a mapping

$$\boldsymbol{\varphi} = (\varphi_1, \varphi_2, \varphi_3) : \mathfrak{S} \to \mathfrak{T}$$

which is C^1 in the sense that each φ_i (as a mapping from $\mathfrak{S}$ into $\mathbb{R}$) has a C^1 extension $\tilde{\varphi}_i$ to some neighborhood of $\mathfrak{S}$. We wish to introduce the concept of quasiconformality of $\boldsymbol{\varphi}$, but to do this we first need to define the *signed area magnification factor* $J(\boldsymbol{\varphi})$ of $\boldsymbol{\varphi}$. Namely, $J(\boldsymbol{\varphi})$ is defined on $\mathfrak{S}$ by

$$(16.72) \qquad J(\boldsymbol{\varphi}) = \boldsymbol{\nu} \cdot \delta(\omega_i \circ \boldsymbol{\varphi}) \times (\delta\varphi_i).$$

This definition of $J(\boldsymbol{\varphi})$ is easily motivated as follows. Let $\mathscr{E}$ be a region of $\mathfrak{S}$ which is such that $\partial\mathscr{E}$ is a simple, smooth curve, and suppose that $\boldsymbol{\varphi}$ is one-to-one with a C^1 inverse in some open subset of $\mathfrak{T}$ which contains $\boldsymbol{\varphi}(\mathscr{E})$. Assuming that the curves $\partial\mathscr{E}$ and $\partial\boldsymbol{\varphi}(\mathscr{E})$ are appropriately oriented, we can apply (16.69) and (16.71) to give

$$A(\boldsymbol{\varphi}(\mathscr{E})) = \int_{\partial\boldsymbol{\varphi}(\mathscr{E})} \omega_i \, dx_i = \pm \int_{\partial\mathscr{E}} \omega_i \circ \boldsymbol{\varphi} \, d\varphi_i = \pm \int_{\mathscr{E}} \boldsymbol{\nu} \cdot (\delta\omega_i \circ \boldsymbol{\varphi}) \times (\delta\varphi_i) \, dA$$

with the + or − sign according as φ is orientation preserving or reversing on $\mathscr{E}$. This identity clearly motivates the definition (16.72).

An important (and intuitively obvious) fact concerning $J(\varphi)$ is that it is independent of coordinates in the following sense: If P, Q are linear isometries of $\mathbb{R}^3$ with

$$\det P = \det Q = 1,$$

and if we define

$$\tilde{\mathfrak{S}} = \{P(X - Y) \mid X \in \mathfrak{S}\}, \ \tilde{\mathfrak{T}} = \{Q(X - \varphi(Y)) \mid X \in \mathfrak{T}\},$$
$$\tilde{\nu} = P \circ \nu \circ P^{-1}, \quad \tilde{\mu} = Q \circ \mu \circ Q^{-1}, \quad \tilde{\omega} = Q \circ \omega \circ Q^{-1}, \quad \tilde{\varphi} = Q \circ \varphi \circ P^{-1},$$

then

$$\nu \cdot (\delta\omega_i \circ \varphi) \times (\delta\varphi_i)(Y) = \tilde{\nu} \cdot (\tilde{\delta}\tilde{\omega}_i \circ \tilde{\varphi}) \times (\tilde{\delta}\tilde{\varphi}_i)(0), \tag{16.73}$$

where $\tilde{\delta}$ denotes the tangential gradient operator on $\tilde{\mathfrak{S}}$. The relation (16.73) is easily checked by first representing the isometries P, Q in terms of orthogonal matrices and then using two elementary facts from linear algebra, namely that if A, B are 3×3 matrices, then $\det AB = \det A \det B$, and if A has rows a, b, c, then $\det A = a \cdot b \times c$. It can also be checked by using (16.70) that

$$\tilde{\mu} \cdot D \times \tilde{\omega} \equiv 1 \quad \text{on } \tilde{\mathfrak{T}}.$$

If the isometries P, Q above are chosen so that

$$\tilde{\nu}(0) = \tilde{\mu}(0) = (0, 0, 1),$$

and if we introduce new coordinates

$$(\zeta, \zeta_3) = (\zeta_1, \zeta_2, \zeta_3) = P(X - Y),$$

then $\tilde{\mathfrak{S}}$ can be represented near 0 in the form

$$\zeta_3 = \tilde{u}(\zeta), \ \zeta \in \mathscr{U},$$

where $\mathscr{U}$ is a neighborhood of $0 \in \mathbb{R}^2$ and $\tilde{u}$ is a $C^2(\mathscr{U})$ function with $D\tilde{u}(0) = 0$. We also then have

$$\tilde{\delta}_j \tilde{\varphi}_3(0) = (0, 0, 1) \cdot \tilde{\delta}_j \tilde{\varphi}(0) = \tilde{\mu}(0) \cdot \tilde{\delta}_j \tilde{\varphi}(0) = 0, \quad j = 1, 2, 3$$

by virtue of the fact that the vectors $\tilde{\delta}_j \tilde{\varphi}(0)$, $j = 1, 2, 3$, are tangent to $\tilde{\mathfrak{T}}$ at 0. That is, we have

$$\tilde{\delta}\tilde{\varphi}_3(0) = 0.$$

Thus if

$$\psi : \mathscr{U} \to \mathbb{R}^2$$

is defined by

$$\psi(\zeta) = [\tilde{\varphi}_1(\zeta, \tilde{u}(\zeta)), \tilde{\varphi}_2(\zeta, \tilde{u}(\zeta))], \quad \zeta \in \mathscr{U},$$

then ψ approximates $\tilde{\varphi}$ near 0 in the sense that

$$|(\psi(\zeta), 0) - \tilde{\varphi}(\zeta, \tilde{u}(\zeta))| = o(|\zeta|) \text{ as } |\zeta| \to 0, \quad \zeta \in \mathscr{U}.$$

Furthermore, using (16.73) together with the definition $\tilde{\delta} = D - \tilde{v}(\tilde{v} \cdot D)$, one now easily checks that

$$(16.74) \qquad J(\varphi)(Y) = D_1\psi_1(0)D_2\psi_2(0) - D_1\psi_2(0)D_2\psi_1(0);$$

that is, $J(\varphi)(Y)$ is just the Jacobian of ψ at 0. Also it follows that

$$(16.75) \qquad |\delta\varphi(Y)|^2 = |\tilde{\delta}\tilde{\varphi}(0)|^2 = |D\psi(0)|^2.$$

In view of (16.74), (16.75) it is now reasonable, by analogy with (12.2), to make the following definition. Namely, the mapping φ is said to be a (K, K')-*quasiconformal* mapping from $\mathfrak{S}$ into $\mathfrak{T}$ if

$$(16.76) \qquad |\delta\varphi(X)|^2 \leqslant 2KJ(\varphi)(X) + K'.$$

at each point $X \in \mathfrak{S}$. Here K, K' are real constants with $K' \geqslant 0$ and $|\delta\varphi(X)|^2 = \sum_{i=1}^{3} |\delta\varphi_i(X)|^2$. We emphasize here that it is not assumed in the above that K is positive (cf. (12.2)). Note that when $K' = 0$, we must have $|K| \geqslant 1$ unless $\delta\varphi \equiv 0$.

Before beginning the proof of the main Hölder continuity result for quasiconformal maps, one further preliminary is needed; namely, if g is an arbitrary C^1 function on $\mathfrak{S}_R$, then

$$(16.77) \qquad \int_{\partial\mathfrak{S}_\rho} g\, ds = D_\rho \int_{\mathfrak{S}_\rho} g|\delta r|\, dA$$

for almost all $\rho \in (0, R)$, where r is the radial distance function relative to Y, defined by

$$r(X) = |X - Y|, \quad X \in \mathbb{R}^3.$$

Formula (16.77) is in fact a special case of the important *co-area formula* (see [FE]). We proceed to prove (16.77). First note that the left side of (16.77) makes sense for almost all $\rho \in (0, R)$ because by Sard's Theorem [SB] we can write

$$(16.78) \qquad \partial\mathfrak{S}_\rho = \bar{\mathfrak{S}}_\rho \cap \{X \in \mathbb{R}^3 \mid |X - Y| = \rho\} = \bigcup_{j=1}^{n(\rho)} \Gamma_\rho^{(j)}$$

for almost all $\rho \in (0, R)$, where the $\Gamma_\rho^{(j)}$ are simple closed C^2 curves and $n(\rho)$ is a positive integer. Actually, Sard's Theorem guarantees that for almost all $\rho \in (0, R)$ the tangential gradient δr vanishes at no point of $\partial\mathfrak{S}_\rho$; the geometric interpretation of this is that the surface $\mathfrak{S}$ and the sphere $\{X \in \mathbb{R}^3 \mid |X - Y| = \rho\}$ intersect *non-tangentially*, and this explains why (16.78) holds. Now let us take some fixed $\rho \in (0, R)$ with $\delta r \neq 0$ on $\partial\mathfrak{S}_\rho$ and let $\varepsilon > 0$ be small enough to ensure $\delta r \neq 0$ on $\bar{\mathscr{G}}$, where $\mathscr{G} = \mathfrak{S}_\rho - \mathfrak{S}_{\rho-\varepsilon}$. We wish now to apply Stokes' Theorem (16.68); the appropriately oriented unit tangent for $\partial\mathscr{G}$ is given by $\mathbf{t} = \nu \times \delta r/|\delta r|$ on $\partial\mathfrak{S}_\rho$ and $\mathbf{t} = -\nu \times \delta r/|\delta r|$ on $\partial\mathfrak{S}_{\rho-\varepsilon}$. Let $\mathbf{F}$ denote a C^1 extension of the vector function $\nu \times \delta r/|\delta r|$ to some open subset of $\mathbb{R}^3$ containing $\bar{\mathscr{G}}$. Then applying (16.68) with

$$\mathbf{v} = \frac{1}{\varepsilon}(r - \rho + \varepsilon) g \mathbf{F},$$

and noting that $\mathbf{v} = 0$ on $\partial\mathfrak{S}_{\rho-\varepsilon}$, we obtain

$$\begin{aligned}
\int_{\partial\mathfrak{S}_\rho} g\, ds &= \int_{\partial\mathscr{G}} \frac{1}{\varepsilon}(r - \rho + \varepsilon) g \mathbf{F} \cdot \mathbf{t}\, ds \\
&= \int_{\mathfrak{S}_\rho - \mathfrak{S}_{\rho-\varepsilon}} \frac{1}{\varepsilon}(r - \rho + \varepsilon)\nu \cdot (g D \times \mathbf{F} + Dg \times \mathbf{F})\, dA \\
&\quad + \frac{1}{\varepsilon} \int_{\mathfrak{S}_\rho - \mathfrak{S}_{\rho-\varepsilon}} g \nu \cdot Dr \times \mathbf{F}\, dA.
\end{aligned}$$

Since

$$\begin{aligned}
\nu \cdot Dr \times \mathbf{F} &= (\nu \times Dr) \cdot (\nu \times \delta r/|\delta r|) \\
&= |\nu \times \delta r|^2/|\delta r| = |\delta r|
\end{aligned}$$

on $\mathfrak{S}_\rho - \mathfrak{S}_{\rho-\varepsilon}$, we then have

$$\begin{aligned}
\int_{\partial\mathfrak{S}_\rho} g\, ds = &\int_{\mathfrak{S}_\rho - \mathfrak{S}_{\rho-\varepsilon}} \frac{1}{\varepsilon}(r - \rho + \varepsilon)\nu \cdot (g D \times \mathbf{F} + Dg \times \mathbf{F})\, dA \\
&+ \frac{1}{\varepsilon}\left\{ \int_{\mathfrak{S}_\rho} g|\delta r|\, dA - \int_{\mathfrak{S}_{\rho-\varepsilon}} g|\delta r|\, dA \right\}.
\end{aligned}$$

and, since

$$0 \leqslant \frac{r-\rho+\varepsilon}{\varepsilon} \leqslant 1$$

on $\mathfrak{S}_\rho - \mathfrak{S}_{\rho-\varepsilon}$, we have (16.77) by letting $\varepsilon \to 0$.

The main Hölder estimate will be a consequence of estimates for the integral $\mathfrak{D}(\rho; Z)$, which is defined for $Z \in \mathfrak{S}$ and $\rho > 0$ by

$$\mathfrak{D}(\rho; Z) = \int_{\mathfrak{S}_\rho(Z)} |\delta\varphi|^2 \, dA.$$

(cf. the Dirichlet integral used in Chapter 12).

The following lemma, which should be compared with inequality (12.8), provides a bound for $\mathfrak{D}(R/2; Y)$. In the statement of the lemma, and subsequently, Λ_1 denotes a constant such that

$$(16.79) \qquad A(\mathfrak{S}_{3R/4}(Y)) \leqslant \Lambda_1 (3R/4)^2.$$

It will be shown in the next section that in the case where $\mathfrak{S}$ is a graph with (K, K')-quasiconformal Gauss map, one can obtain a bound for Λ_1 in terms of K and $K'R^2$.

Lemma 16.13. *Suppose φ is a (K, K')-quasiconformal C^1 mapping from $\mathfrak{S}$ into $\mathfrak{T}$. Then*

$$(16.80) \qquad \mathfrak{D}(R/2; Y) \leqslant C$$

where $C = C(\Lambda_0, K, K'R^2, \Lambda_1)$.

Proof. Let $\Gamma_\rho^{(j)}$, $j = 1, \ldots, n(\rho)$, be as in (16.78), and assume $\Gamma_\rho^{(j)}$ are oriented appropriately for the use of (16.68) over $\mathfrak{S}_\rho$. Then, using the definition (16.72) of $J(\varphi)$ in combination with (16.69), we obtain the identity

$$(16.81) \qquad \int_{\mathfrak{S}_\rho} J(\varphi) \, dA = \sum_{j=1}^{n(\rho)} \int_{\Gamma_\rho^{(j)}} \omega_i \circ \varphi \frac{d\varphi_i}{ds} \, ds$$

for almost all $\rho \in (0, R)$. This identity will play a key role in the proof of the main estimate for $\mathfrak{D}(\rho; Z)$ given in the next theorem. For the present, we simply use Schwarz's inequality in combination with the inequality

$$\left|\frac{d\varphi_i}{ds}\right| \leqslant |\delta\varphi_i|, \quad i = 1, 2, 3,$$

whereupon (16.81), (16.70) and (16.77) imply that for almost all $\rho \in (0, R)$

$$\begin{aligned}\left|\int_{\mathfrak{S}_\rho} J(\varphi)\, dA\right| &\leqslant (\sup_{\Sigma} |\omega|) \int_{\partial\mathfrak{S}_\rho} |\delta\varphi|\, ds \\ &\leqslant \Lambda_0\left(\int_{\partial\mathfrak{S}_\rho} |\delta\varphi|^2\, ds\right)^{1/2}\left(\int_{\partial\mathfrak{S}_\rho} ds\right)^{1/2} \\ &= \Lambda_0\left(D_\rho \int_{\mathfrak{S}_\rho} |\delta r|\, |\delta\varphi|^2\, dA\right)^{1/2}\left(D_\rho \int_{\mathfrak{S}_\rho} |\delta r|\, dA\right)^{1/2}.\end{aligned}$$

Thus, since $|\delta r| \leqslant |Dr| = 1$, we obtain from (16.76) that for almost all $\rho \in (0, 3R/4)$

$$\mathfrak{D}(\rho) \leqslant 2\Lambda_0 |K| (f'(\rho)\mathfrak{D}'(\rho))^{1/2} + \Lambda_1 K' R^2,$$

where we are using $\mathfrak{D}(\rho)$ as an abbreviation for $\mathfrak{D}(\rho; Y)$ and where

$$f(\rho) = A(\mathfrak{S}_\rho(Y)).$$

Note that $f'(\rho)$ and $\mathfrak{D}'(\rho)$ exist for almost all $\rho \in (0, R)$ because $f(\rho)$ and $\mathfrak{D}(\rho)$ are non-decreasing in ρ. Squaring each side of the above inequality, we then obtain

$$\mathfrak{D}^2(\rho) \leqslant 8(\Lambda_0 K)^2 f'(\rho)\mathfrak{D}'(\rho) + 2(K'\Lambda_1 R^2)^2.$$

Now if we let

$$g(\rho) = f(\rho) + \rho R$$

(so that $g'(\rho) \geqslant R$) and let

$$\mathfrak{E}(\rho) = \mathfrak{D}(\rho) + \rho/R,$$

then it is quite a straightforward matter to show that the previous inequality implies an inequality of the form

$$\mathfrak{E}^2(\rho) \leqslant C g'(\rho)\mathfrak{E}'(\rho)$$

for almost all $\rho \in (R/4, R/2)$, where $C = C(\Lambda_0, K, K'R^2, \Lambda_1)$. This can be written

$$-\frac{d}{d\rho}\left(\frac{1}{\mathfrak{E}(\rho)}\right) \geqslant \frac{1}{Cg'(\rho)}. \tag{16.82}$$

Now, by using the Hölder inequality and the monotonicity of g, we have

$$\left(\frac{R}{4}\right)^2=\left(\int_{R/2}^{3R/4} d\rho\right)^2$$

$$\leqslant\left(\int_{R/2}^{3R/4} g'(\rho)\,d\rho\right)\left(\int_{R/2}^{3R/4}\frac{d\rho}{g'(\rho)}\right)$$

$$\leqslant(g(3R/4)-g(R/2))\int_{R/2}^{3R/4}\frac{d\rho}{g'(\rho)},$$

so that

$$\int_{R/2}^{3R/4}\frac{d\rho}{g'(\rho)}\geqslant\frac{(R/4)^2}{g(3R/4)-g(R/2)}$$

$$>\frac{(R/4)^2}{g(3R/4)}.$$

Hence integrating (16.82) over $(R/2, 3R/4)$ and using this last inequality we obtain

$$\frac{1}{\mathfrak{E}(R/2)}-\frac{1}{\mathfrak{E}(3R/4)}\geqslant\frac{(R/4)^2}{Cg(3R/4)},$$

so that

$$\mathfrak{E}(R/2)\leqslant\frac{C}{R^2}\left[A(\mathfrak{S}_{3R/4}(Y))+R^2\right].$$

The desired result (16.80) now follows by using (16.79). □

The next theorem contains the main estimate for $\mathfrak{D}(\rho; Z)$. In the statcment of the theorem, and subsequently, Λ_2 denotes a constant such that

$$\int_{\mathfrak{S}_{R/2}(Y)} H^2\,dA\leqslant\Lambda_2, \tag{16.83}$$

where $H=(\kappa_1+\kappa_2)/2$ is the mean curvature of $\mathfrak{S}$.

Theorem 16.14. *Suppose φ is a (K, K')-quasiconformal C^1 mapping from $\mathfrak{S}$ into $\mathfrak{T}$. Then*

$$\mathfrak{D}(\varrho; Z)\leqslant C(\varrho/R)^\alpha \tag{16.84}$$

for all $Z \in \mathfrak{S}_{R/4}(Y)$ and all $\rho \in (0, R/4)$, where C and α are positive constants depending only on Λ_0, K, $K'R^2$, Λ_1 and Λ_2.

Proof. In the proof we let $\mathfrak{S}_\rho = \mathfrak{S}_\rho(Z)$ and $\mathfrak{D}(\rho) = \mathfrak{D}(\rho; Z)$, where $Z \in \mathfrak{D}_{R/4}(Y)$, and let r denote the radial distance function defined by $r(X) = |X - Z|$, $X \in \mathbb{R}^3$. We take $\rho \in (0, R/4)$ such that δr is never zero on $\partial\mathfrak{S}_\rho$, so that we can assume (16.77) and (16.78). Since the curves $\Gamma_\rho^{(j)}$ are closed, we have

$$\int_{\Gamma^{(j)}} \frac{d\varphi_i}{ds}\, ds = 0; \quad (\Gamma^{(j)} = \Gamma_\rho^{(j)}),$$

hence, if we let X_j denote the initial point (corresponding to arc length $s = 0$) of $\Gamma_\rho^{(j)}$, then the integral on the right hand side of (16.81) can be written

$$\int_{\Gamma^{(j)}} (\omega_i \circ \varphi - \omega_i \circ \varphi(X_j)) \frac{d\varphi_i}{ds}\, ds.$$

Hence (16.81) gives

$$\begin{aligned}
\left| \int_{\mathfrak{S}_\rho} J(\varphi)\, dA \right| &\leq \sum_{j=1}^{n(\rho)} \left\{ \sup_{\Gamma^{(j)}} |\omega \circ \varphi - \omega \circ \varphi(X_j)| \int_{\Gamma^{(j)}} \left| \frac{d\varphi}{ds} \right| ds \right\} \\
&\leq \sum_{j=1}^{n(\rho)} \left\{ \int_{\Gamma^{(j)}} \left| \frac{d\omega \circ \varphi}{ds} \right| ds \int_{\Gamma^{(j)}} \left| \frac{d\varphi}{ds} \right| ds \right\} \\
&\leq \sum_{j=1}^{n(\rho)} \left\{ \int_{\Gamma^{(j)}} |\delta\omega \circ \varphi|\, ds \int_{\Gamma^{(j)}} |\delta\varphi|\, ds \right\} \\
&\leq \sup_{\mathfrak{T}} |D\omega| \sum_{j=1}^{n(\rho)} \left(\int_{\Gamma^{(j)}} |\delta\varphi|\, ds \right)^2 \\
&\leq \Lambda_0 \left(\sum_{j=1}^{n(\rho)} \int_{\Gamma^{(j)}} |\delta\varphi|\, ds \right)^2 \quad \text{by (16.70)}, \\
&= \Lambda_0 \left(\int_{\partial\mathfrak{S}_\rho} |\delta\varphi|\, ds \right)^2 \\
&\leq \Lambda_0 \left(\int_{\partial\mathfrak{S}_\rho} |\delta\varphi|^2 |\delta r|^{-1}\, ds \right) \left(\int_{\partial\mathfrak{S}_\rho} |\delta r|\, ds \right)
\end{aligned}$$

by Schwarz's inequality,

$$= \Lambda_0 \left(D_\rho \int_{\mathfrak{S}_\rho} |\delta\varphi|^2 \, dA \right) \left(D_\rho \int_{\mathfrak{S}_\rho} |\delta r|^2 \, dA \right) \quad \text{by (16.77)},$$

$$\leqslant C\Lambda_0 \rho \mathfrak{D}'(\rho)$$

by (16.36), where $C = C(\Lambda_1, \Lambda_2)$. Therefore by (16.76) we have

$$\mathfrak{D}(\rho) \leqslant C|K|\Lambda_0 \rho \mathfrak{D}'(\rho) + C'K'\rho^2$$

for almost all $\rho \in (0, R/4)$, where $C' = C'(\Lambda_1, \Lambda_2)$ by virtue of (16.34). Now define

$$\mathfrak{G}(\rho) = \mathfrak{D}(\rho) + (\rho/R)^2.$$

It is then not difficult to see that the above inequality implies an inequality of the form

$$\mathfrak{G}(\rho) \leqslant C\rho \mathfrak{G}'(\rho)$$

where $C = C(\Lambda_0, K, K'R^2, \Lambda_1, \Lambda_2)$. This last inequality can be written

$$\frac{d}{d\rho} \log \mathfrak{G}(\rho) \geqslant (C\rho)^{-1},$$

and, since $\mathfrak{G}(\rho)$ in increasing in ρ, we can integrate to obtain

$$\log (\mathfrak{G}(\rho)/\mathfrak{G}(R/4)) \leqslant C^{-1} \log (4\rho/R), \quad \rho \leqslant R/4.$$

Thus

$$\mathfrak{G}(\rho) \leqslant 4^\alpha \mathfrak{G}(R/4)(\rho/R)^\alpha, \quad \rho \leqslant R/4,$$

where $\alpha = C^{-1}$. Since $\mathfrak{S}_{R/4}(Z) \subset \mathfrak{S}_{R/2}(Y)$ we must have

$$\mathfrak{G}(R/4) \leqslant \mathfrak{D}(R/2;\, Y) + 1/16$$

and hence the required estimate follows from (16.80). □

Using the extended Morrey estimate, Lemma 16.4, we can now finally deduce from Theorem 16.14 the Hölder estimate for (K, K')-quasiconformal maps.

Theorem 16.15. *Suppose φ is a (K, K')-quasiconformal C^1 mapping from $\mathfrak{S}$ into $\mathfrak{T}$. Then*

$$\text{(16.85)} \qquad \sup_{X \in \mathfrak{S}^*_\rho(Y)} |\varphi(X) - \varphi(Y)| \leqslant C(\rho/R)^\alpha, \quad \rho \in (0, R/4),$$

where C and α are positive constants depending only on Λ_0, K, $K'R^2$, Λ_1, Λ_2 and $\mathfrak{S}^*_\rho(Y)$ is the component of $\mathfrak{S}_\rho(Y)$ which contains Y.

Proof. Let Z be an arbitrary point of $\mathfrak{S}_{R/4}(Y)$. By the Hölder inequality and the estimates (16.84) and (16.34) we have

$$\int_{\mathfrak{S}_\rho(Z)} |\delta\varphi_i| \, dA \leqslant (CC')^{1/2}\rho(\rho/R)^{\alpha/2}, \quad \rho \in (0, R/4), \ i=1, 2, 3,$$

where C, α are as in (16.84) and $C'=C'(\Lambda_1, \Lambda_2)$. Hence the hypotheses of Lemma 16.4 are satisfied with $K=(CC')^{1/2}$ and $\beta=\alpha/2$. The theorem thus follows. □

16.6. Graphs with Quasiconformal Gauss Map

In this section $\mathfrak{S}$ will denote the graph $\{(x, z) \in \mathbb{R}^3 | x \in \Omega, \ z=u(x)\}$ of a $C^2(\Omega)$ function u, y will denote a fixed point of Ω, and it will be assumed that Ω contains the disc $B_R(Y)=\{x \in \mathbb{R}^2 \mid |x-y|<R\}$. Y will denote the point (y, y_3) $(y_3=u(y))$ of $\mathfrak{S}$ and, as in Section 16.4, $\mathbf{v}$ will denote the upward unit normal function defined on $\Omega \times \mathbb{R}$ by

$$\mathbf{v}(x, z) \equiv \mathbf{v}(x) = \left(-\frac{Du(x)}{v}, \frac{1}{v}\right), \quad v=\sqrt{1+|Du|^2}, \ x \in \Omega, \ z \in \mathbb{R}.$$

The *Gauss map* $\mathbf{G}$ of $\mathfrak{S}$, which maps $\mathfrak{S}$ into the upper hemisphere

$$\mathfrak{T}=\{X=(x_1, x_2, x_3) \in \mathbb{R}^3 \mid |X|=1, \ x_3>0\},$$

is defined by setting

$$\mathbf{G}(x, x_3)=\mathbf{v}(x, x_3) \tag{16.86}$$

at each point $(x, x_3) \in \mathfrak{S}$. The remaining notation and terminology are as in Sections 16.4 and 16.5.

We first want to explicitly obtain the quantities $|\delta\varphi|^2$ and $J(\varphi)$ in the case when $\varphi=\mathbf{G}$. This is quite straightforward if we set up new coordinates as in Section 16.5. In this case the function ψ is defined by

$$\psi(\zeta)=-(1+|D\tilde{u}(\zeta)|^2)^{-1/2} D\tilde{u}(\zeta), \quad \zeta \in \mathscr{U},$$

so that we obtain, from (16.74) and (16.75),

$$J(\mathbf{G})(Y)=D_{11}\tilde{u}(0)D_{22}\tilde{u}(0)-(D_{12}\tilde{u}(0))^2$$

and

$$|\delta \mathbf{G}|^2(Y) = (D_{11}\tilde{u}(0))^2 + 2(D_{12}\tilde{u}(0))^2 + (D_{22}\tilde{u}(0))^2.$$

But then by Section 14.6 we have

$$(16.87) \qquad \begin{aligned} &J(\mathbf{G})(Y) = \kappa_1\kappa_2 \\ &|\delta \mathbf{G}|^2(Y) = \kappa_1^2 + \kappa_2^2, \end{aligned}$$

where κ_1, κ_2 are the principal curvatures of $\mathfrak{S}$ at Y. The product $\kappa_1\kappa_2$ appearing in (16.87) is called the *Gauss curvature* of $\mathfrak{S}$; it is an extremely important geometric invariant in the study of surfaces.

By using the identities (16.87) and recalling the definition (16.76), we now see that $\mathbf{G}$ is (K, K')-quasiconformal if and only if at each point of $\mathfrak{S}$ the principal curvatures κ_1, κ_2 satisfy

$$(16.88) \qquad \kappa_1^2 + \kappa_2^2 \leqslant 2K\kappa_1\kappa_2 + K'.$$

We thus see why the study of quasiconformal maps is relevant to the investigation of equations of mean curvature type; because by squaring (16.63)′ we obtain the inequality

$$\frac{\alpha_1}{\alpha_2}\kappa_1^2 + \frac{\alpha_2}{\alpha_1}\kappa_2^2 = -2\kappa_1\kappa_2 + \frac{\beta^2}{\alpha_1\alpha_2}.$$

That is, by virtue of (16.64)′ and (16.65)′, *we deduce that the Gauss map* $\mathbf{G}$ *of the graph of a solution* u *of* (16.63), (16.64), (16.65), is (K, K')*-quasiconformal with*

$$(16.89) \qquad K = -\gamma, \quad K' = \gamma\mu^2.$$

Thus the results established in this section for graphs with quasiconformal Gauss map are all applicable to the graphs of solutions of (16.63), (16.64), (16.65).

We wish to eventually apply Theorem 16.15 to the Gauss map $\mathbf{G}$, but first we need to discuss appropriate choices for the constants Λ_0, Λ_1 and Λ_2. To begin with, we have already seen in Section 16.5 that we can take

$$\omega(X) = \left(-\frac{x_2}{1+x_3}, \frac{x_1}{1+x_3}, 0\right).$$

One then easily checks that an appropriate choice for the constant Λ_0 is $\Lambda_0 = 4$. Next we notice by Lemma 16.13 and (16.87) we have, provided $\mathbf{G}$ is (K, K')-quasiconformal,

$$\int_{\mathfrak{S}_{R/2}(Y)} (\kappa_1^2 + \kappa_2^2)\, dA \leqslant C,$$

where $C=C(K, K'R^2, \Lambda_1)$. Thus, since

$$\kappa_1^2+\kappa_2^2 \geqslant \tfrac{1}{2}(\kappa_1+\kappa_2)^2 = 2H^2,$$

we can take $\Lambda_2 = C/2$, with C as above. The next lemma shows that we can choose Λ_1 to depend only on K and $K'R^2$.

Lemma 16.16. *Suppose* **G** *is* (K, K')*-quasiconformal. Then*

$$|\mathfrak{S}_{\rho/2}(Z)| \leqslant C\rho^2 \tag{16.90}$$

for each $Z \in \mathfrak{S}$ *and* $\rho>0$ *such that* $\mathfrak{S}_\rho(Z) \subset \mathfrak{S}_R(Y)$, *where* $C=C(K, K'R^2)$.

Proof. The starting point for our proof is the identity (16.44) which, by virtue of formula (14.104), can be written in the form

$$\int_\Omega (\kappa_1^2+\kappa_2^2)\eta \, dx + \int_\Omega (\nu_k D_k \nu_i D_i \eta - 2\nu_k D_k H\eta)\, dx = 0$$

for all $\eta \in C_0^1(\Omega)$. Here we write $H(x)=H(x, u(x))$, $\kappa_i(x)=\kappa_i(x, u(x))$ for $x \in \Omega$. If $\eta \in C_0^2(\Omega)$, we then obtain by integration by parts

$$\int_\Omega (\kappa_1^2+\kappa_2^2-4H^2)\eta \, dx = \int_\Omega (\nu_i \nu_k D_{ik}\eta - 4H\nu_i D_i \eta)\, dx,$$

and hence, as $H=(\kappa_1+\kappa_2)/2$,

$$-2\int_\Omega \kappa_1\kappa_2\eta \, dx = \int_\Omega (\nu_i \nu_k D_{ik}\eta - 2(\kappa_1+\kappa_2)\nu_i D_i\eta)\, dx$$

for all $\eta \in C_0^2(\Omega)$. By (16.88) we then have, replacing η by η^2,

$$\int_\Omega (\kappa_1^2+\kappa_2^2)\eta^2 \, dx \leqslant |K| \int_\Omega (|D^2\eta^2| + 2|\kappa_1+\kappa_2|\, |D\eta^2|)\, dx + K' \int_\Omega \eta^2 \, dx.$$

Since we can write

$$4K(\kappa_1+\kappa_2)\eta|D\eta| \leqslant \tfrac{1}{2}(\kappa_1^2+\kappa_2^2)\eta^2 + (4KD\eta)^2,$$

this gives

$$\tfrac{1}{2}\int_\Omega (\kappa_1^2+\kappa_2^2)\eta^2 \, dx \leqslant \int_\Omega \{C(|D\eta|^2 + \eta|D^2\eta|) + K'\eta^2\}\, dx,$$

where $C=C(K)$. Now let $\rho>0$ and $Z=(z, u(z))$ be such that $B_\rho(z)\subset\Omega$, and let us choose η such that $0\leqslant\eta\leqslant 1$ in Ω, $\eta\equiv 1$ on $B_{\rho/2}(z)$, $\eta\equiv 0$ on $\Omega-B_\rho(z)$, $|D\eta|\leqslant c/\rho$, $|D^2\eta|\leqslant c/\rho^2$ where c is an absolute constant. (It is clear that such a function η exists.) Then, since $K'\leqslant K'R^2/\rho^2$, we obtain

$$\int_\Omega (\kappa_1^2+\kappa_2^2)\eta^2\,dx\leqslant C \tag{16.91}$$

where $C=C(K, K'R^2)$. Consequently, using Hölder's inequality, we have

$$\begin{aligned}\int_\Omega |H\eta|\,dx &\leqslant \left(\int_\Omega (H\eta)^2\,dx\right)^{1/2}|B_\rho(z)|^{1/2}\\ &\leqslant (C\pi)^{1/2}\rho.\end{aligned}$$

The lemma now follows from (16.53). □

From Lemma 16.16, Theorem 16.15 and our previous choices of Λ_0, Λ_1, we can now deduce that if G is (K, K')-quasiconformal then

$$\sup_{X\in\mathfrak{S}^*_\rho(Y)} |\nu(X)-\nu(Y)|\leqslant C(\rho/R)^\alpha, \quad \rho\in(0, R), \tag{16.92}$$

where C and α are positive constants depending only on K, $K'R^2$. Notice that we assert (16.92) for all $\rho\in(0, R)$ rather than $\rho\in(0, R/4)$ as in Theorem 16.15. We can do this because $|\nu|=1$ (which means an inequality of the form (16.92) trivially holds for $\rho\in(R/4, R)$).

The estimate (16.92) can be used to obtain some rather strong regularity results for $\mathfrak{S}$. We first use (16.92) to deduce some facts about local non-parametric representations of $\mathfrak{S}$. Let P be a linear isometry of $\mathbb{R}^3$ such that

$$\tilde\nu(0)=P\nu(y, y_3)=(0, 0, 1)$$

and let

$$\tilde{\mathfrak{S}}=\{(\zeta, \zeta_3)\in\mathbb{R}^3\mid(\zeta, \zeta_3)=P(x-y, x_3-y_3), (x, x_3)\in\mathfrak{S}^*_{\theta R}(Y)\},$$

where $\theta\in(0, 1)$. Since $\mathfrak{S}$ is a C^2 surface we of course know that for *small enough* θ there is a neighborhood $\mathscr{U}$ of $0\in\mathbb{R}^2$ and a $C^2(\mathscr{U})$ function $\tilde u$ with $D\tilde u(0)=0$ and

$$\tilde{\mathfrak{S}}=\operatorname{graph}\tilde u=\{(\zeta, \zeta_3)\in\mathbb{R}^3\mid\zeta\in\mathscr{U}, \zeta_3=\tilde u(\zeta)\}. \tag{16.93}$$

Furthermore, writing

$$\tilde\nu(\zeta)=(1+|D\tilde u(\zeta)|^2)^{-1/2}(-D\tilde u(\zeta), 1), \quad \zeta\in\mathscr{U},$$

we have by (16.92) that

$$|\tilde{\nu}(\zeta)-\tilde{\nu}(0)|\leqslant C\theta^{\alpha}, \quad \zeta\in\mathcal{U},$$

where C, α are as in (16.92). Consequently

$$(1+|D\tilde{u}(\zeta)|^2)^{-1}|D\tilde{u}(\zeta)|^2+[(1+|D\tilde{u}(\zeta)|^2)^{-1/2}-1]^2\leqslant(C\theta^{\alpha})^2, \quad \zeta\in\mathcal{U},$$

which implies

$$(16.94)\qquad |D\tilde{u}(\zeta)|\leqslant[1-(C\theta^{\alpha})^2]^{-1/2}C\theta^{\alpha}<\tfrac{1}{2},$$

provided θ is such that

$$(16.95)\qquad C\theta^{\alpha}<\tfrac{1}{3}.$$

Because of (16.94), we can infer that a representation of the above type holds for any θ satisfying (16.95). This follows from the fact that if $\theta\in(0,1)$ is such that (16.93) and (16.95) hold, then, by using the smoothness of $\mathfrak{S}$ together with (16.94), we can extend $\tilde{u}$ so that a representation of the form (16.93) holds with $\mathfrak{S}^*_{(\theta+\varepsilon)R}(Y)$ $(\varepsilon>0)$ in place of $\mathfrak{S}^*_{\theta R}(Y)$. For later reference we also note that (16.94) implies

$$(16.96)\qquad B_{\theta R/2}(0)\subset\mathcal{U}.$$

We can now prove the non-trivial connectivity result of the following lemma.

Lemma 16.17. *Suppose* $\mathbf{G}$ *is* (K, K')*-quasiconformal. Then there is a constant* $\theta\in(0,1)$, *depending only on* K, $K'R^2$, *such that* $\mathfrak{S}_\rho(Y)$ *is connected for each* $\rho<\theta R$.

Proof. In the proof we will let $C_1, C_2, \ldots$ denote constants depending only on K, $K'R^2$. $\tilde{B}_\sigma$, for $\sigma>0$, will denote the open ball $\{X\in\mathbb{R}^3\mid |X-Y|<\sigma\}$. Let $\theta\in(0,1)$ satisfy (16.95), $\rho=\theta R/2$, $\beta\in(0,\frac{1}{4})$ and define $\mathfrak{G}_\beta$ to be the collection of those components of $\mathfrak{S}_{\rho/2}(Y)$ which intersect the ball $\tilde{B}_{\beta\rho}$. For each $\mathcal{G}\in\mathfrak{G}_\beta$ we can find $Z\in\mathcal{G}\cap\tilde{B}_{\rho/4}$ such that

$$\mathcal{G}\subset\mathfrak{S}^*_\rho(Z),$$

and hence, replacing Y by Z and R by $R/2$ in the discussion preceding the lemma, we see that $\mathcal{G}$ can be represented in the form (16.93), (16.94). Using such a non-parametric representation for each $\mathcal{G}\in\mathfrak{G}_\beta$ and also using the fact that no two elements of $\mathfrak{G}_\beta$ can intersect, it follows that the *union* of all the components $\mathcal{G}\in\mathfrak{G}_\beta$ is contained in a region bounded between two parallel planes π_1, π_2 with

$$(16.97)\qquad \operatorname{dist}(\pi_1,\pi_2)\leqslant C_1(\beta+\theta^{\alpha})\rho$$

where α is as in (16.92).

Our aim now is to show that, for suitable choices of β and θ depending only on K and $K'R^2$, there is only one element (viz. $\mathfrak{S}^*_{\rho/2}(Y)$) in $\mathfrak{G}_\beta$. Suppose that in fact there are two distinct elements $\mathscr{G}_1, \mathscr{G}_2 \in \mathfrak{G}_\beta$. We can clearly choose $\mathscr{G}_1, \mathscr{G}_2$ to be adjacent in the sense that the volume $\mathscr{V}$ enclosed by $\mathscr{G}_1, \mathscr{G}_2$ and $\partial\tilde{B}_{\rho/2}$ contains no other elements $\mathscr{G} \in \mathfrak{G}_\beta$. Thus $\mathscr{V} \cap \tilde{B}_{\beta\rho}$ consists entirely of points above the graph $\mathfrak{S}$ or entirely of points below $\mathfrak{S}$; it is then evident that if the unit normal $\boldsymbol{v}$ points out of (into) $\mathscr{V}$ on $\mathscr{G}_1$, then it also points out of (into) $\mathscr{V}$ on $\mathscr{G}_2$. Furthermore, by (16.97) we have

$$\text{volume } \mathscr{V} = |\mathscr{V}| \leqslant C_2(\beta+\theta^\alpha)\rho^3,$$

$$\text{area } (\overline{\mathscr{V}} \cap \partial\tilde{B}_{\rho/2}) = A(\overline{\mathscr{V}} \cap \partial\tilde{B}_{\rho/2}) \leqslant C_3(\beta+\theta^\alpha)\rho^2. \tag{16.98}$$

An application of the divergence theorem over $\mathscr{V}$ then gives

$$\int_{\mathscr{G}_1} \boldsymbol{v}\cdot\boldsymbol{v}\, dA + \int_{\mathscr{G}_2} \boldsymbol{v}\cdot\boldsymbol{v}\, dA = \pm\left\{ \int_{\mathscr{V}} \operatorname{div} \boldsymbol{v}\, dx\, dx_3 - \int_{\partial\tilde{B}_{\rho/2}\cap\mathscr{V}} \boldsymbol{\eta}\cdot\boldsymbol{v}\, dA \right\},$$

where $\boldsymbol{\eta}$ is the outward unit normal of $\partial\tilde{B}_{\rho/2}$. By (16.11) and (16.98) this gives

$$A(\mathscr{G}_1) + A(\mathscr{G}_2) \leqslant 2\int_{\mathscr{V}} |H(x)|\, dx\, dx_3 + C_3(\beta+\theta^\alpha)\rho^2.$$

Also, by (16.98) and (16.91).

$$\begin{aligned}
\int_{\mathscr{V}} |H(x)|\, dx\, dx_3 &\leqslant \left(\int_{\mathscr{V}} H^2(x)\, dx\, dx_3 \right)^{1/2} |\mathscr{V}|^{1/2} \\
&\leqslant \left(\int_{\tilde{B}_{\rho/2}} H^2(x)\, dx\, dx_3 \right)^{1/2} \{C_2(\beta+\theta^\alpha)\rho^3\}^{1/2} \\
&\leqslant (C_4\rho)^{1/2}\{C_2(\beta+\theta^\alpha)\rho^3\}^{1/2} \\
&\leqslant \sqrt{C_4C_2(\beta+\theta^\alpha)}\rho^2.
\end{aligned}$$

Hence, provided $\beta+\theta^\alpha<1$, we have

$$A(\mathscr{G}_1) + A(\mathscr{G}_2) \leqslant C_5\sqrt{\beta+\theta^\alpha}\rho^2. \tag{16.99}$$

On the other hand, by using a non-parametric representation as in (16.93), (16.94) we infer that

$$A(\mathscr{G}) \geqslant C_6\rho^2 \tag{16.100}$$

for each $\mathscr{G} \in \mathfrak{G}_\beta$; where $C_6 > 0$ is an absolute constant. (Since $\mathfrak{S}^*_{\rho/4}(Z) \subset \mathscr{G}$ we can deduce from (16.94) that (16.100) holds with (for example) $C_6 = \pi/64$). Inequalities (16.99) and (16.100) are clearly contradictory if we choose β, θ small enough (but depending only on K and $K'R^2$). For such a choice of β, θ we thus have

$$\mathfrak{S}_{\beta\rho}(Y) = \mathfrak{S} \cap \tilde{B}_{\beta\rho} = \mathfrak{S}_{\rho/2}(Y) \cap \tilde{B}_{\beta\rho} = \mathfrak{S}^*_{\rho/2}(Y) \cap \tilde{B}_{\beta\rho}.$$

But by using a representation of the form (16.93), (16.94) for $\mathfrak{S}^*_{\rho/2}(Y)$, we clearly have $\mathfrak{S}^*_{\rho/2}(Y) \cap \tilde{B}_{\beta\rho}$ connected. Thus $\mathfrak{S}_{\beta\rho}(Y) = \mathfrak{S}_{\beta\theta R/2}(Y)$ is connected. The lemma follows because β, θ depended only on $K, K'R^2$. □

Because of the above connectivity result we can now replace $\mathfrak{S}^*_\rho(Y)$ in (16.92) by $\mathfrak{S}_\rho(Y)$ for $\rho \leqslant \theta R$. However, since $|\nu| = 1$, an inequality of the form (16.92) is trivial for $\rho > \theta R$. Hence we have the following theorem.

Theorem 16.18. *Suppose* **G** *is* (K, K')*-quasiconformal. Then*

$$\sup_{X \in \mathfrak{S}_\rho(Y)} |\nu(X) - \nu(Y)| \leqslant C(\rho/R)^\alpha, \quad \rho \in (0, R), \tag{16.101}$$

where C and α are positive constants depending only on $K, K'R^2$.

Remarks. (i) The estimate (16.101) implies that

$$|\nu(X) - \nu(\bar{X})| \leqslant 2^\alpha C(|X - \bar{X}|/R)^\alpha \tag{16.102}$$

for all $X, \bar{X} \in \mathfrak{S}_{R/4}(Y)$. This is seen by using (16.101) with $\bar{X}$ in place of Y and with $R/2$ in place of R.

(ii) If $K' = 0$ and $\Omega = \mathbb{R}^2$, then we can let $R \to \infty$ in (16.101), thus showing that $\nu(X) \equiv \nu(Y)$ on $\mathfrak{S}$. That is, we have the following corollary.

Corollary 16.19. *Suppose* **G** *is* $(K, 0)$*-quasiconformal and* $\Omega = \mathbb{R}^2$. *Then u is a linear function.*

Note that Corollary 16.19 can be directly deduced by letting $R \to \infty$ in (16.84), without first proving (16.101) or even (16.92). However, we still require Lemma 16.16 to show that Λ_1 can be chosen to depend only on K.

16.7. Applications to Equations of Mean Curvature Type

Here $\mathfrak{S}$ will denote the graph of a solution u of (16.63), (16.64), (16.65). The remaining terminology will be as in Sections 16.4–16.6.

We first note that, since the Gauss map **G** is automatically (K, K')-quasiconformal with K, K' as in (16.89), we can immediately deduce Theorem 16.12 from Corollary 16.19 above.

We next wish to show that Theorem 16.18 implies a bound for the principal curvatures κ_1, κ_2 of $\mathfrak{S}$, provided a suitable Hölder condition is imposed on the coefficients a^{ij}, b. In order that this condition may be conveniently described, we first extend the matrix $[a^{ij}]$ to be a 3×3 matrix (cf. the procedure of Section 16.4) by defining

$$a^{3i}(x, z, p)=a^{i3}(x, z, p)=\sum_{j=1}^{2} p_j a^{ij}(x, z, p),\ i=1, 2,$$

$$(x, z, p)\in \Omega\times\mathbb{R}\times\mathbb{R}^2,$$

$$a^{33}(x, z, p)=\sum_{i,j=1}^{2} p_i p_j a^{ij}(x, z, p).$$

It will also be convenient to express a^{ij}, b in terms of $X=(x, z)\in\Omega\times\mathbb{R}$ and $\nu=(1+|p|^2)^{-1/2}(-p, 1)$, by defining

$$a_*^{ij}(X, \nu)=a^{ij}(x, z, p),\quad i, j=1, 2, 3,$$
$$b_*(X, \nu)=(1+|p|^2)^{-1/2}b(x, z, p)$$

for all $(x, z, p)\in\Omega\times\mathbb{R}\times\mathbb{R}^2$. Notice that these definitions give a_*^{ij}, b_* on the sets $(\Omega\times\mathbb{R})\times\{\zeta\in\mathbb{R}^3 \mid |\zeta|=1, \zeta_3>0\}$. (In the case when equation (16.63) arises as the non-parametric Euler equation of an elliptic parametric variational problem, we show in Appendix 16.8 that the functions a_*^{ij}, b_* arise quite naturally.) We now assume that the functions a_*^{ij}, b_* satisfy the Hölder conditions

$$\begin{aligned}|a_*^{ij}(X, \nu)-a_*^{ij}(\bar{X}, \bar{\nu})|&\le\mu_1\{(|X-\bar{X}|/R)^\beta+|\nu-\bar{\nu}|^\beta\},\\ |b_*(X, \nu)-b_*(\bar{X}, \bar{\nu})|&\le\mu_2\{(|X-\bar{X}|/R)^\beta+|\nu-\bar{\nu}|^\beta\},\end{aligned}\tag{16.103}$$

for all $X, \bar{X}\in\Omega\times\mathbb{R}$ and all $\nu, \bar{\nu}\in\{\zeta\in\mathbb{R}^3 \mid |\zeta|=1, \zeta_3>0\}$, where μ_1, μ_2 and β are constants with $\beta\in(0, 1)$.

Theorem 16.20. *Suppose* (16.63), (16.64), (16.65) *and* (16.103) *hold. Then if* κ_1, κ_2, *are the principal curvatures of* $\mathfrak{S}$ *at* Y, *we have*

$$\kappa_1^2+\kappa_2^2\le C/R^2,\tag{16.104}$$

where $C=C(\gamma, \mu R, \mu_1, \mu_2 R, \beta)$.

Proof. Choose θ small enough to ensure that $\mathfrak{S}_{\theta R}(Y)$ is connected (Lemma 16.17) and can be represented in the form (16.93) with (16.95) (and hence (16.94)) holding. By applying the discussion of Section 16.4 to both $\mathfrak{S}$ and the transformed surface $\tilde{\mathfrak{S}}$ of (16.93), it then follows that $\tilde{u}$ satisfies an equation of the form

$$\tilde{a}^{ij}(\zeta)D_{ij}\tilde{u}+\tilde{b}(\zeta)=0\tag{16.105}$$

on $\mathscr{U}$, where

$$|\xi|^2 - \frac{(D\tilde{u}\cdot\xi)^2}{1+|D\tilde{u}|^2} \leqslant \tilde{a}^{ij}\xi_i\xi_j \leqslant \gamma\left[|\xi|^2 - \frac{(D\tilde{u}\cdot\xi)^2}{1+|D\tilde{u}|^2}\right]$$

for all $\zeta \in \mathbb{R}^2$, and

$$|\tilde{b}| \leqslant \mu\sqrt{1+|D\tilde{u}|^2}.$$

Consequently, by (16.94), we have for all $\zeta \in B = B_{\theta R/2}(0)$

$$\begin{aligned} &|\xi|^2/2 \leqslant \tilde{a}^{ij}(\zeta)\xi_i\xi_j \leqslant \gamma|\xi|^2, \\ &|\tilde{b}(\zeta)| \leqslant 2\mu. \end{aligned} \tag{16.106}$$

By (16.103), (16.102) and the discussion in Section 16.4, it also follows that we can assume the Hölder estimates

$$\begin{aligned} &|\tilde{a}^{ij}(\zeta) - \tilde{a}^{ij}(\bar{\zeta})| \leqslant C(|\zeta - \bar{\zeta}|/R)^{\alpha\beta}, \quad \zeta, \bar{\zeta} \in B, \quad i, j = 1, 2 \\ &|\tilde{b}(\zeta) - \tilde{b}(\bar{\zeta})| \leqslant C(|\zeta - \bar{\zeta}|/R)^{\alpha\beta}, \quad \zeta, \bar{\zeta} \in B. \end{aligned} \tag{16.107}$$

Furthermore it is clear from (16.94), and the fact that $\tilde{u}(0) = 0$, that

$$\sup_B |\tilde{u}| \leqslant R. \tag{16.108}$$

Then, by using Schauder's interior estimate (Theorem 6.2) in conjunction with (16.106), (16.107) and (16.108), we infer that

$$\begin{aligned} |\tilde{u}|^*_{2,\alpha;B} &\leqslant C\{R + \mu R^2 + R^{2+\alpha\beta}\mu_2 R^{-\alpha\beta}\} \\ &\leqslant CR, \end{aligned}$$

where $C = C(\gamma, \mu R, \mu_1, \mu_2 R, \beta)$. In particular we have

$$\sum_{i,j=1}^{2} |D_{ij}\tilde{u}(0)| \leqslant C/R,$$

whence the theorem follows by (16.87). □

The Hölder estimate (16.101) can also be used to obtain gradient estimates for solutions u of (16.63), (16.64), (16.65). The following theorem deals with the homogeneous case, when no smoothness or continuity restrictions will be imposed on the coefficients.

Theorem 16.21. *Suppose* (16.63), (16.64) *hold*, $b\equiv 0$, *and suppose the functions* $a^{ij}(x, u, Du)$, $i, j=1, 2$ *are measurable on* Ω. *Then*

$$|Du(y)|\leqslant C_1 \exp(C_2 m_R/R), \tag{16.109}$$

where

$$m_R=\sup_{B_R(y)}(u-u(y))$$

and $C_1=C_1(\gamma)$, $C_2=C_2(\gamma)$.

Proof. As in Theorem 16.20 we assume θ is small enough to ensure that $\mathfrak{S}_{\theta R}(Y)$ is connected and that the representation (16.93), (16.94) holds. Notice that since $b\equiv 0$ we can choose θ to depend only on γ. Also, we can here suppose that the matrix of P, $[p_{ij}]$, is chosen such that $p_{32}=0$. Hence we have

$$(1+|Du(x)|^2)^{-1/2}=(1+|D\tilde{u}(\zeta)|^2)^{-1/2}(p_{31}D_1\tilde{u}(\zeta)-p_{33}). \tag{16.110}$$

Now (since $b\equiv 0$), equation (16.105) can be written in the form

$$\alpha D_{11}\tilde{u}+2\beta D_{12}\tilde{u}+D_{22}\tilde{u}=0$$

where $\alpha=\tilde{\alpha}_{11}/\tilde{\alpha}_{22}$ and $\beta=\tilde{\alpha}_{12}/\tilde{\alpha}_{22}$. We now multiply each side of this equation by $p_{31}D_1\varphi$, where $\varphi\in C_0^2(\mathscr{U})$ and integrate over $\mathscr{U}$. Making use of the relations

$$\int_{\mathscr{U}} D_{22}\tilde{u}D_1\varphi\, d\zeta=-\int_{\mathscr{U}} D_2\tilde{u}D_{12}\varphi\, d\zeta=\int_{\mathscr{U}} D_{21}\tilde{u}D_2\varphi\, d\zeta$$

and writing

$$\psi=p_{31}D_1\tilde{u}-p_{33},$$

we then obtain

$$\int_{\mathscr{U}}(\alpha D_1\psi D_1\varphi+2\beta D_2\psi D_1\varphi+D_2\psi D_2\varphi)\, d\zeta=0.$$

That is, ψ is a weak solution of the uniformly elliptic equation

$$D_1(\alpha D_1\psi+2\beta D_2\psi)+D_2(D_2\psi)=0 \quad \text{on } \mathscr{U}.$$

Furthermore, we have $\psi>0$ on $\mathscr{U}$ by (16.110). Hence we can apply the Harnack inequality, Theorem 8.20, to ψ, thus giving

$$\sup_{B_{\theta R/2}(0)}\psi\leqslant C\inf_{B_{\theta R/2}(0)}\psi, \tag{16.111}$$

where $C=C(\gamma)$. But because of (16.110) and (16.94) we know that

$$(1+|Du(x)|^2)^{-1/2} \leqslant \psi(\zeta) \leqslant 2(1+|Du(x)|^2)^{-1/2}$$

on $\mathscr{U}$, and hence defining v on $\Omega \times \mathbb{R}$ by

$$v(x, z)=(1+|Du(x)|^2)^{1/2}, \quad x \in \Omega,\ z \in \mathbb{R},$$

we can deduce from (16.111) that

$$\sup_{X \in \mathfrak{S}_{\theta R/2}} v(X) \leqslant C \inf_{X \in \mathfrak{S}_{\theta R/2}} v(X),$$

where $C=C(\gamma)$. By varying Y it clearly follows that there is a number $\chi \in (0, \frac{1}{2})$, depending only on γ, such that

$$v(X_1) \leqslant Cv(X_2) \tag{16.112}$$

whenever $X_1=(x^{(1)}, u(x^{(1)}))$ and $X_2=(x^{(2)}, u(x^{(2)}))$ are such that $|X_1-X_2| \leqslant \chi R$ and $x^{(1)}, x^{(2)} \in B_{R/2}(y)$. Now let

$$B^+_{\chi R}(y)=\{x \in B_{\chi R} \mid u(x)>u(y)\}$$

and let $Y_1=(y^{(1)}, u(y^{(1)}))$ and $Y_2=(y^{(2)}, u(y^{(2)}))$ be such that $y^{(1)}, y^{(2)} \in \overline{B^+_{\chi R}}(y)$ and

$$|Du(y^{(1)})|=\sup_{B^+_{\chi R}(y)} |Du|,$$

$$|Du(y^{(2)})|=\inf_{B^+_{\chi R}(y)} |Du|.$$

Take a sequence $X_1, \ldots, X_N$ of points in $\mathfrak{S} \cap (\overline{B^+_{\chi R}}(y) \times \mathbb{R})$ such that $|X_{i+1}-X_i|<\chi R$, $i=1, \ldots, N-1$, and such that $X_1=Y_1$, $X_N=Y_2$. Clearly, repeated application of (16.111) implies

$$v(Y_1) \leqslant C^N v(Y_2);$$

that is

$$\sup_{B^+_{\chi R}(y)} \sqrt{1+|Du|^2} \leqslant C^N \inf_{B^+_{\chi R}(y)} \sqrt{1+|Du|^2}.$$

However, it is clear that we can choose N so that

$$N \leqslant C(m_R/R+1),$$

where $C=C(\gamma)$. Hence we obtain

$$\sup_{B^+_{\chi R}(y)} \sqrt{1+|Du|^2} \leqslant \{C_1 \exp(C_2 m_R/R)\} \inf_{B^+_{\chi R}(y)} \sqrt{1+|Du|^2},$$

where $C_1=C_1(\gamma)$, $C_2=C_2(\gamma)$. Finally, Theorem 16.21 follows by using the fact that

$$\inf_{B^+_{\chi R}(y)} |Du| \leqslant \chi^{-1} m_R/R.$$

(See problem 16.5.) □

The Hölder estimate for $\boldsymbol{v}$ can also be used in the non-homogeneous case to deduce a gradient estimate for u, but in this case it is necessary to impose Lipschitz restrictions on the functions a^{ij}_*, b_* introduced above. The interested reader is referred to [SI 4].

16.8. Appendix: Elliptic Parametric Functionals

Let Ω be a bounded domain in $\mathbb{R}^2$ and consider the functional I, defined for C^1 mappings $\mathbf{Y}=(Y_1, Y_2, Y_3)\colon \bar{\Omega} \to \mathbb{R}^3$ by

$$I(\mathbf{Y}) = \int_\Omega G(x, \mathbf{Y}, D_1\mathbf{Y}, D_2\mathbf{Y})\, dx, \tag{16.113}$$

where $G=G(x, X, p)$ is a given continuous function of $(x, X, p) \in \mathbb{R}^2 \times \mathbb{R}^3 \times \mathbb{R}^6$. (Here of course $D_i\mathbf{Y}=(D_iY_1, D_iY_2, D_iY_3)$ for $i=1, 2$). Now let us consider the possibility that I remains invariant under orientation preserving diffeomorphisms of $\mathbb{R}^2$; that is, whenever ψ is a diffeomorphism of $\mathbb{R}^2$ onto itself with positive Jacobian, we would have

$$\int_{\Omega'} G(\zeta, \tilde{\mathbf{Y}}(\zeta), D_1\tilde{\mathbf{Y}}(\zeta), D_2\tilde{\mathbf{Y}}(\zeta))\, d\zeta = \int_\Omega G(x, \mathbf{Y}(x), D_1\mathbf{Y}(x), D_2\mathbf{Y}(x))\, dx,$$

where $\Omega'=\psi(\Omega)$ and $\tilde{\mathbf{Y}}=\mathbf{Y}\circ\psi^{-1}$. A simple computation (cf. [MY 5], p. 349) shows that this would be true for all such diffeomorphisms ψ and domains Ω if and only if there is a real-valued function F on $\mathbb{R}^3 \times \mathbb{R}^3$ such that

$$G(x, X, p) = F(X, P), \quad (x, X, p) \in \mathbb{R}^2 \times \mathbb{R}^3 \times \mathbb{R}^6, \tag{16.114}$$

where

$$P=(p_3p_5-p_2p_6,\, p_1p_6-p_3p_4,\, p_2p_4-p_1p_5),$$

and

(16.115) $\quad F(X, \lambda q)=\lambda F(X, q), \quad (X,q) \in \mathbb{R}^3 \times \mathbb{R}^3, \quad \lambda>0.$

Note in particular that (16.114) implies that $G(x, X, p)$ cannot depend on x; that is, $G(x, X, p)=G(0, X, p)$ for $(x, X, p) \in \mathbb{R}^2 \times \mathbb{R}^3 \times \mathbb{R}^6$. In case $p=(D_1\mathbf{Y}, D_2\mathbf{Y})$, where $\mathbf{Y}$ is a C^1 map from $\bar{\Omega}$ into $\mathbb{R}^3$, P is given by

$$P=(D_1 Y_3 \cdot D_2 Y_2 - D_1 Y_2 \cdot D_2 Y_3,\ D_1 Y_1 \cdot D_2 Y_3 - D_1 Y_3 \cdot D_2 Y_1,\ D_1 Y_2 \cdot D_2 Y_1 - D_1 Y_1 \cdot D_2 Y_2).$$

As is well known, in case $\mathbf{Y}$ is one-to-one and such that the Jacobian matrix $[D_j Y_i(x)]$ has rank 2 for each $x \in \Omega$, this last identity can be written

$$P=\chi \nu,$$

where ν is the unit normal of the embedded surface $\mathfrak{S}=\{\mathbf{Y}(x) \mid x \in \Omega\}$ and χ is the area magnification factor of the mapping $\mathbf{Y}$. Thus, assuming that we orient $\mathfrak{S}$ with unit normal ν such that $\chi>0$, we can write

$$I(\mathbf{Y})=\int_{\mathfrak{S}} F(X, \nu(X))\, dA(X);$$

that is, we can express $I(\mathbf{Y})$ completely in terms of the oriented surface $\mathfrak{S}$ and independently of the particular mapping $\mathbf{Y}$ that is used to represent $\mathfrak{S}$. Through this discussion we are led to consider the functional J, defined for any smooth oriented surface $\mathfrak{S}$ in $\mathbb{R}^3$ having finite area by

(16.116) $$J(\mathfrak{S})=\int_{\mathfrak{S}} F(X, \nu(X))\, dA(X);$$

this functional has the property that $J(\mathfrak{S})=I(\mathbf{Y})$ whenever $\mathbf{Y}$ is a one-to-one C^1 mapping from Ω into $\mathbb{R}^3$ such that $[D_j Y_i]$ has rank 2 at each point of Ω and $\mathfrak{S}=\{\mathbf{Y}(x) \mid x \in \Omega\}$.

If F satisfies (16.115), we call a functional of the form (16.116) a *parametric functional*. The functional J is called elliptic if F is C^2 on $\mathbb{R}^3 \times (\mathbb{R}^3 - \{0\})$ and if the convexity condition

(16.117) $$D_{q_i q_j} F(X, q)\xi_i \xi_j \geq |q|^{-1}|\xi'|^2, \quad \xi' = \xi - \left(\xi \cdot \frac{q}{|q|}\right)\frac{q}{|q|},$$

holds for all $X \in \mathbb{R}^3$, $q \in \mathbb{R}^3 - \{0\}$ and $\xi \in \mathbb{R}^3$. Notice that, up to a scalar factor, (16.117) is the strongest convexity condition possible for F in view of the homogeneity condition (16.115).

If we now consider a *non-parametric* surface $\mathfrak{S}$ given by

$$\mathfrak{S}=\{(x, u(x))\in\mathbb{R}^3 \mid x\in\Omega\},$$

where $u\in C^2(\bar{\Omega})$, then, taking $\boldsymbol{\nu}$ to be the downward unit normal $(Du, -1)/\sqrt{1+|Du|^2}$, we have

$$J(\mathfrak{S})=\int_\Omega F(x, u(x), Du(x), -1)\,dx.$$

Notice that here we have used the relation $dA=\sqrt{1+|Du|^2}\,dx$. The expression on the right can be considered as a non-parametric functional, defined for any $u\in C^2(\bar{\Omega})$. The Euler–Lagrange equation for this non-parametric functional is

$$\sum_{i=1}^{2} D_i[D_{q_i}F(x, u, Du, -1)]-D_{X_3}F(x, u, Du, -1)=0.$$

By using the chain rule and the homogeneity condition (16.115), one can easily check that this equation can be written in the form

$$Qu=a^{ij}(x, u, Du)D_{ij}u+b(x, u, Du)=0,$$

where

$$a^{ij}(x, u, Du)=vD_{q_iq_j}F(x, u, Du, -1), \quad i, j=1, 2, \tag{16.118}$$

$$b(x, u, Du)=v\sum_{i=1}^{3} D_{q_iX_i}F(x, u, Du, -1), \tag{16.119}$$

with $v=\sqrt{1+|Du|^2}$. By using (16.115), (16.117) it is not difficult to check that (16.64), (16.65) hold with constants γ and μ depending on F. That is, the *non-parametric Euler–Lagrange equation for an elliptic parametric functional is an equation of mean curvature type.*

Finally we wish to point out that the functions a^{ij}_*, b_* introduced in Section 16.7 have a natural interpretation in the present context. In fact one can easily check that in case a^{ij}, b are as in (16.118), (16.119), then a^{ij}_*, b_* are given by

$$a^{ij}_*(X, \boldsymbol{\nu})=D_{q_iq_j}F(X, \boldsymbol{\nu}), \quad i, j=1, 2, 3,$$

$$b_*(X, \boldsymbol{\nu})=\sum_{i=1}^{3} D_{q_iX_i}F(X, \boldsymbol{\nu});$$

furthermore the conditions (16.103) hold automatically with μ_1, μ_2 determined by F, provided that $F\in C^3(\mathbb{R}^3\times(\mathbb{R}^3-\{0\}))$.

Notes

The interior gradient bound, Theorem 16.5, for the minimal surface equation was discovered, in the case of two variables, by Finn [FN 2] and in the general case by Bombieri, De Giorgi and Miranda [BDM]. An interior gradient bound for the general prescribed mean curvature equation (16.1) was first established by Ladyzhenskaya and Ural'tseva [LU 6]. The methods of both papers [BDM] and [LU 6] depended upon an isoperimetric inequality of Federer and Fleming (see [FE]) and a resulting Sobolev inequality (see [MD 1] and [LU 6]). Our derivation of Theorem 16.5, together with the relevant preparatory material in Section 16.1, is adapted from the work of Michael and Simon [MSI] and Trudinger [TR 6, 8]. The key idea of employing methods analogous to classical potential theory was invoked in an (unpublished) simplification by Michael of the Sobolev inequality in [MD 1] and subsequently utilized in [MSI] and [TR 6, 8]. The integration by parts formula, Lemma 16.1, is due to Morrey [MY 3]. The mean value inequality, Lemma 16.2, is due to Michael and Simon [MSI] and its extension, Lemma 16.3, is essentially given in [TR 8]. The extension of Morrey's estimate to hypersurfaces, Lemma 16.4, is given in [SI 4]. The process in Section 16.2, by which we conclude an interior gradient bound from the potential type inequality, Lemma 16.3, is taken from [TR 6, 8].

The existence theorems 16.9 and 16.11, for the cases of smooth boundary data, were established by Serrin [SE 4]. Theorem 16.10 essentially appears in Giaquinta [GI]. Note that the results of Section 16.3 can be approached by the variational method described in Section 10.5. In recent years the associated variational problems have also been studied in the space of functions of bounded variation, (see for example [BG 2], [GE], [GI], [MD 5]). A further approach to the equation of prescribed mean curvature is presented in [TE].

The treatment of equations of mean curvature type in Section 16.4 is based on [SI 4, 6]. The pioneering work on two dimensional equations of mean curvature type was done by Finn [FN 1, 2] who treated the case $a^{ij}(x, z, p) \equiv a^{ij}(p)$ and $b \equiv 0$. Finn called his equations "equations of minimal surface type" and stated the structure conditions for the coefficient matrix somewhat differently (but equivalently) to (16.64). The first gradient estimates were obtained in [FN 1, 2]. For a more specialized class of equations than those considered in [FN 2], (in fact for equations of the form (16.63) with $b \equiv 0$ and a^{ij} as in (16.118) for some F with $F(X, p) \equiv F(0, p)$), refinements (including an inequality like that of Theorem 16.21) were obtained by Jenkins and Serrin [JS 1]. The inequality (16.112) seems to be new. A curvature estimate like that in Theorem 16.20 was originally obtained for the minimal surface equation by Heinz [HE 1] and strengthened by E. Hopf [HO 4] and R. Osserman [OM 1]. Jenkins [JE] and Jenkins and Serrin [JS 1] obtained similar curvature estimates for equations of the form (16.63) in the case when $b \equiv 0$ and when a^{ij} had the form (16.118) for some F with $F(X, p) \equiv F(0, p)$. The estimates in [HO 4], [OR 1], [JE] and [JS 1] were in fact obtained in the stronger form

$$(16.120) \qquad (1 + |Du(y)|^2)(\kappa_1^2 + \kappa_2^2)(y) \leqslant C/R^2.$$

For the special class of equations dealt with in [JE] and [JS 1], the methods of Sections 4 to 7 can be modified to also give an inequality of the form (16.120), as shown in [SI 4]. In the case $b \not\equiv 0$ it can easily be shown that an estimate like (16.120) cannot hold in general. The only curvature estimate for the case $b \not\equiv 0$ prior to those obtained here and in [SI 4], as far as the authors are aware, was the result of Spruck [SP] for the constant mean curvature equation. It should also be mentioned that results for special classes of *parametric* surfaces have been obtained in [OM 1], [JE], [JS 1] and [SI 6].

The general Bernstein type result of Corollary 16.19 settles a question raised by Osserman [OM 2, p. 137]; such a result is well known (and there are many proofs) in the case of the minimal surface equation; (see for example [NT]). For the class of equations dealt with in [JE], Jenkins obtained such a result by letting $R \to \infty$ in (16.120). The question of whether the Bernstein theorem for the minimal surface equation carried over to $\mathbb{R}^n$, $n > 2$, provided a great impetus to the study of higher dimensional minimal surfaces. It was eventually shown to be true in case $n \leqslant 7$ by Simons [SM], who used some ideas of Fleming [FL] and De Giorgi [DG 2]. It was shown to be false in case $n > 7$ by Bombieri, De Giorgi and Giusti [BDG]. Curvature estimates of the type established in Theorem 16.20 were shown also to hold for the minimal surface equation when $n \leqslant 7$ by L. Simon [SI 5], and these imply Bernstein's theorem for $n \leqslant 7$.

For a thorough account of two dimensional minimal surfaces the reader is referred to the book [NT]. For recent developments in the higher dimensional theory, see [SI 7, 8], [GT 5].

Problems

16.1. Show that by replacing n by some number $\alpha < n$ in the choice of the function χ in the derivation of Lemma 16.2 we obtain, instead of (16.27) and (16.40), the relations:

$$(16.121)\quad \left(1-\frac{\alpha}{n}\right)\int_{\mathfrak{S}_R} r^{-\alpha} g\, dA + \frac{\alpha}{n}\int_{\mathfrak{S}_R} r^{-\alpha}(1-|\delta r|^2) g\, dA$$

$$= R^{-\alpha}\int_{\mathfrak{S}_R} g\, dA + \int_{\mathfrak{S}_R} (r^{-\alpha} - R^{-\alpha}) g H \nu\cdot(x-y)\, dA$$

$$-\frac{1}{n}\int_{\mathfrak{S}_R} (r^{-\alpha} - R^{-\alpha})(x-y)\cdot\delta g\, dA$$

for $g \in C^1(\mathscr{U})$;

$$(16.122)\qquad \left(1-\frac{\alpha}{n}\right)\int_{\mathfrak{S}} r^{-\alpha} g\, dA+\frac{\alpha}{n}\int_{\mathfrak{S}} r^{-\alpha}(1-|\delta r|^2) g\, dA$$

$$=\int_{\mathfrak{S}} r^{-\alpha} g H\nu\cdot(x-y)\, dA-\frac{1}{n}\int_{\mathfrak{S}} r^{-\alpha}(x-y)\cdot\delta g\, dA$$

for $g\in C^1(\mathscr{U})$.

16.2. Using (16.40) and (16.121) derive the following Sobolev inequalities for $g\in C_0^1(\mathscr{U})$:

$$(16.123)\qquad \left[\int_{\mathfrak{S}} |g|^p dA\right]^{1/p}\leqslant C(n, p)(\operatorname{diam}\mathscr{U})^{1-n}[A(\mathfrak{S})]^{1/p}\int_{\mathfrak{S}} |\delta g|\, dA$$

for $p<n/(n-1)$, $H\equiv 0$, where

$$C(n, p)=\frac{1}{n\omega_n}\left[\frac{n}{n-(n-1)p}\right]^{1/p}.$$

$$(16.124)\qquad \left[\int_{\mathfrak{S}} |g|^2\, dA\right]^{1/2}\leqslant\frac{1}{\sqrt{\pi}}\int_{\mathfrak{S}} (|\delta g|+|Hg|)\, dA$$

for $n=2$. (Note that (16.123) is established for $p=n/(n-1)$ in [MSI].)

16.3. Let g be a non-negative subharmonic function on a C^2 hypersurface $\mathfrak{S}\subset\mathbb{R}^{n+1}$. Derive the following generalization of the mean value inequality (16.21):

$$g(y)\leqslant\begin{cases}\dfrac{1}{\pi R^2}\displaystyle\int_{\mathfrak{S}_R} g\, dA+\frac{1}{4\pi}\int_{\mathfrak{S}_R} gH^2\, dA & \text{if } n=2\\[2ex] \dfrac{1+C(n)[H_0R+(H_0R)^n]}{\omega_n R^n}\displaystyle\int_{\mathfrak{S}_R} g\, dA & \text{if } n>2,\ H_0=\sup\limits_{\mathfrak{S}}|H|.\end{cases}$$

16.4 Using Corollary (16.6) and the Harnack inequality, Theorem 8.28, show that a solution of the minimal surface equation in $\mathbb{R}^n$ which is bounded above by a linear function must itself be linear.

16.5. Derive the inequality

$$\inf_{B^+_{\chi R}(y)} |Du|\leqslant\chi^{-1} m_R/R$$

used in the proof of Theorem 16.21.

16.6. Using (16.60) show that the constants C_1 and C_2 in Theorem 16.5 and Corollary 16.6 need only depend on n and d^2H_1.

16.7. By adapting the Perron process described in Chapter 2, show that condition (16.60) is sufficient to guarantee the existence of a classical solution of the prescribed mean curvature equation, $\mathfrak{M}u = H(1+|Du|^2)^{3/2}$, in a domain Ω. Show that condition (16.60) with $\varepsilon_0 = 0$ is necessary for the existence of a classical solution. It turns out that this condition is also sufficient [GT 3].

Chapter 17

Fully Nonlinear Equations

In this chapter we consider the solvability of the classical Dirichlet problem for certain types of *fully nonlinear* elliptic equations; that is, nonlinear elliptic equations that are not quasilinear. A general second-order equation, on a domain Ω in $\mathbb{R}^n$, can be written in the form,

$$F[u] = F(x, u, Du, D^2u) = 0, \tag{17.1}$$

where F is a real function on the set $\Gamma = \Omega \times \mathbb{R} \times \mathbb{R}^n \times \mathbb{R}^{n \times n}$, where $\mathbb{R}^{n \times n}$ denotes the $n(n+1)/2$ dimensional space of real symmetric $n \times n$ matrices. We denote points in Γ typically by $\gamma = (x, z, p, r)$ where $x \in \Omega$, $z \in \mathbb{R}$, $p \in \mathbb{R}^n$ and $r \in \mathbb{R}^{n \times n}$. When F is an affine function of the r variables, the equation (17.1) is called quasilinear; otherwise, it is called fully nonlinear. When F is differentiable with respect to the r variables, the following definitions extend those in Chapter 10:

The operator F is *elliptic* in a subset $\mathscr{U}$ of Γ if the matrix $[F_{ij}(\gamma)]$, given by

$$F_{ij}(\gamma) = \frac{\partial F}{\partial r_{ij}}(\gamma), \qquad i, j = 1, \ldots, n,$$

is positive for all $\gamma = (x, z, p, r) \in \mathscr{U}$. Letting $\lambda(\gamma)$, $\Lambda(\gamma)$ denote, respectively, the minimum and maximum eigenvalues of $[F_{ij}(\gamma)]$, we call F *uniformly elliptic* (*strictly elliptic*) in $\mathscr{U}$, if Λ/λ $(1/\lambda)$ is bounded in $\mathscr{U}$. If F is elliptic (uniformly elliptic, strictly elliptic) in the whole set Γ, then we simply say that F is *elliptic* (*uniformly elliptic, strictly elliptic*) in Ω. If $u \in C^2(\Omega)$ and F is elliptic (uniformly elliptic, strictly elliptic) on the range of the mapping $x \mapsto (x, u(x), Du(x), D^2u(x))$, we say that F is *elliptic* (*uniformly elliptic, strictly elliptic*) *with respect to u.*

EXAMPLES

(i) *The Monge-Ampère Equation*:

$$F[u] = \det D^2u - f(x) = 0. \tag{17.2}$$

Here $F_{ij}(\gamma)$ is the cofactor of r_{ij}, and F is elliptic only for positive r. Accordingly, (17.2) will be elliptic only for functions $u \in C^2(\Omega)$ that are uniformly convex at each point of Ω and for such a solution to exist we must also have f positive.

(ii) *The Equation of Prescribed Gauss Curvature*: Let $u \in C^2(\Omega)$ and suppose the graph of u has Gauss curvature $K(x)$ at the point $(x, u(x))$, $x \in \Omega$. It follows (see Section 14.6) that u satisfies the equation

$$F[u] = \det D^2u - K(x)(1 + |Du|^2)^{(n+2)/2} = 0, \tag{17.3}$$

which again will be elliptic only for uniformly convex $u \in C^2(\Omega)$. More generally, examples (i) and (ii) may be combined into the family of equations of Monge-Ampère type,

$$F[u] = \det D^2u - f(x, u, Du) = 0, \tag{17.4}$$

where f is a positive function on $\Omega \times \mathbb{R} \times \mathbb{R}^n$.

(iii) *Pucci's Equations*: For $0 < \alpha \leqslant 1/n$, let $\mathscr{L}_\alpha$ denote the set of linear uniformly elliptic operators of the form

$$Lu = a^{ij}(x)D_{ij}u$$

with bounded measurable coefficients a^{ij} satisfying

$$a^{ij}\xi_i\xi_j \geqslant \alpha|\xi|^2, \qquad \Sigma a^{ii} = 1$$

for all $x \in \Omega$, $\xi \in \mathbb{R}^n$. Maximal and minimal operators M_α and m_α are then defined by

$$M_\alpha[u] = \sup_{L \in \mathscr{L}_\alpha} Lu, \qquad m_\alpha[u] = \inf_{L \in \mathscr{L}_\alpha} Lu. \tag{17.5}$$

The operators M_α, m_α are fully nonlinear and are also related by $M_\alpha[-u] = -m_\alpha[u]$. Furthermore, a simple calculation yields the formulae

$$M_\alpha[u] = \alpha\Delta u + (1 - n\alpha)\mathscr{C}_n(D^2u),$$
$$m_\alpha[u] = \alpha\Delta u + (1 - n\alpha)\mathscr{C}_1(D^2u)$$

where $\mathscr{C}_1(r)$, $\mathscr{C}_n(r)$ are the minimum, maximum eigenvalues of the matrix r. We can then consider the extremal equations

$$M_\alpha[u] = f; \qquad m_\alpha[u] = f \tag{17.6}$$

in the domain Ω, for a given function f. Although the functions $\mathscr{C}_1$, $\mathscr{C}_n$ are not differentiable, the concept of ellipticity readily extends to embrace this situation, the equations (17.6) being in fact uniformly elliptic (see below).

(iv) *The Bellman Equation*: When the family $\mathscr{L}_\alpha$ in the preceding example is replaced by an arbitrary family $\mathscr{L}$ of linear elliptic operators we obtain the Bellman equation for the optimal cost in a stochastic control problem; (see [KV 2]).

Specifically let us consider a family $\mathscr{L}$ indexed by a parameter ν belonging to a set V. Suppose that each $L_\nu \in \mathscr{L}$ has the form

$$L_\nu u = a_\nu^{ij}(x)D_{ij}u + b_\nu^i(x)D_i u + c_\nu(x)u \tag{17.7}$$

where a_ν^{ij}, b_ν^i, c_ν are real functions on Ω for each $i, j = 1, \ldots, n$, $\nu \in V$, and for each $\nu \in V$, let f_ν be a real function on Ω. The Bellman equation now takes the form

$$F[u] = \inf_{\nu \in V} (L_\nu u - f_\nu) = 0. \tag{17.8}$$

There are various ways of extending the concept of ellipticity to nondifferentiable F, for example by monotonicity or by approximation; (see Problem 17.1). For our purposes it suffices to have an extension to functions F which are Lipschitz continuous with respect to the r variables. We then call the operator F *elliptic* in a subset $\mathscr{U}$ of Γ if the matrix $[F_{ij}] = [\partial F/\partial r_{ij}]$ is positive wherever it exists in $\mathscr{U}$, and *uniformly elliptic* in $\mathscr{U}$ if the ratio of maximum to minimum eigenvalues Λ/λ is bounded in $\mathscr{U}$. Note that $[F_{ij}]$ exists for almost all $r \in \mathbb{R}^{n \times n}$. With these definitions, it follows that the Bellman equation (17.8) is elliptic in Ω if for each $x \in \Omega$, $\nu \in V$,

$$\lambda(x)|\xi|^2 \leqslant a_\nu^{ij}(x)\xi_i\xi_j \leqslant \Lambda(x)|\xi|^2 \tag{17.9}$$

for all $\xi \in \mathbb{R}^n$, where λ and Λ are positive functions in Ω. Moreover, the Bellman equation is *uniformly elliptic* in Ω if $\Lambda/\lambda \in L^\infty(\Omega)$.

17.1. Maximum and Comparison Principles

The maximum and comparison principles derived in Chapter 10 for quasilinear equations in general form are readily extended to fully nonlinear equations. We shall establish the following form of the comparison principle.

Theorem 17.1. *Let $u, v \in C^0(\bar{\Omega}) \cap C^2(\Omega)$ satisfy $F[u] \geqslant F[v]$ in Ω, $u \leqslant v$ on $\partial\Omega$, where:*

(i) *the function F is continuously differentiable with respect to the z, p, r variables in Γ;*
(ii) *the operator F is elliptic on all functions of the form $\theta u + (1 - \theta)v$, $0 \leqslant \theta \leqslant 1$;*
(iii) *the function F is non-increasing in z for each $(x, p, r) \in \Omega \times \mathbb{R}^n \times \mathbb{R}^{n \times n}$.*

It then follows that $u \leqslant v$ in Ω.

Proof. Writing

$$w = u - v,$$
$$u_\theta = \theta u + (1-\theta)v,$$
$$a^{ij}(x) = \int_0^1 F_{ij}(x, u_\theta, Du_\theta, D^2 u_\theta)\, d\theta,$$
$$b^i(x) = \int_0^1 F_{p_i}(x, u_\theta, Du_\theta, D^2 u_\theta)\, d\theta,$$
$$c(x) = \int_0^1 F_z(x, u_\theta, Du_\theta, D^2 u_\theta)\, d\theta,$$

we obtain

$$\begin{aligned} Lw &= a^{ij}D_{ij}w + b^i D_i w + cw \\ &= F[u] - F[v] \geqslant 0 \quad \text{in } \Omega. \end{aligned}$$

Furthermore conditions (i), (ii) and (iii) imply that L satisfies the hypotheses of the weak maximum principle (Theorem 3.1), and thus $w \leqslant 0$ in Ω. □

Weaker hypotheses are clearly possible in Theorem 17.1. Also, by virtue of the strong maximum principle (Theorem 3.5), we have that either $u < v$ in Ω or u and v coincide. A uniqueness result for the Dirichlet problem follows immediately from Theorem 17.1.

Corollary 17.2. *Let $u, v \in C^0(\bar{\Omega}) \cap C^2(\Omega)$ satisfy $F[u] = F[v]$ in Ω $u = v$ on $\partial\Omega$ and suppose that conditions* (i) *to* (iii) *in Theorem* 17.1 *hold. Then $u \equiv v$ in Ω.*

Maximum principles, Hölder estimates for solutions and boundary gradient estimates for fully nonlinear elliptic equations may often be inferred directly from the corresponding results for quasilinear equations. If $u \in C^2(\Omega)$, we may write the operator $F[u]$ in the form

$$\begin{aligned} F[u] &= F(x, u, Du, D^2u) - F(x, u, Du, 0) + F(x, u, Du, 0) \\ &= a^{ij}(x, u, Du)D_{ij}u + b(x, u, Du), \end{aligned} \tag{17.10}$$

where now

$$a^{ij}(x, z, p) = \int_0^1 F_{ij}(x, z, p, \theta D^2 u(x))\, d\theta,$$
$$b(x, z, p) = F(x, z, p, 0).$$

In particular, defining

$$\mathscr{E}(x, z, p, r) = F_{ij}(x, z, p, r)p_i p_j,$$
$$\mathscr{E}^* = \mathscr{E}/|p|^2,$$
$$\mathscr{D}(x, z, p, r) = \det F_{ij}(x, z, p, r),$$
$$\mathscr{D}^* = \mathscr{D}^{1/n},$$

we thus obtain from Theorems 10.3 and 10.4 the following theorem.

Theorem 17.3. *Let F be elliptic in Ω and suppose there exist non-negative constants μ_1 and μ_2 such that*

$$\frac{F(x, z, p, 0)\ \mathrm{sign}\ z}{\mathscr{E}^*\ (\text{or } \mathscr{D}^*)} \leqslant \mu_1|p| + \mu_2 \qquad \forall (x, z, p, r) \in \Gamma. \tag{17.11}$$

Then, if $u \in C^0(\bar{\Omega}) \cap C^2(\Omega)$ satisfies $F[u] \geqslant 0$, $(=0)$ in Ω, we have

$$\sup_{\Omega} u(|u|) \leqslant \sup_{\partial\Omega} u^+(|u|) + C\mu_2 \tag{17.12}$$

where $C = C(\mu_1, \mathrm{diam}\ \Omega)$.

For equations of Monge-Ampère type a maximum principle analogous to Theorem 10.5 can be concluded directly from Lemma 9.4.

Theorem 17.4. *Let F be given by (17.4) and suppose there exist non-negative functions $g \in L^1_{\mathrm{loc}}(\mathbb{R}^n)$, $h \in L^1(\Omega)$ and a constant N such that*

$$|f(x, z, p)| \leqslant \frac{h(x)}{g(p)} \qquad \forall x \in \Omega, |z| \geqslant N, p \in \mathbb{R}^n; \tag{17.13}$$

$$\int_{\Omega} h\, dx < \int_{\mathbb{R}^n} g(p)\, dp \equiv g_\infty. \tag{17.14}$$

Then, if $u \in C^0(\bar{\Omega}) \cap C^2(\Omega)$ satisfies $F[u] \geqslant 0$ $(=0)$ in Ω, we have

$$\sup_{\Omega} u(|u|) \leqslant \sup_{\partial\Omega} u^+(|u|) + C\ \mathrm{diam}\ \Omega + N, \tag{17.15}$$

where C depends only on g and h. In particular, if F is given by (17.2), we obtain

$$\sup_{\Omega} u(|u|) \leqslant \sup_{\partial\Omega} u^+(|u|) + \frac{\mathrm{diam}\ \Omega}{(\omega_n)^{1/n}} \left(\int_{\Omega} |f| \right)^{1/n}; \tag{17.16}$$

while if F is given by (17.3), the estimate (17.15) holds with $C = C(n, K_0)$, provided

$$K_0 = \int_{\Omega} |K(x)| < \omega_n. \tag{17.17}$$

To conclude this section we note that the preceding results and their proofs also extend to nondifferentiable functions F. In particular, we obtain as a generalization of the comparison principle (Theorem 17.1):

Theorem 17.5. *Let $u, v \in C^0(\bar{\Omega}) \cap C^2(\Omega)$ satisfy $F[u] \geqslant F[v]$ in Ω, $u \leqslant v$ on $\partial\Omega$, where*

(i) *F is locally uniformly Lipschitz with respect to the z, p, r variables in Γ;*
(ii) *F is elliptic in Ω;*
(iii) *F is non-increasing in z for each $(x, p, r) \in \Omega \times \mathbb{R}^n \times \mathbb{R}^{n\times n}$;*
(iv) *$|F_p|/\lambda$ is locally bounded in Γ.*

It then follows that $u \leqslant v$ in Ω.

As an application of Theorem 17.5 we see that the comparison principle will hold for the Bellman operators (17.8) provided (17.9) holds, $c_\nu \leqslant 0$ for all $\nu \in V$ and $|b_\nu|/\lambda$ is locally bounded in Ω, uniformly in ν. By virtue of the maximum principle for strong solutions (Theorem 9.1), all the results of this section carry over to functions $u, v \in W^{2,n}_{\text{loc}}(\Omega)$.

17.2. The Method of Continuity

The topological methods of Chapter 11 are inadequate for the treatment of fully nonlinear elliptic equations or nonlinear boundary value problems. For these problems, we shall use a nonlinear version of the method of continuity (Theorem 5.2). In principle, the continuity method involves the embedding of the given problem in a family of problems indexed by a bounded closed interval, say $[0, 1]$. The subset S of $[0, 1]$ for which the corresponding problems are solvable is shown to be nonempty, closed and open, and hence it coincides with the whole interval. As in the quasilinear case in Chapter 11, the linear theory is again vital but in the present situation it is applied to the Fréchet derivative of the operator F in order to demonstrate the openness of the solvability set S.

We commence with an abstract functional analytic formulation. Let $\mathfrak{B}_1$ and $\mathfrak{B}_2$ be Banach spaces and F a mapping from an open set $\mathfrak{U} \in \mathfrak{B}_1$ into $\mathfrak{B}_2$. The mapping F is called *Fréchet differentiable* at an element $u \in \mathfrak{B}_1$ if there exists a bounded linear mapping $L: \mathfrak{B}_1 \to \mathfrak{B}_2$ such that

$$\|F[u+h] - F[u] - Lh\|_{\mathfrak{B}_2}/\|h\|_{\mathfrak{B}_1} \to 0 \tag{17.18}$$

as $h \to 0$ in $\mathfrak{B}_1$. The linear mapping L is called the *Fréchet derivative* (or *differential*) of F at u and will be denoted by F_u. When $\mathfrak{B}_1, \mathfrak{B}_2$ are Euclidean spaces, $\mathbb{R}^n$, $\mathbb{R}^m$, the Fréchet derivative coincides with the usual notion of differential, and, moreover, the basic theory for the infinite dimensional case can be modelled on that for the finite dimensional case as usually treated in advanced calculus; (see for example [DI]). In particular, it is evident from (17.18) that the Fréchet differentiability of F at u

implies that F is continuous at u and that the Fréchet derivative F_u is determined uniquely by (17.18). We call F *continuously differentiable* at u if F is Fréchet differentiable in a neighbourhood of u and the resulting mapping

$$v \mapsto F_v \in E(\mathfrak{B}_1, \mathfrak{B}_2)$$

is continuous at u. Here $E(\mathfrak{B}_1, \mathfrak{B}_2)$ denotes the Banach space of bounded linear mappings from $\mathfrak{B}_1$ into $\mathfrak{B}_2$ with norm given by

$$\|L\| = \sup_{\substack{v \in \mathfrak{B}_1 \\ v \neq 0}} \frac{\|Lv\|_{\mathfrak{B}_2}}{\|v\|_{\mathfrak{B}_1}}$$

The chain rule holds for Fréchet differentiation, that is, if $F: \mathfrak{B}_1 \to \mathfrak{B}_2$, $G: \mathfrak{B}_2 \to \mathfrak{B}_3$ are Fréchet differentiable at $u \in \mathfrak{B}_1$, $F[u] \in \mathfrak{B}_2$, respectively, then the composite mapping $G \circ F$ is differentiable at $u \in \mathfrak{B}_1$ and

$$(17.19) \qquad (G \circ F)_u = G_{F[u]} \circ F_u$$

The theorem of the mean also holds in the sense that if $u, v \in \mathfrak{B}_1$, $F: \mathfrak{B}_1 \to \mathfrak{B}_2$ is differentiable on the closed line segment γ joining u and v in $\mathfrak{B}_1$, then

$$(17.20) \qquad \|F[u] - F[v]\|_{\mathfrak{B}_2} \leqslant K\|u - v\|_{\mathfrak{B}_1},$$

where

$$K = \sup_{w \in \gamma} \|F_w\|.$$

With the aid of these basic properties we may deduce an *implicit function theorem* for Fréchet differentiable mappings. Suppose that $\mathfrak{B}_1$, $\mathfrak{B}_2$ and X are Banach spaces and that $G: \mathfrak{B}_1 \times X \to \mathfrak{B}_2$ is Fréchet differentiable at a point (u, σ), $u \in \mathfrak{B}_1$, $\sigma \in X$. The *partial Fréchet derivatives*, $G^1_{(u,\sigma)}$, $G^2_{(u,\sigma)}$ at (u, σ) are the bounded linear mappings from $\mathfrak{B}_1$, X, respectively, into $\mathfrak{B}_2$ defined by

$$G_{(u,\sigma)}(h, k) = G^1_{(u,\sigma)}(h) + G^2_{(u,\sigma)}(k)$$

for $h \in \mathfrak{B}_1$, $k \in X$. We state the implicit function theorem in the following form.

Theorem 17.6. *Let $\mathfrak{B}_1$, $\mathfrak{B}_2$ and X be Banach spaces and G a mapping from an open subset of $\mathfrak{B}_1 \times X$ into $\mathfrak{B}_2$. Let (u_0, σ_0) be a point in $\mathfrak{B}_1 \times X$ satisfying:*

(i) $G[u_0, \sigma_0] = 0$;
(ii) *G is continuously differentiable at (u_0, σ_0);*
(iii) *the partial Fréchet derivative $L = G^1_{(u_0,\sigma_0)}$ is invertible.*

Then there exists a neighbourhood $\mathcal{N}$ of σ_0 in X such that the equation $G[u, \sigma] = 0$, is solvable for each $\sigma \in \mathcal{N}$, with solution $u = u_\sigma \in \mathfrak{B}_1$.

Theorem 17.6 may be proved by reduction to the contraction mapping principle (Theorem 5.1). Indeed the equation, $G[u, \sigma] = 0$, is equivalent to the equation

$$u = Tu \equiv u - L^{-1}G[u, \sigma],$$

and the operator T will, by virtue of (17.19) and (17.20), be a contraction mapping in a closed ball $\bar{B} = B_\delta(u_0)$ in $\mathfrak{B}_1$ for δ sufficiently small and σ sufficiently close to σ_0 in X; (see [DI] for details). One can further show that the mapping $F: X \to \mathfrak{B}_1$ defined by $\sigma \to u_\sigma$ for $\sigma \in \mathcal{N}, u_\sigma \in \bar{B}, G[u_\sigma, \sigma] = 0$ is differentiable at σ_0 with Fréchet derivative

$$F_{\sigma_0} = -L^{-1}G^2_{(u_0, \sigma_0)}.$$

In order to apply Theorem 17.6 we suppose that $\mathfrak{B}_1$ and $\mathfrak{B}_2$ are Banach spaces with F a mapping from an open subset $\mathfrak{U} \subset \mathfrak{B}_1$ into $\mathfrak{B}_2$. Let ψ be a fixed element in $\mathfrak{U}$ and define for $u \in \mathfrak{U}$, $t \in \mathbb{R}$ the mapping $G: \mathfrak{U} \times \mathbb{R} \to \mathfrak{B}_2$ by

$$G[u, t] = F[u] - tF[\psi].$$

Let S and E be the subsets of $[0, 1]$ and $\mathfrak{B}_1$ defined by

$$\begin{aligned} S &= \{t \in [0, 1] \mid G[u, t] = 0 \quad \text{for some } u \in \mathfrak{U}\}, \\ E &= \{u \in \mathfrak{U} \mid G[u, t] = 0 \quad \text{for some } t \in [0, 1]\}. \end{aligned} \tag{17.21}$$

Clearly $1 \in S$, $\psi \in E$ so that S and E are not empty. Let us next suppose that the mapping F is continuously differentiable on E with invertible Fréchet derivative F_u. It follows then from the implicit function theorem (Theorem 17.6) that the set S is open in $[0, 1]$. Consequently we obtain the following version of the method of continuity.

Theorem 17.7. *The equation $F[u] = 0$ is solvable for $u \in \mathfrak{U}$ provided the set S is closed in* $[0, 1]$.

Let us now examine the application of Theorem 17.7 to the Dirichlet problem for fully nonlinear elliptic equations. We assume that the function F in equation (17.1) belongs to $C^{2,\alpha}(\Gamma)$ and for the Banach spaces $\mathfrak{B}_1$ and $\mathfrak{B}_2$ take $\mathfrak{B}_1 = C^{2,\alpha}(\bar{\Omega})$, $\mathfrak{B}_2 = C^\alpha(\bar{\Omega})$ for some $\alpha \in (0, 1)$. Clearly the operator F defined by (17.1) maps $\mathfrak{B}_1$ into $\mathfrak{B}_2$, and furthermore by calculation we see that F has continuous Fréchet derivative F_u given by

$$F_u h = Lh = F_{ij}(x)D_{ij}h + b^i(x)D_i h + c(x)h, \tag{17.22}$$

where

$$F_{ij}(x) = F_{ij}(x, u(x), Du(x), D^2u(x)),$$
$$b^i(x) = F_{p_i}(x, u(x), Du(x), D^2u(x)),$$
$$c(x) = F_z(x, u(x), Du(x), D^2u(x));$$

(see Problem 17.2). We cannot expect the mapping F_u to be invertible on all of $C^{2,\alpha}(\bar{\Omega})$ and we accordingly restrict F to the subspace $\mathfrak{B}_1 = \{u \in C^{2,\alpha}(\bar{\Omega}) \mid u = 0 \text{ on } \partial\Omega\}$, (already used in Section 6.3). The linear operator L will then be invertible for any $u \in C^{2,\alpha}(\bar{\Omega})$ for which L is strictly elliptic and $c \leqslant 0$ in Ω, provided $\partial\Omega \in C^{2,\alpha}$ (Theorem 6.14). By means of Theorem 17.7 the solvability of the Dirichlet problem is now reduced to the establishment of apriori estimates in the space $C^{2,\alpha}(\bar{\Omega})$.

Theorem 17.8. *Let Ω be a bounded domain in $\mathbb{R}^n$ with boundary $\partial\Omega \in C^{2,\alpha}$, $0 < \alpha < 1$, $\mathfrak{U}$ an open subset of the space $C^{2,\alpha}(\bar{\Omega})$ and ϕ a function in $\mathfrak{U}$. Set $E = \{u \in \mathfrak{U} \mid F[u] = \sigma F[\phi]$ for some $\sigma \in [0, 1]$, $u = \phi$ on $\partial\Omega\}$ and suppose that $F \in C^{2,\alpha}(\bar{\Gamma})$ together with*

(i) *F is strictly elliptic in Ω for each $u \in E$;*
(ii) *$F_z(x, u, Du, D^2u) \leqslant 0$ for each $u \in E$;*
(iii) *E is bounded in $C^{2,\alpha}(\bar{\Omega})$;*
(iv) *$\bar{E} \subset \mathfrak{U}$.*

Then the Dirichlet problem, $F[u] = 0$ in Ω, $u = \phi$ on $\partial\Omega$ is solvable in $\mathfrak{U}$.

Proof. We can reduce to the case of zero boundary values by replacing u with $u - \phi$. The mapping $G: \mathfrak{B}_1 \times \mathbb{R} \to \mathfrak{B}_2$ is then defined by taking $\psi \equiv 0$ so that

$$G[u, \sigma] = F[u + \phi] - \sigma F[\phi].$$

It then follows from Theorem 17.7 and the remarks preceding the statement of Theorem 17.8, that the given Dirichlet problem is solvable provided the set S is closed. However the closure of S (and also E) follows readily from the boundedness of E (and hypothesis (iv)) by virtue of the Arzela-Ascoli theorem. □

When specialized to the case of quasilinear equations, Theorem 17.7 is somewhat weaker than Theorem 11.8. For the quasilinear case, estimates in $C^{1,\beta}(\bar{\Omega})$ for some $\beta \in (0, 1)$ will imply, by the Schauder theory, estimates in $C^{2,\alpha}(\bar{\Omega})$. The requirement $F \in C^2(\Gamma)$ can be weakened (Problem 17.3), but in order to assure the desired Fréchet differentiability of F we still need smoother coefficient hypotheses than those of Theorem 11.8.

Theorem 17.8 reduces the solvability of the Dirichlet problem for fully nonlinear elliptic equations to a series of derivative estimates extending Steps I to IV in the existence procedure for the quasilinear case as described in Chapter 11. The fully nonlinear problem is in general more involved since the estimation of second derivatives is required. In the ensuing sections we shall establish second derivative estimates for various types of equations including the examples mentioned at the beginning of this chapter. In certain cases, these estimates will be insufficient for the direct application of Theorem 17.7 and some modifications will be necessary. In particular, the nonsmoothness of the function F in the Bellman equation (17.8) is overcome through approximation. For uniformly elliptic equations we shall take $\mathfrak{U} = C^{2,\alpha}(\bar{\Omega})$ (so that hypothesis (iv) becomes redundant) while for equations of Monge-Ampère type, $\mathfrak{U}$ will be the subset of uniformly convex functions and hypothesis (iv) is ensured by the positivity of the function f. While the family of operators,

$$F[u;\sigma] = F[u] - \sigma F[\phi], \qquad \sigma \in [0, 1]$$

suffices for the applications in this chapter, it evidently can be replaced by any family

$$F[u;\sigma] = F(x, u, Du, D^2u; \sigma)$$

with $F \in C^2(\Gamma \times [0, 1])$, $F[u; 0] = F[u]$, for which the Dirichlet problem, $F[u; 1] = 0$ in Ω, $u = \phi$ on $\partial\Omega$, is solvable.

17.3. Equations in Two Variables

For fully nonlinear equations in two variables, the Hölder gradient estimates in Sections 12.2 and 13.2 enable the a priori estimation required for hypothesis (iii) in Theorem 17.8 to be reduced to estimation in $C^2(\bar{\Omega})$. To show this, we assume that $u \in C^3(\Omega)$ is a solution of (17.1) in Ω and differentiate with respect to the variable x_k, thereby obtaining the equation

$$F_{ij}D_{ijk}u + F_{p_i}D_{ik}u + F_zD_ku + F_{x_k} = 0, \tag{17.23}$$

where the arguments of the partial derivatives F_{ij}, F_{p_i}, F_z, F_{x_k} are x, u, Du, D^2u. Setting $w = D_ku$, $f = F_{x_k}(x, u, Du, D^2u)$, we may write (17.23), in the notation of the preceding section, as

$$Lw = a^{ij}D_{ij}w + b^iD_iw + cw = -f.$$

Consequently, if F is elliptic with respect to u, the first derivatives of u will be solutions of linear elliptic equations in Ω. Accordingly, taking $n = 2$, and letting λ, Λ, μ satisfy

$$\begin{aligned} 0 < \lambda|\xi|^2 \leqslant F_{ij}(x, u, Du, D^2u)\xi_i\xi_j \leqslant \Lambda|\xi|^2, \\ |F_p, F_z, F_x(x, u, Du, D^2u)| \leqslant \lambda\mu, \end{aligned} \tag{17.24}$$

for all nonzero $\xi \in \mathbb{R}^2$ we have from Theorem 12.4 or 13.3:

Theorem 17.9. *Let $u \in C^3(\Omega)$ satisfy $F[u] = 0$ in $\Omega \subset \mathbb{R}^2$, where $F \in C^1(\Gamma)$, F is elliptic with respect to u and (17.24) is satisfied. Then for any $\Omega' \subset\subset \Omega$, we have the estimate*

$$[D^2u]_{\alpha;\Omega'} \leqslant \frac{C}{d^\alpha}\{|D^2u|_{0;\Omega} + \mu d(1 + |Du|_{1;\Omega})\} \tag{17.25}$$

where C and α depend only on Λ/λ, and $d = \operatorname{dist}(\Omega', \partial\Omega)$.

To derive the corresponding global estimate we return initially to the general case $n \geqslant 2$ and assume that $\partial\Omega \in C^3$, $u \in C^3(\Omega) \cap C^2(\bar{\Omega})$ and $u = \phi$ on $\partial\Omega$ where $\phi \in C^3(\bar{\Omega})$. Let us fix a point $x_0 \in \partial\Omega$ and flatten the part of $\partial\Omega$ in a ball $B = B(x_0)$, centered at x_0, by a one-to-one mapping ψ from B onto an open set $D \subset \mathbb{R}^n$ such that

$$\begin{aligned} &\psi(B \cap \Omega) \subset \mathbb{R}^n_+ = \{x \in \mathbb{R}^n \mid x_n > 0\}, \\ &\psi(B \cap \partial\Omega) \subset \partial\mathbb{R}^n_+ = \{x \in \mathbb{R}^n \mid x_n = 0\}, \\ &\psi \in C^3(B), \qquad \psi^{-1} \in C^3(D). \end{aligned}$$

Writing

$$y = \psi(x), \qquad w = D_{y_k}u = \frac{\partial x_l}{\partial y_k} D_l u$$

we then have

$$w = D_{y_k}\phi \quad \text{on } B \cap \partial\Omega, \qquad k = 1, \ldots, n-1$$

and, using (17.23), we obtain in $B \cap \Omega$ the equation

$$\begin{aligned} &F_{ij}D_{ij}w + F_{p_i}D_i w + F_z w \\ &\quad = 2F_{ij}D_i\left(\frac{\partial x_l}{\partial y_k}\right)D_{jl}u + F_{ij}D_{ij}\left(\frac{\partial x_l}{\partial y_k}\right)D_l u + F_{p_i}D_i\left(\frac{\partial x_l}{\partial y_k}\right)D_l u - F_{x_l}\left(\frac{\partial x_l}{\partial y_k}\right). \end{aligned} \tag{17.26}$$

Consequently, if $n = 2$ and the hypotheses of Theorem 17.9 are satisfied, we conclude, from the arguments of Sections 12.1 and 12.2 or 13.1 and 13.2, a Hölder estimate for Dw in a neighbourhood of x_0, for the case $k = 1$. Furthermore, for any $D' \subset\subset D$, $y_0 \in D^+ \cap D'$, $R \leqslant d/3$, where $D^+ = \psi(B \cap \Omega)$, $d = \operatorname{dist}(D' \cap D^+, \partial D)$, we have

$$\int_{B_R \cap D^+} |D_y^2 w|^2\, dy \leqslant CR^{n-2+2\alpha}\{(1 + \mu)(1 + |Du|_{1;\Omega}) + |D^3\phi|_{0;\Omega}\}^2,$$

where $\alpha = \alpha(\Lambda/\lambda, \psi) > 0$ and $C = C(\Lambda/\lambda, \psi, \operatorname{diam} \Omega)$, Using equation (17.26) in the case $k = 2$, we then obtain

$$(17.27) \qquad \int\limits_{B_R \cap D^+} |D_y^3 u|^2 \, dy \leqslant CR^{n-2+2\alpha}\{(1 + \mu)(1 + |Du|_{1;\Omega}) + |D^3\phi|_{0;\Omega}\}^2$$

and the desired global Hölder estimate follows by Morrey's estimate, Theorem 7.19.

Theorem 17.10. *Let $u \in C^3(\Omega) \cap C^2(\overline{\Omega})$ satisfy $F[u] = 0$ in $\Omega \subset \mathbb{R}^2$, $u = \phi$ on $\partial\Omega$ where $F \in C^1(\Gamma)$, F is elliptic with respect to u, $\partial\Omega \in C^3$, $\phi \in C^3(\overline{\Omega})$ and (17.24) is satisfied. Then we have the estimate*

$$(17.28) \qquad [D^2 u]_{\alpha;\Omega} \leqslant C\{(1 + \mu)(1 + |u|_{2;\Omega}) + |\phi|_{3;\Omega}\}$$

where α and C depend only on Λ/λ and Ω.

As a consequence of Theorem 17.10 (and the regularity result, Lemma 17.16) the space $C^{2,\alpha}(\overline{\Omega})$ in hypothesis (iii) of Theorem 17.8 can be replaced by $C^2(\overline{\Omega})$, provided $\partial\Omega \in C^3$ and $\phi \in C^3(\overline{\Omega})$. For certain equations the necessary estimation of second derivatives can be achieved through interpolation, as in the estimation of first derivatives in Chapter 12. Indeed, let us assume the following structure conditions:

$$(17.29) \qquad \begin{aligned} &0 < \lambda|\xi|^2 \leqslant F_{ij}(x, z, p, r)\xi_i\xi_j \leqslant \Lambda|\xi|^2 \\ &|F_p, F_z(x, z, p, r)| \leqslant \mu\lambda, \\ &|F_x(x, z, p, r)| \leqslant \mu\lambda(1 + |p| + |r|), \end{aligned}$$

for all nonzero $\xi \in \mathbb{R}^2$, $(x, z, p, r) \in \Gamma$, where λ is a nonincreasing function of $|z|$, and Λ and μ are nondecreasing functions of $|z|$. The estimates (17.25) and (17.28) will then be valid with $\lambda = \lambda(M)$, $\Lambda = \Lambda(M)$, $\mu = \mu(M)$ where $M = |u|_{0;\Omega}$. Consequently, with the aid of the interpolation inequalities, Lemmas 6.32 and 6.35, we can estimate the norms $|u|^*_{2,\alpha;\Omega}$, $|u|_{2,\alpha;\Omega}$ in terms of M.

Theorem 17.11. *Let $u \in C^3(\Omega)$ satisfy $F[u] = 0$ in $\Omega \subset \mathbb{R}^2$ and assume the structure conditions (17.29). Then we have the interior estimate*

$$(17.30) \qquad |u|^*_{2,\alpha;\Omega} \leqslant C$$

where $\alpha > 0$ depends only on Λ/λ and C depends on Λ/λ, μ and $|u|_{0;\Omega}$. In addition, if $u \in C^3(\overline{\Omega})$, $\partial\Omega \in C^3$ and $u = \phi$ on $\partial\Omega$ for $\phi \in C^3(\overline{\Omega})$, we have the global estimate

$$(17.31) \qquad |u|_{2,\alpha;\Omega} \leqslant C$$

where $\alpha > 0$ depends on Λ/λ and Ω, and C depends on Λ/λ, μ, Ω and $|u|_{0;\Omega}$.

We remark here that the hypotheses of Theorems 17.8, 17.9 and 17.10 can be weakened to permit $F \in C^{0,1}(\Gamma)$, $u \in W^{3,2}(\Omega)$ and $\phi \in W^{3,2}(\Omega) \cap C^{2,\beta}(\overline{\Omega})$, (with the structure conditions (17.24), (17.28) holding almost everywhere in Ω, Γ, respectively). The differentiation of equation (17.1) is accomplished with the aid of a generalization of the chain rule (Theorem 7.8). The structure conditions (17.29) may also be relaxed to correspond with the natural conditions for quasilinear equations; (see Notes).

By combining Theorems 17.3, 17.8 and 17.11, we obtain an existence theorem for the Dirichlet problem.

Theorem 17.12. *Let Ω be a bounded domain in $\mathbb{R}^2$ with boundary $\partial\Omega \in C^3$ and let $\phi \in C^3(\overline{\Omega})$. Suppose that the operator F satisfies $F_z \leqslant 0$ in Γ together with the structure condition* (17.29). *Then the classical Dirichlet problem, $F[u] = 0$, $u = \phi$ on $\partial\Omega$, is uniquely solvable, with solution $u \in C^{2,\alpha}(\overline{\Omega})$ for all $\alpha < 1$.*

Note that for the direct application of Theorem 17.8 we require smoother F; the full strength of Theorem 17.12 then follows by approximation of F and the estimate (17.31). Similar approximation also yields an existence theorem for equations of Bellman-Pucci type which we take up in Section 17.5.

17.4. Hölder Estimates for Second Derivatives

We derive in this section interior Hölder estimates for second derivatives of solutions of fully nonlinear elliptic equations under the key assumption that the function F is a concave (or convex) function of the r variables. This restriction, which was not necessary in the two variable case of the preceding section, still enables us to cover the equations of Monge-Ampère and Bellman-Pucci type. To illustrate the main features of the method, we first consider equations of the special form

$$F(D^2u) = g, \tag{17.32}$$

where $F \in C^2(\mathbb{R}^{n\times n})$, $g \in C^2(\Omega)$ and $u \in C^4(\Omega)$. We suppose:

(i) F is uniformly elliptic with respect to u, so that there exist positive constants λ, Λ such that

$$\lambda|\xi|^2 \leqslant F_{ij}(D^2u)\xi_i\xi_j \leqslant \Lambda|\xi|^2 \tag{17.33}$$

for all $\xi \in \mathbb{R}^n$;

(ii) F is concave with respect to u in the sense that F is a concave function on the range of D^2u.

Let γ be an arbitrary unit vector in $\mathbb{R}^n$ and differentiate equation (17.32) twice in the direction γ. We thus obtain

$$
\begin{aligned}
(17.34)\qquad & F_{ij}D_{ij\gamma}u = D_\gamma g, \\
& F_{ij}D_{ij\gamma\gamma}u + F_{ij,kl}D_{ij\gamma}uD_{kl\gamma}u = D_{\gamma\gamma}g,
\end{aligned}
$$

where

$$[F_{ij,kl}] = \left[\frac{\partial^2 F}{\partial r_{ij}\,\partial r_{kl}}\right]$$

is nonpositive by virtue of the concavity of F. Consequently, the function $w = D_{\gamma\gamma}u$ satisfies the differential inequality

$$(17.35)\qquad F_{ij}D_{ij}w \geqslant D_{\gamma\gamma}g$$

in Ω. We now invoke the weak Harnack inequality from Section 9.7. Let B_R, B_{2R} be concentric balls in Ω of radii R, $2R$, respectively, and set for $s = 1, 2$,

$$M_s = \sup_{B_{sR}} w, \qquad m_s = \inf_{B_{sR}} w.$$

Applying Theorem 9.22 to the function $M_2 - w$, we thus obtain

$$
\begin{aligned}
(17.36)\qquad \left\{R^{-n}\int_{B_R}(M_2 - w)^p\right\}^{1/p} &\leqslant C\{M_2 - M_1 + R\|(D_{\gamma\gamma}g)^-\|_{n;B_{2R}}\} \\
&\leqslant C\{M_2 - M_1 + R^2|D^2g|_{0;\Omega}\}
\end{aligned}
$$

To conclude a Hölder estimate for w from (17.36), we need a corresponding inequality for $-w$, which we obtain by considering (17.32) as a functional relationship between the second derivatives of u. To begin with, using the concavity of F again, we have for any $x, y \in \Omega$,

$$
\begin{aligned}
(17.37)\qquad F_{ij}(D^2u(y))(D_{ij}u(x) - D_{ij}u(y)) &\geqslant F(D^2u(x)) - F(D^2u(y)) \\
&= g(x) - g(y).
\end{aligned}
$$

We now get a relationship between *pure* second derivatives of u by means of the following matrix result (from [MW]).

Lemma 17.13. *Let $S[\lambda, \Lambda]$ denote the set of positive matrices in $\mathbb{R}^{n\times n}$ with eigenvalues lying in the interval $[\lambda, \Lambda]$, where $0 < \lambda < \Lambda$. Then there exists a finite set of unit vectors, $\gamma_1, \ldots, \gamma_N \in \mathbb{R}^n$ and positive numbers $\lambda^* < \Lambda^*$, depending only on n, λ*

and Λ such that any matrix $\mathscr{A} = [a^{ij}] \in S[\lambda, \Lambda]$ can be written in the form

$$\mathscr{A} = \sum_{k=1}^{N} \beta_k \gamma_k \otimes \gamma_k, \quad \text{i.e.} \quad a^{ij} = \sum_{k=1}^{N} \beta_k \gamma_{ki} \gamma_{kj}, \tag{17.38}$$

where $\lambda^* \leqslant \beta_k \leqslant \Lambda^*$, $k = 1, \ldots, N$. Furthermore the directions $\gamma_1, \ldots, \gamma_N$ can be chosen to include the coordinate directions e_i, $i = 1, \ldots, n$, together with the directions $(1/\sqrt{2})(e_i \pm e_j)$, $i < j$, $i, j = 1, \ldots, n$.

The proof of Lemma 17.13 is deferred until the end of this section. Applying Lemma 17.13 to the matrix F_{ij}, we obtain from (17.37) the inequality

$$\sum_{k=1}^{N} \beta_k(w_k(y) - w_k(x)) \leqslant g(y) - g(x) \tag{17.39}$$

where $w_k = D_{\gamma_k \gamma_k} u$ and $\beta_k = \beta_k(y)$ satisfy

$$0 < \lambda^* \leqslant \beta_k \leqslant \Lambda^*,$$

the vectors $\gamma_1, \ldots, \gamma_N$ and numbers λ^*, Λ^* depending only on n, λ, Λ. Setting

$$M_{sk} = \sup_{B_{sR}} w_k, \quad m_{sk} = \inf_{B_{sR}} w_k, \qquad s = 1, 2;\ k = 1, \ldots, N,$$

we have that each of the functions w_k satisfies (17.36) so that by summation over $k \neq l$ for some fixed l, we obtain

$$\begin{aligned}
\left\{R^{-n} \int_{B_R} \left[\sum_{k \neq l} (M_{2k} - w_k)\right]^p\right\}^{1/p} &\leqslant N^{1/p} \sum_{k \neq l} \left\{R^{-n} \int_{B_R} (M_{2k} - w_k)^p\right\}^{1/p} \\
&\leqslant C\left\{\sum_{k \neq l} (M_{2k} - M_{1k}) + R^2 |D^2 g|_{0;\Omega}\right\} \\
&\leqslant C\{\omega(2R) - \omega(R) + R^2 |D^2 g|_{0;\Omega}\}
\end{aligned}$$

where, for $s = 1, 2$,

$$\omega(sR) = \sum_{k=1}^{N} \operatorname*{osc}_{B_{sR}} w_k = \sum_{k=1}^{N} (M_{sk} - m_{sk})$$

By (17.39), we have for $x \in B_{2R}$, $y \in B_R$

$$\beta_l(w_l(y) - w_l(x)) \leqslant g(y) - g(x) + \sum_{k \neq l} \beta_k(w_k(x) - w_k(y))$$

so that

$$w_l(y) - m_{2k} \leqslant \frac{1}{\lambda^*}\left\{3R|Dg|_{0;\Omega} + \Lambda^* \sum_{k \neq l}(M_{2k} - w_k)\right\};$$

and hence

$$\text{(17.40)} \qquad \left\{R^{-n} \int_{B_R} (w_l - m_{2l})^p\right\}^{1/p}$$
$$\leqslant C\{\omega(2R) - \omega(R) + R|Dg|_{0;\Omega} + R^2|D^2g|_{2;\Omega}\},$$

where C again depends only on n, λ, Λ. By setting $w = w_l$ in (17.36), adding this to (17.40) and summing over $l = 1, \ldots, N$, we therefore obtain

$$\omega(2R) \leqslant C\{\omega(2R) - \omega(R) + R|Dg|_{0;\Omega} + R^2|D^2g|_{0;\Omega}\},$$

hence

$$\omega(R) \leqslant \delta\omega(2R) + R|Dg|_{0;\Omega} + R^2|D^2g|_{0;\Omega}$$

for $\delta = 1 - 1/C$. Hölder estimates for the functions w_k, $k = 1, \ldots, N$ now follow from Lemma 8.23, and by using the last assertion of Lemma 17.13, we obtain a Hölder estimate for D^2u: *For any ball $B_{R_0} \subset \Omega$ and $R \leqslant R_0$,*

$$\text{(17.41)} \qquad \underset{B_R}{\operatorname{osc}}\, D^2u \leqslant C\left(\frac{R}{R_0}\right)^\alpha \{\underset{B_{R0}}{\operatorname{osc}}\, D^2u + R_0|Dg|_{0;\Omega} + R_0^2|D^2g|_{0;\Omega}\}$$

where C and α are positive constants depending only on n, λ and Λ.

In terms of the interior Hölder norms the estimate (17.41) may be expressed in the form

$$\text{(17.42)} \qquad |u|^*_{2,\alpha;\Omega} \leqslant C(|u|^*_{2;\Omega} + |g|^*_{2,\Omega})$$

where C, α depend on n, λ and Λ.

Let us now proceed to the general case (17.1) with $F \in C^2(\Gamma)$. Corresponding to conditions (i) and (ii) we suppose:

(i)′ F is uniformly elliptic with respect to u, so that there exist positive constants λ, Λ such that

$$\text{(17.43)} \qquad \lambda|\xi|^2 \leqslant F_{ij}(x, u, Du, D^2u)\xi_i\xi_j \leqslant \Lambda|\xi|^2$$

for all $\xi \in \mathbb{R}^n$;

(ii)′ F is concave with respect to r on the range of D^2u.

Again we differentiate twice with respect to a unit vector γ, to obtain the equations

$$F_{ij}D_{ij\gamma}u + F_{p_i}D_{i\gamma}u + F_zD_\gamma u + \gamma_iF_{x_i} = 0,$$

$$\begin{aligned} &F_{ij}D_{ij\gamma\gamma}u + F_{ij,kl}D_{ij\gamma}uD_{kl\gamma}u + 2F_{ij,p_k}D_{ij\gamma}uD_{k\gamma}u \\ &\quad + 2F_{ij,z}D_{ij\gamma}uD_\gamma u + 2\gamma_kF_{ij,x_k}D_{ij\gamma}u + F_{p_i}D_{i\gamma\gamma}u \\ &\quad + F_{p_ip_j}D_{i\gamma}uD_{j\gamma}u + 2F_{p_iz}D_{i\gamma}uD_\gamma u + 2\gamma_jF_{p_i,x_j}D_{i\gamma}u \\ &\quad + F_zD_{\gamma\gamma}u + F_{zz}(D_\gamma u)^2 + 2\gamma_iF_{zx_i}D_\gamma u + \gamma_i\gamma_jF_{x_ix_j} = 0 \end{aligned} \tag{17.44}$$

Using the concavity of F we now obtain, in place of (17.35), the differential inequality

$$F_{ij}D_{ij\gamma\gamma}u \geqslant -A_{ij\gamma}D_{ij\gamma}u - B_\gamma, \tag{17.45}$$

where

$$\begin{aligned} A_{ij\gamma} &= 2F_{ij,p_k}D_{k\gamma}u + 2F_{ij,z}D_\gamma u + 2\gamma_kF_{ij,x_k} + \gamma_jF_{p_i}, \\ B_\gamma &= F_{p_ip_j}D_{i\gamma}uD_{j\gamma}u + 2F_{p_i}zD_{i\gamma}uD_\gamma u + 2\gamma_jF_{p_i,x_j}D_{i\gamma}u \\ &\quad + F_zD_{\gamma\gamma}u + F_{zz}(D_\gamma u)^2 + 2\gamma_iF_{zx_i}D_\gamma u + \gamma_i\gamma_jF_{x_ix_j}. \end{aligned}$$

The third derivatives of u in (17.45) are handled analogously to the second derivatives of u in the derivation of the Hölder gradient estimate (Theorem 13.6). Let us first choose directions $\gamma_1, \ldots, \gamma_N$ in accordance with the full statement of Lemma 17.13 applied to the matrix $[F_{ij}]$. Set

$$M_2 = \sup_\Omega |D^2u|,$$

$$h_k = \frac{1}{2}\left(1 + \frac{D_{\gamma_k\gamma_k}u}{1 + M_2}\right), \qquad k = 1, \ldots, N,$$

(if necessary we replace Ω by a subdomain to ensure the finiteness of M_2). We obtain from (17.45)

$$-F_{ij}D_{ij}h_k \leqslant C(A_0|D^3u| + B_0)/(1 + M_2) \tag{17.46}$$

where $C = C(n)$ and

$$A_0 = \sup_\Omega \{|F_{rp}||D^2u| + |F_{rz}||Du| + |F_{rx}| + |F_p|\},$$

$$\begin{aligned} B_0 = \sup_\Omega \{&|F_{pp}||D^2u|^2 + |F_{pz}||D^2u||Du| + |F_{px}||D^2u| \\ &+ |F_z||D^2u| + |F_{zz}||Du|^2 + |F_{zx}||Du| + |F_{xx}|\}, \end{aligned}$$

where $F_{rp} = [F_{ij,p_k}]_{i,j,k=1,\ldots,n}$, etc., are evaluated at (x, u, Du, D^2u). We now multiply (17.46) by h_k and sum from 1 to N to obtain

$$\sum_{k=1}^N F_{ij}D_ih_kD_jh_k - \tfrac{1}{2}F_{ij}D_{ij}v \leqslant C(A_0|D^3u| + B_0)/(1 + M_2), \tag{17.47}$$

where

$$v = \sum_{k=1}^{N} (h_k)^2.$$

By our choice of γ_k, we can estimate

$$\begin{aligned} |D^3u|^2 &= \sum_{i,j,l=1}^{n} |D_{ijl}u|^2 \\ &\leqslant 4(1 + M_2)^2 \sum_{k=1}^{N} |Dh_k|^2, \end{aligned}$$

and by the ellipticity condition (17.43),

$$\sum_{k=1}^{N} F_{ij} D_i h_k D_j h_k \geqslant \lambda \sum_{k=1}^{N} |Dh_k|^2.$$

Consequently, for $\varepsilon \in (0, 1)$ and

$$w = w_k = h_k + \varepsilon v, \qquad k = 1, \ldots, N,$$

we have, by combining (17.46) and (17.47),

$$\varepsilon\lambda \sum_{k=1}^{N} |Dh_k|^2 - \tfrac{1}{2} F_{ij} D_{ij} w \leqslant C\left\{A_0 \left(\sum_{1}^{N} |Dh_k|^2\right)^{1/2} + \frac{B_0}{1 + M_2}\right\};$$

and hence, by the Cauchy inequality,

$$F_{ij} D_{ij} w \geqslant -\lambda \bar{\mu}, \tag{17.48}$$

where

$$\bar{\mu} = \frac{C(n)}{\lambda} \left(\frac{A_0^2}{\lambda\varepsilon} + \frac{B_0}{1 + M_2}\right).$$

We are now ready to apply again the weak Harnack inequality (Theorem 9.22). Let B_R, B_{2R} be concentric balls in Ω' and set, for $s = 1, 2$, $k = 1, \ldots, N$,

$$\begin{aligned} W_{ks} &= \sup_{B_{sR}} w, \\ M_{ks} &= \sup_{B_{sR}} h_k, \\ m_{ks} &= \inf_{B_{sR}} h_k, \\ \omega(sR) &= \sum_{k=1}^{N} \operatorname*{osc}_{B_{sR}} h_k = \sum_{k=1}^{N} (M_{ks} - m_{ks}). \end{aligned}$$

Applying Theorem 9.22 to the functions $W_{k2} - w_k$, we thus obtain

$$(17.49) \qquad \Phi_{p,R}(W_{k2} - w_k) \equiv \left\{\frac{1}{|B_R|}\int_{B_R}(W_{k2} - w_k)^p\right\}^{1/p} \leqslant C\{W_{k2} - W_{k1} + \bar{\mu}R^2\},$$

where p and C are positive constants depending only on n and Λ/λ. Using the inequalities

$$W_{k2} - w_k \geqslant M_{k2} - h_k - 2\varepsilon\omega(2R),$$
$$W_{k2} - w_{k1} \leqslant M_{k2} - M_{k1} + 2\varepsilon\omega(2R),$$

we can conclude from (17.49) a similar inequality for the functions h_k; namely

$$\Phi_{p,R}(M_{k2} - h_k) \leqslant C\{M_{k2} - M_{k1} + \varepsilon\omega(2R) + \bar{\mu}R^2\}.$$

Then summation over $k \neq l$ for some fixed l yields

$$(17.50) \qquad \Phi_{p,r}\left(\sum_{k\neq l}(M_{k2} - h_k)\right) \leqslant N^{1/p}\sum_{k\neq l}(M_{k2} - h_k)$$
$$\leqslant C\{(1+\varepsilon)\,\omega(2R) - \omega(R) + \bar{\mu}R^2\},$$

where $C = C(n, \Lambda/\lambda)$ as before. To compensate for not having the inequality corresponding to (17.48) for the functions $-h_k$, we again resort to equation (17.1), so that using the concavity of F (condition (ii)′), we have

$$\begin{aligned} &F_{ij}(y, u(y), Du(y), D^2u(y))(D_{ij}u(y) - D_{ij}u(x)) \\ &\qquad \leqslant F(y, u(y), Du(y), D^2u(x)) - F(y, u(y), Du(y), D^2u(y)) \\ &\qquad = F(y, u(y), Du(y), D^2u(x)) - F(x, u(x), Du(x), D^2u(x)) \\ &\qquad \leqslant D_0|x - y|, \end{aligned}$$

where

$$\begin{aligned} D_0 = \sup_{x,y\in\Omega} \{&|F_x(y, u(y), Du(y), D^2u(x))| \\ &+ |F_z(y, u(y), Du(y), D^2u(x))|\,|Du(y)| \\ &+ |F_p(y, u(y), Du(y), D^2u(x))|\,|D^2u(y)|\} \end{aligned}$$

Now by Lemma 17.13 and our choice of γ_k,

$$\begin{aligned} &F_{ij}(y, u(y), Du(y), D^2u(y))(D_{ij}u(y) - D_{ij}u(x)) \\ &\qquad = \sum_{k=1}^{N}\beta_k(y)(D_{\gamma_k\gamma_k}u(y) - D_{\gamma_k\gamma_k}u(x)) \\ &\qquad = 2(1 + M_2)\sum_{k=1}^{N}\beta_k(h_k(y) - h_k(x)), \end{aligned}$$

where

$$0 < \lambda^* \leqslant \beta_k \leqslant \Lambda^*, \qquad k = 1, \ldots, N,$$

and λ^*/λ, Λ^*/λ depend only on n and Λ/λ. Consequently, for $x \in B_{2R}$, $y \in B_R$,

$$\sum_{k=1}^{N} \beta_k(h_k(y) - h_k(x)) \leqslant C\lambda\tilde{\mu}R,$$

where

$$\tilde{\mu} = \frac{D_0}{\lambda(1 + M_2)};$$

and hence for fixed l,

$$\begin{aligned} h_l(y) - m_{l2} &\leqslant \frac{1}{\lambda^*}\left\{C\lambda\tilde{\mu}R + \Lambda^* \sum_{k \neq l}(M_{k2} - h_k(y))\right\} \\ &\leqslant C\left\{\tilde{\mu}R + \sum_{k \neq l}(M_{k2} - h_k(y))\right\}, \end{aligned}$$

where $C = C(n, \Lambda/\lambda)$. Therefore by (17.50), we obtain, for $l = 1, \ldots, N$,

$$\phi_{p,R}(h_l - m_{l2}) \leqslant C\{(1 + \varepsilon)\omega(2R) - \omega(R) + \tilde{\mu}R + \bar{\mu}R^2\},$$

where $C = C(n, \Lambda/\lambda)$. By adding the above inequality for $l = k$ to (17.49) and summing over k, we thus obtain

$$\omega(2R) \leqslant C\{(1 + \varepsilon)\omega(2R) - \omega(R) + \tilde{\mu}R + \bar{\mu}R^2;$$

hence

$$\omega(R) \leqslant \delta\omega(2R) + C\{\varepsilon\omega(2R) + \tilde{\mu}R + \bar{\mu}R^2\}$$

for $\delta = 1 - 1/C$. Finally, by fixing $\varepsilon = (1 - \delta)/2C$ we arrive again at the oscillation estimate

$$\omega(R) \leqslant \bar{\delta}\omega(2R) + C(\tilde{\mu}R + \bar{\mu}R^2),$$

where $0 < \bar{\delta} < 1$ and $C, \bar{\delta}$ depend only on n and Λ/λ. The desired Hölder estimates now follow immediately from Lemma 8.23. Namely, *for any ball $B_{R_0} \subset \Omega$ and $R \leqslant R_0$, we have*

$$\operatorname*{osc}_{B_R} D^2u \leqslant C\left(\frac{R}{R_0}\right)^{\alpha}(1 + M_2)(1 + \tilde{\mu}R_0 + \bar{\mu}R_0^2) \tag{17.51}$$

where C and α are positive constants depending only on n and Λ/λ. The resultant interior estimate is formulated in the following theorem.

Theorem 17.14. *Let $u \in C^4(\Omega)$ satisfy $F[u] = 0$ in Ω where $F \in C^2(\Gamma)$, F is elliptic with respect to u and satisfies (i)′ and (ii)′. Then for any $\Omega' \subset\subset \Omega$, we have the estimate*

$$(17.52)\qquad [D^2u]_{\alpha;\Omega'} \leqslant C$$

where α depends only on n, λ and Λ, and C depends in addition on $|u|_{2;\Omega}$, dist $(\Omega', \partial\Omega)$ and the first and second derivatives of F other than F_{rr}.

A more explicit form of the estimate (17.52) is provided by (17.51), which exhibits the dependence of the constant C on $|u|_{2;\Omega}$ and the derivatives of F. Under further concavity assumptions on F various terms in the expressions for A_0 and B_0 in (17.46) may be removed. For example, when F is jointly concave in z, p and r, we may take in (17.45)

$$A_{ij\gamma} = 2\gamma_k F_{ij,x_k} + \gamma_j F_{p_i},$$
$$B_\gamma = 2\gamma_j F_{p_i,x_j} D_{i\gamma}u + F_z D_{\gamma\gamma}u + 2\gamma_i F_{z,x_i} D_\gamma u + \gamma_i\gamma_j F_{x_ix_j}$$

and hence in (17.46)

$$A_0 = \sup_\Omega \{|F_{rx}| + |F_p|\},$$
$$B_0 = \sup_\Omega \{|F_{px}|\,|D^2u| + |F_z|\,|D^2u| + |F_{zx}|\,|Du| + |F_{xx}|\}.$$

Consequently, under the additional structure conditions

$$(17.53)\qquad \begin{aligned} &0 < \lambda|\xi|^2 \leqslant F_{ij}(x, z, p, r)\xi_i\xi_j \leqslant \Lambda|\xi|^2 \\ &|F_p|, |F_z|, |F_{rx}|, |F_{px}|, |F_{zx}| \leqslant \mu\lambda, \\ &|F_x|, |F_{xx}| \leqslant \mu\lambda(1 + |p| + |r|) \end{aligned}$$

for all nonzero $\xi \in \mathbb{R}^n$, $(x, z, p, r) \in \Gamma$, where λ is a nonincreasing function of $|z|$ and Λ and μ are nondecreasing functions of $|z|$, we may conclude from (17.51) the interior estimate, extending (17.42),

$$|u|^*_{2,\alpha} \leqslant C(1 + |u|^*_2),$$

where α depends only on n, Λ/λ, and C depends in addition on μ, diam Ω. With the aid of the interior interpolation inequality (Lemma 6.32), we thus obtain the following interior estimate.

Theorem 17.15. *Let $u \in C^4(\Omega)$ satisfy $F[u] = 0$ in $\Omega \subset \mathbb{R}^n$ and suppose that F is concave (or convex) in z, p, r and satisfies the structure conditions (17.53). Then we have the interior estimate*

$$|u|^*_{2,\alpha;\Omega} \leqslant C \tag{17.54}$$

where $\alpha > 0$ depends only on n and $\Lambda(M)/\lambda(M)$, and C depends in addition on $\mu(M)$, diam Ω *and $M = |u|_{0;\Omega}$.*

We remark that the estimate (17.54) can in fact be proved under more general hypotheses corresponding to the natural conditions for quasilinear equations; (see Notes). We conclude this section with the proof of Lemma 17.13.

Proof of Lemma 17.13. Letting $\mathbb{R}^{n\times n}_+$ denote the cone of positive matrices in $\mathbb{R}^{n\times n}$, we may represent any $\mathscr{A} \in \mathbb{R}^{n\times n}_+$ in the form

$$\mathscr{A} = \sum_{k=1}^{n'} \gamma_k \otimes \gamma_k \tag{17.55}$$

where $n' = n(n+1)/2 = \dim. \mathbb{R}^{n\times n}$, $\gamma_k \in \mathbb{R}^n$, $k = 1, \ldots, n'$, and the dyadic matrices $\gamma_k \otimes \gamma_k = [\gamma_{ki}\gamma_{kj}]$ are linearly independent. To see this we observe that any two matrices in $\mathbb{R}^{n\times n}_+$ are similar and hence in particular each $\mathscr{A} \in \mathbb{R}^{n\times n}_+$ is similar to the matrix $\mathscr{A}_0$ whose diagonal and nondiagonal terms are, respectively, n and 1. But then

$$\mathscr{A}_0 = \sum_{i=1}^{n} e_i \otimes e_i + \sum_{\substack{i,j=1\\ i<j}}^{n} (e_i + e_j) \otimes (e_i + e_j)$$

so that (17.55) follows by an appropriate base change. Consequently, the family of sets of the form

$$U(\gamma_1, \ldots, \gamma_{n'}) = \left\{\sum_{k=1}^{n'} \beta_k \gamma_k \otimes \gamma_k \,\middle|\, \beta_k > 0, k = 1, \ldots, n'\right\},$$

where $\gamma_k \otimes \gamma_k$ are linearly independent, forms an open cover of $S(\lambda, \Lambda) \subset \mathbb{R}^{n\times n}_+$, and since $S(\lambda, \Lambda)$ is compact there exists a finite subcover. Accordingly, there exists a fixed set of unit vectors $\gamma_1, \ldots, \gamma_N$, depending only on λ, Λ and n such that any $\mathscr{A} \in S(\lambda, \Lambda)$ may be written

$$\mathscr{A} = \sum_{k=1}^{N} \beta_k \gamma_k \otimes \gamma_k$$

with $\beta_k \geqslant 0$. To get the assertion of the lemma we observe that at the outset we could have applied the above procedure to the matrix

$$\mathscr{A} - \lambda^* \sum_{k=1}^{N} \gamma_k \otimes \gamma_k \in S(\lambda/2, \Lambda)$$

for sufficiently small λ^* ($\lambda^* = \lambda/2N$ is sufficient). Note that we can take $\Lambda^* = \Lambda$. A similar consideration shows that any particular finite set of unit vectors may be included among the γ_k. □

17.5. Dirichlet Problem for Uniformly Elliptic Equations

We show in this section that the interior derivative estimates of the preceding section in fact suffice to demonstrate the solvability of the Dirichlet problem for certain types of uniformly elliptic equations, including those of Bellman-Pucci type. Since these estimates were established for C^4 solutions, we first need a regularity result to link them with the hypotheses of the method of continuity as asserted in Theorem 17.8.

Lemma 17.16. *Let $u \in C^2(\Omega)$ satisfy $F[u] = 0$ in Ω where F is elliptic with respect to u. Then, if $F \in C^k(\Gamma)$, $k \geqslant 1$, we have $u \in W^{k+2,p}_{\text{loc}}(\Omega)$ for all $p < \infty$; if $F \in C^{k,\alpha}(\Gamma)$, $0 < \alpha < 1$, we have $u \in C^{k+2,\alpha}(\Omega)$.*

Proof. We use a difference quotient argument similar to that in the proof of Theorem 6.17. Let us fix a coordinate vector e_l, $1 \leqslant l \leqslant n$, and write

$$v(x) = u(x + he_l), \qquad h \in \mathbb{R}$$

$$w = \Delta_l^h u = \frac{1}{h}(u - v),$$

$$u_\theta = \theta u + (1 - \theta)v, \qquad 0 \leqslant \theta \leqslant 1,$$

$$a^{ij}(x) = \int_0^1 F_{ij}(x + \theta h, u_\theta, Du_\theta, D^2 u_\theta)\, d\theta,$$

$$b^i(x) = \int_0^1 F_{p_i}(x + \theta h, u_\theta, Du_\theta, D^2 u_\theta)\, d\theta,$$

$$c(x) = \int_0^1 F_z(x + \theta h, u_\theta, Du_\theta, D^2 u_\theta)\, d\theta,$$

$$f(x) = \int_0^1 F_{x_l}(x + \theta h, u_\theta, Du_\theta, D^2 u_\theta)\, d\theta.$$

Then if we fix a subdomain $\Omega' \subset\subset \Omega$ and take sufficiently small h, the difference quotient w will satisfy in Ω' the linear equation

$$Lw = a^{ij}D_{ij}w + b^i D_i w + cw = -f, \tag{17.56}$$

which will also be elliptic with uniformly continuous coefficients in Ω'. The interior L^p estimates (Theorem 9.11) then yield bounds for $\|D^2 w\|_{p;\Omega''}$ for $\Omega'' \subset\subset \Omega'$, independent of h, and hence by Lemma 7.24 we conclude $u \in W^{3,p}_{\text{loc}}(\Omega)$ for any $p < \infty$. By the Sobolev imbedding theorem (Theorem 7.10), it then follows that $u \in C^{2,\alpha}(\Omega)$ for all $\alpha < 1$. Consequently, if $F \in C^{1,\alpha}(\Gamma)$ for some α, $0 < \alpha < 1$, we may apply the Schauder regularity result (Theorem 6.17) to (17.56) to obtain $u \in C^{3,\alpha}(\Omega)$. Note that if at the outset we are given $u \in C^{2,\beta}(\Omega)$, for some $\beta > 0$, there is no need to use the L^p theory for this result. Lemma 17.16 is thus established for $k = 1$. Further interior regularity now follows by a straightforward iterative or "bootstrapping" procedure. □

With the aid of Lemma 17.16 we see that the estimates of Sections 17.3 and 17.4 will hold for $C^2(\Omega)$ solutions of (17.1). In the case of Section 17.4, we should observe that $u \in W^{4,n}_{\text{loc}}(\Omega)$ is sufficient for the given proofs. To offset the lack of global $C^{2,\alpha}(\bar{\Omega})$ estimates, we modify the function F near the boundary $\partial\Omega$ as follows. Let $\{\eta_m\}$ be a sequence of functions in $C^2_0(\Omega)$ satisfying $0 \leqslant \eta \leqslant 1$ in Ω and $\eta_m(x) = 1$ whenever $d(x, \partial\Omega) \geqslant 1/m$, and consider instead of F the operators F_m given by

$$F_m[u] = \eta_m F[u] + (1 - \eta_m)\Delta u. \tag{17.57}$$

If F satisfies the hypotheses of Theorems 17.14 or 17.15, then so also does F_m with structural constants depending possibly on η_m. We therefore obtain interior $C^{2,\alpha}$ estimates for solutions of the equations, $F_m[u] = 0$, of the same form as those given by Theorems 17.14 and 17.15. But near $\partial\Omega$, $F_m[u] = \Delta u$ so that for appropriately smooth boundary data, $C^{2,\alpha}$ estimates near $\partial\Omega$ will follow from the Schauder theory, in particular from Lemma 6.5. Using this procedure we can now establish the following existence result.

Theorem 17.17. *Let Ω be a bounded domain in $\mathbb{R}^n$ satisfying an exterior sphere condition at each boundary point and suppose the function $F \in C^2(\Gamma)$ is concave (or convex) with respect to z, p, r, nonincreasing with respect to z and satisfies the structure conditions* (17.53). *Then the classical Dirichlet problem, $F[u] = 0$ in Ω, $u = \phi$ on $\partial\Omega$ is uniquely solvable in $C^2(\Omega) \cap C^0(\bar{\Omega})$ for any $\phi \in C^0(\partial\Omega)$.*

Proof. Let us first consider the case of smooth boundary data, namely $\partial\Omega \in C^{2,\beta}$, $\phi \in C^{2,\beta}(\bar{\Omega})$ for some $0 < \beta < 1$. In view of the discussion preceding the theorem, we consider the approximating Dirichlet problems,

$$F_m[u] = 0, \quad u = \phi \quad \text{on } \partial\Omega \tag{17.58}$$

where F_m is given by (17.57). In order to apply the method of continuity (Theorem 17.8), we require apriori bounds in $C^{2,\alpha}(\bar{\Omega})$ for some $\alpha > 0$ for solutions of the problems

$$F_m[u] - tF_m[\phi] = 0, \quad u = \phi \quad \text{on } \partial\Omega, \qquad 0 \leqslant t \leqslant 1. \tag{17.59}$$

Now the equations (17.59) continue to satisfy the hypotheses of Theorem 17.14, uniformly with respect to t; and consequently we have, for any solution $u \in C^{2,\beta}(\bar{\Omega})$ and $\Omega' \subset\subset \Omega$, an estimate

$$|u|_{2,\alpha;\Omega'} \leqslant C,$$

where α depends only on n, $\lambda(M)$, $\Lambda(M)$, and C depends in addition on $\mu(M)$, Ω, Ω', ϕ, η_m and $M = |u|_{0;\Omega}$. Since $F_m[u] = \Delta u$ near $\partial\Omega$, we therefore conclude, by Lemma 6.5, the corresponding global estimate with $\Omega' = \Omega$ and C depending in addition on $\partial\Omega$ and β. Next we observe that the conditions, $F_z \leqslant 0$ and (17.53) together imply (17.11), for $\mu_1 = \mu_2 = \mu(0)$, so that by Theorem 17.3, $M = |u|_{0;\Omega}$ is bounded uniformly in t and m. Therefore, by Theorem 17.8, we obtain the existence of a unique solution $u = u_m \in C^{2,\beta}(\bar{\Omega})$ of the Dirichlet problem (17.58). Since $F_m[u] = F[u]$ for $\operatorname{dist}(x, \partial\Omega) \geqslant 1/m$, we then obtain, by the interior estimate (Theorem 17.15), the convergence of a subsequence of $\{u_m\}$ (uniformly together with first and second derivatives on compact subsets of Ω) to a solution $u \in C^{2,\beta}(\Omega)$ of the equation $F[u] = 0$ in Ω. But in view of the representation (17.10) the barrier considerations for quasilinear equations, in particular Theorem 14.15, are applicable to the Dirichlet problems (17.58) and as a result we obtain that $u \in C^{0,1}(\bar{\Omega})$ and $u = \phi$ on $\partial\Omega$. The extension to continuous ϕ and domains Ω satisfying exterior sphere conditions follows similarly. □

By approximation, the condition $F \in C^2(\Gamma)$ can be weakened in the hypotheses of Theorem 17.17, so that the derivatives appearing in the structure conditions (17.53) need exist only in the weak sense; (see Problem 17.4). Furthermore, by extending the above arguments slightly we can encompass the uniformly elliptic Bellman equations (17.8). In fact the necessary modifications can be illustrated more generally. Let $F^1, \ldots, F^m$ be operators of the form (17.1) and let $G \in C^2(\mathbb{R}^m)$ be a concave function whose derivatives satisfy

$$1 \leqslant \sum_{\nu=1}^{m} D_\nu G \leqslant K \tag{17.60}$$

for some constant K. We now define another operator F by

$$F[u] = G(F^1[u], \ldots, F^m[u]).$$

Then if $u \in C^4(\Omega)$ and $F^1, \ldots, F^m$ all satisfy hypotheses (i)′ and (ii)′ in Theorem 17.14, we obtain by differentiation, in place of (17.44), the equation

$$\sum_{\nu=1}^{m} D_\nu G D_\gamma F^\nu = 0,$$

$$\sum_{\nu=1}^{m} D_\nu G D_{\gamma\gamma} F^\nu + \sum_{\nu,\tau=1}^{m} D_{\nu\tau} G D_\gamma F^\nu D_\gamma F^\tau = 0,$$

so that from the concavity of G we have

$$\sum_{\nu=1}^{m} D_\nu G D_{\gamma\gamma} F^\nu \geqslant 0.$$

Consequently the analysis of Section 17.4 carries over with the derivatives DF, D^2F, replaced respectively by $\sum_{\nu=1}^{m} D_\nu G D F^\nu$, $\sum_{\nu=1}^{m} D_\nu G D^2 F^\nu$. Using (17.60) we thus have, in particular, that the estimate of Theorem 17.14 is applicable to the operator F provided the operators F_ν, $\nu = 1, \ldots, m$, satisfy the structural hypotheses and Λ and μ are replaced by $K\Lambda$ and $K\mu$ respectively. The existence result (Theorem 17.17) extends similarly. The equations of Bellman-Pucci type can then be treated by approximation. Namely, let us define for $y \in \mathbb{R}^m$,

$$G_0(y) = \inf_{\nu=1,\ldots,m} y_\nu$$

and for $h > 0$, let G_h be the mollification of G_0 given by

$$G_h(y) = h^{-n} \int_{R^m} \rho\left(\frac{y-\bar{y}}{h}\right) G_0(\bar{y})\, d\bar{y},$$

where ρ is a mollifier on $\mathbb{R}^m$. Since G_0 is concave, it follows readily that G_h is also concave. Furthermore we have

$$\sum_{\nu=1}^{m} D_\nu G_h = 1,$$

so that (17.60) holds with $K = 1$. It follows then that if $F^1, \ldots, F^m$, Ω and ϕ satisfy the hypotheses of Theorem 17.17, the classical Dirichlet problems,

$$F[u] = G_h(F^1[u], \ldots, F^m[u]) = 0, \qquad u = \phi \quad \text{on } \partial\Omega$$

are uniquely solvable with solutions $u = u_h$ satisfying an estimate

$$|u|^*_{2,\alpha;\Omega} \leqslant C$$

where α and C depend only on $n, \lambda, \Lambda, \mu, \phi$ and Ω, together with a modulus of continuity estimate on $\partial\Omega$ depending on the same quantities. By approximation, the result extends to the limiting case $h = 0$ and, since the above estimates are independent of m, also to the case of a countable family of operators. Accordingly we have the following extension of Theorem 17.17.

Theorem 17.18. *Let Ω be a bounded domain in $\mathbb{R}^n$ satisfying an exterior sphere condition at each boundary point and suppose the functions $F^1, F^2, \ldots \in C^2(\Gamma)$, are* *concave with respect to z, p, r, nonincreasing with respect to z, and satisfy uniformly the structure conditions* (17.53). *Then the classical Dirichlet problem,*

$$F[u] = \inf\{F^1[u], F^2[u], \ldots\} = 0 \quad \text{in } \Omega, \qquad u = \phi \quad \text{on } \partial\Omega \tag{17.61}$$

is uniquely solvable in $C^2(\Omega) \cap C^0(\bar{\Omega})$ *for any* $\phi \in C^0(\partial\Omega)$.

We observe that the Bellman equation (17.8) is included in Theorem 17.18 provided the index set V is countable and the operators L_ν and f_ν satisfy the conditions: $a^{ij}_\nu, b^i_\nu, c_\nu, f_\nu \in C^2(\Omega)$ and

$$\begin{aligned} &\lambda|\xi|^2 \leqslant a^{ij}_\nu \xi_i \xi_j \leqslant \Lambda|\xi|^2 \\ &|a^{ij}_\nu|_{2;\Omega}, |b^i_\nu|_{2;\Omega}, |c_\nu|_{2;\Omega}, |f_\nu|_{2;\Omega} \leqslant \mu\lambda \\ &c_\nu \leqslant 0 \end{aligned} \tag{17.62}$$

for all $\xi \in \mathbb{R}^n$, $\nu \in V$, where λ, Λ and μ are positive constants. Moreover it is evident that we can allow certain types of uncountable sets V, for example a separable metric space on which the mappings: $\nu \to a^{ij}_\nu(x), b^i_\nu(x), c_\nu(x), f_\nu(x)$ are continuous for each $x \in \Omega$. In particular the Pucci equations (17.6) are covered provided $f \in C^2(\Omega) \cap L^\infty(\Omega)$.

17.6. Second Derivative Estimates for Equations of Monge-Ampère Type

In this section we turn our attention to equations of form

$$\det D^2u = f(x, u, Du). \tag{17.63}$$

As indicated earlier, equation (17.63) is elliptic only when the Hessian matrix D^2u is positive (or negative) and it is therefore natural to confine our attention to convex solutions u and positive functions f. Writing equation (17.63) in the form

$$F(D^2u) = \log\det D^2u = g(x, u, Du), \tag{17.64}$$

where $g = \log f$, we then have by a calculation,

$$\begin{aligned} F_{ij} &= u^{ij} \\ F_{ij,kl} &= -u^{ik}u^{jl} = -F_{ik}F_{jl} \end{aligned} \tag{17.65}$$

where $[u^{ij}]$ denotes the inverse of D^2u. Consequently the function F is concave on the cone of nonnegative matrices in $\mathbb{R}^{n\times n}$, and the equation (17.64) will be uniformly elliptic on compact subsets of Ω, with respect to any convex $C^2(\Omega)$ solution. If $g \in C^2(\Omega \times \mathbb{R} \times \mathbb{R}^n)$ we can then apply the results of Section 17.4 to get interior Hölder estimates for the second derivatives of solutions and subsequently, through Lemma 17.16, higher order estimates when g is appropriately smooth. We consider now the question of interior and global bounds for the second derivatives of solutions, with approach reminiscent of the treatment of gradient estimates for non-uniformly elliptic equations in Chapter 15.

First let us note from equation (17.44) that any pure second derivative $D_{\gamma\gamma}u$, of a solution of (17.64), satisfies the equation

$$F_{ij}D_{ij\gamma\gamma}u = F_{ik}F_{jl}D_{ij\gamma}uD_{kl\gamma}u + D_{\gamma\gamma}g, \tag{17.66}$$

and since u is convex we have also $D_{\gamma\gamma}u > 0$. To estimate $D_{\gamma\gamma}u$ from above, we take positive functions $\eta \in C^2(\Omega)$, $h \in C^2(\mathbb{R}^n)$ and set

$$w = \eta h(Du)D_{\gamma\gamma}u$$

so that

$$\frac{D_i w}{w} = \frac{D_i\eta}{\eta} + (\log h)_{p_k} D_{ik}u + \frac{D_{i\gamma\gamma}u}{D_{\gamma\gamma}u},$$

$$\begin{aligned}\frac{D_{ij}w}{w} = \frac{D_i w D_j w}{w^2} + \frac{D_{ij}\eta}{\eta} - \frac{D_i\eta D_j\eta}{\eta^2} \\ + (\log h)_{p_k p_l} D_{ik}uD_{jl}u + (\log h)_{p_k} D_{ijk}u \\ + \frac{D_{ij\gamma\gamma}u}{D_{\gamma\gamma}u} - \frac{D_{i\gamma\gamma}uD_{j\gamma\gamma}u}{(D_{\gamma\gamma}u)^2}.\end{aligned}$$

Consequently, using (17.66) we obtain

$$\begin{aligned}(\eta h)^{-1}F_{ij}D_{ij}w \geqslant D_{\gamma\gamma}u\Big\{\frac{F_{ij}D_{ij}\eta}{\eta} - \frac{F_{ij}D_i\eta D_j\eta}{\eta^2} \\ + (\log h)_{p_k p_l}F_{ij}D_{ik}uD_{jl}u + (\log h)_{p_k}F_{ij}D_{ijk}u\Big\} \\ + F_{ik}F_{jl}D_{ij\gamma}uD_{kl\gamma}u - \frac{1}{D_{\gamma\gamma}u}F_{ij}D_{i\gamma\gamma}uD_{j\gamma\gamma}u + D_{\gamma\gamma}g.\end{aligned} \tag{17.67}$$

An obvious candidate for the function h is given by

$$h(p) = e^{\beta|p|^2/2}, \qquad \beta > 0.$$

For then we have

$$(\log h)_{p_k} = \beta p_k, \qquad (\log h)_{p_k p_l} = \beta\delta_{kl},$$

and hence

$$(\log h)_{p_k p_l}F_{ij}D_{ik}uD_{jl}u = \beta F_{ij}D_{ik}uD_{jk}u = \beta\Delta u$$

by (17.65). Next, if we assume the bounds

$$(17.68)\qquad |Dg(x, u, Du)|, \qquad |D^2 g(x, u, Du)| \leqslant \mu,$$

we have

$$\begin{aligned}
D_{\gamma\gamma}u(\log h)_{p_k} F_{ij} D_{ijk}u + D_{\gamma\gamma}g
&= \beta D_k u D_{\gamma\gamma}u(g_{x_k} + g_z D_k u + g_{p_i} D_{ik}u) + g_{\gamma\gamma} + 2g_{\gamma z} D_\gamma u + 2g_{\gamma p_i} D_{i\gamma}u \\
&\quad + g_{zz}(D_\gamma u)^2 + 2g_{zp_i} D_\gamma u D_{i\gamma}u + g_{p_i p_j} D_{ik}u D_{jk}u \\
&\quad + g_z D_{\gamma\gamma}u + g_{p_i} D_{i\gamma\gamma}u \\
&\geqslant g_{p_i}\left(\frac{D_i w}{w} - \frac{D_i \eta}{\eta}\right) D_{\gamma\gamma}u - C\{1 + |D^2u|^2 + \beta(1 + |D^2u|)\}
\end{aligned}$$

where C depends on μ, $\sup_\Omega |Du|$.

In order to handle the other terms in (17.67) we regard $w = w(x, \gamma)$ as a function on $\Omega \times \partial B_1(0)$ and suppose that w takes a maximum value at a point $y \in \Omega$ and direction γ. The derivative $D_{\gamma\gamma}u(y)$ will then be the maximum eigenvalue of the Hessian $D^2u(y)$ and by a rotation of coordinates we can assume that $D^2u(y)$ is in diagonal form with γ a coordinate direction. For global estimates we take $\eta \equiv 1$ so that terms involving η are not present in (17.67). It then follows from (17.65) that

$$\frac{1}{D_{\gamma\gamma}u} F_{ij} D_{i\gamma\gamma}u D_{j\gamma\gamma}u \leqslant F_{ik} F_{jl} D_{ij\gamma}u D_{kl\gamma}u$$

at the point y, so that by choosing β sufficiently large we obtain an estimate for $D_{\gamma\gamma}u(y)$ in terms of n, μ and $|Du|_{0;\Omega}$.

The interior case is more delicate since η cannot be chosen as an arbitrary cutoff function (because of the term $F_{ij}D_{ij}\eta$). In this case we shall assume that u vanishes continuously on $\partial\Omega$ and choose $\eta = -u$ so that $\eta > 0$ in Ω and

$$F_{ij} D_{ij}\eta = -F_{ij} D_{ij}u = -n.$$

Furthermore since $Dw(y) = 0$, we have

$$\begin{aligned}
\frac{F_{ij} D_i \eta D_j \eta}{\eta^2} &= \frac{\sum F_{ii} |D_i \eta|^2}{\eta^2} \\
&= \frac{|D_\gamma u|^2}{u^2 D_{\gamma\gamma}u} + \sum_{i \neq \gamma} F_{ii}\left(\frac{D_{i\gamma\gamma}u}{D_{\gamma\gamma}u} + D_i u D_{ii}u\right)^2 \\
&\leqslant \frac{|D_\gamma u|^2}{u^2 D_{\gamma\gamma}u} + \sum_{i \neq \gamma} F_{ii}\left(\frac{D_{i\gamma\gamma}u}{D_{\gamma\gamma}u}\right)^2 - \frac{2\beta |Du|^2}{u}
\end{aligned}$$

at the point y. Now

$$\begin{aligned}\frac{1}{D_{\gamma\gamma}u}&\left\{\sum_{i\neq\gamma}F_{ii}(D_{i\gamma\gamma}u)^2+F_{ij}D_{i\gamma\gamma}uD_{j\gamma\gamma}u\right\}\\&=\sum_{i\neq\gamma}F_{\gamma\gamma}F_{ii}(D_{i\gamma\gamma}u)^2+\sum_{i=1}^{n}F_{\gamma\gamma}F_{ii}(D_{i\gamma\gamma}u)^2\\&\leqslant\sum_{i,j=1}^{n}F_{ii}F_{jj}(D_{ij\gamma}u)^2\\&=F_{ik}F_{jl}D_{ij\gamma}uD_{kl\gamma}u.\end{aligned}$$

at y, by virtue of our choice of coordinates. Taking account of the above estimates, in the differential inequality (17.67), we obtain

$$D_{\gamma\gamma}u(y)\leqslant C\left(1-\frac{1}{u(y)}\right)$$

where $C=C(n,\mu,|Du|_{0;\Omega})$, and hence

$$\sup_{\Omega} w\leqslant C, \tag{17.69}$$

where $C=C(n,\mu,|u|_{1;\Omega})$. Finally, to estimate $\eta=-u$ from below, we have, by the convexity of u,

$$\frac{u(x)}{\operatorname{dist}(x,\partial\Omega)}\leqslant\frac{\inf u}{\operatorname{diam}\Omega} \tag{17.70}$$

for any $x\in\Omega$. We formulate the resultant second derivative estimates as follows.

Theorem 17.19. *Let $u\in C^2(\Omega)$ be a convex solution of (17.63) in a domain Ω where $f\in C^2(\Omega\times\mathbb{R}\times\mathbb{R}^n)$ is positive in Ω and $g=\log f$ satisfies (17.68). Then if $u\in C^2(\overline{\Omega})$, we have*

$$\sup_{\Omega}|D^2u|\leqslant C, \tag{17.71}$$

where C depends on n, μ, $|u|_{1;\Omega}$ and $\sup_{\partial\Omega}|D^2u|$. If $u\in C^{0,1}(\overline{\Omega})$ and u is constant on $\partial\Omega$, then for any $\Omega'\subset\subset\Omega$, we have

$$\sup_{\Omega'}|D^2u|\leqslant\frac{C}{d_{\Omega'}} \tag{17.72}$$

where C depends on n, μ, $|u|_{1;\Omega}$ and diam Ω, *and* $d_{\Omega'}=\operatorname{dist}(\Omega',\partial\Omega)$.

When the function $f^{1/n}$ is convex with respect to the p variables, the estimate (17.71) may be derived by simpler means and moreover the result extends to more general functions F and solutions that are not necessarily convex (see Problem 17.5).

An estimate for the second derivatives of solutions of equation (17.64) on the boundary $\partial\Omega$ follows readily from equation (17.26). If we assume that Ω is uniformly convex, $\partial\Omega \in C^3$, and $u = \phi$ on $\partial\Omega$ where $\phi \in C^3(\bar{\Omega})$, then the first term on the right-hand side of (17.26) becomes

$$(17.73)\qquad 2F_{ij}D_i\left(\frac{\partial x_l}{\partial y_k}\right)D_{jl}u = 2\delta_{il}D_i\left(\frac{\partial x_l}{\partial y_k}\right) = 2D_i\left(\frac{\partial x_i}{\partial y_k}\right),$$

by virtue of (17.65). An estimate for the derivatives $D_{y_k y_n}u(x_0)$ for $x_0 \in \Omega$ then follows from Theorem 14.4 or Corollary 14.5. The remaining second derivative $D_{y_n y_n}u(x_0)$ is estimated directly from equation (17.64). For, with respect to a principal coordinate system at x_0, we have (taking $\phi \equiv 0$)

$$(17.74)\qquad \det D^2u = |D_nu|^{n-2}\prod_{i=1}^{n-1}\kappa_i\left\{|D_nu|D_{nn}u - \sum_{i=1}^{n-1}\frac{(D_{in}u)^2}{\kappa_i}\right\}$$

where $\kappa_1, \ldots, \kappa_{n-1}$ are the principal curvatures at x_0, and using (17.70), we thus infer a bound for $D_{nn}u(x_0)$. Therefore, we have proved the following global estimate for second derivatives.

Theorem 17.20. *Let $u \in C^3(\bar{\Omega})$ be a convex solution of equation* (17.64) *in Ω, where $f \in C^2(\bar{\Omega} \times \mathbb{R} \times \mathbb{R}^n)$ is positive and $\partial\Omega \in C^3$ is uniformly convex. Then*

$$(17.75)\qquad \sup_{\Omega}|D^2u| \leqslant C$$

where C depends on n, $|u|_{1;\Omega}$, f, $\partial\Omega$, and $u = 0$ on $\partial\Omega$.

We remark here that the proof of Theorem 17.20 also extends to more general functions F, with possibly nonconvex solutions u; (see Problems 17.6, and 17.7).

17.7. Dirichlet Problem for Equations of Monge-Ampère Type

The considerations of the previous section reduce the solvability of the classical Dirichlet problem for equations of Monge-Ampère type to the establishment of C^1 estimates. For equations in two variables, we can then proceed directly through the method of continuity using the global Hölder estimates for second derivatives

(Theorem 17.10). For higher dimensions, there also exist procedures requiring only interior second derivative estimates but these are more complicated than the method employed for the uniformly elliptic case in Section 17.5; (see Notes). In the next section, we shall treat the recently established *global* Hölder estimates for second derivatives that will enable us to align the general case with the two-variable case.

Restrictions on the growth of the function f in (17.63) arise through the consideration of gradient estimates. For a convex function u in a domain Ω we clearly have

$$\sup_{\Omega} |Du| = \sup_{\partial\Omega} |Du| \tag{17.76}$$

so that the estimation of the gradient of convex solutions of (17.63) reduces to estimation on the boundary only. As in the quasilinear case, such estimates are readily obtained by means of barrier constructions. In fact, let us assume the following structure condition,

$$0 \leqslant f(x, z, p) \leqslant \mu(|z|)d_x^{\beta}|p|^{\gamma} \tag{17.77}$$

for all x in some neighbourhood $\mathcal{N}$ of $\partial\Omega$, $z \in \mathbb{R}$, $|p| \geqslant \mu(|z|)$, where $d_x = \operatorname{dist}(x, \partial\Omega)$, μ is nondecreasing and $\beta = \gamma - n - 1 \geqslant 0$. Then we have the following gradient estimate.

Theorem 17.21. *Let $u \in C^0(\bar{\Omega}) \cap C^2(\Omega)$, $\phi \in C^2(\Omega) \cap C^{0,1}(\bar{\Omega})$ be convex functions in the uniformly convex domain Ω, satisfying*

$$\det D^2u = f(x, u, Du) \quad in\ \Omega, \qquad u = \phi \quad on\ \partial\Omega. \tag{17.78}$$

Then we have

$$\sup_{\Omega} |Du| \leqslant C, \tag{17.79}$$

where C depends on n, μ, β, $\mathcal{N}$, Ω, $|u|_{0;\Omega}$ and $|\phi|_{1;\Omega}$.

Proof. Let $B = B_R(y)$ be an enclosing sphere for the domain Ω at the point $x_0 \in \partial\Omega$ and set

$$w = \phi - \psi(d)$$

where $d(x) = \operatorname{dist}(x, \partial B)$ and ψ is given by (14.11), with ν and k to be determined. Using a principal coordinate system for ∂B at x_0, we may then estimate

$$\begin{aligned}\det D^2w &\geqslant \det(-D^2\psi)\\ &= -\psi''\left(\frac{\psi'}{|x-y|}\right)^{n-1}\\ &\geqslant -\psi''\left(\frac{\psi'}{R}\right)^{n-1}.\end{aligned}$$

while from the structure condition (17.77) we have

$$\begin{aligned} f(x, u(x), Dw) &\leqslant 2^\gamma \mu(M) d^\beta (\psi')^\gamma \\ &= 2^\gamma \mu(M)(\psi')^{n+1}(d\psi')^\beta \end{aligned}$$

provided $x \in \mathcal{N}$, $\psi'(d) \geqslant \mu(M) + |D\phi|$, where $M = |u|_{0;\Omega}$. Choosing $\nu = 1 + 2^\gamma R^{n-1}\mu$, so that $d\psi' \leqslant 1$, and then choosing k and a according to (14.14) (with μ replaced by $\mu + |D\phi|$), and such that $\{x \in \Omega | d < a\} \subset \mathcal{N}$, we see that the convex function w will be a lower barrier at x_0 for (17.63) and function u. Consequently, by the comparison principle (Theorem 17.1), we obtain

$$\frac{u(x) - u(x_0)}{|x - x_0|} \geqslant -C$$

for $d \leqslant a$, where C depends on $n, \beta, \mu, |D\phi|_{0;\Omega}, M, \mathcal{N}$ and R. Using the convexity of u, we have

$$\frac{u(x) - u(x_0)}{|x - x_0|} \leqslant |D\phi|_{0;\Omega},$$

for all $x \in \Omega$, $x_0 \in \partial\Omega$, and the estimate (17.79) follows. □

Note that the Monge-Ampère equation (17.2) is covered by Theorem 17.21 for bounded f, while the equation of prescribed Gauss curvature is encompassed only if the curvature K vanishes Lipschitz continuously on $\partial\Omega$. Combining Theorems 17.4, 17.8, 17.10, 17.20 and 17.21, we obtain the following existence theorem for equations of Monge-Ampère type in two variables.

Theorem 17.22. *Let Ω be a uniformly convex domain in $\mathbb{R}^2$ with boundary $\partial\Omega \in C^3$. Suppose that f is a positive function in $C^2(\bar{\Omega} \times \mathbb{R} \times \mathbb{R}^n)$ satisfying $f_z \geqslant 0$, together with the structure conditions* (17.13), (17.14) *and* (17.77) *for $n = 2$, $\beta = 0$. Then the classical Dirichlet problem*

$$\det D^2u = f(x, u, Du) \quad \text{in } \Omega, \qquad u = 0 \quad \text{on } \partial\Omega, \tag{17.80}$$

is solvable with solution $u \in C^{2,\alpha}(\bar{\Omega})$ for all $\alpha < 1$.

Proof. By virtue of Theorem 17.8 it suffices to have a uniform estimate in the space $C^{2,\alpha}(\bar{\Omega})$, for some $\alpha > 0$, of the solutions of the Dirichlet problems,

$$F[u] = \frac{\det D^2u}{f(x, u, Du)} - 1 = \sigma F[\phi], \qquad u = 0 \quad \text{on } \partial\Omega, \quad 0 \leqslant \sigma \leqslant 1,$$

where $\phi \in C^2(\bar{\Omega})$ is uniformly convex and vanishes on $\partial\Omega$; (see Problem 17.8). When the quantity g_∞ in Theorem 17.4 is infinite, such an estimate follows immediately from Theorems 17.4, 17.10, 17.20 and 17.21. Otherwise, we should, by possibly a new choice of ϕ, ensure that $F[\phi] \leqslant 0$; (for example by replacing ϕ with the solution of the Dirichlet problem, $\det D^2u = \inf f$, $u = 0$ on $\partial\Omega$). □

For Monge-Ampère equations in higher dimensions we have the following analogue of Theorem 17.22, as a consequence of the global estimate (Theorem 17.26) in Section 17.8.

Theorem 17.23. *Let Ω be a uniformly convex domain in $\mathbb{R}^n$ with boundary $\partial\Omega \in C^4$. Suppose that f is a positive function in $C^2(\bar{\Omega} \times \mathbb{R} \times \mathbb{R}^n)$ satisfying $f_z \geqslant 0$ together with the structure conditions* (17.13), (17.14) *and* (17.77) *for $\beta = 0$. Then the classical Dirichlet problem* (17.80) *is solvable with solution $u \in C^{3,\alpha}(\bar{\Omega})$ for all $\alpha < 1$.*

We note here that Theorems 17.20, 17.22 and 17.23 extend to general boundary values $\varphi \in C^4(\bar{\Omega})$; (see [IC 2], [CNS]).

Using the interior second-derivative estimates of Theorem 17.20, the above existence theorems can be extended to allow more general functions f.

Theorem 17.24. *Let Ω be a uniformly convex domain in $\mathbb{R}^n$ and suppose that f is a positive function in $C^2(\Omega \times \mathbb{R} \times \mathbb{R}^n)$ satisfying $f_z \geqslant 0$ together with the structure conditions* (17.13), (17.14) *and* (17.77). *Then the classical Dirichlet problem* (17.80) *is solvable with solution $u \in C^{3,\alpha}(\Omega) \cap C^{0,1}(\bar{\Omega})$ for all $\alpha < 1$.*

Proof. Let $\{f_m\}$ be a sequence of bounded functions in $C^2(\Omega \times \mathbb{R} \times \mathbb{R}^n)$ satisfying $0 < f_m \leqslant f$, $f_z \geqslant 0$ and $f_m = f$ for $|z| + |p| \leqslant m$ and let $\{\Omega_l\}$ be an increasing sequence of uniformly convex C^4 subdomains of Ω satisfying $\Omega_l \subset\subset \Omega, \cup\, \Omega_l = \Omega$. By Theorems 17.22 and 17.23, there exists, for each m, a sequence $\{u_{ml}\}$ of uniformly convex solutions of the Dirichlet problems.

$$\det D^2 u_{ml} = f_m(x, u_{ml}, Du_{ml}) \quad \text{in } \Omega_l, \qquad u_{ml} = 0 \quad \text{on } \partial\Omega_l.$$

Using the bounds, Theorems 17.4, 17.21 and the interior estimates, Theorems 17.14, 17.19, we obtain a subsequence converging uniformly, together with its first and second derivatives on compact subsets of Ω, to a solution u_m of the Dirichlet problem

$$\det D^2 u_m = f_m(x, u_m, Du_m) \quad \text{in } \Omega, \qquad u_m = 0 \quad \text{on } \partial\Omega.$$

But by Theorems 17.4, 17.21 again, it follows that for large enough m, $u_m = u$ is a solution of (17.80). □

For the special case of the equation of prescribed Gauss curvature (17.3) we now obtain from Theorem 17.24.

Corollary 17.25. *Let Ω be a uniformly convex domain in $\mathbb{R}^n$ and K a positive function in $C^2(\Omega) \cap C^{0,1}(\bar{\Omega})$, satisfying*

$$(17.81) \qquad K = 0 \quad \textit{on } \partial\Omega, \qquad \int_\Omega K < \omega_n.$$

Then there exists a unique convex function $u \in C^2(\Omega) \cap C^{0,1}(\bar{\Omega})$ *such that* $u = 0$ *on* $\partial\Omega$ *and whose graph has Gauss curvature* $K(x)$ *at each point* $x \in \Omega$.

Both conditions (17.81) in Corollary 17.25 are necessary in some sense. In fact, let us suppose that in the general Monge-Ampère equation (17.4) the function f satisfies

$$(17.82) \quad f(x, z, p) \geqslant \frac{h(x)}{g(p)} \quad \forall (x, z, p) \in \Omega \times \mathbb{R} \times \mathbb{R}^n,$$

where h and g are positive functions in $L^1(\Omega)$, $L^1_{\mathrm{loc}}(\mathbb{R}^n)$, respectively. Then if $u \in C^2(\Omega)$ is a convex solution of (17.4) in a domain Ω, its normal mapping χ coincides with Du and is one-to-one. Consequently, by integration we have

$$\int_{\chi(\Omega)} g(p)\, dp = \int_{\Omega} g(Du) \det D^2 u \geqslant \int_{\Omega} h,$$

and the condition

$$(17.83) \quad \int_{\Omega} h \leqslant g_\infty = \int_{\mathbb{R}^n} g(p)\, dp$$

is thus necessary for the existence of a convex solution u. Furthermore the strict inequality

$$(17.84) \quad \int_{\Omega} h < g_\infty$$

is then necessary for the existence of a solution u whose normal mapping is not all of $\mathbb{R}^n$, in particular a solution $u \in C^{0,1}(\bar{\Omega})$.

Concerning the other condition in (17.81) we remark that by extension of the interior estimate (Theorem 17.19), we can permit arbitrary nonzero boundary values $\phi \in C^2(\bar{\Omega})$ in Corollary 17.25 if and only if $K = 0$ on $\partial\Omega$ [TU]. The conditions (17.81) are analogous to the conditions (16.58), (16.60) for the equation of prescribed *mean* curvature (16.1).

17.8. Global Second Derivative Hölder Estimates

We consider in this section the global analogue, for equations of Monge-Ampère type, of the interior second-derivative Hölder estimates of Theorem 17.14. The methods automatically embrace equations of the general form (17.1) provided we strengthen hypothesis (ii)′ in Theorem 17.14 by requiring

$$-F_{ij,kl}(x, u, Du, D^2u)\xi_{ij}\xi_{kl} \geqslant \lambda_0|\xi|^2 \tag{17.85}$$

for some positive constant λ_0 and all symmetric matrices $\xi \in \mathbb{R}^{n\times n}$. By (17.65) we see that the equations of Monge-Ampère type, when written in the form (17.64), satisfy (17.85) with $\lambda_0 = (\sup_\Omega \mathscr{C}_n)^{-2}$, where $\mathscr{C}_n$ denotes the maximum eigenvalue of D^2u. Since the hypotheses for the global second-derivative Hölder estimation guarantee, through the linear theory, stronger third-derivative estimates, it is convenient to formulate the results accordingly as follows.

Theorem 17.26. *Let Ω be a bounded domain in $\mathbb{R}^n$ with boundary $\partial\Omega \in C^4$ and let $\phi \in C^4(\bar{\Omega})$. Suppose that $u \in C^3(\bar{\Omega}) \cap C^4(\Omega)$ satisfies $F[u] = 0$ in Ω, $u = \phi$ on $\partial\Omega$, where $F \in C^2(\bar{\Gamma})$, F is elliptic with respect to u and satisfies* (i)′, (ii)′ *together with* (17.85). *Then we have the estimate,*

$$\sup_\Omega |D^3u| \leqslant C, \tag{17.86}$$

where C depends on n, λ, Λ, λ_0, $\partial\Omega$, $|\phi|_{4;\Omega}$, $|u|_{2;\Omega}$ and the first and second derivaties of F.

Proof. Letting γ denote a unit vector in $\mathbb{R}^n$, we can, by use of (17.85) in (17.43), improve the inequality (17.44) so that

$$\begin{aligned} F_{ij}D_{ij\gamma\gamma}u &\geqslant -A_{ij\gamma}D_{ij\gamma}u - B_\gamma + \lambda_0|D^2D_\gamma u|^2 \\ &\geqslant -C \end{aligned} \tag{17.87}$$

by Cauchy's inequality, where C depends on n, λ_0 and A_0, B_0 in (17.46). Let us now fix a point $x_0 \in \partial\Omega$ and suppose that $B_R(y)$ is an exterior ball at x_0. The modulus of continuity of the derivatives $D_{\gamma\gamma}u$ can be estimated at the boundary (in terms of their traces on $\partial\Omega$) by standard barrier arguments, as described for example in Remark 3 in Section 6.3. As shown there, the function w in (6.45) given by

$$w(x) = \tau(R^{-\sigma} - |x - y|^{-\sigma})$$

satisfies

$$F_{ij}D_{ij}w \leqslant -1$$

in Ω for sufficiently large τ, σ depending on λ, Λ and R. Consequently, if $\varepsilon \geqslant 0$ and

$$|D_{\gamma\gamma}u(x) - D_{\gamma\gamma}u(x_0)| \leqslant \varepsilon$$

for $x \in \partial\Omega, |x - x_0| < \delta$, we obtain from (17.87) and the classical maximum principle (Theorem 3.1), the estimate

$$(17.88)\qquad \begin{aligned} D_{\gamma\gamma}u(x) - D_{\gamma\gamma}u(x_0) &\leqslant \varepsilon + Cw + 2\sup_{\partial\Omega}|D_{\gamma\gamma}u|\,|x - x_0|^2/\delta^2 \\ &\leqslant \varepsilon + C|x - x_0| \end{aligned}$$

for any $x \in \Omega$, where C depends on $n, \lambda, \Lambda, \lambda_0, \Lambda_0, B_0, |u|_{2;\Omega}$, diam Ω, R and δ. Now choosing directions $\gamma_1, \ldots, \gamma_N$ in accordance with Lemma 17.13 and using equation (17.1) itself, as in the proof of Theorem 17.14, we conclude from (17.88), the estimate

$$(17.89)\qquad |D^2u(x) - D^2u(x_0)| \leqslant \varepsilon + C|x - x_0|$$

for any $x \in \Omega$, where C depends in addition on the quantity D_0 in the proof of Theorem 17.14.

The estimate (17.89) reduces estimation of the modulus of continuity of second derivatives on the boundary $\partial\Omega$ to that of their traces on $\partial\Omega$. A similar result could also have been established by using the weak Harnack inequality at the boundary (Theorem 9.27), and moreover can be proved without requiring the condition (17.85).

To proceed further we use (17.88) again to obtain a one-sided third derivative bound. Without loss of generality we may assume that u vanishes on $\partial\Omega$, and that $\partial\Omega$ is flat in a neighbourhood of $x_0 \in \partial\Omega$, that is, for some $\delta > 0$,

$$B^+ = \Omega \cap B_\delta(x_0) \subset \mathbb{R}^n_+,$$

$$T = \partial\Omega \cap B_\delta(x_0) \subset \partial\mathbb{R}^n_+.$$

This follows since the form of (17.1), together with (17.85) and the hypotheses (i), (ii) in Theorem 17.8, are preserved by replacement of u with $u - \phi$ and by a C^4 coordinate change ψ. The new constants λ, Λ and λ_0 will, of course, depend on ψ (in particular on $D\psi$) as well as their original values. Therefore, restricting γ to tangential directions in $\partial\mathbb{R}^n_+$ so that $D_{\gamma\gamma}u = 0$ on T, we have from (17.88)

$$D_{\gamma\gamma}u(x) - D_{\gamma\gamma}u(x_0) \leqslant C|x - x_0|$$

for $x \in B^+$; and consequently

$$(17.90)\qquad D_{\gamma\gamma n}u(x_0) \geqslant -C,$$

where C depends on $|\phi|_{4;\Omega}$ and ψ, as well as the quantities in (17.88). Setting

$$h = D_n u + k|x|^2$$

it follows that for sufficiently large k ($k \geqslant C$),

$$D_{\gamma\gamma}h(x) \geqslant 0$$

for $x \in T, |x - x_0| \leqslant \delta/2$; that is, the function h has *convex* trace on $\partial\Omega \cap B_{\delta/2}(x_0)$. Furthermore, we see from the differentiated equation (17.23) that h satisfies a uniformly elliptic equation, in particular

$$|F_{ij}D_{ij}h| \leqslant C$$

where C depends on the quantities in the statement of Theorem 17.26. The above properties of the function h are utilized through the following remarkable lemma, whose proof we defer until the end of this section.

Lemma 17.27. *Let $h \in C^2(B^+) \cap C^0(B^+ \cup T)$ satisfy*

$$Lh = a^{ij}D_{ij}h \leqslant f \tag{17.91}$$

in B^+ where L is uniformly elliptic in B^+ and f/λ is bounded. Then if $h|_T$ is convex, we have for any $x, y \in T \cap B_{\delta/2}(x_0)$ and $i = 1, \dots, n-1$, the estimate

$$|D_i h(x) - D_i h(y)| \leqslant \frac{C}{1 + |\log|x - y||} \tag{17.92}$$

where C depends only on n, $\sup \Lambda/\lambda$, $\sup f/\lambda$, δ *and* $\sup |Dh|$.

Lemma 17.27 provides an estimate for the moduli of continuity at x_0 of the second derivatives $D_{in}u$ restricted to T. A similar estimate for the remaining non-tangential second derivative $D_{nn}u$ follows immediately from (17.1) itself. But then the estimate (17.89) yields an estimate for the modulus of continuity of the full Hessian matrix D^2u at the boundary point x_0 in terms of the quantities specified in the statement of Theorem 17.26, which in turn implies an estimate for the moduli of continuity of the principal coefficients of the differentiated equation (17.23). We can then use the L^p theory, in particular Theorem 9.13, to conclude L^p estimates for the third derivatives $D_{ijk}u$, $i, j = 1, \dots, n$, $k = 1, \dots, n-1$, in a neighbourhood of the point x_0 for any $p < \infty$. Similar estimates follow then for the remaining third derivative $D_{nnn}u$ from (17.23) again. Using Morrey's estimate (Theorem 7.19), we thus obtain Hölder estimates for D^2u at x_0 for any exponent $\alpha < 1$. On combination with the interior estimate (Theorem 17.14), as in the proof of Theorem 8.29, we finally infer global estimates in $C^{2,\alpha}(\overline{\Omega})$ for any $\alpha < 1$. To complete the proof of Theorem 17.26 we simply apply the Schauder theory to the differentiated equation (17.23), thereby obtaining global $C^{3,\alpha}(\overline{\Omega})$ estimates for any $\alpha < 1$. □

Proof of Lemma 17.27. Since $h|_T$ is convex the derivatives $D_i h$, $i = 1, \dots, n-1$, exist almost everywhere on T and it suffices to prove (17.92) for such points x, y. The full result for all x, y follows from convexity and continuity.

We may assume $x = 0$ and that after subtraction of a suitable affine function,

$$h(0) = 0 = D_i h(0), \qquad i = 1, \dots, n-1. \tag{17.93}$$

After a rotation of coordinates we may further assume that

$$D_1 h(y) = \alpha \geqslant 0, \qquad D_i h(y) = 0, \qquad i = 2, \ldots, n-1,$$

where $\alpha \leqslant \sup |Dh| \leqslant C$. In the following we shall use the same letter C to denote constants depending only on the constants in the statement of the lemma. We wish to prove

$$\alpha \leqslant \frac{C}{|\log |y||} \tag{17.94}$$

when $|y|$ is sufficiently small, from which (17.92) will follow. Since $\alpha = 0$ implies the desired conclusion we may suppose $\alpha > 0$.

Writing $x = (x', x_n) = (x_1, \ldots, x_{n-1}, x_n)$ for $x \in \mathbb{R}^n$, we have from the convexity of h on T that $h(x', 0) \geqslant 0$ and

$$h(x', 0) \geqslant h(y', 0) + \alpha(x_1 - y_1) = \alpha(x_1 - \beta), \tag{17.95}$$

where β is defined by the equality. By taking $x' = 0$ and $x' = y'$ in this relation, one sees that

$$0 \leqslant \beta \leqslant y_1. \tag{17.96}$$

We consider the function

$$\begin{aligned} w(x', x_n) = {} & \frac{\alpha}{2}[(x_1 - \beta)^2 + x_n^2]^{1/2} + \frac{\alpha}{2}(x_1 - \beta) \\ & - \alpha\gamma x_n \log[(x_1 - \beta)^2 + x_n^2]^{1/2} - D(x_n + |x'|^2 - Ex_n^2). \end{aligned} \tag{17.97}$$

After suitable choice of positive constants $\gamma, \varepsilon < 1$, and $D, E > 1$, depending only on the constants of the lemma, it will be seen that w provides a barrier with the properties

$$Lw \geqslant f \geqslant Lh \quad \text{in } B_\varepsilon^+ (= B_\varepsilon^+(0)) \tag{17.98}$$

and

$$w \leqslant h \quad \text{on } \partial B_\varepsilon^+. \tag{17.99}$$

Assuming for the moment that the constants in w can be so chosen, we infer from the maximum principle that

$$w \leqslant h \quad \text{in } B_\varepsilon^+.$$

Then since $h(0) = w(0) = 0$, we have, after setting $x' = 0$ in (17.97), dividing by x_n and letting $x_n \to 0$,

$$D_n w(0) \leqslant \liminf_{x_n \to 0} \frac{h(0, x_n)}{x_n} \leqslant \sup |Dh| \leqslant C,$$

or

$$-\alpha\gamma \log \beta - D \leqslant C.$$

It follows from (17.96) that

$$\alpha \leqslant \frac{C + D}{\gamma |\log \beta|} \leqslant \frac{C + D}{\gamma |\log |y'||},$$

proving (17.94).

It remains to determine the constants in w. Setting $\rho = [(x_1 - \beta)^2 + x_n^2]^{1/2}$, we obtain by direct calculation

$$L\rho \geqslant \frac{\lambda}{\rho}$$

and

$$|L(x_n \log \rho)| \leqslant \frac{6\Lambda}{\rho},$$

so that for $0 < \gamma \leqslant \frac{1}{12} \inf \lambda/\Lambda$ we have

$$L(\rho/2 - \gamma x_n \log \rho) \geqslant 0.$$

By now choosing the constant E sufficiently large ($\geqslant \frac{1}{2} \sup f/\lambda + n \sup \Lambda/\lambda$) and $D \geqslant 1$, we can satisfy (17.98).

To establish (17.99) we observe first that the inequality holds on $x_n = 0$. For

$$\begin{aligned} w(x', 0) &= \alpha(x_1 - \beta)^+ - D|x'|^2 \\ &= \alpha(x_1 - \beta) - D|x'|^2 \leqslant h(x', 0) \quad \text{if } x_1 \geqslant \beta \qquad \text{(by (17.95))} \\ &= -D|x'|^2 \leqslant 0 \leqslant h(x', 0) \quad \text{if } x_1 \leqslant \beta. \end{aligned}$$

On the hemispherical portion S of ∂B_ε^+ we have that $x_n + |x'|^2 - Ex_n^2 \geqslant \frac{1}{2}\varepsilon^2$ provided ε is sufficiently small (depending on the choice of E). Since the other terms in w are bounded by $\sup |Dh|$ if $\varepsilon < 1$, a suitably large value of D makes

$$w \leqslant -C \leqslant \inf h$$

on S, which completes the proof of (17.94).

Finally, we remark that (17.94) implies (17.92) if $|x - y| \leqslant \varepsilon$, while $|Dh| \leqslant C$ yields the same inequality if $|x - y| > \varepsilon$ in $B^+_{\delta/2}$. □

The proof of the above lemma is taken almost directly from [CNS].

An Alternative Approach

The approach to global second derivative Hölder estimates given above is due to Caffarelli, Nirenberg and Spruck [CNS]. However, as we shall indicate now, the crucial Hölder estimates for the mixed tangential normal second derivatives on the boundary also follow readily from Theorem 9.3.1. Furthermore this approach, due to Krylov [KV 5], also yields global regularity for the Bellman equation, (17.8). In fact we have the stronger version of Theorem 17.6.

Theorem 17.26′. *Let Ω be a bounded domain in $\mathbb{R}^n$ with boundary $\partial\Omega \in C^3$ and let $\phi \in C^3(\bar{\Omega})$. Suppose that $u \in C^3(\bar{\Omega}) \cap C^4(\Omega)$ satisfies $F[u] = 0$ in Ω, $u = \phi$ on $\partial\Omega$, where $F \in C^2(\bar{\Gamma})$, F is elliptic with respect to u and satisfies (i)′, (ii)′. Then we have the estimate*

$$(17.86) \qquad |u|_{2,\alpha;\Omega} \leqq C$$

where α depends only on n, λ and Λ and C depends in addition on $|u|_{2;\Omega}$ and the first and second derivatives of F other than F_{rr}.

Proof. The situation is reduced according to the proof of Theorem 17.26 to the consideration of the derivatives $D_{in}u$, $i = 1, \ldots, n-1$ on the flat boundary portion T. But now, instead of considering the differential inequality (17.91) satisfied by the normal derivative $D_n u$ and applying Lemma 17.27, we consider the uniformly elliptic differential equations

$$a^{ij} D_{ij} h = f_k$$

satisfied by the tangential derivatives $h = D_k u$, $k = 1, \ldots, n-1$ in B^+ and apply Theorem 9.31. A global Hölder estimate for $D^2 u$ then follows independently of the moduli of continuity of the principal coefficients of the differentiated equation (17.23). □

We remark here, that by using the particular forms of the estimates (9.68), (17.51), we can avoid the barrier argument at the beginning of the proof of Theorem 17.26; (see Problem 13.1 with u replaced by $(D_1 u, \ldots, D_{n-1} u)$ or [TR 14]). Also the present proof of Theorem 17.26 can be modified slightly so as to remove the restriction (17.85). To see this, we recall that the pure second derivatives, $v = D_{\gamma\gamma} u = \gamma_i \gamma_j D_{ij} u$, $\gamma \in \partial\mathbb{R}^n_+$, satisfy in B^+, uniformly elliptic differential inequalities of the form

$$F_{ij} D_{ij} v + C_{ij} D_{ij\gamma} u \geqq -C$$

where by virtue of the differentiated equation (17.23), we may assume $C_{in} = 0$, $i = 1, \ldots, n$. The coefficients C_{ij} and constant C will be bounded in terms of the quantities in the statement of Theorem 17.26 (excluding λ_0). Now let us regard v as a function of both x and γ in the domain $\Omega^* = B^+ \times \{\gamma \in \partial\mathbb{R}^n_+ \mid |\gamma| < 2\} \subset \mathbb{R}^{2n-1}$ and extend the above operator so that

$$\tilde{L}v \equiv F_{ij} D_{ij} v + \tfrac{1}{2} C_{ij} D_{i\gamma_j} v + K_0 \sum_{i=1}^{n-1} D_{\gamma_i \gamma_i} v \geqq -C$$

in Ω^* and K_0 is chosen so that $\tilde{L}$ is uniformly elliptic in Ω^*. By appropriate extension of the barrier w to Ω^*, for example by defining

$$w(x, \gamma) = w(x) + \tau' |\gamma - \bar{\gamma}|^2$$

for constant τ' and $|\bar{\gamma}| = 1$, we are again able to infer the one-sided third derivative bound (17.90), and the rest of the proof of Theorem 17.26 follows automatically.

As a consequence of Theorem 17.26 we can deduce global smoothness of the solutions of the Dirichlet problems in Theorems 17.17, 17.18 when the boundary data are appropriately smooth. In fact, if F satisfies the structure conditions (17.53), we can, by interpolation, replace the dependence of the constant C in (17.86)$'$ on $|u|_{2;\Omega}$ by $|u|_{0;\Omega}$, and furthermore this estimate will be uniform with respect to the approximation of the Dirichlet problem (17.61) treated in Section 17.5. We thus obtain, for example, in place of Theorem 17.18.

Theorem 17.18$'$. *Let Ω be a bounded domain in $\mathbb{R}^n$ with boundary $\partial\Omega \in C^3$ and let $\phi \in C^3(\bar{\Omega})$. Suppose the functions $F^1, F^2, \ldots \in C^2(\bar{\Gamma})$, are concave with respect to z, p, r, non-increasing with respect to z, and satisfy uniformly the structure conditions* (17.53). *Then the classical Dirichlet problem,* (17.61), *is uniquely solvable with solution $u \in C^{2,\alpha}(\bar{\Omega})$ for some positive α depending only on n, λ and Λ.*

17.9. Nonlinear Boundary Value Problems

The method of continuity, previously applied to the Dirichlet problem in Section 17.2, readily extends to other boundary value problems as well. Even for the quasilinear case, the fixed point methods of Chapter 11 are not appropriate, since it is not in general possible to construct a *compact* operator analogous to the operator T in Sections 11.2, 11.4.

We first consider nonlinear boundary value problems of the form,

$$\begin{aligned} F[u] &= F(x, u, Du, D^2u) = 0 \quad \text{in } \Omega, \\ G[u] &= G(x, u, Du) \qquad\quad\;\; = 0 \quad \text{on } \partial\Omega, \end{aligned} \tag{17.100}$$

where F and G are real functions on the sets $\Gamma = \Omega \times \mathbb{R} \times \mathbb{R}^n \times \mathbb{R}^{n\times n}$ and $\Gamma' = \partial\Omega \times \mathbb{R} \times \mathbb{R}^n$ respectively. The case

$$G(x, z, p) = z - \varphi(x) \tag{17.101}$$

corresponds to the Dirichlet problem, $F[u] = 0$ in Ω, $u = \varphi$ on $\partial\Omega$, previously treated in this work. If $\partial\Omega \in C^1$, $G \in C^1(\Gamma')$, the boundary operator G is called *oblique* if

$$\frac{\partial G}{\partial p}(x, z, p) \cdot \nu(x) > 0 \tag{17.102}$$

for all $(x, z, p) \in \Gamma'$, where ν denotes the outer unit normal to $\partial\Omega$, while if

$$\frac{\partial G}{\partial p}(x, u(x), Du(x)) \cdot \nu(x) > 0 \tag{17.103}$$

for all $x \in \partial\Omega$, for some function $u \in C^1(\overline{\Omega})$ we call G *oblique with respect* to u. To apply the method of continuity we shall assume that $F \in C^{2,\alpha}(\overline{\Gamma})$, $G \in C^{3,\alpha}(\Gamma')$ and take for our Banach spaces,

$$\mathfrak{B}_1 = C^{2,\alpha}(\overline{\Omega}), \mathfrak{B}_2 = C^{\alpha}(\overline{\Omega}) \times C^{1,\alpha}(\partial\Omega)$$

for some $\alpha \in (0, 1)$. We then define a mapping $P: \mathfrak{B}_1 \to \mathfrak{B}_2$ by

$$P[u] = (F[u], G[u]),$$

which has a continuous Fréchet derivative on $\mathfrak{B}_1$ given by

$$P_u h = (Lh, Nh), \qquad h \in \mathfrak{B}_1, \tag{17.104}$$

where L is defined by (17.22) and

$$Nh = \gamma(x)h + \beta^i(x)D_i h$$

with

$$\gamma(x) = G_z(x, u(x), Du(x)),$$

$$\beta^i(x) = G_{p_i}(x, u(x), Du(x)).$$

It follows then, from the Schauder theory for the linear oblique derivative problem in Theorem 6.31, that P_u is boundedly invertible if F is strictly elliptic with respect to u, G is oblique with respect to u, $c(=L1) \leqslant 0$, $\gamma \geqslant 0$ and either $c \not\equiv 0$ in Ω or $\gamma \not\equiv 0$ on $\partial\Omega$. Accordingly we have, by virtue of Theorem 17.7, the following extension of Theorem 17.8.

Theorem 17.28. *Let Ω be a $C^{2,\alpha}$ domain in $\mathbb{R}^n$, $0 < \alpha < 1$, $\mathfrak{U}$ an open subset of the space $C^{2,\alpha}(\overline{\Omega})$ and $\psi \in \mathfrak{U}$. Set*

$$E = \{u \in \mathfrak{U} \mid P[u] = \sigma P[\psi] \quad \textit{for some } \sigma \in [0, 1]\}. \tag{17.105}$$

Suppose that $F \in C^{2,\alpha}(\overline{\Gamma})$, $G \in C^{3,\alpha}(\Gamma')$ and that

(i) *F is strictly elliptic in Ω for each $u \in E$;*
(ii) *G is either of the form (17.101) or oblique on $\partial\Omega$ for each $u \in E$;*

(iii) $F_z(x, u, Du, D^2u) \leqslant 0$, $G_z(x, u, Du) \geqslant 0$ *for each* $u \in E$ *with one of these quantities not vanishing identically;*
(iv) E *is bounded in* $C^{2,\alpha}(\bar{\Omega})$;
(v) $\bar{E} \subset \mathfrak{U}$.

Then the boundary value problem (17.100) *is solvable in* $\mathfrak{U}$.

A remark analogous to that at the end of Section 17.2 is also pertinent here. Namely, the family of boundary value problems in (17.105) may be replaced by more general families, $P[u; \sigma] = 0$, depending smoothly on σ, for which $P[u; 1] = P[u]$ and the equation $P[u; 0] = 0$ is solvable in $\mathfrak{U}$. For each $\sigma \in [0, 1]$ the operator $u \to P[u; \sigma]$ must of course satisfy the same hypotheses as P.

We now show that for quasilinear operators

$$F[u] = Qu = a^{ij}(x, u, Du)D_{ij}u + b(x, u, Du), \tag{17.106}$$

condition (iv) in Theorem 17.28 can be weakened to require only the boundedness of the set E in $C^{1,\delta}(\bar{\Omega})$ for some $\delta > 0$. This has the effect of reducing the existence program for quasilinear equations to the same type of apriori estimation as was required for the Dirichlet problem in Chapter 11. For this reduction we need the following convergence lemma.

Lemma 17.29. *Let F be a quasilinear operator of the form* (17.106) *with coefficients* $a^{ij}, b \in C^{\alpha}(\Omega \times \mathbb{R} \times \mathbb{R}^n)$ *and G a boundary operator of the form* (17.100) *with* $G, G_p \in C^{1,\alpha}(\partial\Omega \times \mathbb{R} \times \mathbb{R}^n)$, $0 < \alpha < 1$. *Assume that F is strictly elliptic in Ω, G is oblique on $\partial\Omega$, and suppose there exist sequences* $\{u_m\} \subset C^{2,\alpha}(\bar{\Omega})$, $\{f_m\} \subset C^{\alpha}(\bar{\Omega})$, $\{\varphi_m\} \subset C^{1,\alpha}(\partial\Omega)$ *such that*

(i) $P[u_m] = (f_m, \varphi_m)$;
(ii) $\{u_m\}$, $\{f_m\}$, $\{\varphi_m\}$ *are uniformly bounded in* $C^{1,\alpha}(\bar{\Omega})$, $C^{\alpha}(\bar{\Omega})$, $C^{1,\alpha}(\partial\Omega)$, *respectively;*
(iii) $\{u_m\}$, $\{f_m\}$, $\{\varphi_m\}$ *converge uniformly to functions* u, f, φ, *respectively.*

Then $u \in C^{2,\alpha}(\bar{\Omega})$ *and* $P[u] = (f, \varphi)$.

Proof. Let $v = u_m - u_k$ where m, k are positive integers. Then v is a solution of the linear problem

$$Lv = g \quad \text{in } \Omega, \qquad Nv = \psi \quad \text{on } \partial\Omega$$

where

$$Lu = \bar{a}^{ij}D_{ij}u, \qquad Nu = \beta^i D_i u,$$

$$\bar{a}^{ij} = a^{ij}(x, u_m, Du_m), \qquad \beta^i = \int_0^1 G_{p_i}(x, u_m, Du_k + t(Du_m - Du_k))\, dt,$$

$$g = f_m - f_k + b(x, u_k, Du_k) - b(x, u_m, Du_m) + (a^{ij}(x, u_k, Du_k) - a^{ij}(x, u_m, Du_m))D_{ij}u_k,$$

$$\psi = \varphi_m - \varphi_k + G(x, u_k, Du_k) - G(x, u_m, Du_m).$$

From the hypotheses of the lemma we estimate

$$|g|_{\alpha\delta} \leqslant \varepsilon(m, k)(1 + K_1 + K_2) + CK_1,$$
$$|\psi|_{1,\alpha\delta} \leqslant \varepsilon(m, k)(1 + K_1 + K_2)$$
$$|\beta|_{\alpha\delta} \leqslant C, [\beta]_{1,\alpha\delta} \leqslant C(1 + K_1 + K_2)$$

where $\varepsilon \to 0$ as $m, k \to \infty$, C is independent of m, k and

$$K_1 = \max\{|D^2u_m|_0, |D^2u_k|_0\}, \qquad K_2 = \max\{[D^2u_m]_{\alpha\delta}, [D^2u_k]_{\alpha\delta}\}.$$

Using the estimate of Theorem 6.30, as modified in Problem 6.11, we thus obtain

$$[v]_{2,\alpha\delta} \leqslant C(1 + K_1) + \varepsilon(m, k)K_2$$

and hence by the interpolation inequality, Theorem 6.35, we have

$$[v]_{2,\alpha\delta} \leqslant C + (\tfrac{1}{4} + \varepsilon)K_2, \tag{17.107}$$

where again C is independent of m, k and $\varepsilon \to 0$ as $m, k \to \infty$. But (17.107) implies the boundedness of $[u_m]_{2,\alpha\delta}$. For suppose not. Then for some m, k

$$K_2 = [u_m]_{2,\alpha\delta} > 4[u_k]_{2,\alpha\delta} \geqslant 4C, \qquad \varepsilon(m, k) < \tfrac{1}{4}. \tag{17.108}$$

Hence

$$[v]_{2,\alpha\delta} \geqslant \tfrac{3}{4}K_2,$$

so that by (17.107) we have

$$\tfrac{3}{4}K_2 \leqslant C + \tfrac{1}{2}K_2,$$

that is, $K_2 \leqslant 4C$, contradicting (17.108). Thus $[u_m]_{2,\alpha\delta}$ is bounded independently of m and hence $u \in C^{2,\alpha\delta}(\bar{\Omega})$ with $P[u] = (f, \varphi)$. Finally, by an argument similar to the above we infer also $u \in C^{2,\alpha}(\bar{\Omega})$. □

Combining Lemma 17.29 with Theorem 17.7 we now conclude a version of Theorem 17.28 appropriate for quasilinear operators.

Theorem 17.30. *Let Ω be a $C^{2,\alpha}$ domain in $\mathbb{R}^n$, $0 < \alpha < 1$, and suppose that Q is strictly elliptic in Ω with $a^{ij}, b \in C^2(\bar{\Omega} \times \mathbb{R} \times \mathbb{R}^n)$, and $G \in C^3(\partial\Omega \times \mathbb{R} \times \mathbb{R}^n)$ is oblique on $\partial\Omega$. Suppose also that $a^{ij}_z = 0, b_z \leqslant 0, G_z \geqslant 0$ with either $b_z(x, u(x), Du(x)) \not\equiv 0$ or $G_z(x, u(x), Du(x)) \not\equiv 0$ for each $u \in C^{2,\alpha}(\bar{\Omega})$. Then if for some function $\psi \in C^{2,\alpha}(\bar{\Omega})$, the set*

$$E = \{u \in C^{2,\alpha}(\bar{\Omega}) \,|\, Qu = \sigma Q\psi \quad \text{in } \Omega, \quad G[u] = \sigma G[\psi] \quad \text{on } \partial\Omega \quad \text{for some } \sigma \in [0, 1]\} \tag{17.109}$$

is bounded in $C^{1,\delta}(\overline{\Omega})$ *for some* $\delta > 0$, *the boundary value problem* $Qu = 0$ *in* Ω, $G[u] = 0$ *on* $\partial\Omega$ *is uniquely solvable in* $C^{2,\alpha}(\overline{\Omega})$.

As mentioned previously other families of problems may be used in (17.109). For example, we could consider a family analogous to that used for the Dirichlet problem in Theorem 11.4, namely

$$\begin{aligned} Q_\sigma u &= a^{ij}(x, u, Du)D_{ij}u + \sigma b(x, u, Du) = 0 \quad \text{in } \Omega, \\ G_\sigma u &= \sigma G(x, u, Du) + (1 - \sigma)u = 0 \quad \text{on } \partial\Omega. \end{aligned} \tag{17.110}$$

A typical oblique boundary value problem is the *conormal derivative* boundary value problem for an elliptic divergence structure equation, which takes the form

$$\begin{aligned} Qu &= \operatorname{div} A(x, u, Du) + B(x, u, Du) = 0 \quad \text{in } \Omega \\ G[u] &= A(x, u, Du) \cdot \nu(x) + \varphi(x, u) = 0 \quad \text{on } \partial\Omega. \end{aligned} \tag{17.111}$$

The problem of determining a capillary surface with prescribed contact angle, mentioned in Chapter 10, provides an interesting special case of a conormal derivative boundary value problem. For this example and also for uniformly elliptic Q satisfying natural conditions as in Theorem 15.11, the relevant apriori estimates have been proved, so that appropriate existence theorems may be inferred from Theorem 17.30. The reader is referred to the literature, for example [LU 4], [UR], [GE 3], [LB 2, 3] for further details.

Notes

Fully nonlinear equations in two variables were treated by various authors, including Lewy [LW 1], Nirenberg [NI 1], Pogorelov [PG 1] and Heinz [HE 2], with most attention being devoted to equations of Monge-Ampère type and related geometric problems such as the Minkowski problem to determine a convex hypersurface through prescription of its Gauss curvature. The extremal operators of Pucci were introduced in [PU 2] and the associated Dirichlet problems also solved in the two variable case.

The higher dimensional Monge-Ampère equations were first solved in a generalized sense by Aleksandrov [AL 1] and Bakelman [BA 1] using polyhedral approximation, with further development in [BA 2, 4], which contains a generalized version of the main existence theorem (Theorem 17.24). A generalized solution of the Monge-Ampère equation (17.2) can be defined as a convex function whose normal mapping is absolutely continuous with density f. The interior regularity of generalized solutions (under sufficiently smooth boundary conditions) was established by Pogorelov [PG 2, 3, 4, 5] and Cheng and Yau [CY 1, 2], essential ingredients of their proof being the interior second-derivative bounds of Pogorelov and the interior third-derivative bounds of Calabi [CL]. These authors thus obtained Theorem 17.24 for the Monge-Ampère equation (17.2). Our treatment of second-derivative estimates in Section 17.6 uses the Pogorelov method [PG 1, 5] (but incorporates a suggestion of L. M. Simon concerning the function h).

The main impetus for studying fully nonlinear uniformly elliptic equations arose through stochastic control theory, as the Bellman equation (17.8) is satisfied by a (sufficiently smooth) optimal cost function in a control problem associated with a system of stochastic differential equations. The first significant treatment of this equation by Krylov used probability methods and is described in his book [KV 2]. Partial differential equation techniques were subsequently developed by: Brezis and Evans (who derived $C^{2,\alpha}$ estimates for the *two* operator case); Evans and Friedman [EF] for the constant coefficient case (see also [LD]); and P. L. Lions [LP 4], Evans and Lions [EL 1] for the general uniformly elliptic case. In the papers [LP 4], [EL 1], the existence of a strong solution of the Dirichlet problem is established under conditions (17.62) using a clever method based on approximation by an elliptic system and apriori $C^{1,1}(\overline{\Omega})$ bounds. A full treatment of the (possibly degenerate) Bellman equation and connections with stochastic control theory and the Bellman dynamic programming method is given by P.-L. Lions in [LP 8, 9] and discussed in [LP 10]; (see also [LP 7] for the first-order Hamilton-Jacobi equation).

The theory of classical solutions of fully nonlinear elliptic equations advanced substantially with the discovery of interior second-derivative Hölder estimates by Evans [EV 2, 3] and Krylov [KV 4] who basically proved the results of Sections 17.4 and 17.5. These are treated in Sections 17.4, 17.5 following simplifications of Trudinger [TR 13]. While we have for expedience in Sections 17.3 and 17.4 deduced first- and second-derivative bounds through interpolation, such results hold under more general structure conditions analogous to those of the quasilinear theory [TR 13, 14]. In particular, we may replace the derivative conditions in (17.29) and (17.53) by

$$(17.29)' \qquad |F(x, z, p, 0)| \leqslant \mu\lambda(1 + |p|^2),$$

$$|F_x|, |F_z|, |F_p| \leqslant \tilde{\mu}\lambda(1 + |r|);$$

$$(17.53)' \qquad (1 + |p|)|F_p|, |F_z|, |F_x| \leqslant \mu\lambda(1 + |p|^2 + |r|),$$

$$|F_{rx}|, |F_{rz}|, |F_{rp}| \leqslant \tilde{\mu}\lambda,$$

$$|F_{xx}|, |F_{xz}|, |F_{xp}|, |F_{zz}|, |F_{zp}|, |F_{pp}| \leqslant \tilde{\mu}\lambda(1 + |r|)$$

where μ is nondecreasing in $|z|$, $\tilde{\mu}$ is nondecreasing in $|z| + |p|$, (and the concavity of F with respect to p and z in (17.53) is dropped).

Recently the Monge-Ampère equation in higher dimensions has flourished again. A partial differential equations approach to the classical Dirichlet problem was developed by P.-L. Lions [LP 5, 6] which, on combination with the earlier considerations of Bakelman, yielded Theorem 17.24 for $C^0(\overline{\Omega}) \cap C^2(\Omega)$ solutions. Lions' method was based on approximation by problems defined over $\mathbb{R}^n$. In a similar vein, Cheng and Yau [CY 3, 4] proposed a method based on approximation by problems with infinite boundary values. Also, a probabilistic approach was given by Krylov [KV 3]. However the global regularity, which had been an impediment to the direct application of the method of continuity, was finally settled by Caffarelli, Nirenberg and Spruck [CNS] and Krylov [KV 5, 6], who discovered Theorems 17.26 and 17.26' respectively, and consequently established Theorem

17.23 in its full strength for the Monge-Ampère equation (17.2). The paper [CNS] also reduces the solvability of the Dirichlet problem for general Monge-Ampère equations to the existence of globally smooth subsolutions, from which the case $\beta = 0$, $\gamma = n$ of Theorem 17.23 follows readily. The work of Caffarelli, Nirenberg and Spruck [CNS] and also that of Ivochkina [IC 2, 3] embraced the case of general boundary values $\phi \in C^4(\bar{\Omega})$. In [TU], Theorem 17.24 is extended to boundary data $\partial\Omega \in C^{1,1}$, $\phi \in C^{1,1}(\bar{\Omega})$.

The existence of $C^{2,\alpha}(\bar{\Omega})$ solutions of the Bellman equation, Theorem 17.18′, is due to Krylov [KV 5]. The idea, used in Section 17.8, of invoking the directions γ as variables occurs in Krylov's treatment of interior second derivative estimates in [KV 4]. Theorem 17.18′ may also be extended to cover structure conditions such as (17.29)′, (17.53)′ above; (see [TR 13, 14], [CKNS]).

The reduction of nonlinear boundary value problems to apriori estimation in Theorem 17.30 is due to Fiorenza [FI 2]; (see also Ladyzhenskaya and Ural'tseva [LU 4]). In our treatment we have adopted some simplification by Lieberman [LB 2] in the proof of Lemma 17.29, although the main thrust of Lieberman's work is to replace the method of continuity by another functional analytic approach which permits a weaker hypothesis than condition (iii) in Theorem 17.28. Oblique boundary value problems for fully *nonlinear* equations have been treated recently by Lions and Trudinger [LPT], [TR 14] for Bellman equations, Lieberman and Trudinger [LBT], [TR 14] for general nonlinear boundary conditions and by Lions, Trudinger and Urbas [LTU] for equations of Monge Ampère type.

Problems

17.1. Prove that if the operator (17.1) is elliptic at a point $\gamma \in \Gamma$, then the function F is strictly increasing with respect to r at γ. That is, there exists a positive number ε such that

$$F(x, z, p, r) < F(x, z, p, r + \eta) \tag{17.112}$$

for all nonzero, positive semidefinite matrices $\eta \in \mathbb{R}^{n \times n}$ with $|\eta| < \varepsilon$.

17.2. Prove that the operator F given by (17.1) is Fréchet differentiable as a mapping from $C^{2,\alpha}(\bar{\Omega})$ into $C^{0,\alpha}(\bar{\Omega})$, for any $\alpha \leqslant 1$, if the function $F \in C^{2,\alpha}(\bar{\Gamma})$.

17.3. More generally than Problem 17.2, prove that F is Fréchet differentiable as a mapping from $C^{2,\alpha}(\bar{\Omega})$ into $C^{0,\alpha\gamma}(\bar{\Omega})$ for any $\alpha, \beta \leqslant 1$, $\gamma < \alpha\beta$, if the function F is differentiable with respect to z, p and r with $F, F_z, F_p, F_r \in C^\beta(\bar{\Gamma})$.

17.4. Show that Theorem 17.17 remains valid if the condition $F \in C^2(\Gamma)$ is replaced by the existence of the derivatives in (17.53) as weak derivatives in the sense of Chapter 7.

17.5. Let $u \in C^2(\overline{\Omega}) \cap C^4(\Omega)$ be a subharmonic solution of the equation

$$F(D^2u) = g(x) \tag{17.113}$$

in a domain $\Omega \subset \mathbb{R}^n$, where $F \in C^2(\mathbb{R}^{n \times n})$, $g \in C^2(\overline{\Omega})$, F is elliptic with respect to u, tr $[F_{ij}(D^2u)] \geqslant 1$ and F is concave with respect to D^2u. Prove that

$$\sup_{\Omega} |D^2u| \leqslant C(\sup_{\partial\Omega} |D^2u| + \sup_{\Omega} |D^2g|) \tag{17.114}$$

where $C = C(n)$.

17.6. (a) Let $F \in C^1(\mathbb{R}^{n \times n})$ be invariant under orthogonal transformations. Show that for any $r \in \mathbb{R}^{n \times n}$, the matrices $F'(r)$ and r commute.

(b) Suppose that the conditions of Problem 17.5 are strengthened so that F is invariant under orthogonal transformations and concave for $r \geqslant 0$, with det $[F_{ij}(D^2u)] \geqslant 1$. Suppose also that the domain Ω is uniformly convex with boundary $\partial\Omega \in C^3$ and that $u = 0$ on $\partial\Omega$. Show that if the solution u is convex, then

$$\sup_{\partial\Omega} |D^2u| \leqslant C \tag{17.115}$$

where C depends on n, $|g|_{1;\Omega}$, $|u|_{1;\Omega}$ and $\partial\Omega$.

17.7. (a) Let $F \in C^1(\mathbb{R}^{n \times n})$ be invariant under orthogonal transformations and consider for fixed $k, l \in \{1, \ldots, n\}$ the polar coordinate transformation

$$\begin{aligned} x_k &= a + r\sin\theta, \\ x_l &= b - r\cos\theta, \end{aligned} \tag{17.116}$$

where a, b are constants. If $u \in C^3(\Omega)$ satisfies the equation (17.113) in Ω, show that

$$F_{ij}D_{ij}\left(\frac{\partial u}{\partial\theta}\right) = \frac{\partial g}{\partial\theta} \tag{17.117}$$

in $\Omega \cap \{r > 0\}$.

(b) Let Ω be a C^3 domain in $\mathbb{R}^n$. Suppose that $0 \in \partial\Omega$ and that the inner normal to $\partial\Omega$ at 0 is directed along the x_n axis. If $u \in C^1(\overline{\Omega})$ and $u = 0$ on $\partial\Omega$, show that in a neighbourhood of 0, either

$$\left|\frac{\partial u}{\partial\theta}\right| \quad \text{or} \quad \left|\frac{\partial u}{\partial x_k}\right| \leqslant C|x_k|^2$$

where θ is given by (17.116) with appropriate choice of a, b and $l = n$, and C depends on $\partial\Omega$ and $|u|_{1;\Omega}$. Accordingly show that in Problem 17.6, the assumption of convexity of u can be dropped. (Cf. [IC 1].)

17.8. Let Ω be a C^2 uniformly convex domain in $\mathbb{R}^n$. Show the existence of a uniformly convex function $u \in C^2(\overline{\Omega})$ vanishing on $\partial\Omega$.

17.9. Show that the equation

$$(17.118) \qquad F[u] = (\Delta u)^2 - |D^2 u|^2 = f(x)$$

is elliptic with respect to any solution u (or its negative) if the function f is positive. Using the results of Problems 17.5, 17.7 together with Theorems 17.14, 17.26, show that the classical Dirichlet problem for (17.118) is solvable for any uniformly convex domain $\Omega \in C^3$, zero boundary values and positive $f \in C^2(\overline{\Omega})$. More generally we note here that this result extends to domains Ω whose boundary has positive *mean* curvature.

17.10. Let $F \in C^1(\Omega \times \mathbb{R} \times \mathbb{R}^n \times \mathbb{R}^{n \times n})$ satisfy

$$F(x_0, 0, 0, 0) = 0, \qquad F_r(x_0, 0, 0, 0) > 0$$

at some point $x_0 \in \Omega$. Using the implicit function (Theorem 17.6), show that in some neighbourhood $\mathcal{N}$ of x_0, there exists a $C^2(\mathcal{N})$ solution u of equation (17.1) satisfying $u(x_0) = Du(x_0) = 0$. (Hint: Taking $x_0 = 0$, $t \in \mathbb{R}$, make the change of variables $x = ty$, $u = t^2 v$.)

Bibliography

Adams, R. A.
[AD] Sobolev Spaces. New York: Academic Press 1975.

Agmon, S.
[AG] Lectures on Elliptic Boundary Value Problems. Princeton, N. J.: Van Nostrand 1965.

Agmon, S., A. Douglis, and L. Nirenberg
[ADN 1] Estimates near the boundary for solutions of elliptic partial differential equations satisfying general boundary conditions. I. Comm. Pure Appl. Math. **12**, 623–727 (1959).
[ADN 2] Estimates near the boundary for solutions of elliptic partial differential equations satisfying general boundary conditions. II. Comm. Pure Appl. Math. **17**, 35–92 (1964).

Aleksandrov, A. D.
[AL 1] Dirichlet's problem for the equation Det $\|Z_{ij}\| = \phi$. Vestnik Leningrad Univ. **13**, no. 1, 5–24 (1958) [Russian].
[AL 2] Certain estimates for the Dirichlet problem. Dokl. Akad. Nauk. SSSR **134**, 1001–1004 (1960) [Russian]. English Translation in Soviet Math. Dokl. **1**, 1151–1154 (1960).
[AL 3] Uniqueness conditions and estimates for the solution of the Dirichlet problem. Vestnik Leningrad Univ. **18**, no. 3, 5–29 (1963) [Russian]. English Translation in Amer. Math. Soc. Transl. (2) **68**, 89–119 (1968).
[AL 4] Majorization of solutions of second-order linear equations. Vestnik Leningrad Univ. **21**, no. 1, 5–25 (1966) [Russian]. English Translation in Amer. Math. Soc. Transl. (2) **68**, 120–143 (1968).
[AL 5] Majorants of solutions and uniqueness conditions for elliptic equations. Vestnik Leningrad Univ. **21**, no. 7, 5–20 (1966) [Russian]. English Translation in Amer. Math. Soc. Transl. (2) **68**, 144–161 (1968).
[AL 6] The impossibility of general estimates for solutions and of uniqueness conditions for linear equations with norms weaker than in L_n. Vestnik Leningrad Univ. **21**, no. 13, 5–10 (1966) [Russian]. English Translation in Amer. Math. Soc. Transl. (2) **68**, 162–168 (1968).

Alkhutov, Yu. A.
[AK] Regularity of boundary points relative to the Dirichlet problem for second order elliptic equations. Mat. Zametki **30**, 333–342 (1981) [Russian]. English Translation in Math. Notes **30**, 655–661 (1982).

Allard, W.
[AA] On the first variation of a varifold. Ann. of Math. (2) **95**, 417–491 (1972).

Almgren, F. J.
[AM] Some interior regularity theorems for minimal surfaces and an extension of Bernstein's theorem. Ann. of Math. (2) **84**, 277–292 (1966).

Aubin, T.
[AU 1] Équations du type Monge-Ampère sur les variétés kählériennes compactes. C.R. Acad. Sci. Paris **283**, 119–121 (1976).
[AU 2] Problèmes isopérimètriques et espaces de Sobolev. J. Differential Geometry **11**, 573–598 (1976).
[AU 3] Équations du type Monge-Ampère sur les varietes kählériennes compactes. Bull. Sci. Math. **102**, 63–95 (1978).
[AU 4] Équations de Monge-Ampère réelles. J. Funct. Anal. **41**, 354–377 (1981).

Bakel'man, I. YA.
[BA 1] Generalized solutions of the Monge-Ampère equations. Dokl. Akad. Nauk. SSSR **114**, 1143–1145 (1957) [Russian].

[BA 2] The Dirichlet problem for equations of Monge-Ampère type and their n-dimensional analogues. Dokl. Akad. Nauk SSSR **126**, 923-926 (1959) [Russian].
[BA 3] Theory of quasilinear elliptic equations. Sibirsk. Mat. Ž. **2**, 179-186 (1961) [Russian].
[BA 4] Geometrical methods for solving elliptic equations. Moscow: Izdat. Nauka 1965 [Russian].
[BA 5] Mean curvature and quasilinear elliptic equations. Sibirsk. Mat. Ž. **9**, 1014-1040 (1968) [Russian]. English Translation in Siberian Math. J. **9**, 752-771 (1968).
[BA 6] Geometric problems in quasilinear elliptic equations. Uspehi Mat. Nauk. **25**, no. 3, 49-112 (1970) [Russian]. English Translation in Russian Math. Surveys **25**, no. 3, 45-109 (1970).
[BA 7] The Dirichlet problem for the elliptic n-dimensional Monge-Ampère equations and related problems in the theory of quasilinear equations, Proceedings of Seminar on Monge-Ampère Equations and Related Topics, Istituto Nazionale di Alta Matematica, Rome 1-78 (1982).

Bernstein, S.
[BE 1] Sur la généralisation du problème de Dirichlet. I. Math. Ann. **62**, 253-271 (1906).
[BE 2] Méthode générale pour la résolution du problème de Dirichlet. C. R. Acad. Sci. Paris **144**, 1025-1027 (1907).
[BE 3] Sur la généralisation du problème de Dirichlet II. Math. Ann. **69**, 82-136 (1910).
[BE 4] Conditions nécessaires et suffisantes pour la possibilité du problème de Dirichlet. C. R. Acad. Sci. Paris **150**, 514-515 (1910).
[BE 5] Sur les surfaces définies au moyen de leur courbure moyenne et totale. Ann. Sci. École Norm. Sup. **27**, 233-256 (1910).
[BE 6] Sur les équations du calcul des variations. Ann. Sci. École Norm. Sup. **29**, 431-485 (1912).

Bers, L., and L. Nirenberg
[BN] On linear and non-linear elliptic boundary value problems in the plane. In: Convegno Internazionale sulle Equazioni Lineari alle Derivate Parziali, Trieste, pp. 141-167. Rome: Edizioni Cremonese 1955.

Bers, L., and M. Schechter
[BS] Elliptic equations. In: Partial Differential Equations, pp. 131-299. New York: Interscience 1964.

Bliss, G. A.
[BL] An integral inequality. J. London Math. Soc. **5**, 40-46 (1930).

Bombieri, E.
[BM] Variational problems and elliptic equations. In: Proceedings of the International Congress of Mathematicians, Vancouver 1974, Vol. 1, 53-63. Vancouver: Canadian Mathematical Congress 1975.

Bombieri, E., E. De Giorgi, and E. Giusti
[BDG] Minimal cones and the Bernstein problem. Invent. Math. **7**, 243-268 (1969).

Bombieri, E., E. De Giorgi, and M. Miranda
[BDM] Una maggiorazione a priori relativa alle ipersuperfici minimali non parametriche. Arch. Rational Mech. Anal. **32**, 255-267 (1969).

Bombieri, E., and E. Giusti
[BG 1] Harnack's inequality for elliptic differential equations on minimal surfaces. Invent. Math. **15**, 24-46 (1972).
[BG 2] Local estimates for the gradient of non-parametric surfaces of prescribed mean curvature. Comm. Pure Appl. Math. **26**, 381-394 (1973).

Bony, J. M.
[BY] Principe du maximum dans les espaces de Sobolev. C. R. Acad. Sci. Paris **265**, 333-336 (1967).

Bouligand, G., G. Giraud, and P. Delens
[BGD] Le problème de la derivée oblique en théorie du potentiel. Actualités Scientifiques et Industrielles 219. Paris: Hermann 1935.

Brandt, A.
[BR 1] Interior estimates for second-order elliptic differential (or finite-difference) equations via the maximum principle. Israel J. Math. **7**, 95-121 (1969).
[BR 2] Interior Schauder estimates for parabolic differential- (or difference-) equations via the maximum principle. Israel J. Math. **7**, 254-262 (1969).

Brezis, H., and L. C. Evans

[BV] A variational inequality approach to the Bellman-Dirichlet equation for two elliptic operators. Arch. Rational Mech. Anal. **71**, 1-13 (1979).

Browder, F. E.

[BW 1] Strongly elliptic systems of differential equations. In: Contributions to the Theory of Partial Differential Equations, pp. 15-51. Princeton, N.J.: Princeton University Press 1954.

[BW 2] On the regularity properties of solution of elliptic differential equations. Comm. Pure Appl. Math. **9**, 351-361 (1956).

[BW 3] Apriori estimates for solutions of elliptic boundary value problems, I, II, III. Neder. Akad. Wetensch. Indag. Math. **22**, 149-159, 160-169 (1960), **23**, 404-410 (1961).

[BW 4] Problèmes non-linéaires. Séminaire de Mathématiques Supérieures, No. 15 (Été, 1965). Montreal, Que.: Les Presses de l'Université de Montreal 1966.

[BW 5] Existence theorems for nonlinear partial differential equations. In: Proceedings of Symposia in Pure Mathematics, Volume XVI; pp. 1-60. Providence, R.I.: American Mathematical Society 1970.

Caccioppoli, R.

[CA 1] Sulle equazioni ellittiche a derivate parziali con *n* variabili indipendenti. Atti Accad. Naz. Lincei Rend. Cl. Sci. Fis. Mat. Natur. (6) **19**, 83-89 (1934).

[CA 2] Sulle equazioni ellittiche a derivate parziali con due variabili indipendenti, e sui problemi regolari di calcolo delle variazioni. I. Atti Accad. Naz. Lincei Rend. Cl. Sci. Fis. Mat. Natur. (6) **22**, 305-310 (1935).

[CA 3] Sulle equazioni ellittiche a derivate parziali con due variabili indipendenti, e sui problemi regolari di calcolo delle variazioni. II. Atti Accad. Naz. Lincei Rend. Cl. Sci. Fis. Mat. Natur. (6) **22**, 376-379 (1935).

[CA 4] Limitazioni integrali per le soluzioni di un'equazione lineare ellittica a derivate parziali. Giorn. Mat. Battaglini (4) **4** (80), 186-212 (1951).

Caffarelli, L., L. Nirenberg, and J. Spruck

[CNS] The Dirichlet problem for nonlinear second order elliptic equations, I. Monge-Ampère equations. Comm. Pure Appl. Math. **37**, 369-402 (1984).

Caffarelli, L., J. Kohn, L. Nirenberg, and J. Spruck

[CKNS] The Dirichlet problem for nonlinear second-order elliptic equations II. Comm. Pure Appl. Math. **38**, 209-252 (1985).

Calabi, E.

[CL] Improper affine hyperspheres of convex type and a generalization of a theorem by K. Jörgens. Michigan Math. J. **5**, 105-126 (1958).

Calderon, A. P., and A. Zygmund

[CZ] On the existence of certain singular integrals. Acta Math. **88**, 85-139 (1952).

Campanato, S.

[CM 1] Proprietà di inclusione per spazi di Morrey. Ricerche Mat. **12**, 67-86 (1963).

[CM 2] Equazioni ellittiche del II⁰ ordine e spazi $\mathscr{L}^{(2,\lambda)}$. Ann. Mat. Pura Appl. (4) **69**, 321-381 (1965).

[CM 3] Sistemi ellittici in forma divergenza. Regolarità all' interno. Quaderni della Scuola Norm. Sup. di Pisa (1980).

Campanato, S., and G. Stampacchia

[CS] Sulle maggiorazioni in L^p nella teoria delle equazioni ellittiche. Boll. Un. Mat. Ital. (3) **20**, 393-399 (1965).

Cheng, S. Y., and S. T. Yau

[CY 1] On the regularity of the Monge-Ampère equation det $(\partial^2 u/\partial x_i\, \partial x_j) = F(x, u)$. Comm. Pure Appl. Math. **30**, 41-68 (1977).

[CY 2] On the regularity of the solution of the *n*-dimensional Minowski problem. Comm. Pure Appl. Math. **19**, 495-516 (1976).

[CY 3] On the existence of a complete Kähler metric on non-compact complex manifolds and the regularity of Fefferman's equation. Comm. Pure Appl. Math. **33**, 507-544 (1980).

[CY 4] The real Monge-Ampère equation and affine flat structures. Proc. 1980 Beijing Symposium on Differential Geometry and Differential Equations. Vol. I, 339-370 (1982). Editors, S. Chern, W. T. Wu.

Chicco, M.

[CI 1] Principio di massimo forte per sottosoluzioni di equazioni ellittiche di tipo variazionale. Boll. Un. Mat. Ital. (3) **22**, 368-372 (1967).

[Cl 2] Semicontinuità delle sottosoluzioni di equazioni ellittiche di tipo variazionale. Boll. Un. Mat. Ital. (4) **1**, 548–553 (1968).
[Cl 3] Principio di massimo per soluzioni di problemi al contorno misti per equazioni ellittiche di tipo variazionale. Boll. Un. Mat. Ital. (4) **3**, 384–394 (1970).
[Cl 4] Solvability of the Dirichlet problem in $H^{2,p}(\Omega)$ for a class of linear second order elliptic partial differential equations. Boll. Un. Mat. Ital. (4) **4**, 374–387 (1971).
[Cl 5] Sulle equazioni ellittiche del secondo ordine a coefficienti continui. Ann. Mat. Pura Appl. (4) **88**, 123–133 (1971).

Cordes, H. O.
[CO 1] Über die erste Randwertaufgabe bei quasilinearen Differentialgleichungen zweiter Ordnung in mehr als zwei Variablen. Math. Ann. **131**, 278–312 (1956).
[CO 2] Zero order a priori estimates for solutions of elliptic differential equations. In: Proceedings of Symposia in Pure Mathematics, Volume IV, pp. 157–166. Providence, R.I.: American Mathematical Society 1961.

Courant, R., and D. Hilbert
[CH] Methods of Mathematical Physics. Volumes I, II. New York: Interscience 1953, 1962.

De Giorgi, E.
[DG 1] Sulla differenziabilità e l'analiticità delle estremali degli integrali multipli regolari. Mem. Accad. Sci. Torino Cl. Sci. Fis. Mat. Natur. (3) **3**, 25–43 (1957).
[DG 2] Una estensione del teorema di Bernstein. Ann. Scuola Norm. Sup. Pisa (3) **19**, 79–85 (1965).

Delanoë, P.
[DE 1] Équations du type Monge-Ampère sur les variétés Riemanniennes compactes I. J. Funct. Anal. **40**, 358–386 (1981).
[DE 2] Équations du type Monge-Ampère sur les variétés Riemanniennes compactes II. J. Funct. Anal. **41**, 341–353 (1981).

Dieudonné, J.
[DI] Foundations of Modern Analysis. New York: Academic Press, 1960.

Douglas, J., T. Dupont, and J. Serrin
[DDS] Uniqueness and comparison theorems for nonlinear elliptic equations in divergence form. Arch. Rational Mech. Anal. **42**, 157–168 (1971).

Douglis, A., and L. Nirenberg
[DN] Interior estimates for elliptic systems of partial differential equations. Comm. Pure Appl. Math. **8**, 503–538 (1955).

Dunford, N., and J. T. Schwartz
[DS] Linear Operators, Part I. New York: Interscience 1958.

Edmunds, D. E.
[ED] Quasilinear second order elliptic and parabolic equations. Bull. London Math. Soc. **2**, 5–28 (1970).

Edwards, R. E.
[EW] Functional Analysis: Theory and Applications. New York: Holt, Rinehart and Winston 1965.

Egorov, Yu. V., and V. A. Kondrat'ev
[EK] The oblique derivative problem. Mat. Sb. (N.S.) **78** (120), 148–176 (1969) [Russian]. English Translation in Math USSR Sb. **7**, 139–169 (1969).

Emmer, M.
[EM] Esistenza, unicità e regolarità nelle superfice di equilibrio nei capillari. Ann. Univ. Ferrara **18**, 79–94 (1973).

Evans, L. C.
[EV 1] A convergence theorem for solutions of non-linear second order elliptic equations. Indiana University Math. J. **27**, 875–887 (1978).
[EV 2] Classical solutions of fully nonlinear, convex, second order elliptic equations. Comm. Pure Appl. Math. **25**, 333–363 (1982).
[EV 3] Classical solutions of the Hamilton-Jacobi Bellman equation for uniformly elliptic operators. Trans. Amer. Math. Soc. **275**, 245–255 (1983).
[EV 4] Some estimates for nondivergence structure, second order equations. Trans. Amer. Math. Soc. **287**, 701–712 (1985).

Evans, L. C., and A. Friedman
[EF] Optimal stochastic switching and the Dirichlet problem for the Bellman equation. Trans. Amer. Math. Soc. **253**, 365–389 (1979).

Evans, L. C., and P. L. Lions
[EL 1] Résolution des équations de Hamilton-Jacobi-Bellman pour des opérateurs uniformément elliptiques.C. R. Acad. Sci. Paris, **290**, 1049–1052 (1980).
[EL 2] Fully non-linear second order elliptic equations with large zeroth order coefficient. Ann. Inst. Fourier (Grenoble) **31**, fasc. 2, 175–191 (1981).

Fabes, E. B., C. E. Kenig, and R. P. Serapioni
[FKS] The local regularity of solutions of degenerate elliptic equations. Comm. Part. Diff. Equats. **7**, 77–116 (1982).

Fabes, E. B., and D. W. Stroock
[FS] The L^p-integrability of Green's functions and fundamental solutions for elliptic and parabolic equations. Duke Math. J. **51**, 997–1016 (1984).

Federer, H.
[FE] Geometric Measure Theory. Berlin–Heidelberg–New York: Springer-Verlag 1969.

Fefferman, C., and E. Stein
[FS] H^p spaces of several variables. Acta Math. **129**, 137–193 (1972).

Finn, R.
[FN 1] A property of minimal surfaces. Proc. Nat. Acad. Sci. U.S.A., **39**, 197–201 (1953).
[FN 2] On equations of minimal surface type. Ann. of Math. (2) **60**, 397–416 (1954).
[FN 3] New estimates for equations of minimal surface type. Arch. Rational Mech. Anal. **14**, 337–375 (1963).
[FN 4] Remarks relevant to minimal surfaces, and to surfaces of prescribed mean curvature. J. Analyse Math. **14**, 139–160 (1965).

Finn, R., and D. Gilbarg
[FG 1] Asymptotic behavior and uniqueness of plane subsonic flows. Comm. Pure Appl. Math. **10**, 23–63 (1957).
[FG 2] Three-dimensional subsonic flows, and asymptotic estimates for elliptic partial differential equations. Acta Math. **98**, 265–296 (1957).

Finn, R., and J. Serrin
[FS] On the Hölder continuity of quasi-conformal and elliptic mappings. Trans. Amer. Math. Soc. **89**, 1–15 (1958).

Fiorenza, R.
[FI 1] Sui problemi di derivata obliqua per le equazioni ellittiche. Ricerche Mat. **8**, 83–110 (1959).
[FI 2] Sui problemi di derivata obliqua per le equazioni ellittiche quasi lineari. Ricerche Mat. **15**, 74–108 (1966).

Fleming, W.
[FL] On the oriented Plateau problem. Rend. Circ. Mat. Palermo (2) **11**, 69–90 (1962).

Friedman, A.
[FR] Partial Differential Equations. New York: Holt, Rinehart and Winston 1969.

Friedrichs, K. O.
[FD 1] The identity of weak and strong extensions of differential operators. Trans. Amer. Math. Soc. **55**, 132–151 (1944).
[FD 2] On the differentiability of the solutions of linear elliptic differential equations. Comm. Pure Appl. Math. **6**, 299–326 (1953).

Gårding, L.
[GA] Dirichlet's problem for linear elliptic partial differential equations. Math. Scand. **1**, 55–72 (1953).

Gariepy, R., and W. P. Ziemer
[GZ 1] Behaviour at the boundary of solutions of quasilinear elliptic equations. Arch. Rational Mech. Anal. **56**, 372–384 (1974).
[GZ 2] A regularity condition at the boundary for solutions of quasilinear elliptic equations. Arch. Rational Mech. Anal. **67**, 25–39 (1977).

Gerhardt, C.
[GE 1] Existence, regularity, and boundary behaviour of generalized surfaces of prescribed mean curvature. Math. Z. **139**, 173–198 (1974).
[GE 2] On the regularity of solutions to variational problems in BV (Ω). Math. Z. **149**, 281–286 (1976).

[GE 3] Global regularity of the solutions to the capillarity problem. Ann. Scuola Norm. Sup. Pisa (4) **3**, 157–175 (1976).

[GE 4] Boundary value problems for surfaces of prescribed mean curvature. J. Math. Pures et Appl. **58**, 75–109 (1979).

Giaquinta, M.

[GI 1] On the Dirichlet problem for surfaces of prescribed mean curvature. Manuscripta Math. **12**, 73–86 (1974).

[GI 2] Multiple Integrals in the Calculus of Variations and Nonlinear Elliptic Systems, Annals of Math. Studies 105, Princeton Univ. Press, Princeton, 1983.

Gilbarg, D.

[GL 1] Some local properties of elliptic equations. In: Proceedings of Symposia in Pure Mathematics, Volume IV, pp. 127–141. Providence, R.I.: American Mathematical Society 1961.

[GL 2] Boundary value problems for nonlinear elliptic equations in *n* variables. In: Nonlinear Problems, pp. 151–159. Madison, Wis.: University of Wisconsin Press 1963.

Gilbarg, D., and L. Hörmander

[GH] Intermediate Schauder estimates. Arch. Rational Mech. Anal. **74**, 297–318 (1980).

Gilbarg, D., and J. Serrin

[GS] On isolated singularities of solutions of second order elliptic differential equations. J. Analyse Math. **4**, 309–340 (1955/56).

Giraud, G.

[GR 1] Sur le problème de Dirichlet généralisé (deuxième mémoire). Ann. Sci. École Norm. Sup. (3) **46**, 131–245 (1929).

[GR 2] Généralisation des problèmes sur les opérations du type elliptiques. Bull. Sci. Math. (2) **56**, 248–272, 281–312, 316–352 (1932).

[GR 3] Sur certains problèmes non linéaires de Neumann et sur certains problèmes non linéaires mixtes. Ann. Sci. École Norm. Sup. (3) **49**, 1–104, 245–309 (1932).

[GR 4] Équations à integrales principales d'ordre quelconque. Ann. Sci. Ecole Norm. Sup. **53** (36) 1–40 (1936).

Giusti, E.

[GT 1] Superfici cartesiane di area minima. Rend. Sem. Mat. Fis. Milano **40**, 3–21 (1970).

[GT 2] Boundary value problems for non-parametric surfaces of prescribed mean curvature. Ann. Scuola Norm. Sup. Pisa (4) **3**, 501–548 (1976).

[GT 3] On the equation of surfaces of prescribed mean curvature. Invent. Math. **46**, 111–137 (1978).

[GT 4] Equazioni ellittiche del secondo ordine. Bologna: Pitagora Editrice 1978.

[GT 5] Minimal Surfaces and Functions of Bounded Variation. Boston: Birkhäuser, 1984.

Greco, D.

[GC] Nuove formole integrali di maggiorazione per le soluzioni di un'equazione lineare di tipo ellittico ed applicazioni alla teoria del potenziale. Ricerche Mat. **5**, 126–149 (1956).

Günter, N. M.

[GU] Potential Theory and its Applications to Basic Problems of Mathematical Physics. New York: Frederick Ungar Publishing Co. 1967.

Gustin, N.

[GN] A bilinear integral identity for harmonic functions. Amer. J. Math. **70**, 2, 2–220 (1948).

Hardy, G. H., and J. E. Littlewood

[HL] Some properties of fractional integrals. I. Math. Z. **27**, 565–606 (1928).

Hardy, G. H., J. E. Littlewood, and G. Polya

[HLP] Inequalities. 2nd ed., Cambridge: Cambridge University Press 1952.

Hartman, P.

[HA] On the bounded slope condition. Pacific J. Math. **18**, 495–511 (1966).

Hartman, P., and L. Nirenberg

[HN] On spherical image maps whose Jacobians do not change sign. Amer. J. Math. **81**, 901–920 (1959).

Hartman, P., and G. Stampacchia

[HS] On some non-linear elliptic differential-functional equations. Acta Math. **115**, 271–310 (1966).

Heinz, E.

[HE 1] Über die Lösungen der Minimalflächengleichung. Nachr. Akad. Wiss. Göttingen. Math.-Phys. Kl. **IIa**, 51–56 (1952).

[HE 2] On elliptic Monge-Ampère equations and Weyl's imbedding problem. Analyse Math. **7**, 1–52 (1959).

[HE 3] Interior gradient estimates for surfaces $z = f(x, y)$ with prescribed mean curvature. J. Differential Geometry **5**, 149–157 (1971).

Helms, L. L.

[HL] Introduction to potential theory. New York: Wiley-Interscience 1969.

Hervé, R.-M.

[HR] Recherches axiomatiques sur la théorie des fonctions surharmoniques et du potentiel. Ann. Inst. Fourier (Grenoble) **12**, 415–571 (1962).

Hervé, R.-M., and M. Hervé

[HH] Les fonctions surharmoniques associées à un opérateur elliptique du second ordre à coefficients discontinus. Ann. Inst. Fourier (Grenoble) **19**, fasc. 1, 305–359 (1969).

Hilbert, D.

[HI] Über das Dirichletsche Prinzip. Jber. Deutsch. Math.-Verein. **8**, 184–188 (1900).

Hildebrandt, S.

[HD] Maximum principles for minimal surfaces and for surfaces of continuous mean curvature. Math. Z. **128**, 253–269 (1972).

Hopf, E.

[HO 1] Elementare Bemerkungen über die Lösungen partieller Differentialgleichungen zweiter Ordnung vom elliptischen Typus. Sitz. Ber. Preuss. Akad. Wissensch. Berlin, Math.-Phys. Kl. **19**, 147–152 (1927).

[HO 2] Zum analytischen Charakter der Lösungen regulärer zweidimensionaler Variationsprobleme. Math. Z. **30**, 404–413 (1929).

[HO 3] Über den funktionalen, insbesondere den analytischen Charakter der Lösungen elliptischer Differentialgleichungen zweiter Ordnung. Math. Z. **34**, 194–233 (1932).

[HO 4] On S. Bernstein's theorem on surfaces $z(x, y)$ of nonpositive curvature. Proc. Amer. Math. Soc. **1**, 80–85 (1950).

[HO 5] A remark on linear elliptic differential equations of second order. Proc. Amer. Math. Soc. **3**, 791–793 (1952).

[HO 6] On an inequality for minimal surfaces $z = z(x, y)$. J. Rational Mech. Anal. **2**, 519–522 (1953).

Hörmander, L.

[HM 1] Pseudo-differential operators and non-elliptic boundary problems. Ann. of Math. (2) **83**, 129–209 (1966).

[HM2] Hypoelliptic second order differential equations. Acta Math. **119**, 147–161 (1967).

[HM 3] The boundary problems of physical geodesy. Arch. Rational Mech. Anal. **62**, 1–52 (1976).

Ivanov, A. V.

[IV 1] Interior estimates of the first derivatives of the solutions of quasi-linear nonuniformly elliptic and nonuniformly parabolic equations of general form. Zap. Naučn. Sem. Leningrad. Otdel. Mat. Inst. Steklov. (LOMI) **14**, 24–47 (1969) [Russian]. English Translation in Sem. Math. V. A. Steklov Math. Inst. Leningrad **14**, 9–21 (1971).

[IV 2] The Dirichlet problem for second-order quasi-linear nonuniformly elliptic equations. Zap. Naučn. Sem. Leningrad. Otdel. Mat. Inst. Steklov. (LOMI) **19**, 79–94 (1970) [Russian]. English Translation in Sem. Math. V. A. Steklov Math. Inst. Leningrad **19**, 43–52 (1972).

[IV 3] A priori estimates for solutions of nonlinear second order elliptic equations. Zap. Naučn. Sem. Leningrad, Otdel. Mat. Inst. Steklov. (LOMI) **59**, 31–59 (1976) [Russian]. English translation in J. Soviet Math. **10**, 217–240 (1978).

[IV 4] A priori estimates of the second derivatives of solutions to nonlinear second order equations on the boundary of the domain. Zap. Naučn. Sem. Leningrad. Otdel. Mat. Inst. Steklov. (LOMI) **69**, 65–76 (1977) [Russian]. English translation in J. Soviet Math. **10**, 44–53 (1978).

Ivočkina, N. M.

[IC 1] The integral method of barrier functions and the Dirichlet problem for equations with operators of Monge-Ampère type. Mat. Sb. (N. S.) **112**, 193–206 (1980) [Russian]. English translation in Math. USSR Sb. **40**, 179–192 (1981).

[IC 2] An a priori estimate of $\|u\|^2_{C(\bar\Omega)}$ for convex solutions of Monge-Ampère equations, Zap. Nauchn. Sem. Leningr. Otdel. Mat. Inst. Steklova (LOMI) **96**, 69–79 (1980).

[IC 3] Classical solvability of the Dirichlet problem for the Monge-Ampère equation, Zap. Nauchn. Sem. Leningr. Otdel. Mat. Inst. Steklova (LOMI) **131**, 72–79 (1983).

Ivočkina, N. M., and A. P. Oskolkov
[IO] Nonlocal estimates for the first derivatives of solutions of the first boundary problem for certain classes of nonuniformly elliptic and nonuniformly parabolic equations and systems. Trudy Mat. Inst. Steklov **110**, 65–101 (1970) [Russian]. English Translation in Proc. Steklov Inst. Math. **110**, 72–115 (1970).

Jenkins, H.
[JE] On 2-dimensional variational problems in parametric form. Arch. Rational Mech. Anal. **8**, 181–206 (1961).

Jenkins, H., and J. Serrin
[JS 1] Variational problems of minimal surface type I. Arch. Rational Mech. Anal. **12**, 185–212 (1963).
[JS 2] The Dirichlet problem for the minimal surface equation in higher dimensions. J. Reine Angew. Math. **229**, 170–187 (1968).

John, F., and L. Nirenberg
[JN] On functions of bounded mean oscillation. Comm. Pure Appl. Math. **14**, 415–426 (1961).

Kamynin L. I., and B. N. Himčenko
[KH] Investigations on the maximum principle. Dokl. Akad. Nauk SSSR **240**, 774–777 (1978) [Russian]. English Translation in Soviet Math. Dokl. **19**, 677–681 (1978).

Kellogg, O. D.
[KE 1] On the derivatives of harmonic functions on the boundary. Trans. Amer. Math. Soc. **33**, 486–510 (1931).
[KE 2] Foundations of Potential Theory. New York: Dover 1954.

Kinderlehrer, D., and G. Stampacchia
[KST] An introduction to variational inequalities and their applications. New York: Academic Press (1980).

Kohn, J. J.
[KJ] Pseudo-differential operators and hypoellipticity. In: Proceedings of Symposia in Pure Mathematics, Volume XXIII, pp. 61–69. Providence, R. I.: American Mathematical Society (1973).

Kondrachov, V. I.
[KN] Sur certaines propriétés des fonctions dans l'espace L^p. C. R. (Doklady) Acad. Sci. URSS (N. S.) **48**, 535–538 (1945).

Korn, A.
[KR 1] Über Minimalflächen, deren Randkurven wenig von ebenen Kurven abweichen. Abhandl. Königl. Preuss. Akad. Wiss., Berlin 1909; Anhang, Abh. 2.
[KR 2] Zwei Anwendungen der Methode der sukzessiven Annäherungen. Schwarz Festschrift, Berlin 1914, 215–229.

Koševel, A. E.
[KO] On boundedness in L^p of derivatives of solutions of elliptic differential equations. Mat. Sb. (N. S.) **38** (80), 359–372 (1956) [Russian].

Krylov, N. V.
[KV 1] On the first boundary value problem for second order elliptic equations, Differents. Uravn. **3**, 315–325 (1967) [Russian]. English Translation in Differential Equations **3**, 158–164 (1967).
[KV 2] Controlled diffusion processes. Berlin–Heidelberg–New York: Springer-Verlag (1980).
[KV 3] On control of a diffusion process up to the time of first exit from a region. Izvestia Akad. Nauk. SSSR **45**, 1029–1048 (1981) [Russian]. English Translation in Math. USSR Izv. **19**, 297–313 (1982).
[KV 4] Boundedly inhomogeneous elliptic and parabolic equations. Izvestia Akad. Nauk. SSSR **46**, 487–523 (1982) [Russian]. English Translation in Math. USSR Izv. **20** (1983).
[KV 5] Boundedly inhomogeneous elliptic and parabolic equations in a domain. Izvestia Akad. Nauk. SSSR **47**, 75–108 (1983) [Russian]. English translation in Math. USSR Izv. **22**, 67–97 (1984).
[KV 6] On degenerate nonlinear elliptic equations. Mat. Sb. (N.S.) **120**, 311–330 (1983) [Russian]. English translation in Math. USSR Sbornik **48**, 307–326 (1984).

Krylov, N. V. and M. V. Safonov
[KS 1] An estimate of the probability that a diffusion process hits a set of positive measure. Dokl. Akad. Nauk. SSSR 245, 253–255 (1979) [Russian]. English translation in Soviet Math. Dokl. **20**, 253–255 (1979).
[KS 2] Certain properties of solutions of parabolic equations with measurable coefficients. Izvestia Akad. Nauk. SSSR **40**, 161–175 (1980) [Russian]. English translation in Math. USSR Izv. **161**, 151–164 (1981).

Ladyzhenskaya, O. A.
[LA] Solution of the first boundary problem in the large for quasi-linear parabolic equations. Trudy Moskov. Mat. Obšč. 7, 149–177 (1958) [Russian].

Ladyzhenskaya, O. A., and N. N. Ural'tseva
[LU 1] On the smoothness of weak solutions of quasilinear equations in several variables and of variational problems. Comm. Pure Appl. Math. **14**, 481–495 (1961).
[LU 2] Quasilinear elliptic equations and variational problems with several independent variables. Uspehi Mat. Nauk **16**, no. 1, 19–90 (1961) [Russian]. English Translation in Russian Math. Surveys **16**, no. 1, 17–92 (1961).
[LU 3] On Hölder continuity of solutions and their derivatives of linear and quasilinear elliptic and parabolic equations. Trudy Mat. Inst. Steklov. **73**, 172–220 (1964) [Russian]. English Translation in Amer. Math. Soc. Transl. (2) **61**, 207–269 (1967).
[LU 4] Linear and Quasilinear Elliptic Equations. Moscow: Izdat. "Nauka" 1964 [Russian]. English Translation: New York: Academic Press 1968. 2nd Russian ed. 1973.
[LU 5] Global estimates of the first derivatives of the solutions of quasi-linear elliptic and parabolic equations. Zap. Naučn. Sem. Leningrad. Otdel. Mat. Inst. Steklov. (LOMI) **14**, 127–155 (1969) [Russian]. English Translation in Sem. Math. V. A. Steklov Math. Inst. Leningrad **14**, 63–77 (1971).
[LU 6] Local estimates for gradients of solutions of non-uniformly elliptic and parabolic equations. Comm. Pure Appl. Math. **23**, 677–703 (1970).
[LU 7] Hölder estimates for solutions of second order quasilinear elliptic equations in general form. Zap. Nauchn. Sem. Leningr. Otdel. Mat. Inst. Steklova (LOMI) **96**, 161–168 (1980) [Russian]. English translation in J. Soviet Math. **21**, 762–768 (1983).
[LU 8] On Sounds for max $|u_x|$ for solutions of quasilinear elliptic and parabolic equations and existence theorems. Zap. Nauchn. Sem. Leningr. Otdel. Mat. Inst. Steklova (LOMI) **138**, 90–107 (1984).

Landis, E. M.
[LN 1] Harnack's inequality for second order elliptic equations of Cordes type. Dokl. Akad. Nauk SSSR **179**, 1272–1275 (1968) [Russian]. English Translation in Soviet Math. Dokl. **9**, 540–543 (1968).
[LN 2] Second Order Equations of Elliptic and Parabolic Types. Moscow: Nauka, 1971 [Russian].

Landkof, N. S.
[LK] Foundations of Modern Potential Theory. New York–Heidelberg–Berlin: Springer-Verlag (1972).

Lax, P. D.
[LX] On Cauchy's problem for hyperbolic equations and the differentiability of solutions of elliptic equations. Comm. Pure Appl. Maths. **8**, 615–633 (1955).

Lax, P. D., and A. M. Milgram
[LM] Parabolic equations. In: Contributions to the Theory of Partial Differential Equations, pp. 167–190. Princeton, N.J.: Princeton University Press 1954.

Lebesgue, H.
[LE] Sur le problème de Dirichlet. Rend. Circ. Mat. Palermo **24**, 371–402 (1907).

Leray, J.
[LR] Discussion d'un problème de Dirichlet. J. Math. Pures Appl. **18**, 249–284 (1939).

Leray, J., et J.-L. Lions
[LL] Quelques résultats de Višik sur les problèmes elliptiques non linéaires par les méthodes de Minty-Browder. Bull. Soc. Math. France **93**, 97–107 (1965).

Leray, J., et J. Schauder
[LS] Topologie et équations fonctionelles. Ann. Sci. École Norm. Sup. **51**, 45–78 (1934).

Lewy, H.
[LW 1] A priori limitations for solutions of Monge-Ampère equations I, II. Trans. Amer. Math. Soc. **37**, 417–434 (1935); **41**, 365–374 (1937).

Lewy, H., and G. Stampacchia
[LST] On existence and smoothness of solutions of some noncoercive variational inequalities. Arch. Rational Mech. Anal. **41**, 241–253 (1971).

Lichtenstein, L.
[LC] Neuere Entwicklung der Theorie partieller Differentialgleichungen zweiter Ordnung vom elliptischen Typus. In: Enc. d. Math. Wissensch. 2.3.2, 1277–1334 (1924). Leipzig: B. G. Teubner 1923–27.

Lieberman, G. M.
[LB 1] The quasilinear Dirichlet problem with decreased regularity at the boundary. Comm. Partial Differential Equations **6**, 437–497 (1981).
[LB 2] Solvability of quasilinear elliptic equations with nonlinear boundary conditions. Trans. Amer. Math. Soc. **273**, 753–765 (1982).
[LB 3] The nonlinear oblique derivative problem for quasilinear elliptic equations. Journal of Nonlinear Analysis **8**, 49–65 (1984).
[LB 4] The conormal derivative problem for elliptic equations of variational type. J. Differential Equations **49**, 218–257 (1983).

Lieberman, G. M., and N. S. Trudinger
[LT] Nonlinear oblique boundary value problems for nonlinear elliptic equations. Trans. Amer. Math. Soc. **296**, 509–546 (1986).

Lions, P.-L.
[LP 1] Problèmes elliptiques du 2ème ordre non sous forme divergence. Proc. Roy. Soc. Edinburgh **84A**, 263–271 (1979).
[LP 2] A remark on some elliptic second order problems. Boll. Un. Mat. Ital. (5) **17A**, 267–270 (1980).
[LP 3] Résolution de problèmes elliptiques quasilineaires. Arch. Rational Mech. Anal. **74**, 335–353 (1980).
[LP 4] Résolution analytique des problèmes de Bellman-Dirichlet. Acta Math., **146**, 151–166 (1981).
[LP 5] Generalized solutions of Hamilton-Jacobi equations. London: Pitman (1982).
[LP 6] Sur les équations de Monge-Ampère 1. Manuscripta Math. **41**, 1–44 (1983).
[LP 7] Sur les équations de Monge-Ampère 2. Arch. Rational Mech. Anal. (1984).
[LP 8] Optimal control of diffusion processes and Hamilton-Jacobi Bellman equations I, II. Comm. Part. Diff. Equats. **8**, 1101–1174, 1229–1276 (1983).
[LP 9] Optimal control of diffusion processes and Hamilton-Jacobi Bellman equations III. In: Collège de France Seminar IV, London: Pitman (1984).
[LP 10] Hamilton-Jacobi-Bellman equations and the optimal control of stochastic systems. Proc. Int. Cong. Math. Warsaw, 1983.

Lions, P.-L., and J. L. Menaldi
[LD] Problèmes de Bellman avec les contrôles dans les coefficients de plus haut degré. C.R. Acad. Sci. Paris **287**, 409–412 (1978).

Lions, P.-L., and N.S. Trudinger
[LPT] Linear oblique derivative problems for the uniformly elliptic Hamilton-Jacobi Bellman equation. Math. Zeit. **191**, 1–15 (1986).

Lions P.-L., N. S. Trudinger and J. I. E. Urbas
[LTU] The Neumann problem for equations of Monge-Ampère type. Comm. Pure Appl. Math. **39**, 539–563 (1986).

Littman, W., G. Stampacchia, and H. F. Weinberger
[LSW] Regular points for elliptic equations with discontinuous coefficients. Ann. Scuola Norm. Sup. Pisa (3) **17**, 43–77 (1963).

Mamedov, I. T.
[MV 1] On an a priori estimate of the Hölder norm of solutions of quasilinear parabolic equations with discontinuous coefficients. Dokl. Akad. Nauk. SSSR **252**, (1980) [Russian]. English translation in Soviet Math. Dokl. **21**, 872–875 (1980).
[MV 2] Behavior near the boundary of solutions of degenerate second order elliptic equations. Mat. Zametki **30**, 343–352 (1981) [Russian]. English Translation in Math. Notes **30**, 661–666 (1981).

Marcinkiewicz, J.
[MA] Sur l'interpolation d'opérations. C. R. Acad. Sci. Paris **208**, 1272–1273 (1939).

Maz'ya, V. G.
[MZ] Behavior near the boundary of the solution of the Dirichlet problem for a second order elliptic equation in divergence form. Mat. Zametki **2**, 209–220 (1967) [Russian]. English Translation in Math. Notes **2**, 610–619 (1967).

Melin, A., and J. Sjöstrand
[MSJ] Fourier integral operators with complex phase functions and parametrics for an interior boundary value problem. Comm. Part. Diff. Equats. **1**, 313–400 (1976).

Meyers, N. G.
[ME 1] An L^p-estimate for the gradient of solutions of second order elliptic divergence equations. Ann. Scuola Norm. Sup. Pisa (3) **17**, 189–206 (1963).
[ME 2] An example of non-uniqueness in the theory of quasilinear elliptic equations of second order. Arch. Rational Mech. Anal. **14**, 177–179 (1963).

Meyers, N. G., and J. Serrin
[MS 1] The exterior Dirichlet problem for second order elliptic partial differential equations. J. Math. Mech. **9**, 513–538 (1960).
[MS 2] $H = W$. Proc. Nat. Acad. Sci. U.S.A. **51**, 1055–1056 (1964).

Michael, J. H.
[MI 1] A general theory for linear elliptic partial differential equations. J. Differential Equations **23**, 1–29 (1977).
[MI 2] Barriers for uniformly elliptic equations and the exterior cone condition. J. Math. Anal. and Appl. **79**, 203–217 (1981).

Michael, J. H., and L. M. Simon
[MSI] Sobolev and mean-value inequalities on generalized submanifolds of $\mathbf{R}^n$. Comm. Pure Appl. Math. **26**, 361–379 (1973).

Miller, K.
[ML 1] Barriers on cones for uniformly elliptic operators. Ann. Mat. Pura Appl. (4) **76**, 93–105 (1967).
[ML 2] Exceptional boundary points for the nondivergence equation which are regular for the Laplace equation and vice-versa. Ann. Scuola Norm. Sup. Pisa (3) **22**, 315–330 (1968).
[ML 3] Extremal barriers on cones with Phragmèn-Lindelöf theorems and other applications. Ann. Mat. Pura Appl. (4) **90**, 297–329 (1971).
[ML 4] Nonequivalence of regular boundary points for the Laplace and nondivergence equations, even with continuous coefficients. Ann. Scuola Norm. Sup. Pisa (3) **24**, 159–163 (1970).

Miranda, C.
[MR 1] Sul problema misto per le equazioni lineari ellittiche. Ann. Mat. Pura Appl. (4) **39**, 279–303 (1955).
[MR 2] Partial Differential Equations of Elliptic Type. 2nd ed., Berlin-Heidelberg-New York: Springer-Verlag 1970.

Miranda, M.
[MD 1] Diseguaglianze di Sobolev sulle ipersuperfici minimali. Rend. Sem. Math. Univ. Padova **38**, 69–79 (1967).
[MD 2] Una maggiorazione integrale per le curvature delle ipersuperfici minimale. Rend. Sem. Mat. Univ. Padova **38**, 91–107 (1967).
[MD 3] Comportamento delle successioni convergenti di frontiere minimali. Rend. Sem. Mat. Univ. Padova **38**, 238–257 (1967).
[MD 4] Un principio di massimo forte per le frontiere minimali e una sua applicazione alla risoluzione del problema al contorno per l'equazione delle superfici di area minima. Rend. Sem. Mat. Univ. Padova **45**, 355–366 (1971).
[MD 5] Dirichlet problem with L^1 data for the non-homogeneous minimal surface equation. Indiana Univ. Math. J. **24**, 227–241 (1974).

Morrey, C. B., Jr.
[MY 1] On the solutions of quasi-linear elliptic partial differential equations. Trans. Amer. Math. Soc. **43**, 126–166 (1938).
[MY 2] Functions of several variables and absolute continuity, II. Duke Math. J. **6**, 187–215 (1940).

[MY 3] Multiple integral problems in the calculus of variations and related topics. Univ. California Publ. Math. (N. S.) **1**, 1–130 (1943).
[MY 4] Second order elliptic equations in several variables and Hölder continuity. Math. Z. **72**, 146–164 (1959).
[MY 5] Multiple Integrals in the Calculus of Variations. Berlin-Heidelberg-New York: Springer-Verlag 1966.

Moser, J.
[MJ 1] A new proof of de Giorgi's theorem concerning the regularity problem for elliptic differential equations. Comm. Pure Appl. Math. **13**, 457–468 (1960).
[MJ 2] On Harnack's theorem for elliptic differential equations. Comm. Pure Appl. Math. **14**, 577–591 (1961).

Motzkin, T., and W. Wasow
[MW] On the approximations of linear elliptic differential equations by difference equations with positive coefficients. J. Math. and Phys. **31**, 253–259 (1952).

Nadirašvili, N. S.
[ND] A lemma on the inner derivative, and the uniqueness of the solution of the second boundary value problem for second order elliptic equations. Dokl. Akad. Nauk SSSR **261**, 804–808 (1981) [Russian]. English Translation in Soviet Math. Dokl. **24**, 598–601 (1981).

Nash, J.
[NA] Continuity of solutions of parabolic and elliptic equations. Amer. J. Math. **80**, 931–954 (1958).

Nečas, J.
[NE] Les méthodes directes en théorie des équations elliptiques. Prague: Éditeurs Academia 1967.

Neumann, J. von
[NU] Über einen Hilfssatz der Variationsrechnung. Abh. Math. Sem. Univ. Hamburg **8**, 28–31 (1931).

Nirenberg, L.
[NI 1] On nonlinear elliptic partial differential equations and Hölder continuity. Comm. Pure Appl. Math. **6**, 103–156 (1953).
[NI 2] Remarks on strongly elliptic partial differential equations. Comm. Pure Appl. Math. **8**, 649–675 (1955).
[NI 3] On elliptic partial differential equations. Ann. Scuola Norm. Sup. Pisa (3) **13**, 115–162 (1959).
[NI 4] Elementary remarks on surfaces with curvature of fixed sign. In: Proceedings of Symposia in Pure Mathematics, Volume III, pp. 181–185. Providence, R.I.: American Mathematical Society 1961.

Nitsche, J. C. C.
[NT] Vorlesungen über Minimalflächen. Berlin-Heidelberg-New York: Springer-Verlag 1975.

Novrusov, A. A.
[NO 1] Regularity of boundary points for elliptic equations with continuous coefficients. Vestn. Mosk. Univ. Ser. Mat. Mekh., no. 6, 18–25 (1971) [Russian].
[NO 2] The modulus of continuity of the solution of the Dirichlet problem at a regular boundary point. Mat. Zametki **12**, 67–72 (1972) [Russian]. English translation in Math. Notes 12, no. 1, 472–475 (1973).
[NO 3] Estimates of the Hölder norm of solutions to quasilinear elliptic equations with discontinuous coefficients. Dokl. Akad. Nauk. SSSR **253**, 31–33 (1980) [Russian]. English translation in Soviet Math. Dokl. **22**, 25–28 (1980).
[NO 4] Regularity of the boundary points for second order degenerate linear elliptic equations. Mat. Zametki **30**, 353–362 (1981) [Russian]. English Translation in Math. Notes **30**, 666–671 (1982).

Oddson, J. K.
[OD] On the boundary point principle for elliptic equations in the plane. Bull. Amer. Math. Soc. **74**, 666–670 (1968).

Oleinik, O. A.
[OL] On properties of solutions of certain boundary problems for equations of elliptic type. Mat. Sb. (N. S.) **30**, 695–702 (1952) [Russian].

Oleinik, O. A., and Radkevich, E. V.
[OR] Second order equations with non-negative characteristic form. (Mathematical Analysis 1969). Moscow: Itogi Nauki 1971 [Russian]. English Translation: Providence, R.I., Amer. Math. Soc. 1973.

Oskolkov, A. P.
[OS 1] On the solution of a boundary value problem for linear elliptic equations in an unbounded domain. Vestnik Leningrad. Univ. **16**, no. 7, 38–50 (1961) [Russian].
[OS 2] A priori estimates of first derivatives of solutions of the Dirichlet problem for nonuniformly elliptic differential equations. Trudy Mat. Inst. Steklov **102**, 105–127 (1967) [Russian]. English Translation in Proc. Steklov Inst. Math. **102**, 119–144 (1967).
Osserman, R.
[OM 1] On the Gauss curvature of minimal surfaces. Trans. Amer. Math. Soc. **96**, 115–128 (1960).
[OM 2] A Survey of Minimal Surfaces. New York: Van Nostrand 1969.
Panejah, B. P.
[PN] On the theory of solvability of a problem with oblique derivative. Mat. Sb. (N. S.) **114** (156), 226–268 (1981) [Russian]. English Translation in Math. USSR Sb. **42**, 197–235 (1981).
Pascali, D.
[PA] Operatori Neliniari. Bucharest: Editura Academiei Republicii Socialiste România 1974.
Perron, O.
[PE] Eine neue Behandlung der Randwertaufgabe für $\Delta u = 0$. Math. Z. **18**, 42–54 (1923).
Piccinini, L. C., and S. Spagnola
[PS] On the Hölder continuity of solutions of second order elliptic equations in two variables. Ann. Scuola Norm. Sup. Pisa (3) **26**, 391–402 (1972).
Pogorelov, A. V.
[PG 1] Monge-Ampère equations of elliptic type. Groningen: Noordhoff (1964)
[PG 2] On a regular solution of the n-dimensional Minowski problem. Dokl. Akad. Nauk SSSR **119**, 785–788 (1971) [Russian]. English Translation in Soviet Math. Dokl. **12**, 1192–1196 (1971).
[PG 3] On the regularity of generalized solutions of the equation $\det(\partial^2 u/\partial x_i \partial x_j) = \phi(x_1, \ldots, x_n) > 0$. Dokl. Akad. Nauk SSSR **200**, 543–547 (1971) [Russian]. English Translation in Soviet Math. Dokl. **12**. 1436–1440 (1971).
[PG 4] The Dirichlet problem for the n-dimensional analogue of the Monge-Ampère equation. Dokl. Akad. Nauk SSSR **201**, 790–793 (1971) [Russian]. English Translation in Soviet Math. Dokl. **12**, 1727–1731 (1971).
[PG 5] The Minkowski multidimensional problem. New York: J. Wiley (1978).
Protter, M. H., and H. F. Weinberger
[PW] Maximum Principles in Differential Equations. Englewood Cliffs, N.J.: Prentice-Hall 1967.
Pucci, C.
[PU 1] Su una limitazione per soluzioni di equazioni ellittiche. Boll. Un. Mat. Ital. (3) **21**, 228–233 (1966).
[PU 2] Operatori ellittici estremanti. Ann. Mat. Pura Appl. (4) **72**, 141–170 (1966).
[PU 3] Limitazioni per soluzioni di equazioni ellittiche. Ann. Mat. Pura Appl. (4) **74**, 15–30 (1966).
Radó, T.
[RA] Geometrische Betrachtungen über zweidimensionale reguläre Variationsprobleme. Acta Litt. Sci. Univ. Szeged **2**, 228–253 (1924–26).
Rellich, R.
[RE] Ein Satz über mittlere Konvergenz. Nachr. Akad. Wiss. Göttingen Math.-Phys. Kl., 30–35 (1930).
Riesz, M.
[RZ] Sur les fonctions conjuguées. Math. Z. **27**, 218–244 (1927).
Rodemich, E.
[RO] The Sobolev inequalities with best possible constants. In: Analysis Seminar at California Institute of Technology, 1966.
Royden, H. L.
[RY] Real Analysis. 2nd ed., London: Macmillan 1970.
Safonov, M. V.
[SF] Harnack inequalities for elliptic equations and Hölder continuity of their solutions. Zap. Naučn. Sem. Leningrad. Otdel. Mat. Inst. Steklov. (LOMI) **96**, 272–287 (1980) [Russian]. English translation in J. Soviet Math. **21**, 851–863 (1983).

Schaefer, H.
[SH] Über die Methode der a priori Schranken. Math. Ann. **129**, 415–416 (1955).

Schauder, J.
[SC 1] Der Fixpunktsatz in Funktionalräumen. Studia Math. **2**, 171–180 (1930).
[SC 2] Über den Zusammenhang zwischen der Eindeutigkeit und Lösbarkeit partieller Differentialgleichungen zweiter Ordnung vom elliptischen Typus. Math. Ann. **106**, 661–721 (1932).
[SC 3] Über das Dirichletsche Problem im Großen für nicht-lineare elliptische Differentialgleichungen. Math. Z. **37**, 623–634, 768 (1933).
[SC 4] Über lineare elliptische Differentialgleichungen zweiter Ordnung. Math. Z. **38**, 257–282 (1934).
[SC 5] Numerische Abschätzungen in elliptischen linearen Differentialgleichungen. Studia Math. **5**, 34–42 (1935).

Schulz, F.
[SZ 1] Über die Beschränktheit der zweiten Ableitungen der Lösungen nichtlinearer elliptischer Differentialgleichungen. Math. Z. **175**, 181–188 (1980).
[SZ 2] A remark on fully nonlinear, concave elliptic equations. Proc. Centre for Math. Anal. Aust. Nat. Univ. **8**, 202–207 (1984).

Serrin, J.
[SE 1] On the Harnack inequality for linear elliptic equations. J. Analyse Math. **4**, 292–308 (1955/56).
[SE 2] Local behavior of solutions of quasi-linear elliptic equations. Acta Math. **111**, 247–302 (1964).
[SE 3] The problem of Dirichlet for quasilinear elliptic differential equations with many independent variables. Philos. Trans. Roy. Soc. London Ser. A **264**, 413–496 (1969).
[SE 4] Gradient estimates for solutions of nonlinear elliptic and parabolic equations. In: Contributions to Nonlinear Functional Analysis, pp. 565–601. New York: Academic Press 1971.
[SE 5] Nonlinear elliptic equations of second order. Lecture notes (unpublished), Symposium on Partial Differential Equations at Berkeley, 1971.

Simon, L. M.
[SI 1] Global estimates of Hölder continuity for a class of divergence-form elliptic equations. Arch. Rational Mech. Anal. **56**, 253–272 (1974).
[SI 2] Boundary regularity for solutions of the non-parametric least area problem. Ann. of Math. **103**, 429–455 (1976).
[SI 3] Interior gradient bounds for non-uniformly elliptic equations. Indiana Univ. Math. J. **25**, 821–855 (1976).
[SI 4] Equations of mean curvature type in 2 independent variables. Pacific J. Math. **69**, 245–268 (1977).
[SI 5] Remarks on curvature estimates for minimal hypersurfaces. Duke Math. J. **43**, 545–553 (1976).
[SI 6] A Hölder estimate for quasiconformal mappings between surfaces in Euclidean space, with application to graphs having quasiconformal Gauss map. Acta Math. (1977).
[SI 7] Survey lectures on minimal submanifolds. In: Seminar on Minimal Submanifolds. Annals of Math. Studies **103**, Princeton (1983).
[SI 8] Lectures on geometric measure theory. Proc. Centre for Mathematical Analysis. Australian Nat. Univ. **3**(1983).

Simons, J.
[SM] Minimal varieties in riemannian manifolds. Ann. of Math. (2) **88**, 62–105 (1968).

Simon, L. M., and J. Spruck
[SS] Existence and regularity of a capillary surface with prescribed contact angle. Arch. Rational Mech. Anal. **61**, 19–34 (1976).

Sjöstrand, J.
[SJ] Operators of principal type with interior boundary conditions. Acta Math. **130**, 1–51 (1973).

Slobodeckii, L. M.
[SL] Estimate in L_p of solutions of elliptic systems. Dokl. Akad. Nauk SSSR **123**, 616–619 (1958) [Russian].

Sobolev, S. L.
[SO 1] On a theorem of functional analysis. Mat. Sb. (N.S.) **4** (46), 471–497 (1938) [Russian]. English Translation in Amer. Math. Soc. Transl. (2) **34**, 39–68 (1963).

[SO 2] Applications of Functional Analysis in Mathematical Physics. Leningrad: Izdat. Leningrad. Gos. Univ. 1950 [Russian]. English Translation: Translations of Mathematical Monographs, Vol. 7. Providence, R.I.: American Mathematical Society 1963.

Spruck, J.

[SP] Gauss curvature estimates for surfaces of constant mean curvature. Comm. Pure Appl. Math. **27**, 547–557 (1974).

Stampacchia, G.

[ST 1] Contributi alla regolarizzazione delle soluzioni dei problemi al contorno per equazioni del secondo ordine ellittiche. Ann. Scuola Norm. Sup. Pisa (3) **12**, 223–245 (1958).

[ST 2] I problemi al contorno per le equazioni differenziali di tipo ellittico. In: Atti VI Congr. Un. Mat. Ital. (Naples, 1959), pp. 21–44. Rome: Edizioni Cremonese 1960.

[ST 3] Problemi al contorno ellittici, con dati discontinui, dotati di soluzioni hölderiane. Ann. Mat. Pura Appl. (4) **51**, 1–37 (1960).

[ST 4] Le problème de Dirichlet pour les équations elliptiques du second ordre à coefficients discontinues. Ann. Inst. Fourier (Grenoble) **15**, fasc. 1, 189–258 (1965).

[ST 5] Équations elliptiques du second ordre à coefficients discontinues. Séminaire de Mathématiques Supérieures, No. 16 (Été, 1965). Montreal, Que.: Les Presses de l'Université de Montréal 1966.

Stein, E.

[SN] Singular integrals and differentiability properties of functions. Princeton: Princeton University Press (1970.

Sternberg, S.

[SB] Lectures on Differential Geometry. Englewood Cliffs, N.J.: Prentice-Hall 1964.

Talenti, G.

[TA 1] Equazione lineari ellittiche in due variabili. Le Matematiche **21**, 339–376 (1966).

[TA 2] Best constant in Sobolev inequality. Ann. Mat. Pura Appl. **110**, 353–372 (1976).

[TA 3] Elliptic equations and rearrangements. Ann. Scuola Norm. Sup. Pisa (4) **3**, 697–718 (1976).

[TA 4] Some estimates of solutions to Monge-Ampère type equations in dimension two. Ann. Scuola Norm. Sup. Pisa (4) **8**, 183–230 (1981).

Temam, R.

[TE] Solutions généralisées de certaines équations du type hypersurface minima. Arch. Rational Mech. Anal. **44**, 121–156 (1971).

Trudinger, N. S.

[TR 1] On Harnack type inequalities and their application to quasilinear elliptic equations. Comm. Pure Appl. Math. **20**, 721–747 (1967).

[TR 2] On imbeddings into Orlicz spaces and some applications. J. Math. Mech. **17**, 473–483 (1967).

[TR 3] Some existence theorems for quasi-linear, non-uniformly elliptic equations in divergence form. J. Math. Mech. **18**, 909–919 (1968/69).

[TR 4] On the regularity of generalized solutions of linear, non-uniformly elliptic equations. Arch. Rational Mech. Anal. **42**, 50–62 (1971).

[TR 5] The boundary gradient estimate for quasilinear elliptic and parabolic differential equations. Indiana Univ. Math. J. **21**, 657–670 (1972).

[TR 6] A new proof of the interior gradient bound for the minimal surface equation in n dimensions. Proc. Nat. Acad. Sci. U.S.A. **69**, 821–823 (1972).

[TR 7] Linear elliptic operators with measurable coefficients. Ann. Scuola Norm. Sup. Pisa (3) **27**, 265–308 (1973).

[TR 8] Gradient estimates and mean curvature. Math. Z. **131**, 165–175 (1973).

[TR 9] A sharp inequality for subharmonic functions on two-dimensional manifolds. Math. Z. **133**, 75–79 (1973).

[TR 10] On the comparison principle for quasilinear divergence structure equations. Arch. Rational Mech. Anal. **57**, 128–133 (1974).

[TR 11] Maximum principles for linear, non-uniformly elliptic operators with measurable coefficients. Math. Z. **156**, 291–301 (1977).

[TR 12] Local estimates for subsolutions and supersolutions of general second order elliptic quasilinear equations. Invent. Math. **61**, 67–79 (1980).

[TR 13] Fully nonlinear, uniformly elliptic equations under natural structure conditions. Trans. Amer. Math. Soc. **28**, 217–231 (1983).

[TR 14] Boundary value problems for fully nonlinear elliptic equations. Proc. Centre for Math. Anal. Aust. Nat. Univ. **8**, 65–83 (1984).

[TR 15] On an interpolation inequality and its application to nonlinear elliptic equations. Proc. Amer. Math. Soc. **95**, 73–78 (1985).

Trudinger, N. S., and J. Urbas

[TU] On the Dirichlet problem for the prescribed Gauss curvature equation, Bull. Australian Math. Soc. **278**, 751–770 (1983).

Ural'tseva, N.

[UR] The solvability of the capillarity problem. Vestnik Leningrad. Univ. no. **19**, 54–64 (1973) [Russian]. English Translation in Vestnik Leningrad Univ. *6*, 363–375 (1979).

Wahl, W. von

[WA] Über quasilineare elliptische Differentialgleichungen in der Ebene. Manuscripta Math. **8**, 59–67 (1973).

Weinberger, H. F.

[WE] Symmetrization in uniformly elliptic problems. In: Studies in Mathematical Analysis and Related Topics, pp. 424–428. Stanford, Calif.: Stanford University Press 1962.

Widman, K.-O.

[WI 1] Inequalities for the Green function and boundary continuity of the gradient of solutions of elliptic differential equations. Math. Scand. **21**, 17–37 (1967).

[WI 2] A quantitative form of the maximum principle for elliptic partial differential equations with coefficients in L_∞. Comm. Pure Appl. Math. **21**, 507–513 (1968).

[WI 3] On the Hölder continuity of solutions of elliptic partial differential equations in two variables with coefficients in L_∞. Comm. Pure Appl. Math. **22**, 669–682 (1969).

Wiener, N.

[WN] The Dirichlet problem. J. Math. and Phys. **3**, 127–146 (1924).

Williams, G. H.

[WL] Existence of solutions for nonlinear obstacle problems. Math. Z. **154**, 51–65 (1977).

Winzell, B.

[WZ 1] The oblique derivative problem, I. Math. Ann. **229**, 267–278 (1977).

[WZ 2] The oblique derivative problem, II. Ark. Mat. **17**, 107–122 (1979).

Yau, S. T.

[YA] On the Ricci curvature of a compact Kähler manifold and the complex Monge-Ampère equation. Comm. Pure Appl. Math. **31**, 339–411 (1978).

Yosida, K.

[YO] Functional Analysis. 4th ed., Berlin-Heidelberg-New York: Springer-Verlag 1974.

Epilogue

This book has been devoted to the theory of elliptic, second order, partial differential equations with emphasis on the Dirichlet problem for *linear* and *quasilinear* equations. Its second edition in 1983 included an introductory chapter on *fully nonlinear* elliptic equations as the Krylov-Safonov Hölder estimates had recently opened up the higher dimensional theory. This was analogous to the role of the De Giorgi-Nash Hölder estimates in the higher dimensional quasilinear theory about a quarter century earlier. It should not be surprising that the fully nonlinear theory, with its rich applications to stochastic optimization and geometry, has blossomed since our second edition appeared.

We comment *brlefly* on some of the main developments.

Viscosity solutions. The notion of *viscosity* solution, introduced for first order equations by Crandall and Lions, ([LP 5], [CL], [CIL]), was extended to second order equations, with dramatic consequences in the wake of a breakthrough by Jensen [JEN] enabling approximation by semi-convex or semi-concave functions. The concept of *viscosity* subsolution relates to that of subharmonic function, as introduced in Sections 2.8 and 6.3. Using the terminology of Chapter 17, if $F \in C^0(\Gamma)$ is monotone increasing with respect to $r \in R^{n\times n}$, we call a function $u \in C^0(\Omega)$ a *viscosity subsolution* (*supersolution*) of equation (17.1) in Ω, if for every point $y \in \Omega$ and function $v \in C^2(\Omega)$ satisfying $u \le v \, (\ge v)$ in Ω and $u(y) = v(y)$, we have $F[v](y) \ge 0$, (≤ 0). It is readily seen that for linear elliptic equations, $Lu = f$, this notion coincides with that of Section 6.3. Moreover, Ishii [IS] showed that the Perron process could be used to infer existence of viscosity solutions of the Dirichlet problem, with the aid of comparison principles extending those of Jensen [JEN]. Various aspects of this theory and its widespread applications are described in the expository works [CIL], [FLS].

Uniformly elliptic equations. The second derivative Hölder estimates in Sections 17.4 and 17.8 were improved by Safonov [SE 2], [SF 4] and Caffarelli [CAF] by perturbation arguments from the special case (17.32). As a by-product, simpler proofs of the Schauder estimates for linear equations were obtained by various authors, including in particular an "L^∞-Campanato" method [SF 4], [KV 9]. Caffarelli also deduced L^p-estimates for second derivatives for $p > n$; (see [CC]). The basic theory is also covered in [KV 7], [TR 16].

Non-uniformly elliptic equations. The Monge-Ampère and Gauss curvature equations are special cases of *Hessian* and *curvature* equations determined by the elementary symmetric functions. The classical Dirichlet problem is treated in the works [CNS 2, 3], [IC 4], [KV 7, 8], [TR 17].

Quasilinear equations. The special treatment of the two dimensional case stems from Morrey's gradient estimate, Theorem 12.4. By showing that the exponent in the Hölder estimate, Corollary 9.24, can be arbitrarily small, Safonov [SF 3], confirmed that this approach is not extendible to higher dimensions.

Finally, we note that Korevaar [KOR] showed that the interior gradient bound for the minimal surface and prescribed mean curvature equations can be deduced from the maximum principle along the lines of Section 15.3. The resultant bound is not as precise as that in Theorem 16.5.

Bibliography

[CC] Cabré, X., and L. Caffarelli, Fully nonlinear elliptic equations. Amer. Math. Soc. Colloquium Publications **43** (1995).

[CAF] Caffarelli, L., Interior a priori estimates for solutions of fully non-linear equations. Ann. Math. **130**, 189–213 (1989).

[CNS 2] Caffarelli, L., L. Nirenberg and J. Spruck, The Dirichlet problem for nonlinear second order equations III, Functions of the eigenvalues of the Hessian. Acta Math. **155**, 261–301 (1985).

[CNS 3] Caffarelli, L., L. Nirenberg and J. Spruck, The Dirichlet problem for nonlinear second order equations V, The Dirichlet problem for Weingarten surfaces. Comm. Pure Appl. Math. **41**, 47–70 (1988).

[CIL] Crandall, M. G., H. Ishii and P.-L. Lions, User's guide to viscosity solutions of second order partial differential equations. Bull. Amer. Math. Soc. **27**, 1–67 (1992).

[CL] Crandall, M. G., and P-L. Lions, Viscosity solutions of Hamilton-Jacobi equations. Trans. Amer. Math. Soc. **277**, 1–42 (1983).

[FLS] Fleming, W. H., and H. M. Soner, Controlled Markov processes and viscosity solutions. New York: Springer 1993.

[IS] Ishii, H., On uniqueness and existence of viscosity solutions of fully nonlinear second order PDE's. Comm. Pure Appl. Math. **42**, 14–45 (1989).

[IC 4] Ivočkina, N. M., The Dirichlet problem for the curvature equation of order m, Algebra; Analiz **2**, 192–217 (1990) [Russian]. English translation: Leningrad Math. J. **2**, 631–654 (1991).

[JEN] Jensen, R., The maximum principle for viscosity solutions of fully nonlinear second order partial differential equations. Arch. Rat. Mech. Anal. **101**, 1–27 (1988).

[KOR] Korevaar, N., An easy proof of the interior gradient bound for solutions to the prescribed mean curvature equation, In: Nonlinear Functional Analysis and its Applications. Proc. Symp. Pure Math. **45**, (2), 81–90. Providence: Amer. Math. Soc. 1986.

[KV 7] Krylov, N. V., Nonlinear elliptic and parabolic equations of the second order. Moscow: Nauka 1985 [Russian]. English translation: Dordrecht: Reidel 1987.

[KV 8] Krylov, N. V., On the general notion of fully nonlinear second-order elliptic equations. Trans. Amer. Math. Soc. **347**, 857–895 (1995).

[KV 9] Krylov, N. V., Lectures on elliptic and parabolic equations in Hölder spaces. Providence: Amer. Math. Soc. 1996.

[SF 2] Safonov, M. V., On the classical solution of Bellman's elliptic equation. Dokl. Akad. Nauk SSSR **278**, 810–813 (1984) [Russian]. English translation: Soviet Math. Dokl. **30**, 482–485 (1984).

[SF 3] Safonov, M. V., Unimprovability of estimates of Hölder constants for solutions of linear elliptic equations with measurable coefficients. Mat. Sb. **132**, 275–288 (1987) [Russian]. English translation: Math. USSR Sbornik **60**, 269–281 (1988).

[SF 4] Safonov, M. V., Nonlinear elliptic equations of second order. Lecture Notes, Univ. Firenze 1991.

[TR 16] Trudinger, N. S., Lectures on nonlinear elliptic equations of the second order. Lectures in Mathematical Sciences, Univ. of Tokyo 1995.

[TR 17] Trudinger, N. S., On the Dirichlet problem for Hessian equations. Acta Math. **175**, 151–164 (1995).

Subject Index

Notation Index

Spaces and their norms and seminorms

Other notation (see also page 9)

Printing: Druckhaus Beltz, Hemsbach
Binding: Buchbinderei Schäffer, Grünstadt

Printing: Druckhaus Beltz, Hemsbach
Binding: Buchbinderei Schäffer, Grünstadt

Grundlehren der mathematischen Wissenschaften

A Series of Comprehensive Studies in Mathematics

A Selection

210. Gihman/Skorohod: The Theory of Stochastic Processes I
211. Comfort/Negrepontis: The Theory of Ultrafilters
212. Switzer: Algebraic Topology – Homotopy and Homology
215. Schaefer: Banach Lattices and Positive Operators
217. Stenström: Rings of Quotients
218. Gihman/Skorohod: The Theory of Stochastic Processes II
219. Duvaut/Lions: Inequalities in Mechanics and Physics
220. Kirillov: Elements of the Theory of Representations
221. Mumford: Algebraic Geometry I: Complex Projective Varieties
222. Lang: Introduction to Modular Forms
223. Bergh/Löfström: Interpolation Spaces. An Introduction
224. Gilbarg/Trudinger: Elliptic Partial Differential Equations of Second Order
225. Schütte: Proof Theory
226. Karoubi: K-Theory. An Introduction
227. Grauert/Remmert: Theorie der Steinschen Räume
228. Segal/Kunze: Integrals and Operators
229. Hasse: Number Theory
230. Klingenberg: Lectures on Closed Geodesics
231. Lang: Elliptic Curves. Diophantine Analysis
232. Gihman/Skorohod: The Theory of Stochastic Processes III
233. Stroock/Varadhan: Multidimensional Diffusion Processes
234. Aigner: Combinatorial Theory
235. Dynkin/Yushkevich: Controlled Markov Processes
236. Grauert/Remmert: Theory of Stein Spaces
237. Köthe: Topological Vector Spaces II
238. Graham/McGehee: Essays in Commutative Harmonic Analysis
239. Elliott: Probabilistic Number Theory I
240. Elliott: Probabilistic Number Theory II
241. Rudin: Function Theory in the Unit Ball of C^n
242. Huppert/Blackburn: Finite Groups II
243. Huppert/Blackburn: Finite Groups III
244. Kubert/Lang: Modular Units
245. Cornfeld/Fomin/Sinai: Ergodic Theory
246. Naimark/Stern: Theory of Group Representations
247. Suzuki: Group Theory I
248. Suzuki: Group Theory II
249. Chung: Lectures from Markov Processes to Brownian Motion
250. Arnold: Geometrical Methods in the Theory of Ordinary Differential Equations
251. Chow/Hale: Methods of Bifurcation Theory
252. Aubin: Nonlinear Analysis on Manifolds. Monge-Ampère Equations
253. Dwork: Lectures on ρ-adic Differential Equations
254. Freitag: Siegelsche Modulfunktionen
255. Lang: Complex Multiplication
256. Hörmander: The Analysis of Linear Partial Differential Operators I
257. Hörmander: The Analysis of Linear Partial Differential Operators II
258. Smoller: Shock Waves and Reaction-Diffusion Equations
259. Duren: Univalent Functions
260. Freidlin/Wentzell: Random Perturbations of Dynamical Systems
261. Bosch/Güntzer/Remmert: Non Archimedian Analysis – A System Approach to Rigid Analytic Geometry
262. Doob: Classical Potential Theory and Its Probabilistic Counterpart

263. Krasnosel'skiĭ/Zabreĭko: Geometrical Methods of Nonlinear Analysis
264. Aubin/Cellina: Differential Inclusions
265. Grauert/Remmert: Coherent Analytic Sheaves
266. de Rham: Differentiable Manifolds
267. Arbarello/Cornalba/Griffiths/Harris: Geometry of Algebraic Curves, Vol. I
268. Arbarello/Cornalba/Griffiths/Harris: Geometry of Algebraic Curves, Vol. II
269. Schapira: Microdifferential Systems in the Complex Domain
270. Scharlau: Quadratic and Hermitian Forms
271. Ellis: Entropy, Large Deviations, and Statistical Mechanics
272. Elliott: Arithmetic Functions and Integer Products
273. Nikol'skiĭ: Treatise on the Shift Operator
274. Hörmander: The Analysis of Linear Partial Differential Operators III
275. Hörmander: The Analysis of Linear Partial Differential Operators IV
276. Ligget: Interacting Particle Systems
277. Fulton/Lang: Riemann-Roch Algebra
278. Barr/Wells: Toposes, Triples and Theories
279. Bishop/Bridges: Constructive Analysis
280. Neukirch: Class Field Theory
281. Chandrasekharan: Elliptic Functions
282. Lelong/Gruman: Entire Functions of Several Complex Variables
283. Kodaira: Complex Manifolds and Deformation of Complex Structures
284. Finn: Equilibrium Capillary Surfaces
285. Burago/Zalgaller: Geometric Inequalities
286. Andrianaov: Quadratic Forms and Hecke Operators
287. Maskit: Kleinian Groups
288. Jacod/Shiryaev: Limit Theorems for Stochastic Processes
289. Manin: Gauge Field Theory and Complex Geometry
290. Conway/Sloane: Sphere Packings, Lattices and Groups
291. Hahn/O'Meara: The Classical Groups and K-Theory
292. Kashiwara/Schapira: Sheaves on Manifolds
293. Revuz/Yor: Continuous Martingales and Brownian Motion
294. Knus: Quadratic and Hermitian Forms over Rings
295. Dierkes/Hildebrandt/Küster/Wohlrab: Minimal Surfaces I
296. Dierkes/Hildebrandt/Küster/Wohlrab: Minimal Surfaces II
297. Pastur/Figotin: Spectra of Random and Almost-Periodic Operators
298. Berline/Getzler/Vergne: Heat Kernels and Dirac Operators
299. Pommerenke: Boundary Behaviour of Conformal Maps
300. Orlik/Terao: Arrangements of Hyperplanes
301. Loday: Cyclic Homology
302. Lange/Birkenhake: Complex Abelian Varieties
303. DeVore/Lorentz: Constructive Approximation
304. Lorentz/v. Golitschek/Makovoz: Construcitve Approximation. Advanced Problems
305. Hiriart-Urruty/Lemaréchal: Convex Analysis and Minimization Algorithms I. Fundamentals
306. Hiriart-Urruty/Lemaréchal: Convex Analysis and Minimization Algorithms II. Advanced Theory and Bundle Methods
307. Schwarz: Quantum Field Theory and Topology
308. Schwarz: Topology for Physicists
309. Adem/Milgram: Cohomology of Finite Groups
310. Giaquinta/Hildebrandt: Calculus of Variations I: The Lagrangian Formalism
311. Giaquinta/Hildebrandt: Calculus of Variations II: The Hamiltonian Formalism
312. Chung/Zhao: From Brownian Motion to Schrödinger's Equation
313. Malliavin: Stochastic Analysis
314. Adams/Hedberg: Function Spaces and Potential Theory
315. Bürgisser/Clausen/Shokrollahi: Algebraic Complexity Theory
316. Saff/Totik: Logarithmic Potentials with External Fields
317. Rockafellar/Wets: Variational Analysis
318. Kobayashi: Hyperbolic Complex Spaces

M. Aigner Combinatorial Theory ISBN 978-3-540-61787-7
A. L. Besse Einstein Manifolds ISBN 978-3-540-74120-6
N. P. Bhatia, G. P. Szegő Stability Theory of Dynamical Systems ISBN 978-3-540-42748-3
J. W. S. Cassels An Introduction to the Geometry of Numbers ISBN 978-3-540-61788-4
R. Courant, F. John Introduction to Calculus and Analysis I ISBN 978-3-540-65058-4
R. Courant, F. John Introduction to Calculus and Analysis II/1 ISBN 978-3-540-66569-4
R. Courant, F. John Introduction to Calculus and Analysis II/2 ISBN 978-3-540-66570-0
P. Dembowski Finite Geometries ISBN 978-3-540-61786-0
A. Dold Lectures on Algebraic Topology ISBN 978-3-540-58660-9
J. L. Doob Classical Potential Theory and Its Probabilistic Counterpart ISBN 978-3-540-41206-9
R. S. Ellis Entropy, Large Deviations, and Statistical Mechanics ISBN 978-3-540-29059-9
H. Federer Geometric Measure Theory ISBN 978-3-540-60656-7
S. Flügge Practical Quantum Mechanics ISBN 978-3-540-65035-5
L. D. Faddeev, L. A. Takhtajan Hamiltonian Methods in the Theory of Solitons ISBN 978-3-540-69843-2
I. I. Gikhman, A. V. Skorokhod The Theory of Stochastic Processes I ISBN 978-3-540-20284-4
I. I. Gikhman, A. V. Skorokhod The Theory of Stochastic Processes II ISBN 978-3-540-20285-1
I. I. Gikhman, A. V. Skorokhod The Theory of Stochastic Processes III ISBN 978-3-540-49940-4
D. Gilbarg, N. S. Trudinger Elliptic Partial Differential Equations of Second Order ISBN 978-3-540-41160-4
H. Grauert, R. Remmert Theory of Stein Spaces ISBN 978-3-540-00373-1
H. Hasse Number Theory ISBN 978-3-540-42749-0
F. Hirzebruch Topological Methods in Algebraic Geometry ISBN 978-3-540-58663-0
L. Hörmander The Analysis of Linear Partial Differential Operators I – Distribution Theory and Fourier Analysis ISBN 978-3-540-00662-6
L. Hörmander The Analysis of Linear Partial Differential Operators II – Differential Operators with Constant Coefficients ISBN 978-3-540-22516-4
L. Hörmander The Analysis of Linear Partial Differential Operators III – Pseudo-Differential Operators ISBN 978-3-540-49937-4
L. Hörmander The Analysis of Linear Partial Differential Operators IV – Fourier Integral Operators ISBN 978-3-642-00117-8
K. Itô, H. P. McKean, Jr. Diffusion Processes and Their Sample Paths ISBN 978-3-540-60629-1
T. Kato Perturbation Theory for Linear Operators ISBN 978-3-540-58661-6
S. Kobayashi Transformation Groups in Differential Geometry ISBN 978-3-540-58659-3
K. Kodaira Complex Manifolds and Deformation of Complex Structures ISBN 978-3-540-22614-7
Th. M. Liggett Interacting Particle Systems ISBN 978-3-540-22617-8
J. Lindenstrauss, L. Tzafriri Classical Banach Spaces I and II ISBN 978-3-540-60628-4
R. C. Lyndon, P. E Schupp Combinatorial Group Theory ISBN 978-3-540-41158-1
S. Mac Lane Homology ISBN 978-3-540-58662-3
C. B. Morrey Jr. Multiple Integrals in the Calculus of Variations ISBN 978-3-540-69915-6
D. Mumford Algebraic Geometry I – Complex Projective Varieties ISBN 978-3-540-58657-9
O. T. O'Meara Introduction to Quadratic Forms ISBN 978-3-540-66564-9
G. Pólya, G. Szegő Problems and Theorems in Analysis I – Series. Integral Calculus. Theory of Functions ISBN 978-3-540-63640-3
G. Pólya, G. Szegő Problems and Theorems in Analysis II – Theory of Functions. Zeros. Polynomials. Determinants. Number Theory. Geometry ISBN 978-3-540-63686-1
W. Rudin Function Theory in the Unit Ball of $\mathbb{C}^n$ ISBN 978-3-540-68272-1
S. Sakai C*-Algebras and W*-Algebras ISBN 978-3-540-63633-5
C. L. Siegel, J. K. Moser Lectures on Celestial Mechanics ISBN 978-3-540-58656-2
T. A. Springer Jordan Algebras and Algebraic Groups ISBN 978-3-540-63632-8
D. W. Stroock, S. R. S. Varadhan Multidimensional Diffusion Processes ISBN 978-3-540-28998-2
R. R. Switzer Algebraic Topology: Homology and Homotopy ISBN 978-3-540-42750-6
A. Weil Basic Number Theory ISBN 978-3-540-58655-5
A. Weil Elliptic Functions According to Eisenstein and Kronecker ISBN 978-3-540-65036-2
K. Yosida Functional Analysis ISBN 978-3-540-58654-8
O. Zariski Algebraic Surfaces ISBN 978-3-540-58658-6